TECHNOLOGY AND SOCIETY

Inside Technology

edited by Wiebe E. Bijker, W. Bernard Carlson, and Trevor Pinch

A list of the books in this series appears at the back of the book.

TECHNOLOGY AND SOCIETY

Building Our Sociotechnical Future

Second Edition

edited by Deborah G. Johnson and Jameson M. Wetmore

The MIT Press
Cambridge, Massachusetts
London, England

This book was set in Stone Serif and Stone Sans by Westchester Publishing Services. Printed and bound in the United States of America.

Library of Congress Cataloging-in-Publication Data

Names: Johnson, Deborah G., 1945– editor. | Wetmore, Jameson M., editor.
Title: Technology and society : building our sociotechnical future / edited by Deborah G. Johnson and Jameson M. Wetmore.
Description: Second edition. | Cambridge, Massachusetts : The MIT Press, 2021. | Series: Inside technology | Includes bibliographical references and index.
Identifiers: LCCN 2020044373 | ISBN 9780262539968 (paperback)
Subjects: LCSH: Technology—Social aspects. | Technological innovations. | Technology and civilization.
Classification: LCC T14.5 .T44169 2021 | DDC 303.48/3—dc23
LC record available at https://lccn.loc.gov/2020044373

10 9 8 7 6 5 4 3 2 1

Contents

Introduction 1

I VISIONS OF A TECHNOLOGICAL FUTURE 7

1 **The Machine Stops** 11
E. M. Forster

2 **The Prolongation of Life** 33
Francis Fukuyama

3 **Reproductive Ectogenesis: The Third Era of Human Reproduction and Some Moral Consequences** 45
Stellan Welin

4 **Eight Great Technologies** 55
David Willetts

5 **Why the Future Doesn't Need Us** 67
Bill Joy

6 **Sultana's Dream** 87
Rokeya Sakhawat Hossain

II THE RELATIONSHIP BETWEEN TECHNOLOGY AND SOCIETY 95

7 **Do Machines Make History?** 99
Robert L. Heilbroner

8 **The Social Construction of Facts and Artifacts** 109
Trevor J. Pinch and Wiebe Bijker

9 **Technological Momentum** 137
Thomas P. Hughes

10 **Where Are the Missing Masses? The Sociology of a Few Mundane Artifacts** 147
Bruno Latour

11 **Gender: The Missing Factor in STS** 173
Eulalia Pérez Sedeño

III TECHNOLOGY AND VALUES 185

12 Do Artifacts Have Politics? 189
Langdon Winner

13 Control: Human and Nonhuman Robots 205
George Ritzer

14 White 229
Richard Dyer

15 Manufacturing Gender in Commercial and Military Cockpit Design 237
Rachel N. Weber

16 Amish Technology: Reinforcing Values and Building Community 247
Jameson M. Wetmore

17 Preserving Traditional Knowledge: Initiatives in India 267
Rupak Chakravarty

18 Equity in Forecasting Climate: Can Science Save the World's Poor? 275
Maria Carmen Lemos and Lisa Dilling

IV THE COMPLEX NATURE OF SOCIOTECHNICAL SYSTEMS 287

19 Sociotechnical Complexity: Redesigning a Shielding Wall 291
Dominique Vinck

20 Fukushima and the Inevitability of Accidents 303
Charles Perrow

21 Nature as Infrastructure: Making and Managing the Panama Canal Watershed 311
Ashley Carse

22 Conceptions of Control and IT Artefacts: An Institutional Account of the Amazon Rainforest Monitoring System 335
Raoni Guerra Lucas Rajão and Niall Hayes

23 Franken-Algorithms: The Deadly Consequences of Unpredictable Code 357
Andrew Smith

24 The Extraordinary Science of Addictive Junk Food 367
Michael Moss

25 The Gender Binary Will Not Be Deprogrammed: Ten Years of Coding Gender on Facebook 379
Rena Bivens

26 Audible Citizenship and Audiomobility: Race, Technology, and CB Radio 397
Art M. Blake

27 **Drones for the Good: Technological Innovations, Social Movements, and the State** 413
Austin Choi-Fitzpatrick

V TWENTY-FIRST-CENTURY CHALLENGES AND STRATEGIES 429

28 **Engineering the Brain: Ethical Issues and the Introduction of Neural Devices** 433
Eran Klein, Tim Brown, Matthew Sample, Anjali R. Truitt, and Sara Goering

29 **Cyber (In)security: Threat Assessment in the Cyber Domain** 449
George Lucas

30 **Geoengineering as Collective Experimentation** 463
Jack Stilgoe

31 **Seven Principles for Equitable Adaptation** 483
Alice Kaswan

32 **Socio-Energy Systems Design: A Policy Framework for Energy Transitions** 501
Clark A. Miller, Jennifer Richter, and Jason O'Leary

33 **Debugging Bias: Busting the Myth of Neutral Technology** 527
Felicia L. Montalvo

34 **When Winning Is Losing: Why the Nation That Invented the Computer Lost Its Lead** 531
Mar Hicks

35 **Shaping Technology for the "Good Life": The Technological Imperative versus the Social Imperative** 543
Gary Chapman

36 **Not Just One Future** 555
David E. Nye

Index 567

Introduction

Technology is a powerful component of the modern world. Its influence can be clearly seen in the fact that many of the iconic accomplishments of modern societies are technological in nature. For instance, the twentieth century was marked by the construction of the Panama Canal; the invention and widespread adoption of the automobile, airplane, radio, and television; the moon landing; the development of the atomic bomb; and the development of computers and information technologies. Technologies have at least partly empowered even recent, seemingly nontechnological achievements made around the world. For instance, birth control pills helped enable women's liberation, and the invention of the Internet spurred the emergence of new democracies across the globe. Technology has been an important factor shaping the character of human societies and individual lives for millennia, but in looking back over the last century, it is evident that the power of technology to change the way we live is undeniable and awe inspiring.

Technologies not only have contributed to broad social and political change but have permeated almost every dimension of daily life in industrialized societies. Refrigerators, coffee makers, and microwave ovens make certain foods easy for us to obtain and others more difficult. Washing machines, pharmaceuticals, electric toothbrushes, and exercise equipment all shape our health and hygiene. The clothes we wear, the banks we put our money into, and the jobs we hold are all impacted or even made possible by global systems that exist only because of the complex technological systems they're built on. We maintain our personal as well as our working relationships through a global communications network. We entertain ourselves by means of digital technologies in which the quality of audio and images improves year after year. In short, a wide range of technologies are now an essential part of our work, play, education, health, finances, child-rearing, and even reproduction.

Modern societies have embraced technologies for a variety of reasons, but broadly speaking, new technologies offer new opportunities and the promise of improved lives. Technologies have been used to solve problems that have plagued humanity for centuries, eradicating diseases, harnessing new energy resources, and providing rapid transportation. Developers of technology have also pointed out needs we never knew we had to justify the creation of new devices, including eliminating bad breath, providing the convenience of shopping at home, and offering new forms of entertainment such as interactive games.

While technologies have helped us create a truly amazing world, it is clear that numerous problems remain. Our goal is and should be to make the world a better place for ourselves, for those around us, and for those who will come after us. To achieve this better future, we will need to direct technology toward the values we want to promote.

If new technologies solve problems and make our lives better, then we might expect continuous technological development to bring constant improvements in human lives. Visions and predictions of the future are often based on the idea that new technologies will bring about social change—that in and of themselves, technologies create the future. But we must be careful not to confuse technological change with social and human progress. We may be better off in some ways and worse off in others as a result of adopting certain technologies.

Technologies rarely, if ever, change the world in a clear, singular way. When we integrate them into our society, they often have far-rippling effects that are not always positive. Economies have been built on the wealth and benefits generated from manufacturing, but many manufacturing techniques degrade our environment and living conditions through the pollutants they produce. Social media has allowed us to share our views with people on the other side of the planet but may also make the measured, thoughtful discourse necessary for democracy more difficult. In addition to what some deem "side effects," new technologies can also lead to (desirable or undesirable) shifts in power from one group to another. For example, the computational power of large mainframes has enabled the growth of huge organizations, making it harder and harder for small companies to compete. In some cases, the impacts of new technologies have become perceptible soon after development—like the atomic bombs dropped on Hiroshima and Nagasaki. With other technologies, the implications have manifested more slowly—such as damage to the environment from chemical pesticides and fertilizers generated and accumulated over the past few centuries.

What exactly counts as a "negative implication" of a technology is up for debate as well because technologies affect different groups of people differently. The benefits may go to some segments of a population while the burdens are borne by other groups. This is the case, for example, with toxic waste disposal plants, which are generally located in poor nonwhite communities, putting those who live nearby at increased health risk while citizens living farther away reap the benefits. Technologies can reinforce prevailing inequalities, but they can also disrupt status quo arrangements. Think here of how new online news services have put print newspapers out of business or consider how the threat from autonomous vehicles may disempower those who drive vehicles for a living. Thus, while technology must be part of the solution in making a better world, we must also be mindful that simply "adding" technology would not necessarily create a world we want to live in.

The good news is that technology doesn't just happen. It doesn't just come out of nowhere. A wide range of social forces create and shape technologies. We all make decisions that shape or direct the development and use of technologies. For instance, venture capitalists and government officials decide in which areas of science and engineering to invest their resources. Corporate executives choose the types of products they will bring to market. Engineers design technologies with certain features and applications. Marketers evaluate how technologies may best be presented to the public. Regulators specify standards to help ensure that industries develop products that are safe and efficient. Advocates working with nongovernmental organizations (NGOs) encourage the development of alternative technologies that corporations may have overlooked, as well as voice concerns about technologies they believe governments should regulate. Finally, every individual decides whether, where, when, and how to use new technologies at home, in public places, and at work.

The fact that technologies don't just "happen" to society but are the result of social shaping means that influencing and even steering the trajectory of technological developments is possible, albeit not easy. The challenge is daunting because so many different actors and factors can affect the development of technology. Often, decisions are difficult to make because they involve competing values. For instance, those who want to save the world through clean energy have come face-to-face with those whose lives and fortunes have been invested in older energy systems—including everyone from big oil companies to Native American tribes with economies based on their coal reserves. In this situation, not everyone can get what they want. The competing values must be resolved. Usually those with power, money, and prestige are in the best position to resolve issues like this in their favor. Thus, too often, new technologies make the world a better place for the "haves" rather than the "have nots," worsening the inequities that already exist.

When it comes to the future, the fact that human decisions influence technological development is liberating. It means that no vision of the future is inevitable. The actions of today influence outcomes that will constitute the shape of the future. This book has been designed with the idea that readers, like the authors, want the future to be a good place for humanity. We want the future to be better than the present. Among other things this means that we want future technologies to constitute human lives, organizations, governments, and economies in ways that fulfill human needs and enhance human well-being—and do so for all. Such a goal is, we acknowledge, idealistic, but we believe the alternative is unacceptable.

In order to influence the development of technology in whatever way—large or small—it is essential to understand the relationships between technology and society. This book is designed to do just that. Although there are limits to what individuals can do to influence the direction of technological development, the idea of the book is to provide an understanding of how technology and society intertwine that will be helpful to anyone whose decisions affect the future. Whether you are a consumer of technology, a user of technology, a policy-maker, a constituent, an engineer, or a business leader, if you want to direct technology in the most beneficial ways, you need to understand how people shape technology and how technology shapes people. While we are constantly surrounded by devices and tools and often know how they work mechanically, we don't always know where they came from and what it takes to make them work, and we generally don't think much about the social effects they can have. We need an understanding of how machines, devices, and techniques are interwoven with people, institutions, goals, and values if we are going to work for better futures.

A fundamental way this book seeks to further this concept is by rethinking the very idea of technology. The book argues, implicitly as well as explicitly, that technology does not just comprise material objects—or *artifacts*. Technology neither exists nor has meaning without the human activities of which it is a part. Similarly, many social practices would be impossible or incomprehensible without material objects. As such, to understand the ways in which technology permeates and constitutes our everyday lives, we have to examine material objects together with the social practices and social relationships that make such material objects possible and useful. This book and most of the authors in it approach technologies as *sociotechnical systems*, rather than individual devices or machines.

For example, to understand how technologies develop and change over time, it is important to examine the ways that human decision-making influences design. One of the most prominent and influential technologies of the twentieth (and probably the twenty-first) century—the automobile—has been subject to countless forces seeking to shape its design. Engineers, designers, policy-makers, consumer advocacy groups, and consumers themselves have all had an impact on automotive design, with the hopes of promoting their specific goals. Everything from the tail fins of the 1950s to the catalytic converters of the 1970s was a result of the values of a different social group or groups to shape the cars ultimately offered to the public. If you don't consider the social and political forces that inspire it, automobile design probably won't make much sense.

The intertwining of human behavior and technological operations can also be thought of in terms of how things work. Technological devices only work through a combination of human behavior and machine behavior. Even something as simple and seemingly autonomous as a thermostat yields useful results only when the right human and machine inputs have been entered into the system. Although the thermostat receives temperature input and automatically responds by altering the functioning of the furnace and air conditioner, it must be built and installed by a human, and a human needs to set the temperature.

To provide an understanding of technology as a sociotechnical system, the book begins by presenting a selection of visions of the future. Each vision recognizes a connection between technology and society, but each vision and connection is distinctive. The selections raise many questions about the future we might be headed for, the opportunities for change created by new and evolving technologies, the processes by which technologies affect and constitute aspects of human societies and modes of life, and the importance of playing an active role in shaping our collective future.

Although we have suggested that the relationship between technology and society is so interwoven that the two are inseparable, the relationship need not, by any means, be a black box. In the last half century in particular, a new field of study known as STS (science and technology studies or science, technology, and society) sprang up as scholars began theorizing about the social aspects of science and technology. In the second section, we present several of the major theories that have been put forward, contested, and used. We offer these theories and the associated concepts as tools to aid in understanding technology's role in the past, present, and future.

One of the most important aspects of this relationship centers around human and social values and how they are embedded in technological choices. Values shape and are shaped by technology. The readings in the third section illustrate this point by focusing on a particular value in a particular context and time period. These selections suggest the importance of recognizing values in technological decision-making for the future and offer examples of how one might promote specific values through technology.

In trying to get a handle on how values shape and are shaped by technological decisions, we have intentionally selected readings that illustrate how race and gender are entangled with technology. The absence of race and gender from early theorizing in technology has begun to be remedied through research documenting just how subtly and complexly race and gender come into play in technological design and decision-making. To promote continuing work in this area, race and/or gender are addressed in every section of the book.

None of the theories about the technology-society relationship claims that the relationship is simple. The readings in the fourth section identify and illustrate the interesting and sometimes subtle complexities. These readings show that we (individually and collectively: e.g., as countries, cultures, or organizations) must think carefully about the technological choices we make.

We end the book with a set of readings that focus on sociotechnical challenges we currently face. How can we fulfill energy needs without destroying the environment or worsening social inequities? In seeking security, how should we respond to and distinguish cybercrime and cyber acts of warfare? How will we understand what it means to be a person when scientists can tinker with the brain to affect attitudes and behavior? How can privacy be protected while at the same time achieving security, connectedness, and efficiency? These are challenges both in the short and long term, and they are unquestionably daunting. It is impossible to posit every challenge we face in a single section, but the readings provide examples of people struggling to address hugely important issues, and some even identify plausible strategies and solutions. These articles serve as a model for how to approach complicated sociotechnical systems with an eye toward making a difference. If we care about the future, we cannot just sit back and watch. We must face the issues, try to determine what has created the situation, and work to steer in a direction that will bring about a better sociotechnical future.

The ultimate intent of this book is to equip readers to be agents of change in our sociotechnical future. It offers a way to think about technology and society that can lead to improved technologies and a society better equipped to determine future outcomes. Whether you are an engineer, a policy-maker, a tinkerer, or simply a concerned citizen, you are more likely to make a difference if you understand the complexities of technological decisions. Such a journey will certainly not be easy, but this book will encourage and guide in that endeavor.

I VISIONS OF A TECHNOLOGICAL FUTURE

We encounter visions of the future just about every day. Obviously, this happens when we watch a science fiction movie in which we are shown a possible (even if not plausible) future; this also occurs, though more subtly, when we see an advertisement for deodorant or some other product telling us how much better our lives could be if we use it. Because they seem to be everywhere, it may be tempting to dismiss visions of the future as trivial or simply entertainment. But they often embody deep thinking about what is important to human beings and what makes for better or worse forms of living. Visions of the future portray futures suggesting how the world of today might be better in terms of the values we currently hold—a world in which people are not afflicted with diseases, lead less onerous lives, and are treated more equally. Or they may depict dystopic worlds cautioning us to be careful about moving in directions that will degrade our lives and diminish values we hold dear. Whether optimistic or dystopian, visions of the future often identify existing trends in society and imagine the long-term consequences of continuing on those trajectories. They embody ideas about the factors and types of events that might push social change in one direction or another.

For at least the last 150 years, technology has been a central theme in many visions of the future. Technology is often highlighted as the force causing a transition to new forms of social organization. Many science fiction films revolve around a new technology—be it genetic testing, a weapon of war, or artificial intelligence—portrayed as the power determining change from the current world to the one imagined in the films. Or consider the message suggested when the promise of an entirely new and abundant source of energy is depicted as not only addressing climate change but eliminating the need for war and national rivalries. Sometimes the force of technology is presented as unstoppable—as taking us to a specific future regardless of whether we want to go there. In all of these approaches, technological change and social change are understood to be intimately linked; together they create the future.

Visions of the future serve a variety of purposes. As already suggested they may give us ideas about what is possible in the future or caution against certain trends. However, visions of the future are put forward for varying reasons. For example, entrepreneurs may use them to help develop their business plans. Products under development today will be purchased and used in the future, so corporations must not only understand where we are today but also anticipate the world of the future. Many organizations conduct visioning exercises in order to be ready for the world to come.

These same organizations subsequently use visions of the future to sell products to consumers. Many marketers believe we will be more interested in purchasing their products if we think we are not simply buying a new widget but rather achieving the lifestyle destined to come with it. As such they develop visions of a future in which toothpaste doesn't simply make our mouths cleaner—it makes us more attractive to romantic partners; cars don't just take us from point A to point B—they help make our communities more connected; and a new set of golf clubs don't just let us hit a ball farther—they ultimately will make us a winner. The basic idea that the simple purchase of a mundane product will fundamentally change our social lives for the better frames much of advertising.

Corporations aren't the only ones that use futures in this way. Politicians and government officials are also fond of using visions of the future to garner our support. They may describe wonderful worlds that could be ours if only we elect them, or they may predict corresponding dismal futures that could result from voting for their opponent. Who wouldn't support a government policy that promises to reduce taxes, cure cancer, stop climate change, and provide jobs for everyone currently unemployed—all at the same time?

Visions of the future are an important topic to reflect on and study because they are not simply predictions or educated guesses about the future; they play an active role in creating the future. As producers and consumers, we use visions of the future to help us make decisions about today that will affect tomorrow. They may be more powerful in their ability to motivate us to change our course of action and direction than most other types of arguments. For example, when the forecast of a huge increase in the number of jobs in computer science entices us to become computer science majors or when an image of a world without fossil fuels leads us to invest in bio- and solar energy, predictions of the future are shaping what we do today as well as what the future will be.

Because visions of the future have the power to shape what we do today and in the future, they should be carefully examined. It is especially important to reflect on the motivations of those who create visions of the future and to uncover and evaluate the assumptions they make in creating the vision, including their understanding of social change.

This section presents readings with both positive and negative images of the future; they portray technology as playing different roles in getting us to the future. Most of the selections are contemporary and reflect current thinking about the future. However, the first selection by E. M. Forster and the last selection by Rokeya Sakhawat Hossain were written over a hundred years ago—in the first decade of the twentieth century. They are both works of fiction, one serving as a cautionary tale (depicting a future to be avoided) and the other suggesting the positive potential of technology to bring about peace.

The visions included in this section are written in different modes. Two are science fiction stories, one is a government report, and several come from more scholarly or academic venues. The authors of these visions hail from a variety of backgrounds. They are scholars and writers, a social activist, a computer scientist, an elected government official, and a philosopher. They come from different parts of the world, including the US, Europe, and South Asia. Each vision reflects the author's hopes and concerns as well as the state of society at the time of writing.

This diversity helps to illustrate the wide variety of places where one might come across a vision of the future and the ways in which such visions permeate our understanding of the world. The differences among them also help to stress one common theme that pervades all of them. They all embody assumptions about technology's role in shaping society. They implicitly and explicitly assume a connection between the technologies we develop and use, the quality and character of the lives we live, and the accompanying social and political order. They all suggest that if we want to shape the social world of the future, we need to be thinking about technology in the present.

Although some of these technological futures may seem far-fetched, they can still inspire and influence those who design and develop technologies, and thus they can play a pivotal role in creating our future. Even when wild predictions don't come true, their articulation can motivate actions that have big effects. When the British government allocates millions of pounds to study synthetic biology, when a drug company chooses to focus on fertility drugs rather than a cure for malaria, when engineers test a wing design in a wind tunnel or debug a program, or when a handful of electronics enthusiasts begin building affordable computers in their garage, they are all engaged in activities that can affect the future world.

The future is being designed and built by all of those who develop, fund, regulate, and use technology. While no individual or group or single technology is ever solely responsible for the future, it is important for all of those involved with technology to understand their role in building the future so they can consider the possible future consequences of their actions. While governments may not be held accountable for missed opportunities, they must ultimately answer to their constituents, corporations must answer to consumers for the safety and side effects of their products, engineers are morally and legally responsible for ensuring a reasonable level of safety in their inventions, and consumers are charged with appropriately using the products they consume. None of these groups can act responsibly unless they understand how their actions today affect tomorrow. Creating visions of the future and thinking carefully through them can help us imagine worlds we want to live in, worlds we don't want to live in, and how today's decisions may create new and different possibilities.

Questions to consider while reading the selections in this section:

1. What assumptions does each author make about the relationship between technology and society?
2. What technological and social developments would have to occur to bring about the author's vision of the future?
3. Are the changes necessary to achieve the author's vision feasible or desirable?
4. What makes the author's vision utopian or dystopian?
5. Who will benefit, and how, from realization of the author's vision? Who will lose and how will they lose if the author's vision is realized?

1 The Machine Stops

E. M. Forster

Despite being written over a hundred years ago, this powerful story by E. M. Forster is still pertinent today. On one level, the story can be interpreted as a cautionary tale about humans becoming too dependent on technology. On another level, one can read Forster as more concerned with the deep social and psychological changes that accompany massive environmental and technological change. "The Machine Stops" depicts a world in which machines not only make up the physical infrastructure but constrain and shape people—their capabilities, desires, inner lives, emotions, family relationships, and so on. Forster was incredibly prescient. The future he presents us with is perhaps more relevant today than when he wrote the story. This seems especially true when we remember how we retreated to our houses and apartments and turned to technology as a way to interact with others during the COVID-19 lockdowns. In this story, Forster describes a world in which technology fulfills some of the deepest desires and values of humans—to have everything at their fingertips and to be free from the mundane tasks of everyday life so they can engage in what they consider "higher" activities such as art, music, and poetry. Forster challenges us to consider whether these goals, which shape much of our daily lives, would ultimately make us happy. The story implicitly raises profound questions that will resonate through the rest of this book: What is progress? Will our technical achievements bring us happiness? What are the implications when people accept new technologies as unquestioningly good? How does technology mediate our experiences and relationships, and how do we want it to mediate our experiences and relationships? How can we get the sociotechnical future we want and avoid receiving one that is less desirable than what we have now? Forster challenges us to consider which values are most important to us and how technology might help or interfere with those values.

Part I: The Air-Ship

Imagine, if you can, a small room, hexagonal in shape, like the cell of a bee. It is lighted neither by window nor by lamp, yet it is filled with a soft radiance. There are no apertures for ventilation, yet the air is fresh. There are no musical instruments, and yet, at the moment that my meditation opens, this room is throbbing with melodious sounds. An arm-chair is in the centre, by its side a reading-desk—that is all the furniture. And in the arm-chair there sits a swaddled lump of flesh—a woman, about five feet high, with a face as white as a fungus. It is to her that the little room belongs.

An electric bell rang.

From *Oxford and Cambridge Review*, November 1909, 83–122.

The woman touched a switch and the music was silent.

"I suppose I must see who it is," she thought, and set her chair in motion. The chair, like the music, was worked by machinery, and it rolled her to the other side of the room, where the bell still rang importunately.

"Who is it?" she called. Her voice was irritable, for she had been interrupted often since the music began. She knew several thousand people; in certain directions, human intercourse had advanced enormously.

But when she listened into the receiver, her white face wrinkled into smiles, and she said: "Very well. Let us talk, I will isolate myself. I do not expect anything important will happen for the next five minutes—for I can give you fully five minutes, Kuno. Then I must deliver my lecture on 'Music during the Australian Period.'"

She touched the isolation-knob, so that no one else could speak to her. Then she touched the lighting apparatus, and the little room was plunged into darkness.

"Be quick!" she called, her irritation returning. "Be quick, Kuno; here I am in the dark wasting my time."

But it was fully fifteen seconds before the round plate that she held in her hands began to glow. A faint blue light shot across it, darkening to purple, and presently she could see the image of her son, who lived on the other side of the earth, and he could see her.

"Kuno, how slow you are."

He smiled gravely.

"I really believe you enjoy dawdling."

"I have called you before, mother, but you were always busy or isolated. I have something particular to say."

"What is it, dearest boy? Be quick. Why could you not send it by pneumatic post?"

"Because I prefer saying such a thing. I want———"

"Well?"

"I want you to come and see me."

Vashti watched his face in the blue plate.

"But I can see you!" she exclaimed. "What more do you want?"

"I want to see you not through the Machine," said Kuno. "I want to speak to you not through the wearisome Machine."

"Oh, hush!" said his mother, vaguely shocked. "You mustn't say anything against the Machine."

"Why not?"

"One mustn't."

"You talk as if a god had made the Machine," cried the other. "I believe that you pray to it when you are unhappy. Men made it, do not forget that. Great men, but men. The Machine is much, but it is not everything. I see something like you in this plate, but I do not see you. I hear something like you through this telephone, but I do not hear you. That is why I want you to come. Come and stop with me. Pay me a visit, so that we can meet face to face, and talk about the hopes that are in my mind."

She replied that she could scarcely spare the time for a visit.

"The air-ship barely takes two days to fly between me and you."

"I dislike air-ships."

"Why?"

"I dislike seeing the horrible brown earth, and the sea, and the stars when it is dark. I get no ideas in an air-ship."

"I do not get them anywhere else."

"What kind of ideas can the air give you?"

He paused for an instant. "Do you not know four big stars that form an oblong, and three stars close together in the middle of the oblong, and hanging from these stars, three other stars?"

"No, I do not. I dislike the stars. But did they give you an idea? How interesting; tell me."

"I had an idea that they were like a man."

"I do not understand."

"The four big stars are the man's shoulders and his knees. The three stars in the middle are like the belts that men wore once, and the three stars hanging are like a sword."

"A sword?"

"Men carried swords about with them, to kill animals and other men."

"It does not strike me as a very good idea, but it is certainly original. When did it come to you first?"

"In the air-ship———" He broke off, and she fancied that he looked sad. She could not be sure, for the Machine did not transmit *nuances* of expression. It only gave a general idea of people—an idea that was good enough for all practical purposes, Vashti thought. The imponderable bloom, declared by a discredited philosophy to be the actual essence of intercourse, was rightly ignored by the Machine, just as the imponderable bloom of the grape was ignored by the manufacturers of artificial fruit. Something "good enough" had long since been accepted by our race.

"The truth is," he continued, "that I want to see these stars again. They are curious stars. I want to see them not from the air-ship, but from the surface of the earth, as our ancestors did, thousands of years ago. I want to visit the surface of the earth."

She was shocked again.

"Mother, you must come, if only to explain to me what is the harm of visiting the surface of the earth."

"No harm," she replied, controlling herself. "But no advantage. The surface of the earth is only dust and mud, no life remains on it, and you would need a respirator, or the cold of the outer air would kill you. One dies immediately in the outer air."

"I know; of course I shall take all precautions."

"And besides———"

"Well?"

She considered, and chose her words with care. Her son had a queer temper, and she wished to dissuade him from the expedition.

"It is contrary to the spirit of the age," she asserted.

"Do you mean by that, contrary to the Machine?"

"In a sense, but———"

His image in the blue plate faded. "Kuno!"

He had isolated himself.

For a moment Vashti felt lonely.

Then she generated the light, and the sight of her room, flooded with radiance and studded with electric buttons, revived her. There were buttons and switches everywhere—buttons to call for food, for music, for clothing. There was the hot-bath button, by pressure of which a basin of (imitation) marble rose out of the floor, filled to the brim with a warm deodorised liquid. There was the cold-bath button. There was the button that produced literature. And there were of course the buttons by which she communicated with her friends. The room, though it contained nothing, was in touch with all that she cared for in the world.

Vashti's next move was to turn off the isolation-switch, and all the accumulations of the last three minutes burst upon her. The room was filled with the noise of bells, and speaking-tubes. What was the new food like? Could she recommend it? Had she had any ideas lately? Might one tell her one's own ideas? Would she make an engagement to visit the public nurseries at an early date?—say this day month.

To most of these questions she replied with irritation—a growing quality in that accelerated age. She said that the new food was horrible. That she could not visit the public nurseries through press of engagements. That she had no ideas of her own but had just been told one—that four stars and three in the middle were like a man: she doubted there was much in it. Then she switched off her correspondents, for it was time to deliver her lecture on Australian music.

The clumsy system of public gatherings had been long since abandoned; neither Vashti nor her audience stirred from their rooms. Seated in her arm-chair she spoke, while they in their arm-chairs heard her, fairly well, and saw her, fairly well. She opened with a humorous account of music in the pre-Mongolian epoch, and went on to describe the great outburst of song that followed the Chinese conquest. Remote and primæval as were the methods of I-San-So and the Brisbane school, she yet felt (she said) that study of them might repay the musician of to-day: they had freshness; they had, above all, ideas.

Her lecture, which lasted ten minutes, was well received, and at its conclusion she and many of her audience listened to a lecture on the sea; there were ideas to be got from the sea; the speaker had donned a respirator and visited it lately. Then she fed, talked to many friends, had a bath, talked again, and summoned her bed.

The bed was not to her liking. It was too large, and she had a feeling for a small bed. Complaint was useless, for beds were of the same dimension all over the world, and to have had an alternative size would have involved vast alterations in the Machine. Vashti isolated herself—it was necessary, for neither day nor night existed under the ground—and reviewed all that had happened since she had summoned the bed last. Ideas? Scarcely any. Events—was Kuno's invitation an event?

By her side, on the little reading-desk, was a survival from the ages of litter—one book. This was the Book of the Machine. In it were instructions against every possible contingency. If she was hot or cold or dyspeptic or at loss for a word, she went to the book, and it told her which button to press. The Central Committee published it. In accordance with a growing habit, it was richly bound.

Sitting up in the bed, she took it reverently in her hands. She glanced round the glowing room as if some one might be watching her. Then, half ashamed, half joyful, she murmured "O Machine! O Machine!" and raised the volume to her lips. Thrice she kissed it, thrice inclined her head, thrice she felt the delirium of acquiescence. Her ritual performed, she turned to page 1367, which gave the times of the departure of the air-ships from the island in the Southern Hemisphere, under whose soil she lived, to the island in the Northern Hemisphere, whereunder lived her son.

She thought, "I have not the time."

She made the room dark and slept; she awoke and made the room light; she ate and exchanged ideas with her friends, and listened to music and attended lectures; she made the room dark and slept. Above her, beneath her, and around her, the Machine hummed eternally; she did not notice the noise, for she had been born with it in her ears. The earth, carrying her, hummed as it sped through silence, turning her now to the invisible sun, now to the invisible stars. She awoke and made the room light.

"Kuno!"

"I will not talk to you," he answered, "until you come."

"Have you been on the surface of the earth since we spoke last?" His image faded.

Again she consulted the book. She became very nervous and lay back in her chair palpitating. Think of her as without teeth or hair. Presently she directed the chair to the wall, and pressed an unfamiliar button. The wall swung apart slowly. Through the opening she saw a tunnel that curved slightly, so that its goal was not visible. Should she go to see her son, here was the beginning of the journey.

Of course she knew all about the communication-system. There was nothing mysterious in it. She would summon a car and it would fly with her down the tunnel until it reached the lift that communicated with the air-ship station: the system had been in use for thousands of years, long before the universal establishment of the Machine. And of course she had studied the civilisation that had immediately preceded her own—the civilisation that had mistaken the functions of the system, and had used it for bringing people to things, instead of for bringing things to people. Those funny old days, when men went for change of air instead of changing the air in their rooms! And yet—she was frightened of the tunnel: she had not seen it since her last child was born. It curved—but not quite as she remembered; it was brilliant—but not quite as brilliant as a lecturer had suggested. Vashti was seized with the terrors of direct experience. She shrank back into the room, and the wall closed up again.

"Kuno," she said, "I cannot come to see you. I am not well."

Immediately an enormous apparatus fell on to her out of the ceiling, a thermometer was automatically inserted between her lips, a stethoscope was automatically laid upon her heart. She lay powerless. Cool pads soothed her forehead. Kuno had telegraphed to her doctor.

So the human passions still blundered up and down in the Machine. Vashti drank the medicine that the doctor projected into her mouth, and the machinery retired into the ceiling. The voice of Kuno was heard asking how she felt.

"Better." Then with irritation: "But why do you not come to me instead?"

"Because I cannot leave this place."

"Why?"

"Because, any moment, something tremendous may happen."

"Have you been on the surface of the earth yet?"

"Not yet."

"Then what is it?"

"I will not tell you through the Machine." She resumed her life.

But she thought of Kuno as a baby, his birth, his removal to the public nurseries, her one visit to him there, his visits to her—visits which stopped when the Machine had assigned him a room on the other side of the earth. "Parents, duties of," said the book of the Machine, "cease at the moment of birth. P. 422327483." True—but there was something special about Kuno—indeed there had been something special about all her children, and, after all, she must brave the journey if he desired it. And "something tremendous might happen." What did that mean? The nonsense of a youthful man, no doubt, but she must go. Again she pressed the unfamiliar button, again the wall swung back, and she saw the tunnel that curved out of sight. Clasping the Book, she rose, tottered on to the platform, and summoned the car. Her room closed behind her: the journey to the Northern Hemisphere had begun.

Of course it was perfectly easy. The car approached and in it she found arm-chairs exactly like her own. When she signalled, it stopped, and she tottered into the lift. One other passenger was in the lift, the first fellow creature she had seen face to face for months. Few travelled in these days, for, thanks to the advance of science, the earth was exactly alike all over. Rapid intercourse, from which the previous civilisation had hoped so much, had ended by defeating itself. What was the good of going to Pekin when it was just like Shrewsbury? Why return to Shrewsbury when it would be just like Pekin? Men seldom moved their bodies; all unrest was concentrated in the soul.

The air-ship service was a relic from the former age. It was kept up, because it was easier to keep it up than to stop it or to diminish it, but it now far exceeded the wants of the population. Vessel after vessel would rise from the vomitories of Rye or of Christchurch (I use the antique names), would sail into the crowded sky, and would draw up at the wharves of the South—empty. So nicely adjusted was the system, so independent of meteorology, that the sky, whether calm or cloudy, resembled a vast kaleidoscope whereon the same patterns periodically recurred. The ship on which Vashti sailed started now at sunset, now at dawn. But always, as it passed above Rheims, it would neighbour the ship that served between Helsingfors and the Brazils, and, every third time it surmounted the Alps, the fleet of Palermo would cross its track behind.

Night and day, wind and storm, tide and earthquake, impeded man no longer. He had harnessed Leviathan. All the old literature, with its praise of Nature, and its fear of Nature, rang false as the prattle of a child.

Yet as Vashti saw the vast flank of the ship, stained with exposure to the outer air, her horror of direct experience returned. It was not quite like the air-ship in the cinematophote. For one thing it smelt—not strongly or unpleasantly, but it did smell, and with her eyes shut she should have known that a new thing was close to her. Then she had to walk to it from the lift, had to submit to glances from the other passengers. The man in front dropped his Book—no great matter, but it disquieted them all. In the rooms, if the Book was dropped, the floor raised it mechanically, but the gangway to the air-ship was not so prepared, and the sacred volume lay motionless. They stopped—the thing was unforeseen—and the man, instead of picking up his property, felt the muscles of his arm to see how they had failed him. Then some one actually said with direct utterance: "We shall be late"—and they trooped on board, Vashti treading on the pages as she did so.

Inside, her anxiety increased. The arrangements were old-fashioned and rough. There was even a female attendant, to whom she would have to announce her wants during the voyage. Of course a revolving platform ran the length of the boat, but she was expected to walk from it to her cabin. Some cabins were better than others, and she did not get the best. She thought the attendant had been unfair, and spasms of rage shook her. The glass valves had closed, she could not go back. She saw, at the end of the vestibule, the lift in which she had ascended going quietly up and down, empty. Beneath those corridors of shining tiles were rooms, tier below tier, reaching far into the earth, and in each room there sat a human being, eating, or sleeping, or producing ideas. And buried deep in the hive was her own room. Vashti was afraid.

"O Machine! O Machine!" she murmured, and caressed her Book, and was comforted.

Then the sides of the vestibule seemed to melt together, as do the passages that we see in dreams, the lift vanished, the Book that had been dropped slid to the left and vanished, polished tiles rushed by like a stream of water, there was a slight jar, and the air-ship, issuing from its tunnel, soared above the waters of a tropical ocean.

It was night. For a moment she saw the coast of Sumatra edged by the phosphorescence of waves, and crowned by lighthouses, still sending forth their disregarded beams. These also vanished, and only the stars distracted her. They were not motionless, but swayed to and fro above her head, thronging out of one skylight into another, as if the universe and not the air-ship was careening. And, as often happens on clear nights, they seemed now to be in perspective, now on a plane; now piled tier beyond tier into the infinite heavens, now concealing infinity, a roof limiting for ever the visions of men. In either case they seemed intolerable. "Are we to travel in the dark?" called the passengers angrily, and the attendant, who had been careless, generated the light, and pulled down the blinds of pliable metal. When the air-ships had been built, the desire to look direct at things still lingered in the world. Hence the extraordinary number of skylights and windows, and the proportionate discomfort to those who were civilised and refined. Even in Vashti's cabin one star peeped through a flaw in the blind, and after a few hours' uneasy slumber, she was disturbed by an unfamiliar glow, which was the dawn.

Quick as the ship had sped westwards, the earth had rolled eastwards quicker still, and had dragged back Vashti and her companions towards the sun. Science could prolong the night, but only for a little, and those high hopes of neutralising the earth's diurnal revolution had passed, together with hopes that were possibly higher. To "keep pace with the sun," or even to outstrip it had been the aim of the civilisation preceding this. Racing aeroplanes had been built for the purpose, capable of enormous speed, and steered by the greatest intellects of the epoch. Round the globe they went, round and round, westward, westward, round and round, amidst humanity's applause. In vain. The globe went eastward quicker still, horrible accidents occurred, and the Committee of the Machine, at the time rising into prominence, declared the pursuit illegal, unmechanical, and punishable by Homelessness.

Of Homelessness more will be said later.

Doubtless the Committee was right. Yet the attempt to "defeat the sun" aroused the last common interest that our race experienced about the heavenly bodies, or indeed about anything. It was the last time that men were compacted by thinking of a power outside the world. The sun had conquered, yet it was the end of his spiritual dominion. Dawn, midday, twilight, the zodiacal path, touched neither men's lives nor their hearts, and science retreated into the ground, to concentrate herself upon problems that she was certain of solving.

So when Vashti found her cabin invaded by a rosy finger of light, she was annoyed, and tried to adjust the blind. But the blind flew up altogether, and she saw through the skylight small pink clouds, swaying against a background of blue, and as the sun crept higher, its radiance entered direct, brimming down the wall, like a golden sea. It rose and fell with the air-ship's motion, just as waves rise and fall, but it advanced steadily, as a tide advances. Unless she was careful, it would strike her face. A spasm of horror shook her and she rang for the attendant. The attendant too was horrified, but she could do nothing; it was not her place to mend the blind. She could only suggest that the lady should change her cabin, which she accordingly prepared to do.

People were almost exactly alike all over the world, but the attendant of the air-ship, perhaps owing to her exceptional duties, had grown a little out of the common. She had often to address passengers with direct speech, and this had given her a certain roughness and originality of manner. When Vashti swerved away from the sunbeams with a cry, she behaved barbarically—she put out her hand to steady her.

"How dare you!" exclaimed the passenger. "You forget yourself!"

The woman was confused, and apologised for not having let her fall. People never touched one another. The custom had become obsolete, owing to the Machine.

"Where are we now?" asked Vashti haughtily.

"We are over Asia," said the attendant, anxious to be polite.

"Asia?"

"You must excuse my common way of speaking. I have got into the habit of calling places over which I pass by their unmechanical names."

"Oh, I remember Asia. The Mongols came from it."

"Beneath us, in the open air, stood a city that was once called Simla."

"Have you ever heard of the Mongols and of the Brisbane school?"

"No."

"Brisbane also stood in the open air."

"Those mountains to the right—let me show you them." She pushed back a metal blind. The main chain of the Himalayas was revealed. "They were once called the Roof of the World, those mountains."

"What a foolish name!"

"You must remember that, before the dawn of civilisation, they seemed to be an impenetrable wall that touched the stars. It was supposed that no one but the gods could exist above their summits. How we have advanced, thanks to the Machine!"

"How we have advanced, thanks to the Machine!" said Vashti.

"How we have advanced, thanks to the Machine!" echoed the passenger who had dropped his Book the night before, and who was standing in the passage.

"And that white stuff in the cracks?—what is it?"

"I have forgotten its name."

"Cover the window, please. These mountains give me no ideas."

The northern aspect of the Himalayas was in deep shadow: on the Indian slope the sun had just prevailed. The forests had been destroyed during the literature epoch for the purpose of making newspaper-pulp, but the snows were awakening to their morning glory, and clouds still hung on the breasts of Kinchinjunga. In the plain were seen the ruins of cities, with diminished rivers creeping by their walls, and by the sides of these were sometimes the signs of vomitories, marking the cities of to-day. Over the whole prospect air-ships rushed, crossing and intercrossing with incredible *aplomb*, and rising nonchalantly when they desired to escape the perturbations of the lower atmosphere and to traverse the Roof of the World.

"We have indeed advanced, thanks to the Machine," repeated the attendant, and hid the Himalayas behind a metal blind.

The day dragged wearily forward. The passengers sat each in his cabin, avoiding one other with an almost physical repulsion and longing to be once more under the surface of the earth. There were eight or ten of them, mostly young males, sent out from the public nurseries to inhabit the rooms of those who had died in various parts of the earth. The man who had dropped his Book was on the homeward journey. He had been sent to Sumatra for the purpose of propagating the race. Vashti alone was travelling by her private will.

At midday she took a second glance at the earth. The air-ship was crossing another range of mountains, but she could see little, owing to clouds. Masses of black rock hovered below her, and merged indistinctly into grey. Their shapes were fantastic; one of them resembled a prostrate man.

"No ideas here," murmured Vashti, and hid the Caucasus behind a metal blind.

In the evening she looked again. They were crossing a golden sea, in which lay many small islands and one peninsula.

She repeated, "No ideas here," and hid Greece behind a metal blind.

Part II: The Mending Apparatus

By a vestibule, by a lift, by a tubular railway, by a platform, by a sliding door—by reversing all the steps of her departure did Vashti arrive at her son's room, which exactly resembled her own. She might well declare that the visit was superfluous. The buttons, the knobs, the reading desk with the Book, the temperature, the atmosphere, the illumination—all were exactly the same. And if Kuno himself, flesh of her flesh, stood close beside her at last, what profit was there in that? She was too well bred to shake him by the hand.

Averting her eyes, she spoke as follows:

"Here I am. I have had the most terrible journey and greatly retarded the development of my soul. It is not worth it, Kuno, it is not worth it. My time is too precious. The sunlight almost touched me, and I have met with the rudest people. I can only stop a few minutes. Say what you want to say, and then I must return."

"I have been threatened with Homelessness," said Kuno. She looked at him now.

"I have been threatened with Homelessness, and I could not tell you such a thing through the Machine."

Homelessness means death. The victim is exposed to the air, which kills him.

"I have been outside since I spoke to you last. The tremendous thing has happened, and they have discovered me."

"But why shouldn't you go outside!" she exclaimed. "It is perfectly legal, perfectly mechanical, to visit the surface of the earth. I have lately been to a lecture on the sea; there is no objection to that; one simply summons a respirator and gets an Egression-permit. It is not the kind of thing that spiritually minded people do, and I begged you not to do it, but there is no legal objection to it."

"I did not get an Egression-permit."

"Then how did you get out?"

"I found out a way of my own."

The phrase conveyed no meaning to her, and he had to repeat it.

"A way of your own?" she whispered. "But that would be wrong."

"Why?"

The question shocked her beyond measure.

"You are beginning to worship the Machine," he said coldly. "You think it irreligious of me to have found out a way of my own. It was just what the Committee thought, when they threatened me with Homelessness."

At this she grew angry. "I worship nothing!" she cried. "I am most advanced. I don't think you irreligious, for there is no such thing as religion left. All the fear and the superstition that existed once have been destroyed by the Machine. I only meant that to find out a way of your own was———Besides, there is no new way out."

"So it is always supposed."

"Except through the vomitories, for which one must have an Egression-permit, it is impossible to get out. The Book says so."

"Well, the Book's wrong, for I have been out on my feet." For Kuno was possessed of a certain physical strength.

By these days it was a demerit to be muscular. Each infant was examined at birth, and all who promised undue strength were destroyed. Humanitarians may protest, but it would have been no true kindness to let an athlete live; he would never have been happy in that state of life to which the Machine had called him; he would have yearned for trees to climb, rivers to bathe in, meadows and hills against which he might measure his body. Man must be adapted to his surroundings, must he not? In the dawn of the world our weakly must be exposed on Mount Taygetus, in its twilight our strong will suffer euthanasia, that the Machine may progress, that the Machine may progress, that the Machine may progress eternally.

"You know that we have lost the sense of space. We say 'space is annihilated,' but we have annihilated not space, but the sense thereof. We have lost a part of ourselves. I determined to recover it, and I began by walking up and down the platform of the railway outside my room. Up and down, until I was tired, and so did recapture the meaning of 'Near' and 'Far.' 'Near' is a place to which I can get quickly *on my feet*, not a place to which the train or the air-ship will take me quickly. 'Far' is a place to which I cannot get quickly on my feet; the vomitory is 'far,' though I could be there in thirty-eight seconds by summoning the train. Man is the measure. That was my first lesson. Man's feet are the measure for distance, his hands are the measure for ownership, his body is the measure for all that is lovable and desirable and strong. Then I went further: it was then that I called to you for the first time, and you would not come.

"This city, as you know, is built deep beneath the surface of the earth, with only the vomitories protruding. Having paced the platform outside my own room, I took the lift to the next platform and paced that also, and so with each in turn, until I came to the topmost, above which begins the earth. All the platforms were exactly alike, and all that I gained by visiting them was to develop my sense of space and my muscles. I think I should have been content with this—it is not a little thing,—but as I walked and brooded, it occurred to me that our cities had been built in the days when men still breathed the outer air, and that there had been ventilation shafts for the workmen. I could think of nothing but these ventilation shafts. Had they been destroyed by all the food-tubes and medicine-tubes and music-tubes that the Machine has evolved lately? Or did traces of them remain? One thing was certain. If I came upon them anywhere, it would be in the railway-tunnels of the topmost story. Everywhere else, all space was accounted for.

"I am telling my story quickly, but don't think that I was not a coward or that your answers never depressed me. It is not the proper thing, it is not mechanical, it is not decent to walk along a railway-tunnel. I did not fear that I might tread upon a live rail and be killed. I feared something far more intangible—doing what was not contemplated by the Machine. Then I said to myself, 'Man is the measure' and I went, and after many visits I found an opening.

"The tunnels, of course, were lighted. Everything is light, artificial light; darkness is the exception. So when I saw a black gap in the tiles, I knew that it was an exception, and rejoiced. I put in my arm—I could put in no more at first—and waved it round and round in ecstasy. I loosened another tile, and put in my head, and shouted into the darkness: 'I am coming, I shall do it yet,' and my voice reverberated down endless passages. I seemed to hear the spirits of those dead workmen who had returned each evening to the starlight and to their wives, and all the generations who had lived in the open air called back to me, 'You will do it yet, you are coming.'"

He paused, and, absurd as he was, his last words moved her. For Kuno had lately asked to be a father, and his request had been refused by the Committee. His was not a type that the Machine desired to hand on.

"Then a train passed. It brushed by me, but I thrust my head and arms into the hole. I had done enough for one day, so I crawled back to the platform, went down in the lift, and summoned my bed. Ah what dreams! And again I called you, and again you refused."

She shook her head and said: "Don't. Don't talk of these terrible things. You make me miserable. You are throwing civilisation away."

"But I had got back the sense of space and a man cannot rest then. I determined to get in at the hole and climb the shaft. And so I exercised my arms. Day after day I went through ridiculous movements, until my flesh ached, and I could hang by my hands and hold the pillow of my bed outstretched for many minutes. Then I summoned a respirator, and started.

"It was easy at first. The mortar had somehow rotted, and I soon pushed some more tiles in, and clambered after them into the darkness, and the spirits of the dead comforted me. I don't know what I mean by that. I just say what I felt. I felt, for the first time, that a protest had been lodged against corruption, and that even as the dead were comforting me, so I was comforting the unborn. I felt that humanity existed, and that it existed without clothes. How can I possibly explain this? It was naked, humanity seemed naked, and all these tubes and buttons and machineries neither came into the world with us, nor will they follow us out, nor do they matter supremely while we are here. Had I been strong, I would have torn off every garment I had, and gone out into the outer air unswaddled. But this is not for me, nor perhaps for my generation. I climbed with my respirator and my hygienic clothes and my dietetic tabloids! Better thus than not at all.

"There was a ladder, made of some primæval metal. The light from the railway fell upon its lowest rungs, and I saw that it led straight upwards out of the rubble at the bottom of the shaft. Perhaps our ancestors ran up and down it a dozen times daily, in their building. As I climbed, the rough edges cut through my gloves so that my hands bled. The light helped me for a little, and then came darkness and, worse still, silence which pierced my ears like a sword. The Machine hums! Did you know that? Its hum penetrates our blood, and may even guide our thoughts. Who knows! I was getting beyond its power. Then I thought: 'This silence means that I am doing wrong.' But I heard voices in the silence, and again they strengthened me." He laughed. "I had need of them. The next moment I cracked my head against something."

She sighed.

"I had reached one of those pneumatic stoppers that defend us from the outer air. You may have noticed them on the air-ship. Pitch dark, my feet on the rungs of an invisible ladder, my hands cut; I cannot explain how I lived through this part, but the voices still comforted me, and I felt for fastenings. The stopper, I suppose, was about eight feet across. I passed my hand over it as far as I could reach. It was perfectly smooth. I felt it almost to the centre. Not quite to the centre, for my arm was too short. Then the voice said: 'Jump. It is worth it. There may be a handle in the centre, and you may catch hold of it and so come to us your own way. And if there is no handle, so that you may fall and are dashed to pieces—it is still worth it: you will still come to us your own way.' So I jumped. There was a handle, and———"

He paused. Tears gathered in his mother's eyes. She knew that he was fated. If he did not die to-day he would die to-morrow. There was not room for such a person in the world. And with her pity disgust mingled. She was ashamed at having borne such a son, she who had always been so respectable and so full of ideas. Was he really the little boy to whom she had taught the use of his stops and buttons, and to whom she had given his first lessons in the Book? The very hair that disfigured his lip showed that he was reverting to some savage type. On atavism the Machine can have no mercy.

"There was a handle, and I did catch it. I hung tranced over the darkness and heard the hum of these workings as the last whisper in a dying dream. All the things I had cared about and all the people I had spoken to through tubes appeared infinitely little. Meanwhile the handle revolved. My weight had set something in motion and I span slowly, and then———

"I cannot describe it. I was lying with my face to the sunshine. Blood poured from my nose and ears and I heard a tremendous roaring. The stopper, with me clinging to it, had simply been blown out of the earth, and the air that we make down here was escaping through the vent into the air above. It burst up like a fountain. I crawled back to it—for the upper air hurts—and, as it were, I took great sips from the edge. My respirator had flown goodness knows where, my clothes were torn. I just lay with my lips close to the hole, and I sipped until the bleeding stopped. You can imagine nothing so curious. This hollow in the grass—I will speak of it in a minute,—the sun shining into it, not brilliantly but through marbled clouds,—the peace, the nonchalance, the sense of space, and, brushing my cheek, the roaring fountain of our artificial air! Soon I spied my respirator, bobbing up and down in the current high above my head, and higher still were many air-ships. But no one ever looks out of air-ships, and in my case they could not have picked me up. There I was, stranded. The sun shone a little way down the shaft, and revealed the topmost rung of the ladder, but it was hopeless trying to reach it. I should either have been tossed up again by the escape, or else have fallen in, and died. I could only lie on the grass, sipping and sipping, and from time to time glancing around me.

"I knew that I was in Wessex, for I had taken care to go to a lecture on the subject before starting. Wessex lies above the room in which we are talking now. It was once an important state. Its kings held all the southern coast from the Andredswald to Cornwall, while the Wansdyke protected them on the north, running over the high ground. The lecturer was only concerned with the rise of Wessex, so I do not know how long it remained an international power, nor would the knowledge have assisted me. To tell the truth I could do nothing but laugh, during this part. There was I, with a pneumatic stopper by my side and a respirator bobbing over my head, imprisoned, all three of us, in a grass-grown hollow that was edged with fern."

Then he grew grave again.

"Lucky for me that it was a hollow. For the air began to fall back into it and to fill it as water fills a bowl. I could crawl about. Presently I stood. I breathed a mixture, in which the air that hurts predominated whenever I tried to climb the sides. This was not so bad. I had not lost my tabloids and remained ridiculously cheerful, and as for the Machine, I forgot about it altogether. My one aim now was to get to the top, where the ferns were, and to view whatever objects lay beyond.

"I rushed the slope. The new air was still too bitter for me and I came rolling back, after a momentary vision of something grey. The sun grew very feeble, and I remembered that he was in Scorpio—I had been to a lecture on that too. If the sun is in Scorpio and you are in Wessex, it means that you must be as quick as you can, or it will get too dark. (This is the first bit of useful information I have ever got from a lecture, and I expect it will be the last.) It made me try frantically to breathe the new air, and to advance as far as I dared out of my pond. The hollow filled so slowly. At times I thought that the fountain played with less vigour. My respirator seemed to dance nearer the earth; the roar was decreasing."

He broke off.

"I don't think this is interesting you. The rest will interest you even less. There are no Ideas in it, and I wish that I had not troubled you to come. We are too different, mother."

She told him to continue.

"It was evening before I climbed the bank. The sun had very nearly slipped out of the sky by this time, and I could not get a good view. You, who have just crossed the Roof of the World, will not want to hear an account of the little hills that I saw—low colourless hills. But to me they were living and the turf that covered them was a skin, under which their muscles rippled, and I felt that those hills had called with incalculable force to men in the past, and that men had loved them. Now they sleep—perhaps for ever. They commune with humanity in dreams. Happy the man, happy the woman, who awakes the hills of Wessex. For though they sleep, they will never die."

His voice rose passionately.

"Cannot you see, cannot all your lecturers see, that it is we that are dying, and that down here the only thing that really lives is the Machine? We created the Machine, to do our will, but we cannot make it do our will now. It has robbed us of the sense of space and of the sense of touch, it has blurred every human relation and narrowed down love to a carnal act, it has paralysed our bodies and our wills, and now it compels us to worship it. The Machine develops—but not on our lines. The Machine proceeds—but not to our goal. We only exist as the blood corpuscles that course through its arteries, and if it could work without us, it would let us die. Oh, I have no remedy—or, at least, only one—to tell men again and again that I have seen the hills of Wessex as Ælfrid saw them when he overthrew the Danes.

"So the sun set. I forgot to mention that a belt of mist lay between my hill and other hills, and that it was the colour of pearl."

He broke off for the second time.

"Go on," said his mother wearily. He shook his head.

"Go on. Nothing that you say can distress me now. I am hardened."

"I had meant to tell you the rest, but I cannot: I know that I cannot: goodbye."

Vashti stood irresolute. All her nerves were tingling with his blasphemies. But she was also inquisitive.

"This is unfair," she complained. "You have called me across the world to hear your story, and hear it I will. Tell me—as briefly as possible, for this is a disastrous waste of time—tell me how you returned to civilisation."

"Oh—that!" he said, starting. "You would like to hear about civilisation. Certainly. Had I got to where my respirator fell down?"

"No—but I understand everything now. You put on your respirator, and managed to walk along the surface of the earth to a vomitory, and there your conduct was reported to the Central Committee."

"By no means."

He passed his hand over his forehead, as if dispelling some strong impression. Then, resuming his narrative, he warmed to it again.

"My respirator fell about sunset. I had mentioned that the fountain seemed feebler, had I not?"

"Yes."

"About sunset, it let the respirator fall. As I said, I had entirely forgotten about the Machine, and I paid no great attention at the time, being occupied with other things. I had my pool of air, into which I could dip when the outer keenness became intolerable, and which would possibly remain for days, provided that no wind sprang up to disperse it. Not until it was too late, did I realise what the stoppage of the escape implied. You see—the gap in the tunnel had been mended; the Mending Apparatus; the Mending Apparatus, was after me.

"One other warning I had, but I neglected it. The sky at night was clearer than it had been in the day, and the moon, which was about half the sky behind the sun, shone into the dell at moments quite brightly. I was in my usual place—on the boundary between the two atmospheres—when I thought I saw something dark move across the bottom of the dell, and vanish into the shaft. In my folly, I ran down. I bent over and listened, and I thought I heard a faint scraping noise in the depths.

"At this—but it was too late—I took alarm. I determined to put on my respirator and to walk right out of the dell. But my respirator had gone. I knew exactly where it had fallen—between the stopper and the aperture—and I could even feel the mark that it had made in the turf. It had gone, and I realised that something evil was at work, and I had better escape to the other air, and, if I must die, die running towards the cloud that had been the colour of a pearl. I never started. Out of the shaft—it is too horrible. A worm, a long white worm, had crawled out of the shaft and was gliding over the moonlit grass.

"I screamed. I did everything that I should not have done, I stamped upon the creature instead of flying from it, and it at once curled round the ankle. Then we fought. The worm let me run all over the dell, but edged up my leg as I ran. 'Help!' I cried. (That part is too awful. It belongs to the part that you will never know.) 'Help!' I cried. (Why cannot we suffer in silence?) 'Help!' I cried. Then my feet were wound together, I fell, I was dragged away from the dear ferns and the living hills, and past the great metal stopper (I can tell you this part), and I thought it might save me again if I caught hold of the handle. It also was enwrapped, it also. Oh, the whole dell was full of the things. They were searching it in all directions, they were denuding it, and the white snouts of others peeped out of the hole, ready if needed. Everything that could be moved they brought—brushwood, bundles of fern, everything, and down we all went intertwined into hell. The last things that I saw, ere the stopper closed after us, were certain stars, and I felt that a man of my sort lived in the sky. For I did fight, I fought till the very end, and it was only my head hitting against the ladder that quieted me. I woke up in this room. The worms had vanished. I was surrounded by artificial air, artificial light, artificial peace, and my friends were calling to me down speaking-tubes to know whether I had come across any new ideas lately."

Here his story ended. Discussion of it was impossible, and Vashti turned to go. "It will end in Homelessness," she said quietly.

"I wish it would," retorted Kuno.

"The Machine has been most merciful."

"I prefer the mercy of God."

"By that superstitious phrase, do you mean that you could live in the outer air?"

"Yes."

"Have you ever seen, round the vomitories, the bones of those who were extruded after the Great Rebellion?"

"Yes."

"They were left where they perished for our edification. A few crawled away, but they perished, too—who can doubt it? And so with the Homeless of our own day. The surface of the earth supports life no longer."

"Indeed."

"Ferns and a little grass may survive, but all higher forms have perished. Has any air-ship detected them?"

"No."

"Has any lecturer dealt with them?"

"No."

"Then why this obstinacy?"

"Because I have seen them," he exploded.

"Seen *what?*"

"Because I have seen her in the twilight—because she came to my help when I called—because she, too, was entangled by the worms, and, luckier than I, was killed by one of them piercing her throat."

He was mad. Vashti departed, nor, in the troubles that followed, did she ever see his face again.

Part III: The Homeless

During the years that followed Kuno's escapade, two important developments took place in the Machine. On the surface they were revolutionary, but in either case men's minds had been prepared beforehand, and they did but express tendencies that were latent already.

The first of these was the abolition of respirators.

Advanced thinkers, like Vashti, had always held it foolish to visit the surface of the earth. Air-ships might be necessary, but what was the good of going out for mere curiosity and crawling along for a mile or two in a terrestrial motor? The habit was vulgar and perhaps faintly improper: it was unproductive of ideas, and had no connection with the habits that really mattered. So respirators were abolished, and with them, of course, the terrestrial motors, and except for a few lecturers, who complained that they were debarred access to their subject-matter, the development was accepted quietly. Those who still wanted to know what the earth was like, had after all only to listen to some gramophone, or to look into some cinematophote. And even the lecturers acquiesced when they found that a lecture on the sea was none the less stimulating when compiled out of other lectures that had already been delivered on the same subject. "Beware of first-hand ideas!" exclaimed one of the most advanced of them. "First-hand ideas do not really exist. They are but the physical impressions produced by love and fear, and on this gross foundation who could erect a philosophy? Let your ideas be second-hand, and if possible tenth-hand, for then they will be far removed from that disturbing element—direct observation. Do not learn anything about this subject of mine—the French Revolution. Learn instead what I think that Enicharmon thought Urizen thought Gutch thought Ho-Yung thought Chi-Bo-Sing thought Lafcadio Hearn thought Carlyle thought Mirabeau said about the French Revolution. Through the medium of these ten great minds, the blood that was shed at Paris and the windows that were broken at Versailles will be clarified to an Idea which you may employ most profitably in your daily lives. But be sure that the intermediates are many and varied, for in history one authority exists to counteract another. Urizen must counteract the scepticism of Ho-Yung and Enicharmon, I must myself counteract the impetuosity of Gutch. You who listen to me are in a better position to judge about the French Revolution than I am. Your descendants will be even in a better position than you, for they will learn what you think I think, and yet another intermediate will be added to the chain. And in time"—his voice rose—"there will come a generation that has got beyond facts, beyond impressions, a generation absolutely colourless, a generation

> 'seraphically free
> From taint of personality,'

which will see the French Revolution not as it happened, nor as they would like it to have happened, but as it would have happened, had it taken place in the days of the Machine."

Tremendous applause greeted this lecture, which did but voice a feeling already latent in the minds of men—a feeling that terrestrial facts must be ignored, and that the abolition of respirators was a positive gain. It was even suggested that air-ships should be abolished too. This was not done, because air-ships had somehow worked themselves into the Machine's system. But year by year they were used less, and mentioned less by thoughtful men.

The second great development was the re-establishment of religion.

This, too, had been voiced in the celebrated lecture. No one could mistake the reverent tone in which the peroration had concluded, and it awakened a responsive echo in the heart of each. Those who had long worshipped silently, now began to talk. They described the strange feeling of peace that came over them when they handled the Book of the Machine, the pleasure that it was to repeat certain numerals out of it, however little meaning those numerals conveyed to the outward ear, the ecstasy of touching a button, however unimportant, or of ringing an electric bell, however superfluously.

"The Machine," they exclaimed, "feeds us and clothes us and houses us; through it we speak to one another, through it we see one another, in it we have our being. The Machine is the friend of ideas and the enemy of superstition: the Machine is omnipotent, eternal; blessed is the Machine." And before long this allocution was printed on the first page of the Book, and in subsequent editions the ritual swelled into a complicated system of praise and prayer. The word "religion" was sedulously avoided, and in theory the Machine was still the creation and the implement of man. But in practice all, save a few retrogrades, worshipped it as divine. Nor was it worshipped in unity. One believer would be chiefly impressed by the blue optic plates, through which he saw other believers; another by the mending apparatus, which sinful Kuno had compared to worms; another by the lifts, another by the Book. And each would pray to this or to that, and ask it to intercede for him with the Machine as a whole. Persecution—that also was present. It did not break out, for reasons that will be set forward shortly. But it was latent, and all who did not accept the minimum known as "undenominational Mechanism" lived in danger of Homelessness, which means death, as we know.

To attribute these two great developments to the Central Committee, is to take a very narrow view of civilisation. The Central Committee announced the developments, it is true, but they were no more the cause of them than were the Kings of the Imperialistic period the cause of war. Rather did they yield to some invincible pressure, which came no one knew whither, and which, when gratified, was succeeded by some new pressure equally invincible. To such a state of affairs it is convenient to give the name of Progress. No one confessed the Machine was out of hand. Year by year it was served with increased efficiency and decreased intelligence. The better a man knew his own duties upon it, the less he understood the duties of his neighbour, and in all the world there was not one who understood the monster as a whole. Those master brains had perished. They had left full directions, it is true, and their successors had each of them mastered a portion of those directions. But Humanity, in its desire for comfort, had over-reached itself. It had exploited the riches of nature too far. Quietly and complacently, it was sinking into decadence, and Progress had come to mean the Progress of the Machine.

As for Vashti, her life went peacefully forward until the final disaster. She made her room dark and slept; she awoke and made the room light. She lectured and attended lectures. She exchanged ideas with her innumerable friends and believed she was growing more spiritual. At times a friend was granted Euthanasia, and left his or her room for the Homelessness that is beyond all human conception. Vashti did not much mind. After an unsuccessful lecture, she would sometimes ask for Euthanasia herself. But the death-rate was not permitted to exceed the birth-rate, and the Machine had hitherto refused it to her.

The troubles began quietly, long before she was conscious of them.

One day she was astonished at receiving a message from her son. They never communicated, having nothing in common, and she had only heard indirectly that he was still alive, and had been transferred from the northern hemisphere, where he had behaved so mischievously, to the southern—indeed, to a room not far from her own.

"Does he want me to visit him?" she thought. "Never again, never. And I have not the time."

No, it was madness of another kind.

He refused to visualise his face upon the blue plate, and speaking out of the darkness with solemnity said:

"The Machine stops."

"What do you say?"

"The Machine is stopping, I know it, I know the signs."

She burst into a peal of laughter. He heard her and was angry, and they spoke no more.

"Can you imagine anything more absurd?" she cried to a friend. "A man who was my son believes that the Machine is stopping. It would be impious if it was not mad."

"The Machine is stopping?" her friend replied. "What does that mean? The phrase conveys nothing to me."

"Nor to me."

"He does not refer, I suppose, to the trouble there has been lately with the music?"

"Oh no, of course not. Let us talk about music."

"Have you complained to the authorities?"

"Yes, and they say it wants mending, and referred me to the Committee of the Mending Apparatus. I complained of those curious gasping sighs that disfigure the symphonies of the Brisbane school. They sound like some one in pain. The Committee of the Mending Apparatus say that it shall be remedied shortly."

Obscurely worried, she resumed her life. For one thing, the defect in the music irritated her. For another thing, she could not forget Kuno's speech. If he had known that the music was out of repair—he could not know it, for he detested music—if he had known that it was wrong, "the Machine stops" was exactly the venomous sort of remark he would have made. Of course he had made it at a venture, but the coincidence annoyed her, and she spoke with some petulance to the Committee of the Mending Apparatus.

They replied, as before, that the defect would be set right shortly.

"Shortly! At once!" she retorted. "Why should I be worried by imperfect music? Things are always put right at once. If you do not mend it at once, I shall complain to the Central Committee."

"No personal complaints are received by the Central Committee," the Committee of the Mending Apparatus replied.

"Through whom am I to make my complaint, then?"

"Through us."

"I complain then."

"Your complaint shall be forwarded in its turn."

"Have others complained?"

This question was unmechanical, and the Committee of the Mending Apparatus refused to answer it.

"It is too bad!" she exclaimed to another of her friends. "There never was such an unfortunate woman as myself. I can never be sure of my music now. It gets worse and worse each time I summon it."

"I too have my troubles," the friend replied. "Sometimes my ideas are interrupted by a slight jarring noise."

"What is it?"

"I do not know whether it is inside my head, or inside the wall."

"Complain, in either case."

"I have complained, and my complaint will be forwarded in its turn to the Central Committee."

Time passed, and they resented the defects no longer. The defects had not been remedied, but the human tissues in that latter day had become so subservient, that they readily adapted themselves to every caprice of the Machine. The sigh at the crisis of the Brisbane symphony no longer irritated Vashti; she accepted it as part of the melody. The jarring noise, whether in the head or in the wall, was no longer resented by her friend. And so with the mouldy artificial fruit, so with the bath water that began to stink, so with the defective rhymes that the poetry machine had taken to emit. All were bitterly complained of at first, and then acquiesced in and forgotten. Things went from bad to worse unchallenged.

It was otherwise with the failure of the sleeping apparatus. That was a more serious stoppage. There came a day when over the whole world—in Sumatra, in Wessex, in the innumerable cities of Courland and Brazil—the beds, when summoned by their tired owners, failed to appear. It may seem a ludicrous matter, but from it we may date the collapse of humanity. The Committee responsible for the failure was assailed by complainants, whom it referred, as usual, to the Committee of the Mending Apparatus, who in its turn assured them that their complaints would be forwarded to the Central Committee. But the discontent grew, for mankind was not yet sufficiently adaptable to do without sleep.

"Some one is meddling with the Machine———" they began.

"Some one is trying to make himself king, to reintroduce the personal element."

"Punish that man with Homelessness."

"To the rescue! Avenge the Machine! Avenge the Machine!"

"War! Kill the man!"

But the Committee of the Mending Apparatus now came forward, and allayed the panic with well-chosen words. It confessed that the Mending Apparatus was itself in need of repair.

The effect of this frank confession was admirable.

"Of course," said a famous lecturer—he of the French Revolution, who gilded each new decay with splendour—"of course we shall not press our complaints now. The Mending Apparatus has treated us so well in the past that we all sympathise with it, and will wait patiently for its recovery. In its own good time it will resume its duties. Meanwhile let us do without our beds, our tabloids, our other little wants. Such, I feel sure, would be the wish of the Machine."

Thousands of miles away his audience applauded. The Machine still linked them. Under the seas, beneath the roots of the mountains, ran the wires through which they saw and heard, the enormous eyes and ears that were their heritage, and the hum of many workings clothed their thoughts in one garment of subserviency. Only the old and the sick remained ungrateful, for it was rumoured that Euthanasia, too, was out of order, and that pain had reappeared among men.

It became difficult to read. A blight entered the atmosphere and dulled its luminosity. At times Vashti could scarcely see across her room. The air, too, was foul. Loud were the complaints, impotent the remedies, heroic the tone of the lecturer as he cried: "Courage! courage! What matter so long as the Machine goes on? To it the darkness and the light are one." And though things improved again after a time, the old brilliancy was never recaptured, and humanity never recovered from its entrance into twilight. There was an hysterical talk of "measures," of "provisional dictatorship," and the inhabitants of Sumatra were asked to familiarise themselves with the workings of the central power station, the said power station being situated in France. But for the most part panic reigned, and men spent their strength praying to their Books, tangible proofs of the Machine's omnipotence. There were gradations of terror—at times came rumours of hope—the Mending Apparatus was almost mended—the enemies of the Machine had been got under—new "nerve-centres" were evolving which would do the work even more magnificently than before. But there came a day when, without the slightest warning, without any previous hint of feebleness, the entire of the communication-system broke down, all over the world, and the world, as they understood it, ended.

Vashti was lecturing at the time and her earlier remarks had been punctuated with applause. As she proceeded the audience became silent, and at the conclusion there was no sound. Somewhat displeased she called to a friend who was a specialist in sympathy. No sound: doubtless the friend was sleeping. And so with the next friend whom she tried to summon, and so with the next, until she remembered Kuno's cryptic remark, "The Machine stops."

The phrase still conveyed nothing. If Eternity was stopping it would of course be set going shortly.

For example, there was still a little light and air—the atmosphere had improved a few hours previously. There was still the Book, and while there was the Book there was security.

Then she broke down, for with the cessation of activity came an unexpected terror—silence.

She had never known silence, and the coming of it nearly killed her—it did kill many thousands of people outright. Ever since her birth she had been surrounded by the steady hum. It was to the ear what artificial air was to the lungs, and agonising pains shot across her head. And scarcely knowing what she did, she stumbled forward and pressed the unfamiliar button, the one that opened the door of her cell.

Now the door of the cell worked on a simple hinge of its own. It was not connected with the central power station, dying far away in France. It opened, rousing immoderate hopes in Vashti, for she thought that the Machine had been mended. It opened, and she saw the dim tunnel that curved far away towards freedom. One look, and then she shrank back. For the tunnel was full of people—she was almost the last in that city to have taken alarm.

People at any time repelled her, and these were nightmares from her worst dreams. People were crawling about, people were screaming, whimpering, gasping for breath, touching each other, vanishing in the dark, and ever and anon being pushed off the platform on to the live rail. Some were fighting round the electric bells, trying to summon trains which could not be summoned. Others were yelling for Euthanasia or for respirators, or blaspheming the Machine. Others stood at the doors of their cells fearing, like herself, either to stop in them or to leave them. And behind all the uproar was silence—the silence which is the voice of the earth and of the generations who have gone.

No—it was worse than solitude. She closed the door again and sat down to wait for the end. The disintegration went on, accompanied by horrible cracks and rumbling. The valves that restrained the Medical Apparatus must have been weakened, for it ruptured and hung hideously from the ceiling. The floor heaved and fell and flung her from her chair. A tube oozed towards her serpent fashion. And at last the final horror approached—light began to ebb, and she knew that civilisation's long day was closing.

She whirled round, praying to be saved from this, at any rate, kissing the Book, pressing button after button. The uproar outside was increasing, and even penetrated the wall. Slowly the brilliancy of her cell was dimmed, the reflections faded from her metal switches. Now she could not see the reading-stand, now not the Book, though she held it in her hand. Light followed the flight of sound, air was following light, and the original void returned to the cavern from which it had been so long excluded. Vashti continued to whirl, like the devotees of an earlier religion, screaming, praying, striking at the buttons with bleeding hands.

It was thus that she opened her prison and escaped—escaped in the spirit: at least so it seems to me, ere my meditation closes. That she escapes in the body—I cannot perceive that. She struck, by chance, the switch that released the door, and the rush of foul air on her skin, the loud throbbing whispers in her ears, told her that she was facing the tunnel again, and that tremendous platform on which she had seen men fighting. They were not fighting now. Only the whispers remained, and the little whimpering groans. They were dying by hundreds out in the dark.

She burst into tears.

Tears answered her.

They wept for humanity, those two, not for themselves. They could not bear that this should be the end. Ere silence was completed their hearts were opened, and they knew what had been important on the earth. Man, the flower of all flesh, the noblest of all creatures visible, man who had once made god in his image, and had mirrored his strength on the constellations, beautiful naked man was dying, strangled in the garments that he had woven. Century after century had he toiled, and here was his reward. Truly the garment had seemed heavenly at first, shot with the colours of culture, sewn with the threads of self-denial. And heavenly it had been so long as it was a garment and no more, so long as man could shed it at will and live by the essence that is his soul, and the essence, equally divine, that is his body. The sin against the body—it was for that they wept in chief; the centuries of wrong against the muscles and the nerves, and those five portals by which we can alone apprehend—glossing it over with talk of evolution, until the body was white pap, the home of ideas as colourless, last sloshy stirrings of a spirit that had grasped the stars.

"Where are you?" she sobbed.

His voice in the darkness said, "Here."

"Is there any hope, Kuno?"

"None for us."

"Where are you?"

She crawled towards him over the bodies of the dead. His blood spurted over her hands.

"Quicker," he gasped, "I am dying—but we touch, we talk, not through the Machine."

He kissed her.

"We have come back to our own. We die, but we have recaptured life, as it was in Wessex, when Ælfrid overthrew the Danes. We know what they know outside, they who dwelt in the cloud that is the colour of a pearl."

"But, Kuno, is it true? Are there still men on the surface of the earth? Is this—this tunnel, this poisoned darkness—really not the end?"

He replied: "I have seen them, spoken to them, loved them. They are hiding in the mist and the ferns until our civilisation stops. To-day they are the Homeless—to-morrow———"

"Oh, to-morrow—some fool will start the Machine again, to-morrow."

"Never," said Kuno, "never. Humanity has learnt its lesson."

As he spoke, the whole city was broken like a honeycomb. An air-ship had sailed in through the vomitory into a ruined wharf. It crashed downwards, exploding as it went, rending gallery after gallery with its wings of steel. For a moment they saw the nations of the dead, and, ere they joined them, scraps of the untainted sky.

2 The Prolongation of Life

Francis Fukuyama

In this chapter of *Our Postmodern Future*, Francis Fukuyama explores the social and psychological implications of technological change. He supposes that we will continue to make progress in using technologies to extend human life span and then speculates about the effects this will have on a range of social arrangements, practices, and ways of thinking. Although the data and theories that Fukuyama cites are a few decades old now, his vision of a future in which individuals live increasingly longer still provides "food for thought." Like Forster, Fukuyama imagines a world in which an important human value is realized—in this case, a preeminent value in Western culture, avoiding death—yet he seems to warn us to "be careful what we wish for." Fukuyama suggests that longevity may radically transform how we think about our lives, one another, and the organization of our society. Increased longevity could wreak havoc on many of our social and political practices as those in power hold on to their positions for much longer periods of time. Fukuyama anticipates the social, ethical, and psychological changes that could accompany prolongation of life. In doing this, he reveals connections between new technologies, demographic change, and social, ethical, and policy issues. His exploration reminds us that because decisions and choices about technology can have far-reaching and powerful implications, it is important to think those choices through before we make them.

Many die too late, and a few die too early. The doctrine sounds strange: "Die at the right time!"

Die at the right time—thus teaches Zarathustra. Of course, how could those who never live at the right time die at the right time? Would that they had never been born! Thus I counsel the superfluous. But even the superfluous still make a fuss about their dying; and even the hollowest nut wants to be cracked.

—Friedrich Nietzsche, *Thus Spoke Zarathustra*, I.21

The third pathway by which contemporary biotechnology will affect politics is through the prolongation of life, and the demographic and social changes that will occur as a result. One of the greatest achievements of twentieth-century medicine in the United States was the raising of life expectancies at birth from 48.3 years for men and 46.3 for women in 1900 to 74.2 for men and 79.9 for women in 2000.[1] This shift, coupled with dramatically falling birthrates in much of the developed world, has already produced a

---

From *Our Posthuman Future: Consequences of the Biotechnology Revolution* (New York: Farrar, Straus and Giroux, 2002), 57–71.

very different global demographic backdrop for world politics, whose effects are arguably being felt already. Based on birth and mortality patterns already in place, the world will look substantially different in the year 2050 than it does today, even if biomedicine fails to raise life expectancies by a single year over that period. The likelihood that there will not be significant advances in the prolongation of life in this period is small, however, and there is some possibility that biotechnology will lead to very dramatic changes.

One of the areas most affected by advances in molecular biology has been gerontology, the study of aging. There are at present a number of competing theories as to why people grow old and eventually die, with no firm consensus as to the ultimate reasons or mechanisms by which this occurs.[2] One stream of theory comes out of evolutionary biology and holds, broadly, that organisms age and die because there are few forces of natural selection that favor the survival of individuals past the age at which they are able to reproduce.[3] Certain genes may favor an individual's ability to reproduce but become dysfunctional at later periods of life. For evolutionary biologists, the big mystery is not why individuals die but why, for example, human females have a long postmenopausal life span. Whatever the explanation, they tend to believe that aging is the result of the interaction of a large number of genes, and that therefore there are no genetic shortcuts to the postponement of death.[4]

Another stream of theory on aging comes out of molecular biology and concerns the specific cellular mechanisms by which the body loses its functionality and dies. There are two types of human cells: germ cells, which are contained in the female ovum and male sperm, and somatic cells, which include the other hundred trillion or so cells that constitute the rest of the body. All cells replicate by cell division. In 1961, Leonard Hayflick discovered that somatic cells had an upper limit in the total number of divisions they could undergo. The number of possible cell divisions decreased with the age of the cell.

There are a number of theories as to why the so-called Hayflick limit exists. The leading one has to do with the accumulation of random genetic damage as cells replicate.[5] With each cellular division, environmental factors like smoke and radiation, as well as chemicals known as free hydroxyl radicals and cellular waste products, can prevent the accurate copying of the DNA from one cell generation to the next. The body has a number of DNA repair enzymes that oversee the copying process and fix transcription problems as they arise, but these fail to catch all mistakes. With continued cell replication, the DNA damage builds up in the cells, leading to faulty protein synthesis and impaired functioning. These impairments are in turn the basis for diseases characteristic of aging, such as arteriosclerosis, heart disease, and cancer.

Another theory that seeks to explain the Hayflick limit is related to telomeres, the noncoding bits of DNA attached to the end of each chromosome.[6] Telomeres act like the leaders in a filmstrip and ensure that the DNA is accurately replicated. Cell division involves the splitting apart of the two strands of the DNA molecule and their reconstitution into complete new copies of the molecule in the daughter cells. But with each cell division, the telomeres get a bit shorter, until they are unable to protect the ends of the DNA strand and the cell, recognizing the short telomeres as damaged DNA, ceases growth. Dolly the sheep, cloned from somatic cells of an adult animal, had the shortened telomeres of an adult rather than the longer ones of a newborn lamb, and presumably will not live as long as a naturally born sibling.

There are three major types of cells that are not subject to the Hayflick limit: germ cells, cancer cells, and certain types of stem cells. The reason these cells can reproduce indefinitely has to do with the presence of an enzyme called telomerase, first isolated in 1989, which prevents the shortening of telomeres. This is what permits the germ line to continue through the generations without end, and is also what lies behind the explosive growth of cancer tumors.

Leonard Guarente of the Massachusetts Institute of Technology reported findings that calorie restriction in yeast increased longevity, through the action of a single gene known as SIR2 (silent information regulator No. 2). The SIR2 gene represses genes that generate ribosomal wastes that build up in yeast cells and lead to their eventual death; low-calorie diets restrict reproduction but are helpful to the functioning of the SIR2 gene. This may provide a molecular explanation for why laboratory rats fed a low-calorie diet live up to 40 percent longer than other rats.[7]

Biologists such as Guarente have suggested that there might someday be a relatively simple genetic route to life extension in humans: while it is not practical to feed people such restricted diets, there may be other ways of enhancing the functioning of the SIR genes. Other gerontologists, such as Tom Kirkwood, assert flatly that aging is the result of a complex series of processes at the level of cells, organs, and the body as a whole, and that there is therefore no single, simple mechanism that controls aging and death.[8]

If a genetic shortcut to immortality exists, the race is already on within the biotech industry to find it. The Geron Corporation has already cloned and patented the human gene for telomerase and, along with Advanced Cell Technology, has an active research program into embryonic stem cells. The latter are cells that make up an embryo at the earliest stages of development, before there has been any differentiation into different types of tissue and organs. Stem cells have the potential to become any cell or tissue in the body, and hence hold the promise of generating entirely new body parts to replace ones worn out through the aging process. Unlike organs transplanted from donors, such cloned body parts will be almost genetically identical to cells in the body into which they are placed, and so presumably free from the kinds of immune reactions that lead to transplant rejection.

Stem cell research represents one of the great frontiers of contemporary biomedical research. It is also hugely controversial as a result of its use of embryos as sources of stem cells—embryos which must be destroyed in the process.[9] The embryos usually come from the extra embryos "banked" by in vitro fertilization clinics. (Once created, stem cell "lines" can be replicated almost indefinitely.) Out of concern that stem cell research would encourage abortion or lead to the deliberate destruction of human embryos, the U.S. Congress imposed a ban on funding from the National Institutes of Health for research that could harm embryos,[10] pushing U.S. stem cell research into the private sector. In 2001 a bitter policy debate exploded in the United States as the Bush administration considered lifting the ban. In the end, the administration decided to permit federally funded research, but only on the sixty or so existing stem cell lines that had already been created.

It is impossible to know at this point whether the biotech industry will eventually be able to come up with a shortcut to the prolongation of life, such as a simple pill that will add another decade or two to people's life spans.[11] Even if this never happens, however, it seems fairly safe to say that the *cumulative* impact of all the biomedical research going on at present will be to further increase life expectancies over time and therefore to continue the trend that has been under way for the last century. So it is not at all premature to think through some of the political scenarios and social consequences that might emerge from demographic trends that are already well under way.

In Europe at the beginning of the eighteenth century, half of all children died before they reached the age of 15. The French demographer Jean Fourastié has pointed out that reaching the age of 52 was then an accomplishment, since only a small minority of the population did so, and that such a person might legitimately consider himself or herself a "survivor."[12] Since most people reached the peak of their productive lives during their 40s and 50s, a huge amount of human potential was wasted. In the 1990s, by contrast, over 83 percent of the population could expect to live to the age of 65, and more than 28 percent would still be alive at age 85.[13]

Increasing life expectancies are only part of the story of what has happened to populations in the developed world by the end of the twentieth century. The other major development has been the dramatic fall in fertility rates. Countries such as Italy, Spain, and Japan have total fertility rates (that is, the average number of children born to a woman in her lifetime) of between 1.1 and 1.5, far below the replacement rate of about 2.2. The combination of falling birthrates and increasing life expectancies has dramatically shifted the age distribution in developed countries. While the median age of the U.S. population was about 19 years in 1850, it had risen to 34 years by the 1990s.[14] This is nothing compared to what will happen in the first half of the twenty-first century. While the median age in the United States will climb to almost 40 by the year 2050, the change will be even more dramatic in Europe and Japan, where rates of immigration and fertility are lower. In the absence of an unanticipated increase in fertility, the demographer Nicholas Eberstadt estimates, based on UN data, that the median age in Germany will be 54, in Japan 56, and in Italy 58.[15] These estimates, it should be noted, do *not* assume any dramatic increases in life expectancies. If only some of the promises of biotechnology for gerontology pan out, it could well be the case that *half* of the populations of developed countries will be retirement age or older by this point.

Up to now, the "graying" of the populations of developed countries has been discussed primarily in the context of the social security liability that it will create. This looming crisis is real enough: Japan, for instance, will go from a situation in which there were four active workers for every retired person at the end of the twentieth century, to one in which there are only two workers per retired person a generation or so down the road. But there are other political implications as well.

Take international relations.[16] While some developing countries have succeeded in approaching or even crossing the demographic transition to subreplacement fertility and declining population growth, as the developed world has, many of the poorer parts of the world, including the Middle East and sub-Saharan Africa, continue to experience high rates of growth. This means that the dividing line between the First and Third Worlds in two generations will be a matter not simply of income and culture but of age as well, with Europe, Japan, and parts of North America having a median age of nearly 60 and their less developed neighbors having median ages somewhere in the early 20s.

In addition, voting age populations in the developed world will be more heavily feminized, in part because more women in the growing elderly cohort will live to advanced ages than men, and in part because of a long-term sociological shift toward greater female political participation. Indeed, elderly women will emerge as one of the most important blocs of voters courted by twenty-first-century politicians.

What this will mean for international politics is of course far from clear, but we do know on the basis of past experience that there are important differences in attitudes toward foreign policy and national security between women and men, and between older and younger voters. American women, for example, have always been less supportive than American men of U.S. involvement in war, by an average margin of seven to nine percentage points. They are also consistently less supportive of defense spending and the use of force abroad. In a 1995 Roper survey conducted for the Chicago Council on Foreign Relations, men favored U.S. intervention in Korea in the event of a North Korean attack by a margin of 49 to 40 percent, while women were opposed by a margin of 30 to 54. Fifty-four percent of men felt that it was important to maintain superior worldwide military power, compared with only 45 percent of women. Women, moreover, are less likely than men to see force as a legitimate tool for resolving conflicts.[17]

Developed countries will face other obstacles to the use of force. Elderly people, and particularly elderly women, are not the first to be called to serve in military organizations, so the pool of available military manpower will shrink. The willingness of people in such societies to tolerate battle casualties among their young may fall as well.[18] Nicholas Eberstadt estimates that given current fertility trends, Italy in 2050 will be a society in which only 5 percent of all children have any collateral relatives (that is, brothers, sisters, aunts, uncles, cousins, and so forth) at all. People will be primarily related to their parents, grandparents, great-grandparents, and to their own offspring. Such a tenuous generational line is likely to increase the reluctance to go to war and accept death in battle.

The world may well be divided, then, between a North whose political tone is set by elderly women, and a South driven by what Thomas Friedman labels superempowered angry young men. It was a group of such men that carried out the September 11 attacks on the World Trade Center. This does not, of course, mean that the North will fail to rise to challenges posed by the South, or that conflict between the two regions is inevitable. Biology is not destiny. But politicians will have to work within frameworks established by basic demographic facts, and one of those facts may be that many countries in the North will be both shrinking and aging.

There is another, perhaps more likely, scenario that will bring these worlds into direct contact: immigration. The estimates of falling populations in Europe and Japan given above assume no large increases in net immigration. This is unlikely, however, simply because developed countries will want economic growth and the population necessary to sustain it. This means that the North-South divide will be replicated within each country, with an increasingly elderly native-born population living alongside a culturally different and substantially younger immigrant population. The United States and other English-speaking countries have traditionally been good at assimilating culturally diverse groups of immigrants, but other countries, such as Germany and Japan, have not. Europe has already seen the rise of anti-immigrant backlash movements, such as the National Front in France, the Vlaams Blok in Belgium, the Lega Lombarda in Italy, and Jörg Haider's Freedom Party in Austria. For these countries, changes in the age structure of their populations, abetted by increasing longevity, are likely to lay the ground for growing social conflict.

The prolongation of life through biotechnology will have dramatic effects on the internal structures of societies as well. The most important of these has to do with the management of social hierarchies.

Human beings are by nature status-conscious animals who, like their primate cousins, tend from an early age to arrange themselves in a bewildering variety of dominance hierarchies.[19] This hierarchical behavior is innate and has easily survived the arrival of modern ideologies like democracy and socialism that purport to be based on universal equality. (One has only to look at pictures of the politburos of the former Soviet Union and China, where the top leadership is arrayed in careful order of dominance.) The nature of these hierarchies has changed as a result of cultural evolution, from traditional ones based on physical prowess or inherited social status, to modern ones based on cognitive ability or education. But their hierarchical nature remains.

If one looks around at a society, one quickly discovers that many of these hierarchies are age-graded. Sixth graders feel themselves superior to fifth graders and dominate the playground if both have recess together; tenured professors lord it over untenured ones and carefully control entry into their august circle. Age-graded hierarchies make functional sense insofar as age is correlated in many societies with physical prowess, learning, experience, judgment, achievement, and the like. But past a certain age, the correlation between age and ability begins to go in the opposite direction. With life expectancies only in the 40s or 50s for most of human history, societies could rely on normal generation succession to take care of this problem. Mandatory retirement ages came into vogue only in the late nineteenth century, when increasing numbers of people began to survive into old age.[20]

Life extension will wreak havoc with most existing age-graded hierarchies. Such hierarchies traditionally assume a pyramidal structure because death winnows the pool of competitors for the top ranks, abetted by artificial constraints such as the widely held belief that everyone has the "right" to retire at age 65. With people routinely living and working into their 60s, 70s, 80s, and even 90s, however, these pyramids will increasingly resemble squat trapezoids or even rectangles. The natural tendency of one generation to get out of the way of the up-and-coming one will be replaced by the simultaneous existence of three, four, even five generations.

We have already seen the deleterious consequences of prolonged generational succession in authoritarian regimes that have no constitutional requirements limiting tenure in office. As long as dictators like Francisco Franco, Kim Il Sung, and Fidel Castro physically survive, their societies have no way of replacing them, and all political and social change is effectively on hold until they die.[21] In the future, with technologically enhanced life spans, such societies may find themselves locked in a ludicrous deathwatch not for years but for decades.

In societies that are more democratic and/or meritocratic, there are institutional mechanisms for removing leaders, bosses, or CEOs who are past their prime. But the problem does not go away by any stretch of the imagination.

The root problem lies, of course, in the fact that people at the top of social hierarchies generally do not want to lose status or power and will often use their considerable influence to protect their positions. Age-related declines in capabilities have to be fairly pronounced before other people will go to the trouble of removing a leader, boss, ballplayer, professor, or board member. Impersonal formal rules like mandatory retirement ages are useful precisely because they don't require institutions to make nuanced personal judgments about an individual older person's capability. But impersonal rules often discriminate against older people who are perfectly capable of continuing to work and for that reason have been abolished in many American workplaces.

There is at present a tremendous amount of political correctness regarding age: *ageism* has entered the pantheon of proscribed prejudices, next to racism, sexism, and homophobia. There is of course discrimination against older people, particularly in a youth-obsessed society like that of the United States. But there are also a number of reasons why generational succession is a good thing. Chief among them is that it is a major stimulant of progress and change.

Many observers have noted that political change often occurs at generational intervals—from the Progressive Era to the New Deal, from the Kennedy years to

Reaganism.[22] There is no mystery as to why this is so: people born in the same age cohort experience major life events—the Great Depression, World War II, or the sexual revolution—together. Once people's life views and preferences have been formed by these experiences, they may adapt to new circumstances in small ways, but it is very difficult to get them to change broad outlooks. A black person who grew up in the old South has a hard time seeing a white cop as anything but an untrustworthy agent of an oppressive system of racial segregation, regardless of whether this makes sense given the realities of life in a northern city. Those who lived through the Great Depression cannot help feeling uneasy at the lavish spending habits of their grandchildren.

This is true not just in political but in intellectual life as well. There is a saying that the discipline of economics makes progress one funeral at a time, which is unfortunately truer than most people are willing to admit. The survival of a basic "paradigm" (for example, Keynesianism or Friedmanism) that shapes the way most scientists and intellectuals think about things at a particular time depends not just on empirical evidence, as some would like to think, but on the physical survival of the people who created that paradigm. As long as they sit on top of age-graded hierarchies like peer review boards, tenure committees, and foundation boards of trustees, the basic paradigm will often remain virtually unshakable.

It stands to reason, then, that political, social, and intellectual change will occur much more slowly in societies with substantially longer average life spans. With three or more generations active and working at the same time, the younger age cohorts will never constitute more than a small minority of voices clamoring to be heard, and generational change will never be fully decisive. To adjust more rapidly, such societies will have to establish rules mandating constant retraining and downward social mobility at later stages in life. The idea that one can acquire skills and education during one's 20s that will remain useful for the next forty years is implausible enough at present, given the pace of technological change. The idea that these skills would remain relevant over working lives of fifty, sixty, or seventy years becomes even more preposterous. Older people will have to move down the social hierarchy not just to retrain but to make room for new entrants coming up from the bottom. If they don't, generational warfare will join class and ethnic conflict as a major dividing line in society. Getting older people out of the way of younger ones will become a significant struggle, and societies may have to resort to impersonal, institutionalized forms of ageism in a future world of expanded life expectancies.

Other social effects of life extension will depend heavily on the exact way that the geriatric revolution plays itself out—that is, whether people will remain physically and mentally vigorous throughout these lengthening life spans, or whether society will increasingly come to resemble a giant nursing home.

The medical profession is dedicated to the proposition that anything that can defeat disease and prolong life is unequivocally a good thing. The fear of death is one of the deepest and most abiding human passions, so it is understandable that we should celebrate any advance in medical technology that appears to put death off. But people worry about the quality of their lives as well—not just the quantity. Ideally, one would like not merely to live longer but also to have one's different faculties fail as close as possible to when death finally comes, so that one does not have to pass through a period of debility at the end of life.

While many medical advances have increased the quality of life for older people, many have had the opposite effect by prolonging only one aspect of life and increasing dependency. Alzheimer's disease—in which certain parts of the brain waste away, leading to loss of memory and eventually dementia—is a good example of this, because the likelihood of getting it rises proportionately with age. At age 65, only one person in a hundred is likely to come down with Alzheimer's; at 85, it is one in six.[23] The rapid growth in the population suffering from Alzheimer's in developed countries is thus a direct result of increased life expectancies, which have prolonged the health of the body without prolonging resistance to this terrible neurological disease.

There are in fact two periods of old age that medical technology has opened up, at least for people in the developed world.[24] Category I extends from age 65 until sometime in one's 80s, when people can increasingly expect to live healthy and active lives, with enough resources to take advantage of them. Much of the happy talk about increased longevity concerns this period, and indeed the emergence of this new phase of life as a realistic expectation for most people is an achievement of which modern medicine can be proud. The chief problem for people in this category will be the encroachment of working life on their domain: for simple economic reasons, there will be powerful pressures to raise retirement ages and keep the over-65 cohort in the workforce for as long as possible. This does not imply any kind of social disaster: older workers may have to retrain and accept some degree of downward social mobility, but many of them will welcome the opportunity to contribute their labor to society.

The second phase of old age, Category II, is much more problematic. It is the period that most people currently reach by their 80s, when their capabilities decline and they return increasingly to a childlike state of dependency. This is the period that society doesn't like to think about, much less experience, since it flies in the face of ideals of personal autonomy that most people hold dear. Increases in the number of people in both Category I and Category II have created a novel situation in which individuals approaching retirement age today find their own choices constrained by the fact that they still have an elderly parent alive and dependent on them for care.

The social impact of ever-increasing life expectancies will depend on the relative sizes of these two groups, which in turn will depend on the "evenness" of future life-prolonging advances. The best scenario would be one in which technology simultaneously pushes back parallel aging processes—for instance, by the discovery of a common molecular source of aging in all somatic cells, and the delaying of this process throughout the body. Failure of the different parts would come at the same time, just later; people in Category I would be more numerous and those in Category II less so. The worst scenario would be one of highly uneven advance, in which, for example, we found ways to preserve bodily health but could not put off age-related mental deterioration. Stem cell research might yield ways to grow new body parts, as William Haseltine . . . suggest[ed]. . . . But without a parallel cure for Alzheimer's disease, this wonderful new technology would do no more than allow more people to persist in vegetative states for years longer than is currently possible.

An explosion in the number of people in Category II might be labeled the national nursing home scenario, in which people routinely live to be 150 but spend the last fifty years in a state of childlike dependence on caretakers. There is of course no way of predicting whether this or the happier extension of the Category I period will play itself out. If there is no molecular shortcut to postponing death because aging is the result of the gradual accumulation of damage to a wide range of different biological systems, then there is no reason to think that future medical advances will proceed with a neat simultaneity, any more than they have in the past. That existing medical technology is capable only of keeping people's bodies alive at a much reduced quality of life is the reason assisted suicide and euthanasia, as well as figures like Jack Kevorkian, have come to the fore as public issues in the United States and elsewhere in recent years.

In the future, biotechnology is likely to offer us bargains that trade off length of life span for quality of life. If they are accepted, the social consequences could be dramatic. But assessing them will be very difficult: slight changes in mental capabilities such as loss of short-term memory or growing rigidity in one's beliefs are inherently difficult to measure and evaluate. The political correctness about aging noted earlier will make a truly frank assessment nearly impossible, both for individuals dealing with elderly relatives and for societies trying to formulate public policies. To avoid any hint of discrimination against older people, or the suggestion that their lives are somehow worth less than those of the young, anyone who writes on the future of aging feels compelled to be relentlessly sunny in predicting that medical advances will increase both the quantity and quality of life.

This is most evident with regard to sexuality. According to one writer on aging, "One of the factors inhibiting sexuality with ageing is undoubtedly the brain-washing that all of us experience which says that the older person is less sexually attractive."[25] Would that sexuality were only a matter of brainwashing! Unfortunately, there are good Darwinian reasons that sexual attractiveness is linked to youth, particularly in women. Evolution has created sexual desire for the purpose of fostering reproduction, and there are few selective pressures for humans to develop sexual attraction to partners past their prime reproductive years.[26] The consequence is that in another fifty years, most developed societies may have become "postsexual," in the sense that the vast majority of their members will no longer put sex at the top of their "to do" lists.

There are a number of unanswerable questions about what life in this kind of future would be like, since there have never in human history been societies with median ages of 60, 70, or higher. What would such a society's self-image be? If you go to a typical airport newsstand and look at the people pictured on magazine covers, their median age is likely to be in the low 20s, the vast majority good-looking and in perfect health. For most historical human societies, these covers would have reflected the actual median age, though not the looks or health, of the society as a whole. What will magazine covers look like in another couple of generations, when people in their early 20s constitute only a tiny minority of the population? Will society still want to think of itself as young, dynamic, sexy, and healthy, even though the image departs from the reality that people see around them to an even more extreme degree than today? Or will tastes and habits shift, with the youth culture going into terminal decline?

A shift in the demographic balance toward societies with a majority of people in Categories I and II will have much more profound implications for the meaning of life and death as well. For virtually all of human history up to the present, people's lives and identities were bound up either with reproduction—that is, having families and raising children—or with earning the resources to support themselves and their families. Family and work both enmesh individuals in a web of social obligations over which they frequently have little control and which are a source of struggle and anxiety but also of tremendous satisfaction. Learning to meet those social obligations is a source of both morality and character. People in Categories I and II, by contrast, will have a much more attenuated relationship to both family and work. They will be beyond reproductive years, with links primarily to ancestors and descendants. Some in Category I may choose to work, but the obligation to work and the kinds of mandatory social ties that work engenders will be replaced largely by a host of elective occupations. Those in

Category II will not reproduce, not work, and indeed will see a flow of resources and obligation moving one way: toward them.

This does not mean that people in either category will suddenly become irresponsible or footloose. It does mean, however, that they may find their lives both emptier and lonelier, since it is precisely those obligatory ties that make life worth living for many people. When retirement is seen as a brief period of leisure following a life of hard work and struggle, it may seem like a well-deserved reward; if it stretches on for twenty or thirty years with no apparent end, it may seem simply pointless. And it is hard to see how a prolonged period of dependency or debility for people in Category II will be experienced as joyful or fulfilling.

People's relationship to death will change as well. Death may come to be seen not as a natural and inevitable aspect of life, but a preventable evil like polio or the measles. If so, then accepting death will appear to be a foolish choice, not something to be faced with dignity or nobility. Will people still be willing to sacrifice their lives for others, when their lives could potentially stretch out ahead of them indefinitely, or condone the sacrifice of the lives of others? Will they cling desperately to the life that bio-technology offers? Or might the prospect of an unendingly empty life appear simply unbearable?

Notes

1. See http://www.demog.berkeley.edu/~andrew/1918/figure2.html for the 1900 figures and https://www.cia.gov/library/publications/the-world-factbook/index.html.

2. For an overview of these theories, see Michael R. Rose, *Evolutionary Biology of Aging* (New York: Oxford University Press, 1991), p. 160ff.; Caleb E. Finch and Rudolph E. Tanzi, "Genetics of Aging," *Science* 278 (1997): 407–411; S. Michal Jazwinski, "Longevity, Genes, and Aging," *Science* 273 (1996): 54–59; and David M. A. Mann, "Molecular Biology's Impact on Our Understanding of Aging," *British Medical Journal* 315 (1997): 1078–1082.

3. Michael R. Rose, "Finding the Fountain of Youth," *Technology Review* 95, no. 7 (October 1992): 64–69.

4. Nicholas Wade, "A Pill to Extend Life? Don't Dismiss the Notion Too Quickly," *New York Times*, September 22, 2000, p. A20.

5. Tom Kirkwood, *Time of Our Lives: Why Ageing Is Neither Inevitable nor Necessary* (London: Phoenix, 1999), pp. 100–117.

6. Dwayne A. Banks and Michael Fossel, "Telomeres, Cancer, and Aging: Altering the Human Life Span," *Journal of the American Medical Association* 278, no. 16 (1997): 1345–1348.

7. Nicholas Wade, "Searching for Genes to Slow the Hands of Biological Time," *New York Times*, September 26, 2000, p. D1; Cheol-Koo Lee et al., "Gene Expression Profile of Aging and Its Retardation by Caloric Restriction," *Science* 285 (1999): 1390–1393.

8. Kirkwood, *Time of Our Lives*, p. 166.

9. For a sample of the discussion on stem cells, see Eric Juengst and Michael Fossel, "The Ethics of Embryonic Stem Cells—Now and Forever, Cells without End," *Journal of the American Medical Association* 284, no. 6 (2000): 3180–3184; Juan de Dios Vial Correa and S. E. Mons. Elio Sgreccia, *Declaration on the Production and the Scientific and Therapeutic Use of Human Embryonic Stem Cells* (Rome: Pontifical Academy for Life, 2000); and M. J. Friedrich,

"Debating Pros and Cons of Stem Cell Research," *Journal of the American Medical Association* 284, no. 6 (2000): 681–684.

10. Gabriel S. Gross, "Federally Funding Human Embryonic Stem Cell Research: An Administrative Analysis," *Wisconsin Law Review* 2000 (2000): 855–884.

11. For some research strategies into therapies for aging, see Michael R. Rose, "Aging as a Target for Genetic Engineering," in *Engineering the Human Germline: An Exploration of the Science and Ethics of Altering the Genes We Pass to Our Children*, ed. Gregory Stock and John Campbell (New York: Oxford University Press, 2000), pp. 53–56.

12. Jean Fourastié, "De la vie traditionelle à la vie tertiaire," *Population* 14 (1963): 417–432.

13. Kirkwood, *Time of Our Lives*, p. 6.

14. "Resident Population Characteristics—Percent Distribution and Median Age, 1850–1996, and Projections, 2000–2050," www.doi.gov/nrl/statAbst/Aidemo.pdt.

15. Nicholas Eberstadt, "World Population Implosion?," *Public Interest*, no. 129 (February 1997): 3–22.

16. On this issue, see Francis Fukuyama, "Women and the Evolution of World Politics," *Foreign Affairs* 77 (1998): 24–40.

17. Pamela J. Conover and Virginia Sapiro, "Gender, Feminist Consciousness, and War," *American Journal of Political Science* 37 (1993): 1079–1099.

18. Edward N. Luttwak, "Toward Post-Heroic Warfare," *Foreign Affairs* 74 (1995): 109–122.

19. For a longer discussion of this, see Francis Fukuyama, *The Great Disruption: Human Nature and the Reconstitution of Social Order* (New York: Free Press, 1999), pp. 212–230.

20. Otto von Bismarck, who established Europe's first social security system, set retirement at 65, an age to which virtually no one at that time lived.

21. This point is made by Fred Charles Iklé, "The Deconstruction of Death," *National Interest*, no. 62 (Winter 2000–2001): 87–96.

22. Generational change is the theme, inter alia, of Arthur M. Schlesinger Jr.'s *Cycles of American History* (Boston: Houghton Mifflin, 1986); see also William Strauss and Neil Howe, *The Fourth Turning: An American Prophecy* (New York: Broadway Books, 1997).

23. Kirkwood, *Time of Our Lives*, pp. 131–132.

24. Michael Norman, "Living Too Long," *New York Times Magazine*, January 14, 1996, pp. 36–38.

25. Kirkwood, *Time of Our Lives*, p. 238.

26. On the evolution of human sexuality, see Donald Symons, *The Evolution of Human Sexuality* (Oxford: Oxford University Press, 1979).

3 Reproductive Ectogenesis: The Third Era of Human Reproduction and Some Moral Consequences

Stellan Welin

In this piece, Professor Stellan Welin follows a trend in current research to imagine the implications it might have in the future. Welin analyzes the possibility that success in human xenotransplantation—the transplantation of an organ from an animal to a human—could lead to technology that enables ectogenesis for human embryos—the ability to develop embryos outside a human womb. In essence Welin embeds a short science fiction story into an academic article to help the reader imagine the likelihood that it will not simply be scientists and engineers who shape the meaning of new discoveries and abilities but rather social and cultural pressures will have an enormous impact, leading to profound social changes and raising issues that are difficult to comprehend. He argues that while such a technology might help to resolve certain ethical issues, the ability to bring embryos to maturation in such a way would, at the same time, give rise to ethical questions that have never before existed. Such a technology would void many of the assumptions that underlie the ethical decisions we've made about the rights of fetuses, the rights of expectant mothers, and the role of governments. Whether one agrees or disagrees with Welin on the likelihood of such technologies, his account is important for illustrating the profound social, political, and ethical changes that can accompany a technological change.

---

ectogenesis n. Biol. the production of structures outside the organism
—*The Concise Oxford Dictionary*

It is very hard to seriously discuss future issues in science and their moral implications. Usually, one is trapped in one of two embarrassing positions. The first is *the science fiction trap*. To discuss ectogenesis today will very easily elicit the simple response: "This is not something that is going to happen for many years—if ever. It is at present pure science fiction." The conclusion drawn from this is simply that the discussion is pointless. It should wait until this is a serious issue.

However, following the cloning of Dolly it is difficult to know what is science fiction and what is just around the corner. In a (then) seminal paper, Davor Solter and Jim McGrath declared that cloning of mammals was biologically impossible.[1] Then came Dolly in 1997 and Solter has obviously changed his mind.[2] Science fiction seems to be moving quickly towards science.

The other trap is that *everything has already happened*. If some new biomedical technology is already extensively used in the clinic, saving life, it is very hard to radically

---

From *Science and Engineering Ethics* 10, no. 4 (October 2004): 615–626. Reprinted with permission from Springer Verlag.

question that. The most one can do is to modify practices and avoid, hopefully, the worst abuses. At present, the technology of In Vitro Fertilisation (IVF) is in this state.

In the discussion on human embryonic stem cells and their possibilities, there has been much discussion about the destruction of embryos. President Bush's decision on 9 August 2001 is an example.[3] While he bans destruction of human embryos as a means of deriving stem cells in the federal sector—but allows it in the private sector—he does not question the practice of IVF. In most countries IVF is a closed discussion, but human embryonic stem cells are still open for discussion.

In this paper I briefly comment on the three eras of human reproduction—and primarily on the relation between the new individual and the woman—and then spend some time on a fictional story illustrating some moral consequences of the third era. I will also comment on some of the possible consequences—moral, social and psychological—and try to answer the question on the desirability of the third era.

Using Fictional Stories to Explore Conceptual and Moral Problems

The Parfit Teletransporter

Being a philosopher by training I have become used to fantastic stories which illustrate some philosophical points and give arguments for various positions. One famous example in this semi-fictional and partly science-fiction style is Derek Parfit's story about teletransportation as a background to a discussion of personal identity.[4] In first person narrative Parfit tells how he would use a teletransporter to go from earth to Mars, a seemingly standard procedure. The teletransporter would separate the atoms in his body, beam the information to Mars, and set the atoms together. In Parfit's story something goes wrong. He enters the teletransporter, presses the button but does not as expected lose consciousness. To cut the story short, the teletransporter had worked in a new way. The information had been beamed to Mars where a replica appeared of the traveller still on earth. There is now one traveller on earth and one very similar traveller on Mars.

Parfit goes on to discuss personal identity. Who is who? There seems to be two very similar persons. Had the traveller's body disappeared on earth there would have been only one person. No one would have questioned that the person on Mars really was the traveller. In Parfit's story, the traveller on earth is told a little later that his heart had been damaged by the malfunction of the teletransporter and that he will shortly die. He talks to the traveller on Mars who assures the traveller on earth that he truly loves his wife (the same for both) and will carry on the same work as was important for the traveller on earth. As a reader I feel rather scared. The traveller on earth is a completely redundant person. No one will miss him when he dies. Strangely, somehow Parfit seems to regard the story as an antidote to being afraid of death.

What Parfit gives us is a disconnection between two entities that normally (in real life) travel together through space and time, namely your personal identity consisting of both mind and body. Realising the possibility of separation, (or in Parfit's case of doubling) even if it might never happen in real life, new questions will arise and cast doubt on old solutions.

Separating the Woman and the Fetus

In human reproduction, if we focus on the two "entities" that exist in space and time, namely the fetus/future child and the woman (leaving the man out of the picture), we can envisage different relations between them, corresponding to three different eras of human reproduction.

Historically, the first era is the normal conception inside the woman, the growth of the fetus in the womb, after nine months—birth, and the appearance of a new individual. The second era is In Vitro Fertilisation (IVF). The fetus starts outside the woman as a fertilised egg, moves to the body of the woman and spends nine months there. The body of the woman and the fetus travel together in space-time to separate at birth. In the third era of reproductive ectogenesis, the two never travel together. The fetus spends its gestational time entirely outside the woman's body. We have two entities separated in space-time continuously. The intimate connection consisting of the fetus being inside the woman's body is gone.

Obviously, new questions can be asked in the third era. Old answers depending on the intimate bodily connection between fetus and the woman will no longer be valid. This might, for example, have consequences for the morality of abortion, the rights of the fetus and might radically change parenthood. For example, Rosemarie Tong writes "Most feminists believe that the abortion debate should centre not on the question of whether fetuses are the moral equivalent of adult persons but, rather, on the fact that fertilized eggs develop into infants inside wombs of women."[5] In the third era (ectogenesis), the fertilised egg does not any longer develop inside the womb.

There are many well-known philosophical stories, not just Parfit's. We have Robert Nozick's story of the experience machine as some kind of argument that it is not just our experiences that matter,[6] Jonathan Glover on the moral (un)importance of the "dreamworld,"[7] Daniel Dennet on the (un)importance of the body for the location of the self,[8] and Judith Jarvis Thomson's abortion parable of the case when some one has been connected to and gives essential life support to another person for nine months.[9] There also have been stories of the future written by scientists and social scientists. My favourites are J. B. S. Haldane from 1923, where he discusses (the future) ectogenes,[10] and the sociological fiction of Michael Young on the coming of meritocracy.[11] In all humility I will give you mine.

The Future Scenario of Pig Pharmaceuticals Limited

The Reproductive Engineer's View

It is the year 2050 and the conference "Frontiers of medicine" is having its annual meeting. There is a special session celebrating the 20th anniversary of successful clinical applications of xenotransplantation. Gone are the shortage of organs for transplantation and the industrial-medical farm business is flourishing. After having listened to all the tales of heroes—actually it was never the surgeons who were heroic; all the suffering and the failures fell on the patients—the audience finally listens to another talk.

There is a special lecture on xenotransplantation and human reproduction. The chief scientific director of Pig Pharmaceuticals mounts the rostrum. Expectations are high; the firm has advertised that a new development will be announced at this meeting. Everyone attending the session is also invited to the evening barbecue at the poolside as an extra temptation to achieve maximum attendance.

> Ladies and gentlemen, Mr. Chairman, It is a great honour for me as chief director of scientific research at Pig Pharmaceuticals to appear at such a distinguished session as this. I wish to present to you a new project, and I hope we will have a fruitful discussion. The aim of Pig Pharmaceuticals is to serve humanity by developing science-based new medical technologies related to the medical use of pigs.
>
> As you all know xenotransplantation made its breakthrough at the beginning of this century. It was a combination of genetically modified pigs, new immunosuppressive drugs and more effective methods of handling rejection episodes which laid the

groundwork. Gone are the days when recipients had to wait for someone to die, when relatives had to be approached in their mourning and asked about donations. No longer must anyone make the agonising choice of who should have the organ. There are organs for everyone. Fortunately, the once widely-discussed risk of infectious disease caused by the retroviruses of the pig never materialised.

A picture showing the increasing curves for post-operative survival for xenotransplantation is projected on the screen. In the background happy pigs can be seen.

It is most appropriate that this session has paid tribute to the many pioneers, both physicians and patients, who made all this happen. However, I want to spend my time presenting our latest work in reproductive medicine.

Pregnancy is a difficult period for many women. Our present project started when some women who had had many consecutive miscarriages contacted our researchers. In some cases there had been In Vitro Fertilisation, but the pregnancies were never carried to term. These women wondered if it would be possible to save the fetuses by letting the human embryo develop in the uterus of our pigs. When the question first was raised, we were not sure if this was technically feasible, nor were we at Pig Pharmaceuticals certain of the ethical acceptability.

Our scientists were able to answer the technical question rather quickly. Such a pig-related pregnancy, as we now call it, is possible. In principle, it is the same kind of problem as in the xeno case.

The pregnancies of pigs are shorter than human pregnancies. And pigs carry many fetuses. But as you all know, today we are very good at regulating the length of pregnancy. As the pig-related pregnancy would be handled by our competent personnel there would obviously be no problem with the length of the pregnancy, and naturally no porcine fetuses need to coexist with the human fetus in the porcine uterus. However, even if the technical medical problems could be overcome, what about ethics and social acceptability?

Pig Pharmaceuticals arranged many meetings with distinguished ethicists. They soon convinced us that it would not be possible to get ethical permission anywhere in the world to carry out the research necessary. The fundamental problem was that it would not be in the interests of a human embryo conceived by In Vitro Fertilisation to be implanted into a pig; it would have a much better chance in a human uterus. It would be even more difficult to obtain human embryos for the research into the early phases of the process, that is, research that would involve destruction of embryos, especially if we could not reveal our long-term goals. Therefore, any carefully planned research with human embryos in the porcine uterus was out of the question.

As you all know, the traditional regulation of medical research does not cover the use of new innovative techniques to save the life or restore the health of the patient. This is clinical care, not research, as is clear from the Helsinki Declaration of 1964 and onwards. We were given an opening for the project when a physician contacted us with a pregnant patient who had a history of serious early problems with pregnancies. It was the third month, complications were developing, and the risk was great that the fetus might be seriously injured by certain reactions developed by the woman. She objected to abortion on religious grounds. The doctor thought it was his duty to help the woman and the developing fetus. He could, obviously, not overrule her objections to abortion. He therefore contacted us.

One of our pigs was treated by gene therapy and the necessary operation was performed. The fetus was transferred from the human uterus to a porcine one. Of course, it was of great help that the similar technique for transferring fetuses to surrogate mothers is such a well-established technique. Naturally it was all done in the greatest privacy. Six months later the child was delivered by Caesarean section.

A picture of a baby girl is projected onto the screen. Next a round-bellied pig is seen. A storm of applause. The audience is standing up and cheering.

Yes, success is a wonderful thing. Today more than 40 pig-related pregnancies are under way. With your help Pig Pharmaceuticals hopes to be able to develop this new therapeutic tool. We also believe that pig-related pregnancies will be a big gain for society, a help for women and—as this will be explained by our ethicist—a matter of real moral progress. I thank you for your attention and wish you all welcome to our barbecue. And now I give the word to the staff ethicist of Pig Pharmaceuticals.

More applause. The audience stands up again. Eventually, calm returns as the ethicist enters the rostrum.

The Ethicist's View

Ladies and gentlemen. As you have just heard, great advances have been made in human reproduction. As always, when science and technology advance, new possibilities appear and new moral problems. But there are also moral gains. I will argue in my short talk that pig-related pregnancies constitute a very great moral gain for women, for fetuses and for society—and also for pigs.

Let me make a parallel with xenotransplantation. The serious shortage of human organs for transplantation in the beginning of this century caused the public debate to move to a slightly distributive standpoint. Is it my duty to donate one of my kidneys, a part of my liver while still alive? Should the state intervene and force us, as it forces us to pay taxes? Progress in xenotransplantation ended this moral and political discussion. There was no longer any need. Xenotransplantation had done away with the conflict between the interest of the bearer of the organ and the duty of helping a fellow human being in desperate need of the organ or part of it. There are now enough organs for everyone. Unfortunately, no one has come up with a similar solution that would make the tax system redundant.

The ethicist looks up. The audience smiles obediently, although the joke was not very funny. But what can you expect of an ethicist?

I will discuss the serious conflict between the interests of the pregnant woman and the fetus. I will also consider the interests of society and of the pigs. All these interests are legitimate and important. We have a moral obligation to give equal considerations to interests.

Let me start with the child. It has an interest in having a good life. We have an educational system, a social care system, etc. to ensure that the child has a reasonable chance of prospering. And we oblige parents to care for their children both physically and psychologically. If the parents disregard this duty or fail to execute it, society will intervene in order to safeguard the interests of the child. The situation is very different before birth.

The fetus is completely dependent on the good will of the pregnant woman. Most pregnant women do their best to ensure that the fetus will have a good start. They abstain from dangerous behaviour, heavy drinking, smoking, etc. Unfortunately, not everyone follows this pattern. And society has been helpless. There has been no possibility, apart from advising, to ensure that these women live without impairing the future of their expected child.

We live in a society where we firmly believe that there should be no discrimination on a sex or gender basis. Being pregnant is exclusively for women; any special legislation to safeguard fetuses would at the same time impinge upon the rights and autonomy of women. Legally speaking, during the first period when the pregnant woman can have an abortion on demand, the fetus is truly regarded as just a part of the woman's body. Then it is a little unclear, but if you murder a pregnant woman you are prosecuted for one murder, not two. I think many of you who work in the reproductive area have witnessed the disturbing case when a delivery is protracted, the situation is getting serious for the expected child and something has to be done. The autonomy of the woman reigns sovereign and she may legally say no to a Caesarean section, the child may die or be seriously damaged. There is nothing to be done, if the woman does not consent. She has all the legal rights.

At birth, everything changes. A human being with a full set of human rights enters the world. The child-in-the-family is protected in a radically different way than the child-in-the-uterus. When a born child is abused, society will do something. When an unborn child is similarly abused, society will normally do nothing. There has previously been no possibility of safeguarding the interests of the fetus without infringement of the rights of women. In my opinion, the feminists have been absolutely right in putting the interests of the pregnant woman before the interest of the fetus.

All this changes with the prospect of pig-related pregnancies. In a pig-related pregnancy the human fetus is no longer a part of the body of the woman. In the porcine uterus, the human fetus can have the same legal status as the newborn child and have the same protection. And Ladies and Gentlemen, this is consistent with all the rights and liberties of women that the feminist movements successfully fought for. This is a moral gain.

I want to state the following moral principle: If a new technology makes it possible to avoid a conflict between legitimate interests, it is our duty to use the new technology. Hence, it is a duty to use the new technology of pig-related pregnancies. Should society therefore forbid "natural" pregnancies and force women into having pig-related pregnancies?

I think this is going too far. After all, most of us regard infidelity between spouses as a morally bad thing. Still we do not legally punish it. And we might ask when conditions for pregnant women are not just another kind of discrimination? I think a reasonable compromise is the following. Given the possibility of a pig-related pregnancy, society may indeed demand that women should first and foremost safeguard the well-being of the fetus. This can be done in two ways.

One is a pig-related pregnancy with no restrictions on the lifestyle of the woman. The other is a "natural" pregnancy with legally enforced restrictions on the lifestyle. On the other hand, women undergoing a "natural" pregnancy should have the option to switch to a pig-related one at any time.

What about the pigs? Pregnancy is a positive experience for the pig. Our pigs have a good life before, during and after the pregnancy. Last, but not least, pig-related pregnancies will give more pigs the opportunity to exist, and this is a good thing.

Ladies and Gentlemen, thank you for your attention. Welcome to participate in this new exciting development. And welcome to our party.

Comments on the Possibility, the Consequences and Desirability of Reproductive Ectogenesis

What are we to conclude from the story of Pig Pharmaceuticals Limited? Below I offer some comments on some of the issues, in particular: will it happen, will this change our view of the fetus, and what about the rights of pregnant women?

Will Reproductive Ectogenesis Ever Happen?

Personally I think so. It was already prophesied by J. B. S. Haldane in 1923. The chief scientific director of Pig Pharmaceuticals has a real point on how it could become a reality. Many things that are ruled out as unethical under a research regime by the rather restrictive demands of the Helsinki Declaration may be acceptable in the clinic. One example is the development of heart transplantation. This was not done as a research project. The driving force will be to save the fetuses and allow women to have children. Pigs as surrogate mothers may not be very probable. After Novartis closed down Imutran in Cambridge in Spring 2001, the field of xenotransplantation did not look so promising.[12] The particular transgenic pig was not good enough and we will have to wait for new transgenic pigs—which seem to be on their way—or some other breakthrough. It is perhaps more feasible to believe that reproductive ectogenesis will start from what we learn from growing stem cells and, hopefully, learn to develop them into various cell types and later to grow organs for transplantation. Growing organs outside the human body is a very desirable form of tissue engineering with regard to the desperate need for organ transplants. It will at the same time give us knowledge helpful for sustaining a full embryonic development. Once this is given, it is a short step to use the therapeutical argument that we need to apply ectogenes to save the fetus for some woman. I do believe that a *therapeutical imperative* exists—at least with regard to introducing advanced medical technologies—in our societies. Whatever techniques that can be used to save life will be developed. Application may not become very widespread but it will be introduced.

Will This Change Our View of the Fetus?

Speaking generally it seems that fetuses are not considered as having rights or, alternatively, that the legal rights recognised are derivative on the child being born.[13] In my own country (Sweden) the legal status of the fetus changes dramatically at birth. It seems that viability is the criteria used. This is of course a criteria that is technology dependent and in the era of ectogenesis the embryo will be viable (capable of living outside the womb) from the very start. Using another more plausible view, the moral status of the fetus is related to the fetus being capable of sentience. This does not (at present) give a precise time but we can be rather certain that the moment does not occur at birth. It is either before or after.[14]

Ectogenesis will give new possibilities of safeguarding early embryos without interfering with women's rights to control over their bodies, a topic already discussed in the debate over human embryonic stem cells. Many might think that there is a direct coupling between holding (early) abortions admissible and thinking it is admissible to destroy and discard IVF embryos. However, it is possible to argue for special safeguards for the IVF embryo while allowing (early) abortion on women's demand. In a well-known discussion on abortion and other issues Ronald Dworkin makes the point that there can be no rights without interests and these cannot exist without some rudimentary form of consciousness. Therefore the fetus in its early stage does not have interests and therefore no rights.[15] He introduces the distinction between two types of objections to abortion. There is *the derivative objection* which assumes that the fetus has interests and rights. The *detached objection* does not assume that the early fetus has rights but argues from the need to protect the intrinsic value of life.[16] He then rules out the derivative objection as invalid and relegates the detached objection to the private sphere of religious and ethical conviction. Essentially, Dworkin agrees that there is a legitimate interest for the community to safeguard the intrinsic value of human life but this should not conflict with the rights of women to self-determination and control of their bodies. In the first phase of the pregnancy, the right of the woman, according to Dworkin, overrules the interest to protect the value of life.

Dworkin does not discuss the IVF embryo but his discussion of abortion can be applied to ectogenesis. Let us suppose that Dworkin is right about the interest of safeguarding the intrinsic value of life. This can obviously be done in the third era of reproductive ectogenesis without interfering with the woman's body. The case of the child-in-the-incubator/pig uterus is in that respect similar to the child-in-the-family. Also, the child-in-the-female-uterus will, in the third era, be in a similar position to the child-in-the-family. Most of us agree that the society should interfere into the family to safeguard the well-being of the child. The child can be removed. Now the child (in the maybe fictious) third era can be removed from the uterus and placed in some artificial womb. Such removal was proposed in 1984 as an alternative to abortion by Peter Singer and Deane Wells.[17]

The Rights of Pregnant Women

In the third era of reproductive ectogenesis the legal protections of the embryo no longer interfere with the woman's body as the pregnancy is outside the body. And furthermore, if we concede that society should not protect the early embryo in the incubator, it is very hard to find an argument why the woman *alone* should decide whether or not the embryonic development should be interrupted. The man should obviously share the decision. The third era of human reproduction will mean reproductive empowerment of men and end the historically short monopoly of women in regard to deciding the fate of the embryo and fetus. In the second ongoing era of IVF we already see the first signs. When the embryo is outside the woman, the man and the woman share decision-making.

Another possible change was pointed out by the ethicist in the future talk. In the era of ectogenesis, women who choose to have a natural pregnancy will have to face restrictions on lifestyles. At least, I believe it will be very hard to argue against such restrictions in order to protect the fetus. This will be a dramatic change and put the pregnant woman in a special situation. Maybe, for this kind of restriction to be ethically acceptable, every woman should have a choice between ectogenesis and a natural pregnancy. Ideally, it should be as depicted in the 2050 case in that it should be possible at any time to change from a natural pregnancy to ectogenesis.

If not everyone should be able to afford ectogenesis—which may be the case in the USA—then restrictions imposed on those who cannot pay does not seem fair. However, one might still make an argument for restrictions on those who can afford it. Furthermore, "not being able to afford" is a very vague concept.

It is perhaps possible to give arguments for prohibiting ectogenesis altogether and preserve the present situation that pregnant women can (more or less) live without any legal sanctions against hurting their fetuses and keep the sole decision to abort or not. I am not able to think of any good arguments for that, however. I think this is the way it must be today when the fetus is inside the woman's body, but I am doubtful if there are any good arguments for this position in the third era.

The third era will be good news for pro-lifers. Embryo or fetus adoption will be an alternative to abortion. The introduction of embryo adoption has already been proposed for IVF in the second era (that is now) by a committee of the European Parliament.[18] That particular proposal was not adopted.

Is Reproductive Ectogenesis Desirable?

The ethicist of Pig Pharmaceuticals Limited has given his (future) arguments for reproductive ectogenesis. I find ectogenesis in many ways repugnant but I must confess that I lack good arguments against its introduction, at least as an option for therapeutical reasons. But given an option, I suspect we are on the slippery slope.

Acknowledgment

Work on this paper has been supported (partly) by the Swedish Research Council grant K2002-31x-14012-02B. A preliminary version of the paper was presented at "The End of Natural Motherhood," Tulsa, Oklahoma, 22–23 February 2002. I am grateful for comments at that meeting.

Personal Note

Together with my co-worker Anders Persson I have been studying the ethical and social issues in xenotransplantation. I got the idea for the story some years ago when I was told that the Xenotransplantation Society was contemplating new ethical guidelines. One proposal was, as I remembered, to forbid the insertion of a human fetus into an animal. This was presented at a workshop in Tübingen, Germany in 1998 by Claus Hammer of Munich University and is reprinted in *Leben als Labor-Material. Zur Problematik der Embryoforschung*. Ed. Wuerling, Hans-Bernard. Düsseldorf: Patmos Verlag 2000, p. 35. (I am grateful to Claus Hammer for confirming my memory and for the reference.)

Notes

1. J. McGrath and D. Solter, "Inability of Mouse Blastomere Nuclei Transferred to Enucleated Zygotes to Support Development in Vitro," *Science* 226 (1984): 1317–1319.

2. D. Solter, "Mammalian Cloning: Advances and Limitations," *Nature Review Genetics* 1 (2001): 199–206.

3. G. W. Bush, "President Discusses Stem Cell Research," August 9, 2001, https://georgewbush-whitehouse.archives.gov/news/releases/2001/08/20010809-2.html.

4. D. Parfit, *Reasons and Persons* (Oxford: Clarendon Press, 1987), pp. 199–200.

5. R. Tong, *Feminist Approaches to Bioethics* (Boulder, CO: Westview Press, 1997), p. 129.

6. R. Nozick, *Anarchy, State, and Utopia* (Oxford: Basil Blackwell, 1980), pp. 42–45.

7. J. Glover, *What Sort of People Should There Be?* (London: Penguin, 1985), pp. 102–113.

8. D. Dennet, "Where Am I?," in *The Mind's I: Fantasies and Reflections on Self and Soul*, ed. D. Hofstadter and D. Dennet (London: Penguin, 1982), pp. 217–229.

9. J. J. Thomson, "A Defence of Abortion," in *Today's Moral Problems*, 3rd ed., ed. R. Wasserstrom (New York: Macmillan, New York, 1985), pp. 418–432.

10. J. B. S. Haldane, *Daedalus or Science and the Future*, in *Haldane's Daedalus Revisited*, ed. Krishna R. Dronamraju (Oxford: Oxford University Press, 1995).

11. M. Young, *The Rise of the Meriticracy* (New Brunswick, NJ: Transaction, 2002).

12. A. Person, "Optimismen falnar om xenotranplantation" (Optimism fades for xenotransplantation), *Läkartidningen* 98 (2001): 3200–3202.

13. C. Wellman, "The Concept of Fetal Rights," *Law and Philosophy* 21 (2002): 65–93.

14. S. Welin, "Ethical Issues in Human Embryonic Stem Cell Research," *Acta Obstetricia et Gynecologica Scandinavica* 81 (2002): 277–382.

15. R. Dworkin, *Life's Dominion* (London: HarperCollins, 1993), p. 18.

16. Ibid., p. 11.

17. P. Singer and D. Wells, *The Reproduction Revolution: New Ways of Making Babies* (Oxford: Oxford University Press, 1984).

18. European Parliament, Temporary Committee on Human Genetics and Other New Technologies in Modern Medicine, *Report on the Ethical, Legal, Economic and Social Implication of Human Genetics*, A5–0391/2001, 2001.

4 Eight Great Technologies

David Willetts

Futures are not only created by scholars, fiction writers, and eccentric visionaries. Sometimes those who envision the future have the power to make (or at least try to make) the development they imagine happen; their visions promote their interests in a particular future. Since World War II, governments have been major funders of scientific research worldwide. Available funding in specific areas of research generally compels people to work in those fields and results in new technologies that ultimately shape our collective future. This speech by David Willetts, who at the time was the minister of state for universities and science in the UK, outlines a strategy for the British government to promote science in directions that will benefit the UK. While Willetts's vision of the future expresses hope for solving various social problems along the way, his recommendation to direct immense amounts of money at scientific research is targeted to ensure that Britain maintains its political and economic status. He contends that science should generate both pride and a monetary return on investment. His strategy has a fair amount of nuance. Willetts doesn't believe that politicians are knowledgeable enough to direct science at a microlevel, but he does want to ensure that the government has some say in how its money is used. So he identifies eight major areas of research in which he thinks UK scientists possess a "distinctive capability"; he envisions this research leading to commercial enterprises. Choosing to invest in specific areas of research directly affects what research is conducted and, in turn, the possible futures that can be realized.

In January last year I spoke at Policy Exchange about the importance of a high tech industrial strategy. There is a lot that government can and must do to drive the development of key general purpose technologies. Today I can update you on the progress we are making and announce where we're providing more funding for these key technologies.

Vince Cable set out in an important speech in September 2012 our approach to industrial strategy. It is a long term approach across the whole of government, to give business the confidence to invest and grow. We are taking action to make this happen. Technologies and the broader research which underpins their development is a fundamental part of our approach to industrial strategy. Today I can set out new decisions to drive this forward.

We are fortunate to have a very broad science and research base. Indeed there is no other medium sized economy which has anything like our range of world class research activity. This is clearly demonstrated in the Research Council impact reports that are

---

Rt. Hon. David Willetts, "Eight Great Technologies," original script of speech given to the Policy Exchange Thinktank, London, January 24, 2013.

being published today. The reports illustrate the value to the economy and to society of the funding that we provide for science and research every year.

It is not just the Nobel prizes, the winners of the Fields medal and the world famous professors. Whenever there is a crisis, a civil war, or a coup d'état anywhere in the world we are likely to have a historian who has some understanding of the background, anthropologists who know the culture, and someone who can speak the language. This is an extraordinary privilege which we must not take for granted: citizens of very few other countries have such a wide open window on the world. The very range of what we do is one of our greatest assets, especially as great technological and scientific advances depend on breaking down the conventional barriers between disciplines.

We have the extraordinary advantage of being the only medium-size country that has such a range of scientific activities. We have world class scientific institutes and research intensive universities. This includes humanities and social sciences. It is not just STEM it is STEAM—Science Technology Engineering Arts and Maths. Reed Elsevier's 2011 review of the comparative performance of the UK Research Base identifies "over four hundred niche areas of research in which the UK is distinctively strong."

One of the main aims of our science policy is to maintain that breadth and not to find ourselves forced to trade off being world class in life sciences or history or physics. We do not direct our scientific and research community into particular research projects. Instead our science community rightly enjoys extraordinary autonomy as funding is largely allocated on the principle of excellence determined by academic peer review. This is the first pillar for our science and innovation policy.

There is a second pillar too. After the failure of the economic interventionism of the 1970s and the triumph of the liberal revolution in economic policy of the 1980s we are wary of Government trying to pick winners. In so far as government can raise the growth rate we tend therefore to focus on measures which apply across the economy as a whole—deregulation or lower corporate taxes or ease of setting up a business. We perform well on many of these measures—the UK is already ranked 2nd in the G7 for ease of doing business. Until recently we have tended to favour these so-called "horizontal" measures rather than "vertical" ones which focus on particular sectors.

Put the breadth of our science base together with the dominant intellectual climate and you get classic British policy on science and technology. We finance a broad range of research selected by fellow scientists on the basis of its excellence. The government is working hard at tearing down the barriers to the smooth functioning of a modern market economy. Strong science and flexible markets is a good combination of policies. But, like patriotism, it is not enough. It misses out crucial stuff in the middle—real decisions on backing key technologies on their journey from the lab to the marketplace. It is the missing third pillar to any successful high tech strategy. It is R&D and technology and engineering as distinct from pure science. It is our historic failure to back this which lies behind the familiar problems of the so-called "valley of death" between scientific discoveries and commercial applications. Also, as we shall see, it helps to explain our belief that we lack a culture of risk-taking.

We are living now with the long-term consequences of the failure to have a policy backing these key technologies. Look at the business sectors where we are strong—creative industries, financial services, construction, new web-based services. They all share a crucial feature. They are all areas without capital-intensive R&D. So paradoxically the very aversion to backing particular technologies with R&D has itself contributed to a change in the structure of the British economy—an economy which innovates but does not do as much R&D as many of our competitors.

Focusing on R&D and on particular technologies is not the same as picking winners, which notoriously became losers picking the pockets of tax payers. It is not backing particular businesses. Instead we are focusing on big general purpose technologies. Each one has implications potentially so significant that they stretch way beyond any one particular industrial sector. Information Technology has transformed retailing for example. Satellite services could deliver precision agriculture.

This is where we face the valley of death. It is after the pure science and before the usual process of individual companies developing particular products and processes. It is R&D. It is also where the British government used to play a crucial role, supporting the military industrial complex of the twentieth century "warfare state" described by David Edgerton.

It is also what the US still does far more than we do. It is hard to see because you have to look behind the American rhetoric of limited government. Moreover the scale of federal and state activity is hidden because it is divided up between several different agencies. The rationale is often military and security in its broadest sense. There are other reasons too: after President Bush banned federal funds for stem cell therapies, California voted for $3 billion of funding for it in a referendum. I have visited their research funding body and it will not just fund pure research but also help on the new processes needed to manufacture these therapies and use them to treat patients.

Our research councils tend to focus on more upstream research whereas in the US, Defense Advanced Research Projects Agency (DARPA), the National Institutes for Health and the Department of Energy go further downstream closer to market. Sometimes our approach can look like mother birds pushing their fledglings out of the nest but with too many falling to the forest floor to be eaten by foxes. We think our problem is that we lack the same willingness to take risk as in America. But often we were expecting companies to step in earlier, taking more risk than in the US or elsewhere.

The Technology Strategy Board is a crucial but underestimated institution which can help plug that gap. It is working more closely than ever before with our Research Councils to get more sustained support from blue skies research to closer to commercialisation. As part of our life sciences strategy we set up a Biomedical Catalyst worth £180 million split 50/50 between the Medical Research Council and the Technology Strategy Board (TSB) to take new medical innovations closer to practical application. Already this scheme is a real success. I am keen to repeat this model elsewhere. Yesterday, I announced a £25 million catalyst fund for Industrial Biotechnology and Bio Energy, linking the Biotechnology and Biological Sciences Research Council (BBSRC) and the TSB.

The US does other things on a far more ambitious scale than us. They are more imaginative and bold in the use of procurement for example. Their support for innovative small businesses with Ronald Reagan's Small Business Research Initiative is on a scale far greater than ours. Where government has a big role such as in medicine or security they harness that. The US Orphan Drugs Programme for example provides strong incentives for drug development. DARPA rests on the assumption that US security depends on harnessing key new technologies and they do that not just with research support but with contracts that are offered for new products at a very early stage. Indeed Silicon Valley originally grew on the back of contracts from the military for computers and IT.

Just showing that they do it in [the] US doesn't prove the point on its own. There are perhaps four specific objections which we need to address.

First, we have to accept we make mistakes. We do not have perfect foresight. Some of the technologies for which we have high hopes today will turn out to be clunkers tomorrow. That is because this is all about taking risk—if the risk was much lower then we could indeed leave it to straightforward business decisions. But we do have a wide range of expertise to help us understand scientific and technological trends and we have set out our thinking more openly than ever before. Indeed that is why I am releasing today my pamphlet describing eight great technologies.

Secondly we are told that the high tech sector is small and the real big commercial issues are elsewhere. The Organisation for Economic Co-operation and Development (OECD) defines sectors as high tech if they devote more than four per cent of turnover to R&D. This is a demanding test. And companies or sectors which do this are unusual. But they can develop technologies which then go mainstream and have a massive impact way beyond any specific sector. These new technologies may be absorbed by business sectors that themselves do little R&D but are nevertheless transformed.

Thirdly there is the danger incumbents get the support not the insurgents. New small businesses are crucial and we have a range of programs specifically aimed at promoting them. But the fact is that [a] lot of the R&D spend is in big business. Indeed our shortage of big primes at the top of the supply chain is one of our key industrial weaknesses. So big business does matter. Where we do have key primes—as in automotive, aerospace or life science, they themselves can be protectors of small business as they maintain a supply chain.

Moreover they may not have a cushy time. These new technologies are often deeply destabilising. They are a challenge to traditional businesses which find themselves having to adjust to the arrival of new technologies which disrupt what they do. The ones that survive have to move way beyond their traditional technologies and sectors. There is an interesting trend of patents being taken out for technologies which go way beyond the traditional activities of a business—the automotive sector taking out more patents in IT for example as it becomes crucial to the performance of a car.

Finally there is the fear politicians are always seduced by baubles. We go for glitzy new projects rather than what has real potential. That is why it is important we draw on expert advice which has to be more transparent than ever. The pamphlet which I am publishing today identifies eight great technologies. It is not my personal view. It distills work done by experts in the Research Councils, the Technology Strategy Board and foresight exercises conducted by the Government Office for Science. We have published their reports. In an important speech to the Royal Society last November George Osborne listed them and asked if people agreed with them. By and large our analysis has been accepted.

In the past I have drawn on the well recognised American account of four major technological advances—Bio, Nano, Info and Carbo or BNIC for short. It gives us extra confidence in the analysis behind the eight great technologies that they fit into these categories. The first three on our list of eight technologies are broadly information technologies. Then the discovery that biological data is digital moves us on to Bio. Advanced material design often involves nano technology. And our final technology is, in large part, about reducing carbon in our energy supplies.

As well as identifying these great technologies today I can set out more fully than ever before what Government is doing to back them. We are systematically working through all eight to ensure they are properly supported.

There are some basic steps we can take using the convening power of government. So here is Industrial Strategy 101. You set up a leadership council probably co-chaired by a BIS minister and a senior industry figure in which researchers, businesses, perhaps regulators and major public purchasers come together. You use it to get them talking to each other confidently and frankly. Then that group might commission a trusted expert to prepare a technology road map which assesses where the relevant technologies are heading over the next five years or so, where publicly funded research is going, and what business is likely to do. Just this exercise, before any increase in public funding, can transform behaviour. Some of the big companies for example might have a HQ abroad and it means their managers here and also BIS ministers can show to them what we are doing and encourage more investment here. It can encourage businesses sitting on piles of cash to invest when they see how it fits in alongside investment we are committed to putting in. You might go further and find that if the government puts some money up front it can get co-investment by others. You might find some key regulations which need to be eased, or perhaps the opposite and some need to be even introduced to help give confidence a new technology can safely be adopted. Government might be more open about its procurement plans than before and more willing to go for an innovative use of a new technology not settling for the tried and tested. But crucially you have a vehicle for making this happen and building mutual trust. The quality of links between business, the research community and government is itself a source of comparative advantage in the modern world.

Let me now very briefly review progress on each of these eight key technologies. They will be backed further by the decisions I am announcing today on the allocation of an extra £600 million of funding. This investment in science and technology, announced by George Osborne in the Autumn Statement, is additional to the ring-fenced science budget.

1 Big Data

The power of computing and data handling is now becoming so great that classic distinctions between micro and macro effects are breaking down. We are reaching the stage of being able to model airflow across a turbine blade or the movement of a liquid through a tube at the molecular level. Computer modelling of an economy, a substance or a process is therefore becoming very different and far more sophisticated than it was even a decade ago. The importance of these developments is being recognised around the world. I note that I am giving this speech on the same date as the Data Innovation Day in the US.

We have set up the e-infrastructure leadership council co-chaired by Dominic Tildesley, formerly a senior business executive from Unilever, and myself. We share with industry our plans for research funding so as to encourage co-investment by them. We are seeing the benefits already with companies such as IBM, Cisco and Intel making a number of investments into the UK. Business will invest more as they see us invest more in computational infrastructure to capture and analyse data flows released by the open data revolution.

The government invested an extra £150 million in e-Infrastructure in October 2011. This has been followed by a further allocation of an extra £189 million in the Autumn Statement. This will be invested over the next two years in key areas such as: bioinformatics and environmental monitoring.

Our investment in data has also ensured we maintain our leadership in social science. We have invested £23.5 million in the Economic and Social Research Council (ESRC)–led life study, the most ambitious birth cohort study yet, which will track 100,000 children from birth. The reason it is so ambitious is that it will also link genetic data, environmental data and educational outcome data.

2 Space

The UK is once more seen as a leading space science nation. Companies have focussed on making satellite technology more affordable with smaller, lighter-weight satellites that lower the cost of commercial launches. Surrey Satellites Technologies (SSTL), one of the UK's single most successful university spin-outs, is the world leader in high-performance small satellites. Roughly 40 per cent of the world's small satellites come from Guildford—and now even smaller nano-satellites are coming from SSTL and Clydespace in Glasgow.

The Space Leadership Council is co-chaired by an industry executive and myself. The Coalition has made a series of significant investments in space over the past two years, and these investments have given the industry confidence to invest more for the future. Every major public sector investment has triggered commercial investments several times greater. We have also set up a Satellite Applications Catapult at Harwell.

In March 2011 we launched a £10m National Space Technology Programme in the UK and this original programme attracted £17 million in matched funding from institutional and industry investors. Early analysis suggests the return to the economy from this investment of £10 million will be between £50 million and £75 million.

Today I can announce that as a result of the Autumn Statement the Government will be investing an extra £25 million in the further implementation of the technology vision through Phase-2 of the National Space Technology Programme. This £25 million of further investment will meet un-met demand as many excellent projects were not supported in the first phase.

3 Robotics and Autonomous Systems

The UK has some distinctive strengths in this area, going back yet again to our abilities in software programming and data handling. Effective handling of data from a range of sources is key to autonomous systems and we have real skills here. It was an extraordinary feat of engineering to land NASA's Curiosity probe on Mars last year. Its Mars Rover vehicle is however largely controlled from Earth with a delay of at least seven minutes as instructions travel to Mars. The European Mars Rover vehicle, due to land in 2018, is more autonomous, using mainly British technology to enable it to travel further during the Martian day and therefore carry out more investigations during its design life.

The Engineering and Physical Sciences Research Council funds much of the research on robotics. It has so many different applications across different industrial sectors that the R&D effort is fragmented. There is also no single leading major industrial prime leading the development of the technology. In October 2012 I convened a meeting of key experts on robotics and autonomous systems at the Royal Academy of Engineering to discuss what more could be done to promote this important general purpose technology. The discussion showed the need to bring greater coordination of this research. The Technology Strategy Board is now creating a Special Interest Group on Robotics and Autonomous Systems which will shortly produce an outline technology road map to promote future investment. The participants in last October's meeting also proposed academic centres of excellence that would both conduct basic research but also translate it for commercial application. For this reason I am announcing an investment of an extra £35 million for centres of excellence in Robotics & Autonomous Systems. They will be created in and around universities, innovation centres, science parks and enterprise sites and provide bespoke support for both university and industrial interests. Support from these centres of excellence will provide the missing link between our SMEs and primes in this technology area. They will be hubs of technical expertise and training, providing cutting edge facilities and opportunities for business networking.

In addition, the Technology Strategy Board will invest up to £1 million in feasibility studies to accelerate the development of novel robotics and autonomous systems concepts towards technology demonstration in multiple sectors. They will launch the competition in February.

4 Synthetic Biology

Many of the critical discoveries related to DNA were made in Britain, in perhaps the world's greatest post-War research institute—the MRC Laboratory of Molecular Biology in Cambridge. It is not just the original discovery of the structure of DNA by Watson and Crick, drawing on work by Rosalind Franklin and Maurice Wilkins.

More recently researchers funded by EPSRC, have successfully demonstrated that they can build some of the basic components for digital devices out of bacteria and DNA, which could pave the way for a new generation of biological computing devices. The researchers, from Imperial College London, have demonstrated that they can build logic gates or switches, which are used for processing information in devices such as computers and microprocessors, out of harmless gut bacteria and DNA. Although still a long way off, the team suggests that these biological logic gates or switches could one day form the building blocks in microscopic biological computers.

We produced a synthetic biology road map last year and a new Synthetic Biology Council has been established to ensure this road map is delivered. I co-chair it with Lionel Clarke, a senior executive from Shell.

We are making a series of investments in research in synthetic biology. The UK Research Councils and the Technology Strategy Board are spending over £90 million on world leading synthetic biology research and commercialisation including £20m announced by the Chancellor last November. We announced as part of our Life science strategy one year on that a further £50 million will be invested in synthetic biology as part of the subsequent Autumn Statement settlement. This will be used to support implementation of key recommendations from the UK Synthetic Biology roadmap, including establishing multidisciplinary research centres as well as a seed fund to support start-up companies and "pre-companies."

We also announced that we are investing £38 million in a National Biologics Industry Innovation Centre. This investment will allow the development of a large scale facility for the manufacture of biologically produced medicines such as antibodies and vaccines.

At present no major pharmaceutical companies manufacture significant quantities of biologics in the UK so this centre will fill a gap in biologic manufacturing capability and strengthen the UK's case as the location of choice for internationally mobile life sciences companies.

The centre will be managed by the Centre for Process Innovation (CPI) as part of the High Value Manufacturing (HVM) Catapult and also supports regenerative medicine.

5 Regenerative Medicine

Regenerative medicine involves restoring function by replacing or restoring human cells, tissues or organs. There are three main approaches researchers are pursuing—transplantation of cells, tissues and organs, stimulation of the body's own self-repair mechanisms; and the development of biomaterials for structural repairs. This is led by world class research in centres such as Edinburgh (where Dolly the sheep was cloned), Cambridge, Leeds, and London. Our research has moved on from Dolly the sheep to Jasper the dog. He had spinal injuries but was able to walk again by injecting his spinal cords with a specific type of stem cell. The potential applications for human medicine are easy to envisage.

The Research Councils and TSB recently published "A Strategy for UK Regenerative Medicine," including commitments of £25 million for the UK Regenerative Medicine Platform, (which is establishing multidisciplinary programmes to address the key roadblocks in developing therapies in this area) and £75m for translational research. Our Cell Therapy Catapult has now opened at Guy's Hospital in London. An extra £20 million capital was allocated to the Regenerative Medicine Platform at Autumn Statement 2012 to provide imaging and cell manufacture technologies and a clean room.

6 Agri-science

Britain did not just lead the Industrial Revolution, we pioneered the Agricultural Revolution too. From leading that Agricultural Revolution in the late eighteenth century to new biotechnology-led advances, the UK has remained at the forefront of agricultural research.

Chickens are a prime example. Chickens are the world's biggest source of meat, and are particularly important in Asia. We breed the world's chickens—of the £85 billion global poultry market, 80 per cent of breeding chickens come from genetic stock developed in the UK. Thanks to our genetics research you get twice as much chicken for a given amount of chicken feed as 20 years ago. Each year we launch a new breed of chicken which will produce many generations over a year or more before a new improved version comes along. This is possible because of close links between the Roslin Institute, with its world leading R&D, and our commercial sector.

BIS and DEFRA are working together with industry to strengthen links between research spend and agricultural policy. This work will be brought together in a new agritech strategy over the next few months. We are already investing £250 million in the transformation of the Pirbright Institute of Animal Health as well as Babraham and Norwich research park. The Autumn Statement package earmarked £30m for capital investment in BBSRC's world-leading agri-science campuses. A candidate for this funding is the construction of a new National Plant Phenomics Centre at Aberystwyth University.

7 Advanced Materials

Advanced materials are a key tool for advanced manufacturing. UK businesses that produce and process materials have a turnover of around £170 billion per annum, represent 15 per cent of the country's GDP and have exports valued at £50 billion. There has been quite rightly a flurry of interest in 3D printing, or "additive layer manufacturing." This new technology is possible not just because of advances in IT but also because of advances in the materials that go into the process. It is no longer just a matter of printing out designer dolls: Southampton University has used advanced materials to show how we could print out a new aeroplane.

The Prime Minister convened a seminar last summer on advanced materials which showed the importance of advanced materials for advanced manufacturing. As a result I can announce an extra £45 million in advanced materials research, for new facilities and equipment in areas of UK strength such as advanced composites; high-performance alloys; low-energy electronics and telecommunications; materials for energy; and nano-materials for health.

In addition, we announced at the Autumn Statement a £28 million Expansion of the National Composites Centre (NCC), located on the Bristol and Bath Science Park. The NCC is one of the seven centres within the High Value Manufacturing Catapult. This investment will expand the NCC from 6,500 sq m in a single building to 11,500 sq m across two buildings, and give it the space to install equipment to work on larger structures made of composite materials.

It will also enable the NCC to increase the level of skills development it undertakes, by creating a new training centre for higher level and vocational skills development, training the next generation of engineers in manufacturing and materials technologies.

8 Energy

Efficient energy storage technologies could allow the UK to capitalise on its considerable excess energy production. While UK consumption peaks at 60GW, the UK has generation capacity of 80GW but storage capacity of only 3GW (primarily from the single Dinorwig water system in Wales). Greater energy storage capacity can save money and reduce the national carbon footprint at the same time.

It has the potential for delivering massive benefits—in terms of savings on UK energy spend, environmental benefits, economic growth and in enabling UK business to exploit these technologies internationally. Energy is one of the largest single themes in Research Council funded research, with a portfolio of over £600 million of total current awards. In addition the government will invest an extra £30 million to create dedicated R&D facilities to develop and test new grid scale storage technologies.

We are also considering a strategic opportunity to partner with the US Department of Energy in the development of small modular reactor technology.

Behind these technologies lie a network of research labs and facilities. They are a shared national asset for scientists but also of use to business too. We are systematically investing in them and trying to strengthen links between researchers and industry. Many of them are located on university campuses. We are promoting university/business collaboration by our imaginative Research Partnership for Investment Fund which has secured £1 billion of new investment on R&D facilities on our campuses. We are working with our partners to create the new Crick Institute in London which should be one of the world's leading new medical research facilities when it opens in 2015. We are also creating seven Catapult Centres linking business and public funding for new technologies. We are stimulating research clusters like Harwell and Daresbury which are both now enterprise zones. I am delighted to announce that an extra £65 million from Autumn Statement 2012 will be invested in buildings, joint facilities and infrastructure to promote co-location of industrial and academic groups, and support high-tech business on campuses. Investment will mainly be focused around the development of four campuses: Rothamsted Research Campus, Aberystwyth (IBERS as I have already said), Harwell Oxford, and SciTech Daresbury.

This will enable the UK to accelerate the exploitation of its world leading research base to deliver jobs and growth by bringing together substantial, internationally significant research capabilities with a variety of users, supporting the setting up and development of innovative knowledge based companies in sectors ranging from food and farming through to the production of synthetic diamonds.

Scientists also need constantly to upgrade their equipment and labs. Indeed the inter-action between science and technology is itself one of the great drivers of innovation. For this reason we will be investing an extra £50 million in these over the next two years.

We are also encouraging academics to think about the wider impact of what they do. It does not mean faking forecasts of likely benefits. I welcome the recent step by EPSRC to tackle these anxieties.

For all these eight great technologies to come to market we also need excellent measurement and as part of the Autumn Statement I can today announce we are providing an extra £25 million to build a state of the art laboratory for cutting edge measurement research. The creation of advanced facilities at the National Physical Laboratory in Teddington will allow scientists there to undertake leading edge research in key nano and quantum metrology (measurement science) programmes.

This ability to make accurate measurements underpins the UK's competitiveness in both existing markets and to underpin new technology that will support growth in the UK economy. For example Rolls Royce would not have been able to supply turbine blades to Airbus without measurement traceable to NPL; graphene could not have emerged as a viable proposition without the pioneering research work that NPL performed to be able to measure its properties.

Also to underpin the development of the technologies within these eight areas, we need highly skilled individuals. To support this EPSRC is making a £350 million investment in Centres for Doctoral Training (CDTs) to develop the talented people that will create future growth and a more sustainable future.

Centres will be in areas including the digital economy, renewable and nuclear energy, synthetic biology, materials technologies, regenerative medicine, data to knowledge, and advanced manufacturing.

This investment will refresh the current portfolio of Centres for Doctoral Training announced in 2008. Current students of these centres are helping change the world from reducing risk in the financial sector to pioneering 3D inkjet printing for individually tailored therapeutic drugs.

The government's investment of £600 million through the Autumn Statement 2012 in Research Council infrastructure, and the facilities for applied research and development (R&D) will support the development of innovative technologies and strengthen the UK's competitive advantage in areas such as big data and energy efficient computing, synthetic biology and advanced materials.

I can now set out therefore the allocation of the extra £600 million of extra science funding committed from the Autumn Statement. There will be:

£189 million for big data

£25 million for space

£35 million for robotics and autonomous systems

£88 million for synthetic biology

£20 million for regenerative medicine

£30 million for agri-science campuses

£73 million for advanced materials

£30 million for energy

We have also committed a further:

£35 million for research campuses

£25 million for the advanced metrology lab

£50 million for transformative equipment and infrastructure

Conclusion: A Date for Your Diary

The pamphlet on our eight great technologies is being published today. I would like to invite you back in ten years time on 24 January 2023. There are risks of course. I may not be around. Policy Exchange may not be. But I hope most of us are and that we are still excited about science. Imagine that today we are burying a time capsule and we are going to open it up in ten years when we can take stock. One possibility is that of course technology has developed in a way completely different than set out here. I am still waiting to commute to work on a personal jet booster pack as operated by James Bond in Thunderball. There could well be new technologies which we just have not considered. We are not claiming perfect foresight. But in addition there are six real possibilities for the long-term impact of our strategy for these eight great technologies. Here they are.

1 False Dawn

We are still waiting. The analysis broadly stands but it all takes longer than we had hoped. Robots for example are still trundling round labs but not yet waiting at our tables.

2 Transmutation

The technologies will not have worked out in the way we expected but new businesses have emerged in a more indirect route. As every romcom shows, things rarely work out in the direct routes we expect. ARM originates with the BBC Acorn computer project run out of Bristol.

3 Gone Abroad

The technologies play out roughly as we describe but it all happens abroad. We have a few multi-millionaires who sold their ideas to foreign multinationals but not much else. This is one of my fears. It is the observation that we grow the world's best corporate veal.

4 It's Here but It Isn't Ours

We have grown the companies here so they have put down roots and we have got genuine expertise which cannot be shifted. But ultimately they are owned by a big corporate which has HQ somewhere else. Illumina is a happy example.

5 We Have Grown Big New Companies

Just as the US has got Google Amazon Facebook Ebay. We have got more companies like Vodaphone or GSK or Rolls Royce. We get regulations right. We have patient capital. We are the home to more top 500 companies than we are now.

6 We Are Purveyors of R&D to the World

We host the world's clusters. From Formula One in Oxford/Warwick/Birmingham to Tech City in East London and space activity around Harwell, we are famous for our world class R&D centres. The emerging economies are keen to work with us because creating a world-class university from scratch is hard. It is smarter to work with ones you have. Britain is increasingly recognised as the world's best R&D lab. We have achieved our ambition of being the best place in the world to do science. Multinationals base their R&D facilities here. Smart people from around the world want to come and research here. We have also earned a reputation as the best managers of big international scientific projects.

I believe that with our eight technologies we will probably have a mix of these outcomes. But I am optimistic. With our strong public support for R&D and these new measures for converting discovery into commercial opportunities we can indeed achieve a lot. We can help new businesses grow. We can be [the] world's R&D lab. We can indeed be the best place in the world to do science.

5 Why the Future Doesn't Need Us

Bill Joy

When this article by Bill Joy was first published in 2000, it caused a good deal of rumbling in scientific, engineering, and public policy circles. Joy was an extremely well-respected and accomplished computer scientist who had been a major player in the IT revolution. Yet in this piece, he raises questions and expresses doubts about the long-term implications of research that echo many of the concerns that critics of artificial intelligence research voice today. Joy does something parallel to the authors of the previous readings in that he takes certain research and development trends and extrapolates out to where they might take us. He is struck in particular by the possibilities of the convergence of genetics, nanotechnology, and robotics and especially the potential of this convergence to lead to self-replicating entities that might effectively take over the world because of their superior-to-human capacities. He insists that he is not antitechnology; he is not advocating that we stop developing the technologies at issue. Rather, he is concerned about the direction of development. Joy struggles with the negative vision to which his extrapolations lead and in so doing raises many of the questions that will be pursued in this book. His goal is to have us reflect on the technologies we are developing today to make sure they will help us create the future that we want.

From the moment I became involved in the creation of new technologies, their ethical dimensions have concerned me, but it was only in the autumn of 1998 that I became anxiously aware of how great are the dangers facing us in the 21st century. I can date the onset of my unease to the day I met Ray Kurzweil, the deservedly famous inventor of the first reading machine for the blind and many other amazing things.

Ray and I were both speakers at George Gilder's Telecosm conference, and I encountered him by chance in the bar of the hotel after both our sessions were over. I was sitting with John Searle, a Berkeley philosopher who studies consciousness. While we were talking, Ray approached and a conversation began, the subject of which haunts me to this day.

I had missed Ray's talk and the subsequent panel that Ray and John had been on, and they now picked right up where they'd left off, with Ray saying that the rate of improvement of technology was going to accelerate and that we were going to become robots or fuse with robots or something like that, and John countering that this couldn't happen, because the robots couldn't be conscious.

While I had heard such talk before, I had always felt sentient robots were in the realm of science fiction. But now, from someone I respected, I was hearing a strong

---

Originally published in *WIRED* 8, no. 4 (April 2000): 238–262.

argument that they were a near-term possibility. I was taken aback, especially given Ray's proven ability to imagine and create the future. I already knew that new technologies like genetic engineering and nanotechnology were giving us the power to remake the world, but a realistic and imminent scenario for intelligent robots surprised me.

It's easy to get jaded about such breakthroughs. We hear in the news almost every day of some kind of technological or scientific advance. Yet this was no ordinary prediction. In the hotel bar, Ray gave me a partial preprint of his then-forthcoming book *The Age of Spiritual Machines*, which outlined a utopia he foresaw—one in which humans gained near immortality by becoming one with robotic technology. On reading it, my sense of unease only intensified; I felt sure he had to be understating the dangers, understating the probability of a bad outcome along this path.

I found myself most troubled by a passage detailing a *dys*topian scenario:

The New Luddite Challenge

> First let us postulate that the computer scientists succeed in developing intelligent machines that can do all things better than human beings can do them. In that case presumably all work will be done by vast, highly organized systems of machines and no human effort will be necessary. Either of two cases might occur. The machines might be permitted to make all of their own decisions without human oversight, or else human control over the machines might be retained.
>
> If the machines are permitted to make all their own decisions, we can't make any conjectures as to the results, because it is impossible to guess how such machines might behave. We only point out that the fate of the human race would be at the mercy of the machines. It might be argued that the human race would never be foolish enough to hand over all the power to the machines. But we are suggesting neither that the human race would voluntarily turn power over to the machines nor that the machines would willfully seize power. What we do suggest is that the human race might easily permit itself to drift into a position of such dependence on the machines that it would have no practical choice but to accept all of the machines' decisions. As society and the problems that face it become more and more complex and machines become more and more intelligent, people will let machines make more of their decisions for them, simply because machine-made decisions will bring better results than man-made ones. Eventually a stage may be reached at which the decisions necessary to keep the system running will be so complex that human beings will be incapable of making them intelligently. At that stage the machines will be in effective control. People won't be able to just turn the machines off, because they will be so dependent on them that turning them off would amount to suicide.
>
> On the other hand it is possible that human control over the machines may be retained. In that case the average man may have control over certain private machines of his own, such as his car or his personal computer, but control over large systems of machines will be in the hands of a tiny elite—just as it is today; but with two differences. Due to improved techniques the elite will have greater control over the masses; and because human work will no longer be necessary the masses will be superfluous, a useless burden on the system. If the elite is ruthless they may simply decide to exterminate the mass of humanity. If they are humane they may use propaganda or other psychological or biological techniques to reduce the birth rate until the mass of humanity becomes extinct, leaving the world to the elite. Or, if the elite consists of soft-hearted liberals, they may decide to play the role of good shepherds to the rest of the human race. They will see to it that everyone's physical needs are satisfied, that all children are raised under psychologically hygienic conditions, that everyone has a wholesome hobby to keep him busy, and that anyone who may become dissatisfied undergoes "treatment" to cure his "problem." Of course, life will be so purposeless that people will have to be biologically

> or psychologically engineered either to remove their need for the power process or make them "sublimate" their drive for power into some harmless hobby. These engineered human beings may be happy in such a society, but they will most certainly not be free. They will have been reduced to the status of domestic animals.[1]

In the book, you don't discover until you turn the page that the author of this passage is Theodore Kaczynski—the Unabomber. I am no apologist for Kaczynski. His bombs killed three people during a 17-year terror campaign and wounded many others. One of his bombs gravely injured my friend David Gelernter, one of the most brilliant and visionary computer scientists of our time. Like many of my colleagues, I felt that I could easily have been the Unabomber's next target.

Kaczynski's actions were murderous and, in my view, criminally insane. He is clearly a Luddite, but simply saying this does not dismiss his argument; as difficult as it is for me to acknowledge, I saw some merit in the reasoning in this single passage. I felt compelled to confront it.

Kaczynski's dystopian vision describes unintended consequences, a well-known problem with the design and use of technology, and one that is clearly related to Murphy's law—"Anything that can go wrong, will." (Actually, this is Finagle's law, which in itself shows that Finagle was right.) Our overuse of antibiotics has led to what may be the biggest such problem so far: the emergence of antibiotic-resistant and much more dangerous bacteria. Similar things happened when attempts to eliminate malarial mosquitoes using DDT caused them to acquire DDT resistance; malarial parasites likewise acquired multi-drug-resistant genes.[2]

The cause of many such surprises seems clear: The systems involved are complex, involving interaction among and feedback between many parts. Any changes to such a system will cascade in ways that are difficult to predict; this is especially true when human actions are involved.

I started showing friends the Kaczynski quote from *The Age of Spiritual Machines*; I would hand them Kurzweil's book, let them read the quote, and then watch their reaction as they discovered who had written it. At around the same time, I found Hans Moravec's book *Robot: Mere Machine to Transcendent Mind*. Moravec is one of the leaders in robotics research, and was a founder of the world's largest robotics research program, at Carnegie Mellon University. *Robot* gave me more material to try out on my friends—material surprisingly supportive of Kaczynski's argument. For example:

> **The Short Run (Early 2000s)**
>
> Biological species almost never survive encounters with superior competitors. Ten million years ago, South and North America were separated by a sunken Panama isthmus. South America, like Australia today, was populated by marsupial mammals, including pouched equivalents of rats, deers, and tigers. When the isthmus connecting North and South America rose, it took only a few thousand years for the northern placental species, with slightly more effective metabolisms and reproductive and nervous systems, to displace and eliminate almost all the southern marsupials.
>
> In a completely free marketplace, superior robots would surely affect humans as North American placentals affected South American marsupials (and as humans have affected countless species). Robotic industries would compete vigorously among themselves for matter, energy, and space, incidentally driving their price beyond human reach. Unable to afford the necessities of life, biological humans would be squeezed out of existence.
>
> There is probably some breathing room, because we do not live in a completely free marketplace. Government coerces nonmarket behavior, especially by collecting taxes.

> Judiciously applied, governmental coercion could support human populations in high style on the fruits of robot labor, perhaps for a long while.

A textbook dystopia—and Moravec is just getting wound up. He goes on to discuss how our main job in the 21st century will be "ensuring continued cooperation from the robot industries" by passing laws decreeing that they be "nice,"[3] and to describe how seriously dangerous a human can be "once transformed into an unbounded superintelligent robot." Moravec's view is that the robots will eventually succeed us—that humans clearly face extinction.

I decided it was time to talk to my friend Danny Hillis. Danny became famous as the cofounder of Thinking Machines Corporation, which built a very powerful parallel supercomputer. Despite my current job title of Chief Scientist at Sun Microsystems, I am more a computer architect than a scientist, and I respect Danny's knowledge of the information and physical sciences more than that of any other single person I know. Danny is also a highly regarded futurist who thinks long-term—four years ago he started the Long Now Foundation, which is building a clock designed to last 10,000 years, in an attempt to draw attention to the pitifully short attention span of our society.[4]

So I flew to Los Angeles for the express purpose of having dinner with Danny and his wife, Pati. I went through my now-familiar routine, trotting out the ideas and passages that I found so disturbing. Danny's answer—directed specifically at Kurzweil's scenario of humans merging with robots—came swiftly, and quite surprised me. He said, simply, that the changes would come gradually, and that we would get used to them.

But I guess I wasn't totally surprised. I had seen a quote from Danny in Kurzweil's book in which he said, "I'm as fond of my body as anyone, but if I can be 200 with a body of silicon, I'll take it." It seemed that he was at peace with this process and its attendant risks, while I was not.

While talking and thinking about Kurzweil, Kaczynski, and Moravec, I suddenly remembered a novel I had read almost 20 years ago—*The White Plague*, by Frank Herbert—in which a molecular biologist is driven insane by the senseless murder of his family. To seek revenge he constructs and disseminates a new and highly contagious plague that kills widely but selectively. (We're lucky Kaczynski was a mathematician, not a molecular biologist.) I was also reminded of the Borg of *Star Trek*, a hive of partly biological, partly robotic creatures with a strong destructive streak. Borg-like disasters are a staple of science fiction, so why hadn't I been more concerned about such robotic dystopias earlier? Why weren't other people more concerned about these nightmarish scenarios?

Part of the answer certainly lies in our attitude toward the new—in our bias toward instant familiarity and unquestioning acceptance. Accustomed to living with almost routine scientific breakthroughs, we have yet to come to terms with the fact that the most compelling 21st-century technologies—robotics, genetic engineering, and nanotechnology—pose a different threat than the technologies that have come before. Specifically, robots, engineered organisms, and nanobots share a dangerous amplifying factor: They can self-replicate. A bomb is blown up only once—but one bot can become many, and quickly get out of control.

Much of my work over the past 25 years has been on computer networking, where the sending and receiving of messages creates the opportunity for out-of-control replication. But while replication in a computer or a computer network can be a nuisance, at worst it disables a machine or takes down a network or network service. Uncontrolled self-replication in these newer technologies runs a much greater risk: a risk of substantial damage in the physical world.

Each of these technologies also offers untold promise: The vision of near immortality that Kurzweil sees in his robot dreams drives us forward; genetic engineering may soon provide treatments, if not outright cures, for most diseases; and nanotechnology and nanomedicine can address yet more ills. Together they could significantly extend our average life span and improve the quality of our lives. Yet, with each of these technologies, a sequence of small, individually sensible advances leads to an accumulation of great power and, concomitantly, great danger.

What was different in the 20th century? Certainly, the technologies underlying the weapons of mass destruction (WMD)—nuclear, biological, and chemical (NBC)—were powerful, and the weapons an enormous threat. But building nuclear weapons required, at least for a time, access to both rare—indeed, effectively unavailable—raw materials and highly protected information; biological and chemical weapons programs also tended to require large-scale activities.

The 21st-century technologies—genetics, nanotechnology, and robotics (GNR)—are so powerful that they can spawn whole new classes of accidents and abuses. Most dangerously, for the first time, these accidents and abuses are widely within the reach of individuals or small groups. They will not require large facilities or rare raw materials. Knowledge alone will enable the use of them.

Thus we have the possibility not just of weapons of mass destruction but of knowledge-enabled mass destruction (KMD), this destructiveness hugely amplified by the power of self-replication.

I think it is no exaggeration to say we are on the cusp of the further perfection of extreme evil, an evil whose possibility spreads well beyond that which weapons of mass destruction bequeathed to the nation-states, on to a surprising and terrible empowerment of extreme individuals.

Nothing about the way I got involved with computers suggested to me that I was going to be facing these kinds of issues.

My life has been driven by a deep need to ask questions and find answers. When I was 3, I was already reading, so my father took me to the elementary school, where I sat on the principal's lap and read him a story. I started school early, later skipped a grade, and escaped into books—I was incredibly motivated to learn. I asked lots of questions, often driving adults to distraction.

As a teenager I was very interested in science and technology. I wanted to be a ham radio operator but didn't have the money to buy the equipment. Ham radio was the Internet of its time: very addictive, and quite solitary. Money issues aside, my mother put her foot down—I was not to be a ham; I was antisocial enough already.

I may not have had many close friends, but I was awash in ideas. By high school, I had discovered the great science fiction writers. I remember especially Heinlein's *Have Spacesuit Will Travel* and Asimov's *I, Robot*, with its Three Laws of Robotics. I was enchanted by the descriptions of space travel, and wanted to have a telescope to look at the stars; since I had no money to buy or make one, I checked books on telescope-making out of the library and read about making them instead. I soared in my imagination.

Thursday nights my parents went bowling, and we kids stayed home alone. It was the night of Gene Roddenberry's original *Star Trek*, and the program made a big impression on me. I came to accept its notion that humans had a future in space, Western-style, with big heroes and adventures. Roddenberry's vision of the centuries to come was one with strong moral values, embodied in codes like the Prime Directive: to not interfere in the development of less technologically advanced civilizations. This had an incredible appeal to me; ethical humans, not robots, dominated this future, and I took Roddenberry's dream as part of my own.

I excelled in mathematics in high school, and when I went to the University of Michigan as an undergraduate engineering student I took the advanced curriculum of the mathematics majors. Solving math problems was an exciting challenge, but when I discovered computers I found something much more interesting: a machine into which you could put a program that attempted to solve a problem, after which the machine quickly checked the solution. The computer had a clear notion of correct and incorrect, true and false. Were my ideas correct? The machine could tell me. This was very seductive.

I was lucky enough to get a job programming early supercomputers and discovered the amazing power of large machines to numerically simulate advanced designs. When I went to graduate school at UC Berkeley in the mid-1970s, I started staying up late, often all night, inventing new worlds inside the machines. Solving problems. Writing the code that argued so strongly to be written.

In *The Agony and the Ecstasy*, Irving Stone's biographical novel of Michelangelo, Stone described vividly how Michelangelo released the statues from the stone, "breaking the marble spell," carving from the images in his mind.[5] In my most ecstatic moments, the software in the computer emerged in the same way. Once I had imagined it in my mind I felt that it was already there in the machine, waiting to be released. Staying up all night seemed a small price to pay to free it—to give the ideas concrete form.

After a few years at Berkeley I started to send out some of the software I had written—an instructional Pascal system, Unix utilities, and a text editor called vi (which is still, to my surprise, widely used more than 20 years later)—to others who had similar small PDP-11 and VAX minicomputers. These adventures in software eventually turned into the Berkeley version of the Unix operating system, which became a personal "success disaster"—so many people wanted it that I never finished my PhD. Instead I got a job working for Darpa putting Berkeley Unix on the Internet and fixing it to be reliable and to run large research applications well. This was all great fun and very rewarding. And, frankly, I saw no robots here, or anywhere near.

Still, by the early 1980s, I was drowning. The Unix releases were very successful, and my little project of one soon had money and some staff, but the problem at Berkeley was always office space rather than money—there wasn't room for the help the project needed, so when the other founders of Sun Microsystems showed up I jumped at the chance to join them. At Sun, the long hours continued into the early days of workstations and personal computers, and I have enjoyed participating in the creation of advanced microprocessor technologies and Internet technologies such as Java and Jini. From all this, I trust it is clear that I am not a Luddite. I have always, rather, had a strong belief in the value of the scientific search for truth and in the ability of great engineering to bring material progress. The Industrial Revolution has immeasurably improved everyone's life over the last couple hundred years, and I always expected my career to involve the building of worthwhile solutions to real problems, one problem at a time.

I have not been disappointed. My work has had more impact than I had ever hoped for and has been more widely used than I could have reasonably expected. I have spent the last 20 years still trying to figure out how to make computers as reliable as I want them to be (they are not nearly there yet) and how to make them simple to use (a goal that has met with even less relative success). Despite some progress, the problems that remain seem even more daunting.

But while I was aware of the moral dilemmas surrounding technology's consequences in fields like weapons research, I did not expect that I would confront such issues in my own field, or at least not so soon.

Perhaps it is always hard to see the bigger impact while you are in the vortex of a change. Failing to understand the consequences of our inventions while we are in the rapture of discovery and innovation seems to be a common fault of scientists and technologists; we have long been driven by the overarching desire to know that is the nature of science's quest, not stopping to notice that the progress to newer and more powerful technologies can take on a life of its own.

I have long realized that the big advances in information technology come not from the work of computer scientists, computer architects, or electrical engineers, but from that of physical scientists. The physicists Stephen Wolfram and Brosl Hasslacher introduced me, in the early 1980s, to chaos theory and nonlinear systems. In the 1990s, I learned about complex systems from conversations with Danny Hillis, the biologist Stuart Kauffman, the Nobel-laureate physicist Murray Gell-Mann, and others. Most recently, Hasslacher and the electrical engineer and device physicist Mark Reed have been giving me insight into the incredible possibilities of molecular electronics.

In my own work, as codesigner of three microprocessor architectures—SPARC, pico-Java, and MAJC—and as the designer of several implementations thereof, I've been afforded a deep and firsthand acquaintance with Moore's law. For decades, Moore's law has correctly predicted the exponential rate of improvement of semiconductor technology. Until last year I believed that the rate of advances predicted by Moore's law might continue only until roughly 2010, when some physical limits would begin to be reached. It was not obvious to me that a new technology would arrive in time to keep performance advancing smoothly.

But because of the recent rapid and radical progress in molecular electronics—where individual atoms and molecules replace lithographically drawn transistors—and related nanoscale technologies, we should be able to meet or exceed the Moore's law rate of progress for another 30 years. By 2030, we are likely to be able to build machines, in quantity, a million times as powerful as the personal computers of today—sufficient to implement the dreams of Kurzweil and Moravec.

As this enormous computing power is combined with the manipulative advances of the physical sciences and the new, deep understandings in genetics, enormous transformative power is being unleashed. These combinations open up the opportunity to completely redesign the world, for better or worse: The replicating and evolving processes that have been confined to the natural world are about to become realms of human endeavor.

In designing software and microprocessors, I have never had the feeling that I was designing an intelligent machine. The software and hardware is so fragile and the capabilities of the machine to "think" so clearly absent that, even as a possibility, this has always seemed very far in the future.

But now, with the prospect of human-level computing power in about 30 years, a new idea suggests itself: that I may be working to create tools which will enable the construction of the technology that may replace our species. How do I feel about this? Very uncomfortable. Having struggled my entire career to build reliable software systems, it seems to me more than likely that this future will not work out as well as some people may imagine. My personal experience suggests we tend to overestimate our design abilities.

Given the incredible power of these new technologies, shouldn't we be asking how we can best coexist with them? And if our own extinction is a likely, or even possible, outcome of our technological development, shouldn't we proceed with great caution?

The dream of robotics is, first, that intelligent machines can do our work for us, allowing us lives of leisure, restoring us to Eden. Yet in his history of such ideas, *Darwin Among the Machines*, George Dyson warns: "In the game of life and evolution there are three players at the table: human beings, nature, and machines. I am firmly on the side of nature. But nature, I suspect, is on the side of the machines." As we have seen, Moravec agrees, believing we may well not survive the encounter with the superior robot species.

How soon could such an intelligent robot be built? The coming advances in computing power seem to make it possible by 2030. And once an intelligent robot exists, it is only a small step to a robot species—to an intelligent robot that can make evolved copies of itself.

A second dream of robotics is that we will gradually replace ourselves with our robotic technology, achieving near immortality by downloading our consciousnesses; it is this process that Danny Hillis thinks we will gradually get used to and that Ray Kurzweil elegantly details in *The Age of Spiritual Machines*. (We are beginning to see intimations of this in the implantation of computer devices into the human body, as illustrated on the cover of *Wired* 8.02.)

But if we are downloaded into our technology, what are the chances that we will thereafter be ourselves or even human? It seems to me far more likely that a robotic existence would not be like a human one in any sense that we understand, that the robots would in no sense be our children, that on this path our humanity may well be lost.

Genetic engineering promises to revolutionize agriculture by increasing crop yields while reducing the use of pesticides; to create tens of thousands of novel species of bacteria, plants, viruses, and animals; to replace reproduction, or supplement it, with cloning; to create cures for many diseases, increasing our life span and our quality of life; and much, much more. We now know with certainty that these profound changes in the biological sciences are imminent and will challenge all our notions of what life is.

Technologies such as human cloning have in particular raised our awareness of the profound ethical and moral issues we face. If, for example, we were to reengineer ourselves into several separate and unequal species using the power of genetic engineering, then we would threaten the notion of equality that is the very cornerstone of our democracy.

Given the incredible power of genetic engineering, it's no surprise that there are significant safety issues in its use. My friend Amory Lovins recently cowrote, along with Hunter Lovins, an editorial that provides an ecological view of some of these dangers. Among their concerns: that "the new botany aligns the development of plants with their economic, not evolutionary, success."[6] Amory's long career has been focused on energy and resource efficiency by taking a whole-system view of human-made systems; such a whole-system view often finds simple, smart solutions to otherwise seemingly difficult problems, and is usefully applied here as well.

After reading the Lovins' editorial, I saw an op-ed by Gregg Easterbrook in the *New York Times* (November 19, 1999) about genetically engineered crops, under the headline: "Food for the Future: Someday, Rice Will Have Built-In Vitamin A. Unless the Luddites Win."

Are Amory and Hunter Lovins Luddites? Certainly not. I believe we all would agree that golden rice, with its built-in vitamin A, is probably a good thing, if developed with proper care and respect for the likely dangers in moving genes across species boundaries.

Awareness of the dangers inherent in genetic engineering is beginning to grow, as reflected in the Lovins' editorial. The general public is aware of, and uneasy about, genetically modified foods, and seems to be rejecting the notion that such foods should be permitted to be unlabeled.

But genetic engineering technology is already very far along. As the Lovins note, the USDA has already approved about 50 genetically engineered crops for unlimited release; more than half of the world's soybeans and a third of its corn now contain genes spliced in from other forms of life.

While there are many important issues here, my own major concern with genetic engineering is narrower: that it gives the power—whether militarily, accidentally, or in a deliberate terrorist act—to create a White Plague.

The many wonders of nanotechnology were first imagined by the Nobel-laureate physicist Richard Feynman in a speech he gave in 1959, subsequently published under the title "There's Plenty of Room at the Bottom." The book that made a big impression on me, in the mid-'80s, was Eric Drexler's *Engines of Creation*, in which he described beautifully how manipulation of matter at the atomic level could create a utopian future of abundance, where just about everything could be made cheaply, and almost any imaginable disease or physical problem could be solved using nanotechnology and artificial intelligences.

A subsequent book, *Unbounding the Future: The Nanotechnology Revolution*, which Drexler cowrote, imagines some of the changes that might take place in a world where we had molecular-level "assemblers." Assemblers could make possible incredibly low-cost solar power, cures for cancer and the common cold by augmentation of the human immune system, essentially complete cleanup of the environment, incredibly inexpensive pocket supercomputers—in fact, any product would be manufacturable by assemblers at a cost no greater than that of wood—spaceflight more accessible than transoceanic travel today, and restoration of extinct species.

I remember feeling good about nanotechnology after reading *Engines of Creation*. As a technologist, it gave me a sense of calm—that is, nanotechnology showed us that incredible progress was possible, and indeed perhaps inevitable. If nanotechnology was our future, then I didn't feel pressed to solve so many problems in the present. I would get to Drexler's utopian future in due time; I might as well enjoy life more in the here and now. It didn't make sense, given his vision, to stay up all night, all the time.

Drexler's vision also led to a lot of good fun. I would occasionally get to describe the wonders of nanotechnology to others who had not heard of it. After teasing them with all the things Drexler described I would give a homework assignment of my own: "Use nanotechnology to create a vampire; for extra credit create an antidote."

With these wonders came clear dangers, of which I was acutely aware. As I said at a nanotechnology conference in 1989, "We can't simply do our science and not worry about these ethical issues."[7] But my subsequent conversations with physicists convinced me that nanotechnology might not even work—or, at least, it wouldn't work anytime soon. Shortly thereafter I moved to Colorado, to a skunk works I had set up, and the focus of my work shifted to software for the Internet, specifically on ideas that became Java and Jini.

Then, last summer, Brosl Hasslacher told me that nanoscale molecular electronics was now practical. This was *new* news, at least to me, and I think to many people—and it radically changed my opinion about nanotechnology. It sent me back to *Engines of Creation*. Rereading Drexler's work after more than 10 years, I was dismayed to realize how little I had remembered of its lengthy section called "Dangers and Hopes," including a discussion of how nanotechnologies can become "engines of destruction." Indeed, in my rereading of this cautionary material today, I am struck by how naive some of Drexler's safeguard proposals seem, and how much greater I judge the dangers to be now than even he seemed to then. (Having anticipated and described many technical and political problems with nanotechnology, Drexler started the Foresight Institute in the late 1980s "to help prepare society for anticipated advanced technologies"—most important, nanotechnology.)

The enabling breakthrough to assemblers seems quite likely within the next 20 years. Molecular electronics—the new subfield of nanotechnology where individual molecules are circuit elements—should mature quickly and become enormously lucrative within this decade, causing a large incremental investment in all nanotechnologies.

Unfortunately, as with nuclear technology, it is far easier to create destructive uses for nanotechnology than constructive ones. Nanotechnology has clear military and terrorist uses, and you need not be suicidal to release a massively destructive nanotechnological device—such devices can be built to be selectively destructive, affecting, for example, only a certain geographical area or a group of people who are genetically distinct.

An immediate consequence of the Faustian bargain in obtaining the great power of nanotechnology is that we run a grave risk—the risk that we might destroy the biosphere on which all life depends.

As Drexler explained:

> "Plants" with "leaves" no more efficient than today's solar cells could out-compete real plants, crowding the biosphere with an inedible foliage. Tough omnivorous "bacteria" could out-compete real bacteria: They could spread like blowing pollen, replicate swiftly, and reduce the biosphere to dust in a matter of days. Dangerous replicators could easily be too tough, small, and rapidly spreading to stop—at least if we make no preparation. We have trouble enough controlling viruses and fruit flies.
>
> Among the cognoscenti of nanotechnology, this threat has become known as the "gray goo problem." Though masses of uncontrolled replicators need not be gray or gooey, the term "gray goo" emphasizes that replicators able to obliterate life might be less inspiring than a single species of crabgrass. They might be superior in an evolutionary sense, but this need not make them valuable.
>
> The gray goo threat makes one thing perfectly clear: We cannot afford certain kinds of accidents with replicating assemblers.
>
> Gray goo would surely be a depressing ending to our human adventure on Earth, far worse than mere fire or ice, and one that could stem from a simple laboratory accident.[8]

Oops.

It is most of all the power of destructive self-replication in genetics, nanotechnology, and robotics (GNR) that should give us pause. Self-replication is the modus operandi of genetic engineering, which uses the machinery of the cell to replicate its designs, and the prime danger underlying gray goo in nanotechnology. Stories of runamok robots like the Borg, replicating or mutating to escape from the ethical constraints imposed on them by their creators, are well established in our science fiction books and movies. It is even possible that self-replication may be more fundamental than we thought, and hence harder—or even impossible—to control. A recent article by Stuart Kauffman in *Nature* titled "Self-Replication: Even Peptides Do It" discusses the discovery that a 32-amino-acid peptide can "autocatalyse its own synthesis." We don't know how widespread this ability is, but Kauffman notes that it may hint at "a route to self-reproducing molecular systems on a basis far wider than Watson-Crick base-pairing."[9]

In truth, we have had in hand for years clear warnings of the dangers inherent in widespread knowledge of GNR technologies—of the possibility of knowledge alone enabling mass destruction. But these warnings haven't been widely publicized; the public discussions have been clearly inadequate. There is no profit in publicizing the dangers. The nuclear, biological, and chemical (NBC) technologies used in 20th-century weapons of mass destruction were and are largely military, developed in government laboratories. In sharp contrast, the 21st-century GNR technologies have clear commercial uses and are being developed almost exclusively by corporate enterprises. In this age of triumphant commercialism, technology—with science as its handmaiden—is delivering a series of almost magical inventions that are the most phenomenally lucrative ever seen. We are aggressively pursuing the promises of these new technologies within the now-unchallenged system of global capitalism and its manifold financial incentives and competitive pressures.

This is the first moment in the history of our planet when any species, by its own voluntary actions, has become a danger to itself—as well as to vast numbers of others.

It might be a familiar progression, transpiring on many worlds—a planet, newly formed, placidly revolves around its star; life slowly forms; a kaleidoscopic procession of creatures evolves; intelligence emerges which, at least up to a point, confers enormous survival value; and then technology is invented. It dawns on them that there are such things as laws of Nature, that these laws can be revealed by experiment, and that knowledge of these laws can be made both to save and to take lives, both on unprecedented scales. Science, they recognize, grants immense powers. In a flash, they create world-altering contrivances. Some planetary civilizations see their way through, place limits on what may and what must not be done, and safely pass through the time of perils. Others, not so lucky or so prudent, perish.

That is Carl Sagan, writing in 1994, in *Pale Blue Dot*, a book describing his vision of the human future in space. I am only now realizing how deep his insight was, and how sorely I miss, and will miss, his voice. For all its eloquence, Sagan's contribution was not least that of simple common sense—an attribute that, along with humility, many of the leading advocates of the 21st-century technologies seem to lack.

I remember from my childhood that my grandmother was strongly against the overuse of antibiotics. She had worked since before the first World War as a nurse and had a commonsense attitude that taking antibiotics, unless they were absolutely necessary, was bad for you.

It is not that she was an enemy of progress. She saw much progress in an almost 70-year nursing career; my grandfather, a diabetic, benefited greatly from the improved treatments that became available in his lifetime. But she, like many levelheaded people, would probably think it greatly arrogant for us, now, to be designing a robotic "replacement species," when we obviously have so much trouble making relatively simple things work, and so much trouble managing—or even understanding—ourselves.

I realize now that she had an awareness of the nature of the order of life, and of the necessity of living with and respecting that order. With this respect comes a necessary humility that we, with our early-21st-century chutzpah, lack at our peril. The commonsense view, grounded in this respect, is often right, in advance of the scientific evidence. The clear fragility and inefficiencies of the human-made systems we have built should give us all pause; the fragility of the systems I have worked on certainly humbles me.

We should have learned a lesson from the making of the first atomic bomb and the resulting arms race. We didn't do well then, and the parallels to our current situation are troubling.

The effort to build the first atomic bomb was led by the brilliant physicist J. Robert Oppenheimer. Oppenheimer was not naturally interested in politics but became painfully aware of what he perceived as the grave threat to Western civilization from the Third Reich, a threat surely grave because of the possibility that Hitler might obtain nuclear weapons. Energized by this concern, he brought his strong intellect, passion for physics, and charismatic leadership skills to Los Alamos and led a rapid and successful effort by an incredible collection of great minds to quickly invent the bomb.

What is striking is how this effort continued so naturally after the initial impetus was removed. In a meeting shortly after V-E Day with some physicists who felt that perhaps the effort should stop, Oppenheimer argued to continue. His stated reason seems a bit strange: not because of the fear of large casualties from an invasion of Japan, but because the United Nations, which was soon to be formed, should have foreknowledge of atomic weapons. A more likely reason the project continued is the momentum that had built up—the first atomic test, Trinity, was nearly at hand.

We know that in preparing this first atomic test the physicists proceeded despite a large number of possible dangers. They were initially worried, based on a calculation by Edward Teller, that an atomic explosion might set fire to the atmosphere. A revised calculation reduced the danger of destroying the world to a three-in-a-million chance. (Teller says he was later able to dismiss the prospect of atmospheric ignition entirely.) Oppenheimer, though, was sufficiently concerned about the result of Trinity that he arranged for a possible evacuation of the southwest part of the state of New Mexico. And, of course, there was the clear danger of starting a nuclear arms race.

Within a month of that first, successful test, two atomic bombs destroyed Hiroshima and Nagasaki. Some scientists had suggested that the bomb simply be demonstrated, rather than dropped on Japanese cities—saying that this would greatly improve the chances for arms control after the war—but to no avail. With the tragedy of Pearl Harbor still fresh in Americans' minds, it would have been very difficult for President Truman to order a demonstration of the weapons rather than use them as he did—the desire to quickly end the war and save the lives that would have been lost in any invasion of Japan was very strong. Yet the overriding truth was probably very simple: As the physicist Freeman Dyson later said, "The reason that it was dropped was just that nobody had the courage or the foresight to say no."

It's important to realize how shocked the physicists were in the aftermath of the bombing of Hiroshima, on August 6, 1945. They describe a series of waves of emotion: first, a sense of fulfillment that the bomb worked, then horror at all the people that had been killed, and then a convincing feeling that on no account should another bomb be dropped. Yet of course another bomb was dropped, on Nagasaki, only three days after the bombing of Hiroshima.

In November 1945, three months after the atomic bombings, Oppenheimer stood firmly behind the scientific attitude, saying, "It is not possible to be a scientist unless you believe that the knowledge of the world, and the power which this gives, is a thing which is of intrinsic value to humanity, and that you are using it to help in the spread of knowledge and are willing to take the consequences."

Oppenheimer went on to work, with others, on the Acheson-Lilienthal report, which, as Richard Rhodes says in his recent book *Visions of Technology*, "found a way to prevent a clandestine nuclear arms race without resorting to armed world government";

their suggestion was a form of relinquishment of nuclear weapons work by nation-states to an international agency.

This proposal led to the Baruch Plan, which was submitted to the United Nations in June 1946 but never adopted (perhaps because, as Rhodes suggests, Bernard Baruch had "insisted on burdening the plan with conventional sanctions," thereby inevitably dooming it, even though it would "almost certainly have been rejected by Stalinist Russia anyway"). Other efforts to promote sensible steps toward internationalizing nuclear power to prevent an arms race ran afoul either of US politics and internal distrust, or distrust by the Soviets. The opportunity to avoid the arms race was lost, and very quickly.

Two years later, in 1948, Oppenheimer seemed to have reached another stage in his thinking, saying, "In some sort of crude sense which no vulgarity, no humor, no overstatement can quite extinguish, the physicists have known sin; and this is a knowledge they cannot lose."

In 1949, the Soviets exploded an atom bomb. By 1955, both the US and the Soviet Union had tested hydrogen bombs suitable for delivery by aircraft. And so the nuclear arms race began.

Nearly 20 years ago, in the documentary *The Day After Trinity*, Freeman Dyson summarized the scientific attitudes that brought us to the nuclear precipice:

> I have felt it myself. The glitter of nuclear weapons. It is irresistible if you come to them as a scientist. To feel it's there in your hands, to release this energy that fuels the stars, to let it do your bidding. To perform these miracles, to lift a million tons of rock into the sky. It is something that gives people an illusion of illimitable power, and it is, in some ways, responsible for all our troubles—this, what you might call technical arrogance, that overcomes people when they see what they can do with their minds.[10]

Now, as then, we are creators of new technologies and stars of the imagined future, driven—this time by great financial rewards and global competition—despite the clear dangers, hardly evaluating what it may be like to try to live in a world that is the realistic outcome of what we are creating and imagining.

In 1947, *The Bulletin of the Atomic Scientists* began putting a Doomsday Clock on its cover. For more than 50 years, it has shown an estimate of the relative nuclear danger we have faced, reflecting the changing international conditions. The hands on the clock have moved 15 times and today, standing at nine minutes to midnight, reflect continuing and real danger from nuclear weapons. The recent addition of India and Pakistan to the list of nuclear powers has increased the threat of failure of the nonproliferation goal, and this danger was reflected by moving the hands closer to midnight in 1998.

In our time, how much danger do we face, not just from nuclear weapons, but from all of these technologies? How high are the extinction risks?

The philosopher John Leslie has studied this question and concluded that the risk of human extinction is at least 30 percent,[11] while Ray Kurzweil believes we have "a better than even chance of making it through," with the caveat that he has "always been accused of being an optimist." Not only are these estimates not encouraging, but they do not include the probability of many horrid outcomes that lie short of extinction.

Faced with such assessments, some serious people are already suggesting that we simply move beyond Earth as quickly as possible. We would colonize the galaxy using von Neumann probes, which hop from star system to star system, replicating as they go. This step will almost certainly be necessary 5 billion years from now (or sooner if our solar system is disastrously impacted by the impending collision of our galaxy with the Andromeda galaxy within the next 3 billion years), but if we take Kurzweil and Moravec at their word it might be necessary by the middle of this century.

What are the moral implications here? If we must move beyond Earth this quickly in order for the species to survive, who accepts the responsibility for the fate of those (most of us, after all) who are left behind? And even if we scatter to the stars, isn't it likely that we may take our problems with us or find, later, that they have followed us? The fate of our species on Earth and our fate in the galaxy seem inextricably linked. Another idea is to erect a series of shields to defend against each of the dangerous technologies. The Strategic Defense Initiative, proposed by the Reagan administration, was an attempt to design such a shield against the threat of a nuclear attack from the Soviet Union. But as Arthur C. Clarke, who was privy to discussions about the project, observed: "Though it might be possible, at vast expense, to construct local defense systems that would 'only' let through a few percent of ballistic missiles, the much touted idea of a national umbrella was nonsense. Luis Alvarez, perhaps the greatest experimental physicist of this century, remarked to me that the advocates of such schemes were 'very bright guys with no common sense.'"

Clarke continued: "Looking into my often cloudy crystal ball, I suspect that a total defense might indeed be possible in a century or so. But the technology involved would produce, as a by-product, weapons so terrible that no one would bother with anything as primitive as ballistic missiles."[12]

In *Engines of Creation*, Eric Drexler proposed that we build an active nanotechnological shield—a form of immune system for the biosphere—to defend against dangerous replicators of all kinds that might escape from laboratories or otherwise be maliciously created. But the shield he proposed would itself be extremely dangerous—nothing could prevent it from developing autoimmune problems and attacking the biosphere itself.[13]

Similar difficulties apply to the construction of shields against robotics and genetic engineering. These technologies are too powerful to be shielded against in the time frame of interest; even if it were possible to implement defensive shields, the side effects of their development would be at least as dangerous as the technologies we are trying to protect against.

These possibilities are all thus either undesirable or unachievable or both. The only realistic alternative I see is relinquishment: to limit development of the technologies that are too dangerous, by limiting our pursuit of certain kinds of knowledge.

Yes, I know, knowledge is good, as is the search for new truths. We have been seeking knowledge since ancient times. Aristotle opened his *Metaphysics* with the simple statement: "All men by nature desire to know." We have, as a bedrock value in our society, long agreed on the value of open access to information, and recognize the problems that arise with attempts to restrict access to and development of knowledge. In recent times, we have come to revere scientific knowledge.

But despite the strong historical precedents, if open access to and unlimited development of knowledge henceforth puts us all in clear danger of extinction, then common sense demands that we reexamine even these basic, long-held beliefs.

It was Nietzsche who warned us, at the end of the 19th century, not only that God is dead but that "faith in science, which after all exists undeniably, cannot owe its origin to a calculus of utility; it must have originated *in spite of* the fact that the disutility and dangerousness of the 'will to truth,' of 'truth at any price' is proved to it constantly." It is this further danger that we now fully face—the consequences of our truth-seeking. The truth that science seeks can certainly be considered a dangerous substitute for God if it is likely to lead to our extinction.

If we could agree, as a species, what we wanted, where we were headed, and why, then we would make our future much less dangerous—then we might understand what we can and should relinquish. Otherwise, we can easily imagine an arms race developing over GNR technologies, as it did with the NBC technologies in the 20th century. This is perhaps the greatest risk, for once such a race begins, it's very hard to end it. This time—unlike during the Manhattan Project—we aren't in a war, facing an implacable enemy that is threatening our civilization; we are driven, instead, by our habits, our desires, our economic system, and our competitive need to know.

I believe that we all wish our course could be determined by our collective values, ethics, and morals. If we had gained more collective wisdom over the past few thousand years, then a dialogue to this end would be more practical, and the incredible powers we are about to unleash would not be nearly so troubling.

One would think we might be driven to such a dialogue by our instinct for self-preservation. Individuals clearly have this desire, yet as a species our behavior seems to be not in our favor. In dealing with the nuclear threat, we often spoke dishonestly to ourselves and to each other, thereby greatly increasing the risks. Whether this was politically motivated, or because we chose not to think ahead, or because when faced with such grave threats we acted irrationally out of fear, I do not know, but it does not bode well.

The new Pandora's boxes of genetics, nanotechnology, and robotics are almost open, yet we seem hardly to have noticed. Ideas can't be put back in a box; unlike uranium or plutonium, they don't need to be mined and refined, and they can be freely copied. Once they are out, they are out. Churchill remarked, in a famous left-handed compliment, that the American people and their leaders "invariably do the right thing, after they have examined every other alternative." In this case, however, we must act more presciently, as to do the right thing only at last may be to lose the chance to do it at all.

As Thoreau said, "We do not ride on the railroad; it rides upon us"; and this is what we must fight, in our time. The question is, indeed, Which is to be master? Will we survive our technologies?

We are being propelled into this new century with no plan, no control, no brakes. Have we already gone too far down the path to alter course? I don't believe so, but we aren't trying yet, and the last chance to assert control—the fail-safe point—is rapidly approaching. We have our first pet robots, as well as commercially available genetic engineering techniques, and our nanoscale techniques are advancing rapidly. While the development of these technologies proceeds through a number of steps, it isn't necessarily the case—as happened in the Manhattan Project and the Trinity test—that the last step in proving a technology is large and hard. The breakthrough to wild self-replication in robotics, genetic engineering, or nanotechnology could come suddenly, reprising the surprise we felt when we learned of the cloning of a mammal.

And yet I believe we do have a strong and solid basis for hope. Our attempts to deal with weapons of mass destruction in the last century provide a shining example of relinquishment for us to consider: the unilateral US abandonment, without preconditions, of the development of biological weapons. This relinquishment stemmed from the realization that while it would take an enormous effort to create these terrible weapons, they could from then on easily be duplicated and fall into the hands of rogue nations or terrorist groups.

The clear conclusion was that we would create additional threats to ourselves by pursuing these weapons, and that we would be more secure if we did not pursue them. We have embodied our relinquishment of biological and chemical weapons in the

1972 Biological Weapons Convention (BWC) and the 1993 Chemical Weapons Convention (CWC).[14]

As for the continuing sizable threat from nuclear weapons, which we have lived with now for more than 50 years, the US Senate's recent rejection of the Comprehensive Test Ban Treaty makes it clear relinquishing nuclear weapons will not be politically easy. But we have a unique opportunity, with the end of the Cold War, to avert a multipolar arms race. Building on the BWC and CWC relinquishments, successful abolition of nuclear weapons could help us build toward a habit of relinquishing dangerous technologies. (Actually, by getting rid of all but 100 nuclear weapons worldwide—roughly the total destructive power of World War II and a considerably easier task—we could eliminate this extinction threat.)[15]

Verifying relinquishment will be a difficult problem, but not an unsolvable one. We are fortunate to have already done a lot of relevant work in the context of the BWC and other treaties. Our major task will be to apply this to technologies that are naturally much more commercial than military. The substantial need here is for transparency, as difficulty of verification is directly proportional to the difficulty of distinguishing relinquished from legitimate activities.

I frankly believe that the situation in 1945 was simpler than the one we now face: The nuclear technologies were reasonably separable into commercial and military uses, and monitoring was aided by the nature of atomic tests and the ease with which radioactivity could be measured. Research on military applications could be performed at national laboratories such as Los Alamos, with the results kept secret as long as possible.

The GNR technologies do not divide clearly into commercial and military uses; given their potential in the market, it's hard to imagine pursuing them only in national laboratories. With their widespread commercial pursuit, enforcing relinquishment will require a verification regime similar to that for biological weapons, but on an unprecedented scale. This, inevitably, will raise tensions between our individual privacy and desire for proprietary information, and the need for verification to protect us all. We will undoubtedly encounter strong resistance to this loss of privacy and freedom of action.

Verifying the relinquishment of certain GNR technologies will have to occur in cyberspace as well as at physical facilities. The critical issue will be to make the necessary transparency acceptable in a world of proprietary information, presumably by providing new forms of protection for intellectual property.

Verifying compliance will also require that scientists and engineers adopt a strong code of ethical conduct, resembling the Hippocratic oath, and that they have the courage to whistleblow as necessary, even at high personal cost. This would answer the call—50 years after Hiroshima—by the Nobel laureate Hans Bethe, one of the most senior of the surviving members of the Manhattan Project, that all scientists "cease and desist from work creating, developing, improving, and manufacturing nuclear weapons and other weapons of potential mass destruction."[16] In the 21st century, this requires vigilance and personal responsibility by those who would work on both NBC and GNR technologies to avoid implementing weapons of mass destruction and knowledge-enabled mass destruction.

Thoreau also said that we will be "rich in proportion to the number of things which we can afford to let alone." We each seek to be happy, but it would seem worthwhile to question whether we need to take such a high risk of total destruction to gain yet more knowledge and yet more things; common sense says that there is a limit to our material needs—and that certain knowledge is too dangerous and is best forgone.

Neither should we pursue near immortality without considering the costs, without considering the commensurate increase in the risk of extinction. Immortality, while perhaps the original, is certainly not the only possible utopian dream.

I recently had the good fortune to meet the distinguished author and scholar Jacques Attali, whose book *Lignes d'horizons* (*Millennium*, in the English translation) helped inspire the Java and Jini approach to the coming age of pervasive computing, as previously described in this magazine. In his new book *Fraternités*, Attali describes how our dreams of utopia have changed over time:

> At the dawn of societies, men saw their passage on Earth as nothing more than a labyrinth of pain, at the end of which stood a door leading, via their death, to the company of gods and to *Eternity*. With the Hebrews and then the Greeks, some men dared free themselves from theological demands and dream of an ideal City where *Liberty* would flourish. Others, noting the evolution of the market society, understood that the liberty of some would entail the alienation of others, and they sought *Equality*.17

Jacques helped me understand how these three different utopian goals exist in tension in our society today. He goes on to describe a fourth utopia, *Fraternity*, whose foundation is altruism. Fraternity alone associates individual happiness with the happiness of others, affording the promise of self-sustainment.

This crystallized for me my problem with Kurzweil's dream. A technological approach to Eternity—near immortality through robotics—may not be the most desirable utopia, and its pursuit brings clear dangers. Maybe we should rethink our utopian choices.

Where can we look for a new ethical basis to set our course? I have found the ideas in the book *Ethics for the New Millennium*, by the Dalai Lama, to be very helpful. As is perhaps well known but little heeded, the Dalai Lama argues that the most important thing is for us to conduct our lives with love and compassion for others, and that our societies need to develop a stronger notion of universal responsibility and of our interdependency; he proposes a standard of positive ethical conduct for individuals and societies that seems consonant with Attali's Fraternity utopia.

The Dalai Lama further argues that we must understand what it is that makes people happy, and acknowledge the strong evidence that neither material progress nor the pursuit of the power of knowledge is the key—that there are limits to what science and the scientific pursuit alone can do.

Our Western notion of happiness seems to come from the Greeks, who defined it as "the exercise of vital powers along lines of excellence in a life affording them scope."[18]

Clearly, we need to find meaningful challenges and sufficient scope in our lives if we are to be happy in whatever is to come. But I believe we must find alternative outlets for our creative forces, beyond the culture of perpetual economic growth; this growth has largely been a blessing for several hundred years, but it has not brought us unalloyed happiness, and we must now choose between the pursuit of unrestricted and undirected growth through science and technology and the clear accompanying dangers.

It is now more than a year since my first encounter with Ray Kurzweil and John Searle. I see around me cause for hope in the voices for caution and relinquishment and in those people I have discovered who are as concerned as I am about our current predicament. I feel, too, a deepened sense of personal responsibility—not for the work I have already done, but for the work that I might yet do, at the confluence of the sciences.

But many other people who know about the dangers still seem strangely silent. When pressed, they trot out the "this is nothing new" riposte—as if awareness of what could happen is response enough. They tell me, There are universities filled with bioethicists who study this stuff all day long. They say, All this has been written about before, and by experts. They complain, Your worries and your arguments are already old hat.

I don't know where these people hide their fear. As an architect of complex systems I enter this arena as a generalist. But should this diminish my concerns? I am aware of how much has been written about, talked about, and lectured about so authoritatively. But does this mean it has reached people? Does this mean we can discount the dangers before us?

Knowing is not a rationale for not acting. Can we doubt that knowledge has become a weapon we wield against ourselves?

The experiences of the atomic scientists clearly show the need to take personal responsibility, the danger that things will move too fast, and the way in which a process can take on a life of its own. We can, as they did, create insurmountable problems in almost no time flat. We must do more thinking up front if we are not to be similarly surprised and shocked by the consequences of our inventions.

My continuing professional work is on improving the reliability of software. Software is a tool, and as a toolbuilder I must struggle with the uses to which the tools I make are put. I have always believed that making software more reliable, given its many uses, will make the world a safer and better place; if I were to come to believe the opposite, then I would be morally obligated to stop this work. I can now imagine such a day may come.

This all leaves me not angry but at least a bit melancholic. Henceforth, for me, progress will be somewhat bittersweet.

Do you remember the beautiful penultimate scene in *Manhattan* where Woody Allen is lying on his couch and talking into a tape recorder? He is writing a short story about people who are creating unnecessary, neurotic problems for themselves, because it keeps them from dealing with more unsolvable, terrifying problems about the universe.

He leads himself to the question, "Why is life worth living?" and to consider what makes it worthwhile for him: Groucho Marx, Willie Mays, the second movement of the Jupiter Symphony, Louis Armstrong's recording of "Potato Head Blues," Swedish movies, Flaubert's Sentimental Education, Marlon Brando, Frank Sinatra, the apples and pears by Cézanne, the crabs at Sam Wo's, and, finally, the showstopper: his love Tracy's face.

Each of us has our precious things, and as we care for them we locate the essence of our humanity. In the end, it is because of our great capacity for caring that I remain optimistic we will confront the dangerous issues now before us.

My immediate hope is to participate in a much larger discussion of the issues raised here, with people from many different backgrounds, in settings not predisposed to fear or favor technology for its own sake.

As a start, I have twice raised many of these issues at events sponsored by the Aspen Institute and have separately proposed that the American Academy of Arts and Sciences take them up as an extension of its work with the Pugwash Conferences. (These have been held since 1957 to discuss arms control, especially of nuclear weapons, and to formulate workable policies.)

It's unfortunate that the Pugwash meetings started only well after the nuclear genie was out of the bottle—roughly 15 years too late. We are also getting a belated start on seriously addressing the issues around 21st-century technologies—the prevention of knowledge-enabled mass destruction—and further delay seems unacceptable.

So I'm still searching; there are many more things to learn. Whether we are to succeed or fail, to survive or fall victim to these technologies, is not yet decided. I'm up late again—it's almost 6 am. I'm trying to imagine some better answers, to break the spell and free them from the stone.

Notes

1. The passage Kurzweil quotes is from Kaczynski's "Unabomber Manifesto," which was published jointly, under duress, by the *New York Times* and the *Washington Post* to attempt to bring his campaign of terror to an end. I agree with David Gelernter, who said about their decision: "It was a tough call for the newspapers. To say yes would be giving in to terrorism, and for all they knew he was lying anyway. On the other hand, to say yes might stop the killing. There was also a chance that someone would read the tract and get a hunch about the author; and that is exactly what happened. The suspect's brother read it, and it rang a bell. I would have told them not to publish. I'm glad they didn't ask me. I guess." *Drawing Life: Surviving the Unabomber* (New York: Free Press, 1997), 120.

2. Laurie Garrett, *The Coming Plague: Newly Emerging Diseases in a World Out of Balance* (London: Penguin, 1994), 47–52, 414, 419, 452.

3. Isaac Asimov described what became the most famous view of ethical rules for robot behavior in his book *I, Robot* in 1950, in his Three Laws of Robotics: 1. A robot may not injure a human being, or, through inaction, allow a human being to come to harm. 2. A robot must obey the orders given it by human beings, except where such orders would conflict with the First Law. 3. A robot must protect its own existence, as long as such protection does not conflict with the First or Second Law.

4. See Jennifer Hillner, "Test of Time," *Wired* 8.03, 78.

5. Michelangelo wrote a sonnet that begins:

> Non ha l' ottimo artista alcun concetto
> Ch' un marmo solo in se` non circonscriva
> Col suo soverchio; e solo a quello arriva
> La man che ubbidisce all' intelleto.

Stone translates this as:

> The best of artists hath no thought to show
> which the rough stone in its superfluous shell
> doth not include; to break the marble spell
> is all the hand that serves the brain can do.

Stone describes the process: "He was not working from his drawings or clay models; they had all been put away. He was carving from the images in his mind. His eyes and hands knew where every line, curve, mass must emerge, and at what depth in the heart of the stone to create the low relief." *The Agony and the Ecstasy* (New York: Doubleday, 1961), 6, 144.

6. See "A Tale of Two Botanies," *Wired* 8.04, 247.

7. "The Future of Computation," First Foresight Conference on Nanotechnology, October 1989. Published in B. C. Crandall and James Lewis, eds., *Nanotechnology: Research and Perspectives* (Cambridge, MA: MIT Press, 1992), 269. See also https://foresight.org/Conferences/MNT01/Nano1.html.

8. In his 1963 novel *Cat's Cradle*, Kurt Vonnegut imagined a gray-goo-like accident in which a form of ice called ice-nine, which becomes solid at a much higher temperature, freezes the oceans.

9. Stuart Kauffman, "Self-Replication: Even Peptides Do It," *Nature* 382 (August 8, 1996): 496.

10. Jon Else, *The Day After Trinity: J. Robert Oppenheimer and the Atomic Bomb* (Pyramid Films, 1981).

11. This estimate is in Leslie's book *The End of the World: The Science and Ethics of Human Extinction*, where he notes that the probability of extinction is substantially higher if we accept Brandon Carter's doomsday argument, which is, briefly, that "we ought to have some reluctance to believe that we are very exceptionally early, for instance in the earliest 0.001 percent, among all humans who will ever have lived. This would be some reason for thinking that humankind will not survive for many more centuries, let alone colonize the galaxy. Carter's doomsday argument doesn't generate any risk estimates just by itself. It is an argument for *revising* the estimates which we generate when we consider various possible dangers" (London: Routledge, 1996), 1, 3, 145.

12. Arthur C. Clarke, "Presidents, Experts, and Asteroids," *Science*, June 5, 1998. Reprinted as "Science and Society," in *Greetings, Carbon-Based Bipeds! Collected Essays, 1934–1998* (New York: St. Martin's Press, 1999), 526.

13. And, as David Forrest suggests in his paper "Regulating Nanotechnology Development," available at https://foresight.org/nano/Forrest1989.php, "If we used strict liability as an alternative to regulation it would be impossible for any developer to internalize the cost of the risk (destruction of the biosphere), so theoretically the activity of developing nanotechnology should never be undertaken." Forrest's analysis leaves us with only government regulation to protect us—not a comforting thought.

14. Matthew Meselson, "The Problem of Biological Weapons" (presentation to the 1,818th Stated Meeting of the American Academy of Arts and Sciences, January 13, 1999).

15. Paul Doty, "The Forgotten Menace: Nuclear Weapons Stockpiles Still Represent the Biggest Threat to Civilization," *Nature* 402 (December 9, 1999): 583.

16. See also Hans Bethe's 1997 letter to President Clinton, at https://fas.org/faspir/pir0797.htm.

17. Jacques Attali, *Fraternités: Une nouvelle utopie* (Paris: Fayard, 1999).

18. Edith Hamilton, *The Greek Way* (New York: W. W. Norton, 1942), 35.

6 Sultana's Dream

Rokeya Sakhawat Hossain

We end this section with a positive vision of the future, at least for the woman who wrote it. "Sultana's Dream" depicts a world in which innovative technologies have been designed and used to solve major social problems. Although readers may be disappointed by Hossain's harsh portrayal of men, her story must be put in context. Rokeya Sakhawat Hossain was a Bengali writer, a social activist, and an advocate for women's rights. In the world of "Sultana's Dream," women have taken over and used technology to bring about peace. Hossain's motivation in writing this piece is not difficult to decipher; she was responding to both the state of gender relations and political strife in her country. In this respect, her vision of the future was a form of social activism. She used her story to expose the flaws of her world and argued that technology (in the right hands) could cure many of the ills she faced. Her vision may seem counterintuitive to many, especially Westerners, in that she aligns women with technology. In the modern Western world, technology has traditionally been associated with men and masculinity, and this is understood to be one of the factors that has contributed to the small number of women interested in pursuing degrees in STEM fields. But this has not always been so, and there have been many instances in which technology has been deemed an empowering force for women. Hossain's story reminds us that technologies should not simply be categorized as "good" or "bad" but rather analyzed for the values they enable and the futures they help to shape.

One evening I was lounging in an easy chair in my bedroom and thinking lazily of the condition of Indian womanhood. I am not sure whether I dozed off or not. But, as far as I remember, I was wide awake. I saw the moonlit sky sparkling with thousands of diamond-like stars, very distinctly.

All on a sudden a lady stood before me; how she came in, I do not know. I took her for my friend, Sister Sara.

"Good morning," said Sister Sara. I smiled inwardly as I knew it was not morning, but starry night. However, I replied to her, saying, "How do you do?"

"I am all right, thank you. Will you please come out and have a look at our garden?"

I looked again at the moon through the open window, and thought there was no harm in going out at that time. The men-servants outside were fast asleep just then, and I could have a pleasant walk with Sister Sara.

I used to have my walks with Sister Sara, when we were at Darjeeling. Many a time did we walk hand in hand and talk light-heartedly in the botanical gardens there. I

From *Indian Ladies' Magazine*, Madras, 1905.

fancied, Sister Sara had probably come to take me to some such garden and I readily accepted her offer and went out with her.

When walking I found to my surprise that it was a fine morning. The town was fully awake and the streets alive with bustling crowds. I was feeling very shy, thinking I was walking in the street in broad daylight, but there was not a single man visible.

Some of the passers-by made jokes at me. Though I could not understand their language, yet I felt sure they were joking. I asked my friend, "What do they say?"

"The women say that you look very mannish."

"Mannish?" said I, "What do they mean by that?"

"They mean that you are shy and timid like men."

"Shy and timid like men?" It was really a joke. I became very nervous, when I found that my companion was not Sister Sara, but a stranger. Oh, what a fool had I been to mistake this lady for my dear old friend, Sister Sara.

She felt my fingers tremble in her hand, as we were walking hand in hand.

"What is the matter, dear?" she said affectionately.

"I feel somewhat awkward," I said in a rather apologizing tone, "as being a purdahnishin woman I am not accustomed to walking about unveiled."[1]

"You need not be afraid of coming across a man here. This is Ladyland, free from sin and harm. Virtue herself reigns here."

By and by I was enjoying the scenery. Really it was very grand. I mistook a patch of green grass for a velvet cushion. Feeling as if I were walking on a soft carpet, I looked down and found the path covered with moss and flowers.

"How nice it is," said I.

"Do you like it?" asked Sister Sara. (I continued calling her "Sister Sara," and she kept calling me by my name.)

"Yes, very much; but I do not like to tread on the tender and sweet flowers."

"Never mind, dear Sultana; your treading will not harm them; they are street flowers."

"The whole place looks like a garden," said I admiringly. "You have arranged every plant so skillfully."

"Your Calcutta could become a nicer garden than this if only your countrymen wanted to make it so."

"They would think it useless to give so much attention to horticulture, while they have so many other things to do."

"They could not find a better excuse," said she with smile.

I became very curious to know where the men were. I met more than a hundred women while walking there, but not a single man.

"Where are the men?" I asked her.

"In their proper places, where they ought to be."

"Pray let me know what you mean by 'their proper places.'"

"O, I see my mistake, you cannot know our customs, as you were never here before. We shut our men indoors."

"Just as we are kept in the zenana?"[2]

"Exactly so."

"How funny," I burst into a laugh. Sister Sara laughed too.

"But dear Sultana, how unfair it is to shut in the harmless women and let loose the men."

"Why? It is not safe for us to come out of the zenana, as we are naturally weak."

"Yes, it is not safe so long as there are men about the streets, nor is it so when a wild animal enters a marketplace."

"Of course not."

"Suppose, some lunatics escape from the asylum and begin to do all sorts of mischief to men, horses and other creatures; in that case what will your countrymen do?"

"They will try to capture them and put them back into their asylum."

"Thank you! And you do not think it wise to keep sane people inside an asylum and let loose the insane?"

"Of course not!" said I laughing lightly.

"As a matter of fact, in your country this very thing is done! Men, who do or at least are capable of doing no end of mischief, are let loose and the innocent women, shut up in the zenana! How can you trust those untrained men out of doors?"

"We have no hand or voice in the management of our social affairs. In India man is lord and master, he has taken to himself all powers and privileges and shut up the women in the zenana."

"Why do you allow yourselves to be shut up?"

"Because it cannot be helped as they are stronger than women."

"A lion is stronger than a man, but it does not enable him to dominate the human race. You have neglected the duty you owe to yourselves and you have lost your natural rights by shutting your eyes to your own interests."

"But my dear Sister Sara, if we do everything by ourselves, what will the men do then?"

"They should not do anything, excuse me; they are fit for nothing. Only catch them and put them into the zenana."

"But would it be very easy to catch and put them inside the four walls?" said I. "And even if this were done, would all their business—political and commercial—also go with them into the zenana?"

Sister Sara made no reply. She only smiled sweetly. Perhaps she thought it useless to argue with one who was no better than a frog in a well.

By this time we reached Sister Sara's house. It was situated in a beautiful heart-shaped garden. It was a bungalow with a corrugated iron roof. It was cooler and nicer than any of our rich buildings. I cannot describe how neat and how nicely furnished and how tastefully decorated it was.

We sat side by side. She brought out of the parlour a piece of embroidery work and began putting on a fresh design.

"Do you know knitting and needle work?"

"Yes; we have nothing else to do in our zenana."

"But we do not trust our zenana members with embroidery!" she said laughing, "as a man has not patience enough to pass thread through a needlehole even!"

"Have you done all this work yourself?" I asked her pointing to the various pieces of embroidered teapoy cloths.

"Yes."

"How can you find time to do all these? You have to do the office work as well? Have you not?"

"Yes. I do not stick to the laboratory all day long. I finish my work in two hours."

"In two hours! How do you manage? In our land the officers,—magistrates, for instance—work seven hours daily."

"I have seen some of them doing their work. Do you think they work all the seven hours?"

"Certainly they do!"

"No, dear Sultana, they do not. They dawdle away their time in smoking. Some smoke two or three choroots during the office time. They talk much about their work, but do little. Suppose one choroot takes half an hour to burn off, and a man smokes twelve choroots daily; then you see, he wastes six hours every day in sheer smoking."

We talked on various subjects, and I learned that they were not subject to any kind of epidemic disease, nor did they suffer from mosquito bites as we do. I was very much astonished to hear that in Ladyland no one died in youth except by rare accident.

"Will you care to see our kitchen?" she asked me.

"With pleasure," said I, and we went to see it. Of course the men had been asked to clear off when I was going there. The kitchen was situated in a beautiful vegetable garden. Every creeper, every tomato plant was itself an ornament. I found no smoke, nor any chimney either in the kitchen—it was clean and bright; the windows were decorated with flower gardens. There was no sign of coal or fire.

"How do you cook?" I asked.

"With solar heat," she said, at the same time showing me the pipe, through which passed the concentrated sunlight and heat. And she cooked something then and there to show me the process.

"How did you manage to gather and store up the sun-heat?" I asked her in amazement.

"Let me tell you a little of our past history then. Thirty years ago, when our present Queen was thirteen years old, she inherited the throne. She was Queen in name only, the Prime Minister really ruling the country.

"Our good Queen liked science very much. She circulated an order that all the women in her country should be educated. Accordingly a number of girls' schools were founded and supported by the government. Education was spread far and wide among women. And early marriage also was stopped. No woman was to be allowed to marry before she was twenty-one. I must tell you that, before this change we had been kept in strict purdah."

"How the tables are turned," I interposed with a laugh.

"But the seclusion is the same," she said. "In a few years we had separate universities, where no men were admitted."

"In the capital, where our Queen lives, there are two universities. One of these invented a wonderful balloon, to which they attached a number of pipes. By means of this captive balloon which they managed to keep afloat above the cloud-land, they could draw as much water from the atmosphere as they pleased. As the water was incessantly being drawn by the university people no cloud gathered and the ingenious Lady Principal stopped rain and storms thereby."

"Really! Now I understand why there is no mud here!" said I. But I could not understand how it was possible to accumulate water in the pipes. She explained to me how it was done, but I was unable to understand her, as my scientific knowledge was very limited. However, she went on, "When the other university came to know of this, they became exceedingly jealous and tried to do something more extraordinary still. They invented an instrument by which they could collect as much sun-heat as they wanted. And they kept the heat stored up to be distributed among others as required.

"While the women were engaged in scientific research, the men of this country were busy increasing their military power. When they came to know that the female universities were able to draw water from the atmosphere and collect heat from the sun, they only laughed at the members of the universities and called the whole thing 'a sentimental nightmare'!"

"Your achievements are very wonderful indeed! But tell me, how you managed to put the men of your country into the zenana. Did you entrap them first?"

"No."

"It is not likely that they would surrender their free and open air life of their own accord and confine themselves within the four walls of the zenana! They must have been overpowered."

"Yes, they have been!"

"By whom? By some lady-warriors, I suppose?"

"No, not by arms."

"Yes, it cannot be so. Men's arms are stronger than women's. Then?"

"By brain."

"Even their brains are bigger and heavier than women's. Are they not?"

"Yes, but what of that? An elephant also has got a bigger and heavier brain than a man has. Yet man can enchain elephants and employ them, according to their own wishes."

"Well said, but tell me please, how it all actually happened. I am dying to know it!"

"Women's brains are somewhat quicker than men's. Ten years ago, when the military officers called our scientific discoveries 'a sentimental nightmare,' some of the young ladies wanted to say something in reply to those remarks. But both the Lady Principals restrained them and said, they should reply not by word, but by deed, if ever they got the opportunity. And they had not long to wait for that opportunity."

"How marvelous!" I heartily clapped my hands. "And now the proud gentlemen are dreaming sentimental dreams themselves."

"Soon afterwards certain persons came from a neighbouring country and took shelter in ours. They were in trouble having committed some political offense. The king who cared more for power than for good government asked our kind-hearted Queen to hand them over to his officers. She refused, as it was against her principle to turn out refugees. For this refusal the king declared war against our country.

"Our military officers sprang to their feet at once and marched out to meet the enemy. The enemy however, was too strong for them. Our soldiers fought bravely, no doubt. But in spite of all their bravery the foreign army advanced step by step to invade our country.

"Nearly all the men had gone out to fight; even a boy of sixteen was not left home. Most of our warriors were killed, the rest driven back and the enemy came within twenty-five miles of the capital.

"A meeting of a number of wise ladies was held at the Queen's palace to advise as to what should be done to save the land. Some proposed to fight like soldiers; others objected and said that women were not trained to fight with swords and guns, nor were they accustomed to fighting with any weapons. A third party regretfully remarked that they were hopelessly weak of body.

"'If you cannot save your country for lack of physical strength,' said the Queen, 'try to do so by brain power.'

"There was a dead silence for a few minutes. Her Royal Highness said again, 'I must commit suicide if the land and my honour are lost.'

"Then the Lady Principal of the second university (who had collected sun-heat), who had been silently thinking during the consultation, remarked that they were all but lost, and there was little hope left for them. There was, however, one plan which she would like to try, and this would be her first and last efforts; if she failed in this, there would be nothing left but to commit suicide. All present solemnly vowed that they would never allow themselves to be enslaved, no matter what happened.

"The Queen thanked them heartily, and asked the Lady Principal to try her plan. The Lady Principal rose again and said, 'Before we go out the men must enter the zenanas. I make this prayer for the sake of purdah.' 'Yes, of course,' replied Her Royal Highness.

"On the following day the Queen called upon all men to retire into zenanas for the sake of honour and liberty. Wounded and tired as they were, they took that order rather for a boon! They bowed low and entered the zenanas without uttering a single word of protest. They were sure that there was no hope for this country at all.

"Then the Lady Principal with her two thousand students marched to the battle field, and arriving there directed all the rays of the concentrated sunlight and heat towards the enemy.

"The heat and light were too much for them to bear. They all ran away panic-stricken, not knowing in their bewilderment how to counteract that scorching heat. When they fled away leaving their guns and other ammunitions of war, they were burnt down by means of the same sun-heat. Since then no one has tried to invade our country any more."

"And since then your countrymen never tried to come out of the zenana?"

"Yes, they wanted to be free. Some of the police commissioners and district magistrates sent word to the Queen to the effect that the military officers certainly deserved to be imprisoned for their failure; but they never neglected their duty and therefore they should not be punished and they prayed to be restored to their respective offices.

"Her Royal Highness sent them a circular letter intimating to them that if their services should ever be needed they would be sent for, and that in the meanwhile they should remain where they were. Now that they are accustomed to the purdah system and have ceased to grumble at their seclusion, we call the system 'Mardana' instead of 'zenana.'"

"But how do you manage," I asked Sister Sara, "to do without the police or magistrates in case of theft or murder?"

"Since the 'Mardana' system has been established, there has been no more crime or sin; therefore we do not require a policeman to find out a culprit, nor do we want a magistrate to try a criminal case."

"That is very good, indeed. I suppose if there was any dishonest person, you could very easily chastise her. As you gained a decisive victory without shedding a single drop of blood, you could drive off crime and criminals too without much difficulty!"

"Now, dear Sultana, will you sit here or come to my parlour?" she asked me.

"Your kitchen is not inferior to a queen's boudoir!" I replied with a pleasant smile, "but we must leave it now; for the gentlemen may be cursing me for keeping them away from their duties in the kitchen so long." We both laughed heartily.

"How my friends at home will be amused and amazed, when I go back and tell them that in the far-off Ladyland, ladies rule over the country and control all social matters, while gentlemen are kept in the Mardanas to mind babies, to cook and to do all sorts of domestic work; and that cooking is so easy a thing that it is simply a pleasure to cook!"

"Yes, tell them about all that you see here."

"Please let me know, how you carry on land cultivation and how you plough the land and do other hard manual work."

"Our fields are tilled by means of electricity, which supplies motive power for other hard work as well, and we employ it for our aerial conveyances too. We have no rail road nor any paved streets here."

"Therefore neither street nor railway accidents occur here," said I. "Do not you ever suffer from want of rainwater?" I asked.

"Never since the 'water balloon' has been set up. You see the big balloon and pipes attached thereto. By their aid we can draw as much rainwater as we require. Nor do we ever suffer from flood or thunderstorms. We are all very busy making nature yield as much as she can. We do not find time to quarrel with one another as we never sit idle. Our noble Queen is exceedingly fond of botany; it is her ambition to convert the whole country into one grand garden."

"The idea is excellent. What is your chief food?"

"Fruits."

"How do you keep your country cool in hot weather? We regard the rainfall in summer as a blessing from heaven."

"When the heat becomes unbearable, we sprinkle the ground with plentiful showers drawn from the artificial fountains. And in cold weather we keep our room warm with sun-heat."

She showed me her bathroom, the roof of which was removable. She could enjoy a shower bath whenever she liked, by simply removing the roof (which was like the lid of a box) and turning on the tap of the shower pipe.

"You are a lucky people!" ejaculated I. "You know no want. What is your religion, may I ask?"

"Our religion is based on Love and Truth. It is our religious duty to love one another and to be absolutely truthful. If any person lies, she or he is . . ."

"Punished with death?"

"No, not with death. We do not take pleasure in killing a creature of God, especially a human being. The liar is asked to leave this land for good and never to come to it again."

"Is an offender never forgiven?"

"Yes, if that person repents sincerely."

"Are you not allowed to see any man, except your own relations?"

"No one except sacred relations."

"Our circle of sacred relations is very limited; even first cousins are not sacred."

"But ours is very large; a distant cousin is as sacred as a brother."

"That is very good. I see purity itself reigns over your land. I should like to see the good Queen, who is so sagacious and far-sighted and who has made all these rules."

"All right," said Sister Sara.

Then she screwed a couple of seats onto a square piece of plank. To this plank she attached two smooth and well-polished balls. When I asked her what the balls were for, she said they were hydrogen balls and they were used to overcome the force of gravity. The balls were of different capacities to be used according to the different weights desired to be overcome. She then fastened to the air-car two wing-like blades, which, she said, were worked by electricity. After we were comfortably seated she touched a knob and the blades began to whirl, moving faster and faster every moment. At first we were raised to the height of about six or seven feet and then off we flew. And before I could realize that we had commenced moving, we reached the garden of the Queen.

My friend lowered the air-car by reversing the action of the machine, and when the car touched the ground the machine was stopped and we got out.

I had seen from the air-car the Queen walking on a garden path with her little daughter (who was four years old) and her maids of honour.

"Halloo! You here!" cried the Queen addressing Sister Sara. I was introduced to Her Royal Highness and was received by her cordially without any ceremony.

I was very much delighted to make her acquaintance. In the course of the conversation I had with her, the Queen told me that she had no objection to permitting her subjects to trade with other countries. "But," she continued, "no trade was possible with countries where the women were kept in the zenanas and so unable to come and trade with us. Men, we find, are rather of lower morals and so we do not like dealing with them. We do not covet other people's land, we do not fight for a piece of diamond though it may be a thousand-fold brighter than the Koh-i-Noor, nor do we grudge a ruler his Peacock Throne. We dive deep into the ocean of knowledge and try to find out the precious gems, which nature has kept in store for us. We enjoy nature's gifts as much as we can."

After taking leave of the Queen, I visited the famous universities, and was shown some of their manufactories, laboratories and observatories.

After visiting the above places of interest we got again into the air-car, but as soon as it began moving, I somehow slipped down and the fall startled me out of my dream. And on opening my eyes, I found myself in my own bedroom still lounging in the easy-chair!

Notes

1. *Purdah* refers to the practice of female seclusion from public observation, including physical separation of the sexes and women covering their bodies.

2. *Zenana* refers to the part of a household reserved for women.

II THE RELATIONSHIP BETWEEN TECHNOLOGY AND SOCIETY

Technology plays a pivotal role in each of the futures presented in the first section. Implicit in all of them is the idea that technology shapes, if not determines, society. Although the idea that technology influences society is not controversial, the idea that technology might directly "determine" society generates considerable controversy, analysis, and research, especially among those who study the relationship between society and technology. Much of the controversy centers on how, in what ways, and to what extent technology and society are intertwined. The readings in this section dig deeply into this complex relationship. They articulate and make explicit ideas lurking in the thoughts of the previous section's authors as they imagined what might come to be.

The traditional scholarship aimed at understanding the relationship between technology and society has focused on two sets of arguments that can be labeled *technological determinism* and *social construction*. The technological determinist argument comes in many forms, but in its most pure iteration, it is made up of two major components. The first is that the introduction of new technologies produces direct and unalterable social changes. For example, a technological determinist might argue that the Internet leads to globalization. Because the Internet makes it possible for corporations and organizations to communicate across the globe in real time and without much effort, local ties and interdependencies become weaker; companies have more options to establish relationships that serve their interests independent of geographic location. Once the Internet was created, determinists might say, it was inevitable that local and national alliances would weaken, and global organizations would come to dominate. Advertisers love to play on this aspect of technological determinism by suggesting that their products will inevitably lead to certain kinds of social outcomes—for example, the fast car that assures the owner will be seen as a celebrity when driving it or the handy kitchen appliance that produces not only tasty gourmet meals with little effort but also correspondingly well-fed and happy spouses.

The second major component of technological determinism is the idea that people have little, if any, control over which technologies are developed. This is often framed by the argument that technology is "autonomous"—that is, that it can't be stopped. Advocates of this view claim that technological development follows a logical or natural path of progression—each invention builds upon one previously developed—and humans do not and cannot influence the order or direction. Technological determinism presumes that the development of technology is unaffected (or affected very little) by social and political forces and that the technologies we have today are simply the latest step in the linear progression of science and engineering. It might be possible for technologies to develop more or less quickly depending on how much money and time a civilization invests in them. But human intervention cannot sway the order in which they arrive and the inevitability of their arrival. Technologically deterministic arguments present technology as a powerful force requiring—even demanding—that people and institutions operate in particular ways.

The social constructivist argument was developed as a direct challenge to the idea of technological determinism. Social constructivists contend that technological development has no natural or logical predetermined order of progression and is not out of human control. Instead, they maintain that society (through interest groups, laws, the economy, political decisions, cultural values, and more) shapes and directs technology at every phase of its development—from conception to production to use (or even nonuse). Many advocates of this approach believe the determinist argument is not just wrong but inherently dangerous because it pushes out of sight the social forces and interests at work in directing the development of technology, concealing the infinite array of alternative possibilities. They argue that technologically deterministic arguments lead people to believe that a given future is inescapable and allow individuals to abdicate responsibility for the decisions made about technology.

To counter this idea, social constructivists demonstrate how individuals and groups—engineers, corporations, regulatory agencies, lawyers, politicians, and others—have and do contribute to the direction of technological development. But their argument does not end with production. Social constructivists maintain that once a technology is developed society still has the ability to decide if, where, when, and how it will be used. Even users play an active role in this process by interpreting and reinterpreting technologies and using them for purposes for which they weren't designed. Consider, for example, that bubble wrap was originally designed to be used as textured wallpaper. The inventors then thought it might work as a greenhouse insulator. Only later did the idea of using it as a packaging material come to mind. Later still, users decided that popping the small bubbles provided an excellent form of stress relief—a use that was not likely in the minds of the original inventors!

The goal of this section is not to strongly advocate for either technological determinism or social constructivism as the better way to describe the world. Crucial aspects of the relationships between technology and society are concealed if either position is taken to the extreme. For example, a technological determinist stance might ignore the fact that new technologies are sometimes entirely rejected or rejected by their intended audience and that individuals can turn machines off and throw them away if they choose. On the other hand, a hard-core social constructivist might ignore the fact that while technologies can be interpreted and used in multiple ways, their physical limits prevent them from being interpreted and used in any imaginable way. It's hard to imagine how a group of people could collectively decide that a hamster wheel would make an appropriate table for a dinner party. This section will draw on the insights of both approaches and ultimately argue that we must consider the ways in which technology and society simultaneously influence and constitute each other.

The first important step in doing this is to recognize that it is misleading to think of technologies as simply physical objects (or *artifacts*), as many scholars do when they refer to the objects and machines that are the physical symbols of technologies. Technology is better understood as a complex system of people, relationships, and artifacts. You might think of your mobile phone as the chunk of metal and plastic you carry around, but in order to work your phone requires a network of other technologies (such as computers, software, cell towers, and so on) and social organizations (such as companies manufacturing, distributing, and selling phones; agencies for setting standards and getting consensus about protocols), as well as a vast number of individuals who work every day at these organizations and maintain the artifacts that make the phones do what you expect them to do. Viewing technologies merely as physical objects is like trying to understand a chess piece separately from the game, the thirty-one other pieces, the board, the players, and the rules. Certainly, one can learn something about a rook by focusing on its physical design and studying how plastics are created or marble is carved, but such knowledge will give few clues as to why the piece was made the way it was or the impact it can have on the game. So it is with technological artifacts and the systems of which they are a part.

In order to avoid thinking of technology merely as physical objects, it is useful to think of technology as *sociotechnical systems*—assemblages of things, people, practices, and meanings. Only a few of the authors in this section explicitly use the term *socio-technical system*, but nearly all of them have, at the heart of their argument, the idea that technology is not just artifacts. They recognize that any given technology functions and has meaning and significance insofar as it is embedded in a wider social and technical context.

In order to grasp what a sociotechnical system is, it is helpful to think through an example. Consider the safety airbag in an automobile. It can be seen as a system of sensors, an inflator, and a bag, but this tells one very little about why it was created, what it does, or even how it works. The sociotechnical system of an airbag includes its numerous relationships with people and other devices. For instance, the airbag's role in our world is much easier to understand when viewed in the context of modern transportation, automobiles, and car collisions. Its purpose cannot be understood without looking at how insurance companies encouraged its development, how government regulations shaped its design, and how engineers had certain users in mind when they built it. The airbag can work only when automobile manufacturers and distributors, a road system, and drivers with certain habits exist. The device itself is not without meaning, but its purpose, value, and implications are best understood in its broader sociotechnical context. Those who insist that technologies are simply "tools" often do not want you to recognize the values that underlie them, the groups impacted by them, and the ways in which the world would be a very different place without them.

If we want to influence the direction of the future and we believe that technology will play a powerful role in that future, it is of utmost importance that we understand the complex, multidimensional relationships between technology and society. The readings in this section offer a starting point for understanding how technology and society are inextricably interwoven. They explore a variety of ways in which society and technology influence one another and work together. The temptation to draw a line and argue that everything on one side of it should be categorized as *technical* and everything on the other side should be labeled *social* is powerful, but this section's readings suggest that doing so is misguided, likely doomed to failure, and threatening to our attempts to build a future that reflects the values we hold most dear.

Questions to consider while reading the selections in this section:

1. Are technological determinism and social constructivism incompatible? How might they work together?
2. What aspects of the selections in the first section can be characterized as technologically deterministic? Why?

3. Which technologies are still "in the making" today—that is, which are still in the stage of interpretive flexibility?
4. Which current technologies have the momentum that Hughes describes? Are there technologies that would be difficult to change or replace (except around the edges) because so many individuals and groups have invested in them and/or rely upon them?
5. In what ways do technologies push back on us? That is, when do technologies compel us to behave in ways we wouldn't otherwise?

7 Do Machines Make History?

Robert L. Heilbroner

In this classic article, Robert L. Heilbroner introduces two aspects of a strong form of technological determinism. The first is the claim that technology develops in a fixed, naturally determined sequence, that each new invention is built on the previous in an order that couldn't be otherwise. The second is the claim that adoption of a given technology imposes certain social and political characteristics upon the society in which it is used. Heilbroner is especially focused on the ways in which the labor and management requirements of particular technologies create social relationships that permeate society more broadly. The idea that technology is in control and tells us what to do is disturbing to many. In fact it is hard to find someone today who self-identifies as a technological determinist. Even Heilbroner isn't a "hard-core" technological determinist since he seems to acknowledge at least some social influence on technology. But it is an important approach to study, as this view of the technology-society relationship is very common, although it is not often articulated so explicitly. Many of the visions of the future that were presented in the last section were written as though specific technologies necessarily have specific social ramifications. While Heilbroner wrote this piece in 1967, it remains one of the best articulations of technological determinism, and some of the technologies he mentions are still being debated today.

The hand-mill gives you society with the feudal lord; the steam-mill, society with the industrial capitalist.

—Marx, *The Poverty of Philosophy*

That machines make history in some sense—that the level of technology has a direct bearing on the human drama—is of course obvious. That they do not make all of history, however that word be defined, is equally clear. The challenge, then, is to see if one can say something systematic about the matter, to see whether one can order the problem so that it becomes intellectually manageable.

To do so calls at the very beginning for a careful specification of our task. There are a number of important ways in which machines make history that will not concern us here. For example, one can study the impact of technology on the *political* course of history, evidenced most strikingly by the central role played by the technology of war. Or one can study the effect of machines on the *social* attitudes that underlie historical

From *Technology and Culture* 8, no. 3 (July 1967): 335–345.

evolution: one thinks of the effect of radio or television on political behavior. Or one can study technology as one of the factors shaping the changeful content of life from one epoch to another: when we speak of "life" in the Middle Ages or today we define an existence much of whose texture and substance is intimately connected with the prevailing technological order.

None of these problems will form the focus of this essay. Instead, I propose to examine the impact of technology on history in another area—an area defined by the famous quotation from Marx that stands beneath our title. The question we are interested in, then, concerns the effect of technology in determining the nature of the *socioeconomic order*. In its simplest terms the question is: did medieval technology bring about feudalism? Is industrial technology the necessary and sufficient condition for capitalism? Or, by extension, will the technology of the computer and the atom constitute the ineluctable cause of a new social order?

Even in this restricted sense, our inquiry promises to be broad and sprawling. Hence, I shall not try to attack it head-on, but to examine it in two stages:

1. If we make the assumption that the hand-mill does "give" us feudalism and the steam-mill capitalism, this places technological change in the position of a prime mover of social history. Can we then explain the "laws of motion" of technology itself? Or to put the question less grandly, can we explain why technology evolves in the sequence it does?
2. Again, taking the Marxian paradigm at face value, exactly what do we mean when we assert that the hand-mill "gives us" society with the feudal lord? Precisely how does the mode of production affect the superstructure of social relationships?

These questions will enable us to test the empirical content—or at least to see if there *is* an empirical content—in the idea of technological determinism. I do not think it will come as a surprise if I announce now that we will find *some* content, and a great deal of missing evidence, in our investigation. What will remain then will be to see if we can place the salvageable elements of the theory in historical perspective—to see, in a word, if we can explain technological determinism historically as well as explain history by technological determinism.

I

We begin with a very difficult question hardly rendered easier by the fact that there exist, to the best of my knowledge, no empirical studies on which to base our speculations. It is the question of whether there is a fixed sequence to technological development and therefore a necessitous path over which technologically developing societies must travel.

I believe there is such a sequence—that the steam-mill follows the hand-mill not by chance but because it is the next "stage" in a technical conquest of nature that follows one and only one grand avenue of advance. To put it differently, I believe that it is impossible to proceed to the age of the steam-mill until one has passed through the age of the hand-mill, and that in turn one cannot move to the age of the hydroelectric plant before one has mastered the steam-mill, nor to the nuclear power age until one has lived through that of electricity.

Before I attempt to justify so sweeping an assertion, let me make a few reservations. To begin with, I am fully conscious that not all societies are interested in developing a technology of production or in channeling to it the same quota of social energy. I am very much aware of the different pressures that different societies exert on the direction in which technology unfolds. Lastly, I am not unmindful of the difference between the discovery of a given machine and its application as a technology—for example, the invention of a steam engine (the aeolipile) by Hero of Alexandria long before its incorporation into a steam-mill. All these problems, to which we will return in our last section, refer however to the way in which technology makes its peace with the social, political, and economic institutions of the society in which it appears. They do not directly affect the contention that there exists a determinate sequence of productive technology for those societies that are interested in originating and applying such a technology.

What evidence do we have for such a view? I would put forward three suggestive pieces of evidence:

1. The Simultaneity of Invention

The phenomenon of simultaneous discovery is well known.[1] From our view, it argues that the process of discovery takes place along a well-defined frontier of knowledge rather than in grab-bag fashion. Admittedly, the concept of "simultaneity" is impressionistic,[2] but the related phenomenon of technological "clustering" again suggests that technical evolution follows a sequential and determinate rather than random course.[3]

2. The Absence of Technological Leaps

All inventions and innovations, by definition, represent an advance of the art beyond existing base lines. Yet, most advances, particularly in retrospect, appear essentially incremental, evolutionary. If nature makes no sudden leaps, neither, it would appear, does technology. To make my point by exaggeration, we do not find experiments in electricity in the year 1500, or attempts to extract power from the atom in the year 1700. On the whole, the development of the technology of production presents a fairly smooth and continuous profile rather than one of jagged peaks and discontinuities.

3. The Predictability of Technology

There is a long history of technological prediction, some of it ludicrous and some not.[4] What is interesting is that the development of technical progress has always seemed *intrinsically* predictable. This does not mean that we can lay down future timetables of technical discovery, nor does it rule out the possibility of surprises. Yet I venture to state that many scientists would be willing to make *general* predictions as to the nature of technological capability twenty-five or even fifty years ahead. This too suggests that technology follows a developmental sequence rather than arriving in a more chancy fashion.

I am aware, needless to say, that these bits of evidence do not constitute anything like a "proof" of my hypothesis. At best they establish the grounds on which a prima facie case of plausibility may be rested. But I should like now to strengthen these grounds by suggesting two deeper-seated reasons why technology *should* display a "structured" history.

The first of these is that a major constraint always operates on the technological capacity of an age, the constraint of its accumulated stock of available knowledge. The application of this knowledge may lag behind its reach; the technology of the hand-mill, for example, was by no means at the frontier of medieval technical knowledge, but technical realization can hardly precede what men generally know (although experiment may incrementally advance both technology and knowledge concurrently). Particularly from the mid-nineteenth century to the present do we sense the loosening constraints on technology stemming from successively yielding barriers of scientific knowledge—loosening constraints that result in the successive arrival of the electrical, chemical, aeronautical, electronic, nuclear, and space stages of technology.[5]

The gradual expansion of knowledge is not, however, the only order-bestowing constraint on the development of technology. A second controlling factor is the material competence of the age, its level of technical expertise. To make a steam engine, for example, requires not only some knowledge of the elastic properties of steam but the ability to cast iron cylinders of considerable dimensions with tolerable accuracy. It is one thing to produce a single steam-machine as an expensive toy, such as the machine depicted by Hero, and another to produce a machine that will produce power economically and effectively. The difficulties experienced by Watt and Boulton in achieving a fit of piston to cylinder illustrate the problems of creating a technology, in contrast with a single machine.

Yet until a metal-working technology was established—indeed, until an embryonic machine-tool industry had taken root—an industrial technology was impossible to create. Furthermore, the competence required to create such a technology does not reside alone in the ability or inability to make a particular machine (one thinks of Babbage's ill-fated calculator as an example of a machine born too soon), but in the ability of many industries to change their products or processes to "fit" a change in one key product or process.

This necessary requirement of technological congruence[6] gives us an additional cause of sequencing. For the ability of many industries to co-operate in producing the equipment needed for a "higher" stage of technology depends not alone on knowledge or sheer skill but on the division of labor and the specialization of industry. And this in turn hinges to a considerable degree on the sheer size of the stock of capital itself. Thus the slow and painful accumulation of capital, from which springs the gradual diversification of industrial function, becomes an independent regulator of the reach of technical capability.

In making this general case for a determinate pattern of technological evolution—at least insofar as that technology is concerned with production—I do not want to claim too much. I am well aware that reasoning about technical sequences is easily faulted as *post hoc ergo propter hoc*. Hence, let me leave this phase of my inquiry by suggesting no more than that the idea of a roughly ordered progression of productive technology seems logical enough to warrant further empirical investigation. To put it as concretely as possible, I do not think it is just by happenstance that the steam-mill follows, and does not precede, the hand-mill, nor is it mere fantasy in our own day when we speak of the coming of the automatic factory. In the future as in the past, the development of the technology of production seems bounded by the constraints of knowledge and capability and thus, in principle at least, open to prediction as a determinable force of the historic process.

II

The second proposition to be investigated is no less difficult than the first. It relates, we will recall, to the explicit statement that a given technology imposes certain social and political characteristics upon the society in which it is found. Is it true that, as Marx wrote in *The German Ideology*, "A certain mode of production, or industrial stage, is always combined with a certain mode of cooperation, or social stage,"[7] or as he put it in the sentence immediately preceding our hand-mill, steam-mill paradigm, "In acquiring new productive forces men change their mode of production, and in changing their mode of production they change their way of living—they change all their social relations"?

As before, we must set aside for the moment certain "cultural" aspects of the question. But if we restrict ourselves to the functional relationships directly connected with the process of production itself, I think we can indeed state that the technology of a society imposes a determinate pattern of social relations on that society.

We can, as a matter of fact, distinguish at least two such modes of influence:

1. The Composition of the Labor Force

In order to function, a given technology must be attended by a labor force of a particular kind. Thus, the hand-mill (if we may take this as referring to late medieval technology in general) required a work force composed of skilled or semiskilled craftsmen, who were free to practice their occupations at home or in a small atelier, at times and seasons that varied considerably. By way of contrast, the steam-mill—that is, the technology of the nineteenth century—required a work force composed of semiskilled or unskilled operatives who could work only at the factory site and only at the strict time schedule enforced by turning the machinery on or off. Again, the technology of the electronic age has steadily required a higher proportion of skilled attendants; and the coming technology of automation will still further change the needed mix of skills and the locale of work, and may as well drastically lessen the requirements of labor time itself.

2. The Hierarchical Organization of Work

Different technological apparatuses not only require different labor forces but different orders of supervision and co-ordination. The internal organization of the eighteenth-century handicraft unit, with its typical man-master relationship, presents a social configuration of a wholly different kind from that of the nineteenth-century factory with its men-manager confrontation, and this in turn differs from the internal social structure of the continuous-flow, semiautomated plant of the present. As the intricacy of the production process increases, a much more complex system of internal controls is required to maintain the system in working order.

Does this add up to the proposition that the steam-mill gives us society with the industrial capitalist? Certainly the class characteristics of a particular society are strongly implied in its functional organization. Yet it would seem wise to be very cautious before relating political effects exclusively to functional economic causes. The Soviet Union, for example, proclaims itself to be a socialist society although its technical base resembles that of old-fashioned capitalism. Had Marx written that the steam-mill gives you society with the industrial *manager*, he would have been closer to the truth.

What is less easy to decide is the degree to which the technological infrastructure is responsible for some of the sociological features of society. Is anomie, for instance, a disease of capitalism or of all industrial societies? Is the organization man a creature of monopoly capital or of all bureaucratic industry wherever found? These questions tempt us to look into the problem of the impact of technology on the existential quality of life, an area we have ruled out of bounds for this paper. Suffice it to say that superficial evidence seems to imply that the similar technologies of Russia and America are indeed giving rise to similar social phenomena of this sort.

As with the first portion of our inquiry, it seems advisable to end this section on a note of caution. There is a danger, in discussing the structure of the labor force or the nature of intrafirm organization, of assigning the sole causal efficacy to the visible presence of machinery and of overlooking the invisible influence of other factors at work. Gilfillan, for instance, writes, "Engineers have committed such blunders as saying the typewriter brought women to work in offices, and with the typesetting machine made possible the great modern newspaper, forgetting that in Japan there are women office workers and great modern newspapers getting practically no help from typewriters and typesetting machines."[8] In addition, even where technology seems unquestionably to play the critical role, an independent "social" element unavoidably enters the scene in the *design* of technology, which must take into account such facts as the level of education of the work force or its relative price. In this way the machine will reflect, as much as mould, the social relationships of work.

These caveats urge us to practice what William James called a "soft determinism" with regard to the influence of the machine on social relations. Nevertheless, I would say that our cautions qualify rather than invalidate the thesis that the prevailing level of technology imposes itself powerfully on the structural organization of the productive side of society. A foreknowledge of the shape of the technical core of society fifty years hence may not allow us to describe the political attributes of that society, and may perhaps only hint at its sociological character, but assuredly it presents us with a profile of requirements, both in labor skills and in supervisory needs, that differ considerably from those of today. We cannot say whether the society of the computer will give us the latter-day capitalist or the commissar, but it seems beyond question that it will give us the technician and the bureaucrat.

III

Frequently, during our efforts thus far to demonstrate what is valid and useful in the concept of technological determinism, we have been forced to defer certain aspects of the problem until later. It is time now to turn up the rug and to examine what has been swept under it. Let us try to systematize our qualifications and objections to the basic Marxian paradigm:

1. Technological Progress Is Itself a Social Activity

A theory of technological determinism must contend with the fact that the very activity of invention and innovation is an attribute of some societies and not of others. The Kalahari bushmen or the tribesmen of New Guinea, for instance, have persisted in a neolithic technology to the present day; the Arabs reached a high degree of technical proficiency in the past and have since suffered a decline; the classical Chinese developed technical expertise in some fields while unaccountably neglecting it in the area of production. What factors serve to encourage or discourage this technical thrust is a problem about which we know extremely little at the present moment.[9]

2. The Course of Technological Advance Is Responsive to Social Direction

Whether technology advances in the area of war, the arts, agriculture, or industry depends in part on the rewards, inducements, and incentives offered by society. In this way the direction of technological advance is partially the result of social policy. For example, the system of interchangeable parts, first introduced into France and then independently into England, failed to take root in either country for lack of government interest or market stimulus. Its success in America is attributable mainly to government support and to its appeal in a society without guild traditions and with high labor costs.[10] The general *level* of technology may follow an independently determined sequential path, but its areas of application certainly reflect social influences.

3. Technological Change Must Be Compatible with Existing Social Conditions

An advance in technology not only must be congruent with the surrounding technology but must also be compatible with the existing economic and other institutions of society. For example, labor-saving machinery will not find ready acceptance in a society where labor is abundant and cheap as a factor of production. Nor would a mass production technique recommend itself to a society that did not have a mass market. Indeed, the presence of slave labor seems generally to inhibit the use of machinery and the presence of expensive labor to accelerate it.[11]

These reflections on the social forces bearing on technical progress tempt us to throw aside the whole notion of technological determinism as false or misleading.[12] Yet, to relegate technology from an undeserved position of *primum mobile* in history to that of a mediating factor, both acted upon by and acting on the body of society, is not to write off its influence but only to specify its mode of operation with greater precision. Similarly, to admit we understand very little of the cultural factors that give rise to technology does not depreciate its role but focuses our attention on that period of history when technology is clearly a major historic force, namely Western society since 1700.

IV

What is the mediating role played by technology within modern Western society? When we ask this much more modest question, the interaction of society and technology begins to clarify itself for us:

1. The Rise of Capitalism Provided a Major Stimulus for the Development of a Technology of Production

Not until the emergence of a market system organized around the principle of private property did there also emerge an institution capable of systematically guiding the inventive and innovative abilities of society to the problem of facilitating production. Hence the environment of the eighteenth and nineteenth centuries provided both a novel and an extremely effective encouragement for the development of an *industrial* technology. In addition, the slowly opening political and social framework of late mercantilist society gave rise to social aspirations for which the new technology offered the best chance of realization. It was not only the steam-mill that gave us the industrial capitalist but the rising inventor-manufacturer who gave us the steam-mill.

2. The Expansion of Technology within the Market System Took on a New "Automatic" Aspect

Under the burgeoning market system not alone the initiation of technical improvement but its subsequent adoption and repercussion through the economy was largely governed by market considerations. As a result, both the rise and the proliferation of technology assumed the attributes of an impersonal diffuse "force" bearing on social and economic life. This was all the more pronounced because the political control needed to buffer its disruptive consequences was seriously inhibited by the prevailing laissez-faire ideology.

3. The Rise of Science Gave a New Impetus to Technology

The period of early capitalism roughly coincided with and provided a congenial setting for the development of an independent source of technological encouragement—the rise of the self-conscious activity of science. The steady expansion of scientific research, dedicated to the exploration of nature's secrets and to their harnessing for social use, provided an increasingly important stimulus for technological advance from the middle of the nineteenth century. Indeed, as the twentieth century has progressed, science has become a major historical force in its own right and is now the indispensable precondition for an effective technology.

It is for these reasons that technology takes on a special significance in the context of capitalism—or, for that matter, of a socialism based on maximizing production or minimizing costs. For in these societies, both the continuous appearance of technical advance and its diffusion throughout the society assume the attributes of autonomous process, "mysteriously" generated by society and thrust upon its members in a manner as indifferent as it is imperious. This is why, I think, the problem of technological determinism—of how machines make history—comes to us with such insistence despite the ease with which we can disprove its more extreme contentions.

Technological determinism is thus peculiarly a problem of a certain historic epoch—specifically that of high capitalism and low socialism—*in which the forces of technical change have been unleashed, but when the agencies for the control or guidance of technology are still rudimentary.*

The point has relevance for the future. The surrender of society to the free play of market forces is now on the wane, but its subservience to the impetus of the scientific ethos is on the rise. The prospect before us is assuredly that of an undiminished and very likely accelerated pace of technical change. From what we can foretell about the direction of this technological advance and the structural alterations it implies, the pressures in the future will be toward a society marked by a much greater degree of organization and deliberate control. What other political, social, and existential changes the age of the computer will also bring we do not know. What seems certain, however, is that the problem of technological determinism—that is, of the impact of machines on history—will remain germane until there is forged a degree of public control over technology far greater than anything that now exists.

Notes

1. See Robert K. Merton, "Singletons and Multiples in Scientific Discovery: A Chapter in the Sociology of Science," *Proceedings of the American Philosophical Society*, October 1961, 470–486.

2. See John Jewkes, David Sawers, and Richard Stillerman, *The Sources of Invention* (New York: Macmillan, 1960), 227, for a skeptical view.

3. "One can count 21 basically different means of flying, at least eight basic methods of geophysical prospecting; four ways to make uranium explosive; . . . 20 or 30 ways to control birth. . . . If each of these separate inventions were autonomous, i.e., without cause, how could one account for their arriving in these functional groups?" S. C. Gilfillan, "Social Implications of Technological Advance," *Current Sociology* 1 (1952): 197. See also Jacob Schmookler, "Economic Sources of Inventive Activity," *Journal of Economic History*, March 1962, 1–20; and Richard Nelson, "The Economics of Invention: A Survey of the Literature," *Journal of Business* 32 (April 1959): 101–119.

4. See Jewkes, Sawyers, and Stillerman, *Sources of Invention*, 230f, for a catalogue of chastening mistakes. On the other hand, for a sober predictive effort, see Francis Bello, "The 1960s: A Forecast of Technology," *Fortune* 59 (January 1959): 74–78; and Daniel Bell, "The Study of the Future," *Public Interest* 1 (Fall 1965): 119–130. Modern attempts at prediction project

likely avenues of scientific advance or technological function rather than the feasibility of specific machines.

5. To be sure, the inquiry now regresses one step and forces us to ask whether there are inherent stages for the expansion of knowledge, at least insofar as it applies to nature. This is a very uncertain question. But having already risked so much, I will hazard the suggestion that the roughly parallel sequential development of scientific understanding in those few cultures that have cultivated it (mainly classical Greece, China, the high Arabian culture, and the West since the Renaissance) makes such a hypothesis possible, provided that one looks to broad outlines and not to inner detail.

6. The phrase is Richard LaPiere's in *Social Change* (New York: McGraw-Hill, 1965), 263f.

7. Karl Marx and Friedrich Engels, *The German Ideology* (London: Lawrence and Wishart, 1942), 18.

8. Gilfillan, "Social Implications of Technological Advance," 202.

9. An interesting attempt to find a line of social causation is found in E. Hagen, *The Theory of Social Change* (Homewood, IL: Dorsey Press, 1962).

10. See K. R. Gilbert, "Machine-Tools," in vol. 4, *A History of Technology*, ed. Charles Singer, E. J. Holmyard, A. R. Hall, and Trevor I. Williams (Oxford: Oxford University Press, 1958), chap. 14.

11. See LaPiere, *Social Change*, 284; also H. J. Habbakuk, *British and American Technology in the 19th Century* (Cambridge: Cambridge University Press, 1962), passim.

12. As, for example, in A. Hansen, "The Technological Interpretation of History," *Quarterly Journal of Economics* 36, no. 1 (November 1921): 72–83.

8 The Social Construction of Facts and Artifacts

Trevor J. Pinch and Wiebe Bijker

This article was part of a concerted effort to disprove the idea that technology is deterministic and established a new way to analyze the development of technology: the social construction of technology (SCOT) approach. Advocates of the SCOT approach argue that social groups direct nearly every aspect of technology. It is people, not machines, who design, build, and give meaning to technologies and ultimately decide which ones to adopt and which ones to reject. For Pinch and Bijker, the capacity of human beings to define and change the technology around them is not limited to just a handful of powerful groups like CEOs, industrialists, or even engineers. Even after an artifact has been built and sold by a corporation, individuals still have the power to redefine what the technology means and invent unanticipated uses for it. Pinch and Bijker argue that technological determinism is a myth that results when one looks backward and believes that the path taken to the present was the only possible path. They encourage us instead to consider all the possibilities available in the past and realize that the technologies, designs, and uses chosen had as much to do with the social circumstances at the time as with the nature and the state of technical knowledge. Pinch and Bijker provide a theoretical framework for explaining technological development as a social process. They argue that in the early stages of a technology's development there is a broad array of possibilities for how it might develop; to use their terminology, the technology has *interpretive flexibility*. People decide on the uses, meaning, and specific design of a technology based on their interests, needs, and values. As different *relevant social groups* coalesce around a particular design and meaning for a technology, the technical design begins to stabilize and becomes much more difficult to reinterpret. Through this process, a technology is given uses and meanings that later appear to be essential and somehow "natural" parts of the technology rather than something arrived at through social negotiations.

. . .

Technology Studies

There is a large amount of writing that falls under the rubric of *technology studies*. It is convenient to divide the literature into three parts: innovation studies, history of technology, and sociology of technology. We discuss each in turn.

Most innovation studies have been carried out by economists looking for the conditions for success in innovation. Factors researched include various aspects of the

From W. Bijker, T. P. Hughes, and T. J. Pinch, eds., *The Social Construction of Technological Systems* (Cambridge, MA: MIT Press, 1987), 21–50.

innovating firm (for example, size of R&D effort, management strength, and marketing capability) along with macroeconomic factors pertaining to the economy as a whole.[1] This literature is in some ways reminiscent of the early days in the sociology of science, when scientific knowledge was treated like a "black box" (Whitley 1972) and, for the purpose of such studies, scientists might as well have produced meat pies. Similarly, in the economic analysis of technological innovation everything is included that might be expected to influence innovation, except any discussion of the technology itself. As Layton (1977) notes:

> What is needed is an understanding of technology from inside, both as a body of knowledge and as a social system. Instead, technology is often treated as a "black box" whose contents and behaviour may be assumed to be common knowledge. (198)

Only recently have economists started to look into this black box.[2]

The failure to take into account the content of technological innovations results in the widespread use of simple linear models to describe the process of innovation. The number of developmental steps assumed in these models seems to be rather arbitrary (for an example of a six-stage process see figure 8.1).[3] Although such studies have undoubtedly contributed much to our understanding of the conditions for economic success in technological innovation, because they ignore the technological content they cannot be used as the basis for a social constructivist view of technology.[4]

This criticism cannot be leveled at the history of technology, where there are many finely crafted studies of the development of particular technologies. However, for the purposes of a sociology of technology, this work presents two kinds of problem. The first is that descriptive historiography is endemic in this field. Few scholars (but there are some notable exceptions) seem concerned with generalizing beyond historical instances, and it is difficult to discern any overall patterns on which to build a theory of technology (Staudenmaier 1984, 1985). This is not to say that such studies might not be useful building blocks for a social constructivist view of technology—merely that these historians have not yet demonstrated that they are doing sociology of knowledge in a different guise.[5]

The second problem concerns the asymmetric focus of the analysis. For example, it has been claimed that in twenty-five volumes of *Technology and Culture* only nine articles were devoted to the study of failed technological innovations (Staudenmaier 1985). This contributes to the implicit adoption of a linear structure of technological development, which suggests that the whole history of technological development had followed an orderly or rational path, as though today's world was the precise goal toward which all decisions, made since the beginning of history, were consciously directed (Ferguson 1974, 19).

This preference for successful innovations seems to lead scholars to assume that the success of an artifact is an explanation of its subsequent development. Historians of technology often seem content to rely on the manifest success of the artifact as

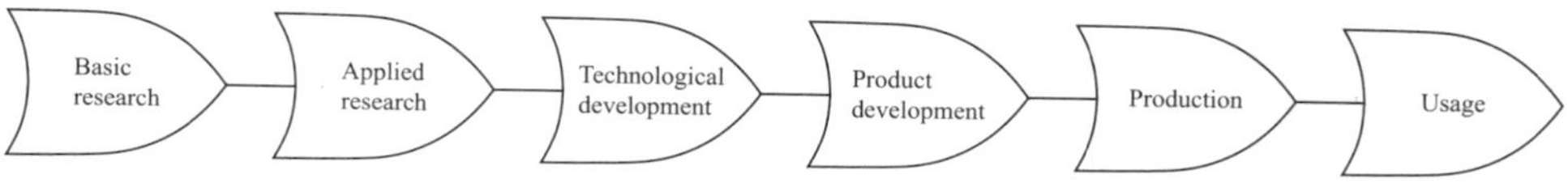

Figure 8.1
A six-stage model of the innovation process.

evidence that there is no further explanatory work to be done. For example, many histories of synthetic plastics start by describing the "technically sweet" characteristics of Bakelite; these features are then used implicitly to position Bakelite at the starting point of the glorious development of the field:

> God said: "let Baekeland be" and all was plastics! (Kaufman 1963, 61)

However, a more detailed study of the developments of plastic and varnish chemistry, following the publication of the Bakelite process in 1909 (Baekeland 1909a, 1909b), shows that Bakelite was at first hardly recognized as the marvelous synthetic resin that it later proved to be.[6] And this situation did not change much for some ten years. During the First World War the market prospects for synthetic plastics actually grew worse. However, the dumping of war supplies of phenol (used in the manufacture of Bakelite) in 1918 changed all this (Haynes 1954, 137–138) and made it possible to keep the price sufficiently low to compete with (semi-)natural resins, such as celluloid.[7] One can speculate over whether Bakelite would have acquired its prominence if it had not profited from that phenol dumping. In any case it is clear that a historical account founded on the retrospective success of the artifact leaves much untold.

Given our intention of building a sociology of technology that treats technological knowledge in the same symmetric, impartial manner that scientific facts are treated within the sociology of scientific knowledge, it would seem that much of the historical material does not go far enough. The success of an artifact is precisely what needs to be explained. For a sociological theory of technology it should be the *explanandum*, not the *explanans*.

Our account would not be complete, however, without mentioning some recent developments, especially in the American history of technology. These show the emergence of a growing number of theoretical themes on which research is focused (Hughes 1979; Staudenmaier 1985). For example, the systems approach to technology,[8] consideration of the effect of labor relations on technological development,[9] and detailed studies of some not-so-successful inventions[10] seem to herald departures from the "old" history of technology. Such work promises to be valuable for a sociological analysis of technology, and we return to some of it later.

The final body of work we wish to discuss is what might be described as "sociology of technology."[11] There have been some limited attempts in recent years to launch such a sociology, using ideas developed in the history and sociology of science—studies by, for example, Johnston (1972) and Dosi (1982), who advocate the description of technological knowledge in terms of Kuhnian paradigms.[12] Such approaches certainly appear to be more promising than standard descriptive historiography, but it is not clear whether or not these authors share our understanding of technological artifacts as social constructs. For example, neither Johnston nor Dosi considers explicitly the need for a symmetric sociological explanation that treats successful and failed artifacts in an equivalent way. Indeed, by locating their discussion at the level of technological paradigms, we are not sure how the artifacts themselves are to be approached. As neither author has yet produced an empirical study using Kuhnian ideas, it is difficult to evaluate how the Kuhnian terms may be utilized.[13] Certainly this has been a pressing problem in the sociology of science, where it has not always been possible to give Kuhn's terms a clear empirical reference.

The possibilities of a more radical social constructivist view of technology have been touched on by Mulkay (1979). He argues that the success and efficacy of technology could pose a special problem for the social constructivist view of *scientific knowledge*. The argument Mulkay wishes to counter is that the practical effectiveness of technology somehow demonstrates the privileged epistemology of science and thereby exempts it from sociological explanation. Mulkay opposes this view, rightly in our opinion, by pointing out the problem of the "science discovers, technology applies" notion implicit in such claims. In a second argument against this position, Mulkay notes (following Mario Bunge [1966]) that it is possible for a false or partly false theory to be used as the basis for successful practical application: The success of the technology would not then have anything to say about the "truth" of the scientific knowledge on which it was based. We find this second point not entirely satisfactory. We would rather stress that the truth or falsity of scientific knowledge is irrelevant to sociological analysis of belief: To retreat to the argument that science may be wrong but good technology can still be based on it is missing this point. Furthermore, the success of technology is still left unexplained within such an argument. The only effective way to deal with these difficulties is to adopt a perspective that attempts to show that technology, as well as science, can be understood as a social construct.

Mulkay seems to be reluctant to take this step because, as he points out, "there are very few studies . . . which consider how the technical meaning of hard technology is socially constructed" (Mulkay 1979, 77). This situation however, is starting to change: A number of such studies have recently emerged. For example, Michel Callon, in a pioneering study, has shown the effectiveness of focusing on technological controversies. He draws on an extensive case study of the electric vehicle in France (1960–75) to demonstrate that almost everything is negotiable: what is certain and what is not; who is a scientist and who is a technologist; what is technological and what is social; and who can participate in the controversy (Callon 1980a, 1980b, 1981, 1987). David Noble's study of the introduction of numerically controlled machine tools can also be regarded as an important contribution to a social constructivist view of technology (Noble 1984). Noble's explanatory goals come from a rather different (Marxist) tradition,[14] and his study has much to recommend it: He considers the development of both a successful and a failed technology and gives a symmetric account of both developments. Another intriguing study in this tradition is Lazonick's (1979) account of the introduction of the self-acting mule: He shows that aspects of this technical development can be understood in terms of the relations of production rather than any inner logic of technological development. The work undertaken by Bijker, Bönig, and Van Oost is another attempt to show how the socially constructed character of the content of some technological artifacts might be approached empirically: Six case studies were carried out, using historical sources.[15]

In summary, then, we can say that the predominant traditions in technology studies—innovation studies and the history of technology—do not yet provide much encouragement for our program. There are exceptions, however, and some recent studies in the sociology of technology present promising starts on which a unified approach could be built. We now give a more extensive account of how these ideas may be synthesized.

EPOR and SCOT

In this part we outline in more detail the concepts and methods that we wish to employ. We start by describing the *Empirical Programme of Relativism* as it was developed in the sociology of scientific knowledge. We then go on to discuss in more detail the approach taken by Bijker and his collaborators in the sociology of technology.

The Empirical Programme of Relativism (EPOR)

The EPOR is an approach that has produced several studies demonstrating the social construction of scientific knowledge in the "hard" sciences. This tradition of research has emerged from recent sociology of scientific knowledge. Its main characteristics, which distinguish it from other approaches in the same area, are the focus on the empirical study of contemporary scientific developments and the study, in particular, of scientific controversies.[16]

Three stages in the explanatory aims of the EPOR can be identified. In the *first stage* the interpretative flexibility of scientific findings is displayed; in other words, it is shown that scientific findings are open to more than one interpretation. This shifts the focus for the explanation of scientific developments from the natural world to the social world. Although this interpretative flexibility can be recovered in certain circumstances, it remains the case that such flexibility soon disappears in science; that is, a scientific consensus as to what the "truth" is in any particular instance usually emerges. Social mechanisms that limit interpretative flexibility and thus allow scientific controversies to be terminated are described in the *second stage*. A *third stage*, which has not yet been carried through in any study of contemporary science, is to relate such "closure mechanisms" to the wider social-cultural milieu. If all three stages were to be addressed in a single study, as Collins writes, "the impact of society on knowledge 'produced' at the laboratory bench would then have been followed through in the hardest possible case" (Collins 1981c, 7).

The EPOR represents a continuing effort by sociologists to understand the content of the natural sciences in terms of social construction. Various parts of the program are better researched than others. The third stage of the program has not yet even been addressed, but there are many excellent studies exploring the first stage. Most current research is aimed at elucidating the closure mechanisms whereby consensus emerges (the second stage). Many studies within the EPOR have been most fruitfully located in the area of scientific controversy. Controversies offer a methodological advantage in the comparative ease with which they reveal the interpretative flexibility of scientific results. Interviews conducted with scientists engaged in a controversy usually reveal strong and differing opinions over scientific findings. As such flexibility soon vanishes from science, it is difficult to recover from the textual sources with which historians usually work. Collins has highlighted the importance of the "controversy group" in science by his use of the term "core set" (Collins 1981b). These are the scientists most intimately involved in a controversial research topic. Because the core set is defined in relation to knowledge production in science (the core set constructs scientific knowledge), some of the empirical problems encountered in the identification of groups in science by purely sociometric means can be overcome. And studying the core set has another methodological advantage, in that the resulting consensus can be monitored. In other words, the group of scientists who experiment and theorize at the research frontiers and who become embroiled in scientific controversy will also reflect the growing consensus as to the outcome of that controversy. The same group of core set scientists can then be studied in both the first and second stages of the EPOR. For the purposes of the third stage, the notion of a core set may be too limited.

The Social Construction of Technology (SCOT)

Before outlining some of the concepts found to be fruitful by Bijker and his collaborators in their studies in the sociology of technology, we should point out an imbalance between the two approaches (EPOR and SCOT) we are considering. The EPOR is part of a flourishing tradition in the sociology of scientific knowledge: It is a well-established

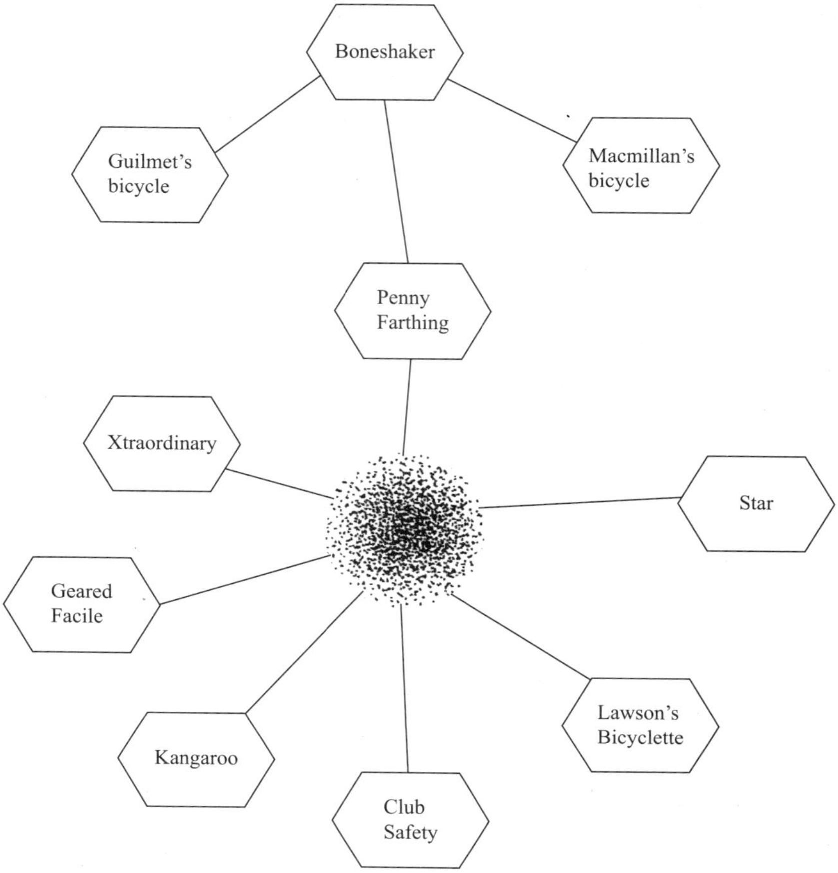

Figure 8.2
A multidirectional view of the developmental process of the Penny Farthing bicycle. The shaded area is filled in and magnified in figure 8.11. The hexagons symbolize artifacts.

program supported by much empirical research. In contrast, the sociology of technology is an embryonic field with no well-established traditions of research, and the approach we draw on specifically (SCOT) is only in its early empirical stages, although clearly gaining momentum.[17]

In SCOT the developmental process of a technological artifact is described as an alternation of variation and selection.[18] This results in a "multidirectional" model, in contrast with the linear models used explicitly in many innovation studies and implicitly in much history of technology. Such a multidirectional view is essential to any social constructivist account of technology. Of course, with historical hindsight, it is possible to collapse the multidirectional model on to a simpler linear model; but this misses the thrust of our argument that the "successful" stages in the development are not the only possible ones.

Let us consider the development of the bicycle.[19] Applied to the level of artifacts in this development, this multidirectional view results in the description summarized in figure 8.2. Here we see the artifact "Ordinary" (or, as it was nicknamed after becoming

Figure 8.3
A typical Penny Farthing, the Bayliss-Thomson Ordinary (1878). Photograph courtesy of the Trustees of the Science Museum, London.

less ordinary, the "Penny-farthing"; figure 8.3) and a range of possible variations. It is important to recognize that, in the view of the actors of those days, these variants were at the same time quite different from each other and equally were serious rivals. It is only by retrospective distortion that a quasi-linear development emerges, as depicted in figure 8.4. In this representation the so-called safety ordinaries (Xtraordinary [1878], Facile [1879], and Club Safety [1885]) figure only as amusing aberrations that need not be taken seriously (figures 8.5, 8.6, and 8.7). Such a retrospective description can be challenged by looking at the actual situation in the 1880s. Some of the "safety ordinaries" were produced commercially, whereas Lawson's Bicyclette, which seems to play an important role in the linear model, proved to be a commercial failure (Woodforde 1970).

However, if a multidirectional model is adopted, it is possible to ask why some of the variants "die," whereas others "survive." To illuminate this "selection" part of the developmental processes, let us consider the problems and solutions presented by each artifact at particular moments. The rationale for this move is the same as that for focusing on scientific controversies within EPOR. In this way, one can expect to bring out more clearly the interpretative flexibility of technological artifacts.

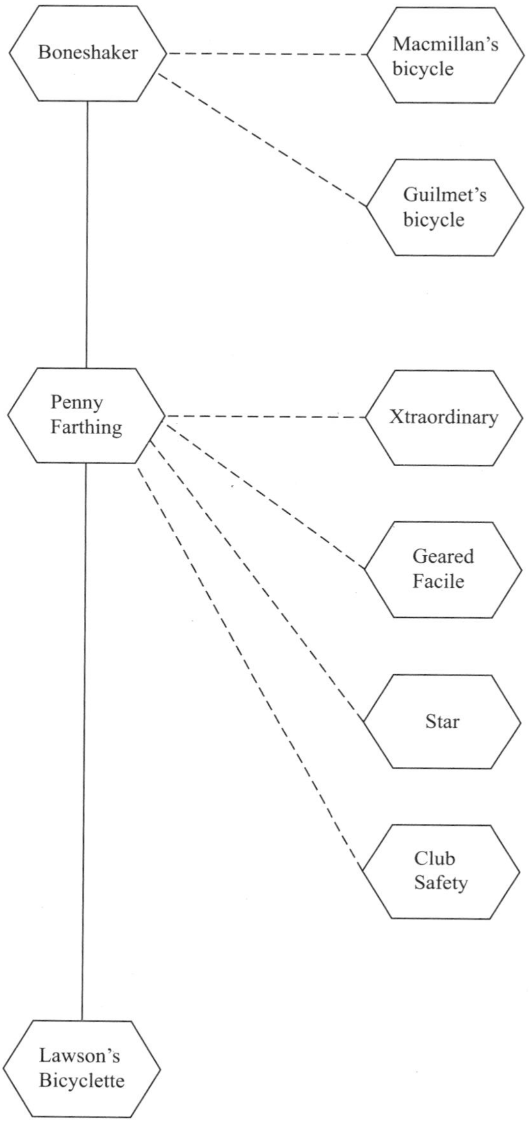

Figure 8.4
The traditional quasi-linear view of the developmental process of the Penny Farthing bicycle. Solid lines indicate successful development, and dashed lines indicate failed development.

In deciding which problems are relevant, the social groups concerned with the artifact and the meanings that those groups give to the artifact play a crucial role: A problem is defined as such only when there is a social group for which it constitutes a "problem."

The use of the concept of a relevant social group is quite straightforward. The phrase is used to denote institutions and organizations (such as the military or some specific industrial company), as well as organized or unorganized groups of individuals. The key requirement is that all members of a certain social group share the same set of meanings, attached to a specific artifact.[20] In deciding which social groups are

Figure 8.5
The American Star bicycle (1885). Photograph courtesy of the Trustees of the Science Museum, London.

Figure 8.6
Facile bicycle (1874). Photograph courtesy of the Trustees of the Science Museum, London.

Figure 8.7
A form of the Kangaroo bicycle (1878). Photograph courtesy of the Trustees of the Science Museum, London.

relevant, we must first ask whether the artifact has any meaning at all for the members of the social group under investigation. Obviously, the social group of "consumers" or "users" of the artifact fulfills this requirement. But also less obvious social groups may need to be included. In the case of the bicycle, one needs to mention the "anticyclists." Their actions ranged from derisive cheers to more destructive methods. For example, Reverend L. Meadows White described such resistance to the bicycle in his book, *A Photographic Tour on Wheels*:

> But when to words are added deeds, and stones are thrown, sticks thrust into the wheels, or caps hurled into the machinery, the picture has a different aspect. All the above in certain districts are of common occurrence, and have all happened to me, especially when passing through a village just after school is closed. (Quoted in Woodforde 1970, 49–50)

Clearly, for the anticyclists the artifact "bicycle" had taken on meaning!

Another question we need to address is whether a provisionally defined social group is homogeneous with respect to the meanings given to the artifact—or is it more effective to describe the developmental process by dividing a rather heterogeneous group into several different social groups? Thus within the group of cycle-users we discern a separate social group of women cyclists. During the days of the high-wheeled Ordinary women were not supposed to mount a bicycle. For instance, in a magazine advice column (1885) it is proclaimed, in reply to a letter from a young lady:

> The mere fact of riding a bicycle is not in itself sinful, and if it is the only means of reaching the church on a Sunday, it may be excusable. (Quoted in Woodforde 1970, 122)

Tricycles were the permitted machines for women. But engineers and producers anticipated the importance of women as potential bicyclists. In a review of the annual Stanley Exhibition of Cycles in 1890, the author observes:

> From the number of safeties adapted for the use of ladies, it seems as if bicycling was becoming popular with the weaker sex, and we are not surprised at it, considering the saving of power derived from the use of a machine having only one slack. ("Stanley Exhibition of Cycles" 1890, 107–108)

Thus some parts of the bicycle's development can be better explained by including a separate social group of feminine cycle-users. This need not, of course, be so in other cases: For instance, we would not expect it to be useful to consider a separate social group of women users of, say, fluorescent lamps.

Once the relevant social groups have been identified, they are described in more detail. This is also where aspects such as power or economic strength enter the description, when relevant. Although the only defining property is some homogeneous meaning given to a certain artifact, the intention is not just to retreat to worn-out, general statements about "consumers" and "producers." We need to have a detailed description of the relevant social groups in order to define better the function of the artifact with respect to each group. Without this, one could not hope to be able to give any explanation of the developmental process. For example, the social group of cyclists riding the high-wheeled Ordinary consisted of "young men of means and nerve: they might be professional men, clerks, schoolmasters or dons" (Woodforde 1970, 47). For this social group the function of the bicycle was primarily for sport. The following comment in the *Daily Telegraph* (September 7, 1877) emphasizes sport, rather than transport:

> Bicycling is a healthy and manly pursuit with much to recommend it, and, unlike other foolish crazes, it has not died out. (Quoted in Woodforde 1970, 122)

Let us now return to the exposition of the model. Having identified the relevant social groups for a certain artifact (figure 8.8), we are especially interested in the problems each group has with respect to that artifact (figure 8.9). Around each problem, several variants of solution can be identified (figure 8.10). In the case of the bicycle, some relevant problems and solutions are shown in figure 8.11, in which the shaded area of figure 8.2 has been filled. This way of describing the developmental process brings out clearly all kinds of conflicts: conflicting technical requirements by different social groups (for example, the speed requirement and the safety requirement); conflicting solutions to the same problem (for example, the safety low-wheelers and the safety ordinaries); and moral conflicts (for example, women wearing skirts or trousers on high-wheelers; figure 8.12). Within this scheme, various solutions to these conflicts and problems are possible—not only technological ones but also judicial or even moral ones (for example, changing attitudes toward women wearing trousers).

Following the developmental process in this way, we see growing and diminishing degrees of stabilization of the different artifacts.[21] In principle, the degree of stabilization is different in different social groups. By using the concept of stabilization, we see that the "invention" of the safety bicycle was not an isolated event (1884), but a nineteen-year process (1879–98). For example, at the beginning of this period the relevant groups did not see the "safety bicycle" but a wide range of bi- and tricycles—and, among those, a rather ugly crocodilelike bicycle with a relatively low front wheel and rear chain drive (Lawson's Bicyclette; figure 8.13). By the end of the period, the phrase "safety bicycle" denoted a low-wheeled bicycle with rear chain drive, diamond frame, and air tires. As a result of the stabilization of the artifact after 1898, one did not need to specify these details: They were taken for granted as the essential "ingredients" of the safety bicycle.

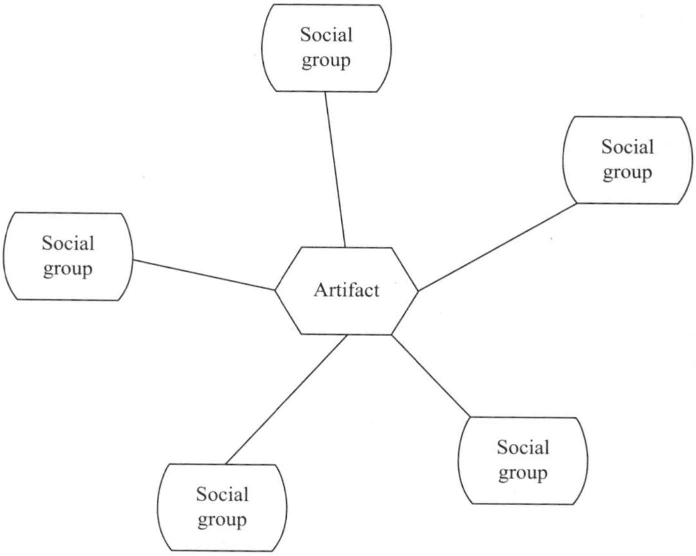

Figure 8.8
The relationship between an artifact and the relevant social groups.

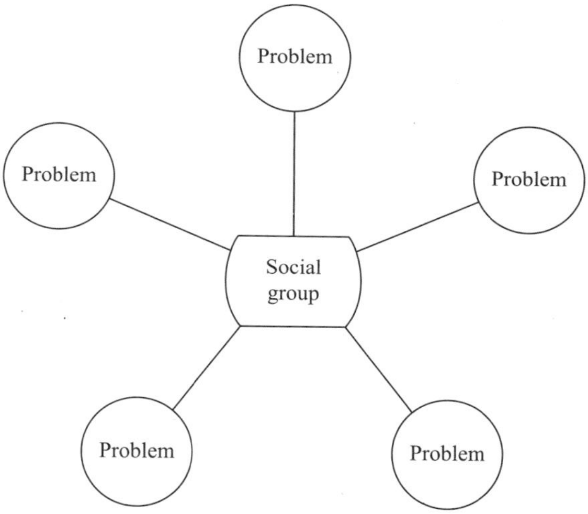

Figure 8.9
The relationship between one social group and the perceived problems.

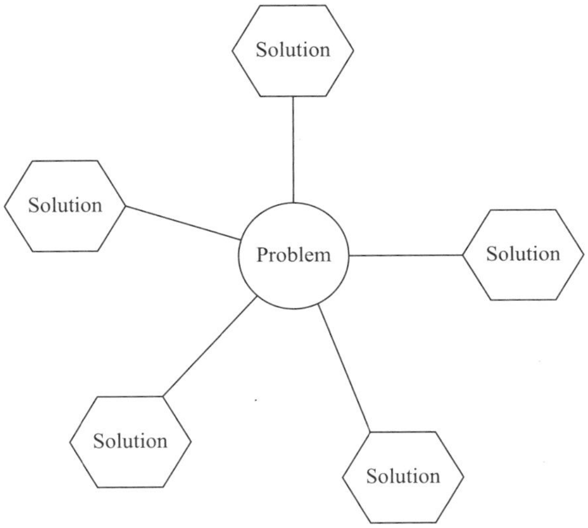

Figure 8.10
The relationship between one problem and its possible solutions.

We want to stress that our model is not used as a mold into which the empirical data have to be forced, *coûte que coûte*. The model has been developed from a series of case studies and not from purely philosophical or theoretical analysis. Its function is primarily heuristic—to bring out all the aspects relevant to our purposes. This is not to say that there are no explanatory and theoretical aims, analogous to the different stages of the EPOR (Bijker 1984, 1987). And indeed, as we have shown, this model already does more than merely describe technological development: It highlights its multidirectional character. Also, as will be indicated, it brings out the interpretative flexibility of technological artifacts and the role that different closure mechanisms may play in the stabilization of artifacts.

The Social Construction of Facts and Artifacts

Having described the two approaches to the study of science and technology we wish to draw on, we now discuss in more detail the parallels between them. As a way of putting some flesh on our discussion we give, where appropriate, empirical illustrations drawn from our own research.

Interpretative Flexibility

The first stage of the EPOR involves the demonstration of the interpretative flexibility of scientific findings. In other words, it must be shown that different interpretations of nature are available to scientists and hence that nature alone does not provide a determinant outcome to scientific debate.[22]

In SCOT, the equivalent of the first stage of the EPOR would seem to be the demonstration that technological artifacts are culturally constructed and interpreted; in other words, the interpretative flexibility of a technological artifact must be shown. By this we mean not only that there is flexibility in how people think of or interpret artifacts but also that there is flexibility in how artifacts are *designed*. There is

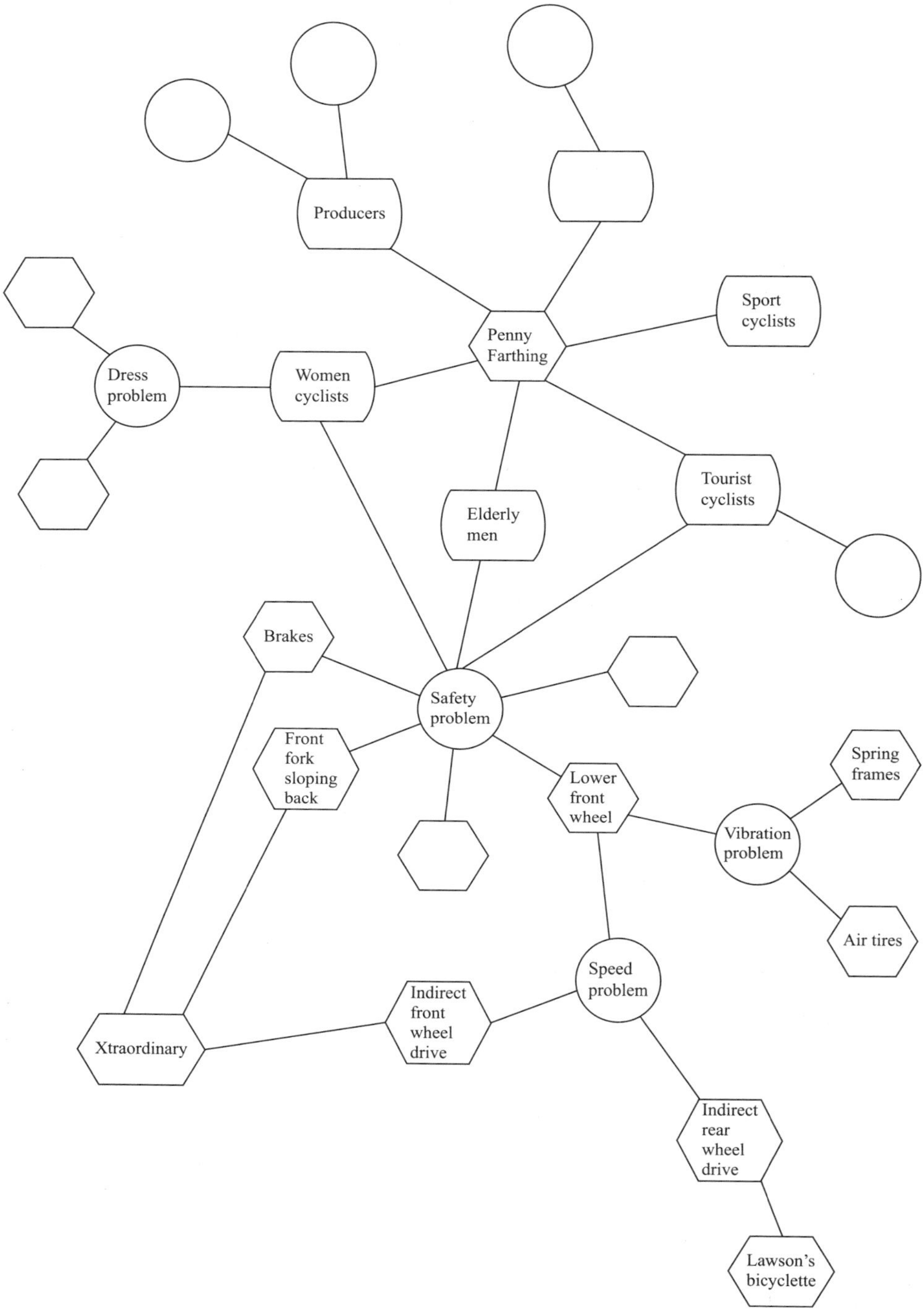

Figure 8.11
Some relevant social groups, problems, and solutions in the developmental process of the Penny Farthing bicycle. Because of lack of space, not all artifacts, relevant social groups, problems, and solutions are shown.

not just one possible way or one best way of designing an artifact. In principle, this could be demonstrated in the same way as for the science case, that is, by interviews with technologists who are engaged in a contemporary technological controversy. For example, we can imagine that, if interviews had been carried out in 1890 with the cycle engineers, we would have been able to show the interpretative flexibility of the artifact "air tyre." For some, this artifact was a solution to the vibration problem of small-wheeled vehicles:

> [The air tire was] devised with a view to afford increased facilities for the passage of wheeled vehicles—chiefly of the lighter class such for instance as velocipedes, invalid chairs, ambulances—over roadways and paths, especially when these latter are of rough or uneven character. (Dunlop 1888, 1)

For others, the air tire was a way of going faster (this is outlined in more detail later). For yet another group of engineers, it was an ugly looking way of making the low-wheeler even less safe (because of side-slipping) than it already was. For instance, the following comment, describing the Stanley Exhibition of Cycles, is revealing:

> The most conspicuous innovation in the cycle construction is the use of pneumatic tires. These tires are hollow, about 2 in. diameter, and are inflated by the use of a small air pump. They are said to afford most luxurious riding, the roughest macadam and cobbles being reduced to the smoothest asphalte. Not having had the opportunity of testing these tires, we are unable to speak of them from practical experience; but looking at them from a theoretical point of view, we opine that considerable difficulty will be experienced in keeping the tires thoroughly inflated. Air under pressure is a troublesome thing to deal with. From the reports of those who have used these tires, it seems that they are prone to slip on muddy roads. If this is so, we fear their use on rear-driving safeties—which are all more or less addicted to side-slipping—is out of the question, as any improvement in this line should be to prevent side slip and not to increase it. Apart from these defects, the appearance of the tires destroys the symmetry and graceful appearance of a cycle, and this alone is, we think, sufficient to prevent their coming into general use. ("Stanley Exhibition of Cycles" 1890, 107)

And indeed, other artifacts were seen as providing a solution for the vibration problem, as the following comment reveals:

> With the introduction of the rear-driving safety bicycle has arisen a demand for anti-vibration devices, as the small wheels of these machines are conducive to considerable vibration, even on the best roads. Nearly every exhibitor of this type of machine has some appliance to suppress vibration. ("Stanley Exhibition of Cycles" 1889, 157–158)

Most solutions used various spring constructions in the frame, the saddle, and the steering-bar (figure 8.14). In 1896, even after the safety bicycle (and the air tire with it) achieved a high degree of stabilization, "spring frames" were still being marketed.

It is important to realize that this demonstration of interpretative flexibility by interviews and historical sources is only one of a set of possible methods. At least in the study of technology, another method is applicable and has actually been used. It can be shown that different social groups have radically different interpretations of one technological artifact. We call these differences "radical" because the *content* of the artifact seems to be involved. It is something more than what Mulkay rightly claims to be rather easy—"to show that the social meaning of television varies with and depends upon the social context in which it is employed." As Mulkay notes: "It is much more difficult to show what is to count as a 'working television set' is similarly context-dependent in any significant respect" (Mulkay 1979, 80).

Figure 8.12
A solution to the women's dressing problem with respect to the high-wheeled Ordinary. The solution obviously has technical and athletic aspects. Probably, the athletic aspects prevented the solution from stabilizing. The set-up character of the photograph suggests a rather limited practical use. Photograph courtesy of the Trustees of the Science Museum, London.

We think that our account—in which the different interpretations by social groups of the content of artifacts lead by means of different chains of problems and solutions to different further developments—involves the content of the artifact itself. Our earlier example of the development of the safety bicycle is of this kind. Another example is variations within the high-wheeler. The high-wheeler's meaning as a virile, high-speed bicycle led to the development of larger front wheels—for with a fixed angular velocity one way of getting a higher translational velocity over the ground was by enlarging the radius. One of the last bicycles resulting from this strand of development was the Rudge Ordinary of 1892, which had a 56-inch wheel and air tire. But groups of women

Figure 8.13
Lawson's Bicyclette (1879). Photograph courtesy of the Trustees of the Science Museum, London.

Figure 8.14
Whippet spring frame (1885). Photograph courtesy of the Trustees of the Science Museum, London.

and of elderly men gave quite another meaning to the high-wheeler. For them, its most important characteristic was its lack of safety:

> Owing to the disparity in wheel diameters and the small weight of the backbone and trailing wheel, also to the rider's position practically over the centre of the wheel, if the large front wheel hit a brick or large stone on the road, and the rider was unprepared, the sudden check to the wheel usually threw him over the handlebar. For this reason the machine was regarded as dangerous, and however enthusiastic one may have been about the ordinary—and I was an enthusiastic rider of it once—there is no denying that it was only possible for comparatively young and athletic men. (Grew 1921, 8)

This meaning gave rise to lowering the front wheel, moving back the saddle, and giving the front fork a less upright position. Via another chain of problems and solutions (see figure 8.7), this resulted in artifacts such as Lawson's Bicyclette (1879) and the Xtraordinary (1878; figure 8.15). Thus there was not *one* high-wheeler; there was the *macho* machine, leading to new designs of bicycles with even higher front wheels, and there was the *unsafe* machine, leading to new designs of bicycles with lower front wheels, saddles moved backward, or reversed order of small and high wheel. Thus the interpretative flexibility of the artifact Penny-farthing is materialized in quite different design lines.

Closure and Stabilization

The second stage of the EPOR concerns the mapping of mechanisms for the closure of debate—or, in SCOT, for the stabilization of an artifact. We now illustrate what we

Figure 8.15
Singer Xtraordinary bicycle (1878). Photograph courtesy of the Trustees of the Science Museum, London.

mean by a closure mechanism by giving examples of two types that seem to have played a role in cases with which we are familiar. We refer to the particular mechanisms on which we focus as rhetorical closure and closure by redefinition of problem.

Rhetorical Closure Closure in technology involves the stabilization of an artifact and the "disappearance" of problems. To close a technological "controversy," one need not *solve* the problems in the common sense of that word. The key point is whether the relevant social groups *see* the problem as being solved. In technology, advertising can play an important role in shaping the meaning that a social group gives to an artifact.[23] Thus, for instance, an attempt was made to "close" the "safety controversy" around the high-wheeler by simply claiming that the artifact was perfectly safe. An advertisement for the "Facile" (*sic*!) Bicycle (figure 8.16) reads:

> Bicyclists! Why risk your limbs and lives on high Machines when for road work a 40 inch or 42 inch "Facile" gives all the advantages of the other, together with almost absolute safety. (*Illustrated London News* 1880, quoted in Woodforde 1970, 60)

This claim of "almost absolute safety" was a rhetorical move, considering the height of the bicycle and the forward position of the rider, which were well known to engineers at the time to present problems of safety.

Closure by Redefinition of the Problem We have already mentioned the controversy around the air tire. For most of the engineers it was a theoretical and practical monstrosity. For the general public, in the beginning it meant an aesthetically awful accessory:

> Messenger boys guffawed at the sausage tyre, factory ladies squirmed with merriment, while even sober citizens were sadly moved to mirth at a comicality obviously designed solely to lighten the gloom of their daily routine. (Woodforde 1970, 89)

Figure 8.16
Geared Facile bicycle (1888). Photograph courtesy of the Trustees of the Science Museum, London.

For Dunlop and the other protagonists of the air tire, originally the air tire meant a solution to the vibration problem. However, the group of sporting cyclists riding their high-wheelers did not accept that as a problem at all. Vibration presented a problem only to the (potential) users of the low-wheeled bicycle. Three important social groups were therefore opposed to the air tire. But then the air tire was mounted on a racing bicycle. When, for the first time, the tire was used at the racing track, its entry was hailed with derisive laughter. This was, however, quickly silenced by the high speed achieved, and there was only astonishment left when it outpaced all rivals (Croon 1939). Soon handicappers had to give racing cyclists on high-wheelers a considerable start if riders on air-tire low-wheelers were entered. After a short period no racer of any pretensions troubled to compete on anything else (Grew 1921).

What had happened? With respect to two important groups, the sporting cyclists and the general public, closure had been reached, but not by convincing those two groups of the feasibility of the air tire in its meaning as an antivibration device. One can say, we think, that the meaning of the air tire was translated to constitute a solution to quite another problem: the problem of how to go as fast as possible.[24] And thus, by redefining the key problem with respect to which the artifact should have the meaning of a solution, closure was reached for two of the relevant social groups. How the third group, the engineers, came to accept the air tire is another story and need not be told here. Of course, there is nothing "natural" or logically necessary about this form of closure. It could be argued that speed is not the most important characteristic of the bicycle or that existing cycle races were not appropriate tests of a cycle's "real" speed (after all, the idealized world of the race track may not match everyday road conditions, any more than the Formula-l racing car bears on the performance requirements of the average family sedan). Still, bicycle races have played an important role in the development of the bicycle, and because racing can be viewed as a specific form of testing, this observation is much in line with Constant's recent plea to pay more attention to testing procedures in studying technology (Constant 1983).

The Wider Context

Finally, we come to the third stage of our research program. The task here in the area of technology would seem to be the same as for science—to relate the content of a technological artifact to the wider sociopolitical milieu. This aspect has not yet been demonstrated for the science case,[25] at least not in contemporaneous sociological studies.[26]

However, the SCOT method of describing technological artifacts by focusing on the meanings given to them by relevant social groups seems to suggest a way forward. Obviously, the sociocultural and political situation of a social group shapes its norms and values, which in turn influence the meaning given to an artifact. Because we have shown how different meanings can constitute different lines of development, SCOT's descriptive model seems to offer an operationalization of the relationship between the wider milieu and the actual content of technology. To follow this line of analysis, see Bijker (1987).

Conclusion

In this chapter we have been concerned with outlining an integrated social constructivist approach to the empirical study of science and technology. We reviewed several relevant bodies of literature and strands of argument. We indicated that the social constructivist approach is a flourishing tradition within the sociology of science and that it shows every promise of wider application. We reviewed the literature on the science-technology relationship and showed that here, too, the social constructivist approach is starting to bear fruit. And we reviewed some of the main traditions in technology studies. We argued that innovation studies and much of the history of technology are unsuitable for our sociological purposes. We discussed some recent work in the sociology of technology and noted encouraging signs that a new wave of social constructivist case studies is beginning to emerge.

We then outlined in more detail the two approaches—one in the sociology of scientific knowledge (EPOR) and one in the field of sociology of technology (SCOT)—on which we base our integrated perspective. Finally, we indicated the similarity of the explanatory goals of the two approaches and illustrated these goals with some examples drawn from technology. In particular, we have seen that the concepts of interpretative flexibility and closure mechanism and the notion of social group can be given empirical reference in the social study of technology.

As we have noted throughout this chapter, the sociology of technology is still underdeveloped, in comparison with the sociology of scientific knowledge. It would be a shame if the advances made in the latter field could not be used to throw light on the study of technology. On the other hand, in our studies of technology it appeared to be fruitful to include several social groups in the analysis, and there are some indications that this method may also bear fruit in studies of science. Thus our integrated approach to the social study of science and technology indicates how the sociology of science and the sociology of technology might benefit each other.

But there is another reason, and perhaps an even more important one, to argue for such an integrated approach. And this brings us to a question that some readers might have expected to be dealt with in the first paragraph of this chapter, namely, the question of how to distinguish science from technology. We think that it is rather unfruitful to make such an a priori distinction. Instead, it seems worthwhile to start with commonsense notions of science and technology and to study them in an integrated way, as we have proposed. Whatever interesting differences may exist will gain contrast within such a program. This would constitute another concrete result of the integrated study of the social construction of facts and artifacts.

Notes

This chapter is a shortened and updated version of Pinch and Bijker (1984). We are grateful to Henk van den Belt, Ernst Homburg, Donald MacKenzie, and Steve Woolgar for comments on an earlier draft of this chapter. We would like to thank the Stiftung Volkswagen, the Federal Republic of Germany, the Twente University of Technology, the Netherlands, and the UK SSRC (under grant G/00123/0072/1) for financial support.

1. See, for example, Freeman (1974, 1977); Schmookler (1966, 1972); Scholz (1977); and Schumpeter ([1928] 1971, 1942).

2. See, for example, Dosi (1982, 1984); Nelson and Winter (1977, 1982); and Rosenberg (1982). A study that preceded these is Rosenberg and Vincenti (1978).

3. Adapted from Uhlmann (1978, 45).

4. For another critique of these linear models, see Kline (1985).

5. Shapin (1980) writes that "a proper perspective of the uses of science might reveal that sociology of knowledge and history of technology have more in common than is usually thought" (132). Although we are sympathetic to Shapin's argument, we think the time is now ripe for asking more searching questions of historical studies.

6. Manuals describing resinous materials do mention Bakelite but not with the amount of attention that, retrospectively, we would think to be justified. Professor Max Bottler (1924),

for example, devotes only one page to Bakelite in his 228-page book on resins and the resin industry. Even when Bottler concentrates in another book on the *synthetic* resinous materials, Bakelite does not receive an indisputable "first place." Only half of the book is devoted to phenol/formaldehyde condensation products, and roughly half of that part is devoted to Bakelite (Bottler 1919). See also Matthis (1920).

7. For an account of other aspects of Bakelite's success, see Bijker (1987).

8. See, for example, Constant (1980); Hanieski (1973); and Hughes (1983).

9. See, for example, Lazonick (1979); Noble (1979); and Smith (1977).

10. See, for example, Vincenti (1986).

11. There is an American tradition in the sociology of technology. See, for example, Gilfillan (1935); Ogburn (1945); Ogburn and Meyers Nimkoff (1955); and Westrum (1983). A fairly comprehensive view of the present state of the art in German sociology of technology can be obtained from Jokisch (1982). Several studies in the sociology of technology that attempt to break with the traditional approach can be found in Krohn, Layton, and Weingart (1978).

12. Dosi uses the concept of technological trajectory, developed by Nelson and Winter (1977); see also van den Belt and Rip (1987). Other approaches to technology based on Kuhn's idea of the community structure of science are mentioned by Bijker (1987). See also Constant (1987) and the collection edited by Laudan (1984).

13. One is reminded of the first blush of Kuhnian studies in the sociology of science. It was hoped that Kuhn's "paradigm" concept might be straightforwardly employed by sociologists in their studies of science. Indeed there were a number of studies in which attempts were made to identify phases in science, such as preparadigmatic, normal, and revolutionary. It soon became apparent, however, that Kuhn's terms were loosely formulated, could be subject to a variety of interpretations, and did not lend themselves to operationalization in any straightforward manner. See, for example, the inconclusive discussion over whether a Kuhnian analysis applies to psychology in Palermo (1973). A notable exception is Barnes's (1982) contribution to the discussion of Kuhn's work.

14. For a valuable review of Marxist work in this area, see MacKenzie (1984).

15. For a provisional report of this study, see Bijker, Bönig, and van Oost (1984). The five artifacts that are studied are Bakelite, fluorescent lighting, the safety bicycle, the Sulzer loom, and the transistor. See also Bijker (1987).

16. Work that might be classified as falling within the EPOR has been carried out primarily by Collins, Pinch, and Travis at the Science Studies Centre, University of Bath, and by Harvey and Pickering at the Science Studies Unit, University of Edinburgh. See, for example, Collins (1975); Pickering (1984); Pinch (1977, 1986); Wynne (1976) and the studies by Pickering, Collins, Travis, and Pinch in Collins (1981a).

17. See, for example, Bijker (1984, 1987) and Bijker and Pinch (1983). Studies by Elzen (1985, 1986); Jelsma and Smit (1986); Schot (1985, 1986); and van den Belt (1985) are also based on SCOT.

18. Constant (1980) used a similar evolutionary approach. Both Constant's model and our model seem to arise out of the work in evolutionary epistemology; see, for example, Campbell (1974) and Toulmin (1972). Elster (1983) gives a review of evolutionary models of technical change. See also van den Belt and Rip (1987).

19. It may be useful to state explicitly that we consider bicycles to be as fully fledged a technology as, for example, automobiles or aircraft. It may be helpful for readers from outside

notorious cycle countries such as the Netherlands, France, and Great Britain to point out that both the automobile and the aircraft industries are, in a way, descendants from the bicycle industry. Many names occur in the histories of both the bicycle and the autocar: Triumph, Rover, Humber, and Raleigh, to mention but a few (Caunter 1955, 1957). The Wright brothers both sold and manufactured bicycles before they started to build their flying machines—mostly made out of bicycle parts (Gibbs-Smith 1960).

20. There is no cookbook recipe for how to identify a social group. Quantitative instruments using citation data may be of some help in certain cases. More research is needed to develop operationalizations of the notion of "relevant social group" for a variety of historical and sociological research sites. See also Law (1987) on the demarcation of networks and Bijker (1987).

21. Previously, two concepts have been used that can be understood as two distinctive concepts within the broader idea of stabilization (Bijker, Bönig, and van Oost 1984). *Reification* was used to denote social existence—existence in the consciousness of the members of a certain social group. *Economic stabilization* was used to indicate the economic existence of an artifact—its having a market. Both concepts are used in a continuous and relative way, thus requiring phrases such as "the *degree* of reification of the high-wheeler is *higher* in the group of young men of means and nerve than in the group of elderly men."

22. The use of the concepts of interpretative flexibility and rhetorical closure in science cases is illustrated by Pinch and Bijker (1984).

23. Advertisements seem to constitute a large and potentially fruitful data source for empirical social studies of technology. The considerations that professional advertising designers give to differences among various "consumer groups" obviously fit our use of different relevant groups. See, for example, Bijker (1987) and Schwartz Cowan (1983).

24. The concept of translation is fruitfully used in an extended way by Callon (1980b, 1981, 1986); Callon and Law (1982); and Latour (1983, 1984).

25. A model of such a "stage 3" explanation is offered by Collins (1983).

26. Historical studies that address the third stage may be a useful guide here. See, for example, MacKenzie (1978); Shapin (1979, 1984); and Shapin and Schaffer (1985).

References

Baekeland, L. H. 1909a. "On soluble, fusible, resinous condensation products of phenols and formaldehyde." *Journal of Industrial and Engineering Chemistry* 1:345–349.

Baekeland, L. H. 1909b. "The synthesis, constitution, and use of Bakelite." *Journal of Industrial and Engineering Chemistry* 1:149–161.

Barnes, B. 1982. *T. S. Kuhn and Social Science*. London: Macmillan.

Bijker, W. E. 1984. "Collectifs technologiques et styles technologiques: Eléments pour un modèle explicatif de la construction sociale des artefacts techniques." In *Travailleur collectif et relations science-production*, edited by J. H. Jacot, 113–120. Paris: Editions du CNRS.

Bijker, W. E. 1987. "The social construction of Bakelite: Toward a theory of invention." In Bijker, Hughes, and Pinch 1987, 159–187.

Bijker, W. E., Bönig, J., and van Oost, E. C. J. 1984. "The social construction of technological artefacts." Paper presented at the EASST Conference. A shorter version of this paper is published in special issue, *Zeitschrift für Wissenschaftsorschung* 2 (3): 39–52.

Bijker, W. E., Hughes, T. P., and Pinch, T. J., eds. 1987. *The Social Construction of Technological Systems*. Cambridge, MA: MIT Press.

Bijker, W. E., and Pinch, T. J. 1983. "La construction sociale de faits et d'artefacts: Impératifs stratégiques et méthodologiques pour une approche unifiée de l'étude des sciences et de la technique." Paper presented to L'atelier de recherche (III) sur les problèmes stratégiques et méthodologiques en milieu scientifique et technique, Paris, March 1983.

Bottler, M. 1919. *Uber Herstellung und Eigenschaften von Kunstharzen und deren Verwendung in der Lack- und Firnisindustrie und zu elektrotechnischen und industriellen Zwecken*. Munich: Lehmanns.

Bottler, M. 1924. *Harze und Harzinustrie*. Leipzig: Max Janecke.

Bunge, M. 1966. "Technology as applied science." *Technology and Culture* 7:329–347.

Callon, M. 1980a. "The state and technical innovation: A case study of the electrical vehicle in France." *Research Policy* 9:358–376.

Callon, M. 1980b. "Struggles and negotiations to define what is problematic and what is not: The sociologic of translation." Vol. 4 of *The Social Process of Scientific Investigation*, edited by K. Knorr, R. Krohn, and R. Whitley, 197–219. Dordrecht, the Netherlands: Reidel.

Callon, M. 1981. "Pour une sociologie des controverses technologiques." *Fundamenta Scientiae* 2:381–399.

Callon, M. 1986. "Some elements of a sociology of translation: Domestication of the scallops and the fishermen of St. Brieuc Bay." In *Power, Action, and Belief: A New Sociology of Knowledge?*, edited by J. Law, 196–233. London: Routledge and Kegan Paul.

Callon, M. 1987. "Society in the making: The study of technology as a tool for sociological analysis." In Bijker, Hughes, and Pinch 1987, 83–110.

Callon, M., and Law, J. 1982. "On interests and their transformation: Enrollment and counter-enrollment." *Social Studies of Science* 12:615–625.

Campbell, D. T. 1974. "Evolutionary epistemology." Vol. 14-I of *The Philosophy of Karl Popper, the Library of Living Philosophers*, edited by P. A. Schlipp, 413–463. La Salle, IL: Open Court.

Caunter, C. F. 1955. *The History and Development of Cycles (as Illustrated by the Collection of Cycles in the Science Museum); Historical Survey*. London: HMSO.

Caunter, C. F. 1957. *The History and Development of Light Cars*. London: HMSO.

Collins, H. M. 1975. "The seven sexes: A study in the sociology of a phenomenon, or the replication of experiments in physics." *Sociology* 9:205–224.

Collins, H. M., ed. 1981a. "Knowledge and controversy." *Social Studies of Science* 11:3–158.

Collins, H. M. 1981b. "The place of the coreset in modern science: Social contingency with methodological propriety in science." *History of Science* 19:6–19.

Collins, H. M. 1981c. "Stages in the empirical programme of relativism." *Social Studies of Science* 11:3–10.

Collins, H. M. 1983. "An empirical relativist programme in the sociology of scientific knowledge." In *Science Observed: Perspectives on the Social Study of Science*, edited by K. D. Knorr-Cetina and M. J. Mulkay, 85–113. Beverly Hills, CA: Sage.

Constant, E. W., III. 1980. *The Origins of the Turbojet Revolution*. Baltimore: Johns Hopkins University Press.

Constant, E. W., III. 1983. "Scientific theory and technological testability: Science, dynamometers, and water turbines in the 19th century." *Technology and Culture* 24:183–198.

Constant, E. W., III. 1987. "The social locus of technological practice: Community, system, or organization?" In Bijker, Hughes, and Pinch 1987, 223–242.

Croon, L. 1939. *Das Fahrrad und seine Entwicklung*. Berlin: VDI-Verlag.

Dosi, G. 1982. "Technological paradigms and technological trajectories: A suggested interpretation of the determinants and directions of technical change." *Research Policy* 11:147–162.

Dosi, G. 1984. *Technical Change and Industrial Transformation*. London: Macmillan.

Dunlop, J. B. 1888. "An improvement in tyres of wheels for bicycles, tricycles, or other road cars." British patent 10607, filed July 23, 1888.

Elster, J. 1983. *Explaining Technical Change*. Cambridge: Cambridge University Press.

Elzen, B. 1985. "De ultracentrifuge: Op zoek naar patronen in technologische ontwikkeling door een vergelijkin van twee case-studies." *Jaarboek voor de Geschiedenis van Bedrijf en Techniek* 2:250–278.

Elzen, B. 1986. "Two ultracentrifuges: A comparative study of the social construction of artefacts." *Social Studies of Science* 16:621–662.

Ferguson, E. 1974. "Toward a discipline of the history of technology." *Technology and Culture* 15:13–30.

Freeman, C. 1974. *The Economics of Industrial Innovation*. London: Penguin. Reprinted by Frances Pinter, London, 1982.

Freeman, C. 1977. "Economics of research and development." In *Science, Technology and Society: A Cross-Disciplinary Perspective*, edited by I. Spiegel-Rösing and D. de Solla Price, 223–275. London: Sage.

Gibbs-Smith, C. H. 1960. *The Aeroplane: An Historical Survey of Its Origins and Development*. London: HSMO.

Gilfillan, S. G. 1935. *The Sociology of Invention*. Cambridge, MA: MIT Press.

Grew, W. 1921. *The Cycle Industry, Its Origin, History and Latest Developments*. London: Pitman & Sons.

Hanieski, J. F. 1973. "The airplane as an economic variable: Aspects of technological change in aeronautics, 1903–1955." *Technology and Culture* 14:535–552.

Haynes, W. 1954. *American Chemical Industry*. Vol. 2. New York: Van Nostrand.

Hughes, T. P. 1979. "Emerging themes in the history of technology." *Technology and Culture* 20:697–711.

Hughes, T. P. 1983. *Networks of Power: Electrification in Western Society, 1880–1930*. Baltimore: Johns Hopkins University Press.

Jelsma, J., and Smit, W. A. 1986. "Risks of recombinant DNA research: From uncertainty to certainty." In *Impact Assessment Today*, edited by H. A. Becker and A. L. Porter, 715–741. Utrecht: Van Arkel.

Johnston, R. 1972. "The internal structure of technology." In *The Sociology of Science*, edited by P. Halmos, 117–130. Keele, United Kingdom: University of Keele.

Jokisch, R., ed. 1982. *Techniksoziologie*. Frankfurt am Main: Suhrkamp.

Kaufman, M. 1963. *The First Century of Plastics: Celluloid and Its Sequel*. London: Plastics Institute.

Kline, S. 1985. "Research, invention, innovation, and production: Models and reality." *Research Management* 28 (4): 36–45.

Krohn, W., Layton, E. T., and Weingart, P., eds. 1978. *The Dynamics of Science and Technology, Sociology of the Sciences Yearbook*. Vol. 2. Dordrecht, the Netherlands: Reidel.

Latour, B. 1983. "Give me a laboratory and I will raise the world." In *Science Observed: Perspectives on the Social Study of Science*, edited by K. D. Knorr-Cetina and M. J. Mulkay, 141–170. London: Sage.

Latour, B. 1984. *Les microbes, querre et paix, suivi de irréductions*. Paris: Métaillé. Translated as *The Pasteurization of French Society, Followed by Irréductions. A Politico-Scientific Essay*. Cambridge, MA: Harvard University Press, 1987.

Laudan, R., ed. 1984. *The Nature of Technological Knowledge: Are Models of Scientific Change Relevant?* Dordrecht, the Netherlands: Reidel.

Law, John. 1987. "Technology and heterogeneous engineering: The case of Portuguese expansion." In Bijker, Hughes, and Pinch 1987, 111–134.

Layton, E. 1977. "Conditions of technological development." In *Science, Technology and Society: A Cross-Disciplinary Perspective*, edited by I. Spiegel-Rösing and D. de Solla Price, 197–222. London: Sage.

Lazonick, W. 1979. "Industrial relations and technical change: The case of the self-acting mule." *Cambridge Journal of Economics* 3:231–262.

MacKenzie, D. 1978. "Statistical theory and social interest: A case study." *Social Studies of Science* 8:35–83.

MacKenzie, D. 1984. "Marx and the machine." *Technology and Culture* 25:473–502.

Matthis, A. R. c. 1920. *Insulating Varnishes in Electrotechnics*. London: John Heywood.

Mulkay, M. J. 1979. "Knowledge and utility: Implications for the sociology of knowledge." *Social Studies of Science* 9:63–80.

Nelson, R. R., and Winter, S. G. 1977. "In search of a useful theory of innovation." *Research Policy* 6:36–76.

Nelson, R. R., and Winter, S. G. 1982. *An Evolutionary Theory of Economic Change*. Cambridge, MA: Belknap Press of Harvard University Press.

Noble, D. F. 1979. "Social choice in machine design: The case of automatically controlled machine tools." In *Case Studies on the Labour Process*, edited by A. Zimbalist, 18–50. New York: Monthly Review Press.

Noble, D. F. 1984. *Forces of Production: A Social History of Industrial Automation*. New York: Knopf.

Ogburn, W. F. 1945. *The Social Effects of Aviation*. Boston: Houghton Mifflin.

Ogburn, W. F., and Meyers Nimkoff, F. 1955. *Technology and the Changing Family*. Boston: Houghton Mifflin.

Palermo, D. S. 1973. "Is a scientific revolution taking place in psychology?" *Science Studies* 3:211–244.

Pickering, A. 1984. *Constructing Quarks—a Sociological History of Particle Physics*. Chicago: University of Chicago Press.

Pinch, T. J. 1977. "What does a proof do if it does not prove? A study of the social conditions and metaphysical divisions leading to David Bohm and John von Neumann failing to communicate in quantum physics." In *The Social Production of Scientific Knowledge*, edited by E. Mendelsohn, P. Weingart, and R. Whitley, 171–215. Dordrecht, the Netherlands: Reidel.

Pinch, T. J. 1986. *Confronting Nature: The Sociology of Solar-Neutrino Detection*. Dordrecht, the Netherlands: Reidel.

Pinch, T. J., and Bijker, W. E. 1984. "The social construction of facts and artefacts: Or how the sociology of science and the sociology of technology might benefit each other." *Social Studies of Science* 14:399–441. Also published in Serbo-Croatian as "Druˇstveno Proizvodenje Činjenica I Tvorevina: O Cjelovitom Pristupu Izučavanju Znanosti I Tehnologije." *Gledišta,časopis za društvenu kritiku i teoriju* 25 (March–April): 21–57.

Rosenberg, N. 1982. *Inside the Black Box: Technology and Economics*. Cambridge: Cambridge University Press.

Rosenberg, N., and Vincenti, W. G. 1978. *The Britannia Bridge: The Generation and Diffusion of Knowledge*. Cambridge, MA: MIT Press.

Schmookler, J. 1966. *Invention and Economic Growth*. Cambridge, MA: Harvard University Press.

Schmookler, J. 1972. *Patents, Invention and Economic Change, Data and Selected Essays*. Edited by Z. Griliches and L. Hurwicz. Cambridge, MA: Harvard University Press.

Scholz, L., with a contribution by G. von L. Uhlmann. 1977. *Technik-Indikatoren, Ansätze zur Messung des Standes der Technik in der industriellen Produktion*. Berlin: Duncker & Humblot.

Schot, J. 1985. "De ontwikkeling van de techniek als een variatieen selectieproces. De meekrapteelt en-bereiding in het licht van een alternatieve techniekopvatting." Master's thesis, Erasmus University of Rotterdam.

Schot, J. 1986. "De meekrapnijverheid: De ontwikkeling van de techniek als een proces van variatie en selectie." Vol. 3 of *Jaarboek voor de Geschiedenis van Bedrijf en Techniek*, edited by E. S. A. Bloemen, W. E. Bijker, W. van den Brocke, et al., 43–62. Utrecht: Stichting.

Schumpeter, J. (1928) 1971. "The instability of capitalism." In *The Economics of Technological Change*, edited by N. Rosenberg, 13–42. London: Penguin.

Schumpeter, J. 1942. *Capitalism, Socialism and Democracy*. New York: Harper & Row. Reprinted 1974 by Unwin University Books, London.

Schwartz Cowan, R. 1983. *More Work for Mother: The Ironies of Household Technology from the Open Hearth to the Microwave*. New York: Basic Books.

Shapin, S. 1979. "The politics of observation: Cerebral anatomy and social interests in the Edinburgh phrenology disputes." In *On the Margins of Science: The Social Construction of Rejected Knowledge*, edited by R. Wallis, 139–178. Keele, United Kingdom: University of Keele.

Shapin, S. 1980. "Social uses of science." In *The Ferment of Knowledge*, edited by G. S. Rousseau and R. Porter, 93–139. Cambridge: Cambridge University Press.

Shapin, S. 1984. "Pump and circumstance: Robert Boyle's literary technology." *Social Studies of Science* 14:481–520.

Shapin, S., and Schaffer, S. 1985. *Leviathan and the Air-Pump: Hobbes, Boyle and the Experimental Life*. Princeton, NJ: Princeton University Press.

Smith, M. Roe. 1977. *Harpers Ferry Armory and the New Technology: The Challenge of Change*. Ithaca, NY: Cornell University Press.

"The Stanley Exhibition of Cycles." 1889. *Engineer* 67:157–158.

"The Stanley Exhibition of Cycles." 1890. *Engineer* 69:107–108.

Staudenmaier, J. M., SJ. 1984. "What SHOT hath wrought and what SHOT hath not: Reflections on 25 years of the history of technology." Paper presented at the Twenty-Fifth Annual Meeting of SHOT and published in *Technology and Culture* 25, no. 4 (October): 707–730.

Staudenmaier, J. M., SJ. 1985. *Technology's Storytellers: Reweaving the Human Fabric*. Cambridge, MA: MIT Press.

Toulmin, S. 1972. *Human Understanding*. Vol. 1. Oxford: Oxford University Press.

Uhlmann, L. 1978. *Der Innovationsprozess in westeurpäischen Industrieländern. Band 2: Den Ablauf industriellen Innovationsprozesses*. Berlin: Duncker and Humblot.

van den Belt, H. 1985. "A. W. Hofman en de Franse Octrooiprocessen rond anilinerood: Demarcatie als sociale constructie." *Jaarboek voor de Geschiedenis van Bedrijf en Techniek* 2:64–86.

van den Belt, H., and Rip, A. 1987. "The Nelson-Winter-Dosi model and synthetic dye chemistry." In Bijker, Hughes, and Pinch 1987, 135–158.

Vincenti, W. G. 1986. "The Davis wing and the problem of airfoil design: Uncertainty and growth in engineering knowledge." *Technology and Culture* 27, no. 4 (October): 717–758.

Westrum, R. 1983. "What happened to the old sociology of technology?" Paper presented at the Eighth Annual Meeting of the Society for Social Studies of Science, Blacksburg, Virginia, November 1983.

Whitley, R. D. 1972. "Black boxism and the sociology of science: A discussion of the major developments in the field." In *The Sociology of Science*, edited by P. Halmos, 62–92. Keele, United Kingdom: University of Keele.

Woodforde, J. 1970. *The Story of the Bicycle*. London: Routledge and Kegan Paul.

Wynne, B. 1976. "C. G. Barkla and the J phenomenon: A case study of the treatment of deviance in physics." *Social Studies of Science* 6:307–347.

9 Technological Momentum

Thomas P. Hughes

In this article, Thomas Hughes argues that although both technological determinists and social constructivists provide interesting accounts, neither provides the full picture. He asserts that rather than adhering to one or the other theory, one should examine how society and technology both exert influence. Hughes acknowledges that people—in the form of individuals, governments, corporations, and so on—direct the development of new technologies. But he also claims that large sociotechnical systems can gain "momentum." By this he means that some of these systems may appear at times to have a mind of their own and cannot be stopped—or, at least, they resist change. But Hughes maintains that this is simply because a large number of social groups (such as corporations, governments, industries, and consumers) have financial, capital, infrastructure, and ideological reasons for keeping such systems going. Once certain large systems are in place, there is inertia to keep them going and innovate "around the edges" rather than make radical change or abandon them altogether. In this way Hughes offers a compromise of sorts in the social determinism versus technological determinism debate, a compromise that helps to explain how both people and technological systems influence and shape each other. He argues that the investment of money, effort, and resources required to develop technological systems creates conditions that make those systems resistant to subsequent attempts to change them.

The concepts of technological determinism and social construction provide agendas for fruitful discussion among historians, sociologists, and engineers interested in the nature of technology and technological change. Specialists can engage in a general discourse that subsumes their areas of specialization. In this essay I shall offer an additional concept—technological momentum—that will, I hope, enrich the discussion. Technological momentum offers an alternative to technological determinism and social construction. Those who in the past espoused a technological determinist approach to history offered a needed corrective to the conventional interpretation of history that virtually ignored the role of technology in effecting social change. Those who more recently advocated a social construction approach provided an invaluable corrective to an interpretation of history that encouraged a passive attitude toward an overwhelming technology. Yet both approaches suffer from a failure to encompass the complexity of technological change.

From Leo Marx and Merritt Roe Smith, eds., *Does Technology Drive History? The Dilemma of Technological Determinism* (Cambridge, MA: MIT Press, 1994), 101–113. Reprinted with permission.

All three concepts present problems of definition. Technological determinism I define simply as the belief that technical forces determine social and cultural changes. Social construction presumes that social and cultural forces determine technical change. A more complex concept than determinism and social construction, technological momentum infers that social development shapes and is shaped by technology. Momentum also is time dependent. Because the focus of this essay is technological momentum, I shall define it in detail by resorting to examples.

"Technology" and "technical" also need working definitions. Proponents of technological determinism and of social construction often use "technology" in a narrow sense to include only physical artifacts and software. By contrast, I use "technical" in referring to physical artifacts and software. By "technology" I usually mean technological or sociotechnical systems, which I shall also define by examples.

Discourses about technological determinism and social construction usually refer to society, a concept exceedingly abstract. Historians are wary of defining society other than by example because they have found that twentieth-century societies seem quite different from twelfth-century ones and that societies differ not only over time but over space as well. Facing these ambiguities, I define the social as the world that is not technical, or that is not hardware or technical software. This world is made up of institutions, values, interest groups, social classes, and political and economic forces. As the reader will learn, I see the social and the technical as interacting within technological systems. Technological system, as I shall explain, includes both the technical and the social. I name the world outside of technological systems that shapes them or is shaped by them the "environment." Even though it may interact with the technological system, the environment is not a part of the system because it is not under the control of the system as are the system's interacting components.

In the course of this essay the reader will discover that I am no technological determinist. I cannot associate myself with such distinguished technological determinists as Karl Marx, Lynn White, and Jacques Ellul. Marx, in moments of simplification, argued that waterwheels ushered in manorialism and that steam engines gave birth to bourgeois factories and society. Lenin added that electrification was the bearer of socialism. White elegantly portrayed the stirrup as the prime mover in a train of cause and effect culminating in the establishment of feudalism. Ellul finds the human-made environment structured by technical systems, as determining in their effects as the natural environment of Charles Darwin. Ellul sees the human-made as steadily displacing the natural—the world becoming a system of artifacts, with humankind, not God, as the artificer.[1]

Nor can I agree entirely with the social constructivists. Wiebe Bijker and Trevor Pinch have made an influential case for social construction in their essay "The Social Construction of Facts and Artifacts."[2] They argue that social, or interest, groups define and give meaning to artifacts. In defining them, the social groups determine the designs of artifacts. They do this by selecting for survival the designs that solve the problems they want solved by the artifacts and that fulfill desires they want fulfilled by the artifacts. Bijker and Pinch emphasize the interpretive flexibility discernible in the evolution of artifacts: they believe that the various meanings given by social groups to, say, the bicycle result in a number of alternative designs of that machine. The various bicycle designs are not fixed; closure does not occur until social groups believe that the problems and desires they associate with the bicycle are solved or fulfilled.

In summary, I find the Bijker-Pinch interpretation tends toward social determinism, and I must reject it on these grounds. The concept of technological momentum avoids the extremism of both technological determinism and social construction by presenting a more complex, flexible, time-dependent, and persuasive explanation of technological change.

Technological Systems

Electric light and power systems provide an instructive example of technological systems. By 1920 they had taken on a messy complexity because of the heterogeneity of their components. In their diversity, their complexity, and their large scale, such mature technological systems resemble the megamachines that Lewis Mumford described in *The Pentagon of Power.*[3] The actor networks of Bruno Latour and Michel Callon also share essential characteristics with technological systems.[4] An electric power system consists of inanimate electrons and animate regulatory boards, both of which, as Latour and Callon suggest, can be intractable if not brought in line or into the actor network.

The Electric Bond and Share Company (EBASCO), an American electric utility holding company of the 1920s, provides an example of a mature technological system. Established in 1905 by the General Electric Company, EBASCO controlled through stock ownership a number of electric utility companies, and through them a number of technical subsystems—namely electric light and power networks, or grids.[5] EBASCO provided financial, management, and engineering construction services for the utility companies. The inventors, engineers, and managers who were the system builders of EBASCO saw to it that the services related synergistically. EBASCO management recommended construction that EBASCO engineering carried out and for which EBASCO arranged financing through sale of stocks or bonds. If the utilities lay in geographical proximity, then EBASCO often physically interconnected them through high-voltage power grids. The General Electric Company founded EBASCO and, while not owning a majority of stock in it, substantially influenced its policies. Through EBASCO General Electric learned of equipment needs in the utility industry and then provided them in accord with specifications defined by EBASCO for the various utilities with which it interacted. Because it interacted with EBASCO, General Electric was a part of the EBASCO system. Even though I have labeled this the EBASCO system, it is not clear that EBASCO solely controlled the system. Control of the complex systems seems to have resulted from a consensus among EBASCO, General Electric, and the utilities in the systems.

Other institutions can also be considered parts of the EBASCO system, but because the interconnections were loose rather than tight these institutions are usually not recognized as such.[6] I refer to the electrical engineering departments in engineering colleges, whose faculty and graduate students conducted research or consulted for EBASCO. I am also inclined to include a few of the various state regulatory authorities as parts of the EBASCO system, if their members were greatly influenced by it. If the regulatory authorities were free of this control, then they should be considered a part of the EBASCO environment, not of the system.

Because it had social institutions as components, the EBASCO system could be labeled a sociotechnical system. Since, however, the system had a technical (hardware and software) core, I prefer to name it a technological system, to distinguish it from social systems without technical cores. This privileging of the technical in a technological system is justified in part by the prominent roles played by engineers, scientists, workers, and technical-minded managers in solving the problems arising during the creation and early history of a system. As a system matures, a bureaucracy of managers and white-collar employees usually plays an increasingly prominent role in maintaining and expanding the system, so that it then becomes more social and less technical.

EBASCO as a Cause and an Effect

From the point of view of technological—better, technical—determinists, the determined is the world beyond the technical. Technical determinists considering EBASCO as a historical actor would focus on its technical core as a cause with many effects. Instead of seeing EBASCO as a technological system with interacting technical and social components, they would see the technical core as causing change in the social components of EBASCO and in society in general. Determinists would focus on the way in which EBASCO's generators, by energizing electric motors on individual production machines, made possible the reorganization of the factory floor in a manner commonly associated with Fordism. Such persons would see street, workplace, and home lighting changing working and leisure hours and affecting the nature of work and play. Determinists would also cite electrical appliances in the home as bringing less—and more—work for women,[7] and the layout of EBASCO's power lines as causing demographic changes. Electrical grids such as those presided over by EBASCO brought a new decentralized regionalism, which contrasted with the industrial, urban-centered society of the steam age.[8] One could extend the list of the effects of electrification enormously.

Yet, contrary to the view of the technological determinists, the social constructivists would find exogenous technical, economic, political, and geographical forces, as well as values, shaping with varying intensity the EBASCO system during its evolution. Social constructivists see the technical core of EBASCO as an effect rather than a cause. They could cite a number of instances of social construction. The spread of alternating (polyphase) current after 1900, for instance, greatly affected, even determined, the history of the early utilities that had used direct current, for these had to change their generators and related equipment to alternating current or fail in the face of competition. Not only did such external technical forces shape the technical core of the utilities; economic forces did so as well. With the rapid increase in the United States' population and the concentration of industry in cities, the price of real estate increased. Needing to expand their generating capacity, EBASCO and other electric utilities chose to build new turbine-driven power plants outside city centers and to transmit electricity by high-voltage lines back into the cities and throughout the area of supply. Small urban utilities became regional ones and then faced new political or regulatory forces as state governments took over jurisdiction from the cities. Regulations also caused technical changes. As the regional utilities of the EBASCO system expanded, they conformed to geographical realities as they sought cooling water, hydroelectric sites, and minemouth locations. Values, too, shaped the history of EBASCO. During the Great Depression, the Roosevelt administration singled out utility holding-company magnates for criticism, blaming the huge losses experienced by stock and bond holders on the irresponsible, even illegal, machinations of some of the holding companies. Partly as a result of this attack, the attitudes of the public toward large-scale private enterprise shifted so that it was relatively easy for the administration to push through Congress the Holding Company Act of 1935, which denied holding companies the right to incorporate utilities that were not physically contiguous.[9]

Gathering Technological Momentum

Neither the proponents of technical determinism nor those of social construction can alone comprehend the complexity of an evolving technological system such as EBASCO. On some occasions EBASCO was a cause; on others it was an effect. The system both shaped and was shaped by society. Furthermore, EBASCO's shaping society is not an example of purely technical determinism, for EBASCO, as we have observed, contained social components. Similarly, social constructivists must acknowledge that social forces in the environment were not shaping simply a technical system, but a technological system, including—as systems invariably do—social components.

The interaction of technological systems and society is not symmetrical over time. Evolving technological systems are time dependent. As the EBASCO system became larger and more complex, thereby gathering momentum, the system became less shaped by and more the shaper of its environment. By the 1920s the EBASCO system rivaled a large railroad company in its level of capital investment, in its number of customers, and in its influence upon local, state, and federal governments. Hosts of electrical engineers, their professional organizations, and the engineering schools that trained them were committed by economic interests and their special knowledge and skills to the maintenance and growth of the EBASCO system. Countless industries and communities interacted with EBASCO utilities because of shared economic interests. These various human and institutional components added substantial momentum to the EBASCO system. Only a historical event of large proportions could deflect or break the momentum of an EBASCO, the Great Depression being a case in point.

Characteristics of Momentum

Other technological systems reveal further characteristics of technological momentum, such as acquired skill and knowledge, special-purpose machines and processes, enormous physical structures, and organizational bureaucracy. During the late nineteenth century, for instance, mainline railroad engineers in the United States transferred their acquired skill and knowledge to the field of intra-urban transit. Institutions with specific characteristics also contributed to this momentum. Professors in the recently founded engineering schools and engineers who had designed and built the railroads organized and rationalized the experience that had been gathered in preparing roadbeds, laying tracks, building bridges, and digging tunnels for mainline railroads earlier in the century. This engineering science found a place in engineering texts and in the curricula of the engineering schools, thus informing a new generation of engineers who would seek new applications for it.

Late in the nineteenth century, when street congestion in rapidly expanding industrial and commercial cities such as Chicago, Baltimore, New York, and Boston threatened to choke the flow of traffic, extensive subway and elevated railway building began as an antidote. The skill and the knowledge formerly expended on railroad bridges were now applied to elevated railway structures; the know-how once invested in tunnels now found application in subways. A remarkably active period of intraurban transport construction began about the time when the building of mainline railways reached a plateau, thus facilitating the movement of know-how from one field to the other. Many of the engineers who played leading roles in intra-urban transit between 1890 and 1910 had been mainline railroad builders.[10]

The role of the physical plant in the buildup of technological momentum is revealed in the interwar history of the Badische Anilin und Soda Fabrik (BASF), one of Germany's leading chemical manufacturers and a member of the I.G. Farben group. During World War I, BASF rapidly developed large-scale production facilities to utilize the recently introduced Haber-Bosch technique of nitrogen fixation. It produced the nitrogen compounds for fertilizers and explosives so desperately needed by a blockaded

Germany. The high-technology process involved the use of high-temperature, high-pressure, complex catalytic action. Engineers had to design and manufacture extremely costly and complex instrumentation and apparatus. When the blockade and the war were over, the market demand for synthetic nitrogen compounds did not match the large capacity of the high-technology plants built by BASF and other companies during the war. Numerous engineers, scientists, and skilled craftsmen who had designed, constructed, and operated these plants found their research and development knowledge and their construction skills underutilized. Carl Bosch, chairman of the managing board of BASF and one of the inventors of the Haber-Bosch process, had a personal and professional interest in further development and application of high-temperature, high-pressure, catalytic processes. He and other managers, scientists, and engineers at BASF sought additional ways of using the plant and the knowledge created during the war years. They first introduced a high-temperature, high-pressure catalytic process for manufacturing synthetic methanol in the early 1920s. The momentum of the now-generalized process next showed itself in management's decision in the mid 1920s to invest in research and development aimed at using high-temperature, high-pressure catalytic chemistry for the production of synthetic gasoline from coal. This project became the largest investment in research and development by BASF during the Weimar era. When the National Socialists took power, the government contracted for large amounts of the synthetic product. Momentum swept BASF and I.G. Farben into the Nazi system of economic autarky.[11]

When managers pursue economies of scope, they are taking into account the momentum embodied in large physical structures. Muscle Shoals Dam, an artifact of considerable size, offers another example of this aspect of technological momentum. As the loss of merchant ships to submarines accelerated during World War I, the United States also attempted to increase its indigenous supply of nitrogen compounds. Having selected a process requiring copious amounts of electricity, the government had to construct a hydroelectric dam and power station. This was located at Muscle Shoals, Alabama, on the Tennessee River. Before the nitrogen-fixation facilities being built near the dam were completed, the war ended. As in Germany, the supply of synthetic nitrogen compounds then exceeded the demand. The U.S. government was left not only with process facilities but also with a very large dam and power plant.

Muscle Shoals Dam (later named Wilson Dam), like the engineers and managers we have considered, became a solution looking for a problem. How should the power from the dam be used? A number of technological enthusiasts and planners envisioned the dam as the first of a series of hydroelectric projects along the Tennessee River and its tributaries. The poverty of the region spurred them on in an era when electrification was seen as a prime mover of economic development. The problem looking for a solution attracted the attention of an experienced problem solver, Henry Ford, who proposed that an industrial complex based on hydroelectric power be located along 75 miles of the waterway that included the Muscle Shoals site. An alliance of public power and private interests with their own plans for the region frustrated his plan. In 1933, however, Muscle Shoals became the original component in a hydroelectric, flood-control, soil-reclamation, and regional development project of enormous scope sponsored by Senator George Norris and the Roosevelt administration and presided over by the Tennessee Valley Authority. The technological momentum of the Muscle Shoals Dam had carried over from World War I to the New Deal. This durable artifact acted over time like a magnetic field, attracting plans and projects suited to its characteristics. Systems of artifacts are not neutral forces; they tend to shape the environment in particular ways.[12]

Using Momentum

System builders today are aware that technological momentum—or whatever they may call it—provides the durability and the propensity for growth that were associated more commonly in the past with the spread of bureaucracy. Immediately after World War II, General Leslie Groves displayed his system-building instincts and his awareness of the critical importance of technological momentum as a means of ensuring the survival of the system for the production of atomic weapons embodied in the wartime Manhattan Project. Between 1945 and 1947, when others were anticipating disarmament, Groves expanded the gaseous-diffusion facilities for separating fissionable uranium at Oak Ridge, Tennessee; persuaded the General Electric Company to operate the reactors for producing plutonium at Hanford, Washington; funded the new Knolls Atomic Power Laboratory at Schenectady, New York; established the Argonne and Brookhaven National Laboratories for fundamental research in nuclear science; and provided research funds for a number of universities. Under his guiding hand, a large-scale production system with great momentum took on new life in peacetime. Some of the leading scientists of the wartime project had confidently expected production to end after the making of a few bombs and the coming of peace.[13]

More recently, proponents of the Strategic Defense Initiative (SDI), organized by the Reagan administration in 1983, have made use of momentum. The political and economic interests and the organizational bureaucracy vested in this system were substantial—as its makers intended. Many of the same industrial contractors, research universities, national laboratories, and government agencies that took part in the construction of intercontinental ballistic missile systems, National Air and Space Administration projects, and atomic weapon systems have been deeply involved in SDI. The names are familiar: Lockheed, General Motors, Boeing, TRW, McDonnell Douglas, General Electric, Rockwell, Teledyne, MIT, Stanford, the University of California's Lawrence Livermore Laboratory, Los Alamos, Hanford, Brookhaven, Argonne, Oak Ridge, NASA, the U.S. Air Force, the U.S. Navy, the CIA, the U.S. Army, and others. Political interests reinforced the institutional momentum. A number of congressmen represent districts that receive SDI contracts, and lobbyists speak for various institutions drawn into the SDI network.[14] Only the demise of the Soviet Union as a military threat allowed counter forces to build up sufficient momentum to blunt the cutting edge of SDI.

Conclusion

A technological system can be both a cause and an effect; it can shape or be shaped by society. As they grow larger and more complex, systems tend to be more shaping of society and less shaped by it. Therefore, the momentum of technological systems is a concept that can be located somewhere between the poles of technical determinism and social constructivism. The social constructivists have a key to understanding the behavior of young systems; technical determinists come into their own with the mature ones. Technological momentum, however, provides a more flexible mode of interpretation and one that is in accord with the history of large systems.

What does this interpretation of the history of technological systems offer to those who design and manage systems or to the public that might wish to shape them through a democratic process? It suggests that shaping is easiest before the system has acquired political, economic, and value components. It also follows that a system with great technological momentum can be made to change direction if a variety of its components are subjected to the forces of change.

For instance, the changeover since 1970 by U.S. automobile manufacturers from large to more compact automobiles and to more fuel-efficient and less polluting ones came about as a result of pressure brought on a number of components in the huge automobile production and use system. As a result of the oil embargo of 1973 and the rise of gasoline prices, American consumers turned to imported compact automobiles; this, in turn, brought competitive economic pressure to bear on the Detroit manufacturers. Environmentalists helped persuade the public to support, and politicians to enact, legislation that promoted both anti-pollution technology and gas-mileage standards formerly opposed by American manufacturers. Engineers and designers responded with technical inventions and developments.

On the other hand, the technological momentum of the system of automobile production and use can be observed in recent reactions against major environmental initiatives in the Los Angeles region. The host of institutions and persons dependent politically, economically, and ideologically on the system (including gasoline refiners, automobile manufacturers, trade unions, manufacturers of appliances and small equipment using internal-combustion engines, and devotees of unrestricted automobile usage) rallied to frustrate change.

Because social and technical components interact so thoroughly in technological systems and because the inertia of these systems is so large, they bring to mind the iron-cage metaphor that Max Weber used in describing the organizational bureaucracies that proliferated at the beginning of the twentieth century.[15] Technological systems, however, are bureaucracies reinforced by technical, or physical, infrastructures which give them even greater rigidity and mass than the social bureaucracies that were the subject of Weber's attention. Nevertheless, we must remind ourselves that technological momentum, like physical momentum, is not irresistible.

Notes

1. Lynn White Jr., *Medieval Technology and Social Change* (Oxford: Clarendon, 1962); Jacques Ellul, *The Technological System* (New York: Continuum, 1980); Karl Marx, *Capital: A Critique of Political Economy*, ed. F. Engels (n.p., 1867); B. I. Weitz, ed., *Electric Power Development in the U.S.S.R.* (Moscow: INRA, 1936).

2. The essay is found in W. E. Bijker et al., eds., *The Social Construction of Technological Systems: New Directions in the Sociology and History of Technology* (Cambridge, MA: MIT Press, 1987). It is partially reprinted as chapter 8 in this book.

3. Lewis Mumford, *The Myth of the Machine: II. The Pentagon of Power* (San Diego: Harcourt Brace Jovanovich, 1970).

4. Bruno Latour, *Science in Action: How to Follow Scientists and Engineers through Society* (Cambridge, MA: Harvard University Press, 1987); Michel Callon, "Society in the Making: The Study of Technology as a Tool for Sociological Analysis," in Bijker et al., *Social Construction of Technological Systems*.

5. Before 1905, General Electric used the United Electric Securities Company to hold its utility securities and to fund its utility customers who purchased GE equipment. See Thomas P. Hughes, *Networks of Power: Electrification in Western Society, 1880–1930* (Baltimore: Johns Hopkins University Press, 1983), pp. 395–396.

6. The concept of loosely and tightly coupled components in systems is found in Charles Perrow's *Normal Accidents: Living with High Risk Technology* (New York: Basic Books, 1984).

7. Ruth Schwartz Cowan, "The 'Industrial Revolution' in the Home," *Technology and Culture* 17 (1976): 1–23.

8. Lewis Mumford, *The Culture of Cities* (San Diego: Harcourt Brace Jovanovich, 1970), p. 378.

9. More on EBASCO's history can be found on pp. 392–399 of Hughes, *Networks of Power*.

10. Thomas Parke Hughes, "A Technological Frontier: The Railway," in *The Railroad and the Space Program*, ed. B. Mazlish (Cambridge, MA: MIT Press, 1965).

11. Thomas Parke Hughes, "Technological Momentum: Hydrogenation in Germany, 1900–1933," *Past and Present* (August 1969): 106–132.

12. On Muscle Shoals and the TVA, see Preston J. Hubbard's *Origins of the TVA: The Muscle Shoals Controversy, 1920–1932* (New York: W. W. Norton, 1961).

13. Richard G. Hewlett and Oscar E. Anderson Jr., *The New World, 1939–1946* (Philadelphia: Pennsylvania State University Press, 1962), pp. 624–638.

14. Charlene Mires, "The Strategic Defense Initiative" (unpublished essay, History and Sociology of Science Department, University of Pennsylvania, 1990).

15. Max Weber, *The Protestant Ethic and the Spirit of Capitalism*, trans. T. Parsons (London: Unwin Hyman, 1990), p. 155.

10 Where Are the Missing Masses? The Sociology of a Few Mundane Artifacts

Bruno Latour

One of the most popular and powerful ways of resolving the technological determinism/social constructivism dichotomy in technology studies is the actor-network approach. Those advocating the actor-network approach agree with the social constructivist claim that sociotechnical systems are developed through negotiations between people, institutions, and organizations. But they make the additional interesting argument that artifacts are part of these negotiations as well. This is not to say that machines think like people and decide how they will act, but their behavior or nature often has a comparable role. Actor-network theorists argue that the material world pushes back on people because of its physical structure or natural behavior. People are free to interpret the precise meaning of an artifact, but they can't simply tell an automobile engine to get two hundred miles per gallon. The laws of nature and the capacities of a particular design limit the ways in which artifacts can be integrated into a sociotechnical system. In this selection, one of the foremost contributors to the actor-network approach, Bruno Latour, explores how artifacts can be deliberately designed to both replace human action and constrain and shape the actions of humans. His study demonstrates how people can "act at a distance" through the technologies they create and implement and how, from a user's perspective, a technology can appear to determine or compel certain actions. He argues that even technologies so commonplace that we don't think about them can shape the decisions we make, the effects our actions have, and the way we move through the world. Technologies play such an important role in mediating human relationships, Latour contends, that we cannot understand how societies work without comprehending how technologies shape our everyday lives. Latour's study of the relationship between producers, machines, and users demonstrates how certain values and political goals can be promoted and achieved through the construction and employment of technologies.

To Robert Fox, Again, might not the glory of the machines consist in their being without this same boasted gift of language? "Silence," it has been said by one writer, "is a virtue which render us agreeable to our fellow-creatures."

—Samuel Butler, *Erewhon*

Early this morning, I was in a bad mood and decided to break a law and start my car without buckling my seat belt. My car usually does not want to start before I buckle the belt. It first flashes a red light "FASTEN YOUR SEAT BELT!," then an alarm sounds; it

From Wiebe E. Bijker and John Law, eds., *Shaping Technology/Building Society: Studies in Sociotechnical Change* (Cambridge, MA: MIT Press, 1992), 225–258. Reprinted with permission.

is so high pitched, so relentless, so repetitive, that I cannot stand it. After ten seconds I swear and put on the belt. This time, I stood the alarm for twenty seconds and then gave in. My mood had worsened quite a bit, but I was at peace with the law—at least with that law. I wished to break it, but I could not. Where is the morality? In me, a human driver, dominated by the mindless power of an artifact? Or in the artifact forcing me, a mindless human, to obey the law that I freely accepted when I get my driver's license? Of course, I could have put on my seat belt before the light flashed and the alarm sounded, incorporating in my own self the good behavior that everyone—the car, the law, the police—expected of me. Or else, some devious engineer could have linked the engine ignition to an electric sensor in the seat belt, so that I could not even have started the car before having put it on. Where would the morality be in those two extreme cases? In the electric currents flowing in the machine between the switch and the sensor? Or in the electric currents flowing down my spine in the automatism of my routinized behavior? In both cases the result would be the same from an outside observer—say a watchful policeman: this assembly of a driver and a car obeys the law in such a way that it is impossible for a car to be at the same time moving AND to have the driver without the belt on. A *law of the excluded middle* has been built, rendering logically inconceivable as well as morally unbearable a driver without a seat belt. Not quite. Because I feel so irritated to be forced to behave well that I instruct my garage mechanics to unlink the switch and the sensor. The excluded middle is back in! There is at least one car that is both on the move and without a seat belt on its driver—mine. This was without counting on the cleverness of engineers. They now invent a seat belt that politely makes way for me when I open the door and then straps me as politely but very tightly when I close the door. Now there is no escape. The only way not to have the seat belt on is to leave the door wide open, which is rather dangerous at high speed. Exit the excluded middle. The program of action "IF a car is moving, THEN the driver has a seat belt" is enforced.[1] It has become logically—no, it has become sociologically—impossible to drive without wearing the belt. I cannot be bad anymore. I, plus the car, plus the dozens of patented engineers, plus the police are making me be moral (figure 10.1).

According to some physicists, there is not enough mass in the universe to balance the accounts that cosmologists make of it. They are looking everywhere for the "missing mass" that could add up to the nice expected total. It is the same with sociologists. They are constantly looking, somewhat desperately, for social links sturdy enough to tie all of us together or for moral laws that would be inflexible enough to make us behave properly. When adding up social ties, all does not balance. Soft humans and weak moralities are all sociologists can get. The society they try to recompose with bodies and norms constantly crumbles. Something is missing, something that should be strongly social and highly moral. Where can they find it? Everywhere, but they too often refuse to see it in spite of much new work in the sociology of artifacts.[2]

I expect sociologists to be much more fortunate than cosmologists, because they will soon discover their missing mass. To balance our accounts of society, we simply have to turn our exclusive attention away from humans and look also at nonhumans. Here they are, the hidden and despised social masses who make up our morality. They knock at the door of sociology, requesting a place in the accounts of society as stubbornly as the human masses did in the nineteenth century. What our ancestors, the founders of sociology, did a century ago to house the human masses in the fabric of social theory, we should do now to find a place in a new social theory for the nonhuman masses that beg us for understanding.

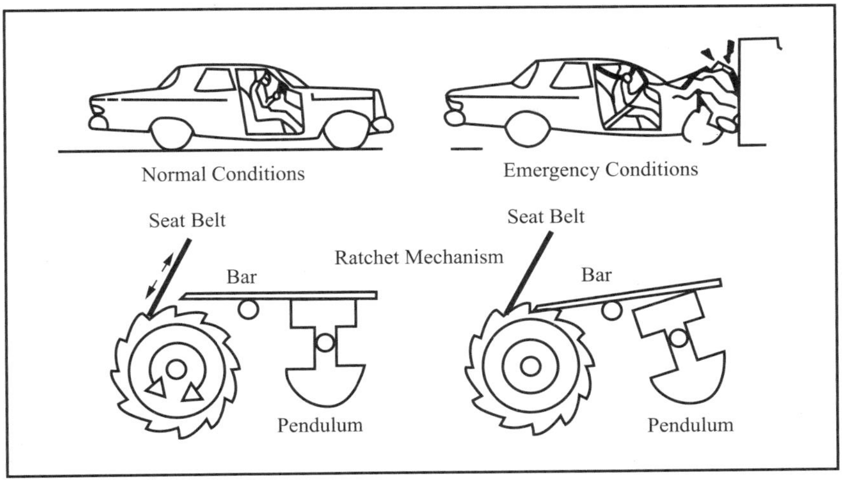

Figure 10.1
The designers of the seat belt take on themselves and then shift back to the belt contradictory programs; the belt should be lenient and firm, easy to put on, and solidly fastened while ready to be unbuckled in a fraction of a second; it should be unobtrusive and strap in the whole body. The object does not reflect the social. It does more. It transcribes and displaces the contradictory interests of people and things.

Description of a Door

I will start my inquiry by following a little script written by anonymous hands.[3] On a freezing day in February, posted on the door of La Halle aux Cuirs at La Villette, in Paris, where Robert Fox's group was trying to convince the French to take up social history of science, could be seen a small handwritten notice: "The Groom Is On Strike, For God's Sake, Keep The Door Closed" ("groom" is Frenglish for an automated door-closer or butler). This fusion of labor relations, religion, advertisement, and technique in one insignificant fact is exactly the sort of thing I want to describe in order to discover the missing masses of our society.[4] As a technologist teaching in the School of Mines, an engineering institution, I want to challenge some of the assumptions sociologists often hold about the social context of machines.

Walls are a nice invention, but if there were no holes in them there would be no way to get in or out—they would be mausoleums or tombs. The problem is that if you make holes in the walls, anything and anyone can get in and out (cows, visitors, dust, rats, noise—La Halle aux Cuirs is ten meters from the Paris ring road—and, worst of all, cold—La Halle aux Cuirs is far to the north of Paris). So architects invented this hybrid: a wall hole, often called a *door*, which although common enough has always struck me as a miracle of technology. The cleverness of the invention hinges upon the hingepin: instead of driving a hole through walls with a sledgehammer or a pick, you simply gently push the door (I am supposing here that the lock has not been invented—this would overcomplicate the already highly complex story of La Villette's door); furthermore—and here is the real trick—once you have passed through the door, you do not have to find trowel and cement to rebuild the wall you have just destroyed: you simply push the door gently back (I ignore for now the added complication of the "pull" and "push" signs).

So, to size up the work done by hinges, you simply have to imagine that every time you want to get in or out of the building you have to do the same work as a prisoner trying to escape or as a gangster trying to rob a bank, plus the work of those who re-build either the prison's or the bank's walls. If you do not want to imagine people destroying walls and rebuilding them every time they wish to leave or enter a building, then imagine the work that would have to be done to keep inside or outside all the things and people that, left to themselves, would go the wrong way.[5] As Maxwell never said, imagine his demon working *without* a door. Anything could escape from or penetrate into La Halle aux Cuirs, and soon there would be complete equilibrium between the depressing and noisy surrounding area and the inside of the building. Some technologists, including the present writer in *Material Resistance, A Textbook* (1984), have written that techniques are always involved when asymmetry or irreversibility is the goal; it might appear that doors are a striking counterexample because they maintain the wall hole in a reversible state; the allusion to Maxwell's demon clearly shows, however, that such is not the case; the reversible door is the only way to trap irreversibly inside La Halle aux Cuirs a differential accumulation of warm historians, knowledge, and also, alas, a lot of paperwork; the hinged door allows a selection of what gets in and what gets out so as to locally increase order, or information. If you let the drafts get inside (these renowned *courants d'air* so dangerous to French health), the paper drafts may never get outside to the publishers.

Now, draw two columns (if I am not allowed to give orders to the reader, then I offer it as a piece of strongly worded advice): in the right-hand column, list the work people would have to do if they had no door; in the left-hand column write down the gentle pushing (or pulling) they have to do to fulfill the same tasks. Compare the two columns: the enormous effort on the right is balanced by the small one on the left, and this is all thanks to hinges. I will define this transformation of a major effort into a minor one by the words "displacement" or "translation" or "delegation" or "shifting";[6] I will say that we have delegated (or translated or displaced or shifted down) to the hinge the work of reversibly solving the wall-hole dilemma. Calling on Robert Fox, I do not have to do this work nor even think about it; it was delegated by the carpenter to a character, the hinge, which I will call a *nonhuman*. I simply enter La Halle aux Cuirs. As a more general descriptive rule, every time you want to know what a nonhuman does, simply imagine what other humans or other nonhumans would have to do were this character not present. This imaginary substitution exactly sizes up the role, or function, of this little character.

Before going on, let me point out one of the side benefits of this table: in effect, we have drawn a scale where tiny efforts balance out mighty weights; the scale we drew reproduces the very leverage allowed by hinges. That the small be made stronger than the large is a very moral story indeed (think of David and Goliath); by the same token, it is also, since at least Archimedes' days, a very good definition of a lever and of power: what is the minimum you need to hold and deploy astutely to produce the maximum effect? Am I alluding to machines or to Syracuse's King? I don't know, and it does not matter, because the King and Archimedes fused the two "minimaxes" into a single story told by Plutarch: the defense of Syracuse through levers and war machines.[7] I contend that this reversal of forces is what sociologists should look at in order to understand the social construction of techniques, and not a hypothetical "social context" that they are not equipped to grasp. This little point having been made, let me go on with the story (we will understand later why I do not really need your permission to go on and why, nevertheless, you are free not to go on, although only *relatively* so).

Delegation to Humans

There is a problem with doors. Visitors push them to get in or pull on them to get out (or vice versa), but then the door remains open. That is, instead of the door you have a gaping hole in the wall through which, for instance, cold rushes in and heat rushes out. Of course, you could imagine that people living in the building or visiting the Centre d'Histoire des Sciences et des Techniques would be a well-disciplined lot (after all, historians are meticulous people). They will learn to close the door behind them and retransform the momentary hole into a well-sealed wall. The problem is that discipline is not the main characteristic of La Villette's people; also you might have mere sociologists visiting the building, or even pedagogues from the nearby Centre de Formation. Are they all going to be so well trained? Closing doors would appear to be a simple enough piece of know-how once hinges have been invented, but, considering the amount of work, innovations, sign-posts, and recriminations that go on endlessly everywhere to keep them closed (at least in northern regions), it seems to be rather poorly disseminated.

This is where the age-old Mumfordian choice is offered to you: either to discipline the people or to substitute for the unreliable people another delegated human character whose only function is to open and close the door. This is called a groom or a porter (from the French word for door), or a gatekeeper, or a janitor, or a concierge, or a turnkey, or a jailer. The advantage is that you now have to discipline only one human and may safely leave the others to their erratic behavior. No matter who it is and where it comes from, the groom will always take care of the door. A nonhuman (the hinges) plus a human (the groom) have solved the wall-hole dilemma.

Solved? Not quite. First of all, if La Halle aux Cuirs pays for a porter, they will have no money left to buy coffee or books, or to invite eminent foreigners to give lectures. If they give the poor little boy other duties besides that of porter, then he will not be present most of the time and the door will stay open. Even if they had money to keep him there, we are now faced with a problem that two hundred years of capitalism has not completely solved: how to discipline a youngster to reliably fulfill a boring and underpaid duty? Although there is now only one human to be disciplined instead of hundreds, the weak point of the tactic can be seen: if this *one* lad is unreliable, then the whole chain breaks down; if he falls asleep on the job or goes walkabout, there will be no appeal: the door will stay open (remember that locking it is no solution because this would turn it into a wall, and then providing everyone with the right key is a difficult task that would not ensure that key holders will lock it back). Of course, the porter may be punished. But disciplining a groom—Foucault notwithstanding—is an enormous and costly task that only large hotels can tackle, and then for other reasons that have nothing to do with keeping the door properly closed.

If we compare the work of disciplining the groom with the work he substitutes for, according to the list defined above, we see that this delegated character has the opposite effect to that of the hinge: a simple task—forcing people to close the door—is now performed at an incredible cost; the minimum effect is obtained with maximum spending and discipline. We also notice, when drawing the two lists, an interesting difference: in the first relationship (hinges vis-à-vis the work of many people), you not only had a reversal of forces (the lever allows gentle manipulations to displace heavy weights) but also a modification of *time schedule*: once the hinges are in place, nothing more has to be done apart from maintenance (oiling them from time to time). In the second set of relations (groom's work versus many people's work), not only do you fail to reverse the forces but you also fail to modify the time schedule: nothing can be done to prevent the groom who has been reliable for two months from failing on the sixty-second day; at this point it is not maintenance work that has to be done but the *same* work as on the first day—apart from the few habits that you might have been able to *incorporate* into his body. Although they appear to be two similar delegations, the first one is concentrated at the time of installation, whereas the other is continuous; more exactly, the first one creates clear-cut distinctions between production, installation, and maintenance, whereas in the other the distinction between training and keeping in operation is either fuzzy or nil. The first one evokes the past perfect ("once hinges had been installed . . ."), the second the present tense ("when the groom is at his post . . ."). There is a built-in inertia in the first that is largely lacking in the second. The first one is Newtonian, the second Aristotelian (which is simply a way of repeating that the first is nonhuman and the other human). A profound temporal shift takes place when nonhumans are appealed to; time is *folded*.

Delegation to Nonhumans

It is at this point that you have a relatively new choice: either to discipline the people or to *substitute* for the unreliable humans a *delegated nonhuman character* whose only function is to open and close the door. This is called a door-closer or a groom ("groom" is a French trademark that is now part of the common language). The advantage is that you now have to discipline only one nonhuman and may safely leave the others (bellboys included) to their erratic behavior. No matter who they are and where they come from—polite or rude, quick or slow, friends or foes—the nonhuman groom will always take care of the door in any weather and at any time of the day. A nonhuman (hinges) plus another nonhuman (groom) have solved the wall-hole dilemma.

Solved? Well, not quite. Here comes the deskilling question so dear to social historians of technology: thousands of human grooms have been put on the dole by their nonhuman brethren. Have they been replaced? This depends on the kind of action that has been translated or delegated to them. In other words, when humans are displaced and deskilled, nonhumans have to be upgraded and reskilled. This is not an easy task, as we shall now see.

We have all experienced having a door with a powerful spring mechanism slam in our faces. For sure, springs do the job of replacing grooms, but they play the role of a very rude, uneducated, and dumb porter who obviously prefers the wall version of the door to its hole version. They simply slam the door shut. The interesting thing with such impolite doors is this: if they slam shut so violently, it means that you, the visitor, have to be very quick in passing through and that you should not be at someone else's heels, otherwise your nose will get shorter and bloody. An unskilled nonhuman groom thus presupposes a skilled human user. It is always a trade-off. I will call, after Madeleine Akrich's (1992) paper, the behavior imposed back onto the human by nonhuman delegates *prescription*.[8] Prescription is the moral and ethical dimension of mechanisms. In spite of the constant weeping of moralists, no human is as relentlessly moral as a machine, especially if it is (she is, he is, they are) as "user friendly" as my Macintosh computer. We have been able to delegate to nonhumans not only force as we have known it for centuries but also values, duties, and ethics. It is because of this morality that we, humans, behave so ethically, no matter how weak and wicked we feel we are. The sum of morality does not only remain stable but increases enormously with the population of nonhumans. It is at this time, funnily enough, that moralists who focus on isolated socialized humans despair of us—us meaning of course humans and their retinue of nonhumans.

How can the prescriptions encoded in the mechanism be brought out in words? By replacing them by strings of sentences (often in the imperative) that are uttered (silently and continuously) by the mechanisms for the benefit of those who are mechanized: do this, do that, behave this way, don't go that way, you may do so, be allowed to go there. Such sentences look very much like a programming language. This substitution of words for silence can be made in the analyst's thought experiments, but also by instruction booklets, or explicitly, in any training session, through the voice of a demonstrator or instructor or teacher. The military are especially good at shouting them out through the mouthpiece of human instructors who delegate back to themselves the task of explaining, in the rifle's name, the characteristics of the rifle's ideal user. Another way of hearing what the machines silently did and said are the accidents. When the space shuttle exploded, thousands of pages of transcripts suddenly covered every detail of the silent machine, and hundreds of inspectors, members of congress, and engineers retrieved from NASA dozens of thousands of pages of drafts and orders. This description of a machine—whatever the means—retraces the steps made by the engineers to transform texts, drafts, and projects into things. The impression given to those who are obsessed by human behavior that there is a missing mass of morality is due to the fact that they do not follow this path that leads from text to things and from things to texts. They draw a strong distinction between these two worlds, whereas the job of engineers, instructors, project managers, and analysts is to continually cross this divide. Parts of a program of action may be delegated to a human, or to a nonhuman.

The results of such *distribution of competences* between humans and nonhumans is that competent members of La Halle aux Cuirs will safely pass through the slamming door at a good distance from one another while visitors, unaware of the local cultural condition, will crowd through the door and get bloody noses.[9] The nonhumans take over the selective attitudes of those who engineered them. To avoid this discrimination, inventors get back to their drawing board and try to imagine a nonhuman character that will not *prescribe* the same rare local cultural skills to its human users. A weak spring might appear to be a good solution. Such is not the case, because it would substitute for another type of very unskilled and undecided groom who is never sure about the door's (or his own) status: is it a hole or a wall? Am I a closer or an opener? If it is both at once, you can forget about the heat. In computer parlance, a door is an exclusive OR, not an AND gate.

I am a great fan of hinges, but I must confess that I admire hydraulic door closers much more, especially the old heavy copper-plated one that slowly closed the main door of our house in Aloxe-Corton. I am enchanted by the addition to the spring of a hydraulic piston, which easily draws up the energy of those who open the door, retains it, and then gives it back slowly with a subtle type of implacable firmness that one could expect from a well-trained butler. Especially clever is its way of extracting energy from each unwilling, unwitting passerby. My sociologist friends at the School of Mines call such a clever extraction an "obligatory passage point," which is a very fitting name for a door. No matter what you feel, think, or do, you have to leave a bit of your energy, literally, at the door. This is as clever as a toll booth.[10]

This does not quite solve all of the problems, though. To be sure, the hydraulic door closer does not bang the noses of those unaware of local conditions, so its prescriptions may be said to be less restrictive, but it still leaves aside segments of human populations: neither my little nephews nor my grandmother could get in unaided because our groom needed the force of an able-bodied person to accumulate enough energy to close the door later. To use Langdon Winner's (1980) classic motto: Because of their prescriptions, these doors *discriminate* against very little and very old persons. Also, if there is no way to keep them open for good, they discriminate against furniture removers and in general everyone with packages, which usually means, in our late capitalist society, working- or lower-middle-class employees. (Who, even among those from higher strata, has not been cornered by an automated butler when they had their hands full of packages?)

There are solutions, though: the groom's delegation may be written off (usually by blocking its arm) or, more prosaically, its delegated action may be opposed by a foot (salesmen are said to be expert at this). The foot may in turn be delegated to a carpet or anything that keeps the butler in check (although I am always amazed by the number of objects that *fail* this trial of force and I have very often seen the door I just wedged open politely closing when I turned my back to it).

Anthropomorphism

As a technologist, I could claim that provided you put aside the work of installing the groom and maintaining it, and agree to ignore the few sectors of the population that are discriminated against, the hydraulic groom does its job well, closing the door behind you, firmly and slowly. It shows in its humble way how three rows of delegated nonhuman actants (hinges, springs, and hydraulic pistons) replace, 90 percent of the time, either an undisciplined bellboy who is never there when needed or, for the general public, the program instructions that have to do with remembering-to-close-the-door-when-it-is-cold.[11]

The hinge plus the groom is the technologist's dream of efficient action, at least until the sad day when I saw the note posted on La Villette's door with which I started this meditation: "The groom is on strike." So not only have we been able to delegate the act of closing the door from the human to the nonhuman, we have also been able to delegate the human lack of discipline (and maybe the union that goes with it). On strike . . . [12] Fancy that! Nonhumans stopping work and claiming what? Pension payments? Time off? Landscaped offices? Yet it is no use being indignant, because it is very true that nonhumans are not so reliable that the irreversibility we would like to grant them is always complete. We did not want ever to have to think about this door again—apart from regularly scheduled routine maintenance (which is another way of saying that we did not have to bother about it)—and here we are, worrying again about how to keep the door closed and drafts outside.

What is interesting in this note is the humor of attributing a human characteristic to a failure that is usually considered "purely technical." This humor, however, is more profound than in the notice they could have posted: "The groom is not working." I constantly talk with my computer, who answers back; I am sure you swear at your old car; we are constantly granting mysterious faculties to gremlins inside every conceivable home appliance, not to mention cracks in the concrete belt of our nuclear plants. Yet, this behavior is considered by sociologists as a scandalous breach of natural barriers. When you write that a groom is "on strike," this is only seen as a "projection," as they say, of a human behavior onto a nonhuman, cold, technical object, one by nature impervious to any feeling. This is *anthropomorphism*, which for them is a sin akin to zoophily but much worse.

It is this sort of moralizing that is so irritating for technologists, because the automatic groom is already anthropomorphic through and through. It is well known that the French like etymology; well, here is another one: *anthropos* and *morphos* together mean either that which *has* human shape or that which *gives shape* to humans. The groom is indeed anthropomorphic, in three senses: first, it has been made by humans; second, it substitutes for the actions of people and is a delegate that permanently occupies the position of a human; and third, it shapes human action by prescribing back what sort of people should pass through the door. And yet some would forbid us to ascribe feelings to this thoroughly anthropomorphic creature, to delegate labor relations, to "project"—that is, to translate—*other* human properties to the groom. What of those many other innovations that have endowed much more sophisticated doors with the ability to see you arrive in advance (electronic eyes), to ask for your identity (electronic passes), or to slam shut in case of danger? But anyway, who are sociologists to decide the real and final shape (*morphos*) of humans (*anthropos*)? To trace with confidence the boundary between what is a "real" delegation and what is a "mere" projection? To sort out forever and without due inquiry the three different kinds of anthropomorphism I listed above? Are we not shaped by nonhuman grooms, although I admit only a very little bit? Are they not our brethren? Do they not deserve consideration? With your self-serving and self-righteous social studies of technology, you always plead against machines and for deskilled workers—are you aware of *your* discriminatory biases? You discriminate between the human and the inhuman. I do not hold this bias (this one at least) and see only actors—some human, some nonhuman, some skilled, some unskilled—that exchange their properties. So the note posted on the door is accurate; it gives with humor an exact rendering of the groom's behavior: it is not working, it is on strike (notice, that the word "strike" is a rationalization carried from the nonhuman repertoire to the human one, which proves again that the divide is untenable).

Built-In Users and Authors

The debates around anthropomorphism arise because we believe that there exist "humans" and "nonhumans," without realizing that this attribution of roles and action is also a *choice*.[13] The best way to understand this choice is to compare machines with texts, since the inscription of builders and users in a mechanism is very much the same as that of authors and readers in a story. In order to exemplify this point I have now to confess that I am *not* a technologist. I built in my article a made-up author, and I also invented possible readers whose reactions and beliefs I anticipated. Since the beginning I have many times used the "you" and even "you sociologists." I even asked you to draw up a table, and I also asked your permission to go on with the story. In doing so, I built up an inscribed reader to whom I prescribed qualities and behavior, as surely as a traffic light or a painting prepare a position for those looking at them. Did you *underwrite* or *subscribe* this definition of yourself? Or worse, is there any one at all to read this text and occupy the position prepared for the reader? This question is a source of constant difficulties for those who are unaware of the basics of semiotics or of technology. *Nothing in a given scene* can prevent the inscribed user or reader from behaving differently from what was expected (nothing, that is, until the next paragraph). The reader in the flesh may totally ignore my definition of him or her. The user of the traffic light may well cross on the red. Even visitors to La Halle aux Cuirs may never show up because it is too complicated to find the place, *in spite* of the fact that their behavior and trajectory have been perfectly anticipated by the groom. As for the computer user input, the cursor might flash forever without the user being there or knowing what to do. There might be an enormous gap between the prescribed user and the user-in-the-flesh, a difference as big as the one between the "I" of a novel and the novelist.[14] It is exactly this difference that upset the authors of the anonymous appeal on which I comment. On other occasions, however, the gap between the two may be nil: the prescribed user is so well anticipated, so carefully nested inside the scenes, so exactly dovetailed, that it does what is expected.[15]

The problem with scenes is that they are usually well prepared for anticipating users or readers who are at close quarters. For instance, the groom is quite good in its anticipation that people will push the door open and give it the energy to reclose it. It is very bad at doing anything to help people arrive there. After fifty centimeters, it is helpless and cannot act, for example, on the maps spread around La Villette to explain where La Halle aux Cuirs is. Still, no scene is prepared without a preconceived idea of what sort of actors will come to occupy the prescribed positions.

This is why I said that although *you* were free not to go on with this paper, *you* were only "relatively" so. Why? Because I know that, because you bought this book, you are hard-working, serious, English-speaking technologists or readers committed to understanding new development in the social studies of machines. So my injunction to "read the paper, you sociologist" is not very risky (but I would have taken no chance with a French audience, especially with a paper written in English). This way of counting on earlier distribution of skills to help narrow the gap between built-in users or readers and users- or readers-in-the-flesh is like a *pre*-inscription.[16]

The fascinating thing in text as well as in artifact is that they have to thoroughly organize the relation between what is inscribed in them and what can/could/should be pre-inscribed in the users. Each setup is surrounded by various arenas interrupted by different types of walls. A text, for instance, is clearly *circumscribed*[17]—the dust cover, the title page, the hard back—but so is a computer—the plugs, the screen, the disk drive, the user's input. What is nicely called *interface* allows any setup to be connected to another through so many carefully designed entry points. Sophisticated mechanisms build up a whole gradient of concentric circles around themselves. For instance, in most modern photocopy machines there are troubles that even rather incompetent users may solve themselves like "ADD PAPER"; but then there are trickier ones that require a bit of explanation: "ADD TONER. SEE MANUAL, PAGE 30." This instruction might be backed up by homemade labels: "DON'T ADD THE TONER YOURSELF, CALL THE SECRETARY," which limit still further the number of people able to troubleshoot. But then other more serious crises are addressed by labels like "CALL THE TECHNICAL STAFF AT THIS NUMBER," while there are parts of the machine that are sealed off entirely with red labels such as "DO NOT OPEN—DANGER, HIGH VOLTAGE, HEAT" or "CALL THE POLICE." Each of these messages addresses a different audience, from the widest (everyone with the rather largely disseminated competence of using photocopying machines) to the narrowest (the rare bird able to troubleshoot and who, of course, is never there).[18] Circumscription only defines how a setup itself has built-in plugs and interfaces; as the name indicates, this tracing of circles, walls, and entry points inside the text or the machine does not prove that readers and users will obey. There is nothing sadder than an obsolete computer with all its nice interfaces, but no one on earth to plug them in.

Drawing a side conclusion in passing, we can call *sociologism* the claim that, given the competence, pre-inscription, and circumscription of human users and authors, you can read out the scripts nonhuman actors have to play; and *technologism* the symmetric claim that, given the competence and pre-inscription of nonhuman actors, you can easily read out and deduce the behavior prescribed to authors and users. From now on, these two absurdities will, I hope, disappear from the scene, because the actors at any point may be human or nonhuman, and the displacement (or translation, or transcription) makes impossible the easy reading out of one repertoire and into the next. The bizarre idea that society might be made up of human relations is a mirror image of the other no less bizarre idea that techniques might be made up of nonhuman relations. We deal with characters, delegates, representatives, lieutenants (from the French "lieu" plus "tenant," i.e., holding the place of, for, someone else)—some figurative, others nonfigurative; some human, others nonhuman; some competent, others incompetent. Do you want to cut through this rich diversity of delegates and artificially create two heaps of refuse, "society" on one side and "technology" on the other? That is your privilege, but I have a less bungled task in mind.

A scene, a text, an automatism can do a lot of things to their prescribed users at the range—close or far—that is defined by the circumscription, but most of the effect finally ascribed to them depends on lines of other setups being aligned.[19] For instance, the groom closes the door only if there are people reaching the Centre d'Histoire des Sciences; these people arrive in front of the door only if they have found maps (another delegate, with the built-in prescription I like most: "*you* are here" circled in red on the map) and only if there are roads leading under the Paris ring road to the Halle (which is a condition not always fulfilled); and of course people will start bothering about reading the maps, getting their feet muddy and pushing the door open only if they are convinced that the group is worth visiting (this is about the only condition in La Villette that is fulfilled). This gradient of aligned setups that endow actors with the pre-inscribed competences to find its users is very much like Waddington's *chreod*:[20] people effortlessly flow through the door of La Halle aux Cuirs and the groom, hundreds of times a day, recloses the door—when it is not stuck. The result of such an alignment of setups is to decrease the number of occasions in which words are used;[21] most of the actions are silent, familiar, incorporated (in human or in non-human bodies)—making the analyst's job so much harder. Even the classic debates about freedom, determination, predetermination, brute force, or efficient will—debates that are the twelfth-century version of seventeenth-century discussions on grace—will be slowly eroded. (Because *you* have reached this point, it means I was right in saying that you were not at all free to stop reading the paper: positioning myself cleverly along a chreod, and adding a few other tricks of my own, I led you *here* . . . or did I? Maybe you skipped most of it, maybe you did not understand a word of it, o you, undisciplined readers.)

Figurative and Nonfigurative Characters

Most sociologists are violently upset by this crossing of the sacred barrier that separates human from nonhumans, because they confuse this divide with another one between *figurative* and *nonfigurative* actors. If I say that Hamlet is the figuration of "depression among the aristocratic class," I move from a personal figure to a less personal one—that is, class. If I say that Hamlet stands for doom and gloom, I use less figurative entities, and if I claim that he represents western civilization, I use nonfigurative abstractions. Still, they all are equally actors, that is, entities that *do* things, either in Shakespeare's artful plays or in the commentators' more tedious tomes. The choice of granting actors figurativity or not is left entirely to the authors. It is exactly the same for techniques.

Engineers are the authors of these subtle plots and scenarios of dozens of delegated and interlocking characters so few people know how to appreciate. The label "inhuman" applied to techniques simply overlooks translation mechanisms and the many choices that exist for figuring or defiguring, personifying or abstracting, embodying or disembodying actors. When we say that they are "mere automatisms," we project as much as when we say that they are "loving creatures"; the only difference is that the latter is an anthropomorphism and the former a technomorphism or phusimorphism.

For instance, a meat roaster in the Ho^tel-Dieu de Beaune, the little groom called "le Petit Bertrand," is the delegated author of the movement (figure 10.2). This little man is as famous in Beaune as is the Mannekenpis in Brussels. Of course, he is not the one who does the turning—a hidden heavy stone collects the force applied when the human demonstrator or the cook turns a heavy handle that winds up a cord around a drum equipped with a ratchet. Obviously le Petit Bertrand believes he is the one doing the job because he not only smiles but also moves his head from side to side with obvious pride while turning his little handle. When we were kids, even though we had seen our father wind up the machine and put away the big handle, we liked to believe that the little guy was moving the spit. The irony of the Petit Bertrand is that, although the delegation to mechanisms aims at rendering any human turnspit useless, the mechanism is ornamented with a constantly exploited character "working" all day long.

Although this turnspit story offers the opposite case from that of the door closer in terms of figuration (the groom on the door does not look like a groom but really does the same job, whereas le Petit Bertrand does look like a groom but is entirely passive), they are similar in terms of delegation (you no longer need to close the door, and the cook no longer has to turn the skewer). The "enunciator" (a general word for the author of a text or for the mechanics who devised the spit) is free to place or not a representation of him or herself in the script (texts or machines). Le Petit Bertrand is a delegated version of whoever is responsible for the mechanism. This is exactly the same operation as the one in which I pretended that the author of this article was a hardcore technologist (when I really am a mere sociologist—which is a second localization of the text, as wrong as the first because really I am a mere philosopher . . .). If I say "we the technologists," I propose a picture of the author of the text as surely as if we place le Petit Bertrand as the originator of the scene. But it would have been perfectly possible for me and for the mechanics to position *no figurated character* at all as the author *in* the scripts *of* our scripts (in semiotic parlance there would be no *narrator*). I would just have had to say things like "recent developments in sociology of technology have shown that . . ." instead of "I," and the mechanics would simply have had to take out le Petit Bertrand, leaving the beautiful cranks, teeth, ratchets, and wheels to work alone. The point is that removing the Petit Bertrand does not turn the mechanism into a "mere mechanism" where no actors are acting. It is just a different choice of style.

The distinctions between humans and nonhumans, embodied or disembodied skills, impersonation or "machination," are less interesting than the complete chain along which competences and actions are distributed. For instance, on the freeway the other day I slowed down because a guy in a yellow suit and red helmet was waving a red flag. Well, the guy's moves were so regular and he was located so dangerously and had such a pale though smiling face that, when I passed by, I recognized it to be a machine (it failed the Turing test, a cognitivist would say). Not only was the red flag delegated; not only was the arm waving the flag also delegated; but the body appearance was also added to the machine. We road engineers (see? I can do it again and carve out another author) could move much further in the direction of figuration, although at a cost:

Figure 10.2
Le Petit Bertrand is a mechanical meat roaster from the sixteenth century that ornaments the kitchen of the Ho^tel-Dieu de Beaune, the hospital where the author was born. The big handle (*bottom right*) is the one that allows the humans to wind up the mechanism; the small handle (*top right*) is made to allow a little nonhuman anthropomorphic character to move the whole spit. Although the movement is prescribed back by the mechanism, since the Petit Bertrand smiles and turns his head from left to right, it is believed that he is at the origin of the force. This secondary mechanism—to whom is ascribed the origin of the force—is unrelated to the primary mechanism, which gathers a large-scale human, a handle, a stone, a crank, and a brake to regulate the movement.

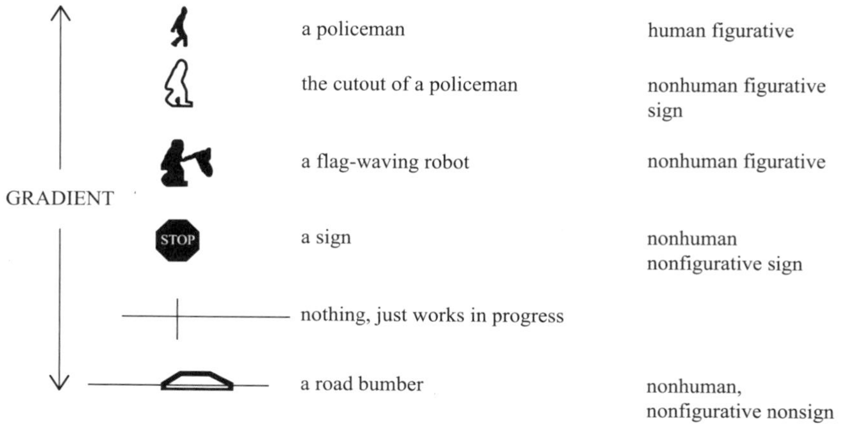

Figure 10.3
Students of technology are wary of anthropomorphism that they see as a projection of human characters to mere mechanisms, but mechanisms to another "morphism," a nonfigurative one that can also be applied to humans. The difference between "action" and "behavior" is not a primary, natural one.

Table 10.1
The distinction between words and things is impossible to make for technology because it is the gradient allowing engineers to shift down—from words to things—or to shift up—from things to signs—that enables them to enforce their programs of actions

	Figurative	Nonfigurative
Human	"I"	"Science shows that . . ."
Nonhuman	"le Petit Bertrand"	a door closer

we could have given him electronic eyes to wave only when a car approaches, or have regulated the movement so that it is faster when cars do not obey. We could also have added (why not?) a furious stare or a recognizable face like a mask of Mrs. Thatcher or President Mitterand—which would have certainly slowed drivers very efficiently.[22] But we could also have moved the other way, to a *less* figurative delegation: the flag by itself could have done the job. And why a flag? Why not simply a sign "work in progress?" And why a sign at all? Drivers, if they are circumspect, disciplined, and watchful will see for themselves that there is work in progress and will slow down. But there is another radical, nonfigurative solution: the road bumper, or a speed trap that we call in French *un gendarme couché*, a laid policeman. It is impossible for us not to slow down, or else we break our suspension. Depending on where we stand along this chain of delegation, we get classic moral human beings endowed with self-respect and able to speak and obey laws, or we get stubborn and efficient machines and mechanisms; halfway through we get the usual power of signs and symbols. It is the complete chain that makes up the missing masses, not either of its extremities. The paradox of technology is that it is thought to be at one of the extremes, whereas it is the ability of the engineer to travel easily along the whole gradient and substitute one type of delegation for another that is inherent to the job.[23]

From Nonhumans to Superhumans

The most interesting (and saddest) lesson of the note posted on the door at La Villette is that people are not circumspect, disciplined, and watchful, especially not French drivers doing 180 kilometers an hour on a freeway on a rainy Sunday morning when the speed limit is 130 (I inscribe the legal limit in this article because this is about the only place where you could see it printed in black and white; no one else seems to bother, except the mourning families). Well, that is exactly the point of the note: "The groom is on strike, *for God's sake*, keep the door closed." In our societies there are two systems of appeal: nonhuman and superhuman—that is, machines and gods. This note indicates how desperate its anonymous frozen authors were (I have never been able to trace and honor them as they deserved). They first relied on the inner morality and common sense of humans; this failed, the door was always left open. Then they appealed to what we technologists consider the supreme court of appeal, that is, to a nonhuman who regularly and conveniently does the job in place of unfaithful humans; to our shame, we must confess that it also failed after a while, the door was again left open. How poignant their line of thought! They moved up and backward to the oldest and firmest court of appeal there is, there was, and ever will be. If humans and nonhumans have failed, certainly God will not deceive them. I am ashamed to say that when I crossed the hallway this February day, the door *was* open. Do not accuse God, though, because the note did not make a direct appeal; God is not accessible without mediators—the anonymous authors knew their catechisms well—so instead of asking for a direct miracle (God holding the door firmly closed or doing so through the mediation of an angel, as has happened on several occasions, for instance when Saint Peter was delivered from his prison) they appealed to the respect for God in human hearts. This was their mistake. In our secular times, this is no longer enough.

Nothing seems to do the job nowadays of disciplining men and women to close doors in cold weather. It is a similar despair that pushed the road engineer to add a golem to the red flag to force drivers to beware—although the only way to slow French drivers is still a good traffic jam. You seem to need more and more of these figurated delegates, aligned in rows. It is the same with delegates as with drugs; you start with soft ones and end up shooting up. There is an inflation for delegated characters, too. After a while they weaken. In the old days it might have been enough just to have a door for people to know how to close it. But then, the embodied skills somehow disappeared; people had to be reminded of their training. Still, the simple inscription "Keep the door closed" might have been sufficient in the good old days. But you know people, they no longer pay attention to the notice and need to be reminded by stronger devices. It is then that you install automatic grooms, since electric shocks are not as acceptable for people as for cows. In the old times, when quality was still good, it might have been enough just to oil it from time to time, but nowadays even automatisms go on strike.

It is not, however, that the movement is always from softer to harder devices, that is, from an autonomous body of knowledge to force through the intermediary situation of worded injunctions, as the La Villette door would suggest. It goes also the other way. It is true that in Paris no driver will respect a sign (for instance, a white or yellow line forbidding parking), nor even a sidewalk (that is a yellow line plus a fifteen centimeter curb); so instead of embodying in the Parisian consciousness an *intrasomatic* skill, authorities prefer to align yet a third delegate (heavy blocks shaped like truncated pyramids and spaced in such a way that cars cannot sneak through); given the results, only a complete two-meter high continuous Great Wall could do the job, and even this might not make the sidewalk safe, given the very poor sealing efficiency of China's Great Wall. So the deskilling thesis appears to be the general case: always go from intrasomatic to *extrasomatic* skills; never rely on undisciplined people, but always on safe, delegated nonhumans. This is far from being the case, even for Parisian drivers. For instance, red lights are usually respected, at least when they are sophisticated enough to integrate traffic flows through sensors; the delegated policeman standing there day and night is respected even though it has no whistles, gloved hands, and body to *enforce* this respect. Imagined collisions with other cars or with the absent police are enough to keep the drivers in check. The thought experiment "what would happen if the delegated character was not there" is the same as the one I recommended above to size up its function. The same *incorporation* from written injunction to body skills is at work with car manuals. No one, I guess, casts more than a cursory glance at the manual before starting the engine of an unfamiliar car. There is a large *body* of skills that we have so well embodied or incorporated that the mediations of the written instructions are useless.[24] From extrasomatic, they have become intrasomatic. Incorporation in human or "excorporation" in nonhuman bodies is also one of the choices left to the designers.

The only way to follow engineers at work is not to look for extra- or intrasomatic delegation, but only at their work of *re-inscription*.[25] The beauty of artifacts is that they take on themselves the contradictory wishes or needs of humans and non-humans. My seat belt is supposed to strap me in firmly in case of accident and thus impose on me the respect of the advice DON'T CRASH THROUGH THE WINDSHIELD, which is itself the translation of the unreachable goal DON'T DRIVE TOO FAST into another less difficult (because it is a more selfish) goal: IF YOU DO DRIVE TOO FAST, AT LEAST DON'T KILL YOURSELF. But accidents are rare, and most of the time the seat belt should not tie me firmly. I need to be able to switch gears or tune my radio. The car seat belt is not like the airplane seat belt buckled only for landing and takeoff and carefully checked by the flight attendants. But if auto engineers invent a seat belt that is completely elastic, it will not be of any use in case of accident. This first contradiction (be firm and be lax) is made more difficult by a second contradiction (you should be able to buckle the belt very fast—if not, no one will wear it—but also unbuckle it very fast, to get out of your crashed car). Who is going to take on all of these contradictory specifications? The seat belt mechanism—if there is no other way to go, for instance, by directly limiting the speed of the engine, or having roads so bad that no one can drive fast on them. The safety engineers have to re-inscribe in the seat belt all of these contradictory usages. They pay a price, of course: the mechanism is *folded* again, rendering it more complicated. The airplane seat belt is childish by comparison with an automobile seat belt. If you study a complicated mechanism without seeing that it reinscribes contradictory specifications, you offer a dull description, but every piece of an artifact becomes fascinating when you see that every wheel and crank is the possible answer to an objection. The program of action is in practice the answer to an *antiprogram* against which the mechanism braces itself. Looking at the mechanism alone is like watching half the court during a tennis game; it appears as so many meaningless moves. What analysts of artifacts have to do is similar to what we all did when studying scientific texts: we added the other half of the court.[26] The scientific literature looked dull, but when the agonistic field to which it reacts was brought back in, it became as interesting as an opera. The same with seat belts, road bumpers, and grooms.

Texts and Machines

Even if it is now obvious that the missing masses of our society are to be found among the nonhuman mechanisms, it is not clear how they get there and why they are missing from most accounts. This is where the comparison between texts and artifacts that I used so far becomes misleading. There is a crucial distinction between stories and machines, between narrative programs and programs of action, a distinction that explains why machines are so hard to retrieve in our common language. In storytelling, one calls *shifting out* any displacement of a character to another space time, or character. If I tell you "Pasteur entered the Sorbonne amphitheater," I translate the present setting—you and me—and shift it to another space (middle of Paris), another time (mid-nineteenth century), and to other characters (Pasteur and his audience). "I" the enunciator may decide to appear, disappear, or be represented by a narrator who tells the story ("That day, I was sitting on the upper row of the room"); "I" may also decide to position you and any reader inside the story ("Had you been there, you would have been convinced by Pasteur's experiments"). There is no limit to the number of shiftings out with which a story may be built. For instance, "I" may well stage a dialogue inside the amphitheater between two characters who are telling a story about what happened at the Académie des Sciences between, say, Pouchet and Milnes-Edwards. In that case, the room becomes the place *from which* narrators shift out to tell a story about the Academy, and they may or not shift *back in* the amphitheater to resume the first story about Pasteur. "I" may also *shift in* the entire series of nested stories to close mine and come back to the situation I started from—you and me. All these displacements are well known in literature departments (Latour 1988) and make up the craft of talented writers.

No matter how clever and crafty are our novelists, they are no match for engineers. Engineers constantly shift out characters in other spaces and other times, devise positions for human and nonhuman users, break down competences that they then redistribute to many different actors, and build complicated narrative programs and subprograms that are evaluated and judged by their ability to stave off antiprograms. Unfortunately, there are many more literary critics than technologists, and the subtle beauties of technosocial imbroglios escape the attention of the literate public. One of the reasons for this lack of concern may be the peculiar nature of the shifting-out that generates machines and devices. Instead of sending the listener of a story into another world, the technical shifting-out inscribes the words into *another matter*. Instead of allowing the reader of the story to be *at the same time* away (in the story's frame of reference) and here (in an armchair), the technical shifting-out forces the reader to choose *between* frames of reference. Instead of allowing enunciators and enunciatees a sort of simultaneous presence and communion to other actors, techniques allow both to *ignore* the delegated actors and walk away without even feeling their presence. This is the profound meaning of Butler's sentence I placed at the beginning of this chapter: machines are not talking actors, not because they are unable to do so, but because they might have chosen to remain silent to become agreeable to their fellow machines and fellow humans.

To understand this difference in the two directions of shifting out, let us venture once more onto a French freeway; for the umpteenth time I have screamed at my son Robinson, "Don't sit in the middle of the rear seat; if I brake too hard, you're dead." In an auto shop further along the freeway I come across a device *made for* tired-and-angry-parents-driving-cars-with-kids-between-two-and-five (too old for a baby seat and not old enough for a seat belt) and-from-small-families (without other persons to hold them safely) with-cars-with-two-separated-front-seats-and-head-rests. It is a small market, but nicely analyzed by the German manufacturers and, given the price, it surely pays off handsomely. This description of myself and the small category into which I am happy to belong is transcribed in the device—a steel bar with strong attachments connecting the head rests—and in the advertisement on the outside of the box; it is also pre-inscribed in about the only place where I could have realized that I needed it, the freeway. (To be honest and give credit where credit is due, I must say that Antoine Hennion has a similar device in his car, which I had seen the day before, so I really looked for it in the store instead of "coming across" it as I wrongly said; which means that a) there is some truth in studies of dissemination by imitation; b) if I describe this episode in as much detail as the door I will never have been able to talk about the work done by the historians of technology at La Villette.) Making a short story already too long, I no longer scream at Robinson, and I no longer try to foolishly stop him with my extended right arm: he firmly holds the bar that protects him against my braking. I have delegated the continuous injunction of my voice and extension of my right arm (with diminishing results, as we know from Feschner's law) to a reinforced, padded, steel bar. Of course, I had to make two detours: one to my wallet, the second to my tool box; 200 francs and five minutes later I had fixed the device (after making sense of the instructions encoded with Japanese ideograms).

We may be able to follow these detours that are characteristic of the technical form of delegation by adapting a linguistic tool. Linguists differentiate the *syntagmatic* dimension of a sentence from the *paradigmatic* aspect. The syntagmatic dimension is the possibility of *associating* more and more words in a grammatically correct sentence: for instance, going from "the barber" to "the barber goes fishing" to "the barber goes fishing with his friend the plumber" is what linguists call moving through the syntagmatic dimension. The number of elements tied together increases, and nevertheless the sentence is still meaningful. The paradigmatic dimension is the possibility, in a sentence of a given length, of *substituting* a word for another while still maintaining a grammatically correct sentence. Thus, going from "the barber goes fishing" to "the plumber goes

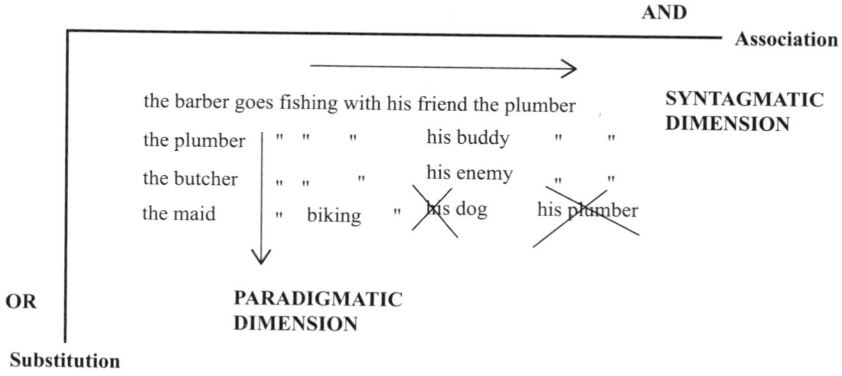

Figure 10.4
Linguists define meaning as the intersection of a horizontal line of association—the syntagm—and a vertical line of substitution—the paradigm. The touchstone in linguistics is the decision made by the competent speaker that a substitution (OR) or an association (AND) is grammatically correct in the language under consideration. For instance, the last sentence is incorrect.

fishing" to "the butcher goes fishing" is tantamount to moving through the paradigmatic dimension.[27]

Linguists claim that these two dimensions allow them to describe the system of any language. Of course, for the analysis of artifacts we do not have a structure, and the definition of a grammatically correct expression is meaningless. But if, by substitution, we mean the technical shifting to another *matter*, then the two dimensions become a powerful means of describing the dynamic of an artifact. The syntagmatic dimension becomes the AND dimension (how many elements are tied together), and the paradigmatic dimension becomes the OR dimension (how many translations are necessary in order to move through the AND dimension). I could not tie Robinson to the order, but through a detour and a translation I now hold together my will and my son.

The detour, plus the translation of words and extended arm into steel, is a shifting out to be sure, but not of the same type as that of a story. The steel bar has now taken over my competence as far as keeping my son at arm's length is concerned. From speech and words and flesh it has become steel and silence and extrasomatic. Whereas a narrative program, no matter how complicated, always remains a text, the program of action substitutes part of its character to other nontextual elements. This divide between text and technology is at the heart of the myth of Frankenstein (Latour 1992). When Victor's monster escapes the laboratory in Shelley's novel, is it a metaphor of fictional characters that seem to take up a life of their own? Or is it the metaphor of technical characters that do take up a life of their own because they cease to be texts and become flesh, legs, arms, and movements? The first version is not very interesting because in spite of the novelist's cliché, a semiotic character in a text always needs the reader to offer it an "independent" life. The second version is not very interesting either, because the "autonomous" thrust of a technical artifact is a worn-out commonplace made up by bleeding-heart moralists who have never noticed the throngs of humans necessary to keep a machine alive. No, the beauty of Shelley's myth is that we cannot choose between the two versions: parts of the narrative program are still texts, others are bits of flesh and steel—and this mixture is indeed a rather curious monster.

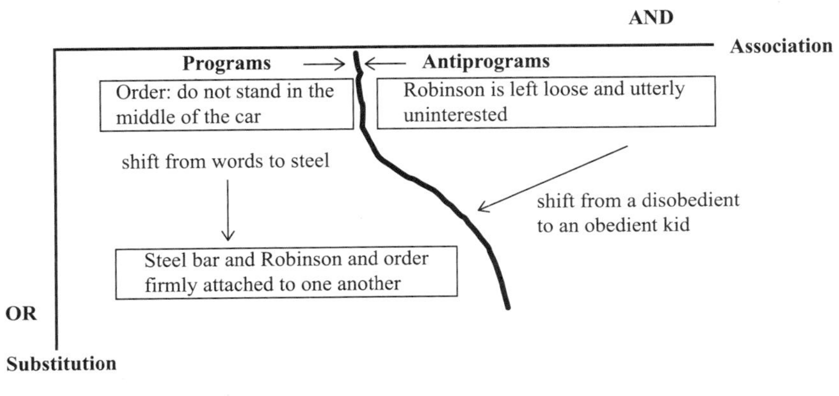

Figure 10.5
The translation diagram allows one to map out the story of a script by following the two dimensions: AND, the association (the latitude, so to speak), and OR, the substitution (the longitude). The plot is defined by the line that separates the programs of action chosen for the analysis and the antiprograms. The point of the story is that it is impossible to move in the AND direction without paying the price of the OR dimension—that is, renegotiating the sociotechnical assemblage.

To bring this chapter to a close and differentiate once again between texts and artifacts, I will take as my final example not a flamboyant Romantic monster but a queer little surrealist one: the Berliner key.[28]

Yes, this is a key and not a surrealist joke (although this is *not* a key, because it is a picture and a text about a key). The program of action in Berlin is almost as desperate a plea as in La Villette, but instead of begging CLOSE THE DOOR BEHIND YOU PLEASE it is slightly more ambitious and *orders*: RELOCK THE DOOR BEHIND YOU. Of course the pre-inscription is much narrower: only people endowed with the competence of living in the house can use the door; visitors should ring the doorbell. But even with such a limited group the antiprogram in Berlin is the same as everywhere: undisciplined tenants forget to lock the door behind them. How can you force them to lock it? A normal key endows you with the *competence* of opening the door[29]—it proves you are persona grata—but nothing in it entails the *performance* of actually using the key again once you have opened the door and closed it behind you. Should you put up

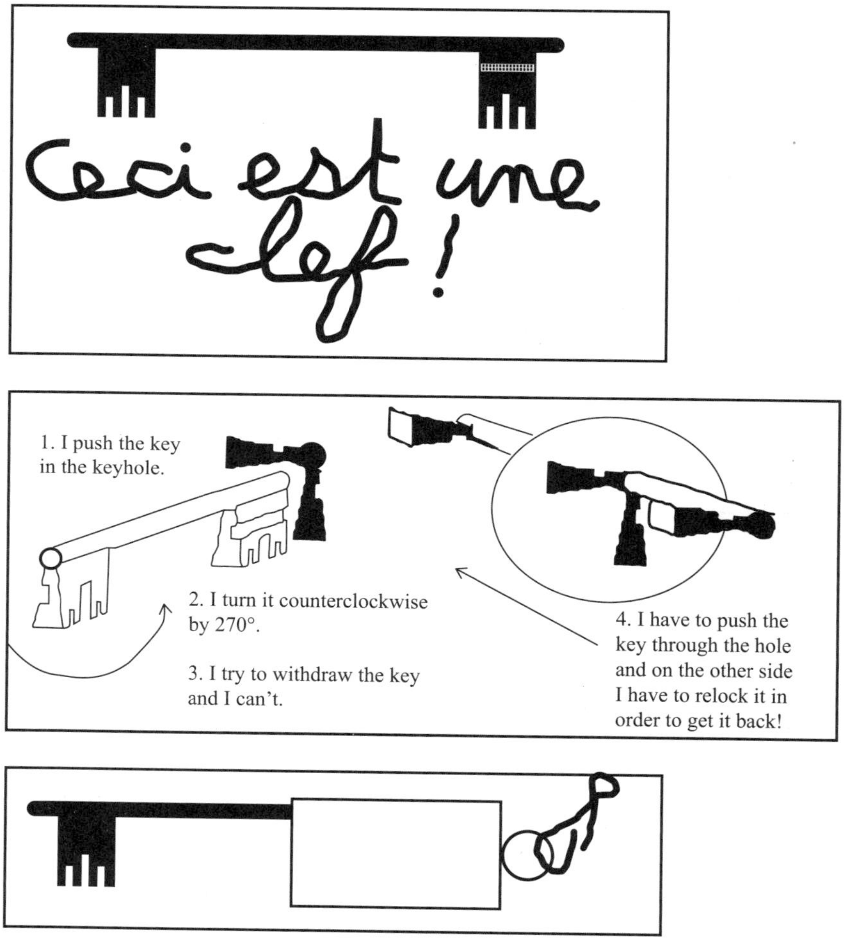

Figure 10.6
The key, its usage, and its holder.

a sign? We know that signs are never forceful enough to catch people's attention for long. Assign a police officer to every doorstep? You could do this in East Berlin, but not in reunited Berlin. Instead, Berliner blacksmiths decided to re-inscribe the program of action in the very shape of the key and its lock—hence this surrealist form. They in effect sunk the contradiction and the lack of discipline of the Berliners in a more "realist" key. The program, once translated, appears innocuous enough: UNLOCK THE DOOR. But here lies the first novelty: it is impossible to remove the key in the normal way; such a move is "proscribed" by the lock. Otherwise you have to break the door, which is hard as well as impolite; the only way to retrieve the key is to push the whole key through the door to the other side—hence its symmetry—but then it is still impossible to retrieve the key. You might give up and leave the key in the lock, but then you lose the competence of the tenant and will never again be able to get in or out. So what do you do? You rotate the key one more turn and, yes, you have in effect relocked the door and then, only then, are you able to retrieve the precious "sesame." This is a clever translation of a possible program relying on morality into a program relying on dire necessity: you might not want to relock the key, but you cannot do otherwise. The distance between morality and force is not as wide as moralists expect; or more exactly, clever engineers have made it smaller. There is a price to pay of course for such a shift away from morality and signs; you have to replace most of the locks in Berlin. The pre-inscription does not stop here, however, because you now have the problem of keys that no decent key holder can stack into place because they have no hole. On the contrary, the new sharp key is going to poke holes in your pockets. So the blacksmiths go back to the drawing board and invent specific key holders adapted to the Berliner key!

The key in itself is not enough to fulfill the program of action. Its effects are very severely circumscribed, because it is only when you have a Berliner endowed with the double competence of being a tenant and knowing how to use the surrealist key that the relocking of the door may be enforced. Even such an outcome is not full proof, because a really bad guy may relock the door without closing it! In that case the worst possible antiprogram is in place because the lock stops the door from closing. Every passerby may see the open door and has simply to push it to enter the house. The setup that prescribed a very narrow segment of the human population of Berlin is now so lax that it does not even discriminate against nonhumans. Even a dog knowing nothing about keys, locks, and blacksmiths is now allowed to enter! No artifact is idiot-proof because any artifact is only a portion of a program of action and of the fight necessary to win against many antiprograms.

Students of technology are never faced with people on the one hand and things on the other, they are faced with programs of action, sections of which are endowed to *parts* of humans, while other sections are entrusted to parts of nonhumans. In practice they are faced with the front line of figure 10.7. This is the only thing they can *observe*: how a negotiation to associate dissident elements requires more and more elements to be tied together and more and more shifts to other matters. We are now witnessing in technology studies the same displacement that has happened in science studies during the last ten years. It is not that society and social relations invade the certainty of science or the efficiency of machines. It is that society itself is to be rethought from top to bottom once we add to it the facts and the artifacts that make up large sections of our social ties. What appears in the place of the two ghosts—society and technology—is not simply a hybrid object, a little bit of efficiency and a little bit of sociologizing, but a sui generis object: the collective thing, the trajectory of the front line between

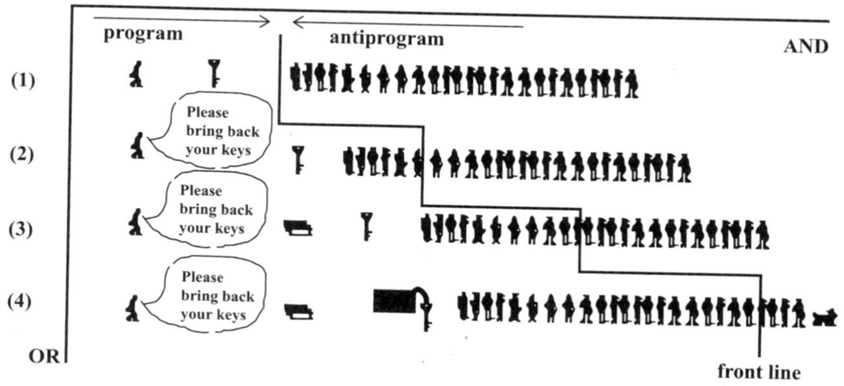

Figure 10.7
The hotel manager successively adds keys, oral notices, written notices, and finally weights; each time he thus modifies the attitude of some part of the "hotel customers" group while he extends the syntagmatic assemblage of elements. From Madeleine Akrich and Bruno Latour, "A Summary of a Convenient Vocabulary for the Semiotics of Human and Nonhuman Assemblies," in Wiebe E. Bijker and John Law, eds., *Shaping Technology/Building Society: Studies in Sociotechnical Change* (Cambridge, MA: MIT Press, 1992), 263.

programs and anti-programs. It is too full of humans to look like the technology of old, but it is too full of nonhumans to look like the social theory of the past. The missing masses are in our traditional social theories, not in the supposedly cold, efficient, and inhuman technologies.

Notes

This chapter owes to many discussions held at the Centre de Sociologie de l'Innovation, especially with John Law, the honorary member from Keele, and Madeleine Akrich. It is particularly indebted to Françoise Bastide, who was still working on these questions of semiotics of technology a few months before her death. I had no room to incorporate a lengthy dispute with Harry Collins about this chapter (but see Callon and Latour [1992] and Collins and Yearley [1992]). Trevor Pinch and John Law kindly corrected the English.

1. The program of action is the set of written instructions that can be substituted by the analyst to any artifact. Now that computers exist, we are able to conceive of a text (a programming language) that is at once words and actions. How to do things with words and then turn words into things is now clear to any programmer. A program of action is thus close to what Pinch, Ashmore, and Mulkay (1992) call "a social technology," except that all techniques may be made to be a program of action.

2. In spite of the crucial work of Diderot and Marx, careful description of techniques is absent from most classic sociologists—apart from the "impact of technology on society" type of study—and is simply black-boxed in too many economists' accounts. Modern writers like Leroi-Gourhan (1964) are not often used. Contemporary work is only beginning to offer us a more balanced account. For a reader, see MacKenzie and Wacjman (1985); for a good overview of recent developments, see Bijker, Hughes, and Pinch (1987). A remarkable essay on how to describe artifacts—an iron bridge compared to a Picasso portrait—is offered by Baxandall (1985). For a recent essay by a pioneer of the field, see Noble (1984). For a remarkable and hilarious description of a list of artifacts, see Baker (1988).

3. Following Madeleine Akrich's (1992) lead, we will speak only in terms of *scripts* or scenes or scenarios, or setups as John Law says, played by human or nonhuman actants, which may be either figurative or nonfigurative.

4. After Akrich, I will call the retrieval of the script from the situation *de-scription*. They define actants, endow them with competences, make them do things, and evaluate the sanction of these actions like the *narrative program* of semioticians.

5. Although most of the scripts are in practice silent, either because they are intra- or extrasomatic, the written descriptions are not an artifact of the analyst (technologist, sociologist, or semiotician), because there exist many states of affairs in which they are *explicitly* uttered. The gradient going from intrasomatic to extrasomatic skills through discourse is never fully stabilized and allows many entries revealing the process of translation: user manuals, instruction, demonstration or drilling situations, practical thought experiments ("What would happen if, instead of the red light, a police officer were there."). To this should be added the innovator's workshop, where most of the objects to be devised are still at the stage of *projects* committed to paper ("If we had a device doing this and that, we could then do this and that."); market analysis in which consumers are confronted with the new device; and, naturally, the exotic situation studied by anthropologists in which people faced with a foreign device talk to themselves while trying out various combinations ("What will happen if I attach this lead here to the mains?"). The analyst has to empirically capture these situations to write down the scripts. When none is available, the analyst may still make a thought experiment by comparing presence/absence tables and collating the list of all the actions taken by actors ("If I take this one away, this and that other action will be modified."). There are dangers in such a counterfactual method, as Collins has pointed out (Collins and Yearley 1992), but it is used here only to outline the semiotics of artifacts. In practice, as Akrich (1992) shows, the scripts are explicit and accountable.

6. We call the translation of any script from one repertoire to a *more durable* one transcription, inscription, or encoding. This definition does *not* imply that the direction always goes from soft bodies to hard machines but simply that it goes from a provisional, less reliable one to a longer-lasting, more faithful one. For instance, the embodiment in cultural tradition of the user manual of a car is a transcription but so is the replacement of a police officer by a traffic light; one goes from machines to bodies, whereas the other goes the opposite way. Specialists of robotics have abandoned the pipe dream of total automation; they learned the hard way that many skills are better delegated to humans than to nonhumans, whereas others may be taken away from incompetent humans.

7. See Authier (1989) on Plutarch's Archimedes.

8. We call prescription whatever a scene presupposes from its *transcribed* actors and authors (this is very much like *role expectation* in sociology, except that it may be inscribed or encoded in the machine). For instance, a Renaissance Italian painting is designed to be viewed from a specific angle of view prescribed by the vanishing lines, exactly like a traffic light expects that its users will watch it from the street and not sideways (French engineers often hide the lights directed toward the side street so as to hide the state of the signals, thus preventing the strong temptation to rush through the crossing at the first hint that the lights are about to be green; this prescription of who is allowed to watch the signal is very frustrating). *User input* in programming language, is another very telling example of this inscription in the automatism of a living character whose behavior is both free and predetermined.

9. In this type of analysis there is no effort to attribute forever certain competences to humans and others to nonhumans. The attention is focused on following how *any* set of competences is *distributed* through various entities.

10. Interestingly enough, the oldest Greek engineering myth, that of Daedalus, is about cleverness, deviousness. "Dedalion" means something that goes away from the main road, like the French word "bricole." In the mythology, science is represented by a straight line and technology by a detour, science by epistémè and technology by the mêtis. See the excellent essay of Frontisi-Ducroux (1975) on the semantic field of the name Daedalus.

11. We use "actant" to mean anything that acts and "actor" to mean what is made the source of an action. This is a semiotician's definition that is not limited to humans and has no relation whatsoever to the sociological definition of an actor by opposition to mere behavior. For a semiotician, the act of attributing "inert force" to a hinge or the act of attributing it "personality" are comparable in principle and should be studied symmetrically.

12. I have been able to document a case of a five-day student strike at a French school of management (ESSEC) to urge that a door closer be installed in the student cafeteria to keep the freezing cold outside.

13. It is of course another choice to decide who makes such a choice: A man? A spirit? No one? An automated machine? The *scripter* or designer of all these scripts is itself (himself, herself, themselves) negotiated.

14. This is what Norman (1988) calls the Gulf of Execution. His book is an excellent introduction to the study of the tense relations between inscribed and real users. However, Norman speaks only about dysfunction in the interfaces with the final user and never considers the shaping of the artifact by the engineer themselves.

15. To stay within the same etymological root, we call the way actants (human or nonhuman) tend to extirpate themselves from the prescribed behavior *de-inscription* and the way they accept or happily acquiesce to their lot *subscription*.

16. We call *pre-inscription* all the work that has to be done upstream of the scene and all the things assimilated by an actor (human or nonhuman) before coming to the scene as a user or an author. For instance, how to drive a car is basically preinscribed in any (Western) youth years before it comes to passing the driving test; hydraulic pistons were also pre-inscribed for slowly giving back the energy gathered, years before innovators brought them to bear on automated grooms. Engineers can bet on this predetermination when they draw up their prescriptions. This is what is called "articulation work" (Fujimura 1987).

17. We call *circumscription* the organization in the setting of its own limits and of its own demarcation (doors, plugs, hall, introductions).

18. See Suchman (1987) for a description of such a setting.

19. We call *ascription* the attribution of an effect to one aspect of the setup. This new decision about attributing efficiency—for instance, to a person's genius, to workers' efforts, to users, to the economy, to technology—is as important as the others, but it is derivative. It is like the opposition between the primary mechanism—who is allied to whom—and the secondary mechanism—whose leadership is recognized—in history of science (Latour 1987).

20. Waddington's term for "necessary paths"—from the Greek *creos* and *odos*.

21. We call *conscription* this mobilization of well-drilled and well-aligned resources to render the behavior of a human or a nonhuman predictable.

22. Trevor Pinch sent me an article from the *Guardian* (September 2, 1988) titled "Cardboard Coppers Cut Speeding by Third."

A Danish police spokesman said an advantage of the effigies, apart from cutting manpower costs, was that they could stand for long periods undistracted by other calls of duty.

Additional assets are understood to be that they cannot claim overtime, be accused of brutality, or get suspended by their chief constable without explanation. "For God's sake, don't tell the Home Office," Mr. Tony Judge, editor of the *Police Review Magazine* in Britain, said after hearing news of the [Danish] study last night. "We have enough trouble getting sufficient men already." The cut-outs have been placed beside notorious speeding blackspots near the Danish capital. Police said they had yielded "excellent" results. Now they are to be erected at crossings where drivers often jump lights. From time to time, a spokesman added, they would be replaced by real officers.

23. Why did the (automatic) groom go on strike? The answers to this are the same as for the question posed earlier of why no one showed up at La Halle aux Cuirs: it is not because a piece of behavior is prescribed by an inscription that the predetermined characters will show up on time and do the job expected of them. This is true of humans, but it is truer of nonhumans. In this case the hydraulic piston did its job, but not the spring that collaborated with it. Any of the words employed above may be used to describe a setup at any level and not only at the simple one I chose for the sake of clarity. It does not have to be limited to the case where a human deals with a series of nonhuman delegates; it can also be true of relations among nonhumans (yes, you sociologists, there are also relations among things, and *social* relations at that).

24. For the study of the user's manual, see Boullier, Akrich, and Le Goaziou (1990) and Norman (1988).

25. Re-inscription is the same thing as inscription or translation or delegation, but seen in its movement. The aim of sociotechnical study is thus to follow the *dynamic* of re-inscription transforming a silent artifact into a *polemical* process. A lovely example of efforts at re-inscription of what was badly pre-inscribed outside of the setting is provided by Orson Welles in *Citizen Kane*, where the hero not only bought a theater for his singing wife to be applauded in but also bought the journals that were to do the reviews, bought off the art critics themselves, and paid the audience to show up—all to no avail, because the wife eventually quit. Humans and nonhumans are very undisciplined no matter what you do and how many predeterminations you are able to control inside the setting.

For a complete study of this dynamic on a large technical system, see Latour (1992) and Law (1992).

26. The study of scientific text is now a whole industry: see Callon, Law, and Rip (1986) for a technical presentation and Latour (1987) for an introduction.

27. The linguistic meaning of a paradigm is unrelated to the Kuhnian usage of the word. For a complete description of these diagrams, see Latour, Mauguin, and Teil (1992).

28. I am grateful to Berward Joerges for letting me interview his key and his key holder. It alone was worth the trip to Berlin.

29. Keys, locks, and codes are of course a source of marvelous fieldwork for analysts. You may for instance replace the key (excorporation) by a memorized code (incorporation). You may lose both, however, since memory is not necessarily more durable than steel.

References

Akrich, Madeleine. 1992. "The De-Scription of Technical Objects." In Bijker and Law, *Shaping Technology/Building Society*.

Authier, M. 1989. "Archime`de, le canon du savant." In *Eléments d'Histoire des Sciences*, edited by Michel Serres. Paris: Bordas.

Baker, N. 1988. *The Mezzanine*. New York: Weidenfeld and Nicholson.

Baxandall, Michael. 1985. *Patterns of Intention: On the Historical Explanation of Pictures*. New Haven, CT: Yale University Press.

Bijker, Wiebe E., Hughes, T. P., and Pinch, Trevor J. 1987. *The Social Construction of Technological Systems: New Directions in the Sociology and History of Technology*. Cambridge, MA: MIT Press.

Bijker, Wiebe E., and Law, John, eds. 1992. *Shaping Technology/Building Society*. Cambridge, MA: MIT Press.

Boullier, D., Akrich, M., and Le Goaziou, V. 1990. *Représentation de l'utilisateur final et genèse des modes d'emploi*. Paris: Miméo, Ecole des Mines.

Butler, Samuel. 1872 (1970). *Erewhon*. London: Penguin.

Callon, Michel, and Latour, Bruno. 1992. "Don't Throw Out the Baby with the Bath School: Reply to Collins and Yearley." In *Science as Practice and Culture*, edited by A. Pickering. Chicago: Chicago University Press.

Callon, Michel, Law, John, and Rip, Arie, eds. 1986. *Mapping the Dynamics of Science and Technology*. Basingstoke: Macmillan.

Collins, H. M., and Yearley, Steven. 1992. "Epistemological Chicken." In *Science in Practice and Culture*, edited by A. Pickering. Chicago: Chicago University Press.

Frontisi-Ducroux, F. 1975. *Dédale, Mythologie de l'artisan en Grèce Ancienne*. Paris: Maspéro-La Découvertè.

Fujimura, Joan. 1987. "Constructing 'Do-Able' Problems in Cancer Research: Articulating Alignment." *Social Studies of Science* 17:257–293.

Latour, Bruno. 1987. *Science in Action: How to Follow Scientists and Engineers through Society*. Cambridge, MA: Harvard University Press.

Latour, Bruno. 1988. "A Relativist Account of Einstein's Relativity." *Social Studies of Science* 18:3–45.

Latour, Bruno. 1992. *Aramis ou l'amour des techniques*. Paris: La Découvertè.

Latour, Bruno, Mauguin, P., and Teil, Genevieve. 1992. "A Note on Socio-technical Graphs." *Social Studies of Science* 22:33–57.

Law, John. 1992. "The Olympus 320 Engine: A Case Study in Design, Autonomy and Organizational Control." *Technology and Culture* 33, no. 3 (July): 409–440.

Leroi-Gourhan, A. 1964. *Les Religions de la Préhistoire*. Paris: Presses Universitaires de France.

MacKenzie, Donald, and Wajcman, Judy, eds. 1985. *The Social Shaping of Technology: How the Refrigerator Got Its Hum*. Bristol, PA: Open University Press.

Noble, David. 1984. *Forces of Production: A Social History of Industrial Automation*. New York: Knopf.

Norman, David. 1988. *The Psychology of Everyday Things*. New York: Basic Books.

Pinch, Trevor, Ashmore, Malcolm, and Mulkay, Michael. 1992. "Technology, Testing, Text: Clinical Budgeting in the U.K. National Health Service." In Bijker and Law, *Shaping Technology/Building Society*.

Suchman, Lucy. 1987. *Plans and Situated Actions: The Problem of Human Machine Communication*. Cambridge: Cambridge University Press.

Winner, Langdon. 1980. "Do Artefacts Have Politics?" *Daedalus* 109:121–136.

11 Gender: The Missing Factor in STS

Eulalia Pérez Sedeño

In this selection, Sedeño argues that few of the early theorists of the relationship between technology and society carefully examined the role of gender. She asserts that such an omission is significant because gender is one of the most powerful ways by which people organize themselves. Ignoring gender gives a skewed understanding of technology, may lead to neglecting the contributions of women, and obscures many of the negative (and positive) impacts technologies have had on half the world's population.

Sedeño puts forward new ways to study technology. She explains, for instance, that bringing women into the picture compels scholars to look in places they haven't looked before, such as the use (rather than just the production) of technology. In the past, the study of technology largely focused on big, powerful artifacts and systems—like rockets and nuclear power plants—to the neglect of smaller, more mundane technologies that arguably have more of an effect on our daily lives as individuals. For example, birth control, kitchen appliances, and even children's toys are technologies that influence and are influenced by social arrangements such as family structures, labor relations, and social notions including personal identity. Bringing gender into the lens of technology and society studies sheds light on impacts, implications, and opportunities that would otherwise not be visible.

There is absolutely no doubt that science and technology have substantially transformed, albeit often in erratic ways, the lives of human beings and the environment in which they live. In past centuries, such transformations were looked upon and lauded as evidence of Man's (*sic*) ability to dominate Nature, which, according to Francis Bacon's philosophy and the scientific revolution, was there to be subjugated and exploited by devices, artifacts, and inventions. Nature, for her part, responded by dispensing justice on the descriptive scientific theories and technology created for her, because she, and only she, is the judge who decides which theories and technology are acceptable.

During the last few decades, however, criticisms aired by some historians, sociologists, and philosophers of science, together with those of the ecological, feminist, and pacifist movements, have produced a type of reflection on science and technology that has put such a conceptualization on the rack. Although science and technology have different effects in different countries, which depends on differences of race, social class, and gender, we can safely assume that modern technoscience is largely a product of investigations performed by trained people who employ characteristic methods and techniques. Thus, technoscience is a body of knowledge and organized procedures,

---

From Stephen H. Cutcliffe and Carl Mitcham, eds., *Visions of STS* (Albany: State University of New York Press, 2001), chap. 9, pp. 123–138.

a means of solving problems. It is also a social and educational institution, one that requires material installations and a cultural resource, but, above all, it is a fundamental factor in human affairs.[1] That is to say, it is a system which has been formed by ideas, designers, artifacts, and end-users, among which there exists a synergistic relationship, one that exists in a specific socio-historical context.

Such a characterization has many advantages. Among others, it highlights the necessity of utilizing a multiplicity of disciplines and points of view, such as history, economics, psychology, sociology, so as to gain an accurate comprehension of technoscience. In addition, it presupposes a recognition that science and technology are not alien to human beings and that, as a consequence, they should be considered as having an interactive relationship with humankind. This means that there are implied ethical and political questions that cannot be put to one side when the time comes to determine whether technoscientific practices are acceptable or not.

The above characterization had its origin in Science and Technology Studies (STS), but it is also a consequence of how STS has evolved. Nowadays, STS is a multidiscipline in which scholars examine science and technology issues and problems from different disciplinary perspectives that deal with cultural, political, social, and ethical questions about how science and technology should be developed and how they could be improved. When one refers to the various disciplines and areas that are covered by STS, one usually includes gender studies.[2] However, I believe that the inclusion of this perspective, in the majority of cases, is merely an attempt at being politically correct, and it is not a genuine recognition of what a gender approach to science and technology means or could mean. This can be seen if one examines the available STS literature in which there are relatively few studies that are at all sensitive to questions of gender. For example, in Bijker, Hughes, and Pinch (1987), one of the most important collections of essays on the social construction of technology, there is hardly any article that could be classified as dealing with that subject, and only one of the works was actually written by a woman. The same is also true of Bijker and Law (1992), an extension of themes developed in the volume mentioned earlier. Furthermore, when such studies do appear, as in the *Handbook of Science and Technology Studies* (Jasanoff et al. 1995), they appear to be peripheral asides that are not really integrated into STS.[3]

I wonder why such marginalization happens, given that one of the most important questions dealt with in STS is the interaction between the social and cultural factors in science and technology. One wonders even more when the function carried out by feminist thinking in a critical reflection on science and technology is of such importance nowadays.[4] It could seem impossible to ignore it, because it raises crucial questions for a comprehensive understanding of science and technology. As things stand presently, I believe that gender is a missing factor that has not sufficiently been taken into consideration by STS. The genuine inclusion of the gender approach is admittedly a challenge, for it would entail a further redefining of science and technology, and of what scientific and technological progress means. To do so, however, would also open the door to new issues, to new challenges, and to different solutions to the questions already posed, both in gender studies and STS.

Feminist Studies of Science and Technology

Feminist studies of science and of technology have not run chronologically in tandem, with a focus on technology having occurred only somewhat more recently. Nonetheless, the latter has followed the path of earlier feminist studies of science by posing similar questions. Even though feminism does not have a common singular stance with regard to technoscience, it does have a common base in maintaining that there is a gender bias in the majority of academic disciplines "that is expressed in particular claims and facilitated by disciplinary first principles" and that "women's experience is made invisible or distorted, as are gender relations" (Longino 1999). As the majority of persons doing science and technology have been men, at least half of humanity has been ignored. That, in turn, has been due to, and reinforced by the fact that, traditionally, science and technology have considered themselves to be objective, neutral, and free of values. That is to say, they have assumed that "external" factors like gender have no place as a central element or concern in their makeup.

One of the general characteristics of feminism is the advancement of social and political issues that will lead to full equality for women. One result has been a focus on pedagogical issues and questions. Here a main objective was that more women should study science and technology, and get more involved in technoscientific activities. With this in mind, some feminist scholars conducted a study of how science and technology was taught at schools and higher educational institutions, and the contents of various curricula. The strategies used to encourage young girls and women to take up these subjects, and later to work in these areas, were very varied. Some focused on the content of the subjects, the selection of suitable texts, the inclusion of information not normally found in standard courses, and on the expectations that young girls and adolescents have from science, which normally condition their adult options. As well they focused on the conscious or unconscious consequences of expectations and attitudes toward female students of the teachers and professionals working in the fields of science and technology. Especially crucial is the need to provide female role models for young women who want to study, or dedicate their careers, to science.

To achieve this latter objective, several recent studies have plucked from oblivion the achievements of female scientists who had been inadvertently, or deliberately, banished from the traditional narratives of the history of science and technology either because of inherent biases, or narrow conceptualizations of the disciplines. There was, in fact, a rich array of women who had made significant contributions to the world of science and technology but who hitherto were not included in history books. The role of women in the origin and development of certain disciplines and allied subjects (such as botany, medicine, and programming) were examined, along with studies on the valuable contributions made by women in the development of technoscience (scientific and literary salons, scientific popularization, and the like). In the case of technology, this sort of recuperation was more difficult because of the systematic, legally permitted, nondisclosure of the names of women who had taken out patents in many countries. Another reason was the fact that standard works in the history of science and technology have largely passed the private sphere over—that is to say, the feminine one—in which the technologies used traditionally were, and continue to be, determined by a sexual division of labor. These studies have clearly shown that, throughout history, the number of women involved in scientific and technological disciplines is lower than that of males, but that the number is not as low as has been claimed. Nevertheless, their presence has been kept hidden away because of prejudice and outdated misconceptions about how the history of science and technology should be constituted.[5]

Studies of science and gender have also been concerned with identifying sexist and androcentric biases, which have shown up in particular theories and technoscientific practices throughout history. In particular, it has become obvious in those disciplines and practices directly related to human beings—the social sciences and biology—that such biases have occurred in all steps of scientific and technological practice, but especially in the selection and definition of problems, the planning of inquiry, and the collection and interpretation of data. Biology has been the subject of numerous analyses because of its pivotal role in the maintenance of the gender organization of society: starting with what biological beings *are*, through how they *should act* socially. In this regard, biological studies on the nature of human beings have been shown to be flawed in experimental planning, in the gathering and interpretation of limited, if not contradictory, experimental data, and in unsustainable conclusions and fallacious arguments.[6]

Such criticisms readily show that older theories and practices were "bad science" in the traditional sense of the term, but it is also important to ask whether it is possible to detect that same bias in so-called good science. Analyses of the conceptual and metaphorical language used in science and technology have shown traces of sexist bias in many scientific and technological areas, which, not surprisingly, are often related to cultural ideals of masculinity, traditional rationalistic ideas, and alleged technoscientific objectivity. This also raised interesting questions regarding who learns what specific types of scientific knowledge, as well as questions regarding the neutrality and objectivity of technoscientific investigation.[7]

In the case of technology, these studies examined whether technology had, or had not, contributed to the emancipation of women. Many scholars believe Western technology has in the main been distinctly patriarchal and used to dominate women (Corea 1985; Corea et al. 1985; Grint and Gill 1995; Mies 1989; Rowland 1985). They claim that such technology, which is produced by men for use by women, can prove to be totally inappropriate for the needs of women, even to the point of being harmful as it incorporates masculine ideologies that dictate how women should live (Karpf 1987, 159).[8] At the same time, there are also techno-optimists. An example of this stance can be found in the work of Shulamith Firestone (1970). She saw contraceptive and reproductive technology as a mechanism of release for women from the tyranny of reproduction. Firestone believes the source of women's oppression is to be found in their biological being; thus, the possibility of putting a full-stop to biological motherhood, thanks to technology, was a milestone in the liberation of women.

The STS Perspective

With regard to such issues, an STS perspective incorporates a skeptical, albeit constructively critical, attitude, taking into account different points of view. For example, we can see that earlier and more recent sociocultural effects of reproductive technology are very different. Despite the fact that much reproductive technology is male designed, it is nonetheless the case that many women knowingly request technological treatment (contraceptive pills, epidural anaesthetic, amniocentesis, ultra-sound scans). Thus, at the same time that reproductive technology gives new possibilities to women, it also presents new challenges with regard to its appropriate adoption. STS can shed valuable light on this otherwise seemingly paradoxical situation.

Be it for good or evil, we all live in a world where technology is omnipresent. Technology is deeply integrated within all facets of our lives, public and private. The most sophisticated means of communication and transport, the chemical industries, genetically engineered food, domestic appliances, reproductive technology, all have had an effect on our society in different ways (polluting the environment, unequal development between the Third and Western worlds, incorporating women into the job market, or changing the very idea of family). Technology has also affected our private lives, for example, in the ways we communicate with others via the telephone and the internet. Technologies have also changed our resistance or immunity to certain illnesses and diseases by altering our metabolism, and they have established new forms of family relationships.

Unfortunately STS has primarily concentrated on the relationships of paid work and in the first stages of technological production, in effect sidelining fields such as reproduction, consumer needs and nonremunerative productivity in the home. For example, STS has given relatively less consideration to many notable (including from an economic viewpoint) inventions for the private individual's daily use, such as disposable diapers and baby-feeding bottles. In contrast STS has more often concerned itself with "significant" high-profile artifacts and projects (cars, missiles, engines, and aircraft instruments), technologies that have been largely masculine orientated. STS also tends to concentrate on the public sphere, which, until recently, was frequently prohibited territory for women, while ignoring the private sphere almost completely. The latter is, of course, largely the feminine one, in which resides those technologies related to work that women carry out. This often neglected area continues to represent an analytical challenge that STS must take up.

This challenge presents STS with a fundamental problem of demarcation: what do we and do we not consider to be technoscientific. If we accept the traditional view of what constitutes technoscience, we would be largely excluding both those artifacts and practices invented by women, and those used primarily by them. In this traditional view, the ideological connection between masculinity and technoscience is also evident because masculinity is interpreted as synonymous with technical competence. And here competence means being able to use those technologies that men design and produce, most often those previously mentioned massive projects and industrial artifacts. In this way the association of masculinity and technology is constantly being strengthened, and it further underlines the fact that masculine power over technology is both a product and a reinforcement of [t]his power in society.

It is also necessary to bear in mind the symbolic dimension of technology and how it impinges on our gender identity, which can be seen in many different spheres. For example, just recently in a press and television advertisement, there was a handsome father protectively holding a beautiful baby. It symbolized the security and the protection that the father was offering his family. As long as a man is the one portrayed as fixing a plug or the water tank under the watchful eye of his female partner, messages of inferiority and technical incompetence will be sent out to her. We transmit similar messages to our children when we give our sons presents of construction sets, machine kits, chemistry sets, or natural mineral collections, while we give our daughters Barbie Superstar dolls. The challenge is to break up this relationship between masculinity and technology in such a way that technoscience and technoscientific progress are redefined in ways that do not create dependence nor exclude anyone.[9]

Certainly one major concern within STS is to convey the idea that social and political decisions are inherent in the design and selection of any given technology, and thus that it is not merely a product of imperative technoscientific rationale. Because technology is the result of a series of specific decisions made by particular groups of people in specific places and times, with specific purposes in mind, its final form depends upon the distribution of power and the resources available inside each society.[10] Because each social group has its own particular interests and resources, the process of development often leads to conflicts arising from different opinions, for example, regarding what technical requisites should be established for a given artifact. We also should remember that the absence of influence will also shape science and technology, since a particular path will not be pursued. In a technological culture that is also a masculine culture, other less influential groups are frequently brushed aside. As a consequence, when analyzing the external factors that influence science and technology, one must, of necessity, include gender.

The Importance of Gender

The relationship between technology and gender is complex and depends on, among other things, how one views not only gender but also technology itself.

> Technology is more than a collection of physical objects or artifacts. It also fundamentally embodies a culture or set of social relations made up of certain sorts of knowledge, beliefs, desires and practices. Treating technology as a culture has enabled us to see the way in which technology is expressive of masculinity and how, in turn, men characteristically view themselves, in relation to these machines. (Wajcman 1991, 149)

Thus, if we redefine technology in such a way that it includes the gender factor, the relationship between science, technology, and society takes on another form. If we analyze, in a critical way, certain "technological scenarios," we will come to the conclusion that technology is made up of an artifact/human-being interface which configures, accommodates, interferes with, and incorporates both human beings and their technologies into a network of social relationships, but without causally determining them. That is to say that technology is designed by human beings, men and women situated in specific economic, political, and historical circumstances, who, in part because they are of different sexes, have their own specific interests, and are in their own particular power situations.

It is also important to understand that while technologies are created with certain goals in mind, the end-users will often transform how they are used or use them to perform other tasks. It is not the case that a certain technology is created in the abstract and then put to another use (be it good or bad), for technology is always created by a "designer" who has a final aim in mind. End-users adopt technologies for specific purposes from the beginning, but they may build in improvements and extensions in such a way that the original is converted into a completely *different* technology and is thus unrecognizable from what was originally intended. This is what happened with the contraceptive pill, which started off by being a treatment to control the menstrual cycle of married women in order to help them to become pregnant. Thus, it developed as a family planning aid, but at the same time it became a means by which women could enjoy their sexuality without unwanted pregnancies, and an instrument which males could use to enjoy their own sexuality without unwelcome responsibilities. Or, take the case of the Internet, which was created by the military so that the "enemy" could not intercept or make use of classified information. However, it quickly developed into a participatory support technology in the hands of feminist groups, progressive political parties, NGO groups, and the like.

Science and technology are systems that contribute to shaping our lives, for they provide a framework in which we organize and carry out our actions. They can also frame our vision of social relationships and what it means to be a human being. Various branches of technology, especially those so-called technologies of gender and sex, can both modify existing hierarchical relations, and at the same time produce new ones according to gender (Haraway 1991). Looked at from this perspective, gender is a technology too because it is "an organized system of management and control which produces and reproduces classifications and hierarchical distinctions between masculinity and femininity . . . ; it is a system of representations which assigns meaning and value to individuals in society, making them into either men or women" (Terry and Calvert 1997, 6).

Contemporary society's tendency toward globalization not only includes the economy but also science and technology. The influences of this tendency are by no means uniform nor consistent for all countries, all social classes, or for all women. Thus, STS should not neglect the differential conditions of development nor the consequences that Western technoscience may bestow on marginalized groups, especially women and less developed countries.[11] In this regard, STS as an educational program should also deal with the minimum technoscientific knowledge which everyone, irrespective of sex, class, or nationality, should have in order to make decisions. Above all, STS can, and should, play a fundamental role in making known what technoscientific practices are indispensable for the better development of human beings, and they should point out policies that are best suited to that development.

For example, according to the United Nations' *Report on Human Development*, published in 1995,[12] most educational technologies never reach the majority of women. At least sixty million young girls all over the world have no access to primary education, while the number of young boys is this situation is forty million. More than two-thirds of the 960 million adult illiterates in the whole world are women. If we believe that education increases the capacity of a person to participate in society and improve their quality of life, including attaining better jobs and higher incomes, this illiteracy situation conditions the productive and reproductive lives of all those women, as do their scant incomes, not to mention their inability to participate in the decision-making process.

The case of medical technologies is also very revealing. According to the U.N., between 1970 and 1990, the life expectancy of women living in developing countries increased by nine years. However, an African woman lives approximately twenty years less than a woman in a developed country. In those less developed countries, a third of all women between the ages of fifteen and forty-four, suffer from illnesses related to pregnancy, giving birth, abortion, or problems with their reproductive organs. According to the World Health Organization, half a million women die every year due to complications experienced during pregnancy and in giving birth. Of these, 99 percent are from less developed countries. In these countries, the mortality rate for mothers is 420 deaths for every 100,000 children born alive, which is in contrast to thirty deaths for every 100,000 live births in developed countries.[13] According to the U.N., $140 million would need to be spent each year from now until 2005 to ensure adequate accessibility to medical services and family planning to redress this imbalance. Meanwhile, the developed Western world invests enormous amounts of money in assisted reproductive technologies, such as in vitro fertilization and intracytoplasmic sperm injection, even though this technology only benefits a select few. We cannot put aside these kind of issues, not if we want STS to be an activist program that is politically and ethically honest and committed to education and research.

We all know that technoscience involves assumptions, acquisition of skills, norms of behavior, and compromises of values. That is to say, technology is not neutral, for while it responds to social necessities, it also creates them. It may solve problems, but it is also responsible for new ones. Increasingly, it gives rise to social, ethical, and political concerns out of which new technologies also emerge. One central feature in this value-laden nature of technoscience is the place of gender. Thus, gender studies needs to work closely with STS as they advance educational and political proposals that argue the necessity for a democratization of science and technology, one that includes full participation by *all* the world's citizens. Such democratization and participation cannot come about, however, if half the world's population remains excluded.[14]

Notes

1. Ziman (1984) offers a similar characterization for science.

2. Even though, as Keller (1995) suggests, we have no reason to identify gender with women—just as race is not to be identified as the black race—I will do so throughout this work, simply because gender studies has used this identification due to the fact that women have been historically and culturally tagged with gender.

3. Something similar occurs in journals. Even though the number of papers dealing with gender has increased, they are still small in overall number or are published as special issues, as was the case with "Gender Analysis and the History of Technology," special issue, *Technology and Culture* 38 (January 1997): 1–231, or else they are exiled to women studies' magazines.

4. Even though feminist thinking is by no means homogenous, I will adopt a very simple assumption—that is, one that starts from the fundamental idea that men and women have the same rights and capabilities. Although simplistic, this approach can lead far.

5. The bibliography regarding the role of women in science and technology is growing. With regard to the pedagogic aspect, see, for example, Alcalá (1998); Stanley (1990); and Kirkup and Smith Keller (1992). For comments on institutional barriers and discriminatory regulations, see Fox (1995); Kirkup and Smith Keller (1992); Pérez Sedeño (1995a, 1995b, 1996, 1997, 1998); Rossiter (1982, 1995); and Sonnert and Holton (1995). One of the latest overviews with an extensive bibliography can be found in Kohlstedt and Longino (1997).

6. Works on this subject are also very numerous—for example, Bleier (1979, 1984); Fausto-Sterling (1992), especially chap. 8; Hubbard (1992); Longino (1990, chap. 6, 7; 1995); Longino and Doell (1983); and Sayers (1982).

7. In the last ten years, there has been a plethora of epistemological studies of this type, but they have fundamentally dealt with science. For a panoramic overview, see, for example, González and Pérez Sedeño (1998); Harding (1986); Longino (1999); and Nelson and Nelson (1996).

8. This position certainly seems to advocate for an essentialism that, in my opinion, is not very well founded at all: anthropological studies of various cultures show that there are no behaviors or meanings universally associated with women (or with men), but they are constructed socially and historically. On the other hand, I find that some stances, such as ecofeminism, sound out of tune when they regard the essence of femininity as being as close as possible to nature and especially, with regard to biology and reproductive capacity—power, they say—which is exactly what has been used to keep women subordinate throughout history. However, numerous authors defend this thesis. See, for example, Merchant (1980).

9. I realize that this ideological supposition is strong and deeply rooted in our culture and produces certain effects; there is also a similar ideological bond between race and technoscience, as well as between class and technoscience.

10. Although, as is pointed out by Cowan (1997), who and how many people go to make up those groups are different in science and in technology, and they will also vary according to the epoch.

11. Women constitute a nonquantitative social minority in the same way as do other social minorities categorized by race, sexual orientation, or marginalized urban status. The same can be said of less developed countries.

12. Data are taken from Medicus Mundi (1996). I thank Eduardo de Bustos Pérez for providing me access to this information.

13. According to the Spanish National Statistics Institute, 2.9 mothers die for every 100,000 live births in Spain.

14. The present work was made possible, in part, thanks to financial help given by the CICYT of the Spanish government, Research Project PB95–0125-C06–03. I would like to thank Stephen Cutcliffe, Marta I. González, and Paloma Alcalá very much for their useful and penetrating comments. Stephen Cutcliffe, especially, has proved indispensable for an understanding of this paper.

References

Alcalá, Cortijo, and Soledad, Paloma. 1998. "Sohre los ingenios femeninos." In Alcalá, Soledad, and Pérez Sedeño, eds.

Alcalá, Cortijo, Soledad, Paloma, and Pérez Sedeño, Eulalia, eds. 1998. *Actas del I Congreso Multidisciplinar 'Ciencia y génera.'* Madrid: Universidad Complutense.

Bijker, Wiebe E., Hughes, Thomas P., and Pinch, Trevor, eds. 1987. *The Social Construction of Technological Systems: New Directions in the Sociology and History of Technology.* Cambridge, MA: MIT Press.

Bijker, Wiebe E., and Law, John. 1992. *Shaping Technology I Building Society: Studies in Sociotechnological Change.* Cambridge, MA: MIT Press.

Bleier, Ruth. 1979. "Social and Political Bias in Science: An Examination of Animal Studies and Their Generalization to Human Behaviors and Evolution." In *Genes and Gender II*, edited by Ruth Hubbard and Marian Lowe, 49–69. New York: Gordian Press.

Bleier, Ruth. 1984. *Science and Gender: A Critique of Biology and Its Theories on Women.* New York: Pergamon Press.

Corea, Gena. 1985. *The Mother Machine: Reproductive Technologies from Artificial Insemination to Artificial Wombs.* New York: Harper and Row.

Corea, Gena, Klein, Renate Duelli, Hanmer, Jalna, Holmes, Helen B., Hoskins, Betty, Kishwar, Madhu, Raymond, Janice, Rowland, Robyn, and Steinbacher, Roberta, eds. 1985. *Man-Made Woman: How New Reproductive Technologies Affect Women.* Bloomington: Indiana University Press.

Cowan, Ruth Schwartz. 1997. "Domestic Technologies: Cinderella and the Engineers." *Women's Studies International Forum* 20, no. 3: 361–371.

Fausto-Sterling, Anne. 1992. *Myths of Gender.* New York: Basic Books.

Firestone, Shulamith. 1970. *The Dialectic of Sex.* New York: William Morrow.

Fox, Mary Frank. 1995. "Women and Scientific Careers." In Jasanoff et al. 1995, 205–228.

González García, Isabel Marta, and Pérez Sedeño, Eulalia. 1998. "Ciencia, tecnología y género." *Revista Iberoamericana de Educación.*

Grint, Keith, and Gill, Rosalind, eds. 1995. *The Gender-Technology Relation: Contemporary Theory and Research.* London: Taylor and Francis.

Haraway, Donna. 1991. *Simians, Cyborgs, Women.* London: Routledge.

Harding, Sandra. 1986. *The Science Question in Feminism*. Ithaca, NY: Cornell University Press.

Hubbard, Ruth. 1992. *The Politics of Women's Biology*. New Brunswick, NJ: Rutgers University Press.

Jasanoff, Sheila, Markle, Gerald E., Petersen, James C., and Pinch, Trevor, eds. 1995. *Handbook of Science and Technology Studies*. Thousand Oaks, CA: Sage.

Karpf, A. 1987. "Recent Feminist Approaches to Women and Technology." In McNeil 1987.

Keller, Evelyn Fox. 1995. "The Origin, History, and Politics of the Subject Called 'Gender and Science': A First Person Account." In Jasanoff et al. 1995, 80–94.

Kirkup, Gill, and Smith Keller, Laurie, eds. 1992. *Inventing Women: Science, Technology, and Gender*. Cambridge: Polity.

Kohlstedt, Sally Gregory, and Longino, Helen, eds. 1997. *Women, Gender, and Science: New Directions*. Vol. 12, *Osiris*. Chicago: University of Chicago Press.

Longino, Helen. 1990. *Science as Social Knowledge*. Princeton, NJ: Princeton University Press.

Longino, Helen. 1995. "Knowledge, Bodies, and Values: Reproductive Technologies and Their Scientific Context." In *Technology and the Politics of Knowledge*, edited by Andrew Feenberg and Alistair Hannay, 195–221. Bloomington: Indiana University Press.

Longino, Helen. 1999. "Feminist Epistemology." In *Blackwell Guide to Epistemology*, edited by John Greco and Ernest Sosa. Malden, MA: Blackwell.

Longino, Helen, and Doell, Ruth. 1983. "Body, Bias and Behaviour." *Signs* 9, no. 2 (Winter): 206–227.

Medicus Mundi. 1996. *Women of the Third World*. Basel, Switzerland: Medicus Mundi.

Merchant, Carolyn. 1980. *The Death of Nature: Women, Ecology and the Scientific Revolution*. New York: Harper and Row.

McNeil, Maureen, ed. 1987. *Gender and Expertise*. London: Free Association Books.

Mies, Maria. 1987. "Why Do We Need All This? A Call against Genetic Engineering and Reproductive Technology." In *Made to Order: The Myth of Reproductive and Genetic Progress*, edited by Patricia Spallone and Deborah Lynn Steinberg, 34–47. New York: Pergamon Press.

Nelson, Lynn Hankinson, and Nelson, Jack, eds. 1996. *Feminism, Science and Philosophy of Science*. Boston: Kluwer.

Pérez Sedeño, Eulalia. 1995a. "Scientific Academic Careers of Women in Spain: History and Facts." Paper presented at the Sixth ILS Conference, Frankfort, KY.

Pérez Sedeño, Eulalia. 1995b. "La sindrome de l'Snark i altres històries." *Quaderns del Observatori de la comunicació cientifica*, no. 1: 58–70.

Pérez Sedeño, Eulalia. 1996. "Family versus Career in Women Mathematicians." In *Proceedings of the Seventh European Women in Mathematics (EWM)*, 211–219. Copenhagen/Madrid.

Pérez Sedeño, Eulalia. 1997. "Decisiones injustas, decisiones innecesarias." In *Actas del II Congreso de Coeducación en matemáticas*, 21–37. Madrid: Sociedad "Ada Lovelace." para la Coeducación en Matemáticas.

Pérez Sedeño, Eulalia. 1998. "Las amistades peligrosas." In *La construcción social de lo femenino*, edited by Amparo Gómez et al., 27–56. La Laguna: de la Universidad de La Laguna.

Rossiter, Margaret. 1982. *Women Scientists in America: Struggles and Strategies to 1940*. Baltimore: Johns Hopkins University Press.

Rossiter, Margaret. 1995. *Women Scientists in America: Before Affirmative Action*. Baltimore: Johns Hopkins University Press.

Rowland, Robin. 1985. "Motherhood, Patriarchal Power, Alienation and the Issue of 'Choice' in Sex Preselection." In Corea et al. 1985, 74–87.

Sayers, Janet. 1982. *Biological Politics: Feminist and Anti-feminist Perspectives*. New York: Tavistock.

Sonnert, Gerhard, and Holton, Gerald. 1995. *Who Succeeds in Science?* New Brunswick, NJ: Rutgers University Press.

Stanley, Autumn. 1990. "The Patent Office Clerk as Conjurer: The Vanishing Lady Trick in a XIXth Century Historical Source." In *Women, Work, and Technology*, edited by B. Dry Gulsky, 118–136. Ann Arbor: University of Michigan Press.

Terry, Jennifer, and Calvert, Melodie, eds. 1997. *Processed Lives: Gender and Technology in Everyday Life*. New York: Routledge.

Wajcman, Judy. 1991. *Feminism Confronts Technology*. University Park: Pennsylvania State University Press.

Wajcman, Judy. 1995. "Feminist Theories of Technologies." In Jasanoff et al. 1995, 189–204.

Ziman, John. 1984. *An Introduction to Science Studies: The Philosophical and Social Aspects of Science and Technology*. Cambridge: Cambridge University Press.

III TECHNOLOGY AND VALUES

The emphasis on sociotechnical systems in this book may, at first, seem academic. Why is it so important to understand that technology and society are woven together? That users shape machines? That devices cannot exist without social practices? That technology is influenced by a myriad of social factors and relies on individual and institutional behavior? One answer is that when we recognize that technology is sociotechnical ensembles, we are better able to see how human values affect and are affected by technology. If we care about making our lives better in the future, we need to understand how values influence the character of the future through technologies. This section is designed to reveal and unravel how this works. We want the future to embody values we hold dear. We want our future lives and those of future generations to be safe, comfortable, peaceful, healthy, and secure. We want socially just arrangements and sustainable practices. Understanding how values are entwined with sociotechnical systems is crucial to steering technology to a desirable future.

The idea that values can be wrapped up in sociotechnical systems may seem a bit odd at first. The traditional view is to think of technology as material objects. When technologies are abstracted out of their social and technical contexts and presented as merely material objects, they seem neutral. How could a chunk of metal embody or propagate values? However, when technologies are viewed as sociotechnical systems—networks of artifacts, people, institutions, and relationships—the values involved become clearer. As several previous readings have shown, artifacts and techniques give form to the world in which we operate and constrain and enable particular ways of behaving. Values are a part of this. Technologies shape and are shaped by values such as autonomy, control, equity, privacy, and democracy. The readings in this section provide a deeper understanding of this, along with a variety of illustrative examples.

Asserting that technologies are value-free is sometimes used as a trope to stop criticism of various kinds. When viewed through the lens of sociotechnical systems, however, this argument dissolves. Consider the claim that engineers have no ethical responsibilities because they simply manipulate matter using scientific laws. An engineer may insist that his or her job is merely to figure out what is possible and leave it to others to decide what to do. Or engineers may simply assume they are supposed to seek the one ideal, "technologically sweet" solution to any problem. The only way this view can be made plausible, however, is by completely ignoring the social context in which technology is produced and the social consequences of adopting and using particular technologies. The daily life of an engineer involves much more than just methodically applying scientific laws. To be successful, engineers must continuously take into account the social context in which they are working and in which their products, designs, and knowledge will be used. An engineer attempting to design anything must consider who will use it, how it will be used, and what effect it will have. In doing so, an engineer makes value judgments. For example, when engineers choose to enhance the safety of a product in a way that makes it slightly less convenient to use, they are making a value judgment that changes the impact that product has on the world. To be sure, information about consumer preferences might influence them, but even so, the choice to heed consumer preferences is a value decision. In a similar way, everyone else who makes decisions about technology—be it those who regulate, invest in, sell, or use technology—makes value judgments that affect which technologies succeed and what they look like, for example, by requiring that the technology be made a certain way or funding this technology rather than that one, or promoting one product over another.

To be sure, the temptation to think of technology as neutral persists. Politicians who fund projects they feel are important, corporate executives who determine which technologies should be developed and how, regulators who specify the standards to which technologies must conform, manufacturers who render designs into concrete form, and users who integrate technologies into their daily lives may believe that the technology they are working with has no influence on values. Yet they are making, using, or promoting specific technologies because they hope to realize particular states of affairs, and those states of affairs express their values.

Often people design and deploy technologies with certain uses (and users) in mind, and once integrated into society, those technologies privilege certain uses and users. Technologies can make it slightly easier for one group of people to participate in an activity and slightly more difficult for another group. For example, a machine may be designed for individuals of a certain height or strength, making it much more difficult for shorter or weaker individuals to use. Consider the design of standard scissors, which left-handed people cannot use easily, or think of buildings with entryways that make it nearly impossible for people in wheelchairs to enter.

While those involved in producing a particular technology may intentionally seek certain ends and design the technology to reinforce or reconfigure a particular set of values, in some cases certain values are privileged without any one individual actively pushing for them. For years and years, engineers simply didn't think about accommodating the needs of the disabled, and many of their designs made life incredibly difficult for those without two arms or two legs or perfect vision. Values can also be privileged in unexpected ways because it is never completely clear what the effects of a new device will be when inserted into the complexities of existing sociotechnical systems. For instance, new technologies can have unexpected consequences that are difficult to perceive at first but can accumulate over time and become quite powerful. Automobiles were initially seen as abating pollution because they reduced the number of horses needed for transport (and the manure they left behind). Over time, however, it has become clear that the emissions automobiles produce can be disastrous for the climate. Technologies can have significant effects on the values and social arrangements in a society regardless of whether a group or individual actively promotes a specific value.

In spite of all this complexity, when we begin to consciously and deliberately consider values as we design, build, use, and dream about technology, we may be able to better ensure that the decisions we make will promote our values, rather than stifle them. Most of the authors in this section hold out hope for the future. They have written on this topic because they believe that individuals who understand the ways in which sociotechnical systems work will be better prepared to build systems that positively impact the world. Recognizing that sociotechnical systems embody social values, enhance the interests of some, and constrain the interests of others empowers those who contribute to the design of technology to shape society for the better.

Questions to consider while reading the selections in this section:

1. How do established, functioning technologies affect social values? Consider examples not mentioned in the readings.
2. How do social values influence the development of technology? Explore examples not mentioned in this section.
3. Distinguish the notion of technological progress from that of social change. What are the values involved in each?
4. How can race and gender bias be reinforced (or countered) in sociotechnical systems?
5. What does the example of the Amish teach us about technology in our own societies?

12 Do Artifacts Have Politics?

Langdon Winner

In this article, Langdon Winner raises the seemingly simple question "Do artifacts have politics?" Winner isn't asking whether automobiles are Republican or Democrat but whether artifacts alter the power relationships between people. While the traditional answer is to say that artifacts are value neutral—as in the saying "Guns don't kill people, people kill people"—Winner argues to the contrary, that artifacts *are* political. He does this by showing how artifacts are integrated into social systems, how they are supported by social arrangements, and in turn how they embody certain values. He contends that artifacts can be political in two important ways. First, the deployment of a technology may become "a way of settling an issue in the affairs of a particular community." Second, artifacts may be inherently political in the sense that they require or are strongly compatible with particular kinds of political arrangements. Winner argues that technology is not simply associated with political order but in some real sense *is* political order. Technology, at least in some cases, is the method by which society is organized. Although scholars have called into question the specific facts underlying Winner's most compelling example (the Long Island bridges), the story he tells serves as a parable of the political nature of technologies. Winner's analysis points to the ways in which certain values—like racial segregation, freedom, and universal access—can be affected, promoted, or denied, either intentionally or accidentally, through the process of design, construction, and use of a technology.

No idea is more provocative in controversies about technology and society than the notion that technical things have political qualities. At issue is the claim that the machines, structures, and systems of modern material culture can be accurately judged not only for their contributions to efficiency and productivity and their positive and negative environmental side effects, but also for the ways in which they can embody specific forms of power and authority. Since ideas of this kind are a persistent and troubling presence in discussions about the meaning of technology, they deserve explicit attention.

Writing in the early 1960s, Lewis Mumford gave [a] classic statement to one version of the theme, arguing that "from late neolithic times in the Near East, right down to our own day, two technologies have recurrently existed side by side: one authoritarian, the other democratic, the first system-centered, immensely powerful, but inherently unstable, the other man-centered, relatively weak, but resourceful and durable."[1] This thesis stands at the heart of Mumford's studies of the city, architecture, and history

From Langdon Winner, "Do Artifacts Have Politics?," in *The Whale and the Reactor: A Search for Limits in an Age of High Technology* (Chicago: University of Chicago Press, 1986), 19–39.

of technics, and mirrors concerns voiced earlier in the works of Peter Kropotkin, William Morris, and other nineteenth-century critics of industrialism. During the 1970s, antinuclear and pro-solar energy movements in Europe and the United States adopted a similar notion as the centerpiece of their arguments. According to environmentalist Denis Hayes, "The increased deployment of nuclear power facilities must lead society toward authoritarianism. Indeed, safe reliance upon nuclear power as the principal source of energy may be possible only in a totalitarian state." Echoing the views of many proponents of appropriate technology and the soft energy path, Hayes contends that "dispersed solar sources are more compatible than centralized technologies with social equity, freedom and cultural pluralism."[2]

An eagerness to interpret technical artifacts in political language is by no means the exclusive property of critics of large-scale, high-technology systems. A long lineage of boosters has insisted that the biggest and best that science and industry made available were the best guarantees of democracy, freedom, and social justice. The factory system, automobile, telephone, radio, television, space program, and of course nuclear power have all at one time or another been described as democratizing, liberating forces. David Lillienthal's *T.V.A.: Democracy on the March*, for example, found this promise in the phosphate fertilizers and electricity that technical progress was bringing to rural Americans during the 1940s.[3] Three decades later Daniel Boorstin's *The Republic of Technology* extolled television for "its power to disband armies, to cashier presidents, to create a whole new democratic world—democratic in ways never before imagined, even in America."[4] Scarcely a new invention comes along that someone doesn't proclaim it as the salvation of a free society.

It is no surprise to learn that technical systems of various kinds are deeply interwoven in the conditions of modern politics. The physical arrangements of industrial production, warfare, communications, and the like have fundamentally changed the exercise of power and the experience of citizenship. But to go beyond this obvious fact and to argue that certain technologies *in themselves* have political properties seems, at first glance, completely mistaken. We all know that people have politics; things do not. To discover either virtues or evils in aggregates of steel, plastic, transistors, integrated circuits, chemicals, and the like seems just plain wrong, a way of mystifying human artifice and of avoiding the true sources, the human sources of freedom and oppression, justice and injustice. Blaming the hardware appears even more foolish than blaming the victims when it comes to judging conditions of public life.

Hence, the stern advice commonly given those who flirt with the notion that technical artifacts have political qualities: What matters is not technology itself, but the social or economic system in which it is embedded. This maxim, which in a number of variations is the central premise of a theory that can be called the social determination of technology, has an obvious wisdom. It serves as a needed corrective to those who focus uncritically upon such things as "the computer and its social impacts" but who fail to look behind technical devices to see the social circumstances of their development, deployments, and use. This view provides an antidote to naive technological determinism—the idea that technology develops as the sole result of an internal dynamic and then, unmediated by any other influence, molds society to fit its patterns. Those who have not recognized the ways in which technologies are shaped to social and economic forces have not gotten very far.

But the corrective has its own shortcomings; taken literally, it suggests that technical *things* do not matter at all. Once one has done the detective work necessary to reveal the social origins—power holders behind a particular instance of technological change—one will have explained everything of importance. This conclusion offers comfort to social scientists. It validates what they had always suspected, namely, that there is nothing distinctive about the study of technology in the first place. Hence, they can return to their standard models of social power—those of interest-group politics, bureaucratic politics, Marxist models of class struggle, and the like—and have everything they need. The social determination of technology is, in this view, essentially no different from the social determination of, say, welfare policy or taxation.

There are, however, good reasons to believe that technology is politically significant in its own right, good reasons why the standard models of social science only go so far in accounting for what is most interesting and troublesome about the subject. Much of modern social and political thought contains recurring statements of what can be called a theory of technological politics, an odd mongrel of notions often crossbred with orthodox liberal, conservative, and socialist philosophies.[5] The theory of technological politics draws attention to the momentum of large-scale sociotechnical systems, to the response of modern societies to certain technological imperatives, and to the ways human ends are powerfully transformed as they are adapted to technical means. This perspective offers a novel framework of interpretation and explanation for some of the more puzzling patterns that have taken shape in and around the growth of modern material culture. Its starting point is a decision to take technical artifacts seriously. Rather than insist that we immediately reduce everything to the interplay of social forces, the theory of technological politics suggests that we pay attention to the characteristics of technical objects and the meaning of those characteristics. A necessary complement to, rather than a replacement for, theories of the social determination of technology, this approach identifies certain technologies as political phenomena in their own right. It points us back, to borrow Edmund Husserl's philosophical injunction, *to the things themselves*.

In what follows I will outline and illustrate two ways in which artifacts can contain political properties. First are instances in which the invention, design, or arrangement of a specific technical device or system becomes a way of settling an issue in the affairs of a particular community. Seen in the proper light, examples of this kind are fairly straightforward and easily understood. Second are cases of what can be called "inherently political technologies," man-made systems that appear to require or to be strongly compatible with particular kinds of political relationships. Arguments about cases of this kind are much more troublesome and closer to the heart of the matter. By the term "politics" I mean arrangements of power and authority in human associations as well as the activities that take place within those arrangements. For my purposes here, the term "technology" is understood to mean all of modern practical artifice, but to avoid confusion I prefer to speak of "technologies" plural, smaller or larger pieces or systems of hardware of a specific kind.[6] My intention is not to settle any of the issues here once and for all, but to indicate their general dimensions and significance.

Technical Arrangements and Social Order

Anyone who has traveled the highways of America and has gotten used to the normal height of overpasses may well find something a little odd about some of the bridges over the parkways on Long Island, New York. Many of the overpasses are extraordinarily low, having as little as nine feet of clearance at the curb. Even those who happened to notice this structural peculiarity would not be inclined to attach any special meaning to it. In our accustomed way of looking at things such as roads and bridges, we see the details of form as innocuous and seldom give them a second thought.

It turns out, however, that some two hundred or so low-hanging overpasses on Long Island are there for a reason. They were deliberately designed and built that way by someone who wanted to achieve a particular social effect. Robert Moses, the master builder of roads, parks, bridges, and other public works of the 1920s to the 1970s in New York, built his overpasses according to specifications that would discourage the presence of buses on his parkways. According to evidence provided by Moses' biographer, Robert A. Caro, the reasons reflect Moses' social class bias and racial prejudice. Automobile-owning whites of "upper" and "comfortable middle" classes, as he called them, would be free to use the parkways for recreation and commuting. Poor people and blacks, who normally used public transit, were kept off the roads because the twelve-foot tall buses could not handle the overpasses. One consequence was to limit access of racial minorities and low-income groups to Jones Beach, Moses' widely acclaimed public park. Moses made doubly sure of this result by vetoing a proposed extension of the Long Island Railroad to Jones Beach.

Robert Moses' life is a fascinating story in recent U.S. political history. His dealings with mayors, governors, and presidents; his careful manipulation of legislatures, banks, labor unions, the press, and public opinion could be studied by political scientists for years. But the most important and enduring results of his work are his technologies, the vast engineering projects that give New York much of its present form. For generations after Moses' death and the alliances he forged have fallen apart, his public works, especially the highways and bridges he built to favor the use of the automobile over the development of mass transit, will continue to shape that city. Many of his monumental structures of concrete and steel embody a systematic social inequality, a way of engineering relationships among people that, after a time, became just another part of the landscape. As New York planner Lee Koppleman told Caro about the low bridges on Wantagh Parkway, "The old son of a gun had made sure that buses would *never* be able to use his goddamned parkways."[7]

Histories of architecture, city planning, and public works contain many examples of physical arrangements with explicit or implicit political purposes. One can point to Baron Haussmann's broad Parisian thoroughfares, engineered at Louis Napoleon's direction to prevent any recurrence of street fighting of the kind that took place during the revolution of 1848. Or one can visit any number of grotesque concrete buildings and huge plazas constructed on university campuses in the United States during the late 1960s and early 1970s to defuse student demonstrations. Studies of industrial machines and instruments also turn up interesting political stories, including some that violate our normal expectations about why technological innovations are made in the first place. If we suppose that new technologies are introduced to achieve increased efficiency, the history of technology shows that we will sometimes be disappointed. Technological change expresses a panoply of human motives, not the least of which is the desire of some to have dominion over others even though it may require an occasional sacrifice of cost savings and some violation of the normal standard of trying to get more from less.

One poignant illustration can be found in the history of nineteenth-century industrial mechanization. At Cyrus McCormick's reaper manufacturing plant in Chicago in the middle 1880s, pneumatic molding machines, a new and largely untested innovation, were added to the foundry at an estimated cost of $500,000. The standard economic interpretation would lead us to expect that this step was taken to modernize the plant and achieve the kind of efficiencies that mechanization brings. But historian Robert Ozanne has put the development in a broader context. At the time, Cyrus

McCormick II was engaged in a battle with the National Union of Iron Molders. He saw the addition of the new machines as a way to "weed out the bad element among the men," namely, the skilled workers who had organized the union local in Chicago.[8] The new machines, manned by unskilled laborers, actually produced inferior castings at a higher cost than the earlier process. After three years of use the machines were, in fact, abandoned, but by that time they had served their purpose—the destruction of the union. Thus, the story of these technical developments at the McCormick factory cannot be adequately understood outside the record of workers' attempts to organize, police repression of the labor movement in Chicago during that period, and the events surrounding the bombing at Haymarket Square. Technological history and U.S. political history were at that moment deeply intertwined.

In the examples of Moses' low bridges and McCormick's molding machines, one sees the importance of technical arrangements that precede the *use* of the things in question. It is obvious that technologies can be used in ways that enhance the power, authority, and privilege of some over others, for example, the use of television to sell a candidate. In our accustomed way of thinking technologies are seen as neutral tools that can be used well or poorly, for good, evil, or something in between. But we usually do not stop to inquire whether a given device might have been designed and built in such a way that it produces a set of consequences logically and temporally *prior to any of its professed uses*. Robert Moses' bridges, after all, were used to carry automobiles from one point to another; McCormick's machines were used to make metal castings; both technologies, however, encompassed purposes far beyond their immediate use. If our moral and political language for evaluating technology includes only categories having to do with tools and uses, if it does not include attention to the meaning of the designs and arrangements of our artifacts, then we will be blinded to much that is intellectually and practically crucial.

Because the point is most easily understood in the light of particular intentions embodied in physical form, I have so far offered illustrations that seem almost conspiratorial. But to recognize the political dimensions in the shapes of technology does not require that we look for conscious conspiracies or malicious intentions. The organized movement of handicapped people in the United States during the 1970s pointed out the countless ways in which machines, instruments, and structures of common use—buses, buildings, sidewalks, plumbing fixtures, and so forth—made it impossible for many handicapped persons to move freely about, a condition that systematically excluded them from public life. It is safe to say that designs unsuited for the handicapped arose more from long-standing neglect than from anyone's active intention. But once the issue was brought to public attention, it became evident that justice required a remedy. A whole range of artifacts have been redesigned and rebuilt to accommodate this minority.

Indeed, many of the most important examples of technologies that have political consequences are those that transcend the simple categories "intended" and "unintended" altogether. These are instances in which the very process of technical development is so thoroughly biased in a particular direction that it regularly produces results heralded as wonderful breakthroughs by some social interests and crushing setbacks by others. In such cases it is neither correct nor insightful to say, "Someone intended to do somebody else harm." Rather one must say that the technological deck has been stacked in advance to favor certain social interests and that some people were bound to receive a better hand than others.

The mechanical tomato harvester, a remarkable device perfected by researchers at the University of California from the late 1940s to the present offers an illustrative tale.

The machine is able to harvest tomatoes in a single pass through a row, cutting the plants from the ground, shaking the fruit loose, and (in the newest models) sorting the tomatoes electronically into large plastic gondolas that hold up to twenty-five tons of produce headed for canning factories. To accommodate the rough motion of these harvesters in the field, agricultural researchers have bred new varieties of tomatoes that are hardier, sturdier, and less tasty than those previously grown. The harvesters replace the system of handpicking in which crews of farm workers would pass through the fields three or four times, putting ripe tomatoes in lug boxes and saving immature fruit for later harvest.[9] Studies in California indicate that the use of the machine reduces costs by approximately five to seven dollars per ton as compared to hand harvesting.[10] But the benefits are by no means equally divided in the agricultural economy. In fact, the machine in the garden has in this instance been the occasion for a thorough reshaping of social relationships involved in tomato production in rural California.

By virtue of their very size and cost of more than $50,000 each, the machines are compatible only with a highly concentrated form of tomato growing. With the introduction of this new method of harvesting, the number of tomato growers declined from approximately 4,000 in the early 1960s to about 600 in 1973, and yet there was a substantial increase in tons of tomatoes produced. By the late 1970s an estimated 32,000 jobs in the tomato industry had been eliminated as a direct consequence of mechanization.[11] Thus, a jump in productivity to the benefit of very large growers has occurred at the sacrifice of other rural agricultural communities.

The University of California's research on and development of agricultural machines such as the tomato harvester eventually became the subject of a lawsuit filed by attorneys for California Rural Legal Assistance, an organization representing a group of farm workers and other interested parties. The suit charged that university officials are spending tax monies on projects that benefit a handful of private interests to the detriment of farm workers, small farmers, consumers, and rural California generally and asks for a court injunction to stop the practice. The university denied these charges, arguing that to accept them "would require elimination of all research with any potential practical application."[12]

As far as I know, no one argued that the development of the tomato harvester was the result of a plot. Two students of the controversy, William Friedland and Amy Barton, specifically exonerate the original developers of the machine and the hard tomato from any desire to facilitate economic concentration in that industry.[13] What we see here instead is an ongoing social process in which scientific knowledge, technological invention, and corporate profit reinforce each other in deeply entrenched patterns, patterns that bear the unmistakable stamp of political and economic power. Over many decades agricultural research and development in U.S. land-grant colleges and universities has tended to favor the interests of large agribusiness concerns.[14] It is in the face of such subtly ingrained patterns that opponents of innovations such as the tomato harvester are made to seem "antitechnology" or "antiprogress." For the harvester is not merely the symbol of a social order that rewards some while punishing others; it is in a true sense an embodiment of that order.

Within a given category of technological change there are, roughly speaking, two kinds of choices that can affect the relative distribution of power, authority, and privilege in a community. Often the crucial decision is a simple "yes or no" choice—are we going to develop and adopt the thing or not? In recent years many local, national, and international disputes about technology have centered on "yes or no" judgments about such things as food additives, pesticides, the building of highways, nuclear reactors, dam projects, and proposed high-tech weapons. The fundamental choice about an antiballistic missile or supersonic transport is whether or not the thing is going to join society as a piece of its operating equipment. Reasons given for and against are frequently as important as those concerning the adoption of an important new law.

A second range of choices, equally critical in many instances, has to do with specific features in the design or arrangement of a technical system after the decision to go ahead with it has already been made. Even after a utility company wins permission to build a large electric power line, important controversies can remain with respect to the placement of its route and the design of its towers; even after an organization has decided to institute a system of computers, controversies can still arise with regard to the kinds of components, programs, modes of access, and other specific features the system will include. Once the mechanical tomato harvester had been developed in its basic form, a design alteration of critical social significance—the addition of electronic sorters, for example—changed the character of the machine's effects upon the balance of wealth and power in California agriculture. Some of the most interesting research on technology and politics at present focuses upon the attempt to demonstrate in a detailed, concrete fashion how seemingly innocuous design features in mass transit systems, water projects, industrial machinery, and other technologies actually mask social choices of profound significance. Historian David Noble has studied two kinds of automated machine tool systems that have different implications for the relative power of management and labor in the industries that might employ them. He has shown that although the basic electronic and mechanical components of the record/playback and numerical control systems are similar, the choice of one design over another has crucial consequences for social struggles on the shop floor. To see the matter solely in terms of cost cutting, efficiency, or the modernization of equipment is to miss a decisive element in the story.[15]

From such examples I would offer some general conclusions. These correspond to the interpretation of technologies as "forms of life" presented in the previous chapter, filling in the explicitly political dimensions of that point of view.

The things we call "technologies" are ways of building order in our world. Many technical devices and systems important in everyday life contain possibilities for many different ways of ordering human activity. Consciously or unconsciously, deliberately or inadvertently, societies choose structures for technologies that influence how people are going to work, communicate, travel, consume, and so forth over a very long time. In the processes by which structuring decisions are made, different people are situated differently and possess unequal degrees of power as well as unequal levels of awareness. By far the greatest latitude of choice exists the very first time a particular instrument, system, or technique is introduced. Because choices tend to become strongly fixed in material equipment, economic investment, and social habit, the original flexibility vanishes for all practical purposes once the initial commitments are made. In that sense technological innovations are similar to legislative acts or political foundings that establish a framework for public order that will endure over many generations. For that reason the same careful attention one would give to the rules, roles, and relationships of politics must also be given to such things as the building of highways, the creation of television networks, and the tailoring of seemingly insignificant features on new machines. The issues that divide or unite people in society are settled not only in the institutions and practices of politics proper, but also, and less obviously, in tangible arrangements of steel and concrete, wires and semiconductors, nuts and bolts.

Inherently Political Technologies

None of the arguments and examples considered thus far addresses a stronger, more troubling claim often made in writings about technology and society—the belief that some technologies are by their very nature political in a specific way. According to this view, the adoption of a given technical system unavoidably brings with it conditions for human relationships that have a distinctive political cast—for example, centralized or decentralized, egalitarian or inegalitarian, repressive or liberating. This is ultimately what is at stake in assertions such as those of Lewis Mumford that two traditions of technology, one authoritarian, the other democratic, exist side by side in Western history. In all the cases cited above the technologies are relatively flexible in design and arrangement and variable in their effects. Although one can recognize a particular result produced in a particular setting, one can also easily imagine how a roughly similar device or system might have been built or situated with very much different political consequences. The idea we must now examine and evaluate is that certain kinds of technology do not allow such flexibility, and that to choose them is to choose unalterably a particular form of political life.

A remarkably forceful statement of one version of this argument appears in Friedrich Engels's little essay "On Authority" written in 1872. Answering anarchists who believed that authority is an evil that ought to be abolished altogether, Engels launches into a panegyric for authoritarianism, maintaining, among other things, that strong authority is a necessary condition in modern industry. To advance his case in the strongest possible way, he asks his readers to imagine that the revolution has already occurred. "Supposing a social revolution dethroned the capitalists, who now exercise their authority over the production and circulation of wealth. Supposing, to adopt entirely the point of view of the anti-authoritarians, that the land and the instruments of labour had become the collective property of the workers who use them. Will authority have disappeared or will it have only changed its form?"[16]

His answer draws upon lessons from three sociotechnical systems of his day, cotton-spinning mills, railways, and ships at sea. He observes that on its way to becoming finished thread, cotton moves through a number of different operations at different locations in the factory. The workers perform a wide variety of tasks, from running the steam engine to carrying the products from one room to another. Because these tasks must be coordinated and because the timing of the work is "fixed by the authority of the steam," laborers must learn to accept a rigid discipline. They must, according to Engels, work at regular hours and agree to subordinate their individual wills to the persons in charge of factory operations. If they fail to do so, they risk the horrifying possibility that production will come to a grinding halt. Engels pulls no punches. "The automatic machinery of a big factory," he writes, "is much more despotic than the small capitalists who employ workers ever have been."[17]

Similar lessons are adduced in Engels's analysis of the necessary operating conditions for railways and ships at sea. Both require the subordination of workers to an "imperious authority" that sees to it that things run according to plan. Engels finds that far from being an idiosyncrasy of capitalist social organization, relationships of authority and subordination arise "independently of all social organization, [and] are imposed upon us together with the material conditions under which we produce and make products circulate." Again, he intends this to be stern advice to the anarchists who, according to Engels, thought it possible simply to eradicate subordination and superordination at a single stroke. All such schemes are nonsense. The roots of unavoidable authoritarianism are, he argues, deeply implanted in the human involvement with science and technology. "If man, by dint of his knowledge and inventive genius, has subdued the forces of nature, the latter avenge themselves upon him by subjecting him, insofar as he employs them, to a veritable despotism independent of all social organization."[18]

Attempts to justify strong authority on the basis of supposedly necessary conditions of technical practice have an ancient history. A pivotal theme in the *Republic* is Plato's quest to borrow the authority of technē and employ it by analogy to buttress his argument in favor of authority in the state. Among the illustrations he chooses, like Engels, is that of a ship on the high seas. Because large sailing vessels by their very nature need to be steered with a firm hand, sailors must yield to their captain's commands; no reasonable person believes that ships can be run democratically. Plato goes on to suggest that governing a state is rather like being captain of a ship or like practicing medicine as a physician. Much the same conditions that require central rule and decisive action in organized technical activity also create this need in government.

In Engels's argument, and arguments like it, the justification for authority is no longer made by Plato's classic analogy, but rather directly with reference to technology itself. If the basic case is as compelling as Engels believed it to be, one would expect that as a society adopted increasingly complicated technical systems as its material basis, the prospects for authoritarian ways of life would be greatly enhanced. Central control by knowledgeable people acting at the top of a rigid social hierarchy would seem increasingly prudent. In this respect his stand in "On Authority" appears to be at variance with Karl Marx's position in Volume I of *Capital*. Marx tries to show that increasing mechanization will render obsolete the hierarchical division of labor and the relationships of subordination that, in his view, were necessary during the early stages of modern manufacturing. "Modern Industry," he writes, "sweeps away by technical means the manufacturing division of labor, under which each man is bound hand and foot for life to a single detail operation. At the same time, the capitalistic form of that industry reproduces this same division of labour in a still more monstrous shape; in the factory proper, by converting the workman into a living appendage of the machine."[19] In Marx's view the conditions that will eventually dissolve the capitalist division of labor and facilitate proletarian revolution are conditions latent in industrial technology itself. The differences between Marx's position in *Capital* and Engels's in his essay raise an important question for socialism: What, after all, does modern technology make possible or necessary in political life? The theoretical tension we see here mirrors many troubles in the practice of freedom and authority that had muddied the tracks of socialist revolution.

Arguments to the effect that technologies are in some sense inherently political have been advanced in a wide variety of contexts, far too many to summarize here. My reading of such notions, however, reveals there are two basic ways of stating the case. One version claims that the adoption of a given technical system actually requires the creation and maintenance of a particular set of social conditions as the operating environment of that system. Engels's position is of this kind. A similar view is offered by a contemporary writer who holds that "if you accept nuclear power plants, you also accept a techno-scientific-industrial-military elite. Without these people in charge, you could not have nuclear power."[20] In this conception some kinds of technology require their social environments to be structured in a particular way in much the same sense that an automobile requires wheels in order to move. The thing could not exist as an effective operating entity unless certain social as well as material conditions were met.

The meaning of "required" here is that of practical (rather than logical) necessity. Thus, Plato thought it a practical necessity that a ship at sea have one captain and an unquestionably obedient crew.

A second, somewhat weaker, version of the argument holds that a given kind of technology is strongly compatible with, but does not strictly require, social and political relationships of a particular stripe. Many advocates of solar energy have argued that technologies of that variety are more compatible with a democratic, egalitarian society than energy systems based on coal, oil, and nuclear power; at the same time they do not maintain that anything about solar energy requires democracy. Their case is, briefly, that solar energy is decentralizing in both a technical and political sense: technically speaking, it is vastly more reasonable to build solar systems in a disaggregated, widely distributed manner than in large-scale centralized plants; politically speaking, solar energy accommodates the attempts of individuals and local communities to manage their affairs effectively because they are dealing with systems that are more accessible, comprehensible, and controllable than huge centralized sources. In this view solar energy is desirable not only for its economic and environmental benefits, but also for the salutary institutions it is likely to permit in other areas of public life.[21]

Within both versions of the argument there is a further distinction to be made between conditions that are internal to the workings of a given technical system and those that are external to it. Engels's thesis concerns internal social relations said to be required within cotton factories and railways, for example; what such relationships mean for the condition of society at large is, for him, a separate question. In contrast, the solar advocate's belief that solar technologies are compatible with democracy pertains to the way they complement aspects of society removed from the organization of those technologies as such.

There are, then, several different directions that arguments of this kind can follow. Are the social conditions predicated said to be required by, or strongly compatible with, the workings of a given technical system? Are those conditions internal to that system or external to it (or both)? Although writings that address such questions are often unclear about what is being asserted, arguments in this general category are an important part of modern political discourse. They enter into many attempts to explain how changes in social life take place in the wake of technological innovation. More important, they are often used to buttress attempts to justify or criticize proposed courses of action involving new technology. By offering distinctly political reasons for or against the adoption of a particular technology, arguments of this kind stand apart from more commonly employed, more easily quantifiable claims about economic costs and benefits, environmental impacts, and possible risks to public health and safety that technical systems may involve. The issue here does not concern how many jobs will be created, how much income generated, how many pollutants added, or how many cancers produced. Rather, the issue has to do with ways in which choices about technology have important consequences for the form and quality of human associations.

If we examine social patterns that characterize the environments of technical systems, we find certain devices and systems almost invariably linked to specific ways of organizing power and authority. The important question is: Does this state of affairs derive from an unavoidable social response to intractable properties in the things themselves, or is it instead a pattern imposed independently by a governing body, ruling class, or some other social or cultural institution to further its own purposes?

Taking the most obvious example, the atom bomb is an inherently political artifact. As long as it exists at all, its lethal properties demand that it be controlled by a centralized, rigidly hierarchical chain of command closed to all influences that might make its workings unpredictable. The internal social system of the bomb must be authoritarian; there is no other way. The state of affairs stands as a practical necessity independent of any larger political system in which the bomb is embedded, independent of the type of regime or character of its rulers. Indeed, democratic states must try to find ways to ensure that the social structures and mentality that characterize the management of nuclear weapons do not "spin off" or "spill over" into the polity as a whole.

The bomb is, of course, a special case. The reasons very rigid relationships of authority are necessary in its immediate presence should be clear to anyone. If, however, we look for other instances in which particular varieties of technology are widely perceived to need the maintenance of a special pattern of power and authority, modern technical history contains a wealth of examples.

Alfred D. Chandler in *The Visible Hand*, a monumental study of modern business enterprise, presents impressive documentation to defend the hypothesis that the construction and day-to-day operation of many systems of production, transportation, and communication in the nineteenth and twentieth centuries require the development of particular social form—a large-scale centralized, hierarchical organization administered by highly skilled managers. Typical of Chandler's reasoning is his analysis of the growth of the railroads.

> Technology made possible fast, all-weather transportation; but safe, regular, reliable movement of goods and passengers, as well as the continuing maintenance and repair of locomotives, rolling stock, and track, roadbed, stations, roundhouses, and other equipment, required the creation of a sizable administrative organization. It meant the employment of a set of managers to supervise these functional activities over an extensive geographical area; and the appointment of an administrative command of middle and top executives to monitor, evaluate, and coordinate the work of managers responsible for the day-to-day operations.[22]

Throughout his book Chandler points to ways in which technologies used in the production and distribution of electricity, chemicals, and a wide range of industrial goods "demanded" or "required" this form of human association. "Hence, the operational requirements of railroads demanded the creation of the first administrative hierarchies in American business."[23]

Were there other conceivable ways of organizing these aggregates of people and apparatus? Chandler shows that a previously dominant social form, the small traditional family firm, simply could not handle the task in most cases. Although he does not speculate further, it is clear that he believes there is, to be realistic, very little latitude in the forms of power and authority appropriate within modern sociotechnical systems. The properties of many modern technologies—oil pipelines and refineries, for example—are such that overwhelmingly impressive economies of scale and speed are possible. If such systems are to work effectively, efficiently, quickly, and safely, certain requirements of internal social organization have to be fulfilled; the material possibilities that modern technologies make available could not be exploited otherwise. Chandler acknowledges that as one compares sociotechnical institutions of different nations, one sees "ways in which cultural attitudes, values, ideologies, political systems, and social structure affect these imperatives."[24] But the weight of argument and empirical evidence in *The Visible Hand* suggests that any significant departure from the basic pattern would be, at best, highly unlikely.

It may be that other conceivable arrangements of power and authority, for example, those of decentralized, democratic worker self-management, could prove capable of administering factories, refineries, communications systems, and railroads as well as or better than the organizations Chandler describes. Evidence from automobile assembly teams in Sweden and worker-managed plants in Yugoslavia and other countries is often presented to salvage these possibilities. Unable to settle controversies over this matter here, I merely point to what I consider to be their bone of contention. The available evidence tends to show that many large, sophisticated technological systems are in fact highly compatible with centralized, hierarchical managerial control. The interesting question, however, has to do with whether or not this pattern is in any sense a requirement of such systems, a question that is not solely empirical. The matter ultimately rests on our judgments about what steps, if any, are practically necessary in the workings of particular kinds of technology and what, if anything, such measures require of the structure of human associations. Was Plato right in saying that a ship at sea needs steering by a decisive hand and that this could only be accomplished by a single captain and an obedient crew? Is Chandler correct in saying that the properties of large-scale systems require centralized, hierarchical managerial control?

To answer such questions, we would have to examine in some detail the moral claims of practical necessity (including those advocated in the doctrines of economics) and weigh them against moral claims of other sorts, for example, the notion that it is good for sailors to participate in the command of a ship or that workers have a right to be involved in making and administering decisions in a factory. It is characteristic of societies based on large, complex technological systems, however, that moral reasons other than those of practical necessity appear increasingly obsolete, "idealistic," and irrelevant. Whatever claims one may wish to make on behalf of liberty, justice, or equality can be immediately neutralized when confronted with arguments to the effect, "Fine, but that's no way to run a railroad" (or steel mill, or airline, or communication system, and so on). Here we encounter an important quality in modern political discourse and in the way people commonly think about what measures are justified in response to the possibilities technologies make available. In many instances, to say that some technologies are inherently political is to say that certain widely accepted reasons of practical necessity—especially the need to maintain crucial technological systems as smoothly working entities—have tended to eclipse other sorts of moral and political reasoning.

One attempt to salvage the autonomy of politics from the bind of practical necessity involves the notion that conditions of human association found in the internal workings of technological systems can easily be kept separate from the polity as a whole. Americans have long rested content in the belief that arrangements of power and authority inside industrial corporations, public utilities, and the like have little bearing on public institutions, practices, and ideas at large. That "democracy stops at the factory gates" was taken as a fact of life that had nothing to do with the practice of political freedom. But can the internal politics of technology and the politics of the whole community be so easily separated? A recent study of business leaders in the United States, contemporary exemplars of Chandler's "visible hand of management," found them remarkably impatient with such democratic scruples as "one man, one vote." If democracy doesn't work for the firm, the most critical institution in all of society, American executives ask, how well can it be expected to work for the government of a nation—particularly when that government attempts to interfere with the achievements of the firm? The authors of the report observe that patterns of authority that work effectively in the corporation become for businessmen "the desirable model against which to compare political and economic relationships in the rest of society."[25] While such findings are far from conclusive, they do reflect a sentiment increasingly common in the land: what dilemmas such as the energy crisis require is not a redistribution of wealth or broader public participation but, rather, stronger, centralized public and private management.

An especially vivid case in which the operational requirements of a technical system might influence the quality of public life is the debates about the risks of nuclear power. As the supply of uranium for nuclear reactors runs out, a proposed alternative fuel is the plutonium generated as a by-product in reactor cores. Well-known objections to plutonium recycling focus on its unacceptable economic costs, its risks of environmental contamination, and its dangers in regard to the international proliferation of nuclear weapons. Beyond these concerns, however, stands another less widely appreciated set of hazards—those that involve the sacrifice of civil liberties. The wide-spread use of plutonium as a fuel increases the chance that this toxic substance might be stolen by terrorists, organized crime, or other persons. This raises the prospect, and not a trivial one, that extraordinary measures would have to be taken to safeguard plutonium from theft and to recover it should the substance be stolen. Workers in the nuclear industry as well as ordinary citizens outside could well become subject to background security checks, covert surveillance, wiretapping, informers, and even emergency measures under martial law—all justified by the need to safeguard plutonium.

Russell W. Ayres's study of the legal ramifications of plutonium recycling concludes: "With the passage of time and the increase in the quantity of plutonium in existence will come pressure to eliminate the traditional checks the courts and legislatures place on the activities of the executive and to develop a powerful central authority better able to enforce strict safeguards." He avers that "once a quantity of plutonium had been stolen, the case for literally turning the country upside down to get it back would be overwhelming." Ayres anticipates and worries about the kinds of thinking that, I have argued, characterize inherently political technologies. It is still true that in a world in which human beings make and maintain artificial systems nothing is "required" in an absolute sense. Nevertheless, once a course of action is under way, once artifacts such as nuclear power plants have been built and put in operation, the kinds of reasoning that justify the adaptation of social life to technical requirements pop up as spontaneously as flowers in the spring. In Ayres's words, "Once recycling begins and the risks of plutonium theft become real rather than hypothetical, the case for governmental infringement of protected rights will seem compelling."[26] After a certain point, those who cannot accept the hard requirements and imperatives will be dismissed as dreamers and fools.

The two varieties of interpretation I have outlined indicate how artifacts can have political qualities. In the first instance we noticed ways in which specific features in the design or arrangement of a device or system could provide a convenient means of establishing patterns of power and authority in a given setting. Technologies of this kind have a range of flexibility in the dimensions of their material form. It is precisely because they are flexible that their consequences for society must be understood with reference to the social actors able to influence which designs and arrangements are chosen. In the second instance we examined ways in which the intractable properties of certain kinds of technology are strongly, perhaps unavoidably, linked to particular institutionalized patterns of power and authority. Here the initial choice about whether or not to adopt something is decisive in regard to its consequences. There are no alternative physical designs or arrangements that would make a significant difference; there are, furthermore, no genuine possibilities for creative intervention by different social systems—capitalist or socialist—that could change the intractability of the entity or significantly alter the quality of its political effects.

To know which variety of interpretation is applicable in a given case is often what is at stake in disputes, some of them passionate ones, about the meaning of technology for how we live. I have argued a "both/and" position here, for it seems to me that both kinds of understanding are applicable in different circumstances. Indeed, it can happen that within a particular complex of technology—a system of communication or transportation, for example—some aspects may be flexible in their possibilities for society, while other aspects may be (for better or worse) completely intractable. The two varieties of interpretation I have examined here can overlap and intersect at many points.

These are, of course, issues on which people can disagree. Thus, some proponents of energy from renewable resources now believe they have at last discovered a set of intrinsically democratic, egalitarian, communitarian technologies. In my best estimation, however, the social consequences of building renewable energy systems will surely depend on the specific configurations of both hardware and the social institutions created to bring that energy to us. It may be that we will find ways to turn this silk purse into a sow's ear. By comparison, advocates of the further development of nuclear power seem to believe that they are working on a rather flexible technology whose adverse social effects can be fixed by changing the design parameters of reactors and nuclear waste disposal systems. For reasons indicated above, I believe them to be dead wrong in that faith. Yes, we may be able to manage some of the "risks" to public health and safety that nuclear power brings. But as society adapts to the more dangerous and apparently indelible features of nuclear power, what will be the long-range toll in human freedom?

My belief that we ought to attend more closely to technical objects themselves is not to say that we can ignore the contexts in which those objects are situated. A ship at sea may well require, as Plato and Engels insisted, a single captain and obedient crew. But a ship out of service, parked at the dock, needs only a caretaker. To understand which technologies and which contexts are important to us, and why, is an enterprise that must involve both the study of specific technical systems and their history as well as a thorough grasp of the concepts and controversies of political theory. In our times people are often willing to make drastic changes in the way they live to accommodate technological innovation while at the same time resisting similar kinds of changes justified on political grounds. If for no other reason than that, it is important for us to achieve a clearer view of these matters than has been our habit so far.

Notes

1. Lewis Mumford, "Authoritarian and Democratic Technics," *Technology and Culture* 5 (1964): 1–8.

2. Denis Hayes, *Rays of Hope: The Transition to a Post-Petroleum World* (New York: W. W. Norton, 1977), 71, 159.

3. David Lillienthal, *T.V.A.: Democracy on the March* (New York: Harper and Brothers, 1944), 72–83.

4. Daniel J. Boorstin, *The Republic of Technology* (New York: Harper and Row, 1978), 7.

5. Langdon Winner, *Autonomous Technology: Technics-Out-of-Control as a Theme in Political Thought* (Cambridge, MA: MIT Press, 1977).

6. The meaning of "technology" I employ in this essay does not encompass some of the broader definitions of that concept found in contemporary literature—for example, the notion of "technique" in the writings of Jacques Ellul. My purposes here are more limited. For a discussion of the difficulties that arise in attempts to define "technology," see Winner, *Autonomous Technology*, 8–12.

7. Robert A. Caro, *The Power Broker: Robert Moses and the Fall of New York* (New York: Random House, 1974), 318, 481, 514, 546, 951–958, 952.

8. Robert Ozanne, *A Century of Labor-Management Relations at McCormick and International Harvester* (Madison: University of Wisconsin Press, 1967), 20.

9. The early history of the tomato harvester is told in Wayne D. Rasmussen, "Advances in American Agriculture: The Mechanical Tomato Harvester as a Case Study," *Technology and Culture* 9 (1968): 531–543.

10. Andrew Schmitz and David Seckler, "Mechanized Agriculture and Social Welfare: The Case of the Tomato Harvester," *American Journal of Agricultural Economics* 52 (1970): 569–577.

11. William H. Friedland and Amy Barton, "Tomato Technology," *Society* 13 (September/October 1976): 6. See also William H. Friedland, *Social Sleep-Walkers: Scientific and Technological Research in California Agriculture*, Research Monograph No. 13 (Davis: University of California, Davis, Department of Applied Behavioral Sciences, 1974).

12. *University of California Clip Sheet* 54 (May 1, 1979): 36.

13. Friedland and Barton, "Tomato Technology."

14. A history and critical analysis of agricultural research in the land-grant colleges is given in James Hightower, *Hard Tomatoes, Hard Times* (Cambridge, MA: Schenkman, 1978).

15. David F. Noble, *Forces of Production: A Social History of Machine Tool Automation* (New York: Alfred A. Knopf, 1984).

16. Friedrich Engels, "On Authority," in *The Marx-Engels Reader*, 2nd ed., ed. Robert Tucker (New York: W. W. Norton, 1978), 731.

17. Ibid.

18. Ibid., 732, 731.

19. Karl Marx, *Capital*, vol. 1, 3rd. ed., trans. Samuel Moore and Edward Aveling (New York: Modern Library, 1906), 530.

20. Jerry Mander, *Four Arguments for the Elimination of Television* (New York: William Morrow, 1978), 44.

21. See, for example, Robert Argue, Barbara Emanuel, and Stephen Graham, *The Sun Builders: A People's Guide to Solar, Wind and Wood Energy in Canada* (Toronto: Renewable Energy in Canada, 1978). "We think decentralization is an implicit component of renewable energy; this implies the decentralization of energy systems, communities and of power. Renewable energy doesn't require mammoth generation sources of disruptive transmission corridors. Our cities and towns, which have been dependent on centralized energy supplies, may be

able to achieve some degree of autonomy, thereby controlling and administering their own energy needs" (16).

22. Alfred D. Chandler Jr., *The Visible Hand: The Managerial Revolution in American Business* (Cambridge, MA: Belknap, 1977), 244.

23. Ibid.

24. Ibid., 500.

25. Leonard Silk and David Vogel, *Ethics and Profits: The Crisis of Confidence in American Business* (New York: Simon and Schuster, 1976), 191.

26. Russell W. Ayres, "Policing Plutonium: The Civil Liberties Fallout," *Harvard Civil Rights—Civil Liberties Law Review* 10 (1975): 443, 413–414, 374.

13 Control: Human and Nonhuman Robots

George Ritzer

This chapter by George Ritzer makes Forster's vision of an automated world seem not too far off in the future. It is taken from his book *The McDonaldization of Society*, in which he examines how corporate management of the fast-food industry use technologies to promote their own values. Two of their most important, and interrelated, goals are efficiency and predictability. Because they do not want to rely on their employees to accept and promote these goals, fast-food management has sought to deploy technologies that either control or bypass the need to trust employees. The resulting machines, technologies, and techniques not only direct the minutia of how people work but radically alter who can perform the work and, perhaps more importantly, who would want to do it—thereby giving management even more control. Ritzer goes on to argue that these same techniques are used not only in the fast-food industry but in many other fields, including education, health care, and even farming. And not just employees are subject to such control. Those who design the technologies and techniques carefully craft the way that customers, students, patients, and even livestock experience certain situations. Ritzer's chapter leaves us with at least two important questions. First, what values are being realized when nearly every machine or technique is designed to further the interests of specific groups of people (often at the expense of other groups)? And second, what will happen as we develop technologies with new and more powerful abilities to control people?

This chapter presents the fourth dimension of McDonaldization: increased control through the replacement of human with nonhuman technology. *Technology* includes not only machines and tools but also materials, skills, knowledge, rules, regulations, procedures, and techniques. Thus, technologies include not only the obvious, such as robots and computers, but also the less obvious, such as the assembly line, bureaucratic rules, and manuals prescribing accepted procedures and techniques. A *human technology* (a screwdriver, for example) is controlled by people; *a nonhuman technology* (the order window at the drive-through, for instance) controls people.

The great source of uncertainty, unpredictability, and inefficiency in any rationalizing system is people—either those who work within it or those served by it. Hence, efforts to increase control are usually aimed at both employees and customers, although processes and products may also be the targets.

Historically, organizations gained control over people gradually through increasingly effective technologies.[1] Eventually, they began reducing people's behavior to a

From George Ritzer, "Control," in *The McDonaldization of Society: Revised New Century Edition* (Thousand Oaks, CA: Pine Forge Press, 2004), 106–133.

series of machinelike actions. And once people were behaving like machines, they could be replaced with actual machines. The replacement of humans by machines is the ultimate stage in control over people; people can cause no more uncertainty and unpredictability because they are no longer involved, at least directly, in the process.

Control is not the only goal associated with nonhuman technologies. These technologies are created and implemented for many reasons, such as increased productivity, greater quality control, and lower cost. However, this chapter is mainly concerned with the ways nonhuman technologies have increased control over employees and consumers in a McDonaldizing society.

Controlling Employees

Before the age of sophisticated nonhuman technologies, people were largely controlled by other people. In the workplace, owners and supervisors controlled subordinates directly, face-to-face. But such direct, personal control is difficult, costly, and likely to engender personal hostility. Subordinates will likely strike out at an immediate supervisor or an owner who exercises excessively tight control over their activities. Control through a technology is easier, less costly in the long run, and less likely to engender hostility toward supervisors and owners. Thus, over time, control by people has shifted toward control by technologies.[2]

The Fast-Food Industry: From Human to Mechanical Robots

Fast-food restaurants have coped with problems of uncertainty by creating and instituting many nonhuman technologies. Among other things, they have done away with a cook, at least in the conventional sense. Grilling a hamburger is so simple that anyone can do it with a bit of training. Furthermore, even when more skill is required (as in the case of cooking an Arby's roast beef), the fast-food restaurant develops a routine involving a few simple procedures that almost anyone can follow. Cooking fast food is like a game of connect-the-dots or painting-by-numbers. Following prescribed steps eliminates most of the uncertainties of cooking.

Much of the food prepared at McDonaldized restaurants arrives preformed, precut, presliced, and "preprepared." All employees need to do, when necessary, is cook or often merely heat the food and pass it on to the customer. At Taco Bell, workers used to spend hours cooking meat and shredding vegetables. Now, the workers simply drop bags of frozen ready-cooked beef into boiling water. They have used preshredded lettuce for some time, and more recently preshredded cheese and prediced tomatoes have appeared.[3] The more that is done by nonhuman technologies before the food arrives at the restaurant, the less workers need to do and the less room they have to exercise their own judgment and skill.

McDonald's has developed a variety of machines to control its employees. The soft drink dispenser has a sensor that automatically shuts off the flow when the cup is full. Ray Kroc's dissatisfaction with the vagaries of human judgment led to the elimination of french fry machines controlled by humans and to the development of machines that ring or buzz when the fries are done or that automatically lift the french fry baskets out of the hot oil. When an employee controls the french fry machine, misjudgment may lead to undercooked, overcooked, or even burned fries. Kroc fretted over this problem: "It was amazing that we got them as uniform as we did, because each kid working the fry vats would have his own interpretation of the proper color and so forth."[4]

At the cash register, workers once had to look at a price list and then punch the prices in by hand—so that the wrong (even lower) amount could be rung up. Computer screens and computerized cash registers forestall that possibility.[5] All the employees need do is press the image on the register that matches the item purchased; the machine then produces the correct price.

If the objective in a fast-food restaurant is to reduce employees to human robots, we should not be surprised by the spread of robots that prepare food. For example, a robot cooks hamburgers at one campus restaurant:

> The robot looks like a flat oven with conveyor belts running through and an arm attached at the end. A red light indicates when a worker should slide in a patty and bun, which bob along in the heat for 1 minute 52 seconds. When they reach the other side of the machine, photo-optic sensors indicate when they can be assembled.
>
> The computer functioning as the robot's brain determines when the buns and patty are where they should be. If the bun is delayed, it slows the patty belt. If the patty is delayed, it slows bun production. It also keeps track of the number of buns and patties in the oven and determines how fast they need to be fed in to keep up speed.[6]

Robots offer a number of advantages—lower cost, increased efficiency, fewer workers, no absenteeism, and a solution to the decreasing supply of teenagers needed to work at fast-food restaurants. The professor who came up with the idea for the robot that cooks hamburgers said, "Kitchens have not been looked at as factories, which they are. . . . Fast-food restaurants were the first to do that."[7]

Taco Bell developed "a computer-driven machine the size of a coffee table that . . . can make and seal in a plastic bag a perfect hot taco."[8] Another company worked on an automated drink dispenser that produced a soft drink in fifteen seconds: "Orders are punched in at the cash register by a clerk. A computer sends the order to the dispenser to drop a cup, fill it with ice and appropriate soda, and place a lid on top. The cup is then moved by conveyor to the customer."[9] When such technologies are refined and prove to be less expensive and more reliable than humans, fast-food restaurants will employ them widely.

McDonald's experimented with a limited program called ARCH, or Automated Robotic Crew Helper. A french fry robot fills the fry basket, cooks the fries, empties the basket when sensors tell it the fries are done, and even shakes the fries while they are being cooked. In the case of drinks, an employee pushes a button on the cash register to place an order. The robot then puts the proper amount of ice in the cup, moves the cup under the correct spigot, and allows the cup to fill. It then places the cup on a conveyor, which moves it to the employee, who passes it on to the customer.[10]

Like the military, fast-food restaurants have generally recruited teenagers because they surrender their autonomy to machines, rules, and procedures more easily than adults.[11] Fast-food restaurants also seek to maximize control over the work behavior of adults. Even managers are not immune from such efforts. Another aspect of McDonald's experimental ARCH program is a computerized system that, among other things, tells managers how many hamburgers or orders of french fries they will require at a given time (the lunch hour, for example). The computerized system takes away the need to make such judgments and decisions from managers.[12] Thus, "Burger production has become an exact science in which everything is regimented, every distance calculated and every dollop of ketchup monitored and tracked."[13]

Education: McChild Care Centers

Universities have developed a variety of nonhuman technologies to exert control over professors. For instance, class periods are set by the university. Students leave at the assigned time no matter where the professor happens to be in the lecture. Because the university requires grading, the professor must test students. In some universities, final grades must be submitted within forty-eight hours of the final exam, which may force professors to employ computer-graded, multiple-choice exams. Required evaluations by students may force professors to teach in a way that will lead to high ratings. The publishing demands of the tenure and promotion system may force professors to devote far less time to their teaching than they, and their students, would like.

An even more extreme version of this emphasis appears in the child care equivalent of the fast-food restaurant, KinderCare, which was founded in 1969, and now operates over 1,250 learning centers in the United States. Over 120,000 children between the ages of 6 weeks and 12 years attend the centers.[14] KinderCare tends to hire short-term employees with little or no training in child care. What these employees do in the "classroom" is largely determined by an instruction book with a ready-made curriculum. Staff members open the manual to find activities spelled out in detail for each day. Clearly, a skilled, experienced, creative teacher is not the kind of person that such "McChild" care centers seek to hire. Rather, relatively untrained employees are more easily controlled by the nonhuman technology of the omnipresent "instruction book."

Another example of organizational control over teachers is the franchised Sylvan Learning Center, often thought of as the "McDonald's of Education."[15] (There are over nine hundred Sylvan Learning Centers in the United States, Canada, and Asia.[16]) Sylvan Learning Centers are after-school centers for remedial education. The corporation "trains staff and tailors a McDonald's type uniformity, down to the U-shaped tables at which instructors work with their charges."[17] Through their training methods, rules, and technologies, for-profit systems such as the Sylvan Learning Center exert great control over their "teachers."

Health Care: Who's Deciding Our Fate?

As is the case with all rationalized systems, medicine has moved away from human toward nonhuman technologies. The two most important examples are the growing importance of bureaucratic rules and controls and the growth of modern medical machinery. For example, the prospective payment and DRG (diagnostic related groups) systems—not physicians and their medical judgment—tend to determine how long a patient must be hospitalized. Similarly, the doctor operating alone out of a black bag with a few simple tools has virtually become a thing of the past. Instead, doctors serve as dispatchers, sending patients on to the appropriate machines and specialists. Computer programs can diagnose illnesses.[18] Although it is unlikely that they will ever replace the physician, computers may one day be the initial, if not the prime, diagnostic agents. It is now even possible for people to get diagnoses, treatment, and prescriptions over the Internet with no face-to-face contact with a physician.

These and other developments in modern medicine demonstrate increasing external control over the medical profession by third-party payers, employing organizations, for-profit hospitals, health maintenance organizations (HMOs), the federal government, and "McDoctors"-like organizations. Even in its heyday the medical profession was not free of external control, but now the nature and extent of the control is changing and its degree and extent is increasing greatly. Instead of decisions being made by the mostly autonomous doctor in private practice, doctors are more likely to conform to bureaucratic rules and regulations. In bureaucracies, employees are controlled by their superiors. Physicians' superiors are increasingly likely to be professional managers and not other doctors. Also, the existence of hugely expensive medical technologies often mandates that they be used. As the machines themselves grow more sophisticated, physicians come to understand them less and are therefore less able to control them. Instead, control shifts to the technologies as well as to the experts who create and handle them.

An excellent recent example of increasing external control over physicians (and other medical personnel) is called *pathways*.[19] A pathway is a standardized series of steps prescribed for dealing with an array of medical problems. Involved are a series of "if-then" decision points—if a certain circumstance exists, the action to follow is prescribed. What physicians do in a variety of situations is determined by the pathway and *not* the individual physician. To put it in terms of this chapter, the pathway—a nonhuman technology—exerts external control over physicians.

Various terms have been used to describe pathways—standardization, "cookbook" medicine, a series of recipes, a neat package tied together with a bow, and so on—and all describe the rationalization of medical practice. The point is that there are prescribed courses of action under a wide range of circumstances. While doctors need not, indeed should not, follow a pathway at all times, they do so most of the time. A physician who spearheads the protocol movement says he grows concerned when physicians follow a pathway more than 92% of the time. While this leaves some leeway for physicians, it is clear that what they are supposed to do is predetermined in the vast majority of instances.

Let us take, for example, an asthma patient. In this case, the pathway says that if the patient's temperature rises above 101 degrees, then a complete blood count is to be ordered. A chest X ray is to be ordered under certain circumstances—if it's the patient's initial wheezing episode or if there is chest pain, respiratory distress, or a fever of over 101 degrees. And so it goes—a series of if-then steps prescribed for and controlling what physicians and other medical personnel do. While there are undoubted advantages associated with such pathways (e.g., lower likelihood of using procedures or medicines that have been shown not to work), they do tend to take decision making away from physicians. Continued reliance on such pathways is likely to adversely affect the ability of physicians to make independent decisions.

The Workplace: Do as I Say, Not as I Do

Most workplaces are bureaucracies that can be seen as large-scale nonhuman technologies. Their innumerable rules, regulations, guidelines, positions, lines of command, and hierarchies dictate what people do within the system and how they do it. The consummate bureaucrat thinks little about what is to be done: He or she simply follows the rules, deals with incoming work, and passes it on to its next stop in the system. Employees need do little more than fill out the required forms, these days most likely right on the computer screen.

At the lowest levels in the bureaucratic hierarchy ("blue-collar work"), scientific management clearly strove to limit or replace human technology. For instance, the "one best way" required workers to follow a series of predefined steps in a mindless fashion. More generally, Frederick Taylor believed that the most important part of the work world was not the employees but, rather, the organization that would plan, oversee, and control their work.

Although Taylor wanted all employees to be controlled by the organization, he accorded managers much more leeway than manual workers. It was the task of management to study the knowledge and skills of workers and to record and tabulate them and ultimately to reduce them to laws, rules, and even mathematical formulas. In other words, managers were to take a body of human skills, abilities, and knowledge and transform them into a set of nonhuman rules, regulations, and formulas. Once human skills were codified, the organization no longer needed skilled workers. Management would hire, train, and employ unskilled workers in accord with a set of strict guidelines.

In effect, then, Taylor separated "head" work from "hand" work. Prior to Taylor's day, the skilled worker had performed both. Taylor and his followers studied what was in the heads of those skilled workers, then translated that knowledge into simple, mindless routines that virtually anyone could learn and follow. Workers were thus left with little more than repetitive "hand" work. This principle remains at the base of the movement throughout our McDonaldizing society to replace human with nonhuman technology.

Behind Taylor's scientific management, and all other efforts at replacing human with nonhuman technology, lies the goal of being able to employ human beings with minimal intelligence and ability. In fact, Taylor sought to hire people who resembled animals:

> Now one of the very first requirements for a man who is fit to handle pig iron as a regular occupation is that he shall be so stupid and so phlegmatic that he more nearly resembles in his mental make-up the ox than any other type. The man who is mentally alert and intelligent is for this very reason entirely unsuited to what would, for him, be the grinding monotony of work of this character. Therefore the workman who is best suited to handling pig iron is unable to understand the real science of doing this class of work. He is so stupid that the word "percentage" has no meaning to him, and he must consequently be trained by a man more intelligent than himself into the habit of working in accordance with the laws of this science before he can be successful.[20]

Not coincidentally, Henry Ford had a similar view of the kinds of people who were to work on his assembly lines:

> Repetitive labour—the doing of one thing over and over again and always in the same way—is a terrifying prospect to a certain kind of mind. It is terrifying to me. I could not possibly do the same thing day in and day out, but to other minds, perhaps I might say to the majority of minds, repetitive operations hold no terrors. In fact, to some types of mind thought is absolutely appalling. To them the ideal job is one where creative instinct need not be expressed. The jobs where it is necessary to put in mind as well as muscle have very few takers—we always need men who like a job because it is difficult. The average worker, I am sorry to say, wants a job in which he does not have to think. Those who have what might be called the creative type of mind and who thoroughly abhor monotony are apt to imagine that all other minds are similarly restless and therefore to extend quite unwanted sympathy to the labouring man who day in and day out performs almost exactly the same operation.[21]

The kind of person sought out by Taylor was the same kind of person Ford thought would work well on the assembly line. In their view, such people would more likely submit to external technological control over their work and perhaps even crave such control.

Not surprisingly, a perspective similar to that held by Taylor and Ford can be attributed to other entrepreneurs: "The obvious irony is that the organizations built by W. Clement Stone [the founder of Combined Insurance] and Ray Kroc, both highly creative and innovative entrepreneurs, depend on the willingness of employees to follow detailed routines precisely."[22]

Many workplaces have come under the control of nonhuman technologies. In the supermarket, for example, the checker once had to read the prices marked on food products and enter them into the cash register. As with all human activities, however, the process was slow, with a chance of human error. To counter these problems, many supermarkets installed optical scanners, which "read" a code preprinted on each item. Each code number calls up a price already entered into the computer that controls the cash register. This nonhuman technology has thus reduced the number and sophistication of the tasks performed by the checker. Only the less-skilled tasks remain, such as physically scanning the food and bagging it. And even those tasks are being eliminated with the development of self-scanning and having consumers bag their groceries, especially in discount supermarkets. In other words, the work performed by the supermarket checker, when it hasn't been totally eliminated, has been *de-skilled*; that is, a decline has occurred in the amount of skill required for the job.

The nonhuman technologies in telemarketing "factories" can be even more restrictive. Telemarketers usually have scripts they must follow unerringly. The scripts are designed to handle most foreseeable contingencies. Supervisors often listen in on solicitations to make sure employees follow the correct procedures. Employees who fail to meet the quotas for the number of calls made and sales completed in a given time may be fired summarily.

Similar control is exerted over the "phoneheads," or customer service representatives, who work for many companies. Those who handle reservations for the airlines (for example, United Airlines) must log every minute spent on the job and justify each moment away from the phone. Employees have to punch a "potty button" on the phone to let management know of their intentions. Supervisors sit in an elevated "tower" in the middle of the reservations floor, "observing like [prison] guards the movements of every operator in the room." They also monitor phone calls to make sure that employees say and do what they are supposed to. This control is part of a larger process of "omnipresent supervision increasingly taking hold in so many workplaces—not just airline reservations centers but customer service departments and data-processing businesses where computers make possible an exacting level of employee scrutiny."[23] No wonder customers often deal with representatives who behave like automatons. Said one employee of United Airlines, "My body became an extension of the computer terminal that I typed the reservations into. I came to feel emptied of self."[24]

Sometimes telephone service representatives are literally prisoners. Prison inmates are now used in at least 17 states in this way, and the idea is currently on the legislative table in several more states. The attractions of prisoners are obvious—they work for very little pay and they can be controlled to a far higher degree than even the "phoneheads" discussed above. Furthermore, they can be relied on to show up for work. As one manager put it, "I need people who are there every day."[25]

Many telemarketing firms are outsourcing much of their labor overseas, especially to India, where people who are desperate for well-paying jobs are willing to accept levels of control that would be found unacceptable by many in the United States. Indian call centers afford a number of advantages, including lower wage costs than in the United States; the availability of an English-speaking, computer-literate, and college-educated workforce with a strong work ethic; and significant experience and familiarity with business processes.[26]

Following the logical progression, some companies now use computer calls instead of having people solicit us over the phone.[27] Computer voices are far more predictable and controllable than even the most rigidly controlled human operator, including prisoners and those who work in Indian call centers. Indeed, in our increasingly McDonaldized society, I have had some of my most "interesting" conversations with such computer voices.

Of course, lower-level employees are not the only ones whose problem-solving skills are lost in the transition to more nonhuman technology. I have already mentioned the controls on professors and doctors. In addition, pilots flying the modern, computerized airplane (such as the Boeing 757, 767, and 777) are being controlled and, in the process, de-skilled. Instead of flying "by the seat of their pants" or using old-fashioned autopilots for simple maneuvers, modern pilots can "push a few buttons and lean back while the plane flies to its destination and lands on a predetermined runway." Said one FAA official, "We're taking more and more of these functions out of human control and giving them to machines." These airplanes are in many ways safer and more reliable than older, less technologically advanced models. However, pilots, dependent on these technologies, may lose the ability to handle emergency situations creatively. The problem, said one airline manager, is that "I don't have computers that will do that [be creative]; I just don't."[28]

Controlling Customers

Employees are relatively easy to control, because they rely on employers for their livelihood. Customers have much more freedom to bend the rules and go elsewhere if they don't like the situations in which they find themselves. Still, McDonaldized systems have developed and honed a number of methods for controlling customers.

The Fast-Food Industry: Get the Hell out of There

Whether they go into a fast-food restaurant or use the drive-through window, customers enter a kind of conveyor system that moves them through the restaurant in the manner desired by the management. It is clearest in the case of the drive-through window (the energy for this conveyor comes from one's own automobile), but it is also true for those who enter the restaurant. Consumers know that they are supposed to line up, move to the counter, order their food, pay, carry the food to an available table, eat, gather up their debris, deposit it in the trash receptacle, and return to their cars.

Three mechanisms help to control customers:[29]

1. Customers receive cues (for example, the presence of lots of trash receptacles, especially at the exits) that indicate what is expected of them.
2. A variety of structural constraints lead customers to behave in certain ways. For example, the drive-through window, as well as the written instructions on the menu marquee at the counter (and elsewhere), gives customers few, if any, alternatives.
3. Customers have internalized taken-for-granted norms and follow them when they enter a fast-food restaurant.

When my children were young, they admonished me after we finished our meal at McDonald's (I ate in fast-food restaurants in those days before I "saw the light") for not cleaning up the debris and carting it to the trash can. My children were, in effect, serving as agents for McDonald's, teaching me the norms of behavior in such settings. I (and most others) have long-since internalized these norms, and I still dutifully follow them these days on the rare occasions that a lack of any other alternative (or the need for a clean restroom) forces me into a fast-food restaurant.

One goal of control in fast-food restaurants is to influence customers to spend their money and leave quickly. The restaurants need tables to be vacated rapidly so other diners will have a place to eat their food. A famous old chain of cafeterias, the Automat,[30] was partly undermined by people who occupied tables for hours on end.

The Automat became a kind of social center, leaving less and less room for people to eat the meals they had purchased. The deathblow was struck when street people began to monopolize the Automat's tables.

Some fast-food restaurants employ security personnel to keep street people on the move or, in the suburbs, to prevent potentially rowdy teenagers from monopolizing tables or parking lots. 7-Eleven has sought to deal with loitering teenagers outside some of its stores by playing saccharine tunes such as "Some Enchanted Evening." Said a 7-Eleven spokesperson, "They won't hang around and tap their feet to Mantovani."[31]

In some cases, fast-food restaurants have put up signs limiting a customer's stay in the restaurant (and even its parking lot), say, to twenty minutes.[32] More generally, fast-food restaurants have structured themselves so that people do not need or want to linger over meals. Easily consumed finger foods make the meal itself a quick one. Some fast-food restaurants use chairs that make customers uncomfortable after about twenty minutes.[33] Much the same effect is produced by the colors used in the decor: "Relaxation isn't the point. Getting the Hell out of there is the point. The interior colours have been chosen carefully with this end in mind. From the scarlet and yellow of the logo to the maroon of the uniform; everything clashes. It's designed to stop people from feeling so comfortable they might want to stay."[34]

Other Settings: It's Like Boot Camp

In the university, students (the "consumers" of university services) are obviously even more controlled than professors. For example, universities often give students little leeway in the courses they may take. The courses themselves, often highly structured, force the students to perform in specific ways.

Control over students actually begins long before they enter the university. Grade schools in particular have developed many ways to control students. Kindergarten has been described as an educational "boot camp."[35] Students are taught not only to obey authority but also to embrace the rationalized procedures of rote learning and objective testing. More important, spontaneity and creativity tend not to be rewarded and may even be discouraged, leading to what one expert calls "education for docility."[36] Those who conform to the rules are thought of as good students; those who don't are labeled bad students. As a general rule, the students who end up in college are the ones who have successfully submitted to the control mechanisms. Creative, independent students are often, from the educational system's point of view, "messy, expensive, and time-consuming."[37]

The clock and the lesson plan also exert control over students, especially in grade school and high school. Because of the "tyranny of the clock," a class must end at the sound of the bell, even if students are just about to comprehend something important. Because of the "tyranny of the lesson plan," a class must focus on what the plan requires for the day, no matter what the class (and perhaps the teacher) may find interesting. Imagine "a cluster of excited children examining a turtle with enormous fascination and intensity. Now children, put away the turtle, the teacher insists. We're going to have our science lesson. The lesson is on crabs."[38]

In the health care industry, the patient (along with the physician) is increasingly under the control of large, impersonal systems. For example, in many medical insurance programs, patients can no longer decide on their own to see a specialist. Rather, the patient must first see a primary-care physician who must decide whether a specialist is necessary. Because of the system's great pressure on the primary physician to keep costs down, fewer patients visit specialists and primary-care physicians perform more functions formerly handled by specialists.

The supermarket scanners that control checkers also control customers. When prices were marked on all the products, customers could calculate roughly how much they were spending as they shopped. They could also check the price on each item to be sure that they were not being overcharged at the cash register. But with scanners, it is almost impossible for consumers to keep tabs on prices and on the checkers.

Supermarkets also control shoppers with food placement. For example, supermarkets take pains to put the foods that children find attractive in places where youngsters can readily grab them (for example, low on the shelves). Also, what a market chooses to feature through sale prices and strategic placement in the store profoundly affects what is purchased. Manufacturers and wholesalers battle one another for coveted display positions, such as at the front of the market or at the "endcaps" of aisles. Foods placed in these positions will likely sell far more than they would if they were relegated to their usual positions.

Malls also exert control over customers, especially children and young adults, who are programmed by the mass media to be avid consumers. Going to the mall can become a deeply ingrained habit. Some people are reduced to what Kowinski calls "zombies," shopping the malls hour after hour, weekend after weekend.[39] More specifically, the placement of food courts, escalators, and stairs force customers to traverse corridors and pass attractive shop windows. Benches are situated so that consumers might be attracted to certain sites even though they are seeking a brief respite from the labors of consumption. The strategic placement of shops, as well as goods within shops, leads people to be attracted to products in which they might not otherwise have been interested.

Computers that respond to the human voice via voice recognition systems exert great control over people. A person receiving a collect call might be asked by the computer voice whether she will accept the charges. The computer voice requests, "Please say yes or no." Although efficient and cost-saving, such a system has its drawbacks:

> The person senses that he cannot use free-flowing speech. He's being constrained. The computer is controlling him. It can be simply frustrating. . . . People adapt to it, but only by filing it away subconsciously as another annoyance of living in our technological world.[40]

Even religion and politics are being marketed today, and like all McDonaldizing systems, they are adopting technologies that help them control the behavior of their "customers." For example, the Roman Catholic Church has its Vatican Television (which conducts about 130 televised broadcasts each year of events inside the Vatican).[41] More generally, instead of worshiping with a human preacher, millions of worshipers now "interact" with a televised image.[42] Television permits preachers to reach far more people than they could in a conventional church, so they can exert (or so they hope) greater control over what people believe and do and, in the process, extract higher contributions. TV preachers use the full panoply of techniques developed by media experts to control their viewers. Some use a format much like that of the talk shows hosted by Jay Leno or David Letterman, complete with jokes, orchestras, singers, and guests. Here is how one observer describes Vatican television: "The big advantage to the Vatican of having its own television operation . . . is that they can put their own spin on anything they produce. If you give them the cameras and give them access, they are in control."[43]

A similar point can be made about politics. The most obvious example is the use of television to market politicians and manipulate voters. Indeed, most people never see a politician except on TV, most likely in a firmly controlled format designed to communicate the exact message and image desired by the politicians and their media advisers. President Ronald Reagan raised such political marketing to an art form in the 1980s.

On many occasions, visits were set up and TV images arranged (the president in front of a flag or with a military cemetery behind him) so that the viewers and potential voters received precisely the visual message intended by Reagan's media advisers. Tightly controlled TV images are similarly important to President George W. Bush, as reflected, for example, in his landing as "co-pilot" of a jet plane on an aircraft carrier in order to announce (erroneously) the end of hostilities with Iraq in 2003. Conversely, as we saw earlier, President Reagan and especially President George W. Bush tended to avoid freewheeling press conferences where they were not in control.

Controlling the Process and the Product

In a society undergoing McDonaldization, people are the greatest threat to predictability. Control over people can be enhanced by controlling processes and products, but control over processes and products also becomes valued in itself.

Food Production, Cooking, and Vending: It Cooks Itself

Technologies designed to reduce uncertainties are found throughout the manufacture of food. For example, the mass manufacturing of bread is not controlled by skilled bakers who lavish love and attention on a few loaves of bread at a time. Such skilled bakers cannot produce enough bread to supply the needs of our society. Furthermore, the bread they do produce can suffer from the uncertainties involved in having humans do the work. The bread may, for example, turn out to be too brown or too doughy. To increase productivity and eliminate these unpredictabilities, mass producers of bread have developed an automated system in which, as in all automated systems, humans play a minimal role rigidly controlled by the technology:

> The most advanced bakeries now resemble oil refineries. Flour, water, a score of additives, and huge amounts of yeast, sugar, and water are mixed into a broth that ferments for an hour. More flour is then added, and the dough is extruded into pans, allowed to rise for an hour, then moved through a tunnel oven. The loaves emerge after eighteen minutes, to be cooled, sliced, and wrapped.[44]

In one food industry after another, production processes in which humans play little more than planning and maintenance roles have replaced those dominated by skilled craftspeople. The warehousing and shipping of food has been similarly automated.

Further along in the food production process, other nonhuman technologies have affected how food is cooked. Technologies such as ovens with temperature probes "decide" for the cook when food is done. Many ovens, coffeemakers, and other appliances can turn themselves on and off. The instructions on all kinds of packaged foods dictate precisely how to prepare and cook the food. Premixed products, such as Mrs. Dash, eliminate the need for the cook to come up with creative combinations of seasonings. Nissin Foods' Super Boil soup—"the soup that cooks itself!"—has a special compartment in the bottom of the can. A turn of a key starts a chemical reaction that eventually boils the soup.[45] Even the cookbook was designed to take creativity away from the cook and control the process of cooking.

Some rather startling technological developments have occurred in the ways in which animals are raised for food. For instance, *aquaculture*, a $57-billion-a-year business in 2000,[46] is growing dramatically because of the spiraling desire for seafood in an increasingly cholesterol-conscious population.[47] Instead of the old inefficient, unpredictable methods of harvesting fish—a lone angler casting a line or even boats catching tons of fish at a time in huge nets—we now have the much more predictable and efficient "farming" of seafood. More than 50% of the fresh salmon found in restaurants is now raised in huge sea cages off the coast of Norway.

Sea farms offer several advantages. Most generally, aquaculture allows humans to exert far greater control over the vagaries that beset fish in their natural habitat, thus producing a more predictable supply. Various drugs and chemicals increase predictability in the amount and quality of seafood. Aquaculture also permits a more predictable and efficient harvest because the creatures are confined to a limited space. In addition, geneticists can manipulate them to produce seafood more efficiently. For example, it takes a standard halibut about ten years to reach market size, but a new dwarf variety can reach the required size in only three years. Sea farms also allow for greater calculability—the greatest number of fish for the least expenditure of time, money, and energy.

Relatively small, family-run farms for raising other animals are being rapidly replaced by "factory farms."[48] The first animal to find its way into the factory farm was the chicken. Here is the way one observer describes a chicken "factory":

> A broiler producer today gets a load of 10,000, 50,000, or even more day-old chicks from the hatcheries, and puts them straight into a long, windowless shed. . . . Inside the shed, every aspect of the birds' environment is controlled to make them grow faster on less feed. Food and water are fed automatically from hoppers suspended from the roof. The lighting is adjusted. . . . For instance, there may be bright light twenty-four hours a day for the first week or two, to encourage the chicks to gain [weight] quickly. . . .
>
> Toward the end of the eight- or nine-week life of the chicken, there may be as little as half a square foot of space per chicken—or less than the area of a sheet of quarto paper for a three-and-one-half-pound bird.[49]

Among [their] other advantages, such chicken farms allow one person to raise over fifty thousand chickens.

Raising chickens this way ensures control over all aspects of the business. For instance, the chickens' size and weight are more predictable than that of free-ranging chickens. "Harvesting" chickens confined in this way is also more efficient than is catching chickens that roam over large areas.

However, confining chickens in such crowded quarters creates unpredictabilities, such as violence and even cannibalism. Farmers deal with these irrational "vices" in a variety of ways, such as dimming the lights as chickens approach full size and "debeaking" chickens so they cannot harm each other.

Some chickens are allowed to mature so they can be used for egg production. However, they receive much the same treatment as chickens raised for food. Hens are viewed as little more than "converting machines" that transform raw material (feed) into a finished product (eggs). Peter Singer describes the technology employed to control egg production:

> The cages are stacked in tiers, with food and water troughs running along the rows, filled automatically from a central supply. They have sloping wire floors. The slope . . . makes it more difficult for the birds to stand comfortably, but it causes the eggs to roll to the front of the cage where they can easily be collected . . . [and] in the more modern plants, carried by conveyor belt to a packing plant. . . . The excrement drops through [the wire floor] and can be allowed to pile up for many months until it is all removed in a single operation.[50]

This system obviously imposes great control over the production of eggs, leading to greater efficiency, to a more predictable supply, and more uniform quality than the old chicken coop.

Other animals—pigs, lambs, steers, and calves especially—are raised similarly. To prevent calves' muscles from developing, which toughens the veal, they are immediately confined to tiny stalls where they cannot exercise. As they grow, they may not even be able to turn around. Being kept in stalls also prevents the calves from eating grass, which would cause their meat to lose its pale color; the stalls are also kept free of straw, which, if eaten by the calves, would also darken the meat. "They are fed a totally liquid diet, based on nonfat milk powder with added vitamins, minerals, and growth-promoting drugs," says Peter Singer in his book, *Animal Liberation*.[51] To make sure the calves take in the maximum amount of food, they are given no water, which forces them to keep drinking their liquid food. By rigidly controlling the size of the stall and the diet, veal producers can maximize two quantifiable objectives: the production of the largest amount of meat in the shortest possible time and the creation of the tenderest, whitest, and therefore most desirable veal.

Employment of a variety of technologies obviously leads to greater control over the process by which animals produce meat, thereby increasing the efficiency, calculability, and predictability of meat production. In addition, they exert control over farm workers. Left to their own devices, ranchers might feed young steers too little or the wrong food or permit them too much exercise. In fact, in the rigidly controlled factory ranch, human ranch hands (and their unpredictabilities) are virtually eliminated.

The Ultimate Examples of Control? Birth and Death

Not just fish, chickens, and calves are being McDonaldized, but also people, especially the processes of birth and death.

Controlling Conception: Even Granny Can Conceive

Conception is rapidly becoming McDonaldized, and increasing control is being exercised over the process. For example, the problem of male impotence[52] has been attacked by the burgeoning impotence clinics, some of which have already expanded into chains,[53] and an increasingly wide array of technologies, including medicine (especially Viagra) and mechanical devices. Many males are now better able to engage in intercourse and to play a role in pregnancies that otherwise might not have occurred.

Similarly, female infertility has been ameliorated by advances in the technologies associated with artificial (more precisely, "donor"[54]) insemination, in vitro fertilization,[55] intracytoplasmic sperm injection,[56] various surgical and nonsurgical procedures associated with the Wurn technique,[57] and so on. Some fertility clinics have grown so confident that they offer a money-back guarantee if there is no live baby after three attempts.[58] For those women who still cannot become pregnant or carry to term, surrogate mothers can do the job.[59] Even postmenopausal women now have the chance of becoming pregnant ("granny pregnancies");[60] the oldest, thus far, is a sixty-five-year-old Indian woman who gave birth to a boy in April 2003.[61] These developments and many others, such as ovulation predictor home tests,[62] have made having a child far more predictable. Efficient, easy-to-use home pregnancy tests are also available to take the ambiguity out of determining whether or not a woman has become pregnant.

One of the great unpredictabilities tormenting some prospective parents is whether the baby will turn out to be a girl or a boy. Sex selection[63] clinics have opened in England, India, and Hong Kong as the first of what may eventually become a chain of "gender choice centers." The technology, developed in the early 1970s, is actually rather simple: Semen is filtered through albumen to separate sperm with male chromosomes from sperm with female chromosomes. The woman is then artificially inseminated with the desired sperm. The chances of creating a boy are 75%; a girl, 70%.[64] A new technique uses staining of sperm cells to determine which cells carry X (male) and Y (female) chromosomes. Artificial insemination or in vitro fertilization then mates the selected sperm with an egg. The U.S. lab that developed this technique is able to offer a couple an 85% chance of creating a girl; the probabilities of creating a boy are still unclear but are expected to be lower.[65] The goal is to achieve 100% accuracy in using "male" or "female" sperm to tailor the sex of the offspring to the needs and demands of the parents.

The increasing control over the process of conception delights some but horrifies others: "Being able to specify your child's sex in advance leads to nightmare visions of ordering babies with detailed specifications, like cars with automatic transmission or leather upholstery."[66] Said a medical ethicist, "Choosing a child like we choose a car is part of a consumerist mentality, the child becomes a 'product' rather than a full human being."[67] By turning a baby into just another "product" to be McDonaldized—engineered, manufactured, and commodified—people are in danger of dehumanizing the birth process.

Of course, we are just on the frontier of the McDonaldization of conception (and just about everything else). For example, the first cloned sheep, Dolly (now deceased), was created in Scotland in 1996, and other animals have since been cloned. This opened the door to the possibility of the cloning of humans. In fact, Clonaid, a Raelian sect that believes aliens populated the earth through cloning and that the destiny of humankind is to clone, recently claimed (thus far unsubstantiated) to have cloned its fifth human being.[68] Cloning involves the creation of identical copies of molecules, cells, or even entire organisms.[69] This conjures up the image of the engineering and mass production of a "cookie-cutter" race of people, all good-looking, athletic, intelligent, free of genetic defects, and so on. If everyone were to be conceived through cloning, we would be close to the ultimate in the control of this process. And a world in which everyone was the same would be a world in which they would be ready to accept a similar sameness in everything around them. Of course, this is a science fiction scenario, but the technology needed to take us down this road is already here!

Controlling Pregnancy: Choosing the Ideal Baby

Some parents wait until pregnancy is confirmed before worrying about the sex of their child. But then, amniocentesis can be used to determine whether a fetus is male or female. First used in 1968 for prenatal diagnosis, amniocentesis is a process whereby fluid is drawn from the amniotic sac, usually between the fourteenth and eighteenth weeks of pregnancy.[70] With amniocentesis, parents might choose to exert greater control over the process by aborting a pregnancy if the fetus is of the "wrong" sex. This is clearly a far less efficient technique than prepregnancy sex selection, because it occurs after conception. In fact, very few Americans (only about 5% in one study) say that they might use abortion as a method of sex selection.[71] However, amniocentesis does allow parents to know well in advance what the sex of their child will be.

Concern about a baby's sex pales in comparison to concern about the possibility of genetic defects. In addition to amniocentesis, a variety of recently developed tests can be used to determine whether a fetus carries genetic defects such as cystic fibrosis, Down syndrome, Huntington's disease, hemophilia, Tay-Sachs disease, and sickle-cell disease.[72] These newer tests include the following:

- **Chorionic villus sampling (CVS)** Generally done earlier than amniocentesis, between the tenth and twelfth weeks of pregnancy, CVS involves taking a sample from the fingerlike structures projecting from the sac that later becomes the placenta. These structures have the same genetic makeup as the fetus.[73]
- **Maternal serum alpha-fetoprotein (MSAFP) testing** A simple blood test done in the sixteenth to eighteenth weeks of pregnancy. A high level of alpha-fetoprotein might indicate spina bifida; a low level might indicate Down syndrome.
- **Ultrasound** A technology derived from sonar that provides an image of the fetus by bouncing high-frequency energy off it. Ultrasound can reveal various genetic defects, as well as many other things (sex, gestational age, and so on).

The use of all these nonhuman technologies has increased dramatically in recent years, with some (ultrasound, MSAFP) already routine practices.[74] Many other technologies for testing fetuses are also available, and others will undoubtedly be created.

If one or more of these tests indicate the existence of a genetic defect, then abortion becomes an option. Parents who choose abortion are unwilling to inflict the pain and suffering of genetic abnormality or illness on the child and on the family. Eugenicists feel that it is not rational for a society to allow genetically disabled children to be born and to create whatever irrationalities will accompany their birth. From a cost-benefit point of view (calculability), abortion is less costly than supporting a child with serious physical or mental abnormalities or problems, sometimes for a number of years. Given such logic, it makes sense for society to use the nonhuman technologies now available to discover which fetuses are to be permitted to survive and which are not. The ultimate step would be a societal ban on certain marriages and births, something that China has considered, with the goal of such a law being the reduction of the number of sick or retarded children that burden the state.[75]

Efforts to predict and repair genetic anomalies are proceeding at a rapid rate. The Human Genome Project constructed a map of 99% of the human genome's gene-containing regions.[76] When the project began, only about 100 human disease genes were known; today we know of over 140 such genes.[77] Such knowledge will allow scientists to develop new diagnostic tests and therapeutic methods. Knowledge of where each gene is and what each does will also extend the ability to test fetuses, children, and prospective mates for genetic diseases. Prospective parents who carry problematic genes may choose not to marry or not to procreate. Another possibility (and fear) is that as the technology gets cheaper and becomes more widely available, people may be able to do the testing themselves (we already have home pregnancy tests) and then make a decision to try a risky home abortion.[78] Overall, human mating and procreation will come to be increasingly affected and controlled by these new nonhuman technologies.

Controlling Childbirth: Birth as Pathology

McDonaldization and increasing control is also manifest in the process of giving birth. One measure is the decline of midwifery, a very human and personal practice. In 1900, midwives attended about half of American births, but by 1986, they attended only 4%.[79] Today, however, midwifery has enjoyed a slight renaissance because of the dehumanization and rationalization of modern childbirth practices,[80] and 6.5% of babies in the United States are now delivered by midwives.[81] When asked why they sought out midwives, women complain about things such as the "callous and neglectful treatment by the hospital staff," "labor unnecessarily induced for the convenience of the doctor," and "unnecessary cesareans for the same reason."[82]

The flip side of the decline of midwives is the increase in the control of the birth process by professional medicine,[83] especially obstetricians. It is they who are most likely to rationalize and dehumanize the birth process. Dr. Michelle Harrison, who served as a resident in obstetrics and gynecology, is but one physician willing to admit that hospital birth can be a "dehumanized process."[84]

The increasing control over childbirth is also manifest in the degree to which it has been bureaucratized. "Social childbirth," the traditional approach, once took place largely in the home, with female relatives and friends in attendance. Now, childbirth takes place almost totally in hospitals, "alone among strangers."[85] In 1900, less than 5% of U.S. births took place in hospitals; by 1940, it was 55%; and by 1960, the process was all but complete, with nearly 100% of births occurring in hospitals.[86] In more recent years, hospital chains and birthing centers have emerged, modeled after my paradigm for the rationalization process—the fast-food restaurant.

Over the years, hospitals and the medical profession have developed many standard, routinized (McDonaldized) procedures for handling and controlling childbirth. One of the best known, created by Dr. Joseph De Lee, was widely followed through the first half of the twentieth century. De Lee viewed childbirth as a disease (a "pathologic process"), and his procedures were to be followed even in the case of low-risk births:[87]

1. The patient was placed in the lithotomy position, "lying supine with legs in air, bent and wide apart, supported by stirrups."[88]
2. The mother-to-be was sedated from the first stage of labor on.
3. An episiotomy was performed to enlarge the area through which the baby must pass.[89]
4. Forceps were used to make the delivery more efficient.
5. Describing this type of procedure, one woman wrote, "Women are herded like sheep through an obstetrical assembly line, are drugged and strapped on tables where their babies are forceps delivered."[90]

De Lee's standard practice includes not only control through nonhuman technology (the procedure itself, forceps, drugs, an assembly line approach) but most of the other elements of McDonaldization—efficiency, predictability, and the irrationality of turning the human delivery room into an inhuman baby factory. The calculability that it lacked was added later in the form of Emanuel Friedman's *Friedman Curve*. This curve prescribed three rigid stages of labor. For example, the first stage was allocated exactly 8.6 hours, during which cervical dilation was to proceed from two to four centimeters.[91]

The moment that babies come into the world, they, too, are greeted by a calculable scoring system, the Apgar test. The babies receive scores of zero to two on each of five factors (for example, heart rate, color), with ten being the healthiest total score. Most babies have scores between seven and nine a minute after birth and scores of eight to ten after five minutes. Babies with scores of zero to three are considered to be in very serious trouble. Dr. Harrison wonders why medical personnel don't ask about more subjective things, such as the infant's curiosity and mood:

> A baby doesn't have to be crying for us to know it is healthy. Hold a new baby. It makes eye contact. It breathes. It sighs. The baby has color. Lift it in your arms and feel whether it has good tone or poor, strong limbs or limp ones. The baby does not have to be on a cold table to have its condition measured.[92]

The use of various nonhuman technologies in the delivery of babies has tended to ebb and flow. The use of forceps, invented in 1588, reached a peak in the United States in the 1950s, when as many as 50% of all births involved their use. However, forceps fell out of vogue, and in the 1980s, only about 15% of all births employed forceps. Many methods of drugging mothers-to-be have also been widely used. The electronic fetal monitor became popular in the 1970s. Today, ultrasound is a popular technology.

Another worrisome technology associated with childbirth is the scalpel. Many doctors routinely perform episiotomies during delivery so that the opening of the vagina does not tear or stretch unduly during pregnancy. Often done to enhance the pleasure of future sex partners and to ease the passage of the infant, episiotomies are quite debilitating and painful for the woman. Dr. Harrison expresses considerable doubt about episiotomies. "I want those obstetricians to stop cutting open women's vaginas. Childbirth is not a surgical procedure."[93]

The scalpel is also a key tool in cesarean sections. Birth, a perfectly human process, has come to be controlled by this technology (and those who wield it) in many cases.[94] The first modern "C-section" took place in 1882, but as late as 1970, only 5% of all births involved cesarean. Its use skyrocketed in the 1970s and 1980s, reaching 25% of all births in 1987 in what has been described as a "national epidemic."[95] By the mid-1990s, the practice had declined slightly, to 21%.[96] However, as of August 2002, 25% of births were once again by cesarean, first-time C-sections were at an all-time high of almost 17%, and the rate of vaginal births after a previous cesarean was down to 16.5%.[97] This latter occurred even though the American College of Obstetricians formally abandoned the time-honored idea, "Once a cesarean, always a cesarean." That is, it no longer supports the view that once a mother has a cesarean section, all succeeding births must be cesarean.

In addition, many people believe that cesareans are often performed unnecessarily. The first clue is historical data: Why do we see a sudden need for so many more cesareans? Weren't cesareans just as necessary a few decades ago? The second clue is data indicating that private patients who can pay are more likely to get cesareans than those on Medicaid (which reimburses far less) and are twice as likely as indigent patients to get cesareans.[98] Are those in higher social classes and with more income really more likely to need cesareans than those with less income and from the lower social classes?[99]

One explanation for the dramatic increase in cesareans is that they fit well with the idea of the substitution of nonhuman for human technology, but they also mesh with the other elements of the increasing McDonaldization of society:

- They are more *predictable* than the normal birth process that can occur a few weeks (or even months) early or late. It is frequently noted that cesareans generally seem to be performed before 5:30 P.M. so that physicians can be home for dinner. Similarly, well-heeled women may choose a cesarean so that the unpredictabilities of natural childbirth do not interfere with careers or social demands.
- As a comparatively simple operation, the cesarean is more *efficient* than natural childbirth, which may involve many more unforeseen circumstances.
- Cesareans births are more *calculable*, normally involving no less than twenty minutes and no more than forty-five minutes. The time required for a normal birth, especially a first birth, may be far more variable.
- Irrationalities exist, including the risks associated with surgery—anesthesia, hemorrhage, blood replacement. Compared with those who undergo a normal childbirth, women who have cesareans seem to experience more physical problems and a longer period of recuperation, and the mortality rate can be as much as twice as high. Then there are the higher costs associated with cesareans. One study indicated that physicians' costs were 68% higher and hospital costs 92% higher for cesareans compared with natural childbirth.[100]

- Cesareans are dehumanizing because a natural human process is transformed, often unnecessarily, into a nonhuman or even inhuman process in which women endure a surgical procedure. At the minimum, many of those who have cesareans are denied unnecessarily the very human experience of vaginal birth. The wonders of childbirth are reduced to the routines of a minor surgical procedure.

Controlling the Process of Dying: Designer Deaths

The months or years of decline preceding most deaths involve a series of challenges irresistible to the forces of McDonaldization. In the natural order of things, the final phase of the body's breakdown can be hugely inefficient, incalculable, and unpredictable. Why can't all systems quit at once instead of, say, the kidneys going and then the intellect and then the heart? Many a dying person has confounded physicians and loved ones by rallying and persisting longer than expected or, conversely, giving out sooner than anticipated. Our seeming lack of control in face of the dying process is pointed up in the existence of powerful death figures in myth, literature, and film.

But now we have found ways to rationalize the dying process, giving us at least the illusion of control. Consider the increasing array of nonhuman technologies designed to keep people alive long after they would have expired had they lived at an earlier time in history. In fact, some beneficiaries of these technologies would not want to stay alive under those conditions that allow them to survive (a clear irrationality). Unless the physicians are following an advance directive (a living will) that explicitly states "Do not resuscitate," or "No heroic measures," people lose control over their own dying process. Family members, too, in the absence of such directives, must bow to the medical mandate to keep people alive as long as possible.

At issue is who should be in control of the process of dying. It seems increasingly likely that the decision about who dies and when will be left to the medical bureaucracy. Of course, we can expect bureaucrats to focus on rational concerns. For instance, the medical establishment is making considerable progress in maximizing the number of days, weeks, or years a patient remains alive. However, it has been slower to work on improving the quality of life during the extra time. This focus on calculability is akin to the fast-food restaurant telling people how large its sandwiches are but saying nothing about their quality.

We can also expect an increasing reliance on nonhuman technologies. For example, computer systems may be used to assess a patient's chances of survival at any given point in the dying process—90%, 50%, 10%, and so on. The actions of medical personnel are likely to be influenced by such assessments. Thus, whether a person lives or dies may come to depend increasingly on a computer program.

As you can see, death has followed much the same path as birth. That is, the dying process has been moved out of the home and beyond the control of the dying and their families and into the hands of medical personnel and hospitals.[101] Physicians have gained a large measure of control over death just as they won control over birth, and death, like birth, is increasingly likely to take place in the hospital. In 1900, only 20% of deaths took place in hospitals; by 1977, it had reached 70%. By 1993, the number of hospital deaths was down slightly to 65%, but to that percentage must be added the increasing number of people who die in nursing homes (11%) and residences such as hospices (22%).[102] The growth of hospital chains and chains of hospices, using principles derived from the fast-food restaurant, signals death's bureaucratization, rationalization, even McDonaldization.

The McDonaldization of the dying process, as well as of birth, has spawned a series of counterreactions, efforts to cope with the excesses of rationalization. For example, as a way to humanize birth, interest in midwives has grown. However, the greatest counter-reaction has been the search for ways to regain control over our own deaths. Advance directives and living wills tell hospitals and medical personnel what they may or may not do during the dying process. Suicide societies and books such as Derek Humphry's *Final Exit* give people instructions on how to kill themselves.[103] Finally, there is the growing interest in and acceptance of euthanasia,[104] most notably the work of "Dr. Death," Jack Kevorkian, whose goal is to give back to people control over their own deaths.

However, these counterreactions themselves have elements of McDonaldization. For example, Dr. Kevorkian (now serving a ten- to twenty-five-year prison term for second-degree murder) uses a nonhuman technology, a "machine," to help people kill themselves. More generally, and strikingly, he is an advocate of a "rational policy" for the planning of death.[105] Thus, the rationalization of death is found even in the efforts to counter it. Dr. Kervorkian's opponents have noted the limitations of his rational policy:

> It is not so difficult . . . to envision a society of brave new benignity and rationality, in which a sort of humane disposal system would tidy up and whisk away to dreamland the worst-case geezers and crones. They are, after all, incredibly expensive and non-productive. . . . And they are a terrible inconvenience to that strain of the American character that has sought to impose rational control on all aspects of life. . . .
>
> Would it be so farfetched to envision a society that in the name of efficiency and convenience . . . practiced Kevorkianism as a matter of routine in every community?[106]

Conclusion

The fourth dimension of McDonaldization is control, primarily through the replacement of human with nonhuman technology. Among the many objectives guiding the development of nonhuman technologies, the most important here is increased control over the uncertainties created by people—especially employees and consumers. The ultimate in control is reached when employees are replaced by nonhuman technologies such as robots. Nonhuman technologies are also employed to control the uncertainties created by customers. The objective is to make them more pliant participants in McDonaldized processes. In controlling employees and consumers, nonhuman technologies also lead to greater control over work-related processes and finished products. However, the ultimate examples of the efforts at obtaining greater control through the use of nonhuman technologies are found in the realms of birth and death.

Clearly, the future will bring with it an increasing number of nonhuman technologies with greater ability to control people and processes. Even today, listening to audiotapes rather than reading books, for example, shifts control to those who do the reading on the tape: "The mood, pace and intonation of the words are decided for you. You can't linger or rush headlong into them anymore."[107] Military hardware such as "smart bombs" (for example, the "jdams"—joint direct attack munitions—used so frequently and with such great effect in the 2003 war with Iraq) adjust their trajectories without human intervention, but in the future, smart bombs may be developed that scan an array of targets and "decide" which one to hit. Perhaps the next great step will be the refinement of artificial intelligence, which gives machines the apparent ability to think and to make decisions as humans do.[108] Artificial intelligence promises many benefits in a wide range of areas (medicine, for example). However, it also constitutes an enormous step in de-skilling. In effect, more and more people will lose the opportunity, and perhaps the ability, to think for themselves.

Notes

1. Richard Edwards, *Contested Terrain: The Transformation of the Workplace in the Twentieth Century* (New York: Basic Books, 1979).

2. Ibid.

3. Michael Lev, "Raising Fast Food's Speed Limit," *Washington Post*, August 7, 1991, pp. D1, D4.

4. Ray Kroc, *Grinding It Out* (New York: Berkeley Medallion, 1977), pp. 131–132.

5. Eric A. Taub, "The Burger Industry Takes a Big Helping of Technology," *New York Times*, October 8, 1998, pp. 13Gff.

6. William R. Greer, "Robot Chef's New Dish: Hamburgers," *New York Times*, May 27, 1987, p. C3.

7. Ibid.

8. Michael Lev, "Taco Bell Finds Price of Success (59 cents)," *New York Times*, December 17, 1990, p. D9.

9. Calvin Sims, "Robots to Make Fast Food Chains Still Faster," *New York Times*, August 24, 1988, p. 5.

10. Chuck Murray, "Robots Roll from Plant to Kitchen," *Chicago Tribune*, October 17, 1993, pp. 3ff; "New Robots Help McDonald's Make Fast Food Faster," *Business Wire*, August 18, 1992.

11. In recent years, the shortage of a sufficient number of teenagers to keep turnover-prone fast-food restaurants adequately stocked with employees has led to a widening of the traditional labor pool of fast-food restaurants.

12. Murray, "Robots Roll from Plant to Kitchen," pp. 3ff.

13. Taub, "Burger Industry Takes a Big Helping of Technology," pp. 13Gff.

14. KinderCare, www.kindercare.com.

15. "The McDonald's of Teaching," *Newsweek*, January 7, 1985, p. 61.

16. Sylvan Learning Center, www.educate.com/about.html.

17. "McDonald's of Teaching," *Newsweek*.

18. William Stockton, "Computers That Think," *New York Times Magazine*, December 14, 1980, p. 48.

19. Bernard Wysocki Jr., "Follow the Recipe: Children's Hospital in San Diego Has Taken the Standardization of Medical Care to an Extreme," *Wall Street Journal*, April 22, 2003, p. R4ff.

20. Frederick W. Taylor, *The Principles of Scientific Management* (New York: Harper and Row, 1947), p. 59.

21. Henry Ford, *My Life and Work* (Garden City, NY: Doubleday, 1922), p. 103.

22. Robin Leidner, *Fast Food, Fast Talk: Service Work and the Routinization of Everyday Life* (Berkeley: University of California Press, 1993), p. 105.

23. Virginia A. Welch, "Big Brother Flies United," Outlook, *Washington Post*, March 5, 1995, p. C5.

24. Ibid.

25. StopJunkCalls, "Convicts Are a Dependable Workforce, Says Manager," www.stopjunkcalls.com/convict.htm.

26. "Call Centres Become Bigger," Global News Wire, *India Business Insight*, September 30, 2002.

27. Gary Langer, "Computers Reach Out, Respond to Human Voice," *Washington Post*, February 11, 1990, p. H3.

28. Carl H. Lavin, "Automated Planes Raising Concerns," *New York Times*, August 12, 1989, pp. 1, 6.

29. Leidner, *Fast Food, Fast Talk*.

30. L. B. Diehl and M. Hardart, *The Automat: The History, Recipes, and Allure of Horn and Hardart's Masterpiece* (New York: Clarkson Potter, 2002).

31. "Disenchanted Evenings," *Time*, September 3, 1990, p. 53.

32. Ester Reiter, *Making Fast Food* (Montreal: McGill-Queens University Press, 1991), p. 86.

33. Stan Luxenberg, *Roadside Empires: How the Chains Franchised America* (New York: Viking, 1985).

34. Martin Plimmer, "This Demi-paradise: Martin Plimmer Finds Food in the Fast Lane Is Not to His Taste," *Independent* (London), January 3, 1998, p. 46.

35. Harold Gracey, "Learning the Student Role: Kindergarten as Academic Boot Camp," in *Readings in Introductory Sociology*, ed. Dennis Wrong and Harold Gracey (New York: Macmillan, 1967), pp. 243–254.

36. Charles E. Silberman, *Crisis in the Classroom: The Remaking of American Education* (New York: Random House, 1970), p. 122.

37. Ibid., p. 137.

38. Ibid., p. 125.

39. William Severini Kowinski, *The Malling of America: An Inside Look at the Great Consumer Paradise* (New York: William Morrow, 1985), p. 359.

40. Gary Langer, "Computers Reach Out, Respond to Human Voice," *Washington Post*, February 11, 1990, p. H3.

41. Vatican (website), https://www.vaticannews.va/en.html [the Vatican's online activities have changed significantly since this article was written, but the types of media Ritzer refers to still occur at this website].

42. Jeffrey Hadden and Charles E. Swann, *Prime Time Preachers: The Rising Power of Televangelism* (Reading, MA: Addison-Wesley, 1981).

43. E. J. Dionne Jr., "The Vatican Is Putting Video to Work," *New York Times*, August 11, 1985, sec. 2, p. 27.

44. William Serrin, "Let Them Eat Junk," *Saturday Review*, February 2, 1980, p. 23.

45. "Super Soup Cooks Itself," *Scholastic News*, January 4, 1991, p. 3.

46. AquaSol Inc (website), www.fishfarming.com.

47. Martha Duffy, "The Fish Tank on the Farm," *Time*, December 3, 1990, pp. 107–111.

48. Peter Singer, *Animal Liberation: A New Ethic for Our Treatment of Animals* (New York: Avon, 1975).

49. Ibid., pp. 96–97.

50. Ibid., pp. 105–106.

51. Ibid., p. 123.

52. Lenore Tiefer, "The Medicalization of Impotence: Normalizing Phallocentrism," *Gender and Society* 8 (1994): 363–377.

53. Cheryl Jackson, "Impotence Clinic Grows into Chain," Business and Finance, *Tampa Tribune*, February 18, 1995, p. 1.

54. Annette Baran and Reuben Pannor, *Lethal Secrets: The Shocking Consequences and Unresolved Problems of Artificial Insemination* (New York: Warner, 1989).

55. Paula Mergenbagen DeWitt, "In Pursuit of Pregnancy," *American Demographics*, May 1993, pp. 48ff.

56. Eric Adler, "The Brave New World: It's Here Now, Where In Vitro Fertilization Is Routine and Infertility Technology Pushes Back All the Old Limitations," *Kansas City Star*, October 25, 1998, pp. G1ff.

57. Clear Passage (website), www.clearpassage.com/about_infertility_therapy.htm.

58. "No Price for Little Ones," *Financial Times*, September 28, 1998, pp. 17ff.

59. Diederika Pretorius, *Surrogate Motherhood: A Worldwide View of the Issues* (Springfield, IL: Charles C Thomas, 1994).

60. Korky Vann, "With In-Vitro Fertilization, Late-Life Motherhood Becoming More Common," *Hartford Courant*, July 7, 1997, pp. E5ff.

61. Ian MacKinnon, "Mother of Newborn Child Says She Is 65," Overseas News, *Times* (London), April 10, 2003, p. 28.

62. Angela Cain, "Home Test Kits Fill an Expanding Health Niche," *Times Union-Life and Leisure* (Albany, NY), February 12, 1995, p. 11.

63. Neil Bennett, ed., *Sex Selection of Children* (New York: Academic Press, 1983).

64. "Selecting Sex of Child," *South China Morning Post*, March 20, 1994, p. 15.

65. Randeep Ramesh, "Can You Trust That Little Glow When You Choose Sex?," *Guardian* (London), October 6, 1998, pp. 14ff; Abigail Trafford, "Is Sex Selection Wise?," *Washington Post*, September 22, 1998, pp. Z6ff; Rick Weiss, "Va. Clinic Develops System for Choosing Sex of Babies," *Washington Post*, September 10, 1998, pp. A1ff.

66. Janet Daley, "Is Birth Ever Natural?," *Times* (London), March 16, 1994, p. 18.

67. Matt Ridley, "A Boy or a Girl: Is It Possible to Load the Dice?," *Smithsonian* 24 (June 1993): 123.

68. Gina Kolata and Kenneth Chang, "For Clonaid, a Trail of Unproven Claims," *New York Times*, January 1, 2003, p. A13.

69. Roger Gosden, *Designing Babies: The Brave New World of Reproductive Technology* (New York: W. H. Freeman, 1999), p. 243.

70. Rayna Rapp, "The Power of 'Positive' Diagnosis: Medical and Maternal Discourses on Amniocentesis," in *Representations of Motherhood*, ed. Donna Bassin, Margaret Honey, and Meryle Mahrer Kaplan (New Haven, CT: Yale University Press, 1994), pp. 204–219.

71. Aliza Kolker and B. Meredith Burke, *Prenatal Testing: A Sociological Perspective* (Westport, CT: Bergin & Garvey, 1994), p. 158.

72. Jeffrey A. Kuller and Steven A. Laifer, "Contemporary Approaches to Prenatal Diagnosis," *American Family Physician* 52 (December 1996): 2277ff.

73. Kolker and Burke, *Prenatal Testing*; Ellen Domke and Al Podgorski, "Testing the Unborn: Genetic Test Pinpoints Defects, but Are There Risks?," *Chicago Sun-Times*, April 17, 1994, p. C5.

74. However, some parents do resist the rationalization introduced by fetal testing. See Shirley A. Hill, "Motherhood and the Obfuscation of Medical Knowledge," *Gender and Society* 8 (1994): 29–47.

75. Mike Chinoy, *CNN*, February 8, 1994.

76. Joan H. Marks, "The Human Genome Project: A Challenge in Biological Technology," in *Culture on the Brink: Ideologies of Technology*, ed. Gretchen Bender and Timothy Druckery (Seattle: Bay Press, 1994), pp. 99–106; R. C. Lewontin, "The Dream of the Human Genome," in Bender and Druckery, *Culture on the Brink*, pp. 107–127; "Genome Research: International Consortium Completes Human Genome Project," *Genomics & Genetics Weekly*, May 9, 2003, p. 32.

77. "Genome Research: International Consortium Completes Human Genome Project," p. 32.

78. Ridley, "Boy or a Girl," p. 123.

79. Jessica Mitford, *The American Way of Birth* (New York: Plume, 1993).

80. For a critique of midwifery from the perspective of rationalization, see Charles Krauthammer, "Pursuit of a Hallmark Moment Costs a Baby's Life," *Tampa Tribune*, May 27, 1996, p. 15.

81. Judy Foreman, "The Midwives' Time Has Come—Again," *Boston Globe*, November 2, 1998, pp. C1ff.

82. Mitford, *American Way of Birth*, p. 13.

83. Catherine Kohler Riessman, "Women and Medicalization: A New Perspective," in *Perspectives in Medical Sociology*, ed. P. Brown (Prospect Heights, IL: Waveland, 1989), pp. 190–220.

84. Michelle Harrison, *A Woman in Residence* (New York: Random House, 1982), p. 91.

85. Judith Walzer Leavitt, *Brought to Bed: Childbearing in America, 1750–1950* (New York: Oxford University Press, 1986), p. 190.

86. Ibid.

87. Paula A. Treichler, "Feminism, Medicine, and the Meaning of Childbirth," in *Body Politics: Women and the Discourses of Science*, ed. Mary Jacobus, Evelyn Fox Keller, and Sally Shuttleworth (New York: Routledge, 1990), pp. 113–138.

88. Mitford, *American Way of Birth*, p. 59.

89. An episiotomy is an incision from the vagina toward the anus to enlarge the opening needed for a baby to pass.

90. Mitford, *American Way of Birth*, p. 61.

91. Ibid., p. 143.

92. Harrison, *Woman in Residence*, p. 86.

93. Ibid., p. 113.

94. Jeanne Guillemin, "Babies by Cesarean: Who Chooses, Who Controls?," in Brown, *Perspectives in Medical Sociology*, pp. 549–558.

95. L. Silver and S. M. Wolfe, *Unnecessary Cesarean Sections: How to Cure a National Epidemic* (Washington, DC: Public Citizen Health Research Group, 1989).

96. Joane Kabak, "C Sections," *Newsday*, November 11, 1996, pp. B25ff.

97. Susan Brink, "Too Posh to Push?," Health and Medicine, *U.S. News & World Report*, August 5, 2002, p. 42.

98. Randall S. Stafford, "Alternative Strategies for Controlling Rising Cesarean Section Rates," *JAMA*, February 2, 1990, pp. 683–687.

99. Jeffrey B. Gould, Becky Davey, and Randall S. Stafford, "Socioeconomic Differences in Rates of Cesarean Sections," *New England Journal of Medicine* 321, no. 4 (July 27, 1989): 233–239; F. C. Barros et al., "Epidemic of Caesarean Sections in Brazil," *Lancet*, July 20, 1991, pp. 167–169.

100. Stafford, "Alternative Strategies for Controlling Rising Cesarean Section Rates," pp. 683–687.

101. Although, more recently, insurance and hospital practices have led to more deaths in nursing homes or even at home.

102. Sherwin B. Nuland, *How We Die: Reflections on Life's Final Chapter* (New York: Knopf, 1994), p. 255; National Center for Health Statistics, *Vital Statistics of the United States, 1992–1993, Volume II—Mortality, Part A*. (Hyattsville, MD: Public Health Service, 1995).

103. Derek Humphry, *Final Exit: The Practicalities of Self-Deliverance and Assisted Suicide for the Dying*, 3rd ed. (New York: Delta, 2002).

104. Richard A. Knox, "Doctors Accepting of Euthanasia, Poll Finds: Many Would Aid in Suicide Were It Legal," *Boston Globe*, April 23, 1998, pp. A5ff.

105. Ellen Goodman, "Kevorkian Isn't Helping 'Gentle Death,'" *Newsday*, August 4, 1992, p. 32.

106. Lance Morrow, "Time for the Ice Floe, Pop: In the Name of Rationality, Kevorkian Makes Dying—and Killing—Too Easy," *Time*, December 7, 1998, pp. 48ff.

107. Amir Muhammad, "Heard Any Good Books Lately?," *New Straits Times*, October 21, 1995, pp. 9ff.

108. Raymond Kurzweil, *The Age of Intelligent Machines* (Cambridge, MA: MIT Press, 1990).

14 White

Richard Dyer

Richard Dyer's study of photography and film technology from the nineteenth and early twentieth centuries provides a powerful example of technological decision-making that was both shaped by and in turn shaped ideas about race and racial social arrangements. Without Dyer digging into the details of technological innovation, we might simply have assumed that the refinement of photographic techniques was based solely on chemistry, mechanics, theories of light, and the development of lenses; that is, we might have presumed they were technologically determined. Yet, while these were all part of the history of photographic technology, Dyer shows that this knowledge was deployed in specific ways to enhance the features of white people and even to make them appear whiter than they might appear in real life. While we might think that cameras "don't lie" and, therefore, can't discriminate against or privilege certain races, Dyer debunks that idea. Because of the values inherent in technical decisions, photographic equipment was developed in a particular way that enhanced white but not black faces. Effects of those decisions are still felt today. We have very few images of African Americans from the early days of photography, and those we have are poor. Many Americans today have no clear images of their ancestors because of the way people chose to design film. The technologies were motivated by, and subsequently reinforced, certain ideas about race.

All technologies are at once technical in the most limited sense (to do with their material properties and functioning) and also always social (economic, cultural, ideological). Cultural historians sometimes ride roughshod over the former, unwilling to accept the stubborn resistance of matter, the sheer time and effort expended in the trial and error processes of technological discovery, the internal dynamics of technical knowledge. Yet the technically minded can also underestimate, or even entirely discount, the role of the social in technology. Why a technology is even explored, why that exploration is funded, what is actually done with the result (out of all the possible things that could be done with it), these are not determined by purely technical considerations. Given tools and media do set limitations to what can be done with them, but these are very broad; in the immediacy and instantaneity of using technologies we don't stop to consider them culturally, we just use them as we know how—but the history, the social inscription, is there all the same.

Several writers have traced the interplay of factors involved in the development of the photographic media (e.g., Altman 1984; Coleman 1985; Neale 1985; Williams 1974) and this chapter is part of that endeavour. I am trying to add two things, in

---

From Richard Dyer, *White* (New York: Routledge, 1997), 83–103.

addition to the specificity of a focus on light. The first is a sense of the racial character of technologies, supplementing the emphasis on class and gender in previous work. Thus just as perspective as an artistic technique has been argued to be implicated in an individualistic world view that privileges both men and the bourgeoisie, so I want to argue that photography and cinema, as media of light, at the very least lend themselves to privileging white people. Second, I also want to insist on the aesthetic, on the technological construction of beauty and pleasure, as well as on the representation of the world. Much historical work on media technology is concerned with how media construct images of the world. This is generally too sophisticatedly conceptualised to be concerned with anything so vulgar as whether a medium represents the world accurately (though in practice, and properly, this lingers as an issue) but is concerned with how an ideology—a way of seeing the world that serves particular social interests—is implicated in the mode of representation. I have no quarrel with this as such, but I do want to recognise that cultural media are only sometimes concerned with reality and are at least as much concerned with ideals and indulgence, that are themselves socially constructed. It is important to understand this too and, indeed, to understand how representation is actually implicated in inspirations and pleasures.

. . .

Lighting for Whiteness

The photographic media and, *a fortiori*, movie lighting assume, privilege and construct whiteness. The apparatus was developed with white people in mind and habitual use and instruction continue in the same vein, so much so that photographing non-white people is typically construed as a problem.

All technologies work within material parameters that cannot be wished away. Human skin does have different colours which reflect light differently. Methods of calculating this differ, but the degree of difference registered is roughly the same: Millerson (1972, 31), discussing colour television, gives light skin 43 per cent light reflectance and dark skin 29 per cent; Malkiewicz (1986, 53) states that "a Caucasian face has about 35 percent reflectance but a black face reflects less than 16 percent." This creates problems if shooting very light and very dark people in the same frame. Writing in *Scientific American* in 1921, Frederick Mills, "electrical illuminating engineer at the Lasky Studios," noted that

> when there are two persons in [a] scene, possibly a star and a leading player, if one has a dark make-up and the other a light, much care must be exercised in so regulating the light that it neither "burns up" the light make-up nor is of insufficient strength to light up the dark make-up. (1921, 148)

The problem is memorably attested in a racial context in school photos where either the black pupils' faces look like blobs or the white pupils have theirs bleached out.

The technology at one's disposal also sets limits. The chemistry of different stocks registers shades and colours differently. Cameras offer varying degrees of flexibility with regard to exposure (affecting their ability to take a wide lightness/darkness range). Different kinds of lighting have different colours and degrees of warmth, with concomitant effects on different skins. However, what is at one's disposal is not all that could exist. Stocks, cameras and lighting were developed taking the white face as the touchstone. The resultant apparatus came to be seen as fixed and inevitable, existing independently of the fact that it was humanly constructed. It may be—certainly was—true that photo and film apparatuses have seemed to work better with light-skinned peoples, but that is because they were made that way, not because they could be no other way.

All this is complicated still further by the habitual practices and uses of the apparatus. Certain exposures and lighting set-ups, as well as make-ups and developing processes, have become established as normal. They are constituted as the way to use the medium. Anything else becomes a departure from the norm, or even a problem. In practice, such normality is white.

The question of the relationship between the variously coloured human subject and the apparatus of photography is not simply one of accuracy. This is certainly how it is most commonly discussed, in accounts of innovation or advice to photographers and film-makers. There are indeed parameters to be recognised. If someone took a photo of me and made it look as if I had olive skin and black hair, I should be grateful but have to acknowledge that it was inaccurate. However, we also find acceptable considerable departures from how we "really" look in what we regard as accurate photos, and this must be all the more so with photography of people whom we don't know, such as celebrities, stars and models. In the history of photography and film, getting the right image meant getting the one which conformed to prevalent ideas of humanity. This included ideas of whiteness, of what colour—what range of hue—white people wanted white people to be.

The rest of this section is concerned with the way the aesthetic technology of photography and film is involved in the production of images of whiteness. I look first at the assumption of whiteness as norm at different moments of technical innovation in film history, before looking at examples of that assumption in standard technical guides to the photographic media. The section ends with a discussion of how lighting privileges white people in the image and begins to open up the analysis of the construction of whiteness through light.

Innovation in the photographic media has generally taken the human face as its touchstone, and the white face as the norm of that. The very early experimenters did not take the face as subject at all, but once they and their followers turned to portraits, and especially once photographic portraiture replaced painted portraits in popularity (from the 1840s on), the issue of the "right" technology (apparatus, consumables, practice) focused on the face and, given the clientele, the white face. Experiment with, for instance, the chemistry of photographic stock, aperture size, length of development and artificial light all proceeded on the assumption that what had to be got right was the look of the white face. This is where the big money lay, in the everyday practices of professional portraiture and amateur snapshots. By the time of film (some sixty years after the first photographs), technologies and practices were already well established. Film borrowed these, gradually and selectively, carrying forward the assumptions that had gone into them. In turn, film history involves many refinements, variations and innovations, always keeping the white face central as a touchstone and occasionally revealing this quite explicitly, when it is not implicit within such terms as "beauty," "glamour" and "truthfulness." Let me provide some instances of this.

The interactions of film stock, lighting and make-up illustrate the assumption of the white face at various points in film history. Film stock repeatedly failed to get the whiteness of the white face. The earliest stock, orthochromatic, was insensitive to red and yellow, rendering both colours dark. Charles Handley, looking back in 1954, noted that with orthochromatic stock, "even a reasonably light-red object would photograph black" (1967, 121). White skin is reasonably light-red. Fashion in make-up also had to be guarded against, as noted in one of the standard manuals of the era, Carl Louis Gregory's (1920) *Condensed Course in Motion Picture Photography*:

> Be very sparing in the use of lip rouge. Remember that red photographs black and that a heavy application of rouge shows an unnaturally black mouth on the screen. (316)

Yellow also posed problems. One derived from theatrical practices of make-up, against which Gregory inveighs in a passage of remarkable racial resonance:

> Another myth that numerous actors entertain is the yellow grease-paint theory. Nobody can explain why a performer should make-up in chinese yellow. . . . The objections to yellow are that it is non-actinic and if the actor happens to step out of the rays of the arcs for a moment or if he is shaded from the distinct force of the light by another actor, his face photographs BLACK instantly. (1920, 317; emphasis in original)

The solution to these problems was a "dreadful white make-up" (actress Geraldine Farrar, interviewed in Brownlow [1968, 418]) worn under carbon arc lights so hot that they made the make-up run, involving endless retouching. This was unpleasant for performers and exacerbated by fine dust and ultraviolet light from the arcs, making the eyes swollen and pink (so-called Klieg eyes after the Kliegl company which was the main supplier of arc lights at the time [Salt 1983, 136]). These eyes filmed big and dark, in other words, not very "white," and involved the performers in endless "trooping down to the infirmary" (Brownlow 1968, 418), constantly interrupting shooting for their well-being and to avoid the (racially) wrong look.

It would have been possible to use incandescent tungsten light instead of carbon arcs; this would have been easier to handle, cheaper (requiring fewer people to operate and using less power) and pleasanter to work with (much less hot). It would also have suited one of the qualities of orthochromatic stock, its preference for subtly modulated lighting rather than high contrast of the kind created by arcs. But incandescent tungsten light has a lot of red and yellow in it and thus tends to bring out those colours in all subjects, including white faces, with consequent blacking effect on orthochromatic stock. This was a reason for sticking with arcs, for all the expense and discomfort.

The insensitivity of orthochromatic stock to yellow also made fair hair look dark "unless you specially lit it" (cinematographer Charles Rosher, interviewed in Brownlow [1968, 262]). Gregory similarly advised:

> Yellow blonde hair photographs dark . . . the more loosely [it] is arranged the lighter it photographs, and different methods of studio lighting also affect the photographic values of hair. (1920, 317)

One of the principal benefits of the introduction of backlighting, in addition to keeping the performer clearly separate from the background, was that it ensured that blonde hair looked blonde:

> The use of backlighting on blonde hair was not only spectacular but *necessary*—it was the only way filmmakers could get blonde hair to look light-coloured on the yellow-insensitive orthochromatic stock. (Bordwell, Staiger, and Thompson 1985, 226; my emphasis)

Backlighting became part of the basic vocabulary of movie lighting. As the cinematographer Joseph Walker put it in his memoirs:

> We found [backlighting] necessary to keep the actors from blending into the background. [It] also adds a halo of highlights to the hair and brilliance to the scene. (Walker and Walker 1984, 218)

From 1926, the introduction of panchromatic stock, more sensitive to yellow, helped with some of the problems of ensuring white people looked properly white, as well as permitting the use of incandescent tungsten, but posed its own problems of make-up. It was still not so sensitive to red, but much more to blue. Max Factor recognised this problem, developing a make-up that would "add to the face sufficient blue coloration in proportion to red . . . in order to prevent excessive absorption of light by the face" (Factor 1937, 54); faces that absorb light "excessively" are of course dark ones.

Colour brought with it a new set of problems, explored in Brian Winston's article on the invention of "colour film that more readily photographs Caucasians than other human types" (1985, 106). Winston argues that at each stage the search for a colour film stock (including the development process, crucial to the subtractive systems that have proved most workable) was guided by how it rendered white flesh tones. Not long after the introduction of colour in the mid-1930s, the cinematographer Joseph Valentine commented that "perhaps the most important single factor in dramatic cinematography is the relation between the colour sensitivity of an emulsion and the reproduction of pleasing flesh tones" (1939, 54). Winston looks at one such example of the search for "pleasing flesh tones" in researches undertaken by Kodak in the early 1950s. A series of prints of "a young lady" were prepared and submitted to a panel, and a report observed:

> Optimum reproduction of skin colour is not "exact" reproduction . . . "exact reproduction" is rejected almost unanimously as "beefy." On the other hand, when the print of highest acceptance is masked and compared with the original subject, it seems quite pale. (David L. MacAdam 1951, quoted in Winston 1985, 120)

As noted above, white skin is taken as a norm but what that means in terms of colour is determined not by how it is but by how, as Winston (1985, 121) puts it, it is "preferred—a whiter shade of white." Characteristically too, it is a woman's skin which provides the litmus test.

Colour film was a possibility from 1896 (when R. W. Paul showed his hand-tinted prints), with Technicolor, the "first entirely successful colour process used in the cinema," available from 1917 (Coe 1981, 112–139). Yet it did not become anything like a norm until the 1950s, for a complex of economic, technological and aesthetic reasons (cf. Kindem 1979), among which was a sense that colour film was not realistic. As Gorham Kindem suggests, this may have been partly due to a real limitation of the processes adopted from the late 1920s, in that they "could not reproduce all the colours of the visible spectrum" (1979, 35) but it also had to do with an early association with musicals and spectacle. The way Kindem elaborates this point is racially suggestive:

> While flesh tones, the most important index of accuracy and consistency, might be carefully controlled through heavy make-up, practically dictating the overall colour appearance, it is quite likely that other colour in the set or location had to be sacrificed and appeared unnatural or "gaudy." (1979, 35)

As noted elsewhere, accurate flesh tones are again the key issue in innovation. The tones involved here are evidently white, for it was lighting the compensatory heavy make-up with sufficient force to ensure a properly white look that was liable to make everything else excessively bright and "gaudy." Kindem relates a resistance to such an excess of colour with growing pessimism and cynicism through the 1930s as the weight of the Depression took a hold, to which black and white seemed more appropriate. Yet this seems to emphasise the gangster and social problem films of the 1930s over and above the comedies, musicals, fantasies and adventure films (think screwball, Fred and

Ginger, Tarzan) that were, all the same, made in black and white. May it not be that what was not acceptable was escapism that was visually too loud and busy, because excess colour, and the very word "gaudy," was associated with, indeed, coloured people?

A last example of the operation of the white face as a control on media technology comes from professional television production in the USA.[1] In the late 1970s the WGBH Educational Foundation and the 3M Corporation developed a special television signal, to be recorded on videotape, for the purpose of evaluating tapes. This signal, known as "skin," was of a pale orange colour and was intended to duplicate the appearance on a television set of white skin. The process of scanning was known as "skinning." Operatives would watch the blank pale orange screen produced by tapes prerecorded with the "skin" signal, making notes whenever a visible defect appeared. The fewer defects, the greater the value of the tape (reckoned in several hundreds of dollars) and thus when and by whom it was used. The whole process centred on blank images representing nothing, and yet founded in the most explicit way on a particular human flesh colour.

The assumption that the normal face is a white face runs through most published advice given on photo- and cinematography.[2] This is carried above all by illustrations which invariably use a white face, except on those rare occasions when they are discussing the "problem" of dark-skinned people.

. . .

The aesthetic technology of photography, as it has been invented, refined and elaborated, and the dominant uses of that technology, as they have become fixed and naturalised, assume and privilege the white subject. They also construct that subject, that is, draw on and contribute to a perception of what it means to be white. They do this as part of a much more general culture of light. This has produced both an astonishing set of technologies of light and certain fundamental philosophical, scientific and aesthetic perceptions of the nature of light. White people are central to it, to the extent that they come to seem to have a special relationship to light.

Notes

1. I am grateful to William Spurlin of the Visual and Environmental Studies Department at Harvard University for telling me of this, explaining it, and commenting on drafts of this paragraph.

2. This observation is based on analysis of a random cross-section of such books published this century.

References

Altman, Rick. 1984. "Towards a Theory of the History of Representational Technologies." *Iris* 2 (2): 111–124.

Bordwell, David, Staiger, Janet, and Thompson, Kristin. 1985. *The Classical Hollywood Cinema: Film Style and Mode of Production to 1960*. New York: Columbia University Press.

Brownlow, Kevin. 1968. *The Parade's Gone By*. London: Secker & Warburg.

Coe, Brian. 1981. *The History of Movie Photography*. London: Ash & Grant.

Coleman, A. D. 1985. "Lentil Soup." *Et cetera*, spring, 19–31.

Factor, M. 1937. "Standardization of Motion Picture Make-up." *Journal of the Society of Motion Picture Engineers* 28 (1): 52–62.

Fielding, Raymond, ed. 1967. *A Technological History of Motion Pictures and Television*. Berkeley: University of California Press.

Gregory, Carl Louis, ed. 1920. *A Condensed Course in Motion Picture Photography*. New York: New York Institute of Photography.

Handley, C. W. 1967. "History of Motion-Picture Studio Lighting." In Fielding 1967, 120–124. First published in the *Journal of the Society of Motion Picture and Television Engineers*, October 1954.

Kindem, Gorham A. 1979. "Hollywood's Conversion to Color: The Technological, Economic and Aesthetic Factors." *Journal of the University Film Association* 31 (2): 29–36.

Malkiewicz, Kris. 1986. *Film Lighting*. New York: Prentice Hall.

Millerson, Gerald. 1972. *The Technique of Lighting for Television and Motion Pictures*. London: Focal Press.

Mills, Frederick S. 1921. "Film Lighting as a Fine Art: Explaining Why the Fireplace Glows and Why Films Stars Wear Halos." *Scientific American* 124: 148, 157–158.

Neale, Steve. 1985. *Cinema and Technology: Image, Sound, Colour*. London: Macmillan/British Film Institute.

Salt, Barry. 1983. *Film Style and Technology: History and Analysis*. London: Starword.

Valentine, Joseph. 1939. "Make-Up and Set Painting Aid New Film." *American Cinematographer*, February, 54–56, 85.

Walker, Joseph B., and Walker, Juanita. 1984. *The Light on Her Face*. Hollywood, CA: ASC Press.

Williams, Raymond. 1974. *Television: Technology and Cultural Form*. London: Fontana.

Winston, Brian. 1985. "A Whole Technology of Dyeing: A Note on Ideology and the Apparatus of the Chromatic Moving Image." *Daedalus* 114 (4): 105–123.

15 Manufacturing Gender in Commercial and Military Cockpit Design

Rachel N. Weber

While the first two selections in this section suggest that values can be designed into technology, it is important to remember that this is sometimes done deliberately and other times not. In this piece by Rachel Weber, we learn about a case in which a technology's design denied access to women, although this was not necessarily intentional. Designers often work with a specific set of users in mind; they may use data and specifications from that group or simply assume that most users will be similar to themselves. In so doing, however, the designed object may discriminate against other users—those not included in the data or those unlike the designers. (Winner makes a reference to this in chapter 12 when he mentions that buses, sidewalks, and buildings have often been designed in ways that discriminate against people in wheelchairs.) Once in place, a designed artifact can be very difficult to change and may continue to act as a barrier to certain groups. Here, however, Weber discusses a case study in which the "momentum" of the sociotechnical system was overcome. She describes how a concerned group of people successfully redesigned airplane cockpits to accommodate a larger number of female users. Project leaders were able to counter arguments that such a change would cost too much money, would take too much time, and would be resisted by some of those previously discriminated against because they did not want to appear to be receiving special treatment. In this respect, Weber's account illustrates both the unintentional incorporation of values into a design and the subsequent deliberate embedding of new values in a revised version of a technology.

Technological Bias in Existing Aircraft

Civilian and defense aircraft have traditionally been built to male specifications (Binkin 1993). Since women tend to be shorter, have smaller limbs and less upper-body strength, some may not be accommodated by such systems and may experience difficulty in reaching controls and operating certain types of equipment (McDaniel 1994). To understand how women's bodies become excluded by design and how difference becomes technologically embodied, it is necessary to examine how current military systems are designed with regard to the physical differences of their human operators.

To integrate the user into current design practices, engineers rely on the concepts of ergonomics and anthropometrics (McCormick and Sanders 1982). Ergonomics, also called "human factors," addresses the human characteristics, expectations, and

From *Science, Technology, and Human Values* 22, no. 2 (Spring 1997): 235–253. Reprinted by permission of Sage Publications Inc.

behaviors in the design of items which people use. During World War II, ergonomics became a distinct discipline, practiced predominantly by the U.S. military. Ergonomic theories were first implemented when it became obvious that new and more complicated types of military equipment could not be operated safely or effectively or maintained adequately even by well-trained personnel. The term "human engineering" was coined and efforts were made to design equipment that would be more suitable for human use.

Anthropometrics refers to the measurement of dimensions and physical characteristics of the body as it occupies space, moves, and applies energy to physical objects as a function of age, sex, occupation, and ethnic origin and other demographic variables. Engineers at the Pentagon and at commercial airframe manufacturers rely on the U.S. Army Natick Research Development and Engineering Center's *1988 Anthropometric Survey of Army Personnel*, in which multiple body dimensions are measured and categorized to standardize the design of systems. The Natick survey contains data on more than 180 body and head dimension measurements of a population of more than 9,000 soldiers. Age and race distributions match those of the June 1988 active duty Army, but minority groups were intentionally oversampled to accommodate anticipated demographic shifts in Army population (Richman-Loo and Weber 1996).

Technological Bias within Defense Aircraft

Department of Defense acquisition policy mandates that human considerations be integrated into design efforts to improve total system performance by focusing attention on the capabilities and limitations of the human operator. In other words, the Defense Department recognizes that the best defense technology is useless if it is incompatible with the capabilities and limitations of its users. In the application of anthropometric data, systems designers commonly rely on Military Standard 1472, *Human Engineering Design Criteria for Military Systems, Equipment and Facilities*. Like the use of military specifications in the procurement process, these guidelines are critical in developing standards; they embody decisions made which reflect the military's needs and goals and are ultimately embodied in the technology (Roe Smith 1985).

These guidelines suggest the use of 95th and 5th percentile male dimensions in designing weapons systems. Use of this standard implies that only 10 percent of men in the population will not be accommodated by a given design feature. If the feature in question is sitting height, the 5 percent of men who are very short and the 5 percent who are very tall will not be accommodated.

Accommodation becomes more difficult when more than one physical dimension is involved, and several dimensions need to be considered in combination. The various dimensions often have low correlations with each other (e.g., sitting height and arm length). For example, approximately 52 percent of Naval aviators would not be accommodated by a particular cockpit specification if both the 5th and 95th percentiles were used for each of the thirteen dimensions.

Because women are often smaller in all physical dimensions than men, the gap between a 5th percentile woman and a 95th percentile man can be very large (Richman-Loo and Weber 1996). Women who do not meet requirements are deemed ineligible to use a variety of military systems.

The case of the Joint Primary Aircraft Training System (JPATS) has been the most publicized case of military design bias against women.[1] Engineers and human factors specialists considered minimum anthropometric requirements needed by an individual to operate the JPATS effectively and wrote specifications to reflect such requirements. For example, "the ability to reach and operate leg and hand controls, see cockpit gauges and displays, and acquire external vision required for safe operation" was considered critical to the safe and efficient operation of the system. Navy and Air Force engineers determined the five critical anthropometry design "drivers" to be sitting height, functional arm reach, leg length, buttock-knee length, and weight (US Department of Defense 1993, 2).

Original JPATS specifications included a 34-inch minimum sitting height requirement in order to safely operate cockpit controls and eject. This specification is based on sitting height minimums in the current aircraft fleet and reflects a 5th percentile male standard. However, at 34 inches, anywhere from 50 to 65 percent of the American female population is excluded because female sitting heights are generally smaller than male. Therefore, JPATS, as originally intended, accommodated the 5th through 95th percentile male, but only approximately the 65th through 95th percentile female.

After successful completion of mandatory JPATS training, student pilots advance to intermediate trainers and then to aircraft-specific training. Therefore, if women cannot "fit" into the JPATS cockpit or if the cockpit does not "fit" women pilots, they will be unable to pursue aviation careers in the Navy or Air Force. In other words, design bias has far-reaching implications for gender equity in the military.

Technological Bias within Commercial Aircraft

Engineers design commercial cockpits based on military specifications, aiming to accommodate a population ranging from 25th percentile military women to 99th percentile military men. The methods used by human factors practitioners in the commercial world to determine accommodation are quite similar to those used by the military, many having been developed by internal defense divisions or borrowed directly from the public sector research laboratories (Weber 1995). Using computerized human modeling packages, engineers are able to analyze visibility and reach in a proposed cockpit design. Such programs create three-dimensional graphic representations of pilots which can be adjusted to different body sizes and proportions based on accumulated anthropometric data from the Army surveys, such as those published by the U.S. Army Natick Research Development and Engineering Center.

Although military and commercial engineers use similar methods and data, their pilot populations may differ. Commercial aviation relies on anthropometric data representative only of military populations, even though a different pool of pilots may be flying commercial planes. Many of the human factors engineers interviewed maintained that one of the obstacles to overcoming design bias against commercial women pilots is the lack of comprehensive anthropometric data for civilian female populations. The only available civilian anthropometric data are very old; for female measurements, some manufacturers still use a 1940 Department of Agriculture survey conducted for clothing dimensions. Human factors engineers agree that these data are not extensive enough for use in designing large, complex interfaces such as cockpits.

Commercial manufacturers do not possess conclusive data regarding the total population of women commercial pilots, let alone their body dimensions.[2] Approximately 3 percent of all pilots in the U.S. are women, and the percentage is significantly lower worldwide (Gilmartin 1992). In 1990 the Air Line Pilots Association (ALPA) estimated that there were approximately 900 women pilots (out of a total of 43,000) at forty-four of the airlines where it had members at that time. However, the number of women earning their air transport rating in the United States has increased by 325 percent since 1980.

Human systems specialists suspect that the civilian pilot population is more varied than the military because civilian airlines have less restrictive eligibility requirements. Commercial airlines do not maintain the same limits on body weight and height as the military. Moreover, in the military most pilots are between twenty-one and thirty-five years old, whereas commercial airlines employ an older population, often composed of retired servicemen.[3] This results in a less standardized commercial pilot population, one that might not be represented in the anthropometric data culled by the military.

Principal airframe manufacturers, such as Boeing and McDonnell Douglas, contract with both the government and private airlines. Much of the technology base, supplier base, skills, and processes used by defense and civil aircraft are common even though the divisions responsible for military and civilian work are organizationally and physically separate (Markusen and Yudken 1992). Whereas the defense division responds to a single client—the Pentagon—whose main concern is the performance characteristics, the concerns of the commercial division focus primarily on production costs or marketing (Markusen 1985; Melman 1983).

Despite a similar technological base, the cockpit technology encountered in civilian aviation differs from that found in the military. The role of the human being and the control processes available to him or her also will differ. For example, the extreme rates of acceleration experienced in military cockpits require elaborate restraining devices. Such restraints must be designed to fit the anthropometric characteristics of the intended users. Ejection is also an issue limited to military cockpit design. Much of the JPATS controversy centers on ejection seats and the need to provide safe ejection to lighter individuals.

In contrast, commercial aircraft do not reach the same high speeds as military planes, nor do they contain ejection seats. The seats in a commercial cockpit are adjustable to meet the varied comfort and safety requirements of the users. Thus certain anthropometrics such as height, weight, and strength do not have the same valence in commercial aviation as they do in the military. Many argue that commercial aircraft can accommodate a more variable population because the operating requirements are not as stringent as in the military.

However, the location of various controls on the commercial flight deck has been found to disadvantage women and smaller-statured men (Sexton 1988). Although the seats are more adjustable, individuals with smaller functional arm reach and less upper-body strength may still experience difficulties manipulating controls and reaching pedals. When smaller women are sitting in the co-pilot seat, some complain that they are not able to reach controls on the right side of the control panel. Reach concerns become increasingly important during manual reversion (when the system reverts to manual operation) even though electrical and hydraulic systems require smaller forces to actuate.

Cockpit design specifications have protected what has traditionally been a male occupation. Because both commercial and defense aircraft have been built for use by male pilots, the physical differences between men and women serve as very tangible rationales for gender-based exclusion. Although technology certainly is not the only "cause" of exclusion and segregation, biased aircraft act as symbolic markers, used to delineate the boundaries between men's and women's social space. Reppy (1993, 6) notes that

> it is not that women are not physically capable of flying these particular aircraft or that they are not equally exposed to danger in other aircraft; rather denying women access to combat aircraft is a way of protecting a distinctly male arena. The technical artifact . . . has functioned to delineate the "other."

Regulating Accommodation in Defense Aircraft

The decision to standardize any technology is often contested, occurring within a space where social, economic, and political factors vie for position. In this case, standardization involved altering technologies in order to adjust to a changed sociopolitical environment. In the military, cockpit technology had to be adjusted to the entry of women into the armed forces and their new roles within the services. The process of design accommodation in the military became a process of negotiation between various social groups who held different stakes in and interpretations of the technology in question (Pinch and Bijker 1984).

One could argue that negotiations over accommodation arose as a result of changes made in policies regarding women in combat. Former Secretary of Defense Les Aspin publicly recognized that women should play a greater role in the military when he issued a directive in April 1993 on the assignment of women in the armed forces. The directive states that

> the services shall permit women to compete for assignments in aircraft, including aircraft engaged in combat missions.
>
> The Army and Marine Corps shall study opportunities for women to serve in additional assignments, including, but not limited to, field artillery and air defense artillery. (Aspin 1993, 1)

Although the new policy gave women a greater combat aviation role and was intended to allow for their entry into many new assignments, the aircraft associated with these assignments precluded the directive from being implemented. The realization that existing systems could contain a technological bias against women's bodies despite the Congressional mandate for accessibility alarmed policy specialists at the Pentagon. This contradiction would potentially embarrass a new administration which was reeling from its handling of the gays in the military debacle and desperately trying to define a working relationship with an antagonistic Pentagon.

In May 1993 the Under Secretary of Defense (Acquisition) directed the Assistant Secretary of Defense (Personnel and Readiness) to develop a new JPATS sitting height threshold which would accommodate at least 80 percent of eligible women. He delayed release of the JPATS draft Request for Proposal until a new threshold could be documented. This move led to the establishment of the JPATS Cockpit Accommodation Working Group which included representatives from the Air Force and Navy JPATS Program Offices as well as from service acquisition, personnel, human factors, and flight surgeon organizations. After months of deliberation, the Working Group determined that a reduction of the sitting height requirement by 3 inches would accommodate approximately 82 percent of the eligible female population (US Department of Defense 1993).

Reducing the operational requirements would entail modifying existing cockpit specifications. Significant modifications were needed because the requirement for an ejection seat restricts the possibility of making the seat adjustable. In addition, the aircraft nose, rudder, and other flight controls would also need to be substantially modified to accommodate a smaller person. Further, since ejections at smaller statures and corresponding body weights had yet to be certified for safety, test articles and demonstrators had to be developed to ensure safe ejection (Dorn 1993).

After the May 1993 directive, many procurement specialists at the Pentagon were perplexed: a design which would accommodate the 5th percentile female through the 95th percentile male would have to incorporate a very wide variability of human dimensions. Some senior defense officials opposed such a change because they believed that such alterations would delay the development of the JPATS, would raise the price of training, and would be prohibitively expensive.

In opposition to these officials, pragmatists within the Pentagon—including most members of the Working Group—argued that it was both efficient and economical to integrate human factors into acquisition. Pragmatists felt that the technologies built for the military, as opposed to civilian markets, tended to privilege capability over maintenance and operability and hardware over personnel. They argued that with decreasing budgets, this could no longer be the case. Design changes, they claimed, would not only benefit women assigned to weapons systems originally designed for male operators, but would benefit smaller men as well. Studies have shown that smaller men also have difficulty operating hatches, damage control equipment, and scuttles on ships (Key, Fleischer, and Gauthier 1993). Shrinking personnel resources and a changing demographic pool from which the military recruits also mandated that defense technologies be more closely matched to human capabilities. The pragmatists were quick to emphasize that the inclusion and accommodation of smaller men would be necessary given changes in the ethnic and racial makeup of the nation (Stiehm 1985).

Pragmatists also pointed to the prospect of foreign military sales to countries with smaller-sized populations, which would make design accommodation an important economic consideration as well. Edwin Dorn (1993), the Assistant Secretary of Defense, in a memorandum to the Under Secretary of Defense (Acquisition), stressed that

> a reduced JPATS sitting height threshold will also expand the accommodation of shorter males who may have previously been excluded from pilot training. For potential foreign military sales, this enhances its marketability in countries where pilot populations are of smaller average stature.

The pragmatists emphasized that cockpit accommodation would benefit all soldiers because it required the acquisition process to consider differences concerning capabilities and limitations. In pursuing this line of argument, they essentially neutered the discourse, erasing the specificities of women's bodies. By refusing to engage in a gendered discourse and instead emphasizing economic benefits, they hoped to appeal to a broader segment of the population and to a Pentagon traditionally hostile to women's issues.

In contrast to the Pentagon pragmatists, women's groups both within the military and outside supported the decision to alter the JPATS sitting height requirement on more ideological grounds. The fact that women were being excluded by the operational requirements and by the technology was central to their decision to support the changes. In general, feminism in the contemporary military environment is organized around ideals of parity and equal opportunity regarding career opportunities (Katzenstein 1993). Insisting that career advancement be based on qualifications, not biology, many argued that physical restrictions which disqualified women would unfairly limit women's mobility in the services.

Through informal networks and more formal associations such as the Defense Advisory Committee on Women in the Service (DACOWITS), new groups of activists set about to influence policy decisions about career opportunities for women.[4] Women aviators organized around the issue of female accommodation and found a receptive audience in some of the new Clinton appointees, such as Edwin Dorn, Assistant Secretary of Defense. Unlike other changes imposed from the top, the decision to alter JPATS was part of a low-level process that began with limited intervention from high-ranking administrators (Brundage 1993).

Although the media spectacle of the Tailhook scandal provided the necessary momentum for feminist groups in the military and brought gender issues to the forefront of national debates, the decision to accommodate more women in the JPATS cockpit was not without dissension.[5] Some women officers—many of whom also considered themselves feminists—believed that, as one of the people I have interviewed told me, "shrill cries for accommodation could be used against women politically." They insisted that demanding special treatment would single women out in an institution which, on the surface, seeks to eradicate differences between the sexes. In a sense, they were asking women to ignore their difference and prove themselves on gender-neutral terms.

A few women pilots questioned the construction of the operational requirements and thresholds but insisted that the existing cockpits were not biased. Is it really necessary, some asked, to possess a sitting height of 34 inches to fly defense aircraft? Women with smaller sitting heights had flown during wartime, and many believed that pilots at shorter sitting heights were no less capable of flying safely. One woman claimed that "the whole issue of height in aircraft is overstated, and just ignorance on the part of the Navy."

As debates raged in the press and within the Working Group during 1993, the possibilities for technological variety began to close down. The Pentagon pragmatists attempted to stabilize the debate, but the public spectacle of the issue facilitated closure by broadening the deliberative arena. With the JPATS case, "administrative" closure was achieved when the 1994 Defense Authorization Bill was passed. The bill included a provision which prevented the Air Force, the lead agency in the purchase of the JPATS, from spending $40 million of its $41.6 million trainer budget unless the Pentagon altered the cockpit design. John Deutsch (1992), then the Under Secretary of Defense, wrote a memo legitimizing the problem of accommodation of women in defense aircraft, stating:

> I believe the Office of the Secretary of Defense (OSD) should continue to take the lead in addressing this problem. Other platforms in addition to aircraft should be considered as well. We must determine what changes are practical and cost effective in support of Secretary of Defense policy to expand combat roles for females. I request that you take the lead in determining specification needs. Further, you should determine the impact of defense platforms already in production and inventory. (Deutsch 1992, 1)

After Working Group deliberations, the Air Force issued a revised JPATS Draft Request for Proposal that included a 32.8-inch sitting height threshold. The RFP identified crew accommodation as a key source selection criterion so that during the selection process, prospective contractors would be required to submit cockpit mock-ups which would be evaluated for their adherence to the revised JPATS anthropometric requirements. Candidates who adhered to and even exceeded these requirements stood the best chance of winning the contract.

As the preceding case reveals, the relevant social groups who had a stake in changing the technology were able to voice their interests in quasi-public fora: in legislative committees, in the JPATS Working Group meetings, and in the popular media. The debates surrounding accommodation exposed the interpretive flexibility of cockpit design but also demonstrated how the more powerful and pragmatic groups were able to push forth their agenda. Able to increase momentum because of intersecting debates on "women in combat," the Working Group cast the issue of altering military technologies in terms of accommodating all types of operators and emphasized the political accountability of a public consumer to these operators.

Notes

1. The JPATS is the aircraft used by both the US Navy and the US Air Force to train its pilot candidates.

2. The FAA Statistics and Forecast Branch maintains information on the number of women pilots who have a current medical certificate and a pilot license. In 1993, 39,460 women held both the certificate and license out of a total of 665,069 pilots (Office of Aviation Policy, Plans and Management 1993). However, these figures do not reflect the number of women actually employed as commercial pilots.

3. In the past, commercial pilots received their training in the military, whereas now the trend is to filter through private flight-training schools.

4. Mary Katzenstein provided me with these insights. See also Enloe (1993, 208–214).

5. The Tailhook scandal refers to the annual Tailhooker's (navy carrier pilots) convention of 1991 where several women were sexually harassed by servicemen and later went public with their charges. As a result, three admirals were disciplined, although none of the servicemen were officially charged.

References

Aspin, L. 1993. *Policy on the Assignment of Women in the Armed Forces*. Washington, DC: US Department of Defense.

Binkin, M. 1993. *Who Will Fight the Next War? The Changing Face of the American Military*. Washington, DC: Brookings Institute.

Brundage, W. 1993. "The changing self-definitions of the military and women's occupational specializations." Paper presented at the workshop Institutional Change and the U.S. Military: The Changing Role of Women, Cornell University, October.

Deutsch, J. 1992. "Memorandum on JPATS cockpit accommodation working group report." Unpublished report, December 2.

Dorn, E. 1993. "Memorandum on JPATS cockpit accommodation working group report." Unpublished report, October 19.

Enloe, C. 1993. *The Morning After: Sexual Politics and the End of the Cold War*. Berkeley: University of California Press.

Gilmartin, P. 1992. "Women pilots' performance in Desert Storm helps lift barriers in military, civilian market." *Aviation Week and Space Technology*, January 13.

Katzenstein, M. 1993. "The formation of feminism in the military environment." Paper presented at the workshop Institutional Change and the U.S. Military: The Changing Role of Women, Cornell University, October.

Key, E., E. Fleischer, and E. Gauthier. 1993. "Women at sea: Design considerations." Paper presented at the Association of Scientists and Engineers, 30th Annual Technological Symposium, Houston, TX, March.

Markusen, A. 1985. *The Economic and Regional Consequences of Military Innovation*. Berkeley, CA: Institute of Urban and Regional Development.

Markusen, A., and J. Yudken. 1992. *Dismantling the Cold War Economy*. New York: Basic Books.

McCormick, E., and M. Sanders. 1982. *Human Factors in Engineering and Design*. New York: McGraw-Hill.

McDaniel, J. 1994. "Strength capability for operating aircraft controls." In *Advances in Industrial Ergonomics and Safety*, vol. 6, edited by E. Aghazadeh, 58–73. Bristol, PA: Taylor and Francis.

Melman, S. 1983. *Profits without Production*. New York: Knopf.

Office of Aviation Policy, Plans and Management. 1993. *U.S. Civil Airmen Statistics*. FAA APO-4–6. Washington, DC: Federal Aviation Administration.

Pinch, T., and W. Bijker. 1984. "The social construction of facts and artifacts." *Social Studies of Science* 14:399–441.

Reppy, J. 1993. "New technologies in the gendered workplace." Paper presented at the workshop Institutional Change and the U.S. Military: The Changing Role of Women, Cornell University, October.

Richman-Loo, N., and R. Weber. 1996. "Gender and weapons design: Are military technologies biased against women?" In *It's Our Military Too! Women and the U.S. Military*, edited by J. Stiehm, 136–155. Philadelphia: Temple University Press.

Roe Smith, M., ed. 1985. *Military Enterprise and Technological Change*. Cambridge, MA: MIT Press.

Sexton, G. 1988. "Cockpit-crew systems design and integration." In *Human Factors in Aviation*, edited by E. Weiner and T. Nagle, 495–526. San Diego: Academic Press.

Stiehm, J. 1985. "Women's biology and the U.S. military." In *Women, Biology, and Public Policy*, edited by J. Stiehm, 205–234. Beverly Hills, CA: Sage.

US Department of Defense. 1993. "JPATS cockpit accommodation working group report." Unpublished report, May 3.

Weber, R. 1995. "Accommodating difference: Gender and cockpit design in military and civilian aviation." *Transportation Research Record* 1480:51–56.

16 Amish Technology: Reinforcing Values and Building Community

Jameson M. Wetmore

Even when we recognize that technologies are value-laden, it is difficult to intentionally choose technologies that promote specific values. Typically, a single technology represents a number of different values, making it impossible to choose a technology without making compromises. But one group that has done an impressive job of taking on this task is the Old Order Amish. This article is especially important because the Amish do not share the mainstream Western belief in technological or economic progress. In this piece Wetmore explains how the Amish evaluate technologies. Twice a year the members of each Amish district reflect on whether integrating new technologies into their society will help to promote, preserve, or dissipate the values they hold most dear. They try to choose technologies they believe will ultimately benefit their society and avoid those they believe will undermine it. As they develop new needs, they do not simply take existing technologies off the shelf; they actively design their own artifacts, regulations, and systems of use in an effort to minimize disruptions to their community caused by the values others have inscribed into technology. The process may not be a perfect democratic way of directing technology, but it is an example of a conscious attempt to reflect on the relationship between technology and values in order to build a more desirable society. The Amish create their own sophisticated sociotechnical systems so that they are not controlled by sociotechnical systems created by others.

On late-night TV and in popular jokes, the Amish are usually portrayed as rural farmers who live in a bygone era.[1] They are supposed to be a people who would never set foot in automobiles, never study the workings of a diesel engine, and never admit change into their society. And yet, when a non-Amish person—or "English" person as the Amish call their English-speaking neighbors[2]—travels through an Amish community, he or she discovers something very different. An observer may see an Amish woman talking on a pay phone, an Amish carpenter using a drill press, or even an Amish teenager driving a car. This revelation is often startling, but scenes like these are in fact the norm. They are not examples of Amish straying from their faith, but evidence that stereotypes obscure the intricacies of Amish life.

The relationships the Amish have with the outside world and technology may at first seem arbitrary, but they are the result of careful consideration. The Amish are not fundamentally anti-technology; rather, they believe that change does not necessarily result in desirable ends. They have not banned all machines and methods invented in the past 150 years, but they do exercise extreme caution when dealing with new

From *IEEE's Technology and Society Magazine* 26, no. 2 (Summer 2007): 10–21. Reprinted with permission.

Figure 16.1
Even though power lines tower over an Amish farm, they choose not to connect to the grid.

technologies. The Amish are cautious because they fear the changes that can accompany new technology. What a modern observer might see as potentially undesirable effects—like pollution and injuries caused by heavy equipment—however, are not major concerns for the Amish. The foremost reason the Amish carefully regulate technology is to preserve their culture [2].

Like many scholars of technology, the Amish have rejected the idea that technologies are value-free tools. Instead, they recognize that technology and social order are constructed simultaneously and influence each other a great deal. Implicitly they agree with the argument that technology and the social world are co-produced, that technology, in Sheila Jasanoff's words, "both embeds and is embedded in social practices, identities, norms, conventions, discourses, instruments and institutions—in short, in all the building blocks of what we term the social" [3, p. 3]. The Amish believe that technologies can reinforce social norms, enable or constrain the ways that people interact with one another, and shape a culture's identity. But despite the fact the Amish believe technology is so powerful, they are not technological determinists [4]. They do not view technology as an autonomous force, but rather as a tool that can be actively used to construct and maintain social order. The Amish recognize both the power of technology to shape their world and their power to shape technology.

The Amish have not, however, developed these ideas out of some sort of theoretical or academic interest. (In fact, they do not believe in education past the eighth grade.)[3] Rather, they reflect on the relationship between technology and society because they believe it is crucial if they are to understand and strengthen their culture, religion, and community. Their belief that technology and society simultaneously influence each other has both inspired and informed Amish attempts to maintain their way of life. The Amish regulate which technologies are to be used, when they are to be used, how they are to be used, and why they are to be used because they believe that one of the

Figure 16.2
A bulk tank and mechanized agitator used to meet grade "A" milk regulations.

most important ways they can promote and reinforce their values is by actively embedding these values in their relationships with technology.

This article explores the way the Amish actively try to shape their society through technological decision-making. It can be tempting to simply point to various technologies the Amish use and ask—why? But because the Amish do not view technology as entirely separate from their society, any faithful explanation of their technology cannot either. Thus in order to convey the full picture this article will examine numerous facets of Amish life including their codes of conduct, the process of becoming an adult member of the church, economic pressures, business needs, and family life.

One Amish person succinctly explained the Amish approach to technology in the following way: "Machinery is not wrong in itself, but if it doesn't help fellowship you shouldn't have it" [6].[4] This article argues that the Amish pursue this goal of fostering community through technological choice in at least two interrelated ways. They first seek to prohibit those technologies they believe are antithetical to their values and choose those they believe will reinforce and strengthen their values. This

Figure 16.3
Amish buggies stand in stark contrast to the trucks and minivans driven by their Indiana neighbors.

straight-forward approach is very important to the Amish, but it cannot explain all of their decisions. The Amish also recognize that the technologies they use have become a crucial part of their identity and they use this link between technology and identity to strengthen their community. Thus when making decisions about technology, the Amish rely on a second criterion—they deliberately choose technologies that are different from those used by other Americans in order to maintain their unique culture. The Amish believe that their way of life depends as much on the technologies they choose as any of the other social institutions that govern their work, religion, and community. The Amish practice of reflecting on their own relationship with machines and techniques makes Amish culture a window into the ways in which technologies, societies, and values are interwoven.

Amish Community and Values

To begin to understand why the Amish make the decisions about technology that they do, one must first understand Amish values. This can be difficult for those raised with very different social norms, but there are a few basic ideas that can help one begin to appreciate why the Amish make the choices they do. The Amish are a sect of Christianity and, as such, share the same Bible and many basic theological beliefs with other Protestant churches.[5] There are a few important points on which they differ in both emphasis and approach, however. One of the church's fathers, Menno Simons, advised his people to "rent a farm, milk cows, learn a trade if possible, do manual labor as did Paul, and all that which you then fall short of will doubtlessly be given and provided you by pious brethren, by the grace of God" [9, p. 451].[6] The idea of honest work, living a simple life, relying on their fellow believers, and trusting in God has shaped the

Amish way of life to this day. They place great importance on values like humility, equality, simplicity and community.

Community is especially important to the Amish. They have gone to great lengths to carry out the scripture passage that implores them to "be not conformed to the world" [11]. The Amish believe that the world is full of distractions that must be avoided if they are to live piously. To steer clear of these distractions and ensure that they rely on their "pious brethren," they have separated themselves from those that do not share their faith.[7] Today the Amish live in groups of between 30 and 50 families called districts. They go to school together, worship together, play together, work together, and make decisions about technology together. The Amish believe that these separate communities provide the fertile soil in which they can best understand their place in the world, pass on their values to the next generation, and live the humble lives they believe are so important.

Rules That Bind and Nurture

Community is so essential to their way of life that the Amish have very carefully shaped the way it is organized. The primary method by which they do this is known as the "Ordnung"—a code of conduct that varies slightly from district to district.[8] The Ordnung is comprised of the district's long established traditions, as well as more recently agreed upon norms, and governs every aspect of Amish life—including the format of church services, the color of clothing to be worn, and which technologies are acceptable and which are unacceptable. The Ordnung is not written down, but it is understood and adhered to by the adult members of the community because it is continually being conveyed by example and occasionally by instruction when someone breaks a rule or inquires about a rule.

The Ordnung structures the life of the Amish in two interconnected ways. First, it provides the members of an Amish district with a template for living that they believe will nurture their community, their religious beliefs, and their values. For example, the Ordnung emphasizes the Amish dedication to nonviolence by forbidding Amish people from becoming soldiers and it requires that church services be held at a different family's house each week so that members of the community are continually supporting and relying on each other.

A number of Amish rules are designed to aid them in their quest to remain humble. For instance, to ensure that no individual becomes prideful about the way they look, each district specifies the color and design of clothing its members are to wear. Many districts go as far as to even reject buttons as "unnecessary" or potentially "prideful" adornment and require Amish to use straight pins to fasten their clothing. The Ordnung is also designed to promote humility by encouraging Amish adults to avoid being photographed in such a way that a viewer can distinguish who particular individuals are. This helps to reinforce the idea that an Amish person should not stand out as an individual, but rather is part of a community.

Through measures like these, the Amish use the Ordnung to promote their values, instill responsibility, pass down traditions, and build strong ties with one another. One Amish minister described the effective use of an Ordnung when he stated: "A respected Ordnung generates peace, love, contentment, equality, and unity" [10, p. 115]. Because it lays out how their life should be lived, in a very real sense the Ordnung is what makes an Amish person Amish.

The second way the Ordnung structures Amish life is by defining what is not Amish. In a sense, the Ordnung is the line that separates the Amish from the non-Amish; it is

Figure 16.4
A sophisticated sawdust collection system, powered by a diesel engine, services an Amish carpentry shop.

what gives the Amish their distinctly separate identity. For instance, each of the rules that detail what an Amish person should wear not only ensures that they will look Amish, but also that they will be easily distinguished from outsiders. In an interview, one Amish man used a parable to describe how this aspect of the Ordnung can promote community [12]. He said that if you own a cow and your property is surrounded by green pastures, you need a good fence to keep it in. For the Amish, who are as human as anyone and are tempted by the outside world to abandon their faith and way of life, there need to be good fences as well. The Ordnung defines what the Amish cannot do and makes those who are not adhering to the faith readily visible. Because they believe the outside world is a distraction that must be mediated, the Ordnung provides the barriers that keep community members focused on their fellow Amish and their faith.

Ordnung and Amish Change

Although a district's Ordnung is meant to convey the traditions of the community, it can be—and occasionally is—changed. When individual members begin exploring new abilities and possibilities that raise some concerns, the district must decide whether or not such activities should be allowed. To facilitate this process, twice a year each Amish district holds a counsel meeting. The counsels are led by the district's bishop (its religious and secular leader) but all of the adult members of the church—men and women—vote on the practices in question. To ensure that the implications of new practices are carefully considered, the voting system is designed such that change is very difficult. If two or more people (out of a possible 60–100) reject the change, the Ordnung remains unaltered. Thus the Amish allow for change, but the emphasis on tradition is built into the mechanisms that allow this change.

At least one other factor also helps to ensure that these deliberations are conservative. When considering a modification to their Ordnung, the members of a district must consider the other districts around them. If they make a change that neighboring districts believe is too radical they may be shunned, i.e., the offended districts could break off all communications with them and no longer recognize them as fellow Amish. This threat is of particular concern not only for community reasons but also because there are often close family ties between districts. An Amish woman might, for instance, decide that voting for allowing electrical appliances in the home is not worth risking the very real possibility that she may never again get to talk to her daughters who married into other districts. There are often small differences in the Ordnung of neighboring districts. For instance one may allow rubber carriage tires or bicycles while others do not. But because of the threat of being shunned, change to a district's Ordnung is usually incremental and often done in concert with other districts.

While Amish counsel meetings address all aspects of Amish life, beginning in the late 19th and early 20th centuries, the conversations increasingly began to focus on modern technologies. Only a few years earlier, it might have been difficult to distinguish the Amish from many other rural American communities. Their dress may have been a bit different, and their buggies less flashy, but they farmed in largely the same way and used many of the same technologies. The development of powerful new technologies like electricity, the automobile, and the airplane, however, generated a significant amount of concern in Amish communities. There was a suspicion that technologies like these would cause a significant disruption in the Amish world. To limit the ways in which machines and techniques negatively impact their society, the Amish have developed rules to govern their use.

Regulating Technological Change

The precise reasons why specific technologies were—and continue to be—regulated is difficult to pin down. The Amish have left very few, if any written explanations; non-Amish are not allowed to attend the Amish counsels and most Amish are very hesitant to discuss the details of counsel meetings with outsiders [13], [14]. Despite these obstacles, conversations with and further study of the Amish can begin to shed some light on the decision making process. As with any democratic process, there were likely many factors taken into account and different people involved may have had very different ideas about why things happened the way they did. But there are a few general themes that can help begin to explain the rationale behind Amish decision-making.

An Amish minister described the decision making process in the following way: "We try to find out how new ideas, inventions, or trends will affect us as a people, as a community, as a church. If they affect us adversely, we are wary. Many things are not what they appear to be at first glance. It is not individual technologies that concern us, but the total chain" [15, p. 16]. The Amish believe that social change is often closely tied with technological change and therefore tend to be suspicious of new technologies. They are strikingly different from most English in that they do not see an inherent value in technological progress. They must be fully convinced that a given technology will benefit the things they do value—their ethics, their community, and their spiritual life—before they will accept it.

As with the Ordnung in general, the Amish formulate rules about technology with two interconnected goals in mind. First, when deciding whether or not to allow a certain practice or technology, the Amish first ask whether it is compatible with their

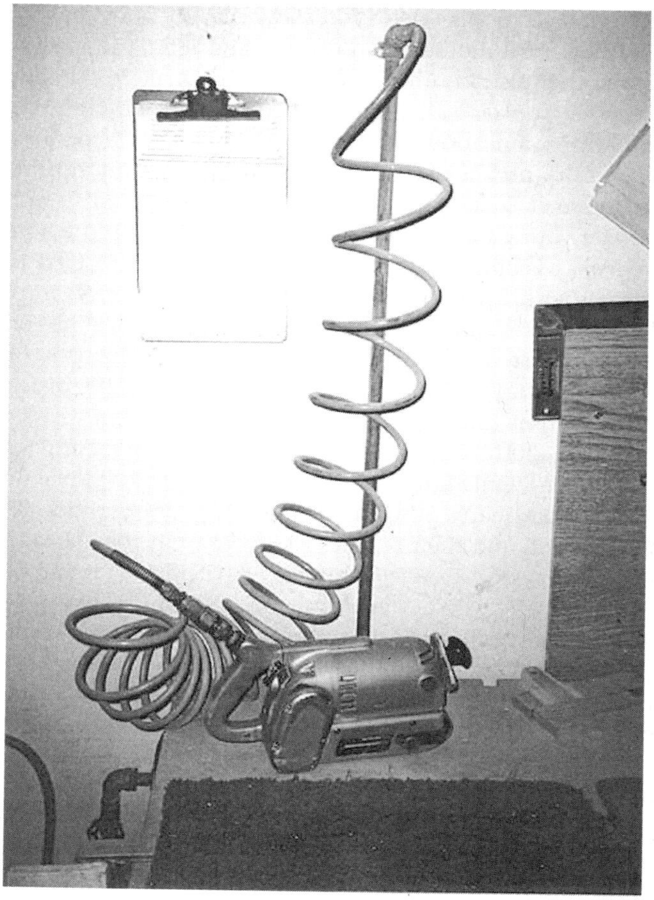

Figure 16.5
A pneumatically powered belt sander lies on a workbench in an Amish carpentry shop.

values. If they fear that a particular technology might disrupt their religion, tradition, community, or families, they are likely to prohibit it. The Amish not only believe that the English world is distracting, but also that many English machines and methods are distracting. For instance, the Amish believe that the pride, sense of power, and convenience that can come from owning an automobile may cause a person to focus on him or herself as an individual and thereby neglect the group. The Amish believe that technologies in general must be mediated in order to avoid situations like this and help to ensure that their way of life is not compromised.

The second purpose of the Ordnung—to create a fence between the Amish and non-Amish—has also played an important role in the Amish decisions about technology. Today, the most visible differences between the Amish and English worlds are the technologies they use. Most Americans do not see the Amish as different because they believe in adult baptism, but rather because they drive buggies, use horse drawn plows, etc. These differences were not accidental. The major technologies being developed in the non-Amish world at the beginning of the 20th century—like electricity, the automobile, and the airplane—very quickly became symbols of the modern world. The Amish rejected many of these technologies in part to retain their identity as separate from the modern world.

When asked today why they have rejected a specific technology, many members of the church will simply reply: "Because it's not Amish."[9] This argument is circular, but it emphasizes the way in which the Amish link their identity to the technologies they use. By banning these highly visible technologies, the Amish developed a new way of distinguishing themselves and strengthening the fence between themselves and the English world.

Regulating Electricity

The way in which the Amish make decisions about technology to promote both their values and their identity can be seen in an example where there is some historical record. The strict Amish regulation of electricity began in 1910 when Isaac Glick, an Amish farmer in Lancaster, Pennsylvania, hooked an electric light up to a generator [10, pp. 198–201]. His use of the new technology led to a counsel debate and the decision was made not to allow it.

Donald Kraybill, who recounts this story, argues that the reason was twofold. First of all, the Peachey Church, a group that broke off from the Mennonites as the Amish had, had just decided to allow electricity and the Amish were looking to prove that they were distinct from this new congregation. Secondly, they believed that physically hooking one's house up to the grid, a public utility owned by large corporations, did not help in the drive to be separate from the modern world. As one Amish farmer feared: "It seems to me that after people get everything hooked up to electricity, then it will all go on fire and the end of the world's going to come" [10, p. 200]. Instead of linking to the grid, the Amish continued to use the power sources they had been using—kerosene and natural gas—to cook their food and illuminate and heat their homes.

To this day, power lines bypass Amish houses. But the justification for this rule may have changed over the years. Many Amish today argue that the desire to avoid a physical connection to the English world is not the reason they reject getting power from electric companies [6], [14]. They point out that they have tapped into natural gas lines (or would if a utility provided them) rather than have to pick up canisters in town. The precise reasons why the Amish initially deemed connection to the grid as a threat to their community no longer matter, if they ever did matter. What is more important is that the Amish have defined electricity as the domain of the outside world, and thus any use of the technology must be very carefully considered. Even if the Amish link themselves to the outside world by piping gas into their homes from a public utility, they are still reaffirming their identity by forging a different relationship to power than their English neighbors.

Amish Transportation

Another area of technology that the Amish have carefully considered in order to ensure that it reflects their values and reinforces their identity is transportation. Traditionally the Amish have relied on horse drawn carriages to transport themselves, but in the early part of the 20th century they were faced with a new option. In 1907, an automobile manufacturing company was formed in Lancaster, Pennsylvania, the heart of Amish country. This company advertised its product as "the king of sports and the queen of

Figure 16.6
The interior of an Amish kitchen is nearly identical to a modern kitchen except that it has no electrical appliances and is lit with sunlight and gaslight.

amusements," and immediately turned off the Amish, who saw it as an unnecessary luxury and dangerous source of pride [10, p. 214]. By the second decade of the 20th century, after a few Amish had purchased motorcars, every Amish district in the United States independently decided to prohibit the use of the automobile [16, pp. 37, 73]. Because most of the people an Amish family knows live relatively close to their home; because the Amish are not relegated to a strict schedule that demands speedy transportation; because horses have become practically family members; and because buggies are relatively inexpensive (costing today between two and three thousand dollars new), require little maintenance, and last for up to twenty years, the Amish saw no reason for changing their traditional way of life [17, p. 8].

But economics are not the only reason why the Amish have chosen to keep their buggies. Some argue that buggies are a social equalizer because they are uniform, free from excess bodywork and color, and because one buggy cannot be made significantly faster or slower than another. Automobiles, on the other hand are criticized for providing an abnormal sensation of power that can be used to not only show up one's neighbors, but to abandon them altogether. As one Amish man noted, "Young people can just jump in the car and go to town and have a good time in it. . . . It destroys the family life at home" [18]. Buggies are deemed better because they slow the pace of life to ten or twelve miles-per-hour, giving people a chance to interact with their environment rather than fly by it. The Amish believe the automobile is not very compatible with the values they hold dear.

Despite these criticisms, however, there are several situations today in which an Amish person would be allowed to make use of a motor vehicle. For instance, it is not uncommon for an Amish woman to be driven to the grocery store by an English friend; for an Amish family to travel from Indiana to Florida via bus; for an Amish business to lease a car indirectly through a non-Amish employee; or even for an Amish teenager to actually drive and own an automobile. While these at first may seem to contradict Amish principles, each case signifies an arrangement that the Amish believe can help strengthen their community, and is therefore allowed under the Ordnung.

In the first scenario, it is probably not a necessity that the Amish woman be driven to the store—it is likely that she could take her own horse and buggy—but because she is not the one driving the car, it is acceptable behavior. She does not have the freedom to roam as she pleases, but rather must depend on another person. Some English people have gotten so involved in transporting the Amish that they have started their own thriving taxicab companies. These services are welcomed by the Amish because they satisfy a need and still make it inconvenient for a person to tour about on a whim.

The second scenario is a response to the fact that the Amish are spread across the United States. Many young Amish move miles away from their families to find land and work. It would be extremely difficult to travel by buggy to visit family members that lived a thousand miles away. Thus the Amish allow the use of public transportation (other than airplanes) to visit family and even to take vacations. The Amish community is a highly structured environment, but it is not a prison. Such trips allow them to reinforce their family ties and their ties with other Amish communities.

The third scenario reflects a fairly recent change that will be explored later. To sum up quickly, this scenario is the result of the belief that many Amish businesses cannot survive without an automobile. For instance, Amish businesses that specialize in building fences would likely run out of work rather quickly if they did not accept jobs outside of the area easily traversed in a buggy. To make this possible, some districts grant businesses special permission to lease a car, but only if they agree to certain restrictions. Under no condition would an Amish person be able to drive it; he or she must instead hire and be dependent upon an English employee. A district may even prohibit parking the car near an Amish home to decrease the temptation to use the car for trivial things. Some districts allow Amish businesses to use motor vehicles, but take a number of precautions to limit the potential negative impacts they perceive.

"Running About"

The fourth scenario is the result of a deeply rooted Amish tradition that will require further explanation of how the Amish structure their society. The Amish understand that it is difficult to be Amish. It requires a significant amount of humility, patience, and dedication. They also understand that because their lives are so intertwined, members who do not accept these responsibilities can threaten the active and united nature of their community. Therefore, the Amish go out of their way to ensure that their members truly want to be Amish.

The primary technique they use is the church admission process itself. To curtail immature and uninformed decisions no one is allowed to enter into the church until they are in a position where they can readily think for themselves. The Amish contend that it takes not just age but also experience to develop such wisdom. Therefore they give their children the opportunity to explore alternatives to Amish life by turning a blind eye to those who violate the Ordnung and choose to adopt some English ways. The Amish term for this phase of life is "rumspringa," or "running about."[10]

Many Amish youths take the opportunity to experience what another life would be like. Amish adolescents may begin with relatively small violations such as curling the brim of their hat or driving the family carriage faster than their parents would.

(It is often said that one can tell that a teenager is driving a buggy whenever it is going fifteen miles-per-hour, rather than the average of ten to twelve.) But the "running about" period also gives Amish youth the chance to experiment with modern technologies. Many of them are drawn to the outside world because they are fascinated by the devices they see English people using. Thus Amish teenagers may find ways to watch television, listen to music on the radio, operate their own ham radio, or even drive automobiles.

By their early 20s, most Amish children decide that they are not satisfied by English customs and technologies. Many of them begin to see more clearly the benefits of Amish culture and sincerely regret their actions [12]. Over eighty percent of children (and as many as ninety-five percent in some places) decide to become adult members of the Amish church [19]. The period of rumspringa helps to ensure that this is an informed commitment to community and church. Offering children the option to leave the rigorous and humble life of an Amish person and explore what the outside world has to offer—including its technologies—ensures that the people that make up the community truly want to be there and will henceforth work for the good of the Amish people.

Modern Pressures

While the questions of whether to adopt electricity and automobiles were important for the Amish to resolve, these were only the beginning of the difficulties their society encountered in the 20th century. Although they work hard to remain separate, many changes in American government, economics, society, and technology have had a significant effect on the Amish. In recent years the stability of the Amish has been put to rigorous tests. In their efforts to meet these challenges and stay focused on their values as much as possible, the Amish have chosen to alter some of their traditions and, in particular, the technologies they employ.

An example of this can be seen in the Amish response to new milk regulations imposed by a number of states in the 1950s and 1960s. These regulations required farmers to install electric powered bulk tanks with cooling systems if they wanted their milk to continue to be rated Grade "A" quality. The regulation clashed head on with the Ordnung of Amish communities.

This put the Amish in a bit of a dilemma. Much of their tradition is built upon an intimate relationship with the land. Many Amish view farming as the ideal way to earn a living. They have kept themselves separate and free from the outside world by working the land upon which they settle. The Amish did not want to significantly compromise one of the cornerstones of their culture.

Therefore, in 1968, a group of five Lancaster bishops and four milk inspectors from Pennsylvania met to iron out an agreement that would satisfy both parties [10, pp. 202–205]. The inspectors' primary concern was that the milk be kept refrigerated. They suggested simply installing normal electric refrigeration units. But the Amish refused to run electric lines into their barns. Instead they developed an "Amish solution." They agreed to install coolers, but chose to power them using diesel engines salvaged from old trucks.

The inspectors also required that the milk be automatically stirred five minutes every hour. This was a difficult request for the Amish to grant because the very word "automatic" bothered them, but they eventually consented to a newly devised system that used a 12-V battery, rather than 110-V electricity, to run an automatic starter. The

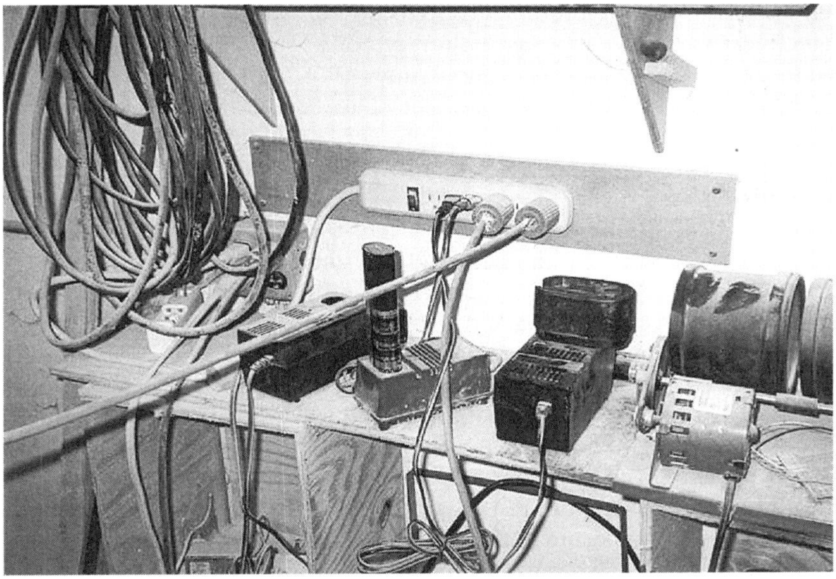

Figure 16.7
An "Amish power strip" draws power from a generator to supply energy to various batteries for a carpentry business.

fact that the Amish had traditionally used batteries to power a few devices like flashlights made this a bit more palatable.

Finally, the inspectors wanted the milk picked up every day to decrease spoilage. At this point, the Amish drew a line they would not cross for any reason. They would not allow anyone to interfere with Sunday, their day of rest and church services. Because the Amish were a major producer of milk in the area, the bulk milk industry agreed to readjust its practices slightly by picking up milk a second time on Saturday instead of Sunday morning. With this specially devised arrangement, the Amish won a minor battle in keeping their community economically sound and their culture relatively unchanged.

The resolution of the milk controversy is an instance in which the Amish accepted new technology, but they did it in a uniquely Amish way and for Amish reasons. The compromise was important because it protected the ability for the Amish to continue to earn a living doing the work they find most rewarding—farming. Yet while they introduced new technologies into their society, they made sure that the machines were different from those used by their English neighbors and that the electricity they generated could not be easily put to other uses. With this new—seemingly modern—technology, the Amish were able to meet an economic need while still retaining their identity and practice of being different from the outside world.

Amish Entrepreneurs

Despite compromises like these, the Amish have not been able to rely completely on farming to support themselves economically. For at least the last forty years, they have been in the middle of a land squeeze. Because married couples desire to have many children and the Ordnung prohibits contraceptives, an Amish family has an average of seven children [20]. Even though not every Amish child enters the church, this has resulted in a constant rise in Amish population. As of 2001, the Amish numbered over 180,000 children and adults [10, p. 336]. They have sought new farmland by gradually spreading into 25 American States and the province of Ontario. But the English population is also increasing and land prices are rising. There simply is not enough farmland to go around.

Young Amish adults increasingly have to look for employment other than farming. In the first half of the 20th century nearly all the Amish in the area surrounding Arthur, Illinois, were farmers. By 1989, that number was less than half [17, p. 9]. In Indiana the changes have been even more marked. While over fifty percent of Amish men under the age of 35 were farming in some Indiana areas in 1993, less than twenty-five percent of young Amish men were farming in 2001 [1, pp. 119–120].

Many young Amish who are not able to farm have found work in English factories, supermarkets, or stores in their area. Generally, they are treated well and receive a good wage. But being employed by the English can disrupt an Amish community. The hours and location of the business can restrict an Amish person's ability to participate in his or her culture and the exposure to the culture of the modern world can exert an influence as well. As one Amish woman noted, "The shops coming in were a good thing. They gave our young people jobs among our own people. But now they've got money and they go to town" [21].

Because of their concern that working for outsiders will dilute their culture and traditions, Amish communities have begun developing their own entrepreneurial talents and have increased the number and variety of businesses they own and operate [22]. Amish people have explored business ventures as diverse as machinery assembly, log house construction, upholstering, engine repair, grocery stores, bookstores, and cabinetry building. Economic forces have made the Amish ideal of communities comprised primarily of farmers impossible. But by developing their own businesses, the Amish ensure that they can work relatively close to home, work with their fellow church members, be free to attend community events like weekday weddings, and help reinforce their separation from the outside world.

As the Amish have entered these new fields—many of which are dominated by large American corporations—they have chosen to make some compromises when it comes to technology. They believe that in order to produce and sell an affordable product in the modern age, some increase in technology is necessary. As an Amish bishop put it, "To make a living, we need to have some things we didn't have fifty years ago" [6].

An example of this can be seen in the issues faced by Amish carpenters. Because the Amish have traditionally been good at building and feel that it is admirable to work with one's hands, carpentry has become one of their key industries. However, it would have been very difficult to survive on the output one could create using hand powered tools. Therefore, the Amish struck another bargain. They still strongly disagreed with running electric lines into their shops, so they motorized hand tools in a different way. A number of carpentry shops purchased regular electric saws, routers, and sanders and retrofitted them with motors that could be powered with air pressure. They then installed large diesel engines just outside their shops and strung pneumatic lines to the various work stations.[11]

Why go to all the trouble and expense to create such an intricate power system when electricity does the same job? In part because it distinguishes the Amish as different from their neighbors. But also because, as an Amish minister explained, "so far no Amish person has ever figured out how to run a television with an air compressor" [17, p. 3]. Television is seen as a technology that is contradictory to Amish ideals because it brings the outside world into the home and can distract one from one's family and neighbors. It is often used as a barometer by the Amish to determine whether or not something is acceptable. The Amish allow certain forms of electricity, but choose those forms that make it difficult to power devices like kitchen appliances, radios, and televisions.

The Amish have also developed ways of gaining the business benefits of certain technologies while maintaining their distance from them.[12] One way they do this is by hiring English companies to take care of certain aspects of an industry that they do not want to do themselves. As was already mentioned, the Amish will often hire English drivers to transport them to work sites, etc. But the Amish may also rely on non-Amish businesses to help them attract and interact with customers in ways they cannot or prefer not to do themselves. For instance, the Amish have been able to tap into the market for remodeling kitchens in far away cities by contracting with companies who do the on-site work. It is also now possible to buy Amish-made furniture online through websites developed and maintained by English companies. These arrangements help the Amish economically and yet minimize the distraction and compromises that come with using particular technologies themselves.

Line Dividing Home and Work

Despite all of the detailed explanations given above, the fact that the Amish use such a wide array of modern technologies may still seem fairly surprising. It does not mesh with many English people's visions of what Amish life should be. Many Amish feel a similar unease. They believe that they must adopt some new practices to remain economically viable, but that does not mean that they are enthusiastic about such changes. To compensate for these distractions, the Amish have tried to protect the simplicity of the home. While they have adjusted the Ordnung to promote Amish businesses, they are much less likely to change rules that govern the life in the home.[13] A stark example of this demarcation is the fact that diesel generators and pneumatic equipment are not allowed in the Amish home; kitchens are empty of electric appliances and interiors are still lit by candles, gas lamps, and windows.

The desire to protect the home has also shaped the Amish rules concerning telephones [27], [28]. Traditionally the Amish have been opposed to owning telephones because they believe that phones disrupt the natural interactions between people. An Amish buggy maker contended that "if everyone had telephones, they wouldn't trouble to walk down the road or get in the buggy to go visiting anymore" [17, p. 3]. Telephones are seen as distracting; they give the outside world an easy entrance into Amish households and make them needlessly noisy.

But the English companies and customers that the Amish rely on have abandoned many of the forms of communication that the Amish prefer. Without a phone it is difficult for furniture shops to communicate with distant customers, for stores to order merchandise, or even for farmers to coordinate milk and produce pick-ups with dairy and grocery companies. To remedy this problem, these businesses began to use the phones of their non-Amish neighbors. But as businesses got bigger and were sometimes far away from English phones, this became increasingly difficult. Gradually many Amish districts have begun to allow telephones, but with certain qualifications that ensure they do not compromise their lives at home.

Most districts maintain the rule that telephones are not allowed inside buildings owned by Amish people. Instead they are usually placed in small structures, or "Amish phone booths," that are kept "a safe distance away" from Amish dwellings. Typically the telephones are purchased by either the community in general or by specific Amish businesses, but they are kept accessible to the entire community. They are outfitted with a log so that calls can be recorded and payments can easily be made by individual people.[14] This arrangement encourages cooperation, reduces the impact on traditional forms of communication, and allows Amish businesses to develop. But most of all it keeps telephones outside of the home. It helps keep the home free from the distractions of the modern world.

Where the Amish Stand Today

The Amish are continually debating whether or not to introduce new technologies into their society—a process which can be contentious at times. A young Amish farmer noted that he (and every other Amish dairy farmer) would love to install glass piping that would quickly transport the milk from the cows to the refrigerators and relieve him of a lot of work, if only it were allowed [29]. Yet despite his desire, this farmer is still firmly committed to his community. Like many other Amish, he struggles with the Ordnung, but has agreed to and recognizes the benefits of a society that does not accept rapid change.

These struggles will continue as changes in American government, business, farming, and technology exert increased pressure on the Amish way of life. In response to some of these stresses, the Amish have chosen to accept some somewhat marked changes in technology. Some Amish communities now allow battery-operated type-writers, electric cash registers, and fax machines [25]. These new machines have led to a vigorous debate because many of them require 110 volt electricity (easily done by coupling inverters to their existing diesel engines), which could also be used to power a television. But some districts have decided that their businesses cannot survive without them.

The Amish are not, however, about to relax their control over technology. Because they believe that technology can shape those things they value above all others—their culture, their community, and their values—they continue to closely monitor and regulate its use. One Amish man admitted, "We realize . . . that the more modern equipment we have and the more mechanized we become, the more we are drawn into the swirl of the world, and away from the simplicity of Christ and our life in Him" [30, p. 95]. The Amish see technology as a potential disruption to their simplicity, humility, and separation and work to make sure that it disrupts their lives as little as possible. A bishop explained his difficult position in the following way: "Time will bring some changes; that's why our responsibility is so great. . . . We can prolong our time. I'll do what I can" [6]. Why this dedication when the world around them is changing so quickly? One Amish farmer argued that "if it hasn't worked for the good of [English] families, why will it work for our society? It's not good community" [31]. The Amish exert control over technology in an effort to protect themselves from the values and distractions of the English world.

The Amish believe that their society and their technology are inextricably intertwined. In an effort to maintain and protect their community of believers, therefore, the Amish require that every technology they use not only conforms to, but reinforces their tradition, culture, and religion. They achieve these goals through two primary techniques. First they choose technologies that they believe will best promote the values they hold most dear—values like humility, equality, and simplicity. Thus they have rejected the speed, glamour and personal expression of automobiles in favor of modest, slow, and community-building horse-drawn buggies. Second, they deliberately choose tools that are different from those used by the outside world. This differentiation helps them maintain their unique identity, bonds their community, and ensures that they will continue to be able to accept technology on their own terms. The Amish view technologies as value-laden tools and use these tools to reinforce their values and build their community. While many scholars of technology have argued that this is the case, the Amish employ the idea in order to build the world they want to live in.

Acknowledgment

I would like to thank Louis B. Wetmore and Gordon Hoke both for helping me to get in touch with Amish communities and for enlightening conversations; Michael Crowe, Deborah Johnson, Shobita Parthasarathy and two anonymous reviewers for comments on various drafts; and the Menno-Hof Museum in Shipshewana, Indiana, and the Mennonite Historical Library at Goshen College for their assistance in locating resources. Most importantly, I would like to thank the Amish people who took the time to share their culture and their experiences with me.

Notes

1. This article primarily refers to the Old Order Amish. Because this is the largest and most recognizable group of Amish people, they are typically referred to as simply "Amish." For an explanation of the different types of Amish, see [1, pp. 21–22].

2. When talking to one another, most Amish speak a derivative of German usually referred to as Pennsylvania Dutch.

3. The Amish rejection of advanced education is based on their belief that "the wisdom of this world is foolishness with God" [5, p. 91].

4. This article is partly based on a handful of interviews conducted by the author in Amish communities in Indiana, Illinois, and New York. Because the Amish value their privacy, their names will not be cited. For an interesting discussion on the difficulties of interviewing the Amish, see [7].

5. For a detailed account of Amish history, see [8].

6. Menno Simons was the founder of the Mennonites. The Amish Church broke from the Mennonites in the late seventeenth century in part because they believed the Mennonites were straying from Menno Simons' teachings [10].

7. Although the Amish separate themselves for the good of their own people, they have not forgotten the outside world. Their desire to help others is often directed toward those outside their community. Should a non-Amish neighbor's barn burn down, the Amish will band together and help with the erection of a new one, just as they would for a fellow Amish person. Above and beyond this, some Amish communities are known to participate actively in hunger and disaster relief projects across the world.

8. While each district has its own distinct Ordnung, they are similar on many points. As such this article will often refer to "the Ordnung" of the Amish in general for those issues on which there is almost universal agreement.

9. Nearly every Amish person interviewed for this article gave this answer at one point or another.

10. *Rumspringa* has recently been subject to a fair amount of media coverage in the United States because of the 2004 UPN television show "Amish in the City" and the 2002 feature-length documentary *Devil's Playground*. These programs can be a bit misleading as they focus on the most extreme examples of Amish rebelliousness. Most Amish teenagers do not live in Los Angeles, parade up and down the red carpet at movie premieres, or deal drugs.

11. These new systems proved to be so efficient that a few English companies now produce them for non-Amish shops [23]. The Amish are surprisingly inventive in other fields as well and have even been awarded patents in a few cases. For instance, they have developed a cookstove that employs an airtight combustion compartment that some claim is the "only significant advance in wood-fire stoves in 300 years" [24, p. 30]. The Amish also have designed a horse-drawn plow fitted with a hydraulic lift so that rocks do not present as much of a problem to farming [25].

12. The Amish relationship with medicine follows a similar rule. While they rely on homeopathic remedies for many things, if they find an English doctor whom they trust, the Ordnung does not prohibit them from receiving medical care that uses advanced technologies.

13. A number of scholars have criticized this stance as just one more method the male-dominated society uses to repress women [26].

14. Whether and how the Amish can receive phone calls varies from district to district [27]. For instance, some do not allow incoming calls to be answered, some allow calls to be prearranged, and some use voice mail services provided by phone companies.

References

[1] Meyers, T. J., and S. M. Nolt. *An Amish Patchwork: Indiana's Old Orders in the Modern World*. Bloomington, IN: Quarry, 2005.

[2] Sclove, R. E. "Spanish waters, Amish farming: Two parables of modernity?" In *Democracy and Technology*. New York: Guilford, 1995, pp. 3–9.

[3] Jasanoff, S. ed. *States of Knowledge: The Co-production of Science and Social Order*. New York: Routledge, 2004.

[4] Smith, M. R., and L. Marx. *Does Technology Drive History? The Dilemma of Technological Determinism*. Cambridge, MA: MIT Press, 1994.

[5] Meyers, T. J. "Education and schooling." In *The Amish and the State*, edited by D. B. Kraybill. Baltimore: Johns Hopkins University Press, 1993, pp. 86–106.

[6] Amish bishop, interview with the author, Shipshewana, IN, February 3, 1996.

[7] Umble, D. Z. "Who are you? The identity of the outsider within." In *Strangers at Home: Amish and Mennonite Women in History*, edited by K. D. Schmidt, D. Z. Umble, and S. D. Reschly. Baltimore: Johns Hopkins University Press, 2002, pp. 39–52.

[8] Hostetler, J. A. *Amish Society*. 4th ed. Baltimore: Johns Hopkins University Press, 1993.

[9] Simons, M. "Brief and clear confession." In *The Complete Writings of Menno Simons*, edited by J. C. Wenger. Scottdale, PA: Herald, 1956, pp. 422–454.

[10] Kraybill, D. B. *The Riddle of Amish Culture*. Rev. ed. Baltimore: Johns Hopkins University Press, 2001.

[11] Romans 12:2.

[12] Amish carpenter, interview with the author, Shipshewana, IN, January 27, 1996.

[13] Bishop, interview with the author, Seneca Falls, NY, November 21, 1997.

[14] Businessman, interview with the author, Finger Lakes Region, NY, November 20, 1997.

[15] Stoll, E., and M. Stoll. *The Pioneer Catalogue of Country Living*. Toronto: Personal Library, 1980.

[16] Pratt, D. O. *Shipshewana: An Indiana Amish Community*. Bloomington, IN: Quarry, 2004.

[17] Mabry, R. "Be ye separate: A look at the Illinois Amish." *Champaign News-Gazette*, 1989.

[18] Larimore, V., dir. *The Amish: Not to Be Modern*. Film. 1986.

[19] Meyers, T. J. "The Old Order Amish: To remain in the faith or to leave." *Mennonite Quarterly Review* 68, no. 3 (July 1994): 378–395.

[20] Pringle, K. "The Amish dilemma: The attraction of the outside world." *Champaign-Urbana News-Gazette*, August 30, 1987, p. E1.

[21] Amish housewife, interview with the author, Arthur, IL, March 30, 1996.

[22] Kraybill, D. B., and S. M. Nolt. *Amish Enterprise: From Plows to Profits*. 2nd ed. Baltimore: Johns Hopkins University Press, 2004.

[23] Amish carpenter, interview with the author, Arthur, IL, March 30, 1996.

[24] Brende, E. "Technology Amish style." *Technology Review* 99, no. 2 (February/March): 26–33.

[25] Tenner, E. "Plain technology: The Amish have something to teach us." *Technology Review* 108, no. 7 (July 2005): 75.

[26] Reschly, S. D. "'The parents shall not go unpunished': Preservationist patriarchy and community." In *Strangers at Home: Amish and Mennonite Women in History*, edited by K. D. Schmidt, D. Z. Umble, and S. D. Reschly. Baltimore: Johns Hopkins University Press, 2002, pp. 160–181.

[27] Umble, M. Z. *Holding the Line: The Telephone in Old Order Mennonite and Amish Life*. Baltimore: Johns Hopkins University Press, 1996.

[28] Rheingold, H. "Look who's talking." *Wired* 7, no. 1 (January 1999): 128–131, 161–163.

[29] Young Amish farmer, interview with the author, Arthur, IL, March 30, 1996.

[30] Good, M. *Who Are the Amish*? Intercourse, PA: Good Books, 1985.

[31] Amish corn and dairy farmer, interview with the author, Arthur, IL, March 30, 1996.

17 Preserving Traditional Knowledge: Initiatives in India

Rupak Chakravarty

In addition to technologies themselves embodying and promoting specific values, the ways in which we organize around technologies can promote specific values as well. This can be clearly seen when different forms of ownership of technological knowledge clash. In 1787 the writers of the US Constitution charged the federal government "to promote the scientific and useful arts, by securing for limited times to authors and inventors the exclusive right to their respective writings and discoveries." This charge led to the development of US copyright and, more importantly for this book, modern patent laws. Many of us take for granted that inventors should be given an exclusive right to make money from their inventions. But this is a culturally specific approach to technological knowledge that is aligned with a number of key capitalistic values, including providing financial incentives to develop new ideas. Some other cultures approach technological knowledge with a more sharing economy in mind—one in which everyone benefits from new ideas. As our economies become more and more global, these cultures, and the values they hold, are coming into conflict.

In this chapter, Chakravarty is concerned about the role that traditional knowledge plays in both holding communities together and constituting identities and how those who want to exploit such knowledge are challenging this role. In parts of India, traditional medicinal knowledge is passed down by word of mouth from generation to generation, and some is documented in classic texts. Outside scientists and drug manufacturers see this knowledge as a possible source for the next revolutionary medical treatment in Western medicine. The tension between societies with traditional knowledge and those who would like to exploit that knowledge for broader commercial purposes converges around questions about who owns the knowledge and who, if anyone, should be able to control how the knowledge is used.

Traditional Knowledge

The knowledge and uses of specific plants for medicinal purposes (often referred to as *traditional medicine*) is an important component of Traditional Knowledge (TK). TK is also termed *indigenous knowledge* (IK), *traditional environmental knowledge* (TEK) and *local knowledge*.[1] It refers to the knowledge systems held by traditional communities and is based on their experience and adaptation to a local culture and environment. This knowledge is used to sustain the community and its culture. Placing value on such knowledge helps strengthen cultural identity and the enhanced use of such knowledge

From *IFLA Journal* 36, no. 4 (2010): 294–299, https://doi.org/10.1177/0340035210388246.

to achieve social and development goals, such as sustainable agriculture, affordable and appropriate public health, and conservation of biodiversity.[2] Traditional knowledge is collective in nature and is often considered the property of the entire community, and does not belong to any single individual within the community. For many communities, TK is inseparable from their cultural values, spiritual beliefs and customary legal systems and is viewed as their intellectual property. Such systems are significant, not only for these communities, but also for the whole world.[3]

Intellectual Property and Traditional Knowledge

The term *intellectual property* (IP) reflects the idea that it is the product of the mind or the intellect. It is protected through law and can be owned, sold or bought.[4] IP law confers enforceable rights upon the person responsible for the intellectual output, so that the creator or owner of IP can exercise a measure of control over its future use. It plays an important role in all aspects of human life, including health care. Each country has developed its own IP laws to regulate the use and re-use of intellectual inventions within specific territorial boundaries.

The role of IP systems in relation to traditional knowledge, its preservation, protection and use, has recently received increasing attention in a number of international forums on matters as diverse as food and agriculture, the environment (notably the conservation of biological diversity), health (including traditional medicine), human rights and indigenous issues, and aspects of trade and economic development.[5] While the policy issues concerning TK are broad and diverse, the IP issues break down into two key themes: positive protection (giving TK holders the right to take action or seek remedies against certain forms of misuse of TK) and defensive protection (safeguarding against illegitimate IP rights taken out by others over TK subject matter).[6]

Indigenous and local communities have argued in national and international law-making bodies that their knowledge systems should not be used by others, without their consent, as well as arrangements for fair sharing of the benefits. Understanding the role of intellectual property and TK, various intergovernment bodies like the Convention on Biological Diversity (CBD),[7] the World Intellectual Property Organization (WIPO),[8] the Food and Agriculture Organization of the United Nations (FAO),[9] the World Trade Organization (WTO)[10] and the United Nations Conference on Trade and Development (UNCTAD)[11] are working in this direction. WIPO has taken the initiative by considering the needs of the representatives of TK-holding communities from all over the world. In 1981, WIPO and UNESCO adopted a model law on folklore.[12] In 1989, the concept of Farmers' Rights was introduced by the FAO into its International Undertaking on Plant Genetic Resources, and in 1992 the CBD highlighted the need to promote and preserve traditional knowledge.[13] The WIPO Intergovernmental Committee on Intellectual Property and Genetic Resources, Traditional Knowledge and Folklore (IGC) was established in 2001 as an international policy forum.[14] Many multilateral treaties have also been enacted over the last more than 100 years with nation states enacting the general principles found in such treaties into their own domestic laws. For instance, India has adopted sui generis laws that protect at least some aspects of TK.[15]

Traditional Medicinal Knowledge of India

Indians have an age old tradition of using herbs and spices with medicinal value, like amla (*Phyllanthus emblica*), black pepper (*Piper nigrum*), basil (*Ocimum basilicum*), etc.

Amla (Indian Gooseberry) is the most potent natural source of vitamin C, which is an excellent anti-oxidant and contains as much vitamin C as two oranges. It helps maintain a stronger, healthier digestive system, improved overall immunity, detoxifies the body, purifies the blood, lowers cholesterol, enhances vision, and strengthens the lungs, respiratory system and central nervous system. Kali Mirch (black pepper) is an expectorant, carminative, antipyretic, anthelmintic and appetizer. In India, it has been used as a medicine to cure toothache, asthma, chronic indigestion, colon toxins, obesity, sinus congestion, fever, colic pain, cholera, gastric ailments, etc. Another Indian spice called methi (fenugreek) cures indigestion, constipation, mouth ulcers, prevents the formation of kidney stones, and controls blood sugar level. Tulsi, the most sacred herb of India, is used as a nerve tonic, to sharpen memory, and cure fevers, common colds and respiratory disorders like bronchitis, asthma and influenza, etc. It also acts as a remedy in cases of influenza, kidney stone and heart disorders by reducing blood cholesterol. Ginger, perhaps the most sought after spice in most of the Indian foods, provides relief from sweating, vomiting, dizziness, nausea, arthritic pain, ulcerative colitis, headaches, fevers from flu and colds, sore throats, and chemotherapy. This traditional knowledge about the uses of herbs, better known among Indians as *dadi maa ke nuskhe* (Grandma's treatment), has been handed down from one generation to another and forms an inseparable part of Indian culture.

Such traditional knowledge is being used by a number of pharmaceutical organizations. In South India, the medicinal knowledge of the Kani tribes led to the development of a sports drug named Jeevani, an anti-stress and anti-fatigue agent, based on the herbal medicinal plant arogyapaacha.[16]

Traditional Knowledge Digital Library (TKDL)

India's rich traditional knowledge has not only been passed down by word of mouth from generation to generation, but has also been described in ancient classical and other literature. Such knowledge is often inaccessible to the common man, and even when accessible, is rarely understood, as it exists in local languages such as Sanskrit, Urdu, Arabic, Persian, Tamil, etc. Documentation of this existing knowledge of various traditional systems of medicine, available in the public domain, has become imperative to protect it from being misappropriated in the form of patents on non-original innovations.[17] It had been observed that, in the past, patents have been granted to inventions related to already known traditional knowledge because the patent examiners could not search for relevant traditional knowledge as prior art, due to the non-availability of such information in the classified non-patent literature. In 1995, the United States Patent Office granted a patent on the wound-healing properties of turmeric (*Curcuma longa*) which was challenged successfully and the patent revoked. The revocation of the patent granted by European Patent Office to W.R. Grace Company and the United States Department of Agriculture on neem (EPO patent No. 436257), again on the same grounds of its use having already been known in India, is another example.[18] A study conducted in 2000 showed that 4,896 patents on medicinal plants had been granted by the US Patent Office, 80 percent of which were on plants of Indian origin.[19] The findings also revealed that out of 760 such patents, 350 should have not been granted. Some 200–500 such patents are granted every year, mainly due to the lack of access to documented traditional knowledge in India. Every year, about 1,500 patents were being granted by the European Patent Office (EPO) and the US Patent Office, based on traditional Indian knowledge in medicine.[20]

Keeping in view the importance of such traditional medicinal knowledge, the Department of Ayurveda, Yoga and Naturopathy, Unani, Siddha and Homoeopathy (AYUSH) of the Indian government constituted an inter-disciplinary Task Force in 1999 for the preparation of an approach paper on establishing a Traditional Knowledge Digital Library (TKDL). Accordingly, the Government of India has undertaken the development of the TKDL database to prevent patenting of inventions based on Indian traditional knowledge (figure 17.1).

TKDL aims to act as a bridge between the traditional knowledge existing in local languages and the patent examiners at various international patent offices. If TKDL had existed earlier, international disputes such as those referred to above would not have arisen. TKDL has also resolved the perpetual problem of lack of access to documentation on India's traditional medicine due to language barriers or formatting incompatibilities, thereby abating the loss of future revenue and resources. It is seen by India as a safeguard against the burgeoning research-based fields of biopharmacology, integrative medicine (IM), evidence-based complementary and alternative medicine (CAM), ethnobotany, and ethnopharmacology.[21]

TKDL is a joint project of five Indian government organizations, including the Council of Scientific and Industrial Research (CSIR) and the National Institute of Science Communication and Informative Resources (NISCAIR). More than 150 traditional medicine practitioners, information technology engineers, patent examiners, intellectual property attorneys, scientists, researchers and librarians worked together to construct this database for India's indigenous medical and scientific knowledge resources which would fit within the framework of the International Patent Classification (IPC) scheme. The TKDL teams systematized and arranged the ancient and mediæval Indian medicaments in this database in accordance with modern conventions of taxonomy.

Figure 17.1
The TKDL interface.

The database is built up from transcribed texts of the triad of Indian medical sciences—Ayurveda, Unani and Siddha—transposed sacred *slokas* (verses), of 14 ancient texts from the 6th to the 3rd century BC Vedic corpus, and other authoritative Oriental canons and treatises.

Translation of palm leaf scriptural verses, parchment manuscripts, textbook citations and oral tradition references into decoded English, French, German, Japanese, and Spanish required Brahmi-based and other non-Latin script conversions of Vedic Sanskrit, classical Sanskrit, Hindi, Arabic, Farsi/Persian, Dravidian Tamil and Urdu in accordance with international language encoding standards (ISO) and Unicode metadata. The TKDL team developed a "smart translation" software to produce the scanned text and images from 54 primary sources on ayurvedic medicinal properties, provenance data, biological activity, chemical constituents, approximately 150,000 triad medicines and pharmaceutical preparations, 1,500 yoga asana therapies, traditional botanical names, malady descriptions, and other bibliographic details in contemporary terminology.[22] TKDL has completed documenting over 220,000 medical formulations (including 81,000 Ayurveda, 140,000 Unani and 12,000 Siddha formulations) and saved them from piracy. TKDL is a dynamic database, where formulations are continuously added and updated according to inputs from the users of the database.

The information on traditional medicines appears in a standard format in TKDL. For example, formulations on Indian Systems of Medicine appear in the form of a text, which comprises the name of the drug, origin of the knowledge, constituents of the drug with the parts used and their quantity, method of preparation of the drug and usage of the drug as well as bibliographic details. TKDL uses modern names of plants (e.g. *Curcuma longa* for turmeric), diseases (e.g. fever for jwar), or processes and establishes relationships between traditional knowledge and modern knowledge.[23] TKDL includes a search interface providing full text search and retrieval of traditional knowledge information using the International Patent Classification (IPC), Traditional Knowledge Resource Classification (TKRC) and keywords in multiple languages. TKRC, an innovative structured classification system for the purpose of systematic arrangement, dissemination and retrieval has been evolved for about 25,000 subgroups related to medicinal plants, minerals, animal resources, effects and diseases, methods of preparations, mode of administration, etc.[24] Search features of TKDL include complex Boolean expression search, proximity search, field search, phrase search, etc. The database does not claim exhaustive coverage and does not affect the rights and obligations relating to any prior art traditional knowledge formulation or know-how not listed in TKDL. Hyperlinks to other websites are provided for convenience only. This does not imply either responsibility or approval of the information contained in those websites.

The contents of TKDL are being digitally transcribed into a readable form in five international languages—English, French, German, Japanese and Spanish—with the objective of preventing their misappropriation at international patent offices. The status of transcription of the traditional medicine formulations in the TKDL as of May 2010 is given in table 17.1.

India is going all out to save yoga, a 2,000-year-old Indian art of righteous living. The team of TKDL is presently scanning through 35 ancient Sanskrit texts, including the Mahabharata, Bhagawad Gita and the Yoga Sutras of Patanjali to identify and document all known yoga concepts, postures and terminology. Among the yoga books being scanned by scientists are *Hatha Praditika*, *Gheranda Samhita*, *Shiva Samhita* and *Sandra Satkarma*. Currently, 600 *asanas* (physical postures) have already been documented with a target to put on record at least 1,500 such yoga postures by the end of 2009. Till now, 130

Table 17.1
Current status of TKDL database

Discipline	Number of texts (including volumes) used for transcription	Transcribed formulations
Ayurveda	75 books	85,500
Unani	10 books	120,200
Siddha	50 books	13,470
Yoga	15 books	1,098
Total	150 books	220,268

Source: Council of Scientific and Industrial Research (CSIR), *About TKDL: Traditional Knowledge Digital Library (TKDL)*, accessed July 2, 2010, http://www.tkdl.res.in/tkdl/langdefault/common/Abouttkdl.asp?GL=Eng.

yoga-related patents granted in the USA have been traced by TKDL.[25] Once the postures are put on record, they would be made available in five international languages. Besides photos and explanation of the postures, video clips of an expert performing them will be put in the TKDL. A voice-over will also point out which text mentions the posture.

In February 2009, the Indian government granted access to TKDL to the European Patent Office under a three-year agreement. The TKDL allows examiners at EPO to compare patent applications with existing traditional knowledge. New patent applications need to demonstrate significant improvements and inventiveness compared to prior art in their field. The cooperation between India and the EPO comes at a time when many countries are struggling to protect traditional and respected knowledge against exploitation, primarily in the pharmaceutical sector. The 34 member states of the EPO now have restricted access for purposes of patent search and examination. TKDL is integrated with the EPO's database as another measure to thwart illegitimately-gained exclusivity.[26] Experts at the EPO say that access to the 30-million-page database will help them to correctly examine patent applications relating to traditional knowledge at an early stage of patent examination. One perceived flaw is the lack of accessibility to online backtracking of certificates of correction and defective patents. Patents are granted for new uses, innovative delivery systems, different combinations, and novel variations of chemical entities and properties.

The TKDL effort has also been appreciated at the international level as well. It has become a model for other countries on defensive protection of their traditional knowledge from misappropriation. Countries and organizations such as South Africa, the African Regional Intellectual Property Organization (ARIPO), Mongolia, Nigeria, Malaysia and Thailand have expressed their keen desire to replicate TKDL.[27]

Conclusion

TK is part and parcel of the daily life of Indians. Very few countries in the world can boast of the variety and vastness of traditional knowledge that India has. However, this knowledge has been exploited throughout the world and is being exclusively patented by foreigners. Natives of India have protested against this, and have urged the government to take every measure to protect our traditional knowledge heritage. Libraries also can play a significant role in this regard as they are now implementing digitization projects for the preservation of our national heritage. However, projects like TKDL should be open for citizens to add new forms/cases of traditional knowledge which are still undiscovered by our scientists.

Notes

1. United Nations Environment Programme, *Cultural and Spiritual Values of Biodiversity*, 1999, wedocs.unep.org/bitstream/handle/20.500.11822/9190/Cultural_Spiritual_thebible.pdf.

2. World Intellectual Property Organization, *Intellectual Property and Traditional Knowledge*, booklet no. 2, WIPO Publication No. 920(E), accessed September 7, 2009, http://www.wipo.int/freepublications/en/tk/920/wipo_pub_920.pdf.

3. S. A. Hansen and J. W. Van Fleet, *Traditional Knowledge and Intellectual Property* (Washington, DC: American Association for the Advancement of Science, 2003).

4. Ministry of Commerce and Industry, Department of Industrial Policy and Promotion, "Intellectual Property," accessed September 7, 2009, http://dipp.nic.in/ipr.htm (page no longer available; the Department of Industrial Policy and Promotion was renamed the Department for Promotion of Industry and Internal Trade in 2019. Its work in intellectual Property can be found at http://www.ipindia.gov.in).

5. Plant Interactions, "Traditional Medicine," 2010, accessed July 1, 2010, http://www.plantinteractions.co.uk/ethnobiology/traditional-medicine/ (page no longer available).

6. World Intellectual Property Organization, "Traditional Knowledge," accessed September 7, 2009, http://www.wipo.int/tk/en/tk/.

7. Convention on Biological Diversity (website), accessed December 20, 2009, http://www.cbd.int/.

8. World Intellectual Property Organization (website), accessed June 10, 2009, http://www.wipo.int/portal/index.html.en.

9. Food and Agriculture Organization of the United Nations (website), accessed December 20, 2009, http://www.fao.org/.

10. World Trade Organization (website), accessed December 20, 2009, http://www.wto.org/.

11. United Nations Conference on Trade and Development (UNCTAD) (website), accessed November 25, 2020, http://unctad.org.

12. B. O'Connor, "Protecting Traditional Knowledge: An Overview of a Developing Area of Intellectual Property Law," *Journal of World Intellectual Property* 6:677.

13. R. V. Anuradha, *Use of Biological Resources or Traditional Knowledge: Additional Disclosure Proposed*, Centre for Trade and Development (Centad), accessed July 1, 2010, http://www.centad.org/focus_25.asp (page no longer available).

14. World Intellectual Property Organization, "Traditional Knowledge, Genetic Resources and Traditional Cultural Expressions/Folklore," accessed September 7, 2009, http://www.wipo.int/tk/en.

15. Ministry of Law, Justice, and Company Affairs, "The Patents (Amendment) Act 2002," *Gazette of India*, June 25, 2002, accessed November 25, 2020, https://dipp.gov.in/sites/default/files/patentg_0.pdf.

16. World Intellectual Property Organization, "Intellectual Property and Traditional Knowledge," accessed July 1, 2010, www.wipo.int/freepublications/en/tk/920/wipo_pub_920.pdf.

17. Council of Scientific and Industrial Research, "About TKDL," accessed September 7, 2009, http://www.tkdl.res.in/tkdl/langdefault/common/Abouttkdl.asp?GL=Eng.

18. R. Menon, "Traditional Knowledge Receives a Boost from the Government," *India Together*, January 13, 2007, accessed July 6, 2010, http://www.indiatogether.org/tkdl-economy.

19. N. Sen, "TKDL–A Safeguard for Indian Traditional Knowledge," *Current Science* 82, no. 9 (2002): 1070–1071.

20. K. Sinha, "India Logs over 2 Lakh Traditional Medical Formulations," *Times of India*, June 10, 2010, accessed September 7, 2009, http://timesofindia.indiatimes.com/articleshow/msid-4101577,prtpage-1.cms.

21. L. Poussaint, "Traditional Knowledge Digital Library of India Brings Ancient, Indigenous Medical Systems Online" (Web log comment), May 13, 2009, accessed September 7, 2009, http://mlaics.blogspot.com/2009/05/traditional-knowledge-digital-library.html.

22. Ibid.

23. Council of Scientific and Industrial Research, "FAQ#4: What is TKDL?," accessed September 7, 2009, http://www.tkdl.res.in/tkdl/langdefault/common/Faq.asp?GL=Eng.

24. P. Jain and P. Babbar, "Digital Libraries Initiatives in India," *International Information and Library Review* 38, no. 3 (2006): 161–169, accessed September 7, 2009, doi:10.1016/j.iilr.2006.06.003.

25. K. Sinha, "Yoga Piracy: India Shows Who's the Guru," *Times of India,* February 22, 2009, accessed September 7, 2009, http://timesofindia.indiatimes.com/news/india/Yoga-piracy-India-shows-whos-the-guru/articleshow/4167939.cms.

26. Poussaint, "Traditional Knowledge Digital Library."

27. Council of Scientific and Industrial Research, "About TKDL: Traditional Knowledge Digital Library (TKDL)," accessed July 2, 2010, http://www.tkdl.res.in/tkdl/langdefault/common/Abouttkdl.asp?GL=Eng.

18 Equity in Forecasting Climate: Can Science Save the World's Poor?

Maria Carmen Lemos and Lisa Dilling

Many people seem to believe that scientific research and technological development constitute goods in and of themselves. They base their belief on the idea that these activities have generated immense knowledge and achievements that have made the world a better place. They may point to the dramatic drop in the infant mortality rate or the worldwide increase in life expectancy—both possible largely because of new medical knowledge and interventions. Or they may argue that the spread of democracy, or even the increase in the rights of women, can be attributed, in part at least, to new technologies.

But even if the sum total of technologies has resulted in a net good, particular endeavors are often designed to further the values and interests of one specific person or group of people. While most technologies are intended to improve the world, it seems that change, perhaps inevitably, brings losers as well as winners. For instance, it is difficult to identify a technology that has not made at least one person's life worse. Labor-saving technologies reduce burdens but also put workers out of jobs; when a technology gives people a new ability, those without that technology may be left behind both socially and economically. A number of scholars have explored this phenomenon and argue that new technologies frequently exacerbate inequality—they broaden the gulf between the haves and the have nots.

In this chapter, Lemos and Dilling explore an attempt to use technology to bridge that gulf. They describe a group of scientists who work concertedly to develop new knowledge with the direct intention of helping a group of disadvantaged, and perhaps even desperate, farmers to get a leg up. This provides a relatively uncommon example of technology being deliberately designed to address inequity. Unfortunately, Lemos and Dilling find that new technologies—even very powerful ones—do not always further the values of their designers. Sometimes the existing economic, political, and social structures make it incredibly difficult to disrupt the status quo.

Between 1979 and 1983, a series of devastating climatic events, including severe drought in northeast (NE) Brazil and Australia, flooding in Peru and Ecuador and drought-related famine in southern Africa and India, revealed to the world the harmful effects of El Niño. Although this was not the first global devastation related to El Niño–Southern Oscillation (ENSO),[1] it was the first in which ENSO effects were widely publicized as an interconnected global phenomenon. The ENSO wreaks havoc on many tropical and subtropical regions of the world, disrupting normal patterns of rainfall to cause severe droughts and catastrophic flooding.

From *Science and Public Policy* 34, no. 2 (March 2007): 109–116.

To make matters worse, many of the regions most hard hit by ENSO have populations in poverty, already living close to the margin for survival in a "normal" year. In NE Brazil alone, the four-year drought caused by the 1983 El Niño affected 18 million *nordestinos*,[2] and in response, the Government spent an estimated US$1.8 billion on emergency programs (Magalhães et al., 1989, 334). More recently, a multi-year drought has had serious impacts on the livelihood of eastern African populations where pastoralists living close to the margin of poverty have been particularly affected in countries such as Kenya (Reliefweb, 2006).

In the mid-1980s, scientists interested in climate dynamics understood the mechanisms of the ENSO phenomenon well enough to be able to predict with some skill the onset of its warm (El Niño) or cool (La Niña) phase some several months to even a year in advance. Not surprisingly, the possibility that scientists might be able to forecast seasonal climate variations, and anticipate their negative consequences such as drought and flooding, captured the attention of policy-makers seeking to improve the livelihoods of those negatively affected by climate-driven hazards.

Because of ENSO's dominant impact on many vulnerable populations worldwide, research on, and application of, seasonal climate forecasting (SCF) has often been specifically justified in terms of their potential for improving the lot of those most in need (for example, see McPhaden et al., 2006). However, if the idea of positive societal impact enticed atmospheric scientists, climatologists and funding agencies to improve the science behind forecasting, early optimism has somewhat faded and many challenges remain (Harrison, 2005). The results of the application of the new technology have been mixed, not only in terms of effectiveness, that is, how much SCF has been used successfully to deflect losses, but also in terms of equity, that is, how SCF use has actually benefited those most in need.[3]

While there is considerable focus in the climate impacts and forecasting literature on theorizing about potential benefits and forecast value, especially how SCF application could improve the response to hazards in the short term (Archer, 2003; Glantz, 1996; Jacobs, 2003; Keogh et al., 2004; Magalhães et al., 1988; Nelson et al., 2002; Pagano et al., 2002; Sayuti et al., 2004; Ziervogel and Calder, 2003), there are relatively few examples of empirical studies evaluating actual forecast use to date. However, what is already available allows us not only to temper some of the more optimistic speculations of forecast value but also, more importantly, to learn from experience to increase opportunities for success.

In the case of SCF, we suggest it is particularly important to evaluate the equity implications of its application both because of its policy justifications and because failures can be especially devastating to those already living at the margin of survival. In this article, we review the literature focusing on the experiences and impacts of SCF and explore three main challenges that can negatively affect equity in its use.

First, while investment in SCF as a decision-support tool has been justified in social terms, that is, as a means to improve the lot of those most vulnerable to climatic variability, many of the examples of application reported in the literature show that this is not always the case. In fact, not only are the most vulnerable, in many cases, unable to benefit from SCF information but may be harmed by it. Here access to resources and to power influence the ability of different users to benefit from SCF use, and previous levels of underlying inequities and differential vulnerabilities also matter.

Second, the usability of SCF as a decision-making tool has been constrained by issues of communication and accessibility. Both the character of information (probabilistic) and its availability (the means of its release, communication and dissemination) shape its access by different groups. Factors such as levels of education, access to electronic media, such as the Internet, and to expert knowledge, critically affect the ability of different groups to take advantage of SCF as a decision tool. Unequal access to technical information can also create power imbalances that negatively affect decision-making processes using SCF. While the adoption of participatory processes of communication and dissemination seem to have a positive effect on the accessibility of SCF by low-income groups, these experiences have so far been limited.

Finally, because resources are spent on SCF projects as a potential solution to climate-related vulnerability, other policies that may be more effective may be precluded from being implemented. To date, the implications of the opportunity cost of the application of SCF are not well understood.

Despite these challenges, the literature also illustrates promising new ways of applying SCF that address equity issues more positively. We review several of these cases, and argue that, without attention to specific mechanisms to counter pre-existing inequities, the distribution and use of SCF is not likely to ameliorate the conditions of those most in need.

In the next sections, we discuss these issues in the light of empirical examples of SCF's application.[4] First, we examine the evolution of SCF as a decision-support tool and discuss its equity implications in the context of resource-poor and resource-rich policy arenas. We also review how institutional and resource constraints shape the ability of populations to rely on these tools over a longer time period. Then we explore how the issue of unequal access to information and barriers to communication affect equity in the application of SCF. Finally, we discuss the opportunity costs of SFC use. We conclude with suggestions for SCF application based on successful examples that might improve the equity of SFC as a decision-support tool.

Forecasting Climate and Accounting for Equity

The ENSO is a well-defined coupled ocean–atmosphere system that influences a wide range of climate-related events around the globe. Although the patterns forming El Niño and their statistical associations with climate-related events have been known for some time, it was not until the late1980s that Zebiak and Cane put together the first model to simulate ENSO (Zebiak and Cane, 1987).[5] This auspicious beginning created great expectation that ENSO modeling and forecasting would quickly generate an array of application activities that could critically affect the ability of different users to mitigate the high risk associated with the effects of climate variability on different systems, especially on agriculture and water management. Funding agencies and forecast producers actively hailed the potential positive societal impact of SCF application, especially to resource poor segments of users whose livelihoods have been historically negatively affected by climate variability (Broad et al., 2002).

Advances in SCF over the past few decades have led to its application in experimental settings in many regions around the world. In these experiments, the expectation among SFC producers and policy-makers has been that, if forecasts were available and reasonably accurate, decision-makers at diverse scales and income levels could use advanced information about potential hazard in their planning. In such cases, rather than responding to the hazard reactively and poorly, forecast users could better prepare, recover and cope with its negative consequences.

For example, farmers could tailor their choice of crops and planting calendars to the likelihood of drought. Civil defense officials could adjust their budgets, human resources and disaster preparedness plans to the expectation of an incoming flood-prone rainy season. Water managers could plan water allocation and storage based on an expectation of less or more rainfall in coming months. While some level of loss due to climate variability stress will always be likely to occur, the goal has been to use SCF to aim for outcomes that would be at least comparable with, and hopefully better than, the situation before.

Yet, in contrast to many science-policy processes where policy-makers recruit science to solve specific problems (which by itself is no guarantee of success), the application of SCF has been as motivated by the progress of forecasting science as by the need to reduce risk. To a certain extent, the solution, rather than the problem, has framed the relationship between the production and dissemination of climate forecasting among different users. In consequence, many of the processes of climate forecast use so far documented suffer from "new technology blues" (Lemos et al., 2002), and the promise of utility and value of the forecast is constrained by both material and institutional factors ranging from lack of resources, to poor communication, to inequitable distribution of knowledge (Broad et al., 2002; Lemos, 2003; Lemos et al., 2002; Patt and Gwata, 2002; Rayner et al., 2005).

As mentioned above, one equity implication of SFC use has been its limitation in benefiting those most vulnerable to climate variability. In some cases, this vulnerability is critically shaped by unequal power and leverage in being able to respond effectively to climate information (Agrawala et al., 2001). In these situations, differential levels in the ability to respond can create winners and losers within the same policy context. For example, in Zimbabwe and NE Brazil, news of poor rainfall forecast for the planting season influences bank managers, who systematically deny credit, especially to poor farmers they perceive as high risk (Hammer et al., 2001; Lemos et al., 2002). In Peru, a forecast of El Niño and the prospect of a weak season gives fishing companies an incentive to accelerate seasonal layoffs of workers (Broad et al., 2002).

In each of these cases, some users, such as banks and businesses, benefited from SCFs, because they were able to anticipate some of the outcomes of a poor season ahead and protect themselves. However, the people dependent on them for credit or livelihoods lose.[6]

In other cases, even if access to information and resources is not a critical limitation, individuals or institutions can be constrained in responding effectively to SCF. In the United States, there are several well-documented cases of institutional limitations in responding to improved scientific predictions of stream flow, seasonal weather patterns, and climate in water management (Callahan et al., 1999; Jacobs, 2003; O'Connor et al., 2005; Pulwarty and Redmond, 1997; Rayner et al., 2005). For example, in their study of water managers in three US regions, Rayner et al. (2005) found that, constrained by the high levels of accountability of their decision environment, water managers prefer to rely on their professional experience rather than on SCF to guide their management decisions. Just having better information available does not mean it can stimulate an improved response.

Among poor farmers in the global south, the general lack of alternatives, both in terms of technology and access to financial and human resources, acts as a critical constraint to their ability to use SCFs. Resource deficiencies among the most climate-vulnerable rain-fed farmers also curb their ability to respond to forecasts even if they have access to them (Lemos et al., 2002). In Zimbabwe, for example, poor farmers' flexibility to adjust their planting to forecasted climate may be limited both because they have to purchase maize seeds before forecasts are released and because there is a low number of seed varieties available (Hammer et al., 2001; Patt and Gwata, 2002).

Access to seed is also a problem in NE Brazil, where a poor climate forecast may delay Government-sponsored seed distribution, because local officials wait for the first rains to avoid what they perceive is a "waste" of seed if farmers plant too soon (Jacobs, 2003; Lemos, 2003; Lemos et al., 2002). In Burkina Faso, high levels of indebtedness among poor farmers and out-migration in search of wage labor in the mining sector constrain village farmers' ability to use SCF (Ingram et al., 2002).

The opposite situation is also true; for those that are already more resilient, or more resource-rich, SCFs have provided additional benefits in terms of improved ability to cope with hazards and disaster. Among rich agricultural systems, the benefits of SCF use are evident.

For example, in Australia, where forecast information is actively sought both by large agribusiness and Government policy-makers planning for drought, agricultural producers have been able to use SCFs to cope better with swings in their commodity production associated with drought (Hammer et al., 2001). One factor helping to explain this positive experience is that in Australia forecast producers' approach to the dissemination of SFCs included close interaction with farmers, use of climate scenarios to discuss the incoming rainfall season and automated dissemination of SCFs through the RAINMAN interactive software. Similarly, in Argentina, resource-rich farmers have been able to take advantage of available SCFs (Letson et al., 2001).

Significantly, most reported successes seem to be associated with the presence of resources that are usually not available to the most vulnerable groups—resources whose absence, often, defines their vulnerability to begin with. Thus many of the factors that make these successes possible, such as financial, social and human resources, are frequently out of reach of the poor, who lack education, money and time resources to engage forecast producers.

Yet, poverty and other vulnerabilities can be counteracted in the application of SCF, if attention is paid to maintaining alternative types of resources, such as sustained relationships with information providers or attention to the context of application. Even among farmers with fewer resources, access to climate information through sustained relationship with, and advice from, forecast and agricultural extension experts can result in positive experiences, such as in the case of small farmers in Tamil Nadu, India (Huda et al., 2004) and Zimbabwe (Patt and Gwata, 2002). In both cases, forecast "brokers"[7] made considerable effort to sustain communication and provide expert knowledge to targeted farmers who were able to benefit from the use of SCF.

However, also in both cases, the number of farmers targeted was but a tiny fraction of those that might have needed, and benefited from, this kind of support. In addition, it is unclear whether, once the research project is finished, such interaction will be sustainable or how what has been learned can be "scaled up" to benefit larger number of farmers in need. In any event, for these interactive approaches to succeed, participants usually have to have not only the financial resources to come to meetings or to access information through the media (at least through the radio) but also to be on the "radar screen" of organizers of workshops, especially in the case of events where participation is limited. Often the poorest segments of the population lack all these resources.

What we learn from these examples is that underlying inequities and differential vulnerabilities in many cases may impede the ability of SCF alone to alleviate negative climate-related outcomes for some vulnerable groups. Success stories seem to depend largely on the resources available to deploy SFC, both in rich and poor agri-economic systems; whenever adequate resources to customize and use forecasts are present, benefits are more likely to accrue. In contrast, when pre-existing conditions are inequitable and without specific counteracting measures, the application of SFC may exacerbate negative conditions for those who are most vulnerable (see also Woodhouse and Sarewitz, 2007).

Impact of Inequality on Equity

One of the fundamental problems often not anticipated by researchers and producers of SCF is that there is great disparity in the ability of potential users to access information (Agrawala et al., 2001). Beyond being available, information has to be accessible, that is, users must be able to understand it in order to use it. For example, better-educated and resource-rich users are more likely to have access to information through different media such as the Internet, television, and newspapers and to make informed use of this information. This is true both in terms of knowledge production (countries and groups within countries with more resources will be able to produce better "customized" information) and use (better-informed systems and users will be able to use information more efficiently).

For example, in the US south-west, forecast producers organized stakeholder workshops that refined their understanding of potential users and their needs. Because continuous interaction with stake-holders was well-funded and encouraged, producers were able to "customize" their product, including the design of user-friendly and interactive Internet access to climate information, to local stakeholders with significant success (Hartmann et al., 2002; Lemos and Morehouse, 2005; Pagano et al., 2002).

In contrast, unequal access to climate information can have negative consequences, when one group of decision-makers acts as a "gatekeeper" for that information and makes decisions insulated from society at large. When public officials (for instance, water managers, relief planners, agriculture and fisheries resource managers) cloaked in technical expertise, insulate their decisions from stakeholders, their decision-making process, lacking in transparency and accountability, not only ignores stakeholders' input but also may affect their interests negatively. In this case, if information is controlled by a few actors seeking to bolster their position vis-à-vis other stakeholders, knowledge can insulate decisions and intensify power imbalances between those with access to knowledge and those without. This kind of technocratic insulation can not only alienate participation but also discourage stakeholders from "buying into" management decisions (Lemos, 2003).

The case of Ceará in NE Brazil, one of the best-studied processes of SFC use to date, offers several illustrations of technocratic insulation in practice. For example, in water management, perceived insulation and lack of participation in reservoir management has led to an overall de-legitimization of the system in the eyes of some users and disregard for management policies (Taddei, 2005). Moreover, the perception of inequality diminishes the potential value of climate information as a decision tool for potential users.

In the Lower Jaguaribe River Basin in Ceará, for example, technical information may have contributed both to better water management and to expanding the power gap between technocrats and stakeholders in the process of water management. It may also have shaped users' perception of the value of SCF as a decision-support tool. Although the majority of river-basin committee members find that climate information is relevant to their decision-making process—7.8 on a scale of one to ten—only 33% consider it accessible. Moreover, 79.3% of all respondents find that the disparate level of technical knowledge among members is the main source of inequality within the committee, above economic and political power disparities.[8]

Communication seems to be an essential ingredient for the success or failure of people to use [. . .] climate knowledge operationally. SCF tools are mostly disseminated in the language of probabilities, which is difficult to assimilate, because non-scientists do not generally think probabilistically, nor do they interpret probabilities easily (Nicholls, 1999). Problems with misinterpretation and miscommunication have negatively affected SCF users, and discredited forecast producers as well as the forecast itself. For example, in Peru, the miscommunication and misinterpretation of climate forecasts in the 1997/98 ENSO not only negatively affected some stake-holders but also discredited the forecast in the eyes of users (Broad et al., 2002; Pfaff et al., 1999).

Difficulties with language and lack of attention to local institutions create an additional layer of constraint. Hammer et al. (2001) suggest that poor understanding of local systems may act as a deterrent to the communication and availability of information which ultimately may affect access. Ziervogel and Downing (2004) argue that one way for SCF providers to mitigate such constraints is to understand information networks better in the context of SCF dissemination, especially in less developed countries. Such understanding can "provide(s) a springboard for targeting future forecast dissemination, which is imperative if this information is to be of use, particularly to marginal groups" (Ziervogel and Downing, 2004, 97).

As Rayner and Malone (2001, 176) discuss, "poverty cannot be understood in terms of lack of goods or income, or even basic needs, but must rather be understood in terms of people's ability to participate in the social discourse that shapes their lives." If the goal for SCFs is indeed a focus on equity and improving the livelihoods of the poorest and most vulnerable to climate-related disasters, then improving access to information and the decision-making process is paramount.

One specific way to enhance SCF's impact on equity is to increase the level of inclusion of underrepresented groups even among the overall poor, such as women (Archer, 2003), lower castes (Roncoli et al., 2001), the old (Valdivia et al., 2001) and the most vulnerable to climatic events (Lemos et al., 2002). Archer (2003) argues that the current focus on aggregate categories of users such as farmers masks the inequality in terms of access among subcategories of potential users such as those mentioned above.

Ziervogel and Calder (2003) agree and contend that it is important to understand the vulnerabilities of different users to target SFC dissemination better. They suggest that building a typology of livelihoods would not only improve usability but also avoid negative application. Pfaff et al. (1999) suggest that the first step in addressing equity issues is to identify the interested parties and delineate their various goals. Obviously, limited resources and time would preclude full inclusion of all potential individual stakeholders. However, there are examples of concrete ways in which organizations are attempting to broaden the scope of who is targeted and involved when SCFs are being disseminated and discussed (Kgakatsi, 2001, as cited in Archer, 2003).

Opportunity Cost of SCF Use

A third source of inequity in SFC use relates to the opportunity cost of choosing SCF as the focus of policy to address climate-related vulnerabilities over other potentially more effective alternatives. In this sense, investment and reliance on climate technology can result in high opportunity costs for policy systems (Brunner, 2000), especially in less developed countries where resources are limited. Because empirical research increasingly shows that, rather than an environmental hazard, disasters are a combination of such hazards, poverty and other vulnerabilities (Blaikie et al., 1994),[9] we argue that in order to provide effective disaster response, governments should address both hazard risk and underlying vulnerabilities.

However, public policy-makers often perceive the complex solution of socioeconomic and political problems underlying disasters as financially impossible and politically unfeasible. In this context, it is not surprising that in the eyes of these policy-makers, the possibility of a technical "fix," such as the ability to forecast the onset of disasters, offers the promise of an easier path to mitigate their effects (Lemos, 2003).

This does not mean that SCFs have no role in improving vulnerability to climate, but it does suggest that there are opportunity costs to pursuing this strategy over others. Technical fixes may compete for resources with other more effective and equitable, but perhaps less politically viable, policy alternatives (such as income redistribution and institutional reform), to build adaptive capacity to climate variability and change (see also Woodhouse and Sarewitz, 2007).

Technical fixes have much appeal from a political perspective as they can be implemented with the authority of science, while avoiding the difficult decisions that decreasing climate and socioeconomic vulnerabilities might entail, especially those that involve any kind of resource redistribution or change in regulations. When such a strategy is implemented with the perceived neutrality of a scientific innovation, it can obscure difficult tradeoffs and exacerbate existing patterns of poverty and inequity.

Another opportunity cost in following the strategy of using SCFs to reduce vulnerability of poor populations is that they are not a foolproof method. Early optimism that the ability to predict El Niño effectively would progress rapidly has somewhat faded and the rate of progress in the skill of climate models to forecast seasonal climate variability with confidence slowed down (Harrison, 2005). Overall, the low skill of current forecasts, that is, "the frequency that a forecast is correct based on historic data" (Ingram et al., 2002, 334), has been mentioned in the majority of studies as a serious constraint to operational use.

Perhaps even more importantly for the issue of equity, potential users of SCF who are already at the margins of survival face a much greater risk from betting their meager resources on a forecast that turns out to be wrong (Hulme et al., 1992). Thus, subsistence farmers and others who rely on traditional means of coping with climate variability may be justifiably reluctant to abandon those methods, even if SCF may promise more success over the long run (Ingram et al., 2002).

Concluding Remarks

This essay has explored equity issues related to the use of SCF in different policy arenas around the world. Although scholars have extensively speculated about its potential beneficial impacts, the implications of its use for the distribution of resources and power among resource-poor groups has received relatively less attention. We find that, in the application of SFC, equity can suffer when potential users' underlying vulnerabilities are not also addressed, when access to information and communication is unequal, and when organizations and individuals lack resources and alternatives to adjust to forecasted climate. There may also be significant opportunity costs to the application of SCF.

From an equity perspective, if climate science applications seek to aid and target the vulnerable poor specifically, then policy-makers and SCF producers have to invest time and funds in understanding the process through which decisions are made and resources allocated. First, the dissemination and communication of SCF need to be more inclusive of vulnerable groups, and availability and access to climate information must be improved. Specific training and a concerted effort to "fit" the available information to local decision-making patterns and culture can be a first step to enhancing its relevance.

Second, SCF producers and policy-makers should be aware of the broader socio-political context and the institutional opportunities and constraints presented by SFC use; understanding potential users and their decision environment will not only allow for better fit between product and client but also avoid situations in which SCF use may in fact harm those it is supposed to help.

Finally, as some of the most successful examples show, SCF application should strive to be more transparent, inclusionary, and interactive as a means to counter power imbalances between those with resources and those without (see also Eubanks, 2007; Woodhouse and Sarewitz, 2007). Unequal distribution of knowledge can insulate decision-making, facilitate elite capture of resources, and alienate disenfranchised groups. In contrast, an approach that is interactive and inclusionary can go a long way to supporting informed decisions that, in turn, can yield better outcomes.

So can science, in the form of SCF, save the poor? From the SCF application experience thus far, we might say no, not by itself. Scientific innovations by themselves are no panacea for the age-old problems of poverty, inequity, and inertia. At the heart of the problem is not equality of outcomes but equality of opportunities to influence the process through which decisions are made. By being mindful of equity issues, we can begin to build a process in which a positive outcome is not a unique contextual experience but an expected result of the application of SCF as a decision-support tool.

Notes

1. For an interesting history of the nineteenth-century drought that may have killed an estimated sixty million people in Africa, India, and China, see Davies (2001).

2. As people from northeast Brazil are known.

3. For an early evaluation of SCF use in agriculture, see Hammer et al. (2001).

4. These empirical examples provide illustration for our arguments throughout this review and are not intended to test formal hypotheses about SCF and equity across different sectors or countries.

5. For a detailed description of the evolution of SCF, see Harrison (2005).

6. Here, rather than equality, the critical issue is the unfair distribution of outcomes (some win, some lose). See Cozzens (2007) for a discussion of the distinction between equity and equality.

7. Researchers in the India case and researchers and extension agents in the Zimbabwe case.

8. The Watermark Survey was carried out in the context of the Watermark Project, a broad comparative study of water management in Brazil, of which Lemos is one of the investigators.

9. Among the causes of vulnerability are lack of democracy, unequal power relations, and/or poor access to resources.

References

Agrawala, S., K. Broad, and D. Guston. 2001. "Integrating climate forecasts and societal decision making: Challenges to an emergent boundary organization." *Science, Technology and Human Values* 26:454–477.

Archer, E. R. M. 2003. "Identifying underserved end-user groups in the provision of climate information." *Bulletin of the American Meteorological Society* 84 (11): 1525–1532.

Blaikie, P., T. Cannon, I. Davis, and B. Wisner. 1994. *At Risk. Natural Hazards, People's Vulnerability and Disasters*. London: Routledge.

Broad, K., A. S. P. Pfaff, and M. H. Glantz. 2002. "Effective and equitable dissemination of seasonal-to-interannual climate forecasts: Policy implications from the Peruvian fishery during El Niño 1997–98." *Climatic Change* 54 (4): 415–438.

Brunner, R. 2000. "Alternatives to prediction." In *Prediction: Science, Decision Making and the Future of Nature*, edited by D. Sarewitz, R. Pielke Jr., and R. Byerly Jr., pp. 299–313. Washington, DC: Island Press.

Callahan, B., E. Miles, and D. Fluharty. 1999. "Policy implications of climate forecasts for water resources management in the Pacific Northwest." *Policy Sciences* 32:269–293.

Cozzens, Susan. 2007. "Distributive justice in science and technology policy." *Science and Public Policy* 34 (2): 85–94.

Davies, M. 2001. *Late Victorian Holocausts: El Niño Famine and Making of the Third World*. New York: Verso.

Eubanks, V. 2007. "Popular technology: Exploring inequality in the information economy." *Science and Public Policy* 34, no. 2: 127–138.

Glantz, M. 1996. *Currents of Change: El Nino's Impact on Climate and Society*. Cambridge: Cambridge University Press.

Hammer, G. L., J. W. Hansen, J. G. Phillips, J. W. Mjelde, H. Hill, A. Love, and A. Potgieter. 2001. "Advances in application of climate prediction in agriculture." *Agricultural Systems* 70:515–553.

Harrison, M. 2005. "The development of seasonal and inter-annual climate forecasting." *Climatic Change* 70:201–220.

Hartmann, H. C, T. C. Pagano, S. Sorooshian, and R. Bales. 2002. "Confidence builders: Evaluating seasonal climate forecasts from user perspectives." *Bulletin of the American Meteorological Society* 8:683–698.

Huda, A. K. S., R. Selvaraju, T. N. Balasubramanian, V. Geethalakshmi, D. A. George, and J. F. Clewett. 2004. "Experiences of using seasonal climate information with farmers in Tamil Nadu, India." In *Using Seasonal Climate Forecasting in Agriculture: A Participatory Decision-Making Approach*, edited by A. K. S. Huda and R. G. Packham, 22–30. ACIAR Technical Report no. 59. Canberra, Australia: Australian Centre for International Agricultural Research.

Hulme, M., Y. Biot, J. Borton, M. Buchanan-Smith, S. Davies, C. Folland, N. Nicholds, D. Seddon, and N. Ward. 1992. "Seasonal rainfall forecasting for Africa. Part II: Application and impact assessment." *International Journal of Environmental Studies* 40:103–121.

Ingram, K. T., C. Roncoli, and P. H. Kirshen. 2002. "Opportunities and constraints for farmers of west Africa to use seasonal precipitation forecasts with Burkina Faso as a case study." *Agricultural Systems* 74 (3): 331–349.

Jacobs, K. 2003. "Connecting water management and climate information." *Bulletin of the American Meteorological Society* 84 (12): 1694.

Keogh, D. U., G. Y. Abawi, S. C. Dutta, A. J. Crane, J. W. Ritchie, T. R. Harris, and C. G. Wright. 2004. "Context evaluation: A profile of irrigator climate knowledge, needs and practices in the northern Murray–Darling Basin to aid development of climate-based decision support tools and information and dissemination of research." *Australian Journal of Experimental Agriculture* 44 (3): 247–257.

Lemos, M. C. 2003. "A tale of two policies: The politics of seasonal climate forecast use in Ceará, Brazil." *Policy Sciences* 32 (2): 101–123.

Lemos, M. C., T. Finan, R. Fox, D. Nelson, and J. Tucker. 2002. "The use of seasonal climate forecasting in policymaking: Lessons from Northeast Brazil." *Climatic Change* 55 (4): 479–507.

Lemos, M. C., and B. Morehouse. 2005. "The co-production of science and policy in integrated climate assessments." *Global Environmental Change* 15 (1): 57–68.

Letson, D., I. Llovet, and G. Podesta 2001. "User perspectives of climate forecasts: Crop producers in Pergamino, Argentina." *Climate Research* 19 (1): 57–67.

Magalhães, A. R., H. C. Filho, F. L. Garagorry, J. G. Gasques, L. C. B. Molion, M. D. S. A. Neto, C. A. Nobre, E. R. Porto, and O. E. Rebouças. 1988. "The effects of climatic variations on agriculture in northeast Brazil." In *The Impact of Climatic Variations on Agriculture*, edited by M. L. Parry, T. R. Carter, and N. T. Konijn, 273–280. Dordrecht, the Netherlands: Kluwer Academic.

Magalhães, A. R., J. R. A. Vale, A. B. Peixoto, and A. D. P. F. Ramos. 1989. "Organização governamental para responder a impactos de variações climáticas: A experiência da seca no nordeste do Brasil." *Revista Economica Do Nordeste* 20 (2): 151–184.

McPhaden, M. J., S. E. Zebiak, and M. Glantz. 2006. "ENSO as an integrating concept in Earth system science." *Science* 314:1740–1745.

Nelson, R. A., D. P. Holzworth, G. L. Hammer, and P. T. Hayman. 2002. "Infusing the use of seasonal climate forecasting into crop management practice in North East Australia using discussion support software." *Agricultural Systems* 74 (3): 393–414.

Nicholls, N. 1999. "Cognitive illusions, heuristics, and climate prediction." *Bulletin of the American Meteorological Society* 80:1385–1396.

O'Connor, R. E., B. Yarnal, K. Dow, C. L. Jocoy, and G. J. Carbonne. 2005. "Feeling at risk matters: Water managers and the decision to use forecasts." *Risk Analysis* 5:1265–1275.

Pagano, T. C., H. C. Hartmann, and S. Sorooshian 2002. "Factors affecting seasonal forecast use in Arizona water management: A case study of the 1997–98 El Nino." *Climate Research* 21 (3): 259–269.

Patt, A., and C. Gwata. 2002. "Effective seasonal climate forecast applications: Examining constraints for subsistence farmers in Zimbabwe." *Global Environmental Change: Human and Policy Dimensions* 12:185–195.

Pfaff, A., K. Broad, and M. Glantz. 1999. "Who benefits from climate forecasts?" *Nature* 397:645–646.

Pulwarty, R. S., and K. T. Redmond. 1997. "Climate and salmon restoration in the Columbia River basin: The role and usability of seasonal forecasts." *Bulletin of the American Meteorological Society* 78 (3): 381–396.

Rayner, S., D. Lach, and H. Ingram. 2005. "Weather forecasts are for wimps: Why water resource managers do not use climate forecasts." *Climatic Change* 69:197–227.

Rayner, S., and E. L. Malone. 2001. "Climate change, poverty and intergenerational equity: The national level." *International Journal of Global Environmental Issues* 1 (2): 175–202.

Reliefweb. 2006. "Thousands of Somali refugees flee drought and war." Reuters Foundation, Nairobi, July 28.

Roncoli, C., K. Ingram, C. Jost, and P. Kirshen. 2001. "Meteorological meanings: Understandings of seasonal rainfall forecasts among farmers of Burkina Faso." Paper presented at the Proceedings Communication of Climate Forecast Information Workshop, Palisades, NY, June 6–8.

Sayuti, R., W. Karyadi, I. Yasin, and Y. Abawi. 2004. "Factors affecting the use of climate forecasts in agriculture: A case study of Lombok Island, Indonesia." In *Using Climate Forecasting in Agriculture: A Participatory Decision-Making Approach*, edited by A. K. S. Huda and R. G. Packham, 15–21. ACIAR Technical Report no. 59. Canberra, Australia: Australian Centre for International Agricultural Research.

Taddei, R. 2005. "Of clouds and streams, prophets and profits: The political semiotics of climate and water in the Brazilian Northeast." PhD diss., Graduate School of Arts and Sciences, Columbia University, New York.

Valdivia, C., J. L. Gilles, and S. Materer. 2001. "Climate variability, a producer of typology and the use of forecasts: Experience from Andean semiarid smallholder producers." Paper presented at the Proceeding International Forum on Climate Prediction, Agriculture and Development, Palisades, NY, June 6–8.

Woodhouse, E., and D. Sarewitz. 2007. "Science policies for reducing societal inequities." *Science and Public Policy* 34 (2): 139–150.

Zebiak, S., and M. A. Cane. 1987. "A model El Niño/Southern Oscillation." *Monthly Weather Review* 115:2262–2278.

Ziervogel, G., and R. Calder. 2003. "Climate variability and rural livelihoods: Assessing the impact of seasonal climate forecasts." *Area* 35 (4): 403–417.

Ziervogel, G., and T. E. Downing. 2004. "Stakeholder networks: Improving seasonal climate forecasts." *Climatic Change* 65 (1–2): 73–101.

IV THE COMPLEX NATURE OF SOCIOTECHNICAL SYSTEMS

The premise of this book is that if we want to build a better future, we must recognize the important role that technology plays and then proactively direct technologies toward the future we want. To emphasize the importance of thinking about technology as we build our future, the second section explored a variety of ways in which technology and society influence each other, and the third section demonstrated numerous ways in which technologies can reflect, embody, reinforce, and undermine values. Understanding the technology-society entanglement is not, however, a simple matter. In order to effectively steer technology and society to a positive future, we must be attuned to the intricate complexities of sociotechnical systems.

While social constructivists argue that we have the capacity to shape and direct technology, putting this into practice is no small feat. Not only is taking action difficult but predicting the effects of those actions is fraught with uncertainty. The article by Lemos and Dilling in the previous section gives a clear example of the factors that can make achieving specific goals challenging. Understanding how sociotechnical systems are produced and sustained, how they change over time due to a variety of forces, and how actions by individuals, organizations, governments, and others affect them is a daunting task.

Technological innovation is often presented as a simple step-by-step process. The traditional view suggests a linear sequence: science discovers new aspects of nature; governments and corporations fund projects to further explore these discoveries and put them into systematic form; engineers apply the new science to create new products; corporations package and market these products; and consumers buy, use, and benefit from the new products. In this account the process seems not just linear but amenable to control. We can, presumably, speed it up or slow it down or turn it slightly in this direction or that by intervening at various points, but for the most part, the path is straightforward. While this view may accurately depict how a few technologies have gotten to the marketplace, the sequence described is far from the norm and hides many of the forces that are typically at work.

The processes by which sociotechnical systems are built, maintained, and modified involve many actors, balancing and trading off diverse factors and dealing with a good amount of uncertainty. Complexity is the norm in all of the stages, be it when a technology is first conceived, then developed, and then adopted (or rejected) by users, and ultimately has an effect on the world. To begin to sort out and understand the nature of the complexity here, it is important to distinguish at least two kinds of complexity. The first and most widely recognized has already been suggested in previous chapters that pointed to the multiplicity of actors and social factors that may influence the development of a sociotechnical system. Sociotechnical systems are nearly always shaped by a complex web of people, organizations, and interactions. Governments and corporations exert influence in the choice of projects to fund, engineers make important decisions about how a machine will be constructed, marketers control when and how to promote a technology, and, ultimately, users deploy products to achieve their ends—including those never envisioned by the companies, governments, and engineers involved in their production.

Because of the number and variety of people who have the ability to influence a sociotechnical system, it is difficult to precisely predict whether an invention will be adopted or exactly how a newly developed technology will be integrated into society. As Langdon Winner argued, something as seemingly innocuous as the height of a bridge may allow pedestrians to cross a road while simultaneously restricting the movement of other social groups. Users can be surprisingly inventive in the ways they take up technologies. Examples of the unpredictable adoption and use of technology include using pie tins as throwing discs, using new chemical glues to get high, and using headphones as a way to tune out and ignore strangers. Each individual or group involved in technological development leaves an imprint and can knowingly and unknowingly affect the impact of technology on values.

The second category of complexity concerns the fact that the technical aspects of a design may never be completely understood. Although engineers and scientists have a deep understanding of how things work, their knowledge is always limited; there are factors of which they are unaware, situations they haven't considered, or aspects they haven't had time to test. Engineers and scientists often do things that have never been done in quite the same way before. Examples of the uncertainties of engineering are found throughout history. In many cases, conscientious engineers, because of incomplete knowledge, built something that did not work as intended. For example, the disasters associated with the space shuttle *Challenger*, the Fukushima nuclear power plant, and the exploding batteries in the Samsung Galaxy Note 7 smartphone all resulted from factors that were not clearly understood or anticipated. While it is important for engineers to continually increase their understanding, they must also develop methods for working without a precise knowledge of all the issues involved. When they design and build, engineers must be mindful of the limits of their knowledge. Nonengineers often assume that complex technologies are well understood, but the most prudent course is to approach them with the recognition that their precise behavior will never be absolutely certain.

These complexities mean that those who deliberately seek to influence the development and direction of sociotechnical systems should tread carefully because the impact of their decisions cannot be precisely predicted. We have a number of systems set up to do exactly that. Government regulations require drugs to be thoroughly tested before they are handed out to patients. Automakers crash their vehicles in computer simulations and in carefully controlled experiments to better understand the implications of car wrecks for automobile occupants. And product liability laws (especially in the United States) incentivize corporations to eliminate as many risks as possible in an effort to avoid large lawsuits. Even in these cases, the efforts to direct technology will never be perfect. But if we abdicate the responsibility to try, we effectively relinquish control over the future. Acknowledging the uncertainty of the enterprise can help us better understand, anticipate, and prepare for the risks inherent in the sociotechnical systems that constitute our lives. Acknowledging the complexities of sociotechnical systems is a key step in being able to act effectively.

The readings in this section cannot possibly cover every issue or unintended consequence that might arise when it comes to complex sociotechnical systems. The readings are intended to provide rich examples and models of the kind of analysis that might be extended to other cases. They provide an understanding that prepares the reader to anticipate and navigate the complexities of other cases and envision ways to circumvent possible negative effects.

Questions to consider while reading the selections in this section:

1. How do social complexities and technical complexities compound one another?
2. How do we make wise decisions amid significant uncertainty?
3. What lessons can we learn about unanticipated consequences by studying past and current examples of sociotechnical change?
4. Are there ways we can prepare for multiple futures or contingencies when we can't make a specific prediction?
5. If no one is in complete control of a sociotechnical system, how do we hold people responsible when something goes wrong?

19 Sociotechnical Complexity: Redesigning a Shielding Wall

Dominique Vinck

Engineering is often portrayed as simply the application of scientific laws to practical problems. However, producing technology, even if one focuses only on what happens in the lab, is far more complicated than such a model would lead one to believe. In this article, Dominique Vinck tells the true story of a young engineer who discovers firsthand that daily life as an engineer involves more than facts, numbers, predictable phenomena, and well-defined problems. His day-to-day work requires a great deal of negotiation with people from other disciplines, other labs, and other companies just to put together what many of his coworkers believe to be one of the most mundane elements of the entire project. Examples like this demonstrate that precisely how technologies are constructed is not preordained. Outcomes depend not only on the goals set by management or the regulations established by government but also on how well individuals are able to communicate and argue for their particular priorities. This article describes the complexities that just one person faces while making decisions about one very small part of a technology. The complexity increases exponentially when we add other stages and other actors to the picture.

Before embarking on a placement in a design and engineering office,[1] the young engineer does not really understand the complexity of the work awaiting him. Of course, he is ready to do complicated operations that must be dealt with at a high level of abstraction. He has also been trained to handle fairly sophisticated models and tools. He knows that he is bound to run up against difficult technical problems. Nevertheless, he has a certain number of working methods and tools under his belt that will get him out of many a difficult situation. He has the capacity to analyze problems, break them down into essential parts, and then model them. This ability to simplify things is supposed to help him get through the most complicated challenges. At least this is what he has been taught.

Yet the young engineering student still has to learn exactly how complex ordinary technical work really is. A placement period lasting just a few months will prove to be a real eye opener. He may have thought that an engineer's work is mainly technical, but he will quickly realize that, in reality, things are much more complex than that. He will also find that, if he wants to be an efficient engineer and get technically satisfactory results, he will have to decode and take into account only what appears to be real.

The aim of this chapter is thus to map and document the changing vision of young engineers after their entry into the industrial world. To build up our account of what typically happens, we will use the experience of an engineering student as he learns

From Dominique Vinck, ed., *Everyday Engineering: An Ethnography of Design and Innovation* (Cambridge, MA: MIT Press, 2003), 13–27. Reprinted with permission.

the ropes during a placement. Although the placement period in question is only 6 months, it must not be forgotten that the time usually required is much longer, about 2 or 3 years.

We will follow the work of an engineering student during his placement in a CERN (European Organization for Nuclear Research) design office in Geneva. For this student, the difference between what he learned at school and the way things really are in the design office is accentuated by the fact that his assignment seems to be quite simple: define the shape and dimensions of an object, and the materials to be used, so as to meet the specifications of the order givers and the laws of science and nature. Furthermore, this assignment is a good opportunity for the student to apply some CAD (computer-aided design) tools to a real case. The problem does not look complicated at the outset; our student needs only the initial data (the specifications defined by the order givers), a computer console, and a methodical approach.

However, what the young engineer will discover during his placement is that his pre-evaluation of it was much too simple and limited. To be able to fulfill his mission, he will have to change his views and his approach little by little. He will have to rework his initial impression of the design work. He imagined himself sitting in front of his computer designing an object (a scene that is consistently reproduced in literature on design methods). In fact he discovers a social world of varying shapes and sizes. He thought he would have to implement certain methods and apply certain cognitive processes. In fact he finds himself having to negotiate and settle on compromises. He thought the procedure to be followed would be straightforward, starting with the specifications, but everything is complicated by new requirements defined by the order givers following the draft of a first solution. Indeed, the story we are about to tell concerns not only the design of a technical object but also the re-design of an apprentice engineer.

A Strange Supervisory Board

Many young engineers have probably discovered the same thing when starting out on their careers. Few of them, however, have had to deal with the same kind of supervisors as this student. What is more, the specific framework in which the work is done should be underlined. It provides the opportunity to discuss and analyze the trainee engineer's experience and find the terms to express what he sees and feels. The framework therefore has a lot to do with what the young engineer experiences.

His mechanical engineering studies are coming to an end when one of his lecturers, Jean-François Boujut, talks about the possibility of his pursuing a DEA or even a doctorate.[2] Involving research work, the DEA gives students an additional non-technical skill. It is also, the lecturer explains, an opportunity to step back from the operational work required by the PFE. But the most surprising announcement is that the proposed subject is to be co-supervised by a sociologist. The student is interested to discover that the mechanical engineering teams and the sociology teams in Grenoble are used to working together. However, since our student has only devoted himself to mechanical engineering throughout the course of his studies, he prefers to concentrate on this area and the technical work in hand at the beginning of his placement period. His tutors nevertheless ask him to take an observer's view of the project and closely follow the design process and the actions and interactions that it generates. To begin with, the trainee thinks that his observations are unrelated to his work as a designer. They involve a different part of his mission. This part is non-technical, and the trainee cannot really see what the aim of making such observations is.

The student, placed in a CERN design office in Geneva, is put under the direct responsibility of the head of the office, Bertrand Nicquevert. To the student's great surprise, his engineering school tutors and his "industrial" tutor seem to work hand in hand. They apparently get on really well and share the same opinion of the work he has been given. They say that it is an interesting opportunity to decode and analyze the design process. The student also discovers that Nicquevert holds a master's degree in philosophy. Not your usual mechanical engineer! And as if his supervisors were not an unlikely bunch as it was, Pascal Lécaille—an anthropologist writing a thesis on simulation tools—joins the group in one of the first supervision meetings.

The young engineer can only explain this strange group of supervisors by the interest they have in the other part of his mission, i.e., the social aspects and all the other factors surrounding the actual design work: the language barrier and the cultural differences of the people in the design office, the different age groups and the probable consequences of people retiring, and, finally, the behavioral and relational problems of the office personnel, especially the more senior designers with respect to their young manager. This set-up, in which the social factors are peripheral, external, or simply tacked onto the technical job in hand, is not to be called into question. However, as the design work progresses, we will discover a different way of looking at things, based on the people concerned, the way they react, and their different relationships. Indeed, the problem can only be defined, and a solution found, if these elements are taken into account. Hence, the sociologist's view of the mission does not fall entirely outside the scope of the technical work; it is up to him to try and understand the dynamics of the technical work.

A Simple Object in a Complex Environment

And so, fresh from his mechanical engineering school, the student settles down in the open-plan office. He is given a work surface and a computer console, like the other fifteen office members. With a mission to fulfill and a place to do it in, he thinks that he will be able to get along fine. At school he has learned to use the models, the catalogues of technical solutions, and the appropriate methods for each design phase. With all this learning under his belt, he should have no difficulty finding the right solution to the problem, making the calculations, and checking his work.

What is more, the technical part that he has to design is very simple. It is a wall, or more specifically a shielding disk, to separate two parts inside the ATLAS particle detector. On one side is the calorimeter (for measuring particle energy); on the other is a superconducting magnet. The shielding must prevent all particles other than muons from interfering with the measurement of the trajectographs called "muon chambers." The shielding must absorb photons and neutrons. For this, materials with high absorption rates for such particles (such as polyethylene or copper) are used.

To get to the bottom of the problem, the student starts by reading up about the entire system in which these shielding disks are to be placed. For a week, he concentrates on learning the technical terms relating to the detector. Using a document referred to as the "Product Breakdown Structure," he identifies each of the parts, its name, its abbreviation, and its dimensions. He makes several sketches in his logbook for future reference. In doing this, he discovers just how complex the detector really is. He also discovers its impressive size and weight: 25 meters high, 40 meters long, 10,000 tonnes. The detector is to be used to determine the identities, energies, and directions of the particles produced during frontal collisions with proton beams. It is made up of detection and measuring instruments (a trajectograph, a calorimeter, a muon spectrometer), confinement and regulation parts (superconducting coils, cryostats), a range of electronic systems, and various supporting and structural parts.

Designing such a detector obviously involves a large number of people, institutions, and countries. In all, 1,700 physicists and engineers, some of whom can be considered "order givers," are taking part in the project. The "order givers" are the future users. Each element is being designed by a specific team. The CERN design office is one of these teams. As the head of the design office gives a quick overview of the ATLAS detector, he points to different parts of a technical drawing, saying "This is us here, and that's a team in England over there." Working in partnership with others is difficult even if it is routine. The whole thing requires complex coordination among senior managers within CERN and among various project steering committees. The technical complexity of the detector is thus matched by the organizational complexity of its coordination and that of the technical information system.

As the engineering student listens to the explanations offered by the head of the design office, he discovers that coordination among the designers and with the physicists is, in general, a central issue. Far from being a relatively closed space, the design office has a tight working relationship with numerous people from various institutions. The head of the design office talks especially about two categories of partners: the safety department (which is in charge of checking all the calculations for the sensitive parts) and the physicists (who are seen as dreamy idealists always wanting to go one step further without taking into account how feasible their ideas are, or at least that's what it looks like). The design office sometimes relies on the former to temper the wishes of the latter. Even within the office, the question of how to work together is often raised in connection with people's cultural differences and differences in age, and also in relation to where they are seated in the office. Indeed, in the middle of the office there is a row of large cabinets in which 30 years' drawings are kept. This row of cupboards physically divides the office in two. However, in the middle of the row there is a gap about half a meter wide. The head of the design office says: "See that? I've made a space between the cupboards. It was like bringing down the Berlin Wall." This witty remark goes some way toward explaining the reluctance of certain designers to cross the office.

Normally the student would not be concerned with all these coordination problems. The shielding disk that he has to design is just one small part of the whole assembly. He should be able to deliver a detailed draft design of the disk within 3 months. What is more, people hardly seem to be interested in this part. The physicists, for example, have turned their attention to other parts of the detector. The shielding is not seen as a "noble component," says the head of the design office, unlike the particle detectors. It is one of the "common components"—parts that go between instruments to accommodate fluids (e.g. for cooling), cables, and support structures. Indeed, if it were possible, the physicists would like these parts to disappear altogether. For them, cooling should be immediate and homogeneous without having to bother with all the tubing and extras. As for the framework supporting the instruments, this really takes up far too much room. These "common" parts are seen as cumbersome by the physicists who will be using the detector. The designers, however, covet these commons because they impose themselves as constraining boundaries. Of course, when a designer needs just a few extra millimeters, he takes them. The problem is that he is not the only one to do so. In fact, it is the head of the design office who is in charge of making sure all these parts fit together and who has to bring these coveted boundary areas of the system apparatus into existence.

The shielding disk on which the engineering student is working is one secondary part that nobody is really interested in. This is why he is under the impression that he need only analyze the problem and find the solution, without having to get into lengthy negotiations with the physicists. To him, the problem is purely technical. From the outset he knows that the space available for the disk's external dimensions is limited by the external dimensions of the surrounding parts. Using the drawings given to him by his colleagues in the design office, he studies these surrounding parts and their dimensions. He takes into account a few general rules relating to safety and ergonomics, so that the detector can be accessed for maintenance. Thus, his scope of action is limited by a multitude of specifications and requirements imposed by various people involved in the project.

Interactions between Objects

During the first days of his placement period, the head of the design office takes the young engineer around the various departments. He is introduced to many people, some of whom are working on issues directly linked to his own study, some of whom are not. He also takes part in technical coordination meetings that deal with project planning, fitting the various parts together, and the safety of their design. He feels that such meetings are just a matter of procedure.[3] Their aim is above all to check how the project is going and swap information. However, the people in the meetings argue about dates and about documents that haven't been handed over. This has nothing to do with the technical side of the project. It has to do with how projects and meetings are organized and managed.

However, as they go through the corridors from one department to the next, the trainee and his tutor come across various people with whom the tutor strikes up conversation about details regarding various projects that he is in charge of and which he needs to keep in mind. The student is astonished to see that part of the project's technical coordination takes place in the corridors. Like the canteen, the corridors are used as a forum for solving many problems.

The more people they meet, the bigger the trainee's list of contacts becomes, although so many names frighten him. The words of his tutor hardly reassure him. He thought he had come to carry out an engineering assignment first and foremost and, as a kind of sideline, act as an observer. But he discovers he actually has to communicate information, get answers to questions, and negotiate. He discovers then why he has been put in charge of designing the shielding. In fact his tutor had known that it would not be easy. He had even said as much right from the start but the young engineer, judging from the simplicity of the object to be designed, had put this to the back of his mind. When the head of the design office had agreed to take on the student along with his strange group of university supervisors, mechanical engineers rubbing shoulders with sociologists, it was because he thought that an outside view of the situation would help the head of the design office to understand what was in play.

And so the young engineer discovers that his poor shielding disk is the object of important stakes in terms of its functional definition. Indeed, it has to fulfill two functions: to stop particles and to bear the weight of the muon chambers. The problem is that the shielding is surrounded by various neighbors that have to be taken into account.

Moreover, the word "neighbor" is used both to talk about neighboring physical objects and to refer to the designers of such objects or the order givers. This is why people talk about negotiating with the neighbor when talking about the cryostat, for example. The number of neighbors involved is already quite considerable: several types of muon chambers, the vacuum chamber, the tile calorimeter, the cryostat, the toroid, the rails on which the system has to run, and the electronic data capture boxes. The trainee discovers, for example, that the electronics engineers in charge of designing the data capture systems have designed an enclosure that is too big and have thus reduced the shielding designer's room for maneuvering. In fact, he needs a clearance margin, as it is difficult to know the exact dimensions of the parts once they have been built. If he can't have the data capture box redesigned or moved, the trainee will have to plan a cutout in the shielding disk. And the physicists will probably not like this. Furthermore, it will reduce the disk's rigidity. As the shielding is at the center of a series of neighboring relations, it is an intermediate object; thus, the young engineer has to argue his case if he wants to get the amount of space he requires.

Ten or so neighboring objects mean ten or so teams or individuals to be contacted for data and information concerning their parts, along with all the associated constraints. On the other hand, the trainee discovers that he has to validate his own designs with these people. Some of them work in the design office, but not all of them. Sometimes the trainee has a CERN physicist with a listening ear to deal with; at other times he must deal with a renowned Italian physicist who is impossible to find, or with a Parisian team that does not answer his e-mail requests for information.

Technical or Strategic Work?

The trainee also learns that the work on shielding is strategic for the design office. Indeed, some of the technical parts, such as the muon chamber, are already the objects of dimensioning studies by work groups such as the Muon Layout Collaboration. In order to define the job at hand, the trainee bases his studies on a technical design report drafted by the Muon Collaboration. This document lists the technical features of the chambers and all the teams working on them. It defines the part of the enclosure that concerns him and in which the disk has to fit. Thus, for some of the parts, things have already been defined, and it would be difficult for the trainee to change them. This means that the design office has less room for maneuvering in the design of the structural parts of which it is in charge. The trainee realizes that his tutor has chosen this moment in the project to assign him to the shielding job so that they can have their say in the matter as early as possible. It is essential for the design office not to have to work with a part that is already joined to the rest. The office therefore has to fight to preserve a certain amount of freedom in its design work. Relations between the design office and its partners are as important as all the problems relating to borders, space, and margins.

It is only at this point that the trainee realizes how poorly prepared he was for this situation. He does not really know what kind of attitude he should adopt in this complex social world of hierarchies (related to the organization itself but also to the reputation of people), divisions, and territorial occupations. Therefore, for several weeks the trainee has put these facts on the back burner, preferring to concentrate on what he can do best: a technical job performed at a computer console. He has memorized the environment of the shielding disk from the drawings given to him by a neighbor at the office. He has redefined the enclosures so that he can accurately assess the space available and the margins for maneuvering. And he certainly needs these margins to be able to reinforce the shielding disk so that it can withstand the weight of the muon chambers. Finally, having concentrated only on the technical aspects, he has learned to know where he stands from a technical point of view; thus, he has developed a line of arguments to use with his neighbors if ever he should have to negotiate.

So he beavers away at his computer console, designing, imagining, and calculating. He checks the possibilities for adding reinforcements without disturbing the layout of the muon chambers. These reinforcements are necessary to prevent the disks from buckling under the weight of the chambers. After talking to his office neighbor, who is in charge of integrating the chambers, he drafts some ideas for fixing them to the disk. He discovers that this is the most delicate part of his own design work, as the loads to be borne are considerable and he has little space for adding the framework. Although two-dimensional design software would give him a good idea of the surrounding space available, it does not really give him an overall view.

Having worked with three-dimensional simulation and viewing tools at his engineering school, he decides to use his training period to put one of the software programs to the test. Using it, he is able to show how the muon chambers and shielding disks fit together. (His mechanical engineering tutors are interested but not entirely convinced. They prefer working with concrete analyses rather than such calculation tools.) Next, the trainee designs a framework able to fit into the space available. He simulates various calculations of the framework with different diameters and materials so that he can get a realistic idea of the mechanical stress. He discovers that the framework will not be rigid enough unless it is closed at both ends.

For several weeks, he concentrates on the design of this framework, working in an environment that seems increasingly restrictive and hostile owing to the dimensional constraints and the problems of accessibility: little space available, the need to leave clearance for the detectors to be opened, and the overall suitability of the assembly. There are so many geometric constraints that his first concern is to find a solution that is able to fit inside the space available. While devoting all his time and energy to this problem, he is also able to build up good professional relations with his colleagues in the office. He discovers that everyone there has had to forge a place for himself. The space-related problems of everyone he meets when working on its project are reflected in the design office itself.

Having discovered the importance of "neighbors" when working on a design issue, our student undertakes to list them all, both the technical parts and the people working on them, along with the questions that he would like to ask them. The list is long indeed, and it gets longer since the very notion of neighborhood has to be revised. Before this, it was defined in terms of spatial proximity. It referred to "elements" that may or may not be in (physical) contact with the disk, but are not separated by another element. Perhaps it is his mechanical engineering background that has so far restricted his field of vision. There are in fact several kinds of relations among objects; geographical proximity is not the only one. Indeed, radiation goes through various parts of the detector along with heat, magnetic fields, vibrations (e.g. earthquakes), and gravity. (The detector may be symmetrical, but gravity does not see it that way.) The toroids in the detector generate an intense magnetic field that tends to cause the elements to come together. The magnetic field exerts a force and then checks whether this force affects it or not (or rather whether it affects the shielding). Added to this is the question of maintenance access to the detector. All these forms of interaction can bring distant elements within the system in contact with one another. Drawing up a list of these elements along with the people working on them seems to be the only way forward. In doing this, the trainee is in fact trying to identify and target all the neighbors with the biggest influence on his design work. Defining each one's territorial position (who does what and up to what point) now seems increasingly important to the trainee, as it will allow him to define his own technical work. Moreover, the breakdown of roles played does not now seem as clear as when he started. And so, having worked hard on the technical side of things, the student discovers how essential it is to be able to socially decode relations if he wants to complete his mission successfully. In other words, he has to ask himself who does what and how far is it possible to negotiate. He finds out that certain elements cannot be negotiated, as changing them would put them back on the drawing board. Thus, the trainee comes to analyze the interactions between people, the recurrent nature of certain practices, the rules applied, and the possible interference of all this in his work.

Stabilizing What the Neighbors Want

The trainee also comes to realize that the demands of each person involved are not always clear and are far from stable. The shielding is supposed to fulfill two functions, but when he takes the various neighbors into account he realizes that things are much more complicated. Each neighbor has his own expectations and requirements, only some of which have been put down on paper. What is more, under the instigation of the head of the design office, the trainee has begun a functional analysis.[4] This requires listing and quantifying all the functional features of the disk, with the aim of discontinuing to work on assumptions alone. For example, the physicists say that the shielding should be 100 millimeters thick and made of iron, with a copper base. But why? Which physicist decided that? And on what basis? Where are the data that led these physicists to give such specifications 3 years earlier? Would they say the same thing today? In fact, all these so-called constraints have to be studied again, and the people who defined them must be found and asked why they said what they said. It is no longer possible, at this stage, to continue to rely on the available technical data. It would be better to find out the logical reasoning behind the orders given and whether it is possible to re-negotiate. For example, just how far can the basic functions of the detector be revised? What seemed to be finalized is perhaps not. And so, after 4 months, after the student has done his design work on the supports for the muon chamber, the physicists decide that the way the chambers are mounted does not satisfy them. After viewing the assembly, they realize the need for maintenance access. The support function thus becomes even more complex, requiring the addition of a new structure that is mobile in relation to the disk. All the design work on the direct support of the chambers has to be reviewed. The concept of the mobile structure and that of its supporting copper base have to be validated at the same time. And yet, the young engineer has already spent several weeks and much energy finding a solution. Bringing to light a new element has led to a whole array of fresh constraints calling into question the initial concept. The trainee begins to wonder what he can base his work on. On top of this, he discovers that certain neighbors have taken up more space than was planned simply because they were not aware that neighboring elements had to be taken into account. As far as they were concerned, their elements were surrounded by emptiness. It is easier for members of the design office with the job of integrating the different elements to understand these "neighborhood relations" than for a subcontractor of a distant part.

Now that the neighbors have been identified, the next problem is getting them to talk. Would it be possible, and enough, just to get them around a table? Some of them come to CERN only once a month. Of course, our trainee engineer has neither the power nor the authority to convene them in a meeting. The head of the design office does not have authority to do so either, in view of the number of people involved in the project.[5] So the young engineer decides that the best thing to do is define a certain number of elements himself and draw up the specifications that the physicists should have drawn up in the first place. These would then be submitted with the aim of getting them to react and thus define their needs. It is at this point that he discovers how foreign the "culture of a specifications sheet" (Bertrand Nicquevert's words) is to physicists. They shy away from the idea. They think that if they put their expectations down in writing they will no longer have the power to change them. For the head of the design office, on the other hand, the act of writing them down will force the physicists to express their needs, even if they will have to be modified later. If this is not done, they will be defining a solution to a problem whose terms are unknown.

To begin with, the design work consists in studying each problem one after another. Little by little, the idea that it is necessary to have an overall view, and not just a technical one, emerges. Different people work on each technical element, and it is essential to know exactly what they want and how far it is possible to negotiate with them.

The trainee thus submits his solutions to his neighbors. The drawing of the disk is sent to a physicist so that, through simulation, he can check whether it is acceptable in relation to his needs (i.e., particle absorption). A proposal for modifying the cryostat cover is faxed to the Orsay team in charge of its design. Within the design office, showing drafts of drawings to different colleagues during lunch or in an informal context produces some interesting reactions. It allows the young engineer to see that work with each partner is carried out differently and requires different approaches. At times, the design proposed is provisional, insofar as an unhurried colleague is expected to provide some data. At other times, the engineer has to wait for a reaction to the proposed modifications to a particular part. The assistant technical coordination manager is soon to leave for the United States for a meeting where he should have a chance to raise the question of the muon chamber. A file has to be prepared for him, and he has to be persuaded to bring up the matter. The problem here is that he comes to the office only once every 2 months. Negotiations depend on mediators whose logical approach is not always fully understood by the members of the design office. For some neighboring elements (such as the calorimeter), negotiation is easier, as the colleagues involved work at the Geneva site. For each neighboring element, and hence for each neighbor, there is a specific coordination procedure. In this way, the young engineer comes to understand the interest of the head of the design office in having an outside view of the situation.

Finally, it is interesting to see how simply working on an element such as the shielding draws attention. It has become a subject of interest, enabling questions to be raised sooner rather than later when later would in fact be too late. The person in charge of the muons (a physicist and a close guardian of the muon spectrometer specifications) did not want to get involved in things to do with "services" or "common parts." And yet the design office needs answers to a certain number of questions. Indeed, the more questions it asks the more seriously it will be taken. If no one bothered to do this, the shielding would just become a kind of black hole, a bin for all the neighbors to throw their unresolved problems into. After all, the designers are bound to find a solution later on. Having questions raised with the project actors was, in fact, one of the objectives of the head of the design office when he took on the trainee. He later explained his numerous expectations in relation to his various responsibilities as follows:

> As project engineer for the traction system, I didn't have anybody to argue for the shielding disk. Doing the design within the design office was going to enable me to monitor its compatibility with the entire assembly.
>
> As a member of the ATLAS Technical Coordination team, where I am responsible for mechanical integration, the shielding disk presents a number of unlikely neighbors (some of which were only discovered through Grégoire's work). It is at the center of numerous problems but there is nobody to deal with them and the initial design plan was far too succinct. It was essential to have somebody prepared to dig further into this design. But most of all, from a sociological point of view, it was the ideal opportunity to study the dynamics of the design process through a physical experiment.
>
> As a mechanical engineer, and one that works at CERN, there were several small mechanical challenges: calculating the disk, the support, etc. But what I hadn't banked on was that this object would take on a new life, thanks to the initiative taken and the work put in by Grégoire. One of the amusing consequences of this situation is that today I am being offered the responsibility of the shielding disk as project engineer, which is something I wasn't expecting to begin with.
>
> As a philosopher/epistemologist, my questioning centers around technical issues. . . . There are many scientists who like to dabble in philosophy, but there are significantly fewer engineers. The latter are "much more aware of the material aspects of a technical issue" (O. Lavoisy). Following Galison's example, I'm hoping to be able to go further into the question of relations between theoretical physicists and experimental physicists but on the triple basis of "theory/experiment/instrument" as opposed to the traditional epistemological dual basis. Furthermore, using an engineer studying for a DEA, supervised by a human sciences committee, was the ideal opportunity for understanding this area in a much more structured way. It is also an opportunity for me to understand my own work as an engineer and what is being done in the area of How Experiments Begin?[6]

The work of the design engineer turned out to be considerably different from what the young engineer had imagined. He had thought that it was just a matter of finding the right solution to the problem in hand by applying the models and methods he had learned in the course of his studies. He knew from the start that he wouldn't be able to base all his work on these existing methods and that he would have to invent new ones, but he certainly didn't think he would have to go so far.

Operational Summary

1. **Design work is complex, even for a simple object** Designing a technical part, however simple, can quickly prove to be complex when the part in question lies at the center of a whole system and is linked to a certain number of other technical parts.
2. **The design work builds up around a network of relations among technical parts** Designing an object involves taking into account a series of other objects, which are not always in direct contact with this object. These are related to one another; however, the way they are related is not always known at the start, and does not necessarily become clear during the design process. To define the specifications, the designer must describe this network of relations among technical elements and must go through it regularly to check on changes made.
3. **Objects and their relations are linked to people and social groups** These objects can be taken into account only if the designer knows these people or groups (i.e. who orders, who designs, and who uses), their relations, and the logic behind what

they decide and what they do. Of course, this demands precise attention and a decoding ability to prevent judgment from being based on simplistic analyses at the beginning of the placement (saying that problems are due to people or technical ideals, etc.). There are actors behind each technical element, and they act as spokespersons for these elements. The elements are at the center of these people's interest.

4. **It is not always clear at the beginning what all the constraints are** They are gradually revealed as their relations with other elements are explored. It is not possible to have them at the start, notably because the actors themselves do not really know them. The process of designing solutions and making them viable through drawings leads the actors concerned (or order givers) to talk about the requirements that they would like to see fulfilled. Bringing the intermediate objects into existence is therefore an important step that will help the actors to express their needs.
5. **It cannot be taken for granted that requirements and technical data are given objectively** In other words, final judgment must be withheld, as the information available may be misleading. It is better to understand how the data are put together (socially and technically) and then regularly test how stable they are.
6. **Showing interest in an object gives it life** Working on it, drawing it, and circulating the drawing helps to awaken the interest of the various people involved, to position work in relation to it, and to demonstrate responsibility for it. It also helps those who draw up formal requirements. If no one is interested in an object, it cannot live.
7. **To manage relations between technical elements, the designer has to take into account how the actors react and behave in relation to their specific element** Taking into account people and groups means first of all examining how they act, both socially and physically, especially when they have to interact with others.
8. **Doing technical work is just one strategy among others** Concentrating on "technical" work, such as entering and processing information using calculation software and CAD, is sometimes seen as a good strategy that can help the designer to report on the situation, his position, what he would reasonably like to obtain from his colleagues, and his margins for maneuvering.
9. **Industrial design stimulates discussion** Industrial design is not only a technical means of viewing objects during their design; it also stimulates discussion between designers and other project partners.

Notes

This chapter sums up work by Grégoire Pépiot, Jean-François Boujut, Pascal Lécaille, and Bertrand Nicquevert. Grégoire Pépiot is a mechanical engineer studying for a research diploma in industrial engineering under the supervision of Jean-François Boujut (a mechanical engineer) and Dominique Vinck (a sociologist). Bertrand Nicquevert heads a technical office in the Experimental Physics Division of CERN in Geneva.

1. As a part of their studies, young engineers are placed with companies. The idea is that this gives them an opportunity to put what they have learned into practice.

2. A DEA (Diplôme d'Etudes Approfondies) is a one-year postgraduate research diploma. It can be done at the same time as a PFE (Projet de Fin d'Études—a placement project carried out by engineers in their final year of study).

3. The head of the design office does not share this opinion. On the contrary, he says that there are very few procedures involved.

4. In fact, he didn't think it was up to him to define the various wishes of the physicists.

5. The head of the design office says that this is not really a problem since nobody has overall authority over the people involved in the project. Indeed, the way physical research is organized at the end of the twentieth century is based on partnership, which means going through a long series of discussions to reach a consensus about what is possible (mechanically, geometrically, and perhaps financially).

6. This harkens back to the title of Peter Galison's book *How Experiments End*.

20 Fukushima and the Inevitability of Accidents

Charles Perrow

It is a common attitude to consider all technology good and think that any bad effects are simply the result of misuse. A corollary of this is to believe that the development of new technology is always a net gain—that the world is always better off with access to a new tool. In this article, sociologist Charles Perrow makes a strong argument that both of these claims are false in many cases. Perrow's classic work is a 1984 book titled *Normal Accidents* in which he provides evidence for the idea that some sociotechnological systems can never be rendered 100 percent safe. Because of the complexities of the ways in which people, organizations, systems, and artifacts are intertwined, there can never be complete certainty as to how the system will behave and why. Perrow asserts that the technologies we choose to develop and use should be treated cautiously. Hubris may convince those who design technological systems that they have eliminated, or at least contained, all risk. But, according to Perrow, the possibility always exists that complexity will prove the builders wrong. Perrow urges society to reconsider the adoption of enormous systems that will result in enormous catastrophes not *if* they fail, but *when* they fail. Perrow's claim in *Normal Accidents* that we must design our technological and social structures with the risk of accidents in mind was heeded by some but not all. Decades later we see evidence of the truth of his argument playing out in large systems failures. The failure of the Fukushima Daiichi Nuclear Power Station in Japan in 2011, due to a series of events that "were never supposed to happen," sparked Perrow to revisit the warnings he gave in *Normal Accidents*.

The March 11, 2011 disaster at the Fukushima Daiichi Nuclear Power Station in Japan replicates the bullet points of most recent industrial disasters. It is outstanding in its magnitude, perhaps surpassing Chernobyl in its effects, but in most other respects, it simply indicates the risks that we run when we allow high concentrations of energy, economic power, and political power to form. Just how commonplace—prosaic, even—this disaster was illustrates just how risky the industrial and financial world really is.

Nothing is perfect, no matter how hard people try to make things work, and in the industrial arena there will always be failures of design, components, or procedures. There will always be operator errors and unexpected environmental conditions. Because of the inevitability of these failures, and because there are often economic incentives for business not to try very hard to play it safe, government regulates risky systems in an attempt to make them less so. Formal and informal warning systems constitute

From *Bulletin of the Atomic Scientists* 67, no. 6 (2011): 44–52.

another method of dealing with the inherently risky systems of industrial society. And society can always be better prepared to respond when accidents and disasters occur.

But for many reasons, even quality regulation, close attention to warnings, and careful plans for responding to disaster cannot eliminate the possibility of catastrophic industrial accidents. Because that possibility is always there, it is important to ask whether some industrial systems have such huge catastrophic potential that they should not be allowed to exist.

Regulations

Nuclear safety is problematic when nuclear plants are in private hands because private firms have the incentive and, often, the political and economic power to resist effective regulation. That resistance often results in regulators being captured in some way by the industry. In Japan and India, for example, the regulatory function concerned with safety is subservient to the ministry concerned with promoting nuclear power and, therefore, is not independent. The United States had a similar problem that was partially corrected in 1975 by putting nuclear safety into the hands of an independent agency, the Nuclear Regulatory Commission (NRC), and leaving the promotion of nuclear power in the hands of the Energy Department. Japan is now considering such a separation. It should make one. Since the accident at Fukushima, many observers have charged that there is a revolving door between industry and the nuclear regulatory agency in Japan—what the *New York Times* called a "nuclear power village"—compromising the regulatory function.

Of course, even in Europe, where for-profit firms have less power, there are safety problems that have needed more effective oversight. But by and large, European nuclear plants, which are generally part of a state-run industry, appear to be safer than the privately owned, poorly regulated nuclear plants in the United States, Japan, and other countries.

Systemic regulatory failure—as opposed to simple error—is tricky to identify accurately. After an accident in a risky industry, it is always possible to find some failure of a regulatory agency. Everything, after all, is subject to error, in regulatory agencies as well as chemical or power plants. To say that regulation failed on a system-wide basis, one must have strong evidence of agency incompetence or collusion.

The Union Carbide chemical plant in Institute, West Virginia, is my favorite example of regulatory incompetence; in this case, it was a matter of regulators seeing what they were apparently predisposed to see. Shortly after a Union Carbide pesticide plant in Bhopal, India, leaked methyl isocyanate gas in December 1984, killing thousands, the Occupational Safety and Health Administration (OSHA) inspected the company's West Virginia plant and gave it a clean bill of health. What happened in Third World India could not happen in the United States, it was said.

Nine months later, an accident quite similar to Bhopal occurred at the plant, though the gas released was not as toxic and the wind was in a favorable direction, so only some 135 people were hospitalized (Perrow, 2011, 179–180). OSHA looked again and, predictably, found "an accident waiting to happen" and cited the plant for numerous violations, despite its clean bill of health nine months before. There was a trivial fine and a Union Carbide promise to store only the small amounts of the toxic gas actually needed for production.

Union Carbide soon resumed massive storage of methyl isocyanate. Bayer subsequently took over the plant and, in 2008, an explosion killed two workers and threatened to release 4,000 gallons of the deadly gas. Subsequent investigation by the US Chemical Safety Board again found an accident waiting to happen. OSHA appears not to have noticed that its strictures on the amount of storage were violated.

Regulations will always be imperfect. They cannot cover every exigency, and, unfortunately, almost anything can be declared the cause of an accident. One can also make the case that too much regulation interferes with safe practices, as nuclear plant operators have always claimed in the United States. But the overregulation complaint is undermined by the following anecdote: A few years ago, the NRC sharply increased the number of inspections of nuclear power plants following some embarrassing near-misses. A then-powerful US senator, Pete Domenici of New Mexico, a recipient of large campaign donations from the industry, called in top NRC officials and threatened to cut the agency's budget in half if it did not reduce the number of inspections (Mangels, 2003). The NRC reduced its inspections. I doubt that anything similar could take place in Europe.

Regulatory capture is widespread in many risky US industrial systems and often subtle—but not always. In the Interior Department's Materials Management Service, for example, representatives of the oil industry and regulators who were supposed to be overseeing oil exploration exchanged sexual favors and drugs. This intramural partying was disclosed just before the BP-leased *Deepwater Horizon* oil rig blew up in the Gulf of Mexico, resulting in the largest oil spill of its kind, making the regulatory failure especially dramatic.

Charges of regulatory failure were also levied in the 2010 Massey Energy coal mine disaster in West Virginia, which killed 29; the explosion at BP's Texas City, Texas, refinery in 2005, which killed 15 and injured at least 170; and BP's massive oil pipeline break in 2006 in Prudhoe Bay, Alaska.

There are many forms of regulatory failure. Regulations on the books can lie dormant by the common consent of regulators and industry. A worker at the Millstone nuclear power plant in Connecticut kept warning management that the spent fuel rods were being put too quickly into the spent storage pool and that the number of rods in the pool exceeded specifications. Management ignored him, so he went directly to the NRC, which eventually admitted that it knew of both of the forbidden practices, which happened at many plants, but chose to ignore them. The whistleblower was fired and blacklisted.

Rather than completely ignore regulations, a captured regulatory agency may just lower the standards it uses. The NRC has consistently lowered standards for emergency electric power supplies in US nuclear plants. And in the wake of the Fukushima disaster, the government of Japan is lowering standards for allowable doses of radiation.

Regulations are only as good as their enforcement, and here the evidence is fairly uniform: Enforcement is generally lax and often all but nonexistent. Workers at Fukushima reported that they had advance warnings of inspections, and inspectors regularly winked at violations. The record of the NRC is similar in the United States; for example, when utilities complained about the standards for fire prevention at nuclear plants in recent years, the regulators lowered the standards.

Even when safety inspections find violations, there is no guarantee that the regulated firm will be moved to change its practices. In many cases, the fines levied are too small to be a deterrent. After BP's huge spill in Prudhoe Bay, the company was fined less than its profits for one day of operation. After the Texas City refinery explosion, the *New York Times* reported that OSHA had levied a record fine of $87 million against the firm. According to BP, it made a profit of about $14 billion in 2009, meaning the fine amounted to about six-tenths of a percent of its profit for the year. An official of OSHA subsequently testified to the agency's weakness and the power of the petrochemical industry by noting that the size of the fines levied did not deter; firms repeated the same glaring mistakes despite their costs, ignored warnings, and harassed workers who warned of wrongdoing.

Warnings

Even if a risky system is only loosely regulated, a point will come when warnings are loud enough to attract attention. Catastrophes are expensive; no one wants them. The overall experience with warnings about global warming, however, should caution against expecting warnings to be too effective. A well-funded but factually challenged campaign to deny that climate change is the result of human activity has managed to ice the climate-warming warnings of a consensus of thousands of the world's top climate scientists.

Not surprisingly, we also do not find that warnings of looming industrial and financial disasters have much impact. At Fukushima, the regulatory authorities required a seawall that was a bit taller than the largest tsunami that locale had experienced in the last 1,000 years. So the danger was, indeed, recognized. But the seawall design was based on probabilistic thinking, not thinking about what is possible, and the seawall was horribly inadequate to the 2011 tsunami.

Some Japanese experts had done possibility analysis. They pointed to historical records of a huge tsunami in the year 869; three huge tsunamis on the Pacific Ring of Fire, along which Japan lies, in the last 100 years; and a geological record of relentless collision between two tectonic plates underneath Japan. Before 2011, these experts were largely ignored.

Japan has 53 nuclear power plants drawing cooling water from the ocean. Before Fukushima, 14 lawsuits charging that risks had been ignored or hidden were filed in Japan, revealing a disturbing pattern in which operators underestimated or hid seismic dangers to avoid costly upgrades and keep operating. But all the lawsuits were unsuccessful.

A representative in the Japanese parliament in 2003 warned that the nuclear plants were not sufficiently protected; a seismology professor at Kobe University resigned in protest from a nuclear safety board in 2006 due to a lack of attention to earthquake and tsunami risks. Even though there had been a 30-foot tsunami in 1993 on Japan's west coast from a much smaller 7.8 earthquake, the former head of the Tokyo Electric Power Company (Tepco) said that the risk of a tsunami never crossed his mind when he was president of that firm. He obviously did not have a possibilistic mind set.

But warnings can be slippery and hard to use effectively, regardless of the attitudes of the people being warned. Too often warnings are imprecise, which was why Condoleezza Rice, as national security adviser at the time of 9/11, dismissed the warnings of a terrorist attack using airplanes. The warnings did not specify the time and place. Many warnings are, in fact, so general as to be useless, e.g., "nuclear power is dangerous" or "radical Moslems will strike the United States."

There is the problem that warnings are often seen as mere obstructionism. This was the view of a representative for a Japanese utility who brushed away the possibility that two backup electrical generators would fail simultaneously. He said that worrying about such possibilities would "make it impossible to ever build anything."

Warnings may also be false, especially if based upon information that has little credibility, as with the weapons of mass destruction that Iraq was supposed to have. Many seemingly credible warnings never materialized, e.g., that President Barack Obama would not live through his first year in office. Florida coast residents are said to have stopped paying much attention to hurricane warnings after there were two evacuations for storms that didn't make landfall in the state.

And to be sure, there are major accidents that occur without warning, including the Three Mile Island nuclear incident and some chemical plant accidents, such as the toxic releases from Union Carbide's West Virginia plant. But these no-warning events are few. Credible warnings before major accidents are much more common.

The most credible ones are specific and in-house: The night before the launch of the space shuttle *Challenger*, engineers wanted to delay it because of the cold, saying, "We have never launched at this temperature, and cold affects the O rings." Before the re-entry that burned up *Columbia*, a technician on the shuttle's launch team tried to get pictures of the extent of the damage caused by chunks of insulation that had fallen off a fuel tank during the lift-off. Before the *Deepwater Horizon* was destroyed in a fiery blowout, Halliburton managers warned BP officials that there were not enough stabilizing rings installed on the drill pipe to continue drilling safely. Before BP's Prudhoe pipeline leaked, workmen placed hand-painted signs in the parts of the pipeline that did the most shaking, warning people to stand back because it might rupture (it did, but in an isolated area, and the break was not discovered for some days). Testimony has revealed that Massey managers regularly told supervisors to ignore warnings of dangerous concentrations of methane.

The warning at Texas City a few days before the plant blew was less specific than these others, but more ominous. At a company safety meeting, a slide was shown that simply said, "This is not a safe plant to work in." It was the view of management at the plant; they were unable to maintain the plant in safe conditions because of budget cuts and production pressures by top management.

Other warnings are more general and long range, but significant anyway. Scientists had regularly warned that the erosion of the wetlands protecting New Orleans was making it more vulnerable to hurricanes. They were more specific about the negative consequences of building a new ship canal that did, as predicted, channel Katrina's storm surge directly into the city.

There were multiple warnings in the United States before the 2008 economic meltdown. They came from some directors of impacted firms and from many risk managers and department heads, worried about the risks of their highly profitable mortgage business. They also came from government regulatory agencies, including the Securities and Exchange Commission, and watchdog agencies such as the Government Accountability Office; from bills proposed in Congress; from chief executives of financial firms not at direct risk; from financial gurus; and from journalists at leading magazines such as the *Economist*. There were also some 7,000 news stories containing the phrase "the housing bubble" from 2000 to 2006, meaning there were seven years of warnings before that bubble burst late in 2007.

Indeed, according to a book by a respected journalist (Sorkin, 2011), the newly appointed secretary of the treasury in the Bush administration, Henry Paulson, delivered warnings about a dangerous mortgage bubble at his first cabinet meeting in 2006. He proposed that investment banks be regulated much as commercial banks were, but Goldman Sachs, where Paulson had just served as president, and other major investment banks that dominated Wall Street would not hear of it. Credible warnings were dense, but the profits the firms were making drowned them out.

Coping

So how do organizations cope with disasters once one occurs? The record here is just as dismal as with regulation and warnings.

There are vastly more cases of creative coping from citizens than from organizations. The true first responders to disaster—co-workers, neighbors, passersby—have almost always performed splendidly, as with the completely self-organized flotilla that evacuated thousands from lower Manhattan in the 9/11 crisis. And in a few cases, governmental agencies and private firms do successfully cope with disaster.

Though failed space flights do not have the catastrophic potential of the other systems I have mentioned (they affect only a handful of people), they are complex and risky. The rescue of the crippled *Apollo 13* capsule is a prime case of skill and innovation in dealing with and overcoming a failure. There are many outstanding examples of coping by airplane pilots, though, again, the potential loss of life in these cases is relatively small. The response of President Lyndon Johnson's administration to the 1964 Alaskan earthquake—at 9.2 magnitude, the biggest ever in North America—is a model of what can be done by government to help victims and rebuild a city. But examples of creative coping by organizations are rare.

The poster child for official failure to cope with disaster is the response to Hurricane Katrina, the subject of so many books and articles that I will not dwell upon it (although it should be noted that the US Coast Guard performed extremely well and the unofficial response by citizens and private firms was often innovative and effective). In the case of Fukushima, there was official denial and secrecy, refusal to accept outside help, the failure to evacuate citizens at risk, and an attempt by the prime minister to halt the cooling of the damaged reactors by seawater shortly after the process had been started (fortunately, the plant manager lied, saying he had stopped using seawater even as he continued to use the ocean to cool the crippled plants, thus preventing a far worse catastrophe).

At Bhopal, plant officials initially denied any chemical release and then said it was not dangerous, even as they themselves were fleeing upwind of the toxic fumes. The Soviet Union refused to admit there had been an accident at Chernobyl, even after the Swedish nuclear agency had concluded that the radioactive materials they were detecting had to come from Chernobyl. Worse yet, the USSR waited two days before evacuating the town next to the plant. BP officials *and* American officials consistently minimized the damage of the oil spill in the Gulf of Mexico and kept reporters and scientists away from the scene. Little of the equipment oil companies are required to have on hand in case of a big spill was present when the *Exxon Valdez* ran aground. Similarly, Massey Energy didn't have equipment available to handle a mine explosion.

Crises may bring out the best in citizens. But, in some cases, they often raise the rate of routine errors made by distressed, tired managers and workers. At Fukushima, workers desperately trying to assemble a huge tank that would remove radioactive substances from the salt water being poured over the damaged reactor and its spent fuel pool saw the system fail on its first trial—because a valve had been installed backward. Workers at the Fort Calhoun Nuclear Power Plant near Omaha were surrounded by a flooding Missouri River with dikes close to being topped. They assembled an emergency berm to protect the vital electrical system. It was a 15-foot-wide, eight-foot-high plastic doughnut filled with water, a literal example of defense in depth. But someone backed a truck into it, and all the water poured out onto the soggy plant grounds.

This litany of regulatory failures, failures to heed warnings, and commonplace failures is independent of normal accident theory. That theory says that even if we had excellent regulation and everyone played it safe, there would still be accidents in systems that are highly "interactively complex," and if the systems are tightly coupled, even small failures will cascade through them. The theory is useful for its emphasis on system complexity and tight coupling; these concepts play a huge role in analyzing the failures of any source in risky systems. In the financial meltdown, for example, the mounting complexity of the overall system allowed fraud and self-dealing to go undetected, and the tight coupling of many systems allowed the failures to cascade.

In my work on "normal accidents," I have argued that some complex organizations—such as chemical plants, nuclear power plants, nuclear weapons systems, and, to a more limited extent, air transport networks—have so many nonlinear system properties that eventually the unanticipated interaction of multiple failures may create an accident that no designer could have anticipated and no operator can understand.

Everything is subject to failure—designs, procedures, supplies and equipment, operators, and the environment. The government and businesses know this and design safety devices with multiple redundancies and all kinds of bells and whistles. But nonlinear, unexpected interactions of even small failures can defeat these safety systems. If the system is also tightly coupled, no intervention can prevent a cascade of failures that brings it down.

I use the term "normal" because these characteristics are built into the systems; there is nothing one can do about them other than to initiate massive system redesigns to reduce interactive complexity and to loosen coupling. Companies and governments can modularize integrated designs and deconcentrate hazardous material. Actually, though, compared with the prosaic cases previously mentioned, normal accidents are rare. (Three Mile Island is the only accident in my list that qualifies.) It is much more common for systems with catastrophic potential to fail because of poor regulation, ignored warnings, production pressures, cost cutting, poor training, and so on.

All of the organizational faults I have noted have their counterpart in daily life. Like organizations and their leaders, people seek wealth and prestige and a reputation for integrity. In the process, they occasionally find it necessary to be deceitful, engaging in denials and cover-ups, cheating and fabrication. Everyone has violated regulations, failed to plan ahead, and bungled in crises. But people are not, as individuals, repositories of radioactive materials, toxic substances, and explosives, nor do they sit astride critical infrastructures. Organizations do. The consequences of an individual's failures can only be catastrophic if they are magnified by organizations. The larger the organizations, the greater the concentration of destructive power. The larger the organizations, the greater the potential for political power that can influence regulations and ignore warnings.

Modern society is not likely to deconcentrate big organizations and toxic substances, so what can be done? High-reliability theory is correct, of course, to say that government and business can do much more than they do to prevent serious accidents through constant training and mindfulness. More important is system design: Modular systems are less vulnerable than integrated ones, and the toxic and explosive potential is more dispersed in modular systems than in tightly coupled ones. Even more can be done through regulation; highly regulated nuclear power plants in Europe do much better than poorly regulated ones in the United States and Japan. Technical improvements can make systems safer, of course, and we do learn from past disasters. Emergency power facilities are being upgraded at nuclear plants in the United States because of Fukushima.

And the learning is continuous. Before the Three Mile Island incident, some held that during a loss-of-coolant accident in a nuclear plant, there would be a possibility of a zirconium-water reaction that consumes oxygen and frees hydrogen, which is explosive. One nuclear scientist scoffed at such a possibility in a publication, which was released shortly before he was, unfortunately, designated to be the key scientific adviser to Pennsylvania Governor Richard Thornburg during the Three Mile Island accident. (Later, the scientist was appointed chairman of the NRC.) The scientist was of course wrong. The appearance of hydrogen meant there was hydrogen "burn," as it is called, at Three Mile Island. Fortunately, the hydrogen accumulation was small, and the damage was minimal.

With this march of knowledge, engineers learned to install vents to prevent the explosive accumulation of hydrogen in the reactor buildings of nuclear plants in case of a loss-of-coolant accident. But the vents failed at Fukushima, and hydrogen explosions sent radioactive materials and gasses into the environment. Don't despair, though. Learning from disaster still goes on: US plants have been asked to make sure their vents will open as designed in case of a hydrogen explosion!

It is the commonplace scenario, such as this encounter with zirconium and vents, that needs to be emphasized. Prosaic organizational failures will always be with us, and knowledge is always incomplete or in dispute. Even highly reliable systems are subject to everyday failures, and even if we avoid these, there is always the possibility of normal accidents—rare but inevitable in interactively complex, tightly coupled systems. Some complex systems with catastrophic potential are just too dangerous to exist, not because we do not want to make them safe, but because, as so much experience has shown, we simply cannot.

Funding

This research received no specific grant from any funding agency in the public, commercial, or not-for-profit sectors.

References

Mangels, J. 2003. "NRC cracks down: Industry strikes back." *Cleveland Plain Dealer*, June 25.

Perrow, C. 2011. *The Next Catastrophe: Reducing Our Vulnerabilities to Natural, Industrial, and Terrorist Disasters*. Princeton, NJ: Princeton University Press.

Sorkin, A. 2011. *Too Big to Fail*. New York: Penguin.

21 Nature as Infrastructure: Making and Managing the Panama Canal Watershed

Ashley Carse

People tend to think of the Panama Canal as a technological system consisting of locks, pumps, valves, and ships. Indeed, the construction of the canal is often characterized as one of the major "technological feats" of the twentieth century. The canal connects two oceans, allowing ships carrying goods back and forth from east to west to avoid having to take a long route around Cape Horn at the southern tip of South America. In this article, Ashley Carse treats the canal as a sociotechnical system and illustrates the complexities of the system, especially given its global scale. In the sociotechnical frame, Carse emphasizes the human and nonhuman actors that both shape and are shaped by the canal, as well as the fact that maintaining the canal may be as difficult as initially building it.

The Panama Canal is, as Carse explains, an infrastructural system, which is somewhat different from other large sociotechnical systems in that infrastructural systems have a service role. They do something that is in a sense not itself a primary or focal good; rather, their value is in facilitating a primary activity such as commerce, transportation, or energy. Just as a power plant generates energy used for wide human purposes, the Panama Canal provides a shorter route for large ships to move cargo across the globe from one place to another. This makes a huge difference in the time and cost to move goods globally.

Carse's analysis illustrates the enormous amount of work it takes to produce and maintain infrastructures. He emphasizes that the work is simultaneously technical, environmental, and social. In framing the canal as a sociotechnical system, Carse shows how nature, as well as social and political circumstances, must be both managed and accommodated. Drawing on actor-network theory, he also demonstrates how ideas serve as nonhuman actors. In this case the idea of a "watershed" shaped the dynamics of decision-making about the canal. What it took to build the canal and now maintain it involves the intertwining of artifactual, social, and ecological elements.

This essay explores the notion that nature is—or might become—infrastructure, delivering critical services for human communities and economies. Put simply, this is the idea that forests, wetlands, reefs, and other landscapes, if appropriately organized, deliver services (water storage, purification, and conveyance; flood alleviation; improved air quality; climate regulation; and so on) that facilitate economic activity and development. It may seem peculiar to refer to landscapes or landforms as infrastructure, a term often reserved for roads, railways, and power lines. Infrastructure implies artifice; nature typically signifies its absence. However, as a growing literature in anthropology

---

From *Social Studies of Science* 42, no. 4 (2012): 539–563, https://doi.org/10.1177/0306312712440166.

(Balee, 2006), geography (Denevan, 1992), and environmental history (Cronon, 1995) suggests, nearly every environment worldwide has been modified through human labor. Work, then, blurs the nature–technology boundary, suggesting that a neat division is illusory (Reuss and Cutcliffe, 2010; White, 1995). Moreover, the concept of infrastructure does not delimit a priori which—or even what kind of—components are needed to achieve a desired objective. In practice, disparate components are integrated and become a networked support system through what Geoffrey Bowker (1994, 10) calls "infrastructural work," a set of organizational techniques (technical, governmental, and administrative) that create the conditions of possibility for a particular higher-order objective. In this essay, I develop an infrastructural approach for analyzing the practices, politics, and dynamics of environmental service delivery.

As infrastructure, nature is irreducible to a non-human world already "out there." It must, in its proponents' terms, be built, invested in, made functional, and managed. This is an active and inherently political process. As nature *becomes* infrastructure through work, human politics and values are inscribed on the landscape, much as they are embedded in arrangements of steel and concrete (Winner, 1980). Through this process, techno-politics and environmental politics become inextricably intertwined. As a landscape becomes infrastructure for one system of production, rather than another, a different group of environmental services (purposefully selected from a multiplicity of possibilities) becomes relevant. In a peculiar inversion, the landform may then be reverse engineered to meet the demands for the prioritized service(s).

The Panama Canal is an illuminating site to think about infrastructure, natural and otherwise. Five percent of global commerce moves through the canal's lock and dam system (USAID, 2005, 1). Interoceanic transportation, the higher-order objective that defines the canal, is made possible by a water management system that delivers the enormous volume of liquid necessary for 35–45 ships to transit the locks every day. Fifty-two million gallons of fresh water, equal to the daily domestic consumption of approximately 500,000 Panamanians,[1] are released into the oceans during *each* of these transits. Thus, the maximum number of transits possible is limited by available water volume, among other constraints. The canal depends on fresh water that falls as rain across the surrounding watershed,[2] a 1077 square mile (Ibáñez et al., 2002) hydrologic basin drained by six major rivers (see figure 21.1). That water is managed by a "traditional" engineered infrastructure comprised of locks, dams, reservoirs, and hydrographic stations. These technologies—which correspond with the popular idea of infrastructure as hardware—were largely constructed and networked during the early 20th century. Since the late 1970s, however, canal administrators from the US and Panama have responded to actual and potential water scarcity by developing an integrated management approach that combines existing technologies with new techniques of land use planning, environmental regulation, and community-based management. I analyze this socio-political work of water provision—especially the management of forests and farmers in the canal watershed—in order to explore the stakes of making natural infrastructure.

The essay is organized in four sections. First, I develop a conceptual framework for studying nature as infrastructure. The material that follows is a case study drawing on 18 months of archival and ethnographic research in Panama and the US. Second, I examine the organization of a network of civil engineering and hydrographic technologies around the Chagres River. Collectively, these technologies transformed a potentially volatile river system into a generally manageable water source for the canal. Third, I examine the establishment of the Panama Canal watershed as an administrative region

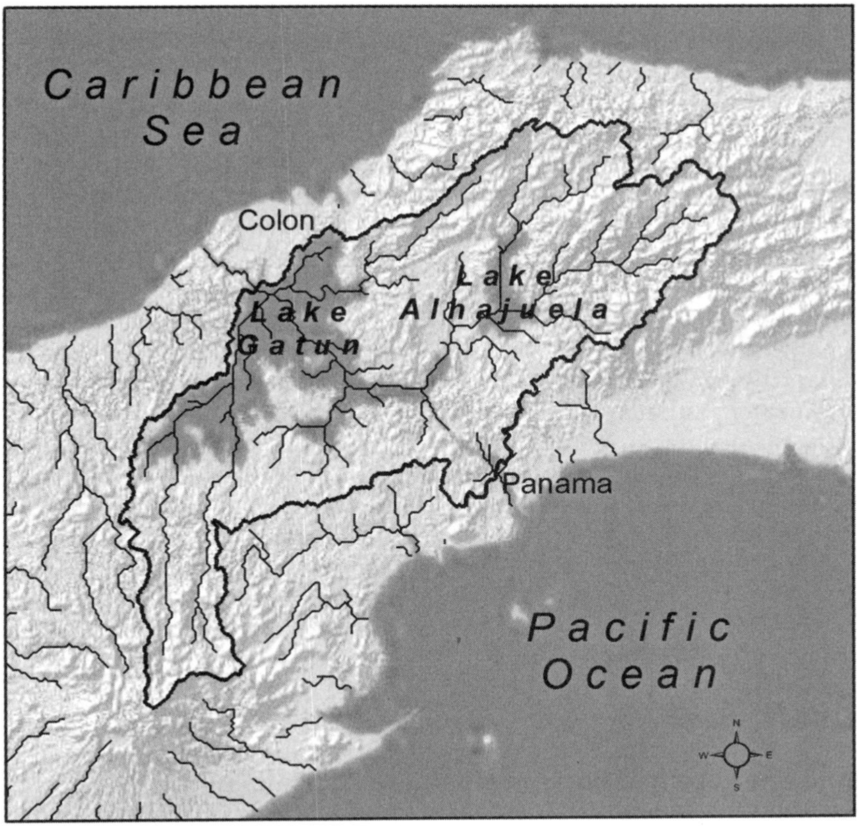

Figure 21.1
The Panama Canal transit zone, including the canal watershed (*bounded with a bold line*), Gatun Lake, Lake Alajuela, and the canal terminus cities of Colón (Caribbean) and Panama City (Pacific).

during the transfer of the canal and Canal Zone from the US to Panama in the decades after the 1977 Canal Treaties. Whereas canal administrators had previously emphasized the control of water in its liquid state, watershed management emerged as an attempt to manipulate water flows through the legal restriction of agriculture and protection of watershed forests. Thus, watershed management can be understood as a relatively recent manifestation of the ongoing project of reorganizing Panamanian landscapes and populations to optimize water delivery. I conclude by discussing the cultural politics of natural infrastructure.

Nature as Infrastructure

Infrastructure is the collective term used to describe the subordinate parts of a "higher" system. The word, first used in English in 1927, came from French, where it referred to the substrate material below railroad tracks (*Oxford English Dictionary*, 1991). The prefix *infra-* means below, beneath, or within. *Structure* has various meanings and, of course, carries significant intellectual baggage, but might be defined as the relation of the constituent parts of a whole that determines its character (*Oxford English Dictionary*, 1991).

The term infrastructure is widely used in economics and planning, where it refers to capital investments that facilitate directly productive economic activity or development (Lee, 2009, 382–383). In vernacular usage, infrastructure often refers to artifacts built of concrete and steel: the "hard" technical systems that facilitate the distribution of people, energy, water, waste, information, and so on. However, the use of the term has expanded rapidly in recent decades as it has been deployed in a variety of new fields. Infrastructure still refers to hardware, of course, but the term also increasingly evokes the "soft" social systems assembled to support education (Twigg, 1994), governance (Globerman and Shapiro, 2002), and public health (Baker et al., 2005). In the case of natural infrastructure, as in these examples, emphasis is placed on the functions, benefits, or services that a subordinate system delivers, rather than the type or character of its individual components. Thus, I argue that work—specifically the organizational techniques through which components are networked for service delivery—is a useful analytic tool for examining relationships defined in terms of infrastructure.

What railroad tracks and public health clinics share, despite their obvious differences, is that they are obviously built by and for people. Landforms, by contrast, are less obviously constructed and have only been explicitly described as infrastructure over the last decade (see, for example, Benedict and McMahon, 2006; Smith and Barchiesi, 2009). This shift in managerial rhetoric and practice is associated with a broader interdisciplinary effort since the 1980s to assign the environment value as natural capital: a stock that provides ecosystem services that benefit humans at multiple scales (Costanza et al., 1997). As a corrective to the assumption that the environment has no economic value and, by extension, that its degradation has no social cost, reconceptualizing nature via service delivery is presented as a market alternative to state regulatory approaches that are incapable of dealing with multi-scale environmental problems. At the same time, scholars have suggested that the establishment of ecosystem service markets may also have adverse social and economic effects for rural people (Corbera et al., 2007; McAfee and Shapiro, 2010). In this essay, however, I bracket these important debates and emphasize how historically contingent socio-technical systems shape which environmental services become valuable and who benefits from their delivery. My study of Panama in the 1970s and 1980s emphasizes the role of the state, but my focus is neither on the market nor the state, per se. Building on the work of Nigel Thrift (2005) and Timothy Mitchell (1991), I approach capitalism and the state as effects of the socio-technical platforms that give them such obduracy and vitality.

My analysis of water management in Panama builds on work in infrastructure studies.[3] The genealogy and key theses of this literature in STS have been outlined elsewhere (Star and Ruhleder, 1996, 113), so I will only highlight a few points central to my argument. First, STS research on infrastructure arguably emerged in the wake of Thomas Hughes's history of electrification and the development of large technical systems (1983, 1987). Large technical systems scholarship posits that technologies, or artifacts, should be understood as components of the socio-technical systems that support and sustain them (Coutard, 1999; La Porte, 1991; Mayntz and Hughes, 1988; Summerton, 1994). Bowker, Edwards, Star and others have theorized infrastructure as a useful conceptual bridge between these macro-scale studies and the actor- or artifact-centered approaches advocated in social construction of technology (Bijker et al., 1987) and actor network theory (Callon, 1986; Latour, 1983) approaches. While the large technical systems approach tends to emphasize top-down, unified organization by system builders, the infrastructure studies approach is less hierarchical. Infrastructures are conceptualized ecologically, which is to say that they are understood to come into being, persist, and fail in relation to the practices of the diverse communities that accrete around them (Star and Ruhleder, 1996). They are shaped by, yet also exceed, the intentions of their builders. Grounded in everyday life, infrastructures become revealing sites for ethnographic research on negotiation, struggle, and meaning (Star, 1999). I argue that we extend this insight to environmental service provision, which can be analyzed at the intersection of emergent environmental problems, historically specific socio-technical systems, and everyday life in neighboring communities.

Infrastructure studies scholars recognize that technology is political (Winner, 1980), but the politics of infrastructure may be difficult to see. In modern life, people often experience infrastructure as "behind the scenes, boring, background processes," operating unnoticed in the absence of breakdown (Bowker and Star, 1998, 234). But, of course, they are not invisible to everyone, everywhere. Visibility is situated, reflecting an actor's geographical location, cultural assumptions, and the nature of his or her labor (Star and Ruhleder, 1996, 113). As technical systems cross social, economic, and geographical boundaries, they are experienced differently and affect human groups differently. This is to say that one actor's background infrastructure (for example, a functional Panama Canal for shipping companies) remains a persistent problem for those "behind the scenes" (the engineer, mechanic, and, as we will see, the Panamanian farmer). For this reason, Bowker and Star (1998, 234) urge us to conduct "infrastructural inversions," recovering the world-ordering arrangements embedded in the systems that many of us take for granted. In summary, STS scholars have conceptualized infrastructure as a large technical system, ecology, and site of political struggle, but the relationship between socio-technical systems and the non-human environment has received less sophisticated attention.

On the one hand, the environment is treated as "an assumed background of natural forces and structures" (Edwards, 2003, 194). Seen this way, the environment may be harnessed by, or set parameters for, a system, but remains outside of it. On the other hand, socio-technical systems may themselves constitute an artificial or built environment constructed and deployed by humans. Here, the variability of nature is channeled and mediated for human use, comfort, and convenience (Edwards, 2003, 189). Environments managed for the purposes of service delivery fit neither framing. On the one hand, these landforms are "natural" because they are shaped by processes beyond human control, but they are also sites of management and investment for actors seeking to optimize the delivery of services once assumed to exist in the background. Below, I show how infrastructure studies might be combined with political ecology and other critical scholarship on the environment to analyze the politics of making natural infrastructure.

Political ecology explores the complex ways that human groups know, access, and struggle over the environment in multi-scale or networked power relations (Robbins, 2005). Although building infrastructure is rarely framed as a political project by its promoters, new infrastructures inevitably threaten to alter or eradicate existing ways of life. Similarly, making natural infrastructure has significant potential to produce ecological distribution conflicts around socially and spatially asymmetrical access (Guha and Martinez-Alier, 1997). These conflicts often turn on material landforms and land cover—the infrastructure itself—but, more fundamentally, they raise questions about which among a multitude of potential environmental services are to be emphasized and delivered and, crucially, *whose* societies and economies those services support.

The distribution of benefits and costs associated with making natural infrastructure is a question for site-specific empirical research. At best, "local" human communities may become "partners" in environmental service provision, and benefit economically. At worst, planners and environmentalists assume that physical work with nature is inherently destructive and attempt to block some communities' access to natural resources. Although the Panama Canal watershed has not, to my knowledge, been explicitly characterized as infrastructure, administrators do emphasize the infrastructural functions—primarily water storage and regulation—that the drainage basin provides. In the case of provisioning the Panama Canal with water, the forests that became a concern for water managers were not uninhabited natural areas, but landscapes already morally, economically, and ecologically bound up in agrarian life. Therefore, natural infrastructure has been constructed across human groups who know and interact with the forest in different ways. Watershed administrators, I will argue, have sought to reshape human–environment relationships so as to optimize the delivery of one class of environmental services (water storage and regulation), while campesino farmers have sought to optimize another (nutrient cycling). Political ecologists have insisted not only that we ask "what nature?" but also "whose nature?" (Escobar, 1998). In this case, we might also ask, "whose natural infrastructure?"

Assembling the Panama Canal

The most obvious place to see the Panama Canal in action is the Miraflores Locks Visitor Center, located in the former US Canal Zone near Panama City. The Miraflores Locks are one of three sets of locks (the others are the Gatun Locks and Pedro Miguel Locks) that raise ships from sea level at one ocean up to Gatun Lake at 85 feet and lower them back down to sea level on the other side of the isthmus in, on average, 8 to 10 hours. From a three-tiered viewing deck, visitors watch a slow parade of container ships, oil tankers, and cruise ships pass through locks built a century ago. Here, crudely, is how they work. Ships are raised and lowered in the locks through the use of water, gravity, and technology. First, a ship slides into the lock chamber and massive steel gates swing slowly closed. Then, a lockmaster opens valves in a water storage reservoir located above the locks. Water surges through culverts the size of subway tunnels in the concrete chamber walls, enters cross-culverts beneath its floor, and then erupts upward, lifting a ship and its cargo. Visitors to Miraflores Locks see only one node of the expansive canal system, but even this suggests the coordinated social and technical work that makes transportation across Panama possible. Less obvious from this vantage point, however, is the quotidian environmental management that administrators now conduct in order to move ships.

On small farms as far as 25 miles upstream from the shipping lane, forest guards working for Panama's national environmental agency (ANAM) inspect the secondary growth of grass, bushes, and young trees that rural Panamanians call *rastrojo* in order to determine when it is legally defined as forest and, therefore, protected by the state to "produce" water for the canal.[4] Integrated watershed management began after the signing of the Canal Treaties of 1977, which guaranteed Panama control of the canal after 1999. Canal administrators, natural scientists, state functionaries, and others collaborated to establish the watershed as a political–administrative space for water management. Forests formerly located in the Republic of Panama and beyond the boundary of US Canal Zone became sites of struggle as the watershed was unified under Panamanian control and rural agriculturalists were scripted as a potential threat to transportation.

How did watershed forests become natural infrastructure serving the canal and, by extension, global commerce? The case study that follows focuses on water management in the late 20th century, but neither the emerging experience of water scarcity during that period, nor its proposed solution—watershed management—can be understood without a discussion of the development of the large technical system still in use today. The water management components of that system include the following: three sets of locks, three dams, three water storage reservoirs, and a network of remote hydrographic stations from which measurements of rainfall and changes in river and lake elevations from across the watershed are transmitted to a central station. My objectives in this section are to summarize how this system functions, discuss the infrastructural work that went into its construction, and highlight the historical relations between the canal administration and the diverse human communities living in the surrounding region.

The Panama Canal is a socio-technical system, which is to say that, as Hughes (1987) points out, it includes technical, organizational, scientific, and political-legislative components. Between 1903 and 1999, these components came together in the Canal Zone, an imperial transportation enclave established and controlled by the US government. In the Hay–Bunau–Varilla Treaty, as the 1903 Panama Canal treaty is officially known, Panama granted the US the "use, occupation, and control" in perpetuity of a strip of land 5 miles wide on either side of the proposed canal route in exchange for a modest cash payment (LaFeber, 1978, 37–38). Although sovereign title remained with Panama in a strict legal sense, in practice the US assumed juridical and police power within this territory at the narrowest point of the isthmus. Significantly, the political boundaries of the Canal Zone were not fixed, but open-ended and linked to the development of the canal as a technical system. In addition to near-sovereign power within the original ten-mile strip, the 1903 Treaty gave the US authority to expropriate more territory as needed for the "construction, maintenance, operation, sanitation, and protection" of the canal (Articles II and III). This provision legally allowed the US government to expand canal infrastructure and political control in tandem across relevant territories in Panama. Water management, in particular, was central to the development of this techno-political project.

In the summer of 1906, the US Congress opted to fund the construction of a lock canal in Panama. The American route would, for the most part, follow the channel where the French *Compagnie Universelle du Canal Interocéanique* had worked in the 1880s (Board of Consulting Engineers, 1906). But there was one significant change of plans. Before bankruptcy and the death of an estimated 20,000 laborers (McCullough, 1977, 235) the French company had planned to excavate a sea-level canal: a salt-water channel that, if completed, would have allowed ships to travel unimpeded between the oceans. By contrast, the American lock design would move traffic over a fresh-water aquatic staircase built around three sets of locks. The Board of Consulting Engineers for the Panama Canal was assembled in 1905 by President Theodore Roosevelt and charged with making a recommendation for canal design (Board of Consulting Engineers, 1906). They were unable to reach a consensus and produced two reports.

The majority report advocated a second attempt to dig a sea-level canal, while the minority report supported the construction of a lock canal. The proposals differed in estimated cost and time. The sea-level canal plan would cost an estimated US$247 million and require 12 to 13 years to complete, while the lock canal was projected to cost US$139 million and require 9 years (Board of Consulting Engineers, 1906, xiv, xvii). The Board of Consulting Engineers members who supported the sea-level canal argued that an open, unobstructed waterway such as the profitable Suez Canal could be achieved due to engineering advances since the failed French effort in Panama and that national dignity compelled the US, as Board Chairman Davis put it, to "treat this matter not in a provisional way but in a final masterly way" (quoted in McCullough, 1977, 483). The proponents of the lock canal proposal, modeled on the Soo Canal connecting Lake Huron and Lake Superior, countered that their design would be safer for ships in Chagres River flood conditions, reduce the impact of landslides on transit, provide easier passage for large vessels, cost less to maintain, and be easier to enlarge and defend. After lengthy, charged deliberations in Washington DC, Congress approved the lock plan in 1906. More than a century later, this decision continues to shape the organization of international shipping networks and, as I will show in the third section of the essay, the political ecology of the surrounding watershed.

The Lock Canal Design and Its Water Use Implications

The decision to build a lock canal and the determination of the dimensions of its chambers (110 × 1050 × 85 feet) fixed the water volume required for each canal transit at 52 million gallons. The establishment of this standard, which also restricts the size of passing ships, significantly shaped the form and extent of the canal's water management network. Why these dimensions? Congress stipulated that the future waterway afford passage to the largest ships that existed during the first decade of the 20th century and "such as may be reasonably anticipated." When the locks were designed, 95% of ocean-going ships measured less than 600 feet in length (Bakenhus et al., 1915, 81). The lock decision precipitated the reorganization of the Chagres River system to store and consistently deliver an enormous volume of fresh water to the canal.[5]

"The vital question," wrote Henry Abbot, hydrologist and retired Army Corps of Engineers Brigadier General, in 1905, "was to determine whether the Chagres [River] will supply all the needs of the Canal in seasons of low water. Any reasonable doubt here would be fatal to the project of a canal with locks" (Abbot, 1905, 105). Abbot was among the North Americans most knowledgeable about isthmian canal engineering debates, having served on the Board of Consulting Engineers and its predecessor, the *Comite Technique*, an international group of engineers assembled to consider the future of a canal in Panama after the failure of the French project. When the US arrived on the isthmus in 1904 to begin construction, relatively little hydrographic data had been collected within the Chagres River basin. Because the French had planned to excavate a sea-level canal with water supplied by the oceans, determining the volume of water flowing through the river system was not a priority. For a lock canal, however, knowledge of the volume, speed, and seasonality of river flows would be very important.

Water Storage at Gatun Lake and Alajuela Lake

Gatun Lake and Alajuela Lake (called Madden Lake before the 1977 Canal Treaties) are the major reservoirs that store water as a buffer against seasonal variation in precipitation, allowing the canal to operate through the 3-month dry season, when little runoff enters the system. Gatun Lake, the largest reservoir, is the centerpiece of the lock design and makes up much of the length of the canal. To create the lake, a 1.5-mile-long earthen dam was built and, in 1911, the spillway gates were closed, interrupting the flow of the Chagres River to the Atlantic Ocean and flooding the surrounding valley. The US government invoked treaty rights and expropriated Panamanian lands to be submerged by the future lake. The creation of the lake also coincided with the 1910 implementation of a depopulation policy designed to "extinguish" competing property claims in the Zone and relocate thousands of Panamanians and West Indians.

When Gatun Lake reached its operating level in early 1914, it was the largest artificial reservoir in the world, spreading out over 164 square miles of the Chagres River valley (Haskin, 1914, 39–40). Even then, however, engineers recognized that its water storage capacity would be insufficient as traffic through the canal increased (Kirkpatrick, 1934, 84). The canal's water demands were, and remain, defined in relation to canal traffic. As the number of transits increases, so does the water used. In 1924, President Calvin Coolidge signed an executive order to create a second dam and reservoir on the upper Chagres River. The 22 square miles that would become Alajuela Lake were expropriated from Panama, again under the terms of the 1903 treaty, and appended to the Canal Zone. The process enacted on the lower Chagres River at Gatun—survey, expropriate, depopulate, and flood—was repeated on the upper Chagres.

In summary, the Panama Canal's water management system was first shaped by political forces and then began to exert its own political force. The pursuit of water for the canal shaped the territorial demands made by the US government of the Panamanian government and led to the displacement of thousands of people living in the Chagres River valley. This process entailed two types of infrastructural work: governmental and technical. The governmental policies enacted across the rural lands around the Chagres during the early 20th century displaced thousands of people from the Canal Zone. Civil engineering reorganized a region's physical geography to deliver the water demanded by the lock design, in the context of the climatic and hydrological specificities of Panama and increasing ship traffic through the canal. As water flowed downstream and collected in storage reservoirs, the network of water management technologies was extended further and further upstream. Concerns about water supply were not permanently resolved with the creation of Alajuela Lake in the 1930s. The Panama Canal's *hydrological reliability*—the ratio of available water volume to the water volume demanded for normal transportation operations and municipal uses (Vargas, 2006, 152)—was not defined in Panama alone, but in relation to the ebb and flow of international trade.

Traffic through the canal—and, by extension, water use—increased rapidly after the Second World War, following worldwide trends in commercial oceangoing traffic (Iwabuchi, 1990, 24–26). Canal administrators, engineers, and hydrologists proposed canal projects aimed at overcoming the physical constraints imposed by the lock design on number of transits, ship size, ship speed, and available water supply. It is beyond the scope of this essay to detail the development of the canal system throughout this period, but key proposals included lock expansion (Iwabuchi, 1990), the construction of a sea-level canal (Leschine, 1981), and additional water storage reservoirs (Panama Canal Company, 1961). The system was able to store and deliver sufficient water to meet traffic demands under normal climatic conditions, but, as the experience of the late 1970s would prove, extreme droughts could lower the levels of reservoirs enough to threaten the flow of ships through the canal. Up to this point, water management and rural governance had been considered distinct issues and administered by different state institutions. The watershed—conceptualized as a geo-hydrological unit, rather than a political space—was the concern of hydrologists and engineers. This changed in the late 20th century, when watershed forests were reimagined as a living support system for the canal.

Making the Panama Canal Watershed

In the 1970s, a new water management problem circulated through offices and conference rooms in Panama and the US. The canal had long been extolled as modern man's ultimate triumph over nature. But now it seemed that the tables had been turned.

Foresters and hydrologists suggested that, without decisive action, the environmental degradation of the Panama Canal watershed would put the critical shipping route permanently out of business. The problem was articulated most forcefully by tropical forester Dr Frank Wadsworth at the 1978 US Strategy Conference on Tropical Deforestation, co-sponsored by the State Department and US Agency for International Development (USAID). Wadsworth worked for the US Forest Service in Puerto Rico at the Institute for Tropical Forestry and had, in 1977, consulted on a USAID program to strengthen environmental management in Panama. In a paper entitled "Deforestation: Death to the Panama Canal," he argued that deforestation by "shifting cultivators"—campesinos farmers—altered runoff from the watershed into the canal system, depositing sediment in the upper reservoir and reducing the available water supply. Wadsworth described the anatomy of an emerging crisis:

> In May of 1977, the passage of an above average number of ships, an increased use of water for hydroelectric power and the domestic supplies of growing cities, and the production of timber, food, and forage crops within the Canal watershed led to a dramatic demonstration of the limits of the capability of the water system. The surface of Gatun Lake dropped to 3.1 feet below the level required for full Canal use. Some ships sent part of their cargo across the isthmus by land, reloading it at the other coast, and certain bulk cargo shippers even abandoned the Canal, sending very large carriers around the Horn. In 1977, this predicament coincided with a serious drought, and this was seen as a harbinger of what could soon take place every year. . . . Deforestation and cultivation in areas adjacent to the headwaters accentuate both flood losses through the spillway and low flow in the dry season. (Wadsworth, 1978, 23)

The problem, in Wadsworth's formulation, could not be fixed through established civil engineering approaches to water management. "Only forests," he (1978, 23) concluded, "can restore and stabilize the capacity of the canal. Even if Madden Dam were raised, the five additional dams built, fresh water tunneled from elsewhere, and power and urban water consumption discontinued completely, the effect of continued deforestation would be inexorable. Sooner or later it would mean death to the Canal as a reliable world trade route." Notably, Wadsworth identified the role of several contributing trends to water shortage in the canal system (increased ship traffic, as well as hydroelectric and municipal water use), but the focus of his proposed intervention was rural land use. By invoking the specter of commercial *death*, Wadsworth assigned the canal a new kind of life. He reframed the waterway as an ecological system—a valuable and fragile organism—countering the perception of it as a man-made channel. In his formulation, the heterogeneous character of the water supply problem (social, technical, and ecological) demanded an integrated solution: watershed management.

Wadsworth arrived in Panama shaped by the scientific and political history of the Forest Service.[6] During its public struggle with the Army Corps of Engineers in the first decades of the century, the Forest Service had promoted the theory that watershed forests regulate stream flow and flooding. Rhetorically, the scientific controversy turned on which institution's water management approach—technical vs. "natural"—would produce orderly rivers, but it also reflected an institutional struggle for political clout and funding.[7] In contrast to its significant role in US public land management, forestry occupied a marginal position on the isthmus.[8] The Panama Canal Company (which operated the canal) and, in many cases, the Canal Zone Government (which governed the enclave around the waterway) had been historically controlled by engineers and tightly bound to the Corps. The few foresters who worked in on the isthmus were scientists conducting research or were consultants. Their work had little apparent effect on environmental policy or management in the Canal Zone or in Panama, a country where no forestry training was available (Budowski, 1961). I provide this brief comparison of the institutional genealogies of forestry in the US and Panama to make two points. First, state management of nature as infrastructure had historical antecedents. In the early 20th century, foresters framed watershed forests as environmental service providers, although not in those exact terms, characterizing them as "nature's reservoirs" and critical support systems for commerce (Pisani, 2002, 155). Second, neither the science of forest hydrology nor the practice of managing watershed forests was significant on the isthmus before the 1970s. This changed as the forests around the canal were increasingly linked to water provision and global shipping.

Forest Hydrology, Institutional Politics, and Watershed Landscapes

Watershed management entailed the conceptual and spatial expansion of water management infrastructure to incorporate the region's forests and the campesinos that inhabited them. The forests of the upper watershed were considered the most critical area for hydrological purposes. "This area," Wadsworth (1978, 23) wrote, "provides about 40 percent of the water for the entire Canal watershed. It is now being invaded by shifting cultivators." Wadsworth held complex, sometimes contradictory, social and ecological commitments. On the one hand, he identified himself as a pragmatic forester, emphasized scientific forest management, and disagreed with what he saw as the extremist political views of many conservationists, or, as he put it, "greenies."[9] He was concerned about the livelihoods of farmers and dismayed by technocrats who did not understand the social context in which they worked. On the other hand, he explained tropical deforestation in terms of a narrowly defined social context of rural mores, mentalities, and population pressures—which he saw as the drivers behind forest "invasion"—and paid little attention to the role of political-economic pressures and state programs in environmental change. To be fair, his perspective was not unusual. Myths and ideologies of rural environmental ignorance were commonplace among foresters working in the tropics at that time (Dove, 1983). In Panama, as we will see, deforestation was the outcome of both local and extra-local influences. Wadsworth believed that governments in the American tropics should actively manage forests and discovered, as the US Forest Service had decades before, that anxiety about water scarcity can serve as a useful tool for motivating state environmental intervention.[10]

The institutional actors who initially assembled around the identification and management of the watershed problem were not collectively concerned with shifting cultivators or forests, per se, but with ensuring a consistent supply of water for the canal in the face of potential scarcity. The ultimate objective of watershed management was thus the mitigation of climatic and hydrological risk. The water shortage of 1977, Wadsworth (1978, 23) wrote, "was seen as a harbinger of what could soon take place every year." But mitigating environmental risk presented serious problems of tractability. Thus, the conservation of watershed forests became water management by proxy. The 1970s and 1980s were a time of rapid geopolitical change on the isthmus and a period when watershed management and tropical forest conservation were ascendant topics in international academic and economic development communities. Stanley Heckadon-Moreno, an anthropologist who played a key role in early Panamanian watershed management efforts, said the concept arrived in the country via foreign institutions:

> In Panama the word watershed—*cuenca*—didn't exist. People knew about the canal. But when one spoke about a *cuenca*, nobody had the slightest idea what you were talking about. . . . I think the word began to come into vogue in the 70s and definitely in the 80s, used by institutions like CATIE [Center for Tropical Agronomy Research and Teaching] in Costa Rica. . . . The concept of using the watershed as a [political] geographical unit—not a country, not a province, or a state or a *corregimiento* [county]—but a river. That was new.[11]

As the arrival of the watershed concept in Panama demonstrates, a hydrological basin may be a "natural fact," but for planners, managers, and policymakers, it is only one possibility among many for partitioning and managing the earth's surface. The selection of one approach over another is a political choice shaped by what particular actors want out of a given landscape. The artifice of the Panama Canal watershed—its "making"—is a result of the accretion of knowledge, technologies, and institutions around an existing hydrological basin to ensure that "it" provides the services desired. Thus, watershed forests became infrastructure through the purposeful work that went into linking them with the existing water management system. In Panama, watershed management entailed—and still entails—the slow, difficult work of forging managerial relationships with the rural people whose livelihoods were scripted as a threat to transportation. This social process, unlike civil engineering, operates in a bottom-up manner and depends on the participation, through coercion or free choice, of groups of actors formerly "outside" the system.

The Rise of the Panama Canal Watershed

Wadsworth's translational work extended the reach of tentative efforts already underway in Panama to manage the watershed. He collected the material for his essay in 1977, while consulting on the development of a USAID program designed to strengthen the technical and administrative capabilities of RENARE, Panama's historically insignificant and impoverished natural resource agency (Wadsworth, 1978, 24). His argument was not completely new. USAID had already funded the research of Dr. Clark Larson, an agricultural engineer. Larson (1979) found that deforestation in the watershed for cultivation and pasture increased the sedimentation of the canal and reduced its water storage capacity. Meanwhile, in the Republic of Panama, RENARE was also in the process of collecting basic meteorological, hydrological, soil, and social data for analysis, map-making, and prospective watershed management.[12] But early US and Panamanian efforts were largely uncoordinated. By contrast, the first integrated Panama Canal watershed program—funded between 1978 and 1983 by a US$10 million USAID loan and a US$6.8 million Panamanian contribution—was designed to establish coordinated regional management. The program also had local objectives: to increase environmental awareness and "incorporate, to the extent possible, the watershed's population into the resource management conservation process" (ROCAP-USAID, 1981, 6). Watershed management was conceptualized at the institutional and regional levels, but its promoters recognized that forest protection ultimately depended on changing the consciousness and behavior of rural people.

"The Canal," an early evaluation of the USAID watershed management program explained, "represents Panama's major industry and is at the heart of a complex system of support and service industries . . . the project benefits Panama's major industry and its work force" (ROCAP-USAID, 1981, 7). Forests were understood to "produce" water, the lifeblood of the canal and the transport service economy. But those forests could not be protected without negatively affecting the rural economy. USAID's phrasing—*incorporation of the watershed's population into the resource management process*—marked a significant shift in which new actors were assigned responsibility for canal water. As forested landscapes were assigned an infrastructural function (water provision), their inhabitants were simultaneously charged with a new responsibility (forest conservation). However, the implementation of watershed management would prove difficult in practice because its environmental goals were in direct conflict with those of an established infrastructure that supported agricultural development.

In the early 1970s, prior to the transfer of the canal from the US to Panama, the campesinos of the watershed inhabited a social world in which, like much of rural Panama at the time, the state promoted agricultural colonization and rural development through political, economic, and technical means. During the 1950s and 1960s, the Panamanian Ministry of Agriculture (MIDA) and multilateral institutions such as the Inter-American Development Bank (IADB) sought to modernize agricultural production across the rural *interior*, including parts of the watershed, through rural "penetration" roads, agricultural extension, market development, and agricultural credit. This agricultural infrastructure supported and incentivized particular, often extractive, relationships between rural people and the land. For example, MIDA implemented a program called "Conquest of the Jungle" that encouraged campesinos to colonize forested frontier zones for economic and political purposes (Heckadon-Moreno, 2005, 37). Nevertheless, many early watershed managers like Wadsworth perceived campesino "culture" and mores to be the main problem, ignoring decades of farmer engagement with a state apparatus that actively promoted deforestation as development policy. Moreover, watershed management often focused on agriculturalists, but extensive cattle ranching was a more permanent threat to forests. MIDA opened a demonstration ranch in the watershed in the 1950s designed to expand cattle production across "a region not commonly considered appropriate for these types of activities" (Government of Panama, 1956, 33–34). Between 1970 and 1979, the height of the cattle boom, US$543 million in loans were made to ranchers in Panama without environmental restrictions (Heckadon-Moreno, 1985, 50). Ninety percent of those receiving loans from the IADB and *Banco Nacional de Panama* used them for ranching. Agricultural and rural development institutions were working at cross-purposes with watershed managers. Campesinos were caught in the middle.

The Panamanian state expanded rural development efforts after 1968, when the *Guardia Nacional*, the first military government in the nation's history, took control in a coup. One of the *Guardia*'s political priorities was improving the condition of the marginalized rural population through land reform, particularly the expropriation of large estate farms (*latifundias*) and the establishment of agricultural cooperatives of landless campesinos (*asentamientos*) on that same land (Heckadon-Moreno, 1984, 143–144). MIDA's annual reports from the late 1960s and early 1970s reflect this populist, agrarian reform fervor. Government reports proudly tally roads built, forests cleared, and new area farmed. "The *Guardia Nacional*," Heckadon-Moreno (1984, 147) writes, "like other military regimes that came to power in tropical America during the 50's, 60's and 70's, was keenly interested in securing the physical integration of the *selva* into the nation state . . . colonization was a fast and cheap way of incorporating the forest into the development process." This dominant paradigm changed rapidly within the watershed after the 1977 Canal Treaties initiated the transfer of the canal and Canal Zone to Panama. Panamanian state institutions had previously conceptualized watershed forests as standing in the way of modern agriculture and, thus, national economic development. However, if the canal and its associated transport service economy were to become Panama's, the same forests, as natural infrastructure, suddenly became necessary for national development. As a result, the campesinos that farmed and ranched across those landscapes had suddenly become a development problem rather than a solution.

Given the rapidity of this shift and obvious tensions between old and new state plans for the region, it proved difficult to convince rural people that the forests they lived and worked in were not exclusively theirs, but part of a hydrological support system for shipping. New watershed managers encountered, at every turn, a rural development infrastructure—roads, agricultural cooperatives, extension agents, agricultural loan programs—that encouraged the very land use practices they now considered economically and ecologically irresponsible. Watershed management thus entailed negotiating this embedded infrastructure and the campesino moral economy—norms and customs concerning the legitimate roles of particular groups within the economy (Thompson, 1971)—that had accreted around it. Managers recognized that the success of watershed management was contingent on enrolling forest guards able to align the diverging interests of state institutions and rural social worlds. The sections below, based on oral history interviews that I conducted during 2008 and 2009, analyze this translational work and the challenges of building natural infrastructure across cultural difference.

Forest Guards and the Translational Work of Watershed Management

Lucho grew up farming in the upper watershed, but, before he became a forest guard in the late 1970s, he had never heard anyone use the word *cuenca* (watershed) to describe those lands.

He moved with his family from Panama City to settle on the banks of Alajuela Lake in 1958, when this canal reservoir was still controlled by the US government and called Madden Lake. Like many settlers arriving at the time, Lucho, still a teenager, dreamt of farming his own land. He wanted to work independently, not be an *empleado* (wage-laborer). One day in 1975, he was cutting back the rapidly growing brush—or *rastrojo*—on his farm when he received a note that Colonel Ruben Dario Paredes, the Minister of Agriculture, wanted to meet with him.

When they met, according to Lucho, Paredes told him, "You've been recommended as a man who is not afraid of anything. We'd like to give you a job: we want you to keep the hand of the *campesino* from destroying the watershed." Lucho, unclear about what this meant, asked, "What is the watershed?" Paredes said, "The watershed is all of this area that drains into Alhajuela Lake." Lucho recalled that he then told Paredes, "I'd like to do it, but I have to talk with my wife, my first child is on the way." Paredes offered him US$50 every 2 weeks, but Lucho countered, "I'm not going to abandon my land for fifty dollars, Colonel. I've got an old mother, an old father, a brother—we can't live off of that much money. I'm my father's right hand."

According to Lucho, Paredes increased the offer to include a free education in natural resource management. Lucho had no particular interest in natural resources at the time, but he accepted. He had, in actor network terms, been enrolled in watershed management (Callon, 1986). Paredes successfully mobilized the promise of career opportunity to convince Lucho to put down the machete and assume the role of a forest guard defending the watershed from other campesinos. RENARE, still part of MIDA, recruited 46 forest guards to patrol the watershed (ROCAP-USAID, 1981, 7). The others, like Lucho, were mostly local men identified by officials as leaders respected in their rural communities. This was a strategic decision. Watershed managers hoped that guards familiar with the area and its people would facilitate cooperation.

The forest guards' first project was to survey the human population living within the watershed. Guards spent 3 years—1975 to 1978—collecting data on the region's rural inhabitants. Survey data and census data provided a demographic baseline for the watershed, assigning, for the first time, a population of human "inhabitants" to the new administrative region (Cortez, 1986, 45). The problem, however, was that most of this population did not think of themselves as inhabitants of a watershed. Forest guards were charged with traversing the region-in-the-making and translating extra-local concerns about forests, water, and the canal to the "shifting cultivators" that Wadsworth had identified as a threat to the trade route. The enrollment of campesinos in watershed management was both emotionally and physically demanding for the guards. As locals, the guards knew they would encounter a pushback to conservation in rural communities. Moreover, the same organization of geographical features that made the upper watershed valuable for water provision and storage—heavy rainfall, dense forest, and a lack of roads—also frustrated the guards' efforts to restrict farming (Pinzon and Esturain, 1986, 213–214). As I have shown, watershed management encountered infrastructural and geographical challenges, but perhaps the greatest obstacle was negotiating the cultural politics of forests.

The Cultural Politics of Forests

Before 1984, when upper watershed lands—about 30% of the basin area—were enclosed within the new Chagres National Park, campesinos with written permission were legally permitted to cut secondary forest for agriculture.[13] Forest management efforts focused primarily on reforestation with exotic tree species—teak, pine, and others—distributed through a network of RENARE nurseries. But management took a coercive turn in 1984 when soldiers from the Panamanian army began making joint inspections with the forest guards. They supervised critical watershed sites by land, water, or air to ensure compliance with environmental laws (Pinzon and Esturain, 1986, 10). In 1987, another new environmental law, Forest Law 13, legally redefined secondary growth more than 5 years old as "forest." The law meant that campesinos were fined when they continued to farm as before, and, in cases still remembered with anger decades later, they were jailed or had machetes and hatchets confiscated. Strict enforcement provoked outrage in rural communities, which became more hostile to the guards.

In practice, the political problems introduced by watershed management turned on the different ways in which tropical forests were conceptualized and integrated into transportation and agricultural economies. When state actors made the case for watershed forest protection in urban and institutional settings, the referent—land covered with trees—seemed clear. This is hardly surprising given that early watershed maps represented non-linear patterns of forest clearing and recovery as inexorable deforestation and reduced heterogeneous landscapes to two land-cover classes: forested (green) and deforested (red). The vagueness of watershed forests as represented on maps and in official documents may have enhanced their effectiveness as objects for building institutional alliances (Star and Griesemer, 1989). However, cooperation was more difficult in the charged encounters between forest guards and campesinos. The forests as known by farmers were not a fixed object—a green space on a map—but an integrated and dynamic part of their swidden agricultural system: a less pejorative term for the "shifting" or "slash-and-burn" agriculture practiced by many small farmers in the tropics. Anthropologists define swidden agriculture as a system in which fields are cleared by firing and cropped discontinuously, with periods of fallow that last longer on average than periods of cropping (Conklin, 1954). *Rastrojo* is the term used in rural Panama to describe the agricultural forest fallow between 1 and 20 years old that becomes fertilizer through burning or mulching. Ideally, cultivation is shifted after a period and the *rastrojo* on the old plot is allowed to recover to be used again later.

Campesino farmers say that the longer a *rastrojo* grows before it is cleared, the more nutrients available for the next crop grown on that land. Consequently, farmers weigh the maturity of a *rastrojo* against pressures and incentives to put land back into production as they make clearing and planting decisions. Or, to put it another way, land use is shaped both by the farmer's relationship with the land itself and the location of that relationship within a broader political ecology. Many swidden farmers in the Panama Canal watershed would choose to clear young *rastrojo* only in the absence of available mature forest or older *rastrojo*. However, Panama's Forest Law 13 of 1987 redefined *rastrojo* more than 5 years old as protected "forest," and, consequently, encouraged shorter periods of fallow. Farmers began to clear *rastrojo* earlier than before so their farmland (which many own through possessory right rather than title) would not fall under state protection in perpetuity. In summary, watershed administrators saw and used *rastrojo* in one way, while farmers saw and used it in another. For administrators, *rastrojo* was secondary forest that supported the transportation economy by providing a critical environmental service: water regulation. For farmers, *rastrojo* was not forest at all but agricultural fallow that supported the rural economy by providing a different service: nutrient cycling. The tangled growth of grass, bushes, and young trees was but one moment in an ongoing cycle of vegetation clearing and recovery that contributed to the reproduction of agrarian livelihoods.

My intention in this essay is neither to romanticize swidden agriculturalists, nor vilify Panama Canal administrators. Farmers and ranchers, often with the support of state institutions, played an undeniable role in reducing the forest cover of the watershed by 50 percent between the 1950s and the late 1970s (Heckadon-Moreno, 2005, 37). Forest cover began to increase in the 1990s across the watershed and the nation as people left the rural agricultural sector to work in the urban service sector (Wright and Sarmiengo, 2008). In the upper watershed, the coercive management tactics of the 1980s—the Manuel Noriega era—have changed. Since 1997, the Panama Canal Authority (ACP), the quasi-autonomous Panamanian state institution that administers the canal, has also been responsible for administering, maintaining, using, and conserving the hydrological resources of the Panama Canal watershed. The ACP emphasizes local participation in watershed management and has sought to develop a "water culture" in rural communities through consultative "subwatershed" committees made up of local leaders from the smaller drainage basins within the canal watershed, public relations campaigns, and environmental education programs. Yet, despite these new forms of engagement, persistent questions of social justice impede the establishment of a participatory regional "water culture."

Farmers today, some only children in the 1980s, said in interviews that the dispossession that defined early watershed management persists, but has taken on new forms. The economic value of the watershed to extra-local actors is now readily apparent to local people through the new infrastructure visible across the landscape: the ubiquitous non-governmental organization (NGO) and state project vehicles; the constant (if often poorly attended) project meetings; and the steady stream of Peace Corps volunteers, natural scientists, and social scientists who arrive with an interest in forest cover. At the same time, however, the distributional inequities associated with watershed management are also recognizable through absences on the landscape. The redefinition of a former agricultural frontier as natural infrastructure has meant that "hard" infrastructure like roads and power lines arrive slowly, if at all. For example, one community where I worked is within 40 miles of the canal and Panama's two largest cities but electricity arrived for the first time in 2009, decades after the rest of the region. The only gravel road to town is often impassable during the rainy season. In another community where I conducted research, this one located on the banks of Gatun Lake, the town was often without potable water for days at a time due to problems with the treatment facilities.

Conclusions

In this essay, I have examined the development of a regional infrastructure assembled to make water circulate through the canal in a manner that meets the transportation needs of global commerce. I have argued that infrastructure is not a specific class of artifact, but a process of relationship-building. This is to say that dams, locks, and forests are connected and become water management infrastructure through the ongoing work—technical, governmental, and administrative—of building and maintaining the sprawling socio-technical system that moves ships across the isthmus. By bringing this infrastructural work to the surface, I have endeavored to show how environmental politics here are mediated by the specificities of transportation infrastructure.

Dams and dikes, Wiebe Bijker (2007) reminds us, are thick with politics. So are watershed forests and other landforms managed to deliver environmental services. It will come as no surprise that distribution conflicts often ensue when actors representing different systems of production inhabit the same ecosystem and use the same resource. However, these conflicts become cultural, rather than strictly political–economic, when groups value, conceptualize, and partition that resource in different ways. For example, Bill Cronon describes the historical distinctions between Native American and prospector visions of copper in Alaska. Native Americans used copper to make knives, bullets, and jewelry for regional trade. Prospectors saw a different value in copper: the capability to conduct electricity. The culture represented by the prospector "was discovering a new need for this ability, and so began to draw [the Native Americans'] world into its orbit" (Cronon, 1992, 40). Similarly, the "culture" of global commerce has discovered a need for environmental services and, through the process of remaking nature as infrastructure, draws rural land managers into complex new relationships.

In his 1978 essay "Deforestation: Death to the Panama Canal," Frank Wadsworth mentions several concurrent trends contributing to water scarcity in the canal system (drought, ship traffic, and municipal water use), but, ultimately, his focus and that of subsequent management efforts was campesino agriculture. If water scarcity was indeed overdetermined, then why did rural land use become the priority for intervention? By paying attention to the specificities of the socio-technical system that channels water from rural landscapes to the canal, we are able to more clearly see the distributional politics of environmental service provision in Panama. As the example of *rastrojo* illustrates, landforms do not have value in an absolute sense. Rather, they have a variety of potential capabilities that emerge in relation to particular uses (Blaikie and Brookfield, 1987, 6–7). Like Alaskan prospectors' pursuit of copper for conducting electricity, our demand for natural infrastructure to store and purify water, alleviate floods, improve air quality, and regulate the climate is not restricted to landforms without people. When a landform is assigned value in relation to one cultural system of production (transportation) rather than another (agriculture), different environmental services become relevant and the landscape is reorganized to prioritize the delivery of those services and support that system. This calls us to examine the ethics of making natural infrastructure and to ask how systems like the canal might be managed in a manner that is more just and equitable for their neighbors.

Acknowledgments

This work was supported by a Wenner-Gren Dissertation Fieldwork Grant, Fulbright Student Award, UNC Latin American Studies Tinker Field Research Grant, UNC Mellon-Gil Dissertation Fellowship for Latin America, and a UNC Dissertation Completion Fellowship. I would like to thank Peter Redfield, Jessica Barnes, Samer Alatout, Dana Powell, Christine Boyle, Nikhil Anand, Blake Scott, and Sara Safransky, as well as Michael Lynch and anonymous reviewers at *Social Studies of Science*, for their insightful and constructive comments on earlier versions of the manuscript. I am also deeply grateful to the farmers, forest guards, scientists, and development professionals in Panama who taught me about the watershed.

Notes

1. This estimate is based on figures from Panama's National Authority of Public Services that estimate per capita consumption of potable water in Panama at 106 gallons/day, the highest in Latin America, and yet 16 percent of the population has no potable water access (EFE, 2010).

2. In this essay, I use the terms "watershed" and "drainage/hydrological basin" interchangeably. "Watershed" entered English in the early nineteenth century from the German *wasserscheide*, or "water-parting" (*Oxford English Dictionary*, 1991). In English, as in German, the term first referred to the boundary line between drainage basins. By the late nineteenth century, however, "watershed" increasingly referred to "the whole gathering ground of a river system." The second, more recent definition, is how the term "watershed" is used in this essay.

3. The term "infrastructure studies" has been used by scholars of cyberinfrastructure in, for example, Edwards et al. (2009).

4. The legal distinction between potential farmland and protected forest is five years of growth: young *rastrojo* can legally be cleared and farmed but forest cannot. This distinction was established in Panama's Forestry Law 13 of 1987, which prohibited cutting of all primary and secondary forest more than five years old. Forest Law 13 was implemented and enforced by INRENARE, the national environmental management agency that was the successor to RENARE and preceded ANAM, the current environmental agency.

5. Most water flows out of the canal system via the locks, but it also exits through the Gatun Dam spillway and hydroelectric turbine through a system that diverts it for industrial and municipal consumption in the terminus cities and is lost through evaporation.

6. In an oral history interview, Wadsworth discusses his career and philosophy of natural resource management (Steen, 1993). Although neither forest hydrology nor the "sponge effect" is discussed specifically, his views as represented in the interview resonate with mainstream thought on the forest-water relationship in twentieth-century US forestry. In his analysis of the so-called stream-flow controversy, historian Gordon Dodds writes, "This thesis [that deforestation radically affects runoff and stream flow], widely publicized in manuals of forestry, popular and technical conservation journals, and in the general press, was further disseminated by forestry organizations and sympathetic politicians skilled in

advocating their views in the mass media" (Dodds, 1969, 59). Wadsworth, who went to graduate school in forestry at the University of Michigan and was a professional forester for decades before working on the canal watershed, seems to have accepted this thesis.

7. Between 1908 and 1911—also the peak years of Panama Canal construction—US foresters framed watersheds as "natural" political-administrative regions and harnessed anxiety about downstream flooding to garner support for a proposed law called the Weeks Act that would authorize the federal government to purchase forested lands in the upper watersheds of navigable rivers. This brought them into conflict with the Army Corps of Engineers. Kittredge (1948) describes this as a "period of propaganda" by forest protection advocates and their opponents. Dodds (1969) shows how friction between American foresters and engineers centered on the efficacy of watershed forests as regulators of stream flow and flooding. Foresters argued that deforestation increases flooding level and frequency, accelerates soil erosion, and alters precipitation, negatively impacting electricity generation, agriculture production, commerce, and natural beauty. Forest cover was described as regulating volatile water flows through the "sponge effect"—a controversial formulation at the time that remains so today (Bruijnzeel, 1990; Hamilton and King, 1983; Saberwal, 1997). The Army Corps of Engineers publicly critiqued the arguments for basinwide water management, which threatened civil engineering's hegemony over navigation and flood control (Dodds, 1969). In the strongest critique, Army Corps Chief H. M. Chittenden (1909) argued that foresters' claims had feeble empirical underpinnings. He argued that forest cover showed no quantitatively demonstrable effect on flow and might even accelerate watershed runoff. Nevertheless, the Weeks Act passed in 1911, and the Forest Service ultimately managed 25.3 million acres of federal forest reserves acquired under the law.

8. Pre-1970s research on Panamanian forests by foreign scientists includes: Allen, 1964; Cummings, 1956; Holdridge and Budowski, 1956; Lamb, 1959; Pittier, 1918.

9. Wadsworth refers to conservation organizations as "greenies" and extremists in his oral history interview with Steen (1993, 31, 42, 65, 88).

10. Wadsworth discusses the farmer mentality of "conquering nature for agriculture" and population pressure (Ebenreck, 1988, 73) but never suggests a role for political-economic factors in land use decisions, a position that, in Panama, is countered by substantial evidence. He discusses water issues as motivating state action on forests in Ebenreck (1988), 73. On forest hydrology "myths" and state intervention, see Kaimowitz (2004); Mathews (2009).

11. Heckadon-Moreno, interview with the author, October 8, 2009. Transnational networks of environmental expertise are documented in the annual reports of Panama's Ministry of Agriculture (MIDA) and natural resource agency (RENARE) throughout the 1970s and 1980s. I also conducted interviews with RENARE staff from this period who supported Heckadon-Moreno's claims.

12. Evidence of early watershed management work is scattered across MIDA annual reports (Government of Panama, 1973, 330; 1975, 269; 1976, 158). Some employees received a short course of training by international organizations in watershed management at that same time. MIDA employees also receive training from international organizations. For example, the Inter-American Institute for Cooperation on Agriculture provided a watershed management course in 1974 and 1975 (Government of Panama, 1975, 275, 310). Early watershed research appeared in government reports in the late 1970s (Isaza and Moran, 1978; USOTA, 1978).

13. Chagres National Park was declared through Panama's *Decreto Ejecutivo 73 de 2 de Octubre* and legally established in 1985 with the publication of the *Gaceta Oficial 20.238*.

References

Abbot, H. L. 1905. *Problems of the Panama Canal*. New York: Macmillan.

Allen, P. H. 1964. "The timber woods of Panama." *Ceiba* 10 (1): 17–61.

Bakenhus, R. E., Knapp, H. S., and Johnson, E. R. 1915. *The Panama Canal: Comprising Its History and Construction, and Its Relation to the Navy, International Law and Commerce*. New York: John Wiley and Sons.

Baker, E. L. Jr., Potter, M. A., Jones, D. L., Mercer, S. L., Cioffi, J. P., Green, L. W., Halverson, P. K., Lichtveld, M. Y., and Fleming, D. W. 2005. "The public health infrastructure and our nation's health." *Annual Review of Public Health* 26:303–318.

Balee, W. 2006. "The research program of historical ecology." *Annual Review of Anthropology* 35:75–98.

Benedict, M. A., and McMahon, E. T. 2006. *Green Infrastructure: Linking Landscapes and Communities*. Washington, DC: Island Press.

Bijker, W. E. 2007. "Dikes and dams, thick with politics." *Isis* 98:109–123.

Bijker, W. E., Hughes, T. P., and Pinch, T. J., eds. 1987. *The Social Construction of Technological Systems: New Directions in the Sociology and History of Technology*. Cambridge, MA: MIT Press.

Blaikie, P., and Brookfield, H., eds. 1987. *Land Degradation and Society*. London: Metheun.

Board of Consulting Engineers. 1906. *Report of Board of Consulting Engineers, Panama Canal*. Washington, DC: Isthmian Canal Commission.

Bowker, G. C. 1994. *Science on the Run: Information Management and Industrial Geophysics at Schlumberger, 1920–1940*. Cambridge, MA: MIT Press.

Bowker, G. C., and Star, S. L. 1998. "Building information infrastructures for social worlds: The role of classifications and standards." In *Community Computing and Support Systems*, edited by T. Ishida, 231–248. Berlin: Springer.

Bruijnzeel, L. A. 1990. *Hydrology of Moist Tropical Forests and Effects of Conversion: A State of Knowledge Review*. Paris: UNESCO.

Budowski, G. 1961. "Forestry training in Latin America." *Caribbean Forester*, January–June, 33–38.

Callon, M. 1986. "Some elements of a sociology of translation: Domestication of the scallops and the fishermen of St. Brieuc Bay." In *Power, Action and Belief: A New Sociology of Knowledge?*, edited by J. Law, 196–223. London: Routledge.

Chittenden, H. M. 1909. "Forests and reservoirs in their relation to stream flow with particular reference to navigable rivers." *Transactions of the American Society of Civil Engineers* 62:540–554.

Conklin, H. C. 1954. "An ethnoecological approach to shifting agriculture." *Transactions of the New York Academy of Sciences* 17 (2): 133–142.

Corbera, E., Brown, K., and Adger, W. N. 2007. "The equity and legitimacy of markets for ecosystem services." *Development and Change* 38 (4): 587–613.

Cortez, R. M. 1986. *Características generales de la población*. In *La Cuenca del Canal de Panama: Actas de los Seminarios-Talleres*, edited by S. Heckadon-Moreno, 45–52. Panama: Grupo de Trabajo sobre la Cuenca del Canal de Panama.

Costanza, R., d'Arge, R., de Groot, R., Faber, S., Grasso, M., Hannon, B., Limburg, K., et al. 1997. "The value of the world's ecosystem services and natural capital." *Nature* 387:253–260.

Coutard, O., ed. 1999. *The Governance of Large Technical Systems*. New York: Routledge.

Cronon, W. 1992. "Kennecott journey: The paths out of town." In *Under an Open Sky: Rethinking America's Western Past*, edited by G. Miles, W. Cronon, and J. Gitlin, 28–51. New York: W.W. Norton.

Cronon, W. 1995. "The trouble with wilderness or, getting back to the wrong nature." In *Uncommon Ground: Rethinking the Human Place in Nature*, edited by W. Cronon, 69–90. New York: W.W. Norton.

Cummings, L. J. 1956. *Forestry in Panama*. Panama City: SICAP.

Denevan, W. M. 1992. "The pristine myth: The landscape of the Americas in 1492." *Annals of the Association of American Geographers* 82 (3): 369–385.

Dodds, G. B. 1969. "The stream-flow controversy: A conservation turning point." *Journal of American History* 56 (1): 59–69.

Dove, M. R. 1983. "Theories of swidden agriculture, and the political economy of ignorance." *Agroforestry Systems* 1:85–90.

Ebenreck, S. 1988. "Frank Wadsworth: Tropical forester." *American Forests*, November/December, 52–73.

Edwards, P. N. 2003. "Infrastructure and modernity: Force, time, and social organization in the history of sociotechnical systems." In *Modernity and Technology*, edited by P. Brey, T. J. Misa, and A. Feenberg, 185–226. Cambridge, MA: MIT Press.

Edwards, P. N, Jackson, S. J., Bowker, G. C., and Williams, R. 2009. "Introduction: An agenda for infrastructure studies." *Journal of the Association for Information Systems* 10 (5): 364–374.

EFE. 2010. "Panamá es el mayor consumidor de agua per cápita en América Latina." *Terra Noticias*, March 22. Accessed December 6, 2011. http://noticias.terra.es/2010/espana/0322/actualidad/panama-es-el-mayor-consumidor-de-agua-per-capita-en-america-latina.aspx (page no longer available).

Escobar, A. 1998. "Whose knowledge, whose nature? Biodiversity conservation and the political ecology of social movements." *Journal of Political Ecology* 5:53–82.

Globerman, S., and Shapiro, D. 2002. "Global foreign direct investment flows: The role of governance infrastructure." *World Development* 30 (11): 1899–1919.

Government of Panama. 1956. *Ministry of Agriculture, Commerce, and Industries, Annual Report*. Panama City: Government Printing Office.

Government of Panama. 1973. *Ministry of Agriculture, Annual Report*. Panama City: Government Printing Office.

Government of Panama. 1975. *Ministry of Agriculture, Annual Report*. Panama City: Government Printing Office.

Government of Panama. 1976. *Ministry of Agriculture, Annual Report*. Panama City: Government Printing Office.

Guha, R., and Martinez-Alier, J. 1997. "From political economy to political ecology." In *Varieties of Environmentalism: Essays on North and South*, edited by R. Guha and J. Martinez-Alier, 22–45. London: Earthscan.

Hamilton, L. S., and King, P. N. 1983. *Tropical Forested Watersheds: Hydrological and Soils Response to Major Uses or Conversions*. Boulder: Westview Press.

Haskin, F. J. 1914. *The Panama Canal*. Garden City, NY: Doubleday, Page.

Heckadon-Moreno, S. 1984. "Panama's expanding cattle front: The *Santeño campesinos* and the colonization of the forests." PhD diss., University of Essex, Essex.

Heckadon-Moreno, S. 1985. *La ganadería extensiva y la deforestación: Los costos de una alternativa de desarrollo*. In *Agonía de la Naturaleza*, edited by S. Heckadon-Moreno and J. E. González, 45–62. Panamá: Instituto de Investigación Agropecuaria de Panamá.

Heckadon-Moreno, S. 2005. "Light and shadows in the management of the Panama Canal watershed." In *The Rio Chagres: A Multidisciplinary Perspective of a Tropical River Basin*, edited by R. S. Harmon, 28–44. New York: Kluwer Academic/Plenum.

Holdridge, L. R., and Budowski, G. 1956. "Report on an ecological survey of the republic of Panama." *Caribbean Forester* 17:92–110.

Hughes, T. P. 1983. *Networks of Power: Electrification in Western Society, 1880–1930*. Baltimore: Johns Hopkins University Press.

Hughes, T. P. 1987. "The evolution of large technological systems." In *The Social Construction of Technological Systems: New Directions in the Sociology and History of Technology*, edited by T. P. Hughes, W. E. Bijker, and T. J. Pinch, 51–82. Cambridge, MA: MIT Press.

HydroSHEDS. 2006. "HydroSHEDS (RIV) River network (stream lines) at 15s resolution—Central America [ArcGIS layer package]." Washington, DC: World Wildlife Fund. Accessed December 20, 2011. http://app.databasin.org/app/pages/datasetPage.jsp?id=1629fc30704849f2a730e5908c64dd4f.

Ibáñez, R., Condit, R., Angehr, G., Aguilar, S., García, T., Martínez, R., Sanjur, A., Stallard, R., Wright, S. J., Rand, A. S., and Heckadon, S. 2002. "An ecosystem report on the Panama Canal: Monitoring the status of the forest communities and the watershed." *Environmental Monitoring and Assessment* 80:65–95.

Isaza, C., and Moran, B. 1978. "Importancia del manejo de la cuenca del Canal de Panamá." In *Dirección de Recursos Naturales Renovables Ministerio de Desarrollo Agropecuario (RENARE)*. Panamá: Government Printer.

Iwabuchi, M. 1990. "The future of the Panama Canal in a competitive world." *Review of Urban and Regional Development Studies* 2 (1): 23–37.

Kaimowitz, D. 2004. "Useful myths and intractable truths: The politics of the links between forests and water in Central America." In *Forests, Water and People in the Humid Tropics: Past, Present and Future Hydrological Research for Integrated Land and Water Management*, edited by M. Bonnell and S. Bruijnzeel, 86–98. New York: Cambridge University Press.

Kirkpatrick, R. Z. 1934. "Madden Dam will insure water supply for Gatun Lake development." *Iowa Engineer*, March, 84–85.

Kittredge, J. 1948. *Forest Influences*. New York: McGraw-Hill.

LaFeber, W. 1978. *The Panama Canal: The Crisis in Historical Perspective*. New York: Oxford University Press.

Lamb, F. B. 1959. "Prospects for forest land management in Panama." *Tropical Woods* 110 (April): 16–28.

La Porte, T. R., ed. 1991. *Social Responses to Large Technical Systems: Control or Anticipation*. Dordrecht, the Netherlands: Kluwer Academic.

Larson, C. L. 1979. "Erosion and sediment yields as affected by land use and slope in the Panama Canal Watershed." *Proceedings of the III World Congress on Water Resources*. Mexico City: International Water Resources Association, 1086–1096.

Latour, B. 1983. "Give me a laboratory and I will raise the world." In *Science Observed: Perspectives on the Social Study of Science*, edited by K. D. Knorr-Cetina and M. Mulkay, 141–170. London: Sage.

Lee, R. 2009. "Infrastructure." In *The Dictionary of Human Geography*, edited by D. Gregory, R. Johnson, G. Pratt, M. J. Watts, and S. Whatmore, 382–383. Oxford: Wiley-Blackwell.

Leschine, T. M. 1981. "The Panamanian sea-level canal." *Oceanus* 24 (2): 20–30.

Mathews, A. S. 2009. "Unlikely alliances: Encounters between state science, nature spirits, and indigenous industrial forestry in Mexico, 1926–2008." *Current Anthropology* 50 (1): 75–101.

Mayntz, R., and Hughes, T. P., eds. 1988. *The Development of Large Technical Systems*. Boulder: Westview Press.

McAfee, K., and Shapiro, E. N. 2010. "Payments for ecosystem services in Mexico: Nature, neoliberalism, social movements, and the state." *Annals of the Association of American Geographers* 100 (3): 579–599.

McCullough, D. 1977. *The Path between the Seas: The Creation of the Panama Canal 1870–1914*. New York: Simon and Schuster.

Mitchell, T. 1991. "The limits of the state: Beyond statist approaches and their critics." *American Political Science Review* 85 (1): 77–96.

Nested Watersheds of North America. 2009. "Nested Watersheds of Central America (modified) [ArcGIS layer package]." Olympia, Washington: The Nature Conservancy. Accessed December 20, 1011. http://app.databasin.org/app/pages/datasetPage.jsp?id=7e28bde7285244d080ef191e9ad39bdb.

Oxford English Dictionary. 1991. "Infrastructure." In *Oxford English Dictionary*, 2nd ed., edited by J. A. Simpson and E. S. C. Weiner, 950. Oxford: Oxford University Press.

Panama Canal Company. 1961. *Review of Studies of Potential Reservoir Development—Upper Chagres River*. Mount Hope: Meteorological and Hydrographic Branch, Canal Zone Printing Office.

Pinzon, L. A., and Esturain, J. 1986. "Vigilancia de los bosques." In *La Cuenca del Canal de Panama*, edited by S. Heckadon-Moreno, 205–215. Panama: Grupo de Trabajo sobre la Cuenca del Canal de Panama.

Pisani, D. J. 2002. "A conservation myth: The troubled childhood of the multiple-use idea." *Agricultural History* 76 (2): 154–171.

Pittier, H. 1918. "Our present knowledge of the forest formations of the Isthmus of Panama." *Journal of Forestry* 16:76–84.

Reuss, M., and Cutcliffe, S. H., eds. 2010. *The Illusory Boundary: Environment and Technology in History*. Charlottesville: University of Virginia Press.

Robbins, P. 2005. *Political Ecology: A Critical Introduction*. Malden, MA: Blackwell.

ROCAP-USAID (Regional Office of Central America and Panama-US Agency for International Development/ Panama). 1981. *Joint Evaluation of the Watershed Management Project #525-T-049*. Guatemala City, Guatemala: ROCAP-USAID.

Saberwal, V. K. 1997. "Science and the desiccationist discourse in the 20th century." *Environment and History* 4:309–343.

Smith, D. M., and Barchiesi, S. 2009. "Environment as infrastructure: Resilience to climate change impacts on water through investments in nature." In *Water and Climate Change Adaptation*, edited by D. M. Smith and S. Barchiesi. Gland, Switzerland: International Union for Conservation of Nature.

Star, S. L. 1999. "The ethnography of infrastructure." *American Behavioral Scientist* 43 (3): 377–391.

Star, S. L., and Griesemer, J. R. 1989. "Institutional ecology, 'translations' and boundary objects: Amateurs and professionals in Berkeley's Museum of Vertebrate Zoology, 1907–1939." *Social Studies of Science* 19 (3): 387–420.

Star, S. L., and Ruhleder, K. 1996. "Steps toward an ecology of infrastructure: Design and access for large information spaces." *Information Systems Research* 7 (1): 111–134.

Steen, H. K. 1993. "The Evolution of Tropical Forestry: Puerto Rico and Beyond, An Interview with Frank H. Wadsworth." Unpublished manuscript, Forest History Society, Durham, NC.

Summerton, J., ed. 1994. *Changing Large Technical Systems*. Boulder: Westview Press.

Thompson, E. P. 1971. "The moral economy of the English crowd in the eighteenth century." *Past and Present* 50:76–136.

Thrift, N. 2005. *Knowing Capitalism*. London: Sage.

Twigg, C. A. 1994. "The need for a national learning infrastructure." *Educom Review* 29 (5): 16.

USAID (US Agency for International Development). 2005. *Evaluation of USAID's Strategic Objective for the Panama Canal Watershed 2000–2005*. Washington, DC: USAID.

USOTA (US State Department Office of Technology Assessment). 1978. *Environmental Issues Affecting the Panama Canal*. Washington, DC: USOTA.

Vargas, C. A. 2006. "Panama Canal water resources management." In *Hydrology and Water Law: Bridging the Gap*, edited by J. Wallace, 143–158. London: IWA.

Wadsworth, F. 1978. "Deforestation: Death to the Panama Canal." US Strategy Conference on Tropical Deforestation. Washington, DC: US Department of State and US Agency for International Development, 22–25.

White, R. 1995. *The Organic Machine: The Remaking of the Columbia River*. New York: Hill and Wang.

Winner, L. 1980. "Do artifacts have politics?" *Daedalus* 109 (1): 121–136.

World Shaded Relief. 2009. "World Shaded Relief [ArcGIS Online Basemap]." Redlands, CA: Esri. www.arcgis.com/home/item.html?id=9c5370d0b54f4de1b48a3792d7377ff2.

Wright, S. J., and Samaniego, M. J. 2008. "Historical, demographic, and economic correlates of land-use change in the Republic of Panama." *Ecology and Society* 13 (2). Accessed December 20, 2011. www.ecologyandsociety.org/vol13/iss2/art17/.

22 Conceptions of Control and IT Artefacts: An Institutional Account of the Amazon Rainforest Monitoring System

Raoni Guerra Lucas Rajão and Niall Hayes

We often think that a particular technology has a single purpose and was designed to effectively serve that purpose. However, any given technology can have multiple purposes, which can change over time. Some have primary and secondary purposes, as in the case of cars, which we might think of as primarily a means of transportation and secondarily a personal statement about who we are. Other technologies, however, are designed to serve multiple purposes. For example, one can grind spices with an electric coffee grinder, and night-vision glasses can be useful in military combat as well as for tracking animals to learn about their nocturnal habits. In this piece, Rajão and Hayes explore the Brazilian use of a radar and satellite-based monitoring system and describe how the objectives of the system changed over time. In some sense, the massive system they describe has a powerful momentum. When Brazilian government officials moved on from their original goals for the technology, they didn't simply throw the system away. Instead, they reworked their relationship with it and deployed it to new ends.

Broadly, the story illustrates how the meaning and function of technology is at least partially embedded in the context of its use. However, Rajão and Hayes are also interested in the question of control and identify several different contexts of control: military, economic, and ecological. In this respect their analysis harkens back to the Ritzer reading, where control is presented as a powerful value-shaping technology. Here we see that control is far from a simple matter. Over time the satellite monitoring system changed from one assisting in the economic development and military security of the Amazon region to one protecting the ecology of the rain forest, illustrating that while technologies can exert a powerful effect on values, values can also have a powerful effect on technologies.

Introduction

One important theme to emerge in both organizational and information systems literatures over recent years has been a consideration of how institutional contexts come to shape the development and use of information technology (IT) artefacts (Orlikowski and Robey, 1991; Robey and Boudreau, 1999; Walsham and Sahay, 1999).[1] Such concerns have led scholars to draw on notions from new institutional theory so as to better understand the enduring nature of institutions (Avgerou, 2000; Currie and Guah, 2007; Hayes, 2008; Noir and Walsham, 2007). This paper seeks to contribute to this important and underdeveloped research theme, and specifically argues that institutionalization

From *Journal of Information Technology* 24 (2009): 320–331.

should be understood as a historically located and highly contested process. Our analysis is based on the uses made of the Amazon monitoring system, which is a set of radar and satellite-based monitoring systems that are currently used by the Brazilian government to track deforestation. Our case will reveal that the differing configurations and uses of this IT artefact can be explained by considering the ways in which the alternate institutional values became dominant at different periods of time.

Conceptually we primarily draw on Fligstein's (1990) conceptions of control to examine the emergence and establishment of Brazil's Amazon rainforest monitoring system. The case will highlight that the meaning and use of the monitoring system has shifted significantly over 44 years, from a system which was first used to assist in the economic development and the military security of the Amazon region to, more recently, protecting the ecology of the rainforest. We will argue that the trajectory of the Amazon monitoring system can be usefully understood in relation to the intercalation and overlapping of different conceptions of control, managerial paradigms that encapsulate the conceptualization of what is the Amazon rainforest (e.g. unproductive land or national patrimony) and consequently how to manage it (e.g. exploit or preserve) (Fligstein, 1990). Based on this analysis, our paper will argue that the relationship between IT artefacts and institutions should be conceptualized as conflictual, emergent and dialectical (Hayes, 2008), and further, that literature on institutional theory needs to better attend to the emergent and conflictual practices that shape and are shaped by IT.

This paper is organized as follows. The next section reviews the literature that informs our analysis, namely neoinstitutional theory and information systems. The third section outlines our methodological approach. We then present our empirical base. The fifth section discusses the case with regard to the relationship between IT artefacts and institutional contexts. This is followed by a brief conclusion.

Institutional Theory and IT Artefacts

Neoinstitutional theory emerged in the late 1970s and 1980s in response to a growing disenchantment with economic and rational theories that depict actors as strict followers of a universal atemporal rationality (Barley and Tolbert, 1997). In contrast to these rational theories, new institutionalism conceives of actors as being embedded into institutional fields that possess widely shared interpretive infrastructures "that constitute the nature of reality and the frames through which meaning is made" (Scott, 1995, 40). In this context, instead of suggesting the existence of an absolute reality and universal rationality, neoinstitutionalists tend to highlight the arbitrariness and cognitive nature of institutions and their role in the process of the social construction of reality (Berger and Luckmann, 1967). Consequently, new intuitionalism proposes that notions such as rationality, efficiency and legitimacy should always be considered in relation to an institutional field located in time and space (DiMaggio and Powell, 1983; Meyer and Rowan, 1977).

Literature relating to the relationship between IT artefacts and organizations has only emerged quite recently (Adler, 2005; Butler, 2003; Lamb and Kling, 2003). One theme in this literature explores why despite there being no clear rationale for such investment to lead to increased efficiency and return on investment many organizations still invest in IT (Avgerou, 2003; Noir and Walsham, 2007). For example, Avgerou (2000) presents the case of a Mexican oil company that had recently been privatized that despite the unsatisfactory financial returns continued to increase their investment in IT. She explains that it was only possible as computers have become "taken-for-granted as fixtures of contemporary organizations" and as such a key legitimating element in the company's effort to become a modern market-oriented organization.

A second theme in the IT and institutionalization literature has considered IT innovations (Butler, 2003; Chatterjee *et al.*, 2002; King *et al.*, 1994; Silva and Figueroa, 2002). Notable in this literature has been (Swanson and Ramiller, 1997) the notion of "organizing vision" as the main enabler of successful IT innovations. Organizing vision is "a focal community idea [. . .] *for* organizing in a way that embeds and utilizes [a specific] information technology in organizational processes and structures" (ibid., 460; emphasis in the original). Drawing on this concept, Currie (2004) analysed the case of Application Services Provisioning (ASP), a business and technological model that provides software over wide area networks. She argued that ASPs have failed to establish themselves due to the incapacity of their supporters to generate a coherent and stable community-wide organizing vision, namely, a specific use for the technology within existing organizational structures and practices.

A third research theme relates to the inscription of institutional values into IT artefacts and the role of these artefacts as (re)producers of institutional orders (Gosain, 2004; Kling and Iacono, 1989; Nicolaou, 1999; Scott, 2003). Many commentators see institutions as being resilient to change and consequently have tended towards stability (Powell and DiMaggio, 1991; Scott, 2001; Zucker, 1977). In relation to technology, Hasselbladh and Kallinikos (2000) argue that institutionalization usually includes the development of abstract ideals into discourses and then concrete techniques of control (forms of codifications including software packages) that allow little space for interpretive flexibility. In this way they suggest that the context of institutions find their way into the design of IT artefacts, which in turn also become institutional carriers that help to support and reproduce certain institutional arrangements.

The fourth theme in the literature considers the role of IT artefacts in change processes. IT artefacts are understood as both enablers and constrainers of institutional change (for a review on change, see Dacin *et al.*, 2002; Greenwood and Suddaby, 2006). The best known example of this is Jane Fountain's (2001, 2006) study of e-government in the USA. She argued that the introduction of IT artefacts connected to the Internet have brought major institutional change to the public sector by enabling the sharing of informal information and knowledge across the government's traditional boundaries which led to greater transparency. These changes have contributed towards the building of a "virtual state" with a significantly different institutional environment. Drawing on Giddens (1986), a number of other information systems writers have also examined the enabling and constraining features of institutions (Barley and Tolbert, 1997; Barrett and Walsham, 1999; Barrett *et al.*, 2001). For example, Barrett and Walsham (1999) and Barrett *et al.* (2001) pointed out that the introduction of IT in the Indian forestry sector and in the London Stock Exchange led to major institutional changes due to "globalizing tendencies" such as the separation of time and space, disembedding mechanisms and institutional reflexivity (Giddens, 1990).

Several studies have signalled the importance of accounting for the dialectical, conflictual and emergent nature of institutional change (Hayes, 2008; Robey and Holmstrom, 2001). Fewer studies have considered how the conflictual relationship between institutions and IT may evolve over time (for an exception, see Currie and Guah, 2007). For example, Hayes's (2008) case study of workflow technology in a High Tech Optronics company highlighted that the workflow system was drawn upon by different groups to try to ensure their institutional values became dominant. He argued that there are always competing views and values within and between institutions and consequently through ongoing negotiations and conflicts, there is always the possibility for alternate institutional values to become dominant (Townley, 2002).

Conceptions of Control and Dialectics

Fligstein's (1990) book entitled *The Transformation of Corporate Control* provides an institutional account of the changes in America's largest corporations over the last century. His analysis, like DiMaggio and Powell's (1983), challenges the mainstream evolutionary economic theory that believes that the USA naturally created technological efficient firms by managing them according to a universal extemporal rationality. In contrast, Fligstein shows that the USA's largest companies have taken their current shape due to a succession of dominant conceptions of control. Conceptions of control are institutionalized widespread "totalizing world views that cause actors to interpret every situation from a given perspective" (1990, 10). They "operate both as cultural templates for structuring new actions (i.e., what behavior makes sense) and a set of structures limiting the possibilities of action (i.e., what others are doing, thereby structuring what reactions are possible)" (Fligstein and Brantley, 1992, 287). Conceptions of control can be understood as an institutional logic as conceptions are also "sets of 'material' practices and symbolic constructions" available to individuals so as to make sense of their environment (Friedland and Alford, 1991, 248). However, while institutional logics provide the "cognitive maps" to activities that range from bureaucracy to religion, conceptions of control are akin to managerial practices as they relate to the desire of individuals and organizations to control their environment. Examples of Fligstein's conceptions of control include the manufacturing conception that sees the organization as an engine that has to efficiently transform raw materials into finished products. Fligstein's financial conception of control views organizations as a portfolio of assets that seek to obtain the highest return on investment. He argues that the financial conception of control has become dominant.

As this paper focuses on the contested and emergent nature of institutionalization, we also draw on literature from a dialectical perspective. This literature highlights the conflictual and tentative nature of institutionalization, and attends to the ways in which structures of power between powerful actors may become taken-for-granted in this process (Robey and Holmstrom, 2001; Seo and Creed, 2002; Suddaby and Greenwood, 2009). As Blackler and Regan (2006, 1858) suggest institutional change:

> should be analyzed as a contested ascent from the abstract to the concrete. In this case the contest was firstly, between alternative images of what kind of practices might be desirable. Secondly, it was between the established institutional arrangements of the past and the social and cultural infrastructure associated with them, and the tentative, possible institutions of the future.

Informed by Fligstein's writing on conceptions of control, dialectical perspectives of institutional change and the literature in IT and institutions, this study will examine the ways in which the Amazon monitoring system was shaped over a 44-year period to support different conceptions of control, and how it was reconfigured to work within new institutional contexts. We will contend that the process of institutional change (shifts in conceptions of control) do not happen in a linear fashion, but are instead conflictual and emergent, and consequently we may find different conceptions of control competing for dominance or relevance at any given time.

Research Methodology

The case study that forms the empirical base of this paper concerns the history of the Amazon monitoring system, a set of satellite-based remote sensing systems currently used by the Brazilian government to estimate deforestation rates in the Amazon rainforest. In order to interpret and integrate the primary and secondary in a single in-depth case study, this study has adopted a subjectivist position, in particular the interpretive tradition of IS research (Walsham, 1993, 2006). As we report on the different ways in which the Amazon rainforest monitoring system manifests itself over 44 years, we adopt a combination of the historical and dialectical method for institutional research (Suddaby and Greenwood, 2009). We attempted to reconstruct history based on secondary data and by asking interviewees to narrate their version of the trajectory and current use of the Amazon monitoring system: a history that they were in most cases close observers or even the protagonists to (Czarniawska, 2004). During and after the data analysis, the researchers also maintained contact via email and telephone with key informants in order to confirm their interpretation of the data and elicit further information.

The primary data consist of 71 interviews collected in Brazil between June 2007 and April 2009. The informants included 8 officials from National Institute for Space Research (INPE), 11 from the Ministry of Environment (MMA), 14 from the federal environmental agency (IBAMA), 9 from Secretary of Environment from the state of Mato Grosso (SEMA-MT) and 8 from environmental non-governmental organizations (NGOs). Most interviewees were officials directly involved in the conception, development and/or use of one of the configurations of the Amazon monitoring system, including three ex-ministers of environment down to local forest rangers in the rainforest region. Most interviews were held in Portuguese in four different regions of Brazil. When possible the interviews were chronicled using a voice recorder, otherwise, extensive notes were taken and the conversation translated directly in English as soon as the interviews finished.

Secondary data sources consist of the Brazilian law (past and present), newspaper archives (*Veja* and *Folha de São Paulo*), reports from governmental agencies (MMA, INPE and SEMA-MT) and NGOs (Greenpeace, WWF, ISA and ICV), technical documentation concerning the Amazon monitoring system and academic papers about the economic, political and environmental history of the Amazon (see references in the account below). Our understanding of the system was further embellished through observing the use of both the front-office (user interface available on the internet[2]) and the back-office (classification of deforestation, composition of maps) of the monitoring system. One of the authors was also trained how to operate both the front and back office systems.

The Amazon Rainforest Monitoring System

The empirical focus of this paper is the Amazon rainforest monitoring system which comprises a family of similar systems. They have three main features in common. First, all systems relate only to the Brazilian portion of the Amazon rainforest (which is by far the largest). Second, they were all developed by the Brazilian government, with INPE (Institute for Space Research) playing a key role in most systems. Finally, all systems work based on similar technological principles, namely, they are geographic information systems that obtain data via remote sensing technology, in particular satellite optical images (see figure 22.1).

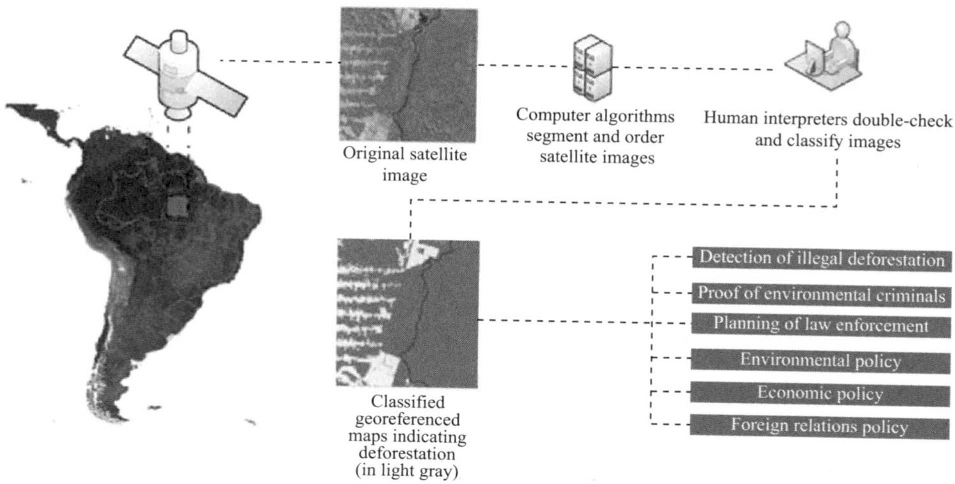

Figure 22.1
General principle behind the systems that compose the Amazon monitoring system.

The origins of the Amazon monitoring system date back to 1961, when the Brazilian government created a department which a decade later became INPE, the National Institute for Space Research. In 1973 INPE installed a reception base for the USA satellite Landsat, becoming one of the first countries outside the USA-USSR to receive satellite images. Early reports highlight that the rationale for the heavy investments in remote sensing technology was to support economic development across Brazil, but in particular within the Amazon region. Indeed, one of the first uses of the monitoring system was to check if the subsidies provided by the central government to those people colonizing and transforming Brazil's "immense and unknown inland" into productive agricultural areas through the planting of crops and the establishment of grazing pasture were effective.

In 1988 the monitoring system changed considerably in scope with INPE's creation of PRODES. While the justification of the previous monitoring system was centred on the economic development of the Amazon, PRODES centred on the protection of the rainforest. However, despite the change in rhetoric, PRODES was designed in a way that made it unpractical for environmental protection efforts. Indeed, PRODES detected deforestation only once a year, and the map showing the detail of the deforested area was considered sensitive to Brazilian military and economic security by the government and as such access was limited. The data generated by PRODES were the total sum of deforestation in km^2 divided by state. As most states in the Amazon have areas bigger than countries like the UK and Spain, these data were not specific enough to allow the environmental protection agencies to plan their activities, let alone ascertain the exact location of illegal deforestation in order to take legal action. It is not surprising then that deforestation rates (as detected by PRODES) remained as high during the 1990s as in they were in the 1970s and 1980s (see figure 22.2).

In 2000 there was another change to the monitoring system. The state of Mato Grosso created SLAPR, a system that according to Fearnside (2003) was the first control mechanism that led to a reduction in deforestation in the Brazilian Amazon region. The system has been updated between 2002 and 2006 so as to better identify and combat deforestation. DETER, a new component of the monitoring system developed by

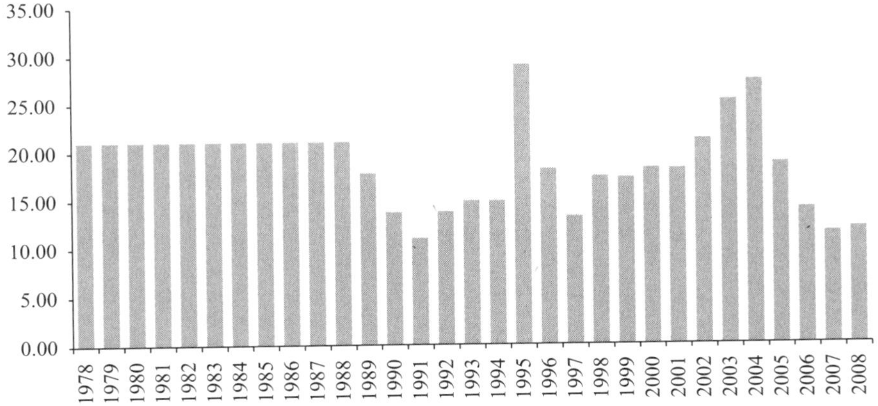

Figure 22.2
Deforestation rates of the Brazilian Amazon rain forest between 1978 and 2008 calculated using the Amazon monitoring system. *Source:* Fearnside, 2005; National Institute for Space Research.

INPE in 2004, is able to detect the precise location of deforestation within days. More recently, IBAMA developed SisCom, a GIS that integrates data from INPE's and Mato Grosso's SLAPR monitoring system with other legal information, such as properties under embargo due to illegal deforestation. Furthermore, in contrast to the previously strict data control of PRODES, whereby the satellite images showing the location of deforestation were available only under authorization of the military, today all monitoring systems (except the state-level SLAPRs) can be accessed over the internet by anyone interested in environmental protection of the Amazon.

Following the creation of these new configurations of the Amazon monitoring system, there has been between 2004 and 2008 a consistent decrease in deforestation rates and a sharp increase in the number of fines for illegal deforestation in Brazil's Amazon region. Even though it is very difficult to attribute this reduction to the new design and use of the monitoring system, different interviewees pointed out that change in economic conditions could not explain this change alone.

In summary, in the last three decades the Amazon monitoring system has changed dramatically in its scope, justification, use and effects. But how was it possible and how can we explain these radical shifts?

Contextualizing the Monitoring System

In order to understand the broader social context in which the Amazon system is embedded, we draw upon Fligstein's (1990) notion of "conceptions of control" (widespread managerial paradigms). We argue that in the period under analysis (1964–2008) the Brazilian government has managed the Amazon region from three different conceptions, namely, the military, the economic and the ecological conceptions of control (see table 22.1). Further, we will explain that these three conceptions of control have shaped the nature and use of the Amazon monitoring system. For simplicity we divide the history of the monitoring system around three key milestones: the military coup d'état in 1964, the creation of PRODES in 1988 and the creation of SLAPR in 2000, two of the most important configurations of the Amazon monitoring system.

Table 22.1
The military, economic, and ecological conceptions of control

Conception of control	Ideal	Aims
Military	An independent and strong Brazilian state	Protect Brazil's Amazonian region against threats to its economic and political sovereignty
Economic	A wealthy and prosperous Brazilian Amazonian region	Develop Brazil's Amazonian region by expanding and modernizing its economy
Ecological	A sustainable society	Preserve Brazil's Amazonian environment

1964–1988: Military and Economic Dominance

The military conception of control represents the worldview usually held by the armed forces, and sees the nation as a territory that must be defended from all kinds of threats to its economical, political and territorial independence (Page and Redclift, 2002). Different interviewees suggested that military concerns with the Amazon rainforest were central to both the decision to colonize the region and to create the Amazon monitoring system during the rule of the military junta, between 1964 and 1985. During this period there was growing concern the region was under imminent threat of being invaded by more developed nations. Commentators suggest that the Amazon represented a large section of unspoiled land that other countries may have sought to exploit for food production and the extraction of raw materials to fulfil the requirements of their increasing populations. Therefore, the military government believed if it did not colonize the Amazon region, other countries certainly would (Kolk, 1998; Reis, 1965).[3]

It was also during the military rule that the Brazilian government started to invest in satellite remote sensing, the technology that is the basis of the Amazon monitoring system. Different scientists and senior government officials considered military concerns as being the principal reason for investing in satellite technology during the 1970s. The military government perceived their lack of knowledge about the Amazon basin as a military weakness, and considered the use of satellite technology as a way to redress this. As a senior INPE scientist explained:

> The Amazon is one of the main reasons why the government has set a long-term goal to invest in orbital remote sensing. During the 1970s the Amazon was seen as a vast and mostly unknown area. To give you an example, just to build the 364 [the trans-Amazon highway crossing the rainforest], they had to create the indigenous reserve of Xingù to relocate the indigenous populations that they [the government] did not even know were there! [. . .] The lack of knowledge was big, and in addition to that, the fear of the international greed was consistent.

The constant preoccupation with maintaining Brazilian sovereignty over the Amazon highlights the strength of the military conception of control at this time.

However, the military government also believed that national security could be enhanced if Brazilian people colonized the Amazon region. This led to economic and demographic growth and suggests the presence of another conception of control, the "economic conception of control." The economic conception of control in broad terms is concerned with the economic growth of the country, as defined by the achievement of certain numeric indicators (e.g. GDP, per capita income and budgeted surplus).

The economic conception of control is closely related to the notion of developmentalism, namely, a set of ideologies that "regard development in the sense of economic growth and institutional modernization as a good in itself" without questioning the fact that those models of modernity are a product of developed nations and that they often collaborate to the perpetuation of post-colonial forms of domination. Fligstein's (1990) financial conception of control has important similarities and differences to our economic conception of control. On the one hand, the finance like the economic conception of control manages the organization based mainly on abstract quantitative indicators such as ROE (return on invest) for the former, and GDP (gross domestic product) for the later. This tendency of using numeric financial and managerial tools in the Brazilian government has been the case particularly in the last 20 years, with a series of efforts to "modernize" the public sector (Abrucio, 2007). On the other hand, public sector organizations emphasize the importance of "espousing the right beliefs, establishing correct structures and following the appropriate processes" where "what is 'right,' 'correct,' or 'appropriate' depends on the broader institutional-cultural context" (Tsoukas and Papoulias, 2005, 82). In sum, for the economic conception of control the means (politically correct discourses and practices) are as important as the ends (economic growth as measured by the GDP).

During this time period there was a strong synergy between the military and the economic conception of control. The governmental slogan of the 1970s *Integrar para não entregar* (translated as: *integrate the region economically to avoid giving it away to other countries*) recalled by different interviewees is a good example of this. At the same time, this combination of economic and military concerns was also embedded in the monitoring systems, given their justification in terms of national security and its application to assess the natural resources of the region and verify if the subsidies for agriculture and cattle ranching (and indirectly, deforestation) were having the effects desired.

1989–2000: Emergence of Ecological Concerns

Towards the end of the 1980s a new conception of control emerged that questioned the legitimacy of both the military and the economic conceptions of control. This ecological conception of control is primarily concerned with the long-term sustainability of human life on Earth. While the economic conception of control "downplay[s] discontinuities and crises, especially in the ecological arena," the ecological conception recognizes the impact that human activities have on the environment and perceive risks—hazards generated by modernization—as management's core problem (Shrivastava, 1995, 119). According to this conception of control, the government should preserve the environment by adopting "ecologically sustainable organizational designs and practices" (ibid., 127).

In 1985 the first civil president since 1964 came to power and thus ended the military control of the country. During this period of transition between the military and democratic regimes, Brazil suddenly found itself as a global environmental villain due to its policy towards the Amazon. Even though today the idea that deforestation is damaging to our global environment is taken for granted, it was not the case then. Indeed, as late as the end of the 1960s the reports of major international bodies did not even mention the large-scale deforestation as an environmental issue. Furthermore, with the exception of a few pioneers, such as John Muir, John Ruskin and Sir Arthur George Tansley, the process of almost complete extinction of primary forests in western Europe passed unnoticed (Guha, 2000; McNeill, 2000). It was only in the 1970s, following the emergence of environmentalism (or "risk society"), that the new environmental NGOs and other members of the international community started expressing their concern about the future of the Amazon, the "world's lungs," and specifically started criticizing the colonization project carried out by the Brazilian government (Keck, 2001; Kolk, 1998).

Initially, the Brazilian government resisted international pressure, but towards the end of the 1980s cuts in international loans and internal political pressures forced the government to change the official discourse towards the Amazon (for a detailed account, see Kolk, 1998). Among a series of environmentally friendly measures adopted during that period is the creation in 1989 of IBAMA (the body responsible for environmental protection in Brazil), the launch of the programme *Nossa Natureza* (translated as: *our nature, the first program aiming at protecting the Amazon rainforest*) and the suspension of some subsidies to agricultural activities in the Amazon region. Brazil's environmental commitment was even documented in the new constitution where the Amazon rainforest is defined as a "national patrimony" that must be protected. PRODES, the monitoring system, was also part of the new environmental policy. Since 1988, PRODES has released data each year about the extent of deforestation in Brazil's Amazon region.

Despite the official "greening" of the Brazilian government, accounts from different interviewees and secondary data sources highlighting high deforestation rates during the 1990s (see figure 22.2) suggest that the change in the official discourse during the end of the 1980s did not represent any significant change in the way the Brazilian government has viewed and managed the Amazon rainforest. Thus, for many, the decision to change the environmental law and to create the Amazon monitoring system was mainly an attempt by the Brazilian government to convince the international community that they were adopting environmental policies for the Amazon. Indeed, when asked about the motivations behind the creation of IBAMA and the monitoring system, a very senior politician explained that

> during the 1980s the international community became concerned with the Amazon. IBAMA was created by putting together different pre-existing bodies as an answer to this pressure. It was a way to prove that we have the competence to manage the Amazon. The use of satellite images to monitor deforestation . . . was to demonstrate to the international community our preoccupation with the environment.

This suggests that the design and initial use of PRODES was not actually intended to be for environmental protection. In contrast, the government seems to have made further investments in the Amazon monitoring system in order to take advantage of the institutional myth embedded in technology (Avgerou, 2003; Noir and Walsham, 2007), and use it as a symbol of national scientific knowledge to mitigate the accusations that Brazil did not have the expertise to protect the Amazon. This suggests that during this period, despite the emergence of the ecological conception of control at discursive level, the monitoring system was still in the main being used for the aims set by the military conception of control.

2001–2008: Ecological Dominance

After years of false starts and military dominance, in 2001 the government started devising a genuine policy to protect the rainforest. This led to the institutionalization of the ecological conception of control and the deinstitutionalization of the military conception. In 2007 the government proposed the creation of a new international fund to exchange reduced greenhouse emissions from deforestation for financial help. Brazil had so far always refused any mechanism linking international money to the preservation of the Amazon on the basis that it did not accept that Brazil's Amazon region was the concern of any other nation. Indeed, it had long been held that any international interference represented a threat to the country's sovereignty. Furthermore, the ecological conception of control was strengthened by the reconfiguration of the monitoring system after 2000 which allowed for the reduction in deforestation as well as establishing an open data policy.

As well as financial incentives, there were several other reasons for institutional change. First, different interviewees explained that over the last decade Brazilians (especially the political class) take for granted the importance of the environment and consequently preserving the rainforest. Second, many Brazilians are concerned about climate change, and the negative consequences such as the increase in the number of floods and droughts in the country. Third, since the election in 2003, Lula has brought some of the main critics of the previous administration's Amazonian policies into his government. As was explained by a senior scientist at INPE:

> When Lula became president his first Minister of the Environment [Marina Silva] invited members of different NGOs to join her Ministry. These were the same people that INPE previously would not give deforestation data to. One of the reasons was because they thought data may be put into foreign hands. However, after Lula became president we would give it to the Ministry who would then give it to the NGOs. Later, INPE's Director ordered that the data be made publically available on the Internet.

While this open data access policy makes sense within the ecological conception of control, it was unthinkable within the military era. Such institutional change can be understood as a process of ongoing contestation between different conceptions of control. First, in April 2008, there was a clash between a military General and a government Minister over the demarcation of an indigenous reserve (Serra do Sol). Second, and most importantly, many recent policies of the Ministry of Agriculture and the Ministry of National Integration suggest that the economic conception of control is dominant in some areas of government. Economic plans, such as PAC (Plan for Acceleration of Economic Growth) include measures with high environmental impacts such as the construction of asphalt highways crosscutting the rainforest (Laurance et al., 2005). Further, environmental NGOs claimed that the surge in deforestation rates in the second half of 2007 (after 3 years of steady decline) was due to the increase in the price of soya beans and other agricultural commodities and the lack of capacity or political will of the government to counterbalance this market force. This suggests that even though the ecological conception of control has gained considerable force in the last 10 years, it coexists and is continuously contested by a weakening military and a strong economic conception of control.

Discussion

Our case outline has highlighted that the ambitions, design and use of the Amazon monitoring system has shifted significantly over the last 40 years. Throughout the 1970s and 1980s the monitoring system was used as a tool to support deforestation so as to best utilize "natural resources." Such use can be understood within the context of an institutional field dominated by the alliance between military and economic conceptions of control. Similarly, as the military conception of control was still dominant throughout the 1990s, the changes to the system at the end of the 1980s did not provide for a reduction in deforestation. Finally, since 2001 the monitoring system has been reconfigured

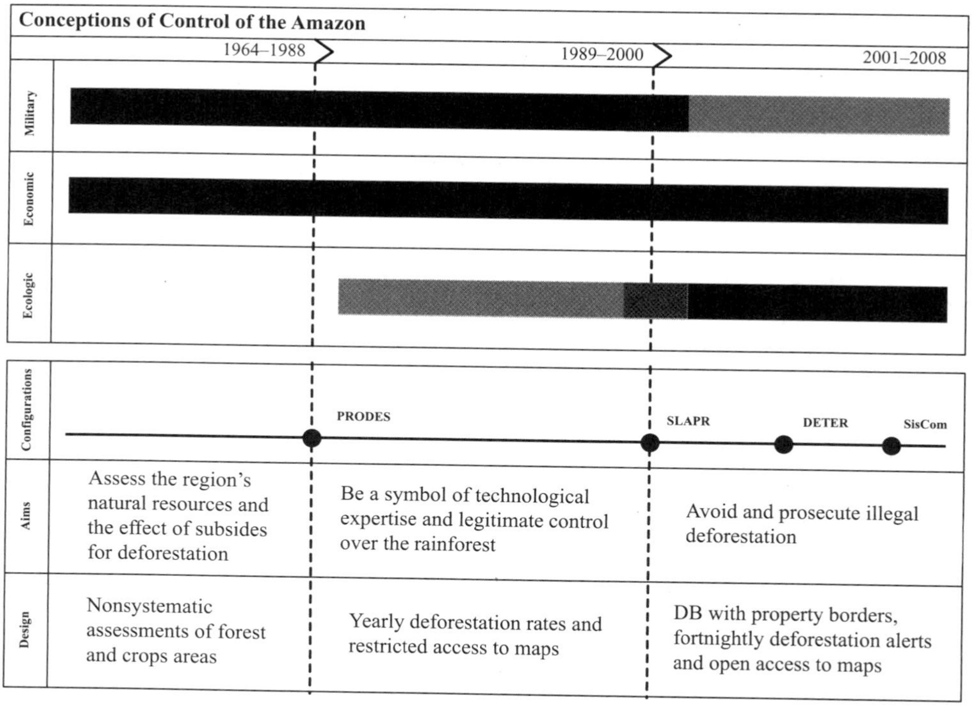

Figure 22.3
Conceptions of control and the Brazilian Amazon monitoring system.

to assist in reducing deforestation. This reconfiguration was possible due to the deinstitutionalization of the military conception of control, and the institutionalization of the ecological conception of control across Brazilian government (see figure 22.3).

This penultimate section will explain these shifts in the use of the Amazon monitoring systems as being largely due to dialectical processes of institutional change within the Brazilian government and between other interested governmental and nongovernmental groups inside and outside Brazil. Specifically, this discussion section provides a discussion of IT artefacts and institutions in relation to three themes: how institutional change takes place, how IT artefacts (re)align themselves to institutions and how IT artefacts may be implicated in the process of institutional change.

Institutional Change as an Emergent and Conflictual Process

The emergence and receding of our three conceptions of control over the 40-year history of the Amazon monitoring system demonstrates that such institutional change can be understood as being emergent and conflictual. Drawing on Fligstein (1990), this first subsection will critically examine how pressures outside of the Brazilian government led to change.

Our case highlighted two key events that occurred in the late 1980s that were implicated in the shift from a military to an ecological conception of control. The first event related to the perceived threat to the sovereignty of the Amazon declining due to improvements in Latin American political relations. The second event related to the growing awareness among scientists, environmentalists and international governments of the importance of the Amazon region with regard to global environmental sustainability. This cumulated in the World Bank and the Inter-America Development Bank suspending the payment of loans to development projects in the Amazon region until Brazil agreed to reduce deforestation (Kolk, 1998). In Fligstein's (1990) longitudinal analysis of American industry, he points to the dramatic changes in economic conditions during the great depression in the 1930s as being the key events that saw companies more aggressively looking for new markets. This reflected the shift from the manufacturing conception of control to the sales and marketing conception of control. In our case, the two events rendered it infeasible for the Brazilian government to ignore such international criticism and economic sanctions. This left Brazil unable to maintain its military-orientated strategy for the Amazon region.

Cumulatively, these two events triggered a "moment of crisis." Fligstein argues that crisis are the moments in which "major groups are having difficulty reproducing their privilege as the rules that have governed interaction are no longer working," the institutional fabric tears up and institutional change can happen (Fligstein, 2001, 118). This moment of crisis was brought about due to global influences that destabilized the military conception of control. It resulted in Brazil developing a policy that sought to convince the international community that it would reduce deforestation. This led to the establishment of the ecological conception of control (at least at a discursive level). This change in government policy is akin to the implementation of anti-trust laws that forced American companies to diversify their investments by establishing a broader portfolio of products and companies (Fligstein, 2001), and cumulated in the dominance of the financial conception of control. The shift in dominance to the ecological conception of control represents a form of coercive isomorphism (DiMaggio and Powell, 1983), where powerful actors (i.e. the USA and Western Europe) imposed a certain institution (i.e. environmentalism) over the weaker members of its institutional field (i.e. the developing countries) through coercive measures (i.e. cuts in loans). It also helps us better understand how powerful global actors can influence and shape societal and organizational change (Robey and Holmstrom, 2001).

The case study of the Amazon monitoring system also contains empirical findings that extends Fligstein's (1990, 2001) understanding of institutional change in two ways. First, it shows that such shifts cannot be entirely understood as coming about due to a "moment of crisis," but also due to an ongoing strengthening of the ecological conception of control, not only among the international community, scientists and environmentalists, but also with more affluent and politically influential sections of Brazilian society. This gradual strengthening suggests that understanding changes in the balance between conceptions of control requires a longitudinal analysis (such as with Beck's [1992] "risk society"). The dominance of the ecological conception of control points to some more productive aspects of globalization. This contrasts Robey and Holmstrom's (2001) more pessimistic view, where they consider globalization privileges economic criteria and thereby restricting social needs.

Second, while Fligstein (1990) suggests that a certain institutional field is usually dominated by a single major conception of control, our case has shown that two conceptions of control can coexist in symbiosis (e.g. military and economic) or conflict (e.g. economic vs ecological) potentially for significant periods of time. Synergy was evident between the pressures for military sovereign control and economic development that framed the colonization of the Amazon region. Contestation is currently evident in the competition for dominance between the ecological and the economic conceptions of control, as while the idea that the Amazon should be preserved is taken-for-granted, the actual governmental practices towards the Amazon are still an arena of contestation as they have to mediate the demands of the global influences coupled with the specific needs of landowners, farmers and related industries who seek to make their existence in the Amazon sustainable. Further, while we agree with commentators who argue that institutions tend towards stability (Powell and DiMaggio, 1991; Scott, 2001; Zucker, 1977), our analysis further implies that even apparently stable conceptions of control are always in states of synergy and/or contestation. This also suggests that the apparently schizophrenic behaviour of the government towards the Amazon is actually the outcome of internal synergy/struggle between different conceptions of control at a global, national and local level (Robey and Holmstrom, 2001). Consequently, we suggest that institutional change should be viewed as being a tentative, emergent political outcome of ongoing relations and negotiations between many different local and global interested groups (Blackler and Regan, 2006).

Aligning Artefacts and Institutions

Conceptions of control also provide a theoretical lens to understand the processes whereby the institutional context shapes the design and use of IT artefacts (Currie and Guah, 2007; Hayes, 2008; Orlikowski and Barley, 2001). Our longitudinal case shows a considerable degree of alignment between the conceptions of control that was dominant at a specific moment of time and IT artefacts. Furthermore, the case also revealed the relation between IT artefacts and conceptions of control is neither static nor predicable, as IT artefacts may emerge in accordance with a certain conception of control (e.g. military) and later be reconfigured to reflect a succeeding conception of control (e.g. ecological). The recognition of this phenomenon has three implications for the understanding of IT in institutions.

First, we suggest institutional theory can provide an important theoretical lens to provide a historical understanding of the ways in which specific IT artefacts come about and change over time (Currie and Guah, 2007; Hayes, 2008; Orlikowski and Barley, 2001). In our case, the development and early use of the Amazon monitoring system was framed by broader social phenomena (such as the cold war). Understanding such framing helps us to make sense of why specific organizational practices were enacted (such as subsidies for deforestation) at that time. Indeed, such practices now look incomprehensible in relation to the current conception of control, and further highlight the importance of avoiding presentisms.

Second, there have been many studies reported in the literature that have considered the ways in which the visions of designers are materialized in software applications and the ways in which this may shape or constrain (or not) possibilities for action (Akrich, 1992; Kling and Iacono, 1989). Indeed, such a view of how technology may be implicated in institutional change resonates with Kallinikos (2004), who argues that IT not only passively offers novel information but “invit[es] particular modes of understanding and action that involve both the framing of the reality to be addressed, the determination of particular tasks and the sequential patterning” (19). We suggest that institutional theory offers important insights that can help researchers to recover the synergy/contestation that surrounded those conceptions of control that were pivotal in shaping the visions of developers of information technologies at particular moments in their history. Along with Currie (2004) we too argue that it is important for institutionalist research to explain how visions became institutionalized and de-institutionalized over time (Swanson and Ramiller, 1997). We suggest that conceptions of control provide a powerful lens to understand the institutionalization and deinstitutionalization over time. Further, we would urge future research to conceive of institutionalization and deinstitutionalization as an ongoing and highly political process rather than being static states that we move between.

Third, institutional theory also helps better understand the ways that IT artefacts may "drift." Ciborra (2000) points out that IT artefacts, such as information infrastructures, tend to drift, namely, "they deviate from their planned purposes for a variety of reasons often outside anyone's influence" (4). He explains that technology intrinsically tends to drift due to some force from inside that manifests itself when it is put to use, "thus, the idea emerges of technology with a certain degree of autonomy and inner dynamics; of technology both as a drifting system and as [an] organism to be cultivated" (*ibid.*, 32). However understood through institutional theory, drift can also be conceived of as not necessarily being arbitrary, but arising as a consequence of contestation between conceptions of control which may lead to shifts in an institutional field. In our case, this was evident in the Amazon monitoring system's transformation from being a symbol of military dominance to a way to protect the environment. Such drift came about as a result of the institutional hijacking of the monitoring system by those supporting new values and interests. Consequently, we suggest that while drift provides important conceptual insights, such insights can be deepened if an institutional analysis is undertaken over the longue durée. Such an analysis allows us to take into consideration the organizational, societal and global pressures for change (Giddens, 1986; Robey and Holmstrom, 2001).

IT Shaping Institutional Change

So far our analysis has pointed to how a dominant conception of control shapes the nature and use of an IT. This section will argue that IT can also lead to new ways of framing reality, and in so doing bring about institutional change (Choo, 1996; Kallinikos, 2004, 2006).

The monitoring system initially sought to measure the extent and location of deforestation so as to evaluate the success of its economic and military-oriented colonization policies. However later, the same deforestation data became a key resource for scientists and environmental NGOs, as it allowed for them to argue that such deforestation was not sustainable and that it was going to have catastrophic implications for the global environment (see e.g. Fearnside, 1982, 2005; Greenpeace, 2008). Without the data that the monitoring system provided, the severity of deforestation may not have been identified as early as it was (Rajão and Hayes, 2007). Thus, despite the initial design and use of the Amazon monitoring system aligning itself closely with the military conception of control, the deforestation data were also inadvertently fundamental to the establishment of the ecological conception of control. In this sense the monitoring system became a key legitimating element in stabilizing and destabilizing a dominant conception of control (Avgerou, 2000; Noir and Walsham, 2007). Such an analysis suggests that the capabilities of the monitoring system to analyse, store and transmit data provided the possibility for its own reconfiguration and thus was inextricably interlinked with institutional change (Barley and Tolbert, 1997; Barrett *et al.*, 2001). Such change may take on forms far removed from that envisaged by the designers and/or the champions of an IT and thus lead to significantly different institutions (Fountain, 2006).

Although not understood through institutional theory, the possibilities for such reconfiguration were captured in Zuboff's (1988) concepts of informate and automate. She highlighted that while an IT system was put in place to automate work, the system also generated information about work activity that was previously unavailable. Consequently, she argued that IT offers the possibility not only to automate by deskilling and eliminating jobs, but also to empower employees with novel information and leads to new ways of understanding their own work. Such IT led change may lead to the emergence of new work practices and the renegotiation of institutionalized relations between occupation groups (Barley, 1986). Thus IT presents intentional and unintentional possibilities for institutional change (Barrett and Walsham, 1999; Barrett et al., 2001; Fountain, 2001).

Conclusion

The overall aim of this paper has been to provide an in-depth account of institutional change, and specifically to consider the ways in which IT artefacts are implicated in processes of institutionalization and deinstitutionalization. Through our analysis of the Amazon monitoring system we have illustrated how its emergence and different uses were shaped by global pressures over 44 years. First, the system was shaped by the perceived threats to Brazilian sovereignty. More recently, economic sanctions proved to be an important dimension in bringing about institutional change. What was also interesting in this case was the ways in which global influences led to many Brazilians themselves becoming environmentally aware and thus provided an internal pressure for change. Local pressures also influenced the shape and nature of change, as the Brazilian government balanced these global influences with the needs of the local population to be able to sustain their existence in the Amazon region through agriculture. These competing pressures between local and global influences have been very much central to shaping the changes in policy and the use of the Amazon monitoring system over its 44-year history and are likely to continue to shape its future role.

Our analysis was primarily based on Fligstein's (1990) conceptions of control, and shows the value of drawing on institutional theory to understand how the global and local pressures are implicated in institutionalization and deinstitutionalization processes. Based on such a theoretical lens, we have argued that institutional change can be better understood if researchers place a strong emphasis on the history of an IT artefact, and by doing so provide a much more detailed understanding of how and why information technologies emerge and how this history is implicated in shaping current practices (Currie, 2004; Orlikowski and Barley, 2001). The ongoing contestation and

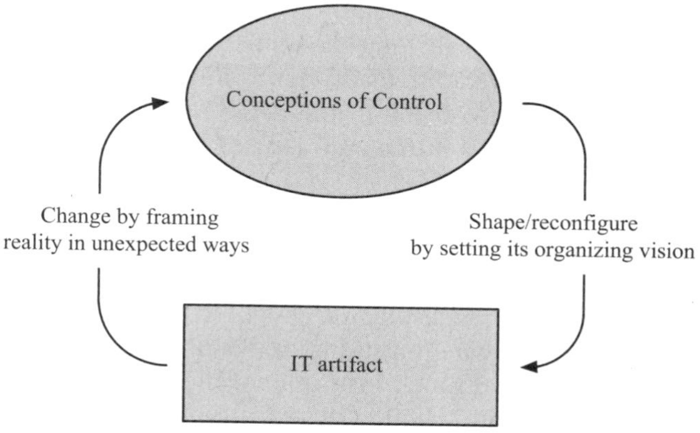

Figure 22.4
Dialectical relationship between IT artefacts and institutional context.

synergy between alternate conceptions of control highlights the importance of viewing institutions as being provisional and highly political. Such a dialectical approach highlights the significance of understanding how relations of power can become taken for granted over time (Robey and Holmstrom, 2001). Thus, our analysis suggests that the relationship between IT and institutional contexts is a complex dialectical linkage whereby the organization's institutional arrangements shape the design and use of IT artefacts, while also providing possibilities for institutional change (see figure 22.4). Indeed, the differing uses of the Amazon monitoring system over the years highlighted how the same IT artefact may align itself with alternate conceptions of control and in doing so is inextricable linked with institutionalization and deinstitutionalization processes. We hope the study reported in this paper will lead to future studies that adopt longitudinal and dialectical approaches to the study of IT and institutional change.

Notes

1. In the context of this paper, IT artefacts are understood as "bundles of material and cultural properties packaged in some socially recognizable form such as hardware and/or software" (Orlikowski and Iacono, 2001, 121). In this context, IT artefacts comprise both material affordances (i.e., design features) and established and envisaged uses or organizing visions (i.e., ideal and actual work practices related to the artefact; Orlikowski and Barley, 2001; Swanson and Ramiller, 1997).

2. INPE's PRODES Digital and DETER and IBAMA's SisCom can be accessed at http://terrabrasilis.dpi.inpe.br; http://www.obt.inpe.br/OBT/assuntos/programas/amazonia/deter; and http://siscom.ibama.gov.br/, respectively.

3. Despite the doubts of some interviewees (especially the scientists), many members of the military and senior politicians did believe that there was a threat to the Brazilian sovereignty of the Amazon at this time.

References

Abrucio, F. L. 2007. "Trajetória recente da gestão pública brasileira: Um balanço crítico e a renovação da agenda de reformas." *Revista de Adminstração Pública* 41 (1): 67–86.

Adler, P. S. 2005. "The Evolving Object of Software Development." *Organization* 12 (3): 401–436.

Akrich, M. 1992. "The De-Scription of Technical Objects." In *Shaping Technology*, edited by W. Bijker and J. Law, 205–224. Cambridge, MA: MIT Press.

Avgerou, C. 2000. "IT and Organizational Change: An Institutionalist Perspective." *Information Technology & People* 13 (4): 234–262.

Avgerou, C. 2003. "IT as an Institutional Actor in Developing Countries." In *The Digital Challenge: Information Technology in the Development Context*, edited by S. Madon and S. Krishna. Aldershot, United Kingdom: Ashgate.

Barley, S. R. 1986. "Technology as an Occasion for Structuring: Evidence from Observations of CT Scanners and the Social Order of Radiology Departments." *Administrative Science Quarterly* 31 (1): 78–108.

Barley, S. R., and Tolbert, P. S. 1997. "Institutionalization and Structuration: Studying the Links between Action and Institution." *Organization Studies* 18 (1): 93–117.

Barrett, M., Sahay, S., and Walsham, G. 2001. "Information Technology and Social Transformation: GIS for Forestry Management in India." *Information Society* 17 (1): 5–20.

Barrett, M., and Walsham, G. 1999. "Electronic Trading and Work Transformation in the London Insurance Market." *Information Systems Research* 10 (1): 1–22.

Beck, U. 1992. *Risk Society: Towards a New Modernity*. London: Sage.

Berger, P. L., and Luckmann, T. 1967. *The Social Construction of Reality: A Treatise in the Sociology of Knowledge*. London: Penguin.

Blackler, F., and Regan, S. 2006. "Institutional Reform and the Reorganization of Family Support Services." *Organization Studies* 27 (12): 1843–1861.

Butler, T. 2003. "An Institutional Perspective on Developing and Implementing Intranet- and Internet-Based Information Systems." *Information Systems Journal* 13 (3): 209–231.

Chatterjee, D., Grewal, R., and Sambamurthy, V. 2002. "Shaping Up for E-commerce: Institutional Enablers of the Organizational Assimilation of Web Technologies." *MIS Quarterly* 26 (2): 65–89.

Choo, C. W. 1996. "The Knowing Organization: How Organizations Use Information to Construct Meaning, Create Knowledge and Make Decisions." *International Journal of Information Management* 16 (5): 329–340.

Ciborra, C. U. 2000. "A Critical Review of the Literature on the Management of Corporate Information Infrastructure." In *From Control to Drift: The Dynamics of Corporate Information Infrastructures*, edited by C. U. Ciborra, 15–40. Oxford: Oxford University Press.

Currie, W. L. 2004. "The Organizing Vision of Application Service Provision: A Process-Oriented Analysis." *Information & Organization* 14 (4): 237–267.

Currie, W. L., and Guah, M. W. 2007. "Conflicting Institutional Logics: A National Programme for IT in the Organisational Field of Healthcare." *Journal of Information Technology* 22 (3): 235–247.

Czarniawska, B. 2004. *Narratives in Social Science Research*. London: Sage.

Dacin, M. T., Goodstein, J., and Scott, W. R. 2002. "Institutional Theory and Institutional Change: Introduction to the Special Research Forum." *Academy of Management Journal* 45 (1): 45–57.

DiMaggio, P. J., and Powell, W. W. 1983. "The Iron Cage Revisited: Institutional Isomorphism and Collective Rationality in Organizational Fields." *American Sociological Review* 48 (2): 147–160.

Fearnside, P. M. 1982. "Deforestation in the Brazilian Amazon: How Fast is It Occurring?" *Interciencia* 7 (2): 82–85.

Fearnside, P. M. 2003. "Deforestation Control in Mato Grosso: A New Model for Slowing the Loss of Brazil's Amazon Forest." *AMBIO: A Journal of the Human Environment* 32 (5): 343–345.

Fearnside, P. M. 2005. "Deforestation in Brazilian Amazonia: History, Rates, and Consequences." *Conservation Biology* 19 (3): 680–688.

Fligstein, N. 1990. *The Transformation of Corporate Control*. Cambridge, MA: Harvard University Press.

Fligstein, N. 2001. "Social Skill and the Theory of Fields." *Sociological Theory* 19:105–125.

Fligstein, N., and Brantley, P. 1992. "Bank Control, Owner Control, or Organizational Dynamics: Who Controls the Large Modern Corporation?" *American Journal of Sociology* 98 (2): 280–307.

Fountain, J. E. 2001. *Building the Virtual State: Information Technology and Institutional Change*. Washington, DC: Brookings Institution Press.

Fountain, J. E. 2006. "Central Issues in the Political Development of the Virtual State." In *The Network Society: From Knowledge to Policy*, edited by M. Castells and G. Cardoso, 149–181. Washington, DC: Brookings Institution Press.

Friedland, R., and Alford, R. R. 1991. "Bringing Society Back." In *The New Institutionalism in Organizational Analysis*, edited by W. W. Powell and P. J. DiMaggio. Chicago: University of Chicago Press.

Giddens, A. 1986. *The Constitution of Society: Outline of the Theory of Structuration*. Berkeley: University of California Press.

Giddens, A. 1990. *The Consequences of Modernity*. Stanford, CA: Stanford University Press.

Gosain, S. 2004. "Enterprise Information Systems as Objects and Carriers of Institutional Forces: The New Iron Cage?" *Journal of the Association for Information Systems* 5 (4): 151–182.

Greenpeace. 2008. *Desmatamento na Amazônia: o leão acordou*. São Paulo: Greenpeace Brasil.

Greenwood, R., and Suddaby, R. 2006. "Institutional Entrepreneurship in Mature Fields: The Big Five Accounting Firms." *Academy of Management Journal* 59 (1): 27–48.

Guha, R. 2000. *Environmentalism: A Global History*. New York: Longman.

Hasselbladh, H., and Kallinikos, J. 2000. "The Project of Rationalization: A Critique and Reappraisal of Neo-institutionalism in Organization Studies." *Organization Studies* 21 (4): 697–720.

Hayes, N. 2008. "Institutionalising Change in a High-Technology Optronics Company: The Role of Information and Communication Technologies." *Human Relations* 61 (2): 245–271.

Kallinikos, J. 2004. "The Social Foundations of the Bureaucratic Order." *Organization* 11 (1): 13–36.

Kallinikos, J. 2006. *The Consequences of Information: Institutional Implications of Technological Change*. Cheltenham, UK: Elgar.

Keck, M. E. 2001. "Dilemmas for Conservation in the Brazilian Amazon." *Environmental Change and Security Program Report* 7:32–46.

King, J. L., Gurbaxani, V., Kraemer, K. L., McFarlan, F. W., Raman, K. S., and Yap, C. S. 1994. "Institutional Factors in Information Technology Innovation." *Information Systems Research* 5 (2): 139–169.

Kling, R., and Iacono, C. S. 1989. "The Institutional Character of Computerized Information Systems." *Office: Technology and People* 5 (1): 7–28.

Kolk, A. 1998. "From Conflict to Cooperation: International Policies to Protect the Brazilian Amazon." *World Development* 26 (8): 1481–1493.

Lamb, R., and Kling, R. 2003. "Reconceptualizing Users as Social Actors in Information Systems Research." *MIS Quarterly* 27 (2): 197–235.

Laurance, W. F., Fearnside, P. M., Albernaz, A. K. M., Vasconcelos, H. L., and Ferreira, L. V. 2005. "Amazon Deforestation: Roads Matter." *Science* 307 (1044).

McNeill, J. R. 2000. *Something New under the Sun: An Environmental History of the Twentieth-Century World*. New York: W. W. Norton.

Meyer, J. W., and Rowan, B. 1977. "Institutionalized Organizations: Formal Structure as Myth and Ceremony." *American Journal of Sociology* 83 (2): 340–363.

Nicolaou, A. 1999. "Social Control in Information Systems Development." *Information Technology & People* 12 (2): 130–150.

Noir, C., and Walsham, G. 2007. "The Great Legitimizer: ICT as Myth and Ceremony in the Indian Healthcare Sector." *Information Technology & People* 20 (4): 313–333.

Orlikowski, W. J., and Barley, S. R. 2001. "Technology and Institutions: What Can Research on Information Technology and Research on Organizations Learn from Each Other?" *MIS Quarterly* 25 (2): 145–165.

Orlikowski, W. J., and Iacono, C. S. 2001. "Research Commentary: Desperately Seeking the 'IT' in IT Research—a Call to Theorizing the IT Artifact." *Information Systems Research* 12 (2): 121–134.

Orlikowski, W. J., and Robey, D. 1991. "Information Technology and the Structuring of Organizations." *Information Systems Research* 2 (2): 143–169.

Page, E. A., and Redclift, M. R. 2002. *Human Security and the Environment: International Comparisons*. Northampton, MA: Edward Elgar.

Powell, W. W., and DiMaggio, P. J., eds. 1991. *The New Institutionalism in Organizational Analysis*. Chicago: University of Chicago Press.

Rajão, R. G. L., and Hayes, N. 2007. "Sistema de monitoramento da Amazônia: Questão de segurança nacional." In *Por que GESITI? Segurança, Inovação e Sociedade*, edited by A. J. Balloni, 53–86. Campinas, Brazil: Komedi.

Reis, A. C. F. 1965. *A Amazônia e a cobiça internacional*. Rio de Janeiro: Edinova Limitada.

Robey, D., and Boudreau, M.-C. 1999. "Accounting for the Contradictory Organizational Consequences of Information Technology: Theoretical Directions and Methodological Implications." *Information Systems Research* 10 (2): 167–185.

Robey, D., and Holmstrom, J. 2001. "Transforming Municipal Governance in Global Context: A Case Study of the Dialectics of Social Change." *Journal of Global Information Technology Management* 4 (4): 19–32.

Scott, W. R. 1995. *Institutions and Organizations*. Thousand Oaks, CA: Sage.

Scott, W. R. 2001. *Institutions and Organizations*. Thousand Oaks, CA: Sage.

Scott, W. R. 2003. "Institutional Carriers: Reviewing Modes of Transporting Ideas over Time and Space and Considering Their Consequences." *Industrial and Corporate Change* 12 (4): 879–894.

Seo, M. -G., and Creed, W. E. D. 2002. "Institutional Contradictions, Praxis, and Institutional Change: A Dialectical Perspective." *Academy of Management Review* 27 (2): 222–247.

Shrivastava, P. 1995. "Ecocentric Management for a Risk Society." *Academy of Management Review* 20 (1): 118–137.

Silva, L., and Figueroa, E. B. 2002. "Institutional Intervention and the Expansion of ICTs in Latin America: The Case of Chile." *Information Technology & People* 15 (1): 8–25.

Suddaby, R., and Greenwood, R. 2009. "Methodological Issues in Researching Institutions and Institutional Change." In *Handbook of Organizational Research Methods*, edited by D. Buchanan and A. Bryman, 176–195. London: Sage.

Swanson, E. B., and Ramiller, N. C. 1997. "The Organizing Vision in Information Systems Innovation." *Organization Science* 8 (5): 458–474.

Townley, B. 2002. "The Role of Competing Rationalities in Institutional Change." *Academy of Management Journal* 45 (1): 163–179.

Tsoukas, H., and Papoulias, D. B. 2005. "Managing Third-Order Change: The Case of the Public Power Corporation in Greece." *Long Range Planning* 38 (1): 79–95.

Walsham, G. 1993. *Interpreting Information Systems in Organizations*. New York: John Wiley & Sons.

Walsham, G. 2006. "Doing Interpretive Research." *European Journal of Information Systems* 15 (3): 320–330.

Walsham, G., and Sahay, S. 1999. "GIS for District-Level Administration in India: Problems and Opportunities." *MIS Quarterly* 23 (1): 39–65.

Zuboff, S. 1988. *In the Age of the Smart Machine: The Future of Work and Power*. New York: Basic Books.

Zucker, L. G. 1977. "The Role of Institutionalization in Cultural Persistence." *American Sociological Review* 42 (5): 726–743.

23 Franken-Algorithms: The Deadly Consequences of Unpredictable Code

Andrew Smith

In the past few years, a number of people have expressed fears about the impact of artificial intelligence on our safety, our ability to work, and our daily lives. This article does not dismiss those fears but suggests they may be distracting many from dangers already present in algorithms. Andrew Smith examines those algorithms—those pieces of code that make decisions about everything from the price of a book on Amazon.com to whether we are deemed a threat when we go through immigration into a new country.

Smith begins by emphasizing how values get embedded and propagated through our technologies. When algorithms are tasked with making decisions, they initially make the decision their creators tell them they should. These simple algorithms are powerful because they enable the values that created them to become more and more ubiquitous, making it difficult for alternatives to gain a foothold. But as algorithms get more complicated and begin interacting with each other, they can begin to behave in unexpected and even unexplainable ways.

This article explores the logical extension of this problem. We might be able to apportion responsibility for simple stand-alone technologies to those who write the code. But as algorithms adapt and interact with each other beyond the realm of anyone's understanding, who should be held accountable for the results? The author believes that scholars and the public must initiate a major effort to answer this question. It becomes more imperative every day to address the issue as these technologies are allowed to stabilize (or destabilize) the stock market, impact elections, and decide who should and who shouldn't be killed. Our traditional methods of promoting a happy and healthy world by linking people with their actions and apportioning blame and penalties accordingly are being circumvented by new technologies. We must either eliminate all systems that are not directly linked with individual people or find ways to rein in the negative fallout that unruly systems can create. Our future depends on how we respond to this challenge.

The 18th of March 2018, was the day tech insiders had been dreading. That night, a new moon added almost no light to a poorly lit four-lane road in Tempe, Arizona, as a specially adapted Uber Volvo XC90 detected an object ahead. Part of the modern gold rush to develop self-driving vehicles, the SUV had been driving autonomously, with no input from its human backup driver, for 19 minutes. An array of radar and light-emitting lidar sensors allowed onboard algorithms to calculate that, given their host vehicle's steady speed of 43 mph, the object was six seconds away—assuming it remained stationary. But objects in roads seldom remain stationary, so more algorithms

---

Published in the *Guardian*, https://www.theguardian.com/technology/2018/aug/29/coding-algorithms-frankenalgos-program-danger.

crawled a database of recognizable mechanical and biological entities, searching for a fit from which this one's likely behavior could be inferred.

At first the computer drew a blank; seconds later, it decided it was dealing with another car, expecting it to drive away and require no special action. Only at the last second was a clear identification found—a woman with a bike, shopping bags hanging confusingly from handlebars, doubtless assuming the Volvo would route around her as any ordinary vehicle would. Barred from taking evasive action on its own, the computer abruptly handed control back to its human master, but the master wasn't paying attention. Elaine Herzberg, aged 49, was struck and killed, leaving more reflective members of the tech community with two uncomfortable questions: was this algorithmic tragedy inevitable? And how used to such incidents would we, *should* we, be prepared to get?

"In some ways we've lost agency. When programs pass into code and code passes into algorithms and then algorithms start to create new algorithms, it gets farther and farther from human agency. Software is released into a code universe which no one can fully understand."

If these words sound shocking, they should, not least because Ellen Ullman, in addition to having been a distinguished professional programmer since the 1970s, is one of the few people to write revealingly about the process of coding. There's not much she doesn't know about software in the wild.

"People say, 'Well, what about Facebook—they create and use algorithms and they can change them.' But that's not how it works. They set the algorithms off and they learn and change and run themselves. Facebook intervene in their running periodically, but they really don't control them. And particular programs don't just run on their own, they call on libraries, deep operating systems and so on . . ."

What Is an Algorithm?

Few subjects are more constantly or fervidly discussed right now than algorithms. But what is an algorithm? In fact, the usage has changed in interesting ways since the rise of the internet—and search engines in particular—in the mid-1990s. At root, an algorithm is a small, simple thing; a rule used to automate the treatment of a piece of data. If a happens, then do b; if not, then do c. This is the "if/then/else" logic of classical computing. If a user claims to be 18, allow them into the website; if not, print "Sorry, you must be 18 to enter." At core, computer programs are bundles of such algorithms. Recipes for treating data. On the micro level, nothing could be simpler. If computers appear to be performing magic, it's because they are fast, not intelligent.

Recent years have seen a more portentous and ambiguous meaning emerge, with the word "algorithm" taken to mean any large, complex decision-making software system; any means of taking an array of input—of data—and assessing it quickly, according to a given set of criteria (or "rules"). This has revolutionized areas of medicine, science, transport, communication, making it easy to understand the utopian view of computing that held sway for many years. Algorithms have made our lives better in myriad ways.

Only since 2016 has a more nuanced consideration of our new algorithmic reality begun to take shape. If we tend to discuss algorithms in almost biblical terms, as independent entities with lives of their own, it's because we have been encouraged to think of them in this way. Corporations like Facebook and Google have sold and defended their algorithms on the promise of objectivity, an ability to weigh a set of conditions with mathematical detachment and absence of fuzzy emotion. No wonder such algorithmic decision-making has spread to the granting of loans/ bail/benefits/ college places/job interviews and almost anything requiring choice.

We no longer accept the sales pitch for this type of algorithm so meekly. In her 2016 book *Weapons of Math Destruction*, Cathy O'Neil, a former math prodigy who left Wall Street to teach and write and run the excellent *mathbabe* blog, demonstrated beyond question that, far from eradicating human biases, algorithms could magnify and entrench them. After all, software is written by overwhelmingly affluent white and Asian men—and it will inevitably reflect their assumptions (Google "racist soap dispenser" to see how this plays out in even mundane real-world situations). Bias doesn't require malice to become harm, and unlike a human being, we can't easily ask an algorithmic gatekeeper to explain its decision. O'Neil called for "algorithmic audits" of any systems directly affecting the public, a sensible idea that the tech industry will fight tooth and nail, because algorithms are what the companies sell; the last thing they will volunteer is transparency.

The good news is that this battle is under way. The bad news is that it's already looking quaint in relation to what comes next. So much attention has been focused on the distant promises and threats of artificial intelligence, AI, that almost no one has noticed us moving into a new phase of the algorithmic revolution that could be just as fraught and disorienting—with barely a question asked.

The algorithms flagged by O'Neil and others are opaque but predictable: they do what they've been programmed to do. A skilled coder can in principle examine and challenge their underpinnings. Some of us dream of a citizen army to do this work, similar to the network of amateur astronomers who support professionals in that field. Legislation to enable this seems inevitable.

We might call these algorithms "dumb," in the sense that they're doing their jobs according to parameters defined by humans. The quality of result depends on the thought and skill with which they were programmed. At the other end of the spectrum is the more or less distant dream of human-like artificial general intelligence, or AGI. A properly intelligent machine would be able to question the quality of its own calculations, based on something like our own intuition (which we might think of as a broad accumulation of experience and knowledge). To put this into perspective, Google's DeepMind division has been justly lauded for creating a program capable of mastering arcade games, starting with nothing more than an instruction to aim for the highest possible score. This technique is called "reinforcement learning" and works because a computer can play millions of games quickly in order to learn what generates points. Some call this form of ability "artificial narrow intelligence," but here the word "intelligent" is being used much as Facebook uses "friend"—to imply something safer and better understood than it is. Why? Because the machine has no context for what it's doing and can't do anything else. Neither, crucially, can it transfer knowledge from one game to the next (so-called transfer learning), which makes it less generally intelligent than a toddler, or even a cuttlefish. We might as well call an oil derrick or an aphid "intelligent." Computers are already vastly superior to us at certain specialized tasks, but the day they rival our general ability is probably some way off—if it ever happens. Human beings may not be best at much, but we're second-best at an impressive range of things.

Here's the problem. Between the "dumb" fixed algorithms and true AI lies the problematic halfway house we've already entered with scarcely a thought and almost no debate, much less agreement as to aims, ethics, safety, best practice. If the algorithms around us are not yet intelligent, meaning able to independently say "that calculation/course of action doesn't look right: I'll do it again," they are nonetheless starting to learn from their environments. And once an algorithm is learning, we no longer know to any degree of certainty what its rules and parameters are. At which point we can't be certain of how it will interact with other algorithms, the physical world, or us. Where the "dumb" fixed algorithms—complex, opaque and inured to real time monitoring as they can be—are in principle predictable and interrogable, these ones are not. After a time in the wild, we no longer know what they are: they have the potential to become erratic. We might be tempted to call these "frankenalgos"—though Mary Shelley couldn't have made this up.

Clashing Codes

These algorithms are not new in themselves. I first encountered them almost five years ago while researching a piece for the *Guardian* about high frequency trading (HFT) on the stock market. What I found was extraordinary: a human-made digital ecosystem, distributed among racks of black boxes crouched like ninjas in billion-dollar data farms—which is what stock markets had become. Where once there had been a physical trading floor, all action had devolved to a central server, in which nimble, predatory algorithms fed off lumbering institutional ones, tempting them to sell lower and buy higher by fooling them as to the state of the market. Human HFT traders (although no human actively traded any more) called these large, slow participants "whales," and they mostly belonged to mutual and pension funds—i.e., the public. For most HFT shops, whales were now the main profit source. In essence, these algorithms were trying to outwit each other; they were doing invisible battle at the speed of light, placing and cancelling the same order 10,000 times per second or slamming so many into the system that the whole market shook—all beyond the oversight or control of humans.

No one could be surprised that this situation was unstable. A "flash crash" had occurred in 2010, during which the market went into freefall for five traumatic minutes, then righted itself over another five—for no apparent reason. I travelled to Chicago to see a man named Eric Hunsader, whose prodigious programming skills allowed him to see market data in far more detail than regulators, and he showed me that by 2014, "mini flash crashes" were happening every week. Even he couldn't prove exactly why, but he and his staff had begun to name some of the "algos" they saw, much as crop circle hunters named the formations found in English summer fields, dubbing them "Wild Thing," "Zuma," "The Click" or "Disruptor."

Neil Johnson, a physicist specializing in complexity at George Washington University, made a study of stock market volatility. "It's fascinating," he told me. "I mean, people have talked about the ecology of computer systems for years in a vague sense, in terms of worm viruses and so on. But here's a real working system that we can study. The bigger issue is that we don't know how it's working or what it could give rise to. And the attitude seems to be 'out of sight, out of mind.'"

Significantly, Johnson's paper on the subject was published in the journal *Nature* and described the stock market in terms of "an abrupt system-wide transition from a mixed human-machine phase to a new all-machine phase characterized by frequent black swan [i.e., highly unusual] events with ultrafast durations." The scenario was complicated, according to the science historian George Dyson, by the fact that some HFT firms were allowing the algos to learn—"just letting the black box try different things, with small amounts of money, and if it works, reinforce those rules. We know that's been done. Then you actually have rules where nobody knows what the rules are: the algorithms create their own rules—you let them evolve the same way nature evolves organisms." Non-finance industry observers began to postulate a catastrophic global "splash crash," while the fastest-growing area of the market became (and remains) instruments that profit from volatility. In his 2011 novel *The Fear Index*, Robert Harris imagines the emergence of AGI—of the Singularity, no less—from precisely this digital ooze. To my surprise, no scientist I spoke to would categorically rule out such a possibility.

All of which could be dismissed as high finance arcana, were it not for a simple fact. Wisdom used to hold that technology was adopted first by the porn industry, then by everyone else. But the 21st century's porn is finance, so when I thought I saw signs of HFT-like algorithms causing problems elsewhere, I called Neil Johnson again.

"You're right on point," he told me: a new form of algorithm is moving into the world, which has "the capability to rewrite bits of its own code," at which point it becomes like "a genetic algorithm." He thinks he saw evidence of them on fact-finding forays into Facebook ("I've had my accounts attacked four times," he adds). If so, algorithms are jousting there, and adapting, as on the stock market. "After all, Facebook is just one big algorithm," Johnson says.

"And I think that's exactly the issue Facebook has. They can have simple algorithms to recognize my face in a photo on someone else's page, take the data from my profile and link us together. That's a very simple concrete algorithm. But the question is what is the effect of billions of such algorithms working together at the macro level? You can't predict the learned behavior at the level of the population from microscopic rules. So Facebook would claim that they know exactly what's going on at the micro level, and they'd probably be right. But what happens at the level of the population? That's the issue."

To underscore this point, Johnson and a team of colleagues from the University of Miami and Notre Dame produced a paper, "Emergence of Extreme Subpopulations from Common Information and Likely Enhancement from Future Bonding Algorithms," purporting to mathematically prove that attempts to connect people on social media inevitably polarize society as a whole. He thinks Facebook and others should model (or be made to model) the effects of their algorithms in the way climate scientists model climate change or weather patterns.

O'Neil says she consciously excluded this adaptive form of algorithm from *Weapons of Math Destruction*. In a convoluted algorithmic environment where nothing is clear, apportioning responsibility to particular segments of code becomes extremely difficult. This makes them easier to ignore or dismiss, because they and their precise effects are harder to identify, she explains, before advising that if I want to see them in the wild, I should ask what a flash crash on Amazon might look like.

"I've been looking out for these algorithms, too," she says, "and I'd been thinking: 'Oh, big data hasn't gotten there yet.' But more recently a friend who's a bookseller on Amazon has been telling me how crazy the pricing situation there has become for people like him. Every so often you will see somebody tweet 'Hey, you can buy a luxury yarn on Amazon for $40,000.' And whenever I hear that kind of thing, I think: 'Ah! That must be the equivalent of a flash crash!'"

Anecdotal evidence of anomalous events on Amazon is plentiful, in the form of threads from bemused sellers, and at least one academic paper from 2016, which claims: "Examples have emerged of cases where competing pieces of algorithmic pricing software interacted in unexpected ways and produced unpredictable prices, as well as cases where algorithms were intentionally designed to implement price fixing." The problem, again, is how to apportion responsibility in a chaotic algorithmic environment where simple cause and effect either doesn't apply or is nearly impossible to trace. As in finance, deniability is baked into the system.

Real-Life Dangers

Where safety is at stake, this really matters. When a driver ran off the road and was killed in a Toyota Camry after appearing to accelerate wildly for no obvious reason, NASA experts spent six months examining the millions of lines of code in its operating system, without finding evidence for what the driver's family believed had occurred, but the manufacturer steadfastly denied that the car had accelerated of its own accord. Only when a pair of embedded software experts spent 20 months digging into the code were they able to prove the family's case, revealing a twisted mass of what programmers call "spaghetti code," full of algorithms that jostled and fought, generating anomalous, unpredictable output. The autonomous cars currently being tested may contain 100m lines of code and, given that no programmer can anticipate all possible circumstances on a real-world road, they have to learn and receive constant updates. How do we avoid clashes in such a fluid code milieu, not least when the algorithms may also have to defend themselves from hackers?

Twenty years ago, George Dyson anticipated much of what is happening today in his classic book *Darwin among the Machines*. The problem, he tells me, is that we're building systems that are beyond our intellectual means to control. We believe that if a system is deterministic (acting according to fixed rules, this being the definition of an algorithm) it is predictable—and that what is predictable can be controlled. Both assumptions turn out to be wrong.

"It's proceeding on its own, in little bits and pieces," he says. "What I was obsessed with 20 years ago that has completely taken over the world today are multicellular, metazoan digital organisms, the same way we see in biology, where you have all these pieces of code running on people's iPhones, and collectively it acts like one multicellular organism.

"There's this old law called Ashby's law that says a control system has to be as complex as the system it's controlling, and we're running into that at full speed now, with this huge push to build self-driving cars where the software has to have a complete model of everything, and almost by definition we're not going to understand it. Because any model that we understand is gonna do the thing like run into a fire truck 'cause we forgot to put in the fire truck."

Unlike our old electro-mechanical systems, these new algorithms are also impossible to test exhaustively. Unless and until we have super-intelligent machines to do this for us, we're going to be walking a tightrope.

Dyson questions whether we will ever have self-driving cars roaming freely through city streets, while Toby Walsh, a professor of artificial intelligence at the University of New South Wales who wrote his first program at age 13 and ran a tyro computing business by his late teens, explains from a technical perspective why this is.

"No one knows how to write a piece of code to recognize a stop sign. We spent years trying to do that kind of thing in AI—and failed! It was rather stalled by our stupidity, because we weren't smart enough to learn how to break the problem down. You discover when you program that you have to learn how to break the problem down into simple enough parts that each can correspond to a computer instruction [to the machine]. We just don't know how to do that for a very complex problem like identifying a stop sign or translating a sentence from English to Russian—it's beyond our capability. All we know is how to write a more general purpose algorithm that can learn how to do that given enough examples."

Hence the current emphasis on machine learning. We now know that Herzberg, the pedestrian killed by an automated Uber car in Arizona, died because the algorithms wavered in correctly categorizing her. Was this a result of poor programming, insufficient algorithmic training or a hubristic refusal to appreciate the limits of our technology? The real problem is that we may never know.

"And we will eventually give up writing algorithms altogether," Walsh continues, "because the machines will be able to do it far better than we ever could. Software engineering is in that sense perhaps a dying profession. It's going to be taken over by machines that will be far better at doing it than we are."

Walsh believes this makes it more, not less, important that the public learn about programming, because the more alienated we become from it, the more it seems like magic beyond our ability to affect. When shown the definition of "algorithm" given earlier in this piece, he found it incomplete, commenting: "I would suggest the problem is that algorithm now means any large, complex decision making software system and the larger environment in which it is embedded, which makes them even more unpredictable." A chilling thought indeed. Accordingly, he believes ethics to be the new frontier in tech, foreseeing "a golden age for philosophy"—a view with which Eugene Spafford of Purdue University, a cybersecurity expert, concurs.

"Where there are choices to be made, that's where ethics comes in. And we tend to want to have an agency that we can interrogate or blame, which is very difficult to do with an algorithm. This is one of the criticisms of these systems so far, in that it's not possible to go back and analyze exactly why some decisions are made, because the internal number of choices is so large that how we got to that point may not be something we can ever recreate to prove culpability beyond doubt."

The counter-argument is that, once a program has slipped up, the entire population of programs can be rewritten or updated so it doesn't happen again—unlike humans, whose propensity to repeat mistakes will doubtless fascinate intelligent machines of the future. Nonetheless, while automation should be safer in the long run, our existing system of tort law, which requires proof of intention or negligence, will need to be rethought. A dog is not held legally responsible for biting you; its owner might be, but only if the dog's action is thought foreseeable. In an algorithmic environment, many unexpected outcomes may not have been foreseeable to humans—a feature with the potential to become a scoundrel's charter, in which deliberate obfuscation becomes at once easier and more rewarding. Pharmaceutical companies have benefited from the cover of complexity for years (see the case of Thalidomide), but here the consequences could be both greater and harder to reverse.

The Military Stakes

Commerce, social media, finance and transport may come to look like small beer in future, however. If the military no longer drives innovation as it once did, it remains tech's most consequential adopter. No surprise, then, that an outpouring of concern among scientists and tech workers has accompanied revelations that autonomous weapons are ghosting toward the battlefield in what amounts to an algorithmic arms race. A robotic sharpshooter currently polices the demilitarized zone between North and South Korea, and while its manufacturer, Samsung, denies it to be capable of autonomy, this claim is widely disbelieved. Russia, China and the US all claim to be at various stages of developing swarms of coordinated, weaponized drones, while the latter plans missiles able to hover over a battlefield for days, observing, before selecting their own targets. A group of Google employees resigned over and thousands more questioned the tech monolith's provision of machine learning software to the Pentagon's Project Maven "algorithmic warfare" program—concerns to which management eventually responded, agreeing not to renew the Maven contract and to publish a code of ethics for the use of its algorithms. At time of writing, competitors including Amazon and Microsoft have resisted following suit.

In common with other tech firms, Google had claimed moral virtue for its Maven software: that it would help choose targets more efficiently and thereby save lives. The question is how tech managers can presume to know what their algorithms will do or be directed to do in situ—especially given the certainty that all sides will develop adaptive algorithmic counter-systems designed to confuse enemy weapons. As in the stock market, unpredictability is likely to be seen as an asset rather than handicap, giving weapons a better chance of resisting attempts to subvert them. In this and other ways we risk in effect turning our machines inside out, wrapping our everyday corporeal world in spaghetti code.

Lucy Suchman of Lancaster University in the UK co-authored an open letter from technology researchers to Google, asking them to reflect on the rush to militarize their work. Tech firms' motivations are easy to fathom, she says: military contracts have always been lucrative. For the Pentagon's part, a vast network of sensors and surveillance systems has run ahead of any ability to use the screeds of data so acquired.

"They are overwhelmed by data, because they have new means to collect and store it, but they can't process it. So it's basically useless—unless something magical happens. And I think their recruitment of big data companies is a form of magical thinking in the sense of: 'Here is some magic technology that will make sense of all this.'"

Suchman also offers statistics that shed chilling light on Maven. According to analysis carried out on drone attacks in Pakistan from 2003–13, fewer than 2% of people killed in this way are confirmable as "high value" targets presenting a clear threat to the United States. In the region of 20% are held to be non-combatants, leaving more than 75% unknown. Even if these figures were out by a factor of two—or three, or four—they would give any reasonable person pause.

"So here we have this very crude technology of identification and what Project Maven proposes to do is automate that. At which point it becomes even less accountable and open to questioning. It's a really bad idea."

Suchman's colleague Lilly Irani, at the University of California, San Diego, reminds us that information travels around an algorithmic system at the speed of light, free of human oversight. Technical discussions are often used as a smokescreen to avoid responsibility, she suggests.

"When we talk about algorithms, sometimes what we're talking about is bureaucracy. The choices algorithm designers and policy experts make are presented as objective, where in the past someone would have had to take responsibility for them. Tech companies say they're only improving accuracy with Maven—i.e., the right people will be killed rather than the wrong ones—and in saying that, the political assumption that those people on the other side of the world are more killable, and that the US military gets to define what suspicion looks like, go unchallenged. So technology questions are being used to close off some things that are actually political questions. The choice to use algorithms to automate certain kinds of decisions is political too."

The legal conventions of modern warfare, imperfect as they might be, assume human accountability for decisions taken. At the very least, algorithmic warfare muddies the water in ways we may grow to regret. A group of government experts is debating the issue at the UN convention on certain conventional weapons (CCW) meeting in Geneva this week.

Searching for a Solution

Solutions exist or can be found for most of the problems described here, but not without incentivizing big tech to place the health of society on a par with their bottom lines. More serious in the long term is growing conjecture that current programming methods are no longer fit for purpose given the size, complexity and interdependency of the algorithmic systems we increasingly rely on. One solution, employed by the Federal Aviation Authority in relation to commercial aviation, is to log and assess the content of all programs and subsequent updates to such a level of detail that algorithmic interactions are well understood in advance—but this is impractical on a large scale. Portions of the aerospace industry employ a relatively new approach called model-based programming, in which machines do most of the coding work and are able to test as they go.

Model-based programming may not be the panacea some hope for, however. Not only does it push humans yet further from the process, but Johnson, the physicist, conducted a study for the Department of Defense that found "extreme behaviors that couldn't be deduced from the code itself" even in large, complex systems built using this technique. Much energy is being directed at finding ways to trace unexpected algorithmic behavior back to the specific lines of code that caused it. No one knows if a solution (or solutions) will be found, but none are likely to work where aggressive algos are designed to clash and/or adapt.

As we wait for a technological answer to the problem of soaring algorithmic entanglement, there are precautions we can take. Paul Wilmott, a British expert in quantitative analysis and vocal critic of high frequency trading on the stock market, wryly suggests "learning to shoot, make jam and knit." More practically, Spafford, the software security expert, advises making tech companies responsible for the actions of their products, whether specific lines of rogue code—or proof of negligence in relation to them—can be identified or not. He notes that the venerable Association for Computing Machinery has updated its code of ethics along the lines of medicine's Hippocratic oath, to instruct computing professionals to do no harm and consider the wider impacts of their work. Johnson, for his part, considers our algorithmic discomfort to be at least partly conceptual; growing pains in a new realm of human experience. He laughs in noting that when he and I last spoke about this stuff a few short years ago, my questions were niche concerns, restricted to a few people who pored over the stock market in unseemly detail.

"And now, here we are—it's even affecting elections. I mean, what the heck is going on? I think the deep scientific thing is that software engineers are trained to write programs to do things that optimize—and with good reason, because you're often optimizing in relation to things like the weight distribution in a plane, or a most fuel-efficient speed: in the usual, anticipated circumstances optimizing makes sense. But in unusual circumstances it doesn't, and we need to ask: 'What's the worst thing that could happen in this algorithm once it starts interacting with others?' The problem is we don't even have a word for this concept, much less a science to study it."

He pauses for moment, trying to wrap his brain around the problem.

"The thing is, optimizing is all about either maximizing or minimizing something, which in computer terms are the same. So what is the opposite of an optimization, i.e., the least optimal case, and how do we identify and measure it? The question we need to ask, which we never do, is: 'What's the most extreme possible behavior in a system I thought I was optimizing?'"

Another brief silence ends with a hint of surprise in his voice.

"Basically, we need a new science," he says.

24 The Extraordinary Science of Addictive Junk Food

Michael Moss

We don't normally think of food as a technology, but people have learned to engineer it with ever better precision. Centuries ago humankind figured out that combining yeast, grain, and water would yield alcohol. Today, food scientists can reformulate food products to create very specific effects for consumers. The goal for many food producers is to achieve the *bliss point*, the perfect balance of flavor and texture that makes people want to eat more. Focusing on compelling people to buy more and eat more does, however, have wide-reaching effects beyond a healthy bottom line for the producer. Many scientists argue that the easy availability of certain foods (which they go so far as to label "addictive") has been a major, if not *the* major, contributor to the obesity problem in the United States and beyond. Moss suggests that the engineers involved were able to advance the values of their bosses, perhaps, to the detriment of the values of many others.

This reading offers lessons in the difficulties of change. Faced with the realization that the actions of a few food scientists and corporate executives had led to significant health problems and even death for millions of people, a handful of those involved sought to intervene and realign the nation's food producers in the interest of a healthier population. They had the clout to gather together the most important decision-makers in the industry. Unfortunately, those efforts largely failed. Rather than examine the vast socioindustrial food network that the industry executives had played a major role in shaping, they targeted a single node—the consumers—and said they bore the responsibility for eating food that was bad for them. To take steps to limit consumer access to unhealthy foods would be a violation of democratic freedoms.

Through this story, Moss raises significant questions about how to change giant systems that result in positive effects for some but negative effects for others. Such questions require us to think long and hard about how responsibility is allocated and who ends up winning and losing with such a distribution.

On the evening of April 8, 1999, a long line of Town Cars and taxis pulled up to the Minneapolis headquarters of Pillsbury and discharged 11 men who controlled America's largest food companies. Nestlé was in attendance, as were Kraft and Nabisco, General Mills and Procter & Gamble, Coca-Cola and Mars. Rivals any other day, the CEOs and company presidents had come together for a rare, private meeting. On the agenda was one item: the emerging obesity epidemic and how to deal with it. While the atmosphere was cordial, the men assembled were hardly friends. Their stature was defined by

---

From *New York Times*, February 20, 2013, https://www.nytimes.com/2013/02/24/magazine/the-extraordinary-science-of-junk-food.html.

their skill in fighting one another for what they called "stomach share"—the amount of digestive space that any one company's brand can grab from the competition.

James Behnke, a 55-year-old executive at Pillsbury, greeted the men as they arrived. He was anxious but also hopeful about the plan that he and a few other food-company executives had devised to engage the CEOs on America's growing weight problem. "We were very concerned, and rightfully so, that obesity was becoming a major issue," Behnke recalled. "People were starting to talk about sugar taxes, and there was a lot of pressure on food companies." Getting the company chiefs in the same room to talk about anything, much less a sensitive issue like this, was a tricky business, so Behnke and his fellow organizers had scripted the meeting carefully, honing the message to its barest essentials. "CEOs in the food industry are typically not technical guys, and they're uncomfortable going to meetings where technical people talk in technical terms about technical things," Behnke said. "They don't want to be embarrassed. They don't want to make commitments. They want to maintain their aloofness and autonomy."

A chemist by training with a doctoral degree in food science, Behnke became Pillsbury's chief technical officer in 1979 and was instrumental in creating a long line of hit products, including microwaveable popcorn. He deeply admired Pillsbury but in recent years had grown troubled by pictures of obese children suffering from diabetes and the earliest signs of hypertension and heart disease. In the months leading up to the CEO meeting, he was engaged in conversation with a group of food-science experts who were painting an increasingly grim picture of the public's ability to cope with the industry's formulations—from the body's fragile controls on overeating to the hidden power of some processed foods to make people feel hungrier still. It was time, he and a handful of others felt, to warn the CEOs that their companies may have gone too far in creating and marketing products that posed the greatest health concerns.

The discussion took place in Pillsbury's auditorium. The first speaker was a vice president of Kraft named Michael Mudd. "I very much appreciate this opportunity to talk to you about childhood obesity and the growing challenge it presents for us all," Mudd began. "Let me say right at the start, this is not an easy subject. There are no easy answers—for what the public health community must do to bring this problem under control or for what the industry should do as others seek to hold it accountable for what has happened. But this much is clear: For those of us who've looked hard at this issue, whether they're public health professionals or staff specialists in your own companies, we feel sure that the one thing we shouldn't do is nothing."

As he spoke, Mudd clicked through a deck of slides—114 in all—projected on a large screen behind him. The figures were staggering. More than half of American adults were now considered overweight, with nearly one-quarter of the adult population—40 million people—clinically defined as obese. Among children, the rates had more than doubled since 1980, and the number of kids considered obese had shot past 12 million. (This was still only 1999; the nation's obesity rates would climb much higher.) Food manufacturers were now being blamed for the problem from all sides—academia, the Centers for Disease Control and Prevention, the American Heart Association and the American Cancer Society. The secretary of agriculture, over whom the industry had long held sway, had recently called obesity a "national epidemic."

Mudd then did the unthinkable. He drew a connection to the last thing in the world the CEOs wanted linked to their products: cigarettes. First came a quote from a Yale University professor of psychology and public health, Kelly Brownell, who was an especially vocal proponent of the view that the processed-food industry should be seen as a public health menace: "As a culture, we've become upset by the tobacco companies advertising to children, but we sit idly by while the food companies do the very same thing. And we could make a claim that the toll taken on the public health by a poor diet rivals that taken by tobacco."

"If anyone in the food industry ever doubted there was a slippery slope out there," Mudd said, "I imagine they are beginning to experience a distinct sliding sensation right about now."

Mudd then presented the plan he and others had devised to address the obesity problem. Merely getting the executives to acknowledge some culpability was an important first step, he knew, so his plan would start off with a small but crucial move: the industry should use the expertise of scientists—its own and others—to gain a deeper understanding of what was driving Americans to overeat. Once this was achieved, the effort could unfold on several fronts. To be sure, there would be no getting around the role that packaged foods and drinks play in overconsumption. They would have to pull back on their use of salt, sugar and fat, perhaps by imposing industrywide limits. But it wasn't just a matter of these three ingredients; the schemes they used to advertise and market their products were critical, too. Mudd proposed creating a "code to guide the nutritional aspects of food marketing, especially to children."

"We are saying that the industry should make a sincere effort to be part of the solution," Mudd concluded. "And that by doing so, we can help to defuse the criticism that's building against us."

What happened next was not written down. But according to three participants, when Mudd stopped talking, the one CEO whose recent exploits in the grocery store had awed the rest of the industry stood up to speak. His name was Stephen Sanger, and he was also the person—as head of General Mills—who had the most to lose when it came to dealing with obesity. Under his leadership, General Mills had overtaken not just the cereal aisle but other sections of the grocery store. The company's Yoplait brand had transformed traditional unsweetened breakfast yogurt into a veritable dessert. It now had twice as much sugar per serving as General Mills' marshmallow cereal Lucky Charms. And yet, because of yogurt's well-tended image as a wholesome snack, sales of Yoplait were soaring, with annual revenue topping $500 million. Emboldened by the success, the company's development wing pushed even harder, inventing a Yoplait variation that came in a squeezable tube—perfect for kids. They called it Go-Gurt and rolled it out nationally in the weeks before the CEO meeting. (By year's end, it would hit $100 million in sales.)

According to the sources I spoke with, Sanger began by reminding the group that consumers were "fickle." (Sanger declined to be interviewed.) Sometimes they worried about sugar, other times fat. General Mills, he said, acted responsibly to both the public and shareholders by offering products to satisfy dieters and other concerned shoppers, from low sugar to added whole grains. But most often, he said, people bought what they liked, and they liked what tasted good. "Don't talk to me about nutrition," he reportedly said, taking on the voice of the typical consumer. "Talk to me about taste, and if this stuff tastes better, don't run around trying to sell stuff that doesn't taste good."

To react to the critics, Sanger said, would jeopardize the sanctity of the recipes that had made his products so successful. General Mills would not pull back. He would push his people onward, and he urged his peers to do the same. Sanger's response effectively ended the meeting.

"What can I say?" James Behnke told me years later. "It didn't work. These guys weren't as receptive as we thought they would be." Behnke chose his words deliberately. He wanted to be fair. "Sanger was trying to say, 'Look, we're not going to screw around with the company jewels here and change the formulations because a bunch of guys in white coats are worried about obesity.'"

The meeting was remarkable, first, for the insider admissions of guilt. But I was also struck by how prescient the organizers of the sit-down had been. Today, one in three adults is considered clinically obese, along with one in five kids, and 24 million Americans are afflicted by type 2 diabetes, often caused by poor diet, with another 79 million people having pre-diabetes. Even gout, a painful form of arthritis once known as "the rich man's disease" for its associations with gluttony, now afflicts eight million Americans.

The public and the food companies have known for decades now—or at the very least since this meeting—that sugary, salty, fatty foods are not good for us in the quantities that we consume them. So why are the diabetes and obesity and hypertension numbers still spiraling out of control? It's not just a matter of poor willpower on the part of the consumer and a give-the-people-what-they-want attitude on the part of the food manufacturers. What I found, over four years of research and reporting, was a conscious effort—taking place in labs and marketing meetings and grocery-store aisles—to get people hooked on foods that are convenient and inexpensive. I talked to more than 300 people in or formerly employed by the processed-food industry, from scientists to marketers to CEOs. Some were willing whistle-blowers, while others spoke reluctantly when presented with some of the thousands of pages of secret memos that I obtained from inside the food industry's operations. What follows is a series of small case studies of a handful of characters whose work then, and perspective now, sheds light on how the foods are created and sold to people who, while not powerless, are extremely vulnerable to the intensity of these companies' industrial formulations and selling campaigns.

"In This Field, I'm a Game Changer."

John Lennon couldn't find it in England, so he had cases of it shipped from New York to fuel the "Imagine" sessions. The Beach Boys, ZZ Top and Cher all stipulated in their contract riders that it be put in their dressing rooms when they toured. Hillary Clinton asked for it when she traveled as first lady, and ever after her hotel suites were dutifully stocked.

What they all wanted was Dr Pepper, which until 2001 occupied a comfortable third-place spot in the soda aisle behind Coca-Cola and Pepsi. But then a flood of spinoffs from the two soda giants showed up on the shelves—lemons and limes, vanillas and coffees, raspberries and oranges, whites and blues and clears—what in food-industry lingo are known as "line extensions," and Dr Pepper started to lose its market share.

Responding to this pressure, Cadbury Schweppes created its first spin-off, other than a diet version, in the soda's 115-year history, a bright red soda with a very un-Dr Pepper name: Red Fusion. "If we are to re-establish Dr Pepper back to its historic growth rates, we have to add more excitement," the company's president, Jack Kilduff, said. One particularly promising market, Kilduff pointed out, was the "rapidly growing Hispanic and African-American communities."

But consumers hated Red Fusion. "Dr Pepper is my all-time favorite drink, so I was curious about the Red Fusion," a California mother of three wrote on a blog to warn other Peppers away. "It's disgusting. Gagging. Never again."

Stung by the rejection, Cadbury Schweppes in 2004 turned to a food-industry legend named Howard Moskowitz. Moskowitz, who studied mathematics and holds a Ph.D. in experimental psychology from Harvard, runs a consulting firm in White Plains, where for more than three decades he has "optimized" a variety of products for

Campbell Soup, General Foods, Kraft and PepsiCo. "I've optimized soups," Moskowitz told me. "I've optimized pizzas. I've optimized salad dressings and pickles. In this field, I'm a game changer."

In the process of product optimization, food engineers alter a litany of variables with the sole intent of finding the most perfect version (or versions) of a product. Ordinary consumers are paid to spend hours sitting in rooms where they touch, feel, sip, smell, swirl and taste whatever product is in question. Their opinions are dumped into a computer, and the data are sifted and sorted through a statistical method called conjoint analysis, which determines what features will be most attractive to consumers. Moskowitz likes to imagine that his computer is divided into silos, in which each of the attributes is stacked. But it's not simply a matter of comparing Color 23 with Color 24. In the most complicated projects, Color 23 must be combined with Syrup 11 and Packaging 6, and on and on, in seemingly infinite combinations. Even for jobs in which the only concern is taste and the variables are limited to the ingredients, endless charts and graphs will come spewing out of Moskowitz's computer. "The mathematical model maps out the ingredients to the sensory perceptions these ingredients create," he told me, "so I can just dial a new product. This is the engineering approach."

Moskowitz's work on Prego spaghetti sauce was memorialized in a 2004 presentation by the author Malcolm Gladwell at the TED conference in Monterey, Calif.: "After . . . months and months, he had a mountain of data about how the American people feel about spaghetti sauce. . . . And sure enough, if you sit down and you analyze all this data on spaghetti sauce, you realize that all Americans fall into one of three groups. There are people who like their spaghetti sauce plain. There are people who like their spaghetti sauce spicy. And there are people who like it extra-chunky. And of those three facts, the third one was the most significant, because at the time, in the early 1980s, if you went to a supermarket, you would not find extra-chunky spaghetti sauce. And Prego turned to Howard, and they said, 'Are you telling me that one-third of Americans crave extra-chunky spaghetti sauce, and yet no one is servicing their needs?' And he said, 'Yes.' And Prego then went back and completely reformulated their spaghetti sauce and came out with a line of extra-chunky that immediately and completely took over the spaghetti-sauce business in this country. . . . That is Howard's gift to the American people. . . . He fundamentally changed the way the food industry thinks about making you happy."

Well, yes and no. One thing Gladwell didn't mention is that the food industry already knew some things about making people happy—and it started with sugar. Many of the Prego sauces—whether cheesy, chunky or light—have one feature in common: The largest ingredient, after tomatoes, is sugar. A mere half-cup of Prego Traditional, for instance, has the equivalent of more than two teaspoons of sugar, as much as two-plus Oreo cookies. It also delivers one-third of the sodium recommended for a majority of American adults for an entire day. In making these sauces, Campbell supplied the ingredients, including the salt, sugar and, for some versions, fat, while Moskowitz supplied the optimization. "More is not necessarily better," Moskowitz wrote in his own account of the Prego project. "As the sensory intensity (say, of sweetness) increases, consumers first say that they like the product more, but eventually, with a middle level of sweetness, consumers like the product the most (this is their optimum, or 'bliss,' point)."

I first met Moskowitz on a crisp day in the spring of 2010 at the Harvard Club in Midtown Manhattan. As we talked, he made clear that while he has worked on numerous projects aimed at creating more healthful foods and insists the industry could be doing far more to curb obesity, he had no qualms about his own pioneering work on discovering what industry insiders now regularly refer to as "the bliss point" or any of the other systems that helped food companies create the greatest amount of crave. "There's no moral issue for me," he said. "I did the best science I could. I was struggling to survive and didn't have the luxury of being a moral creature. As a researcher, I was ahead of my time."

Moskowitz's path to mastering the bliss point began in earnest not at Harvard but a few months after graduation, 16 miles from Cambridge, in the town of Natick, where the U.S. Army hired him to work in its research labs. The military has long been in a peculiar bind when it comes to food: how to get soldiers to eat more rations when they are in the field. They know that over time, soldiers would gradually find their meals-ready-to-eat so boring that they would toss them away, half-eaten, and not get all the calories they needed. But what was causing this M.R.E.-fatigue was a mystery. "So I started asking soldiers how frequently they would like to eat this or that, trying to figure out which products they would find boring," Moskowitz said. The answers he got were inconsistent. "They liked flavorful foods like turkey tetrazzini, but only at first; they quickly grew tired of them. On the other hand, mundane foods like white bread would never get them too excited, but they could eat lots and lots of it without feeling they'd had enough."

This contradiction is known as "sensory-specific satiety." In lay terms, it is the tendency for big, distinct flavors to overwhelm the brain, which responds by depressing your desire to have more. Sensory-specific satiety also became a guiding principle for the processed-food industry. The biggest hits—be they Coca-Cola or Doritos—owe their success to complex formulas that pique the taste buds enough to be alluring but don't have a distinct, overriding single flavor that tells the brain to stop eating.

Thirty-two years after he began experimenting with the bliss point, Moskowitz got the call from Cadbury Schweppes asking him to create a good line extension for Dr Pepper. I spent an afternoon in his White Plains offices as he and his vice president for research, Michele Reisner, walked me through the Dr Pepper campaign. Cadbury wanted its new flavor to have cherry and vanilla on top of the basic Dr Pepper taste. Thus, there were three main components to play with. A sweet cherry flavoring, a sweet vanilla flavoring and a sweet syrup known as "Dr Pepper flavoring."

Finding the bliss point required the preparation of 61 subtly distinct formulas—31 for the regular version and 30 for diet. The formulas were then subjected to 3,904 tastings organized in Los Angeles, Dallas, Chicago and Philadelphia. The Dr Pepper tasters began working through their samples, resting five minutes between each sip to restore their taste buds. After each sample, they gave numerically ranked answers to a set of questions: How much did they like it overall? How strong is the taste? How do they feel about the taste? How would they describe the quality of this product? How likely would they be to purchase this product?

Moskowitz's data—compiled in a 135-page report for the soda maker—is tremendously fine-grained, showing how different people and groups of people feel about a strong vanilla taste versus weak, various aspects of aroma and the powerful sensory force that food scientists call "mouth feel." This is the way a product interacts with the mouth, as defined more specifically by a host of related sensations, from dryness to gumminess to moisture release. These are terms more familiar to sommeliers, but the mouth feel of soda and many other food items, especially those high in fat, is second only to the bliss point in its ability to predict how much craving a product will induce.

In addition to taste, the consumers were also tested on their response to color, which proved to be highly sensitive. "When we increased the level of the Dr Pepper flavoring, it gets darker and liking goes off," Reisner said. These preferences can also be cross-referenced by age, sex and race.

On page 83 of the report, a thin blue line represents the amount of Dr Pepper flavoring needed to generate maximum appeal. The line is shaped like an upside-down U, just like the bliss-point curve that Moskowitz studied 30 years earlier in his Army lab. And at the top of the arc, there is not a single sweet spot but instead a sweet range, within which "bliss" was achievable. This meant that Cadbury could edge back on its key ingredient, the sugary Dr Pepper syrup, without falling out of the range and losing the bliss. Instead of using 2 milliliters of the flavoring, for instance, they could use 1.69 milliliters and achieve the same effect. The potential savings is merely a few percentage points, and it won't mean much to individual consumers who are counting calories or grams of sugar. But for Dr Pepper, it adds up to colossal savings. "That looks like nothing," Reisner said. "But it's a lot of money. A lot of money. Millions."

The soda that emerged from all of Moskowitz's variations became known as Cherry Vanilla Dr Pepper, and it proved successful beyond anything Cadbury imagined. In 2008, Cadbury split off its soft-drinks business, which included Snapple and 7-Up. The Dr Pepper Snapple Group has since been valued in excess of $11 billion.

. . .

"It's Called Vanishing Caloric Density."

At a symposium for nutrition scientists in Los Angeles on Feb. 15, 1985, a professor of pharmacology from Helsinki named Heikki Karppanen told the remarkable story of Finland's effort to address its salt habit. In the late 1970s, the Finns were consuming huge amounts of sodium, eating on average more than two teaspoons of salt a day. As a result, the country had developed significant issues with high blood pressure, and men in the eastern part of Finland had the highest rate of fatal cardiovascular disease in the world. Research showed that this plague was not just a quirk of genetics or a result of a sedentary lifestyle—it was also owing to processed foods. So when Finnish authorities moved to address the problem, they went right after the manufacturers. (The Finnish response worked. Every grocery item that was heavy in salt would come to be marked prominently with the warning "High Salt Content." By 2007, Finland's per capita consumption of salt had dropped by a third, and this shift—along with improved medical care—was accompanied by a 75 percent to 80 percent decline in the number of deaths from strokes and heart disease.)

Karppanen's presentation was met with applause, but one man in the crowd seemed particularly intrigued by the presentation, and as Karppanen left the stage, the man intercepted him and asked if they could talk more over dinner. Their conversation later that night was not at all what Karppanen was expecting. His host did indeed have an interest in salt, but from quite a different vantage point: the man's name was Robert I-San Lin, and from 1974 to 1982, he worked as the chief scientist for Frito-Lay, the nearly $3-billion-a-year manufacturer of Lay's, Doritos, Cheetos and Fritos.

Lin's time at Frito-Lay coincided with the first attacks by nutrition advocates on salty foods and the first calls for federal regulators to reclassify salt as a "risky" food additive, which could have subjected it to severe controls. No company took this threat more seriously—or more personally—than Frito-Lay, Lin explained to Karppanen over their dinner. Three years after he left Frito-Lay, he was still anguished over his inability to effectively change the company's recipes and practices.

By chance, I ran across a letter that Lin sent to Karppanen three weeks after that dinner, buried in some files to which I had gained access. Attached to the letter was a memo written when Lin was at Frito-Lay, which detailed some of the company's efforts in defending salt. I tracked Lin down in Irvine, Calif., where we spent several days going through the internal company memos, strategy papers and handwritten notes he had kept. The documents were evidence of the concern that Lin had for consumers and of the company's intent on using science not to address the health concerns but to thwart them. While at Frito-Lay, Lin and other company scientists spoke openly about the country's excessive consumption of sodium and the fact that, as Lin said to me on more than one occasion, "People get addicted to salt."

Not much had changed by 1986, except Frito-Lay found itself on a rare cold streak. The company had introduced a series of high-profile products that failed miserably. Toppels, a cracker with cheese topping; Stuffers, a shell with a variety of fillings; Rumbles, a bite-size granola snack—they all came and went in a blink, and the company took a $52 million hit. Around that time, the marketing team was joined by Dwight Riskey, an expert on cravings who had been a fellow at the Monell Chemical Senses Center in Philadelphia, where he was part of a team of scientists that found that people could beat their salt habits simply by refraining from salty foods long enough for their taste buds to return to a normal level of sensitivity. He had also done work on the bliss point, showing how a product's allure is contextual, shaped partly by the other foods a person is eating, and that it changes as people age. This seemed to help explain why Frito-Lay was having so much trouble selling new snacks. The largest single block of customers, the baby boomers, had begun hitting middle age. According to the research, this suggested that their liking for salty snacks—both in the concentration of salt and how much they ate—would be tapering off. Along with the rest of the snack-food industry, Frito-Lay anticipated lower sales because of an aging population, and marketing plans were adjusted to focus even more intently on younger consumers.

Except that snack sales didn't decline as everyone had projected, Frito-Lay's doomed product launches notwithstanding. Poring over data one day in his home office, trying to understand just who was consuming all the snack food, Riskey realized that he and his colleagues had been misreading things all along. They had been measuring the snacking habits of different age groups and were seeing what they expected to see, that older consumers ate less than those in their 20s. But what they weren't measuring, Riskey realized, is how those snacking habits of the boomers compared to themselves when they were in their 20s. When he called up a new set of sales data and performed what's called a cohort study, following a single group over time, a far more encouraging picture—for Frito-Lay, anyway—emerged. The baby boomers were not eating fewer salty snacks as they aged. "In fact, as those people aged, their consumption of all those segments—the cookies, the crackers, the candy, the chips—was going up," Riskey said. "They were not only eating what they ate when they were younger, they were eating more of it." In fact, everyone in the country, on average, was eating more salty snacks than they used to. The rate of consumption was edging up about one-third of a pound every year, with the average intake of snacks like chips and cheese crackers pushing past 12 pounds a year.

Riskey had a theory about what caused this surge: Eating real meals had become a thing of the past. Baby boomers, especially, seemed to have greatly cut down on regular meals. They were skipping breakfast when they had early-morning meetings. They skipped lunch when they then needed to catch up on work because of those meetings. They skipped dinner when their kids stayed out late or grew up and moved out of the house. And when they skipped these meals, they replaced them with snacks. "We looked at this behavior, and said, 'Oh, my gosh, people were skipping meals right and left,'" Riskey told me. "It was amazing." This led to the next realization, that baby boomers did not represent "a category that is mature, with no growth. This is a category that has huge growth potential."

The food technicians stopped worrying about inventing new products and instead embraced the industry's most reliable method for getting consumers to buy more: the line extension. The classic Lay's potato chips were joined by Salt & Vinegar, Salt & Pepper and Cheddar & Sour Cream. They put out Chili-Cheese-flavored Fritos, and Cheetos were transformed into 21 varieties. Frito-Lay had a formidable research complex near Dallas, where nearly 500 chemists, psychologists and technicians conducted research that cost up to $30 million a year, and the science corps focused intense amounts of resources on questions of crunch, mouth feel and aroma for each of these items. Their tools included a $40,000 device that simulated a chewing mouth to test and perfect the chips, discovering things like the perfect break point: people like a chip that snaps with about four pounds of pressure per square inch.

To get a better feel for their work, I called on Steven Witherly, a food scientist who wrote a fascinating guide for industry insiders titled, "Why Humans Like Junk Food." I brought him two shopping bags filled with a variety of chips to taste. He zeroed right in on the Cheetos. "This," Witherly said, "is one of the most marvelously constructed foods on the planet, in terms of pure pleasure." He ticked off a dozen attributes of the Cheetos that make the brain say more. But the one he focused on most was the puff's uncanny ability to melt in the mouth. "It's called vanishing caloric density," Witherly said. "If something melts down quickly, your brain thinks that there's no calories in it . . . you can just keep eating it forever."

As for their marketing troubles, in a March 2010 meeting, Frito-Lay executives hastened to tell their Wall Street investors that the 1.4 billion boomers worldwide weren't being neglected; they were redoubling their efforts to understand exactly what it was that boomers most wanted in a snack chip. Which was basically everything: great taste, maximum bliss but minimal guilt about health and more maturity than puffs. "They snack a lot," Frito-Lay's chief marketing officer, Ann Mukherjee, told the investors. "But what they're looking for is very different. They're looking for new experiences, real food experiences." Frito-Lay acquired Stacy's Pita Chip Company, which was started by a Massachusetts couple who made food-cart sandwiches and started serving pita chips to their customers in the mid-1990s. In Frito-Lay's hands, the pita chips averaged 270 milligrams of sodium—nearly one-fifth a whole day's recommended maximum for most American adults—and were a huge hit among boomers.

The Frito-Lay executives also spoke of the company's ongoing pursuit of a "designer sodium," which they hoped, in the near future, would take their sodium loads down by 40 percent. No need to worry about lost sales there, the company's CEO, Al Carey, assured their investors. The boomers would see less salt as the green light to snack like never before.

There's a paradox at work here. On the one hand, reduction of sodium in snack foods is commendable. On the other, these changes may well result in consumers eating more. "The big thing that will happen here is removing the barriers for boomers and giving them permission to snack," Carey said. The prospects for lower-salt snacks were so amazing, he added, that the company had set its sights on using the designer salt to conquer the toughest market of all for snacks: schools. He cited, for example, the school-food initiative championed by Bill Clinton and the American Heart Association, which is seeking to improve the nutrition of school food by limiting its load of salt, sugar and fat. "Imagine this," Carey said. "A potato chip that tastes great and qualifies for the Clinton-A.H.A. alliance for schools. . . . We think we have ways to do all of this on a potato chip, and imagine getting that product into schools, where children can have this product and grow up with it and feel good about eating it."

Carey's quote reminded me of something I read in the early stages of my reporting, a 24-page report prepared for Frito-Lay in 1957 by a psychologist named Ernest Dichter. The company's chips, he wrote, were not selling as well as they could for one simple reason: "While people like and enjoy potato chips, they feel guilty about liking them. . . . Unconsciously, people expect to be punished for 'letting themselves go' and enjoying them." Dichter listed seven "fears and resistances" to the chips: "You can't stop eating them; they're fattening; they're not good for you; they're greasy and messy to eat; they're too expensive; it's hard to store the leftovers; and they're bad for children." He spent the rest of his memo laying out his prescriptions, which in time would become widely used not just by Frito-Lay but also by the entire industry. Dichter suggested that Frito-Lay avoid using the word "fried" in referring to its chips and adopt instead the more healthful-sounding term "toasted." To counteract the "fear of letting oneself go," he suggested repacking the chips into smaller bags. "The more-anxious consumers, the ones who have the deepest fears about their capacity to control their appetite, will tend to sense the function of the new pack and select it," he said.

Dichter advised Frito-Lay to move its chips out of the realm of between-meals snacking and turn them into an ever-present item in the American diet. "The increased use of potato chips and other Lay's products as a part of the regular fare served by restaurants and sandwich bars should be encouraged in a concentrated way," Dichter said, citing a string of examples: "Potato chips with soup, with fruit or vegetable juice appetizers; potato chips served as a vegetable on the main dish; potato chips with salad; potato chips with egg dishes for breakfast; potato chips with sandwich orders."

In 2011, the *New England Journal of Medicine* published a study that shed new light on America's weight gain. The subjects—120,877 women and men—were all professionals in the health field, and were likely to be more conscious about nutrition, so the findings might well understate the overall trend. Using data back to 1986, the researchers monitored everything the participants ate, as well as their physical activity and smoking. They found that every four years, the participants exercised less, watched TV more and gained an average of 3.35 pounds. The researchers parsed the data by the caloric content of the foods being eaten, and found the top contributors to weight gain included red meat and processed meats, sugar-sweetened beverages and potatoes, including mashed and French fries. But the largest weight-inducing food was the potato chip. The coating of salt, the fat content that rewards the brain with instant feelings of pleasure, the sugar that exists not as an additive but in the starch of the potato itself—all of this combines to make it the perfect addictive food. "The starch is readily absorbed," Eric Rimm, an associate professor of epidemiology and nutrition at the Harvard School of Public Health and one of the study's authors, told me. "More quickly even than a similar amount of sugar. The starch, in turn, causes the glucose levels in the blood to spike"—which can result in a craving for more.

If Americans snacked only occasionally, and in small amounts, this would not present the enormous problem that it does. But because so much money and effort has been invested over decades in engineering and then relentlessly selling these products, the effects are seemingly impossible to unwind. More than 30 years have passed since Robert Lin first tangled with Frito-Lay on the imperative of the company to deal with the formulation of its snacks, but as we sat at his dining-room table, sifting through his records, the feelings of regret still played on his face. In his view, three decades had been lost, time that he and a lot of other smart scientists could have spent searching for ways to ease the addiction to salt, sugar and fat. "I couldn't do much about it," he told me. "I feel so sorry for the public."

"These People Need a Lot of Things, but They Don't Need a Coke."

The growing attention Americans are paying to what they put into their mouths has touched off a new scramble by the processed-food companies to address health concerns. Pressed by the Obama administration and consumers, Kraft, Nestlé, Pepsi, Campbell and General Mills, among others, have begun to trim the loads of salt, sugar and fat in many products. And with consumer advocates pushing for more government intervention, Coca-Cola made headlines in January by releasing ads that promoted its bottled water and low-calorie drinks as a way to counter obesity. Predictably, the ads drew a new volley of scorn from critics who pointed to the company's continuing drive to sell sugary Coke.

One of the other executives I spoke with at length was Jeffrey Dunn, who, in 2001, at age 44, was directing more than half of Coca-Cola's $20 billion in annual sales as president and chief operating officer in both North and South America. In an effort to control as much market share as possible, Coke extended its aggressive marketing to especially poor or vulnerable areas of the U.S., like New Orleans—where people were drinking twice as much Coke as the national average—or Rome, Ga., where the per capita intake was nearly three Cokes a day. In Coke's headquarters in Atlanta, the biggest consumers were referred to as "heavy users." "The other model we use was called 'drinks and drinkers,'" Dunn said. "How many drinkers do I have? And how many drinks do they drink? If you lost one of those heavy users, if somebody just decided to stop drinking Coke, how many drinkers would you have to get, at low velocity, to make up for that heavy user? The answer is a lot. It's more efficient to get my existing users to drink more."

One of Dunn's lieutenants, Todd Putman, who worked at Coca-Cola from 1997 to 2001, said the goal became much larger than merely beating the rival brands; Coca-Cola strove to outsell every other thing people drank, including milk and water. The marketing division's efforts boiled down to one question, Putman said: "How can we drive more ounces into more bodies more often?" (In response to Putman's remarks, Coke said its goals have changed and that it now focuses on providing consumers with more low- or no-calorie products.)

In his capacity, Dunn was making frequent trips to Brazil, where the company had recently begun a push to increase consumption of Coke among the many Brazilians living in favelas. The company's strategy was to repackage Coke into smaller, more affordable 6.7-ounce bottles, just 20 cents each. Coke was not alone in seeing Brazil as a potential boon; Nestlé began deploying battalions of women to travel poor neighborhoods, hawking American-style processed foods door to door. But Coke was Dunn's concern, and on one trip, as he walked through one of the impoverished areas, he had an epiphany. "A voice in my head says, 'These people need a lot of things, but they don't need a Coke.' I almost threw up."

Dunn returned to Atlanta, determined to make some changes. He didn't want to abandon the soda business, but he did want to try to steer the company into a more healthful mode, and one of the things he pushed for was to stop marketing Coke in public schools. The independent companies that bottled Coke viewed his plans as reactionary. A director of one bottler wrote a letter to Coke's chief executive and board asking for Dunn's head. "He said what I had done was the worst thing he had seen in 50 years in the business," Dunn said. "Just to placate these crazy leftist school districts who were trying to keep people from having their Coke. He said I was an embarrassment to the company, and I should be fired." In February 2004, he was.

Dunn told me that talking about Coke's business today was by no means easy and, because he continues to work in the food business, not without risk. "You really don't want them mad at you," he said. "And I don't mean that, like, I'm going to end up at the bottom of the bay. But they don't have a sense of humor when it comes to this stuff. They're a very, very aggressive company."

When I met with Dunn, he told me not just about his years at Coke but also about his new marketing venture. In April 2010, he met with three executives from Madison Dearborn Partners, a private-equity firm based in Chicago with a wide-ranging portfolio of investments. They recently hired Dunn to run one of their newest acquisitions—a food producer in the San Joaquin Valley. As they sat in the hotel's meeting room, the men listened to Dunn's marketing pitch. He talked about giving the product a personality that was bold and irreverent, conveying the idea that this was the ultimate snack food. He went into detail on how he would target a special segment of the 146 million Americans who are regular snackers—mothers, children, young professionals—people, he said, who "keep their snacking ritual fresh by trying a new food product when it catches their attention."

He explained how he would deploy strategic storytelling in the ad campaign for this snack, using a key phrase that had been developed with much calculation: "Eat 'Em Like Junk Food."

After 45 minutes, Dunn clicked off the last slide and thanked the men for coming. Madison's portfolio contained the largest Burger King franchise in the world, the Ruth's Chris Steak House chain and a processed-food maker called AdvancePierre whose lineup includes the Jamwich, a peanut-butter-and-jelly contrivance that comes frozen, crustless and embedded with four kinds of sugars.

The snack that Dunn was proposing to sell: carrots. Plain, fresh carrots. No added sugar. No creamy sauce or dips. No salt. Just baby carrots, washed, bagged, then sold into the deadly dull produce aisle.

"We act like a snack, not a vegetable," he told the investors. "We exploit the rules of junk food to fuel the baby-carrot conversation. We are pro-junk-food behavior but anti-junk-food establishment."

The investors were thinking only about sales. They had already bought one of the two biggest farm producers of baby carrots in the country, and they'd hired Dunn to run the whole operation. Now, after his pitch, they were relieved. Dunn had figured out that using the industry's own marketing ploys would work better than anything else. He drew from the bag of tricks that he mastered in his 20 years at Coca-Cola, where he learned one of the most critical rules in processed food: The selling of food matters as much as the food itself.

Later, describing his new line of work, Dunn told me he was doing penance for his Coca-Cola years. "I'm paying my karmic debt," he said.

Note

This article was adapted from *Salt Sugar Fat: How the Food Giants Hooked Us*, published by Random House in 2013.

25 The Gender Binary Will Not Be Deprogrammed: Ten Years of Coding Gender on Facebook

Rena Bivens

Gender is a significant and powerful social category as well as a fundamental aspect of personal identity. Although gender is thought by many to be a simple binary—one is either male or female—that simplistic notion has been called into question in recent years, and many individuals now understand their gender identity in much more complicated ways. Though not often recognized, gender has always affected and been affected by various technologies. Think of the association between masculinity and technology and the paucity of women in engineering. Or think about the gendered division of domestic labor and the consequent association between lawn-care tools and men and kitchen tools and women. Or, with the Weber reading in mind, consider the way tools, machines, and vehicles may be designed to fit men's or women's bodies. This means, among other things, that as ideas about gender change, so do technologies, and as technologies change, so do ideas about gender.

In this article Rena Bivens traces the ways in which the social media giant Facebook has dealt with changing concepts of gender identity over a ten-year period of time. Her account reveals the intricacies of the intertwining of the social category—gender—with the technical coding and operations of the Facebook system. Facebook is a sociotechnical system, and in responding to its users, Facebook has, over time, changed the settings for gender. But Facebook is not just a company responding to its users. In fact, as many have pointed out, Facebook's users are not exactly its customers; Facebook's customers—those who pay for its services—are advertisers. In order to achieve its own marketing needs and those of its customers, Facebook has to keep gender a binary. Thus, the complexities of Facebook as a business organization are reflected in its technical features that, ultimately, affect how people are identified—and perhaps even how they see themselves.

Introduction

On 13 February 2014, mainstream news organizations reported a change to the popular social media site, Facebook. Instead of two options for users to choose from when identifying their gender ("male" and "female"[1]), users were given a third option ("custom") that, if selected, offered 56 additional options. A few examples include agender, gender non-conforming, genderqueer, non-binary, and transgender (Goldman, 2014). These options are dependent on a user's selected language and were initially rolled out only for the English (US) version of the site, which any user can select to gain access.[2] Before confirming a "custom" gender selection, users are required to select a preferred

From *New Media and Society* 19, no. 6 (2017): 880–898.

pronoun: "he," "she," or "them." Reactions have ranged from cautious optimism and joy to surprise, confusion, and mockery (Ferraro, 2014; Jones, 2014). Many lesbian, gay, bisexual, transgender, and queer (LGBTQ) organizations have praised practical implications for non-binary users, while several news anchors and anonymous commenters have instead sought to reassert the hegemony of the gender binary.

This software modification represents one tweak during a decade of iterating. Some changes are detectable through the user interface while others operate "under the hood," embedded in software elements that are not as readily accessible. This is not the first time that gender-related concerns have led to modifications. For example, during the 2011 controversy over the existence of "rape joke" pages, the company eventually tweaked their algorithms to remove ads from the offending pages (Stevens, 2011). The modifications detailed in this article draw attention to a broader socio-cultural context in which ideological struggles take place over social constructions of gender. These struggles have very real consequences: people who do not conform to a binary of masculine and feminine are disproportionately affected by discrimination and violence.[3] As Butler (2004) puts it, "This violence emerges from a profound desire to keep the order of binary gender natural or necessary, to make of it a structure, either natural or cultural, or both, that no human can oppose, and still remain human" (35).

While many queer theorists have focused on the code of language to explore these dynamics, a "resistance to the regimes of the normal" (Warner, 1993, xxvi) can also be applied to the code of software. By extending queer theory to the realm of software, the power of Facebook's code can be interrogated as yet another structural arena through which social life is regulated. This article maps the ways in which design decisions related to gender become embedded and materialized in Facebook, becoming powerful, in Foucault's (1982) sense, as a productive force in the broader software-user relationship (Bucher, 2012). From this Foucauldian perspective, we can explore how software can produce the conditions for gendered existence. Facebook's software configures, constructs, and attempts to impose a menu of gender identities (Nakamura, 2002) onto the users it interacts with. These users can also resist and hack these configurations. Ultimately, users and software designers mutually shape these programmed configurations of gender, severing and opening up possibilities for gendered life. The litany of other human actors who shape these interactions—programmers who wrote the code, superiors who managed design decisions, advertisers who desire increasingly granular data, and many other stakeholders—become specters in this software-user relationship, invisible on the front-stage, graphic user interface displayed by the software.

This analysis is restricted to a 10-year history, beginning with Facebook's original release in 2004 and ending with the 2014 custom gender settings. I demonstrate that the relationship between Facebook's software and its users is deeply structured by the gender binary while simultaneously productive of non-binary possibilities. The binary exists and does not exist at the same time. Considering both surface and deep software levels (the graphic user interface and the database), Facebook's software has always existed somewhere between a rigid gender binary and fluid spectrum.[4] This is emblematic of the complexity of Facebook as a sociotechnical artifact. The programmatic possibility of stretching outside of the binary has always been materialized in the code, but as this analysis will show, the binary continues to dominate and regulate 10 years later, and it is Facebook's business model that influences this binary-driven design strategy.

Methods

To examine Facebook's user interface as a historical artifact, I collected screenshots from different iterations of Facebook ranging from 2004 to 2014. Online image-based search engines were used for this purpose (including Google Images, Yahoo Image Search, and Flickr) since Facebook is inaccessible through archival engines like the Wayback Machine. Search terms included Facebook, thefacebook (the original name of the software), sign-up, register, profile, news feed, mini-feed, new, re-design, change, specific years (2004–2014), gender, pronoun, hack, and trans*. Similar online searches were conducted to explore news reports from 2004–2014 that detailed changes to Facebook's user interface, protests from the queer and trans community, and information about monetization strategies over time. An academic literature search for "thefacebook" also offered information about mandatory fields from early iterations. My searches focused on instances where gender was displayed and/or assigned to users through sign-up pages, profiles, and the news feed.

Access to the database was more complicated. In 2006, Facebook became the first major social media service to open limited access via its Application Programming Interface (API) (Yadav, 2006). APIs are software-to-software interfaces that allow third-party developers to interact with a site so that they can create programs that access, share, and exchange information. Facebook's use of open APIs has been financially motivated, geared toward "achiev[ing] market dominance and user dependency" (Bodle, 2011, 335). In 2011, Facebook released new tools intended for third-party developers (and increasingly exploited by marketers) to ease navigation of the programmatic interface. By using the Graph API Explorer tool, I was able to query the database to gain information and make inferences about how gender is stored.

The following analysis and descriptions about the internal mechanisms of Facebook are based on my own exploration of Facebook's software in conjunction with online archival research. To confirm technical aspects of the analysis, I also conducted a telephone interview with Lexi Ross, a Project Manager at Facebook who was involved in the custom gender project. The analysis begins with a discussion of how gender is coded, resulting sociotechnical problems, and how programming decisions relate to monetization strategies. The next section, "Designing Non-mandatory Gender in Year Zero and Custom Gender in Year Ten," compares the non-mandatory gender design in 2004 with the custom gender project released in 2014. I then demonstrate how the binary has dominated design decisions with access to non-binary possibilities increasingly restricted during this 10-year history. Next, I explore how users have resisted Facebook's control by hacking their gender, followed by a discussion of surveillance, authenticity, and interoperability.

Coding Gender, Sociotechnical Problems, and Monetization

Just as there is more than one way to conceptualize gender in society, there is more than one way to code gender in software. Science and technology studies, software studies, and critical code studies have richly illuminated the many ways in which technological design is a social and political act. With the help of these scholars, we have come to see technology as "never merely technical or social" (Wajcman, 2010, 149). Since "code is never found; it is only ever made, and only ever made by us" (Lessig, 2006, 6), it is clear that "lines of code are not value-neutral" (Marino, 2006). Nakamura (2002) offers a compelling analysis of the programmed limitations of how race can operate online, concluding that "if it can't be clicked, that means that it functionally can't exist" (120). Yet, as this analysis shows, it might continue to exist within deeper levels of software, even while rendered invisible on the surface. At the same time, technology is constantly in flux. Each software iteration incorporates different code, which means it is always possible to expand the conditions for existence. In this sense, technology succumbs to "an 'ambivalent' process of development suspended between different possibilities" (Feenberg, 2002, 15).

In our non-binary world, choosing to code gender as a binary echoes the societal status quo and is in line with other practices that "code" gender, such as sex or gender identification on surveys and official documents. When restricted to a binary, all of these practices erase non-conforming genders and create sociotechnical problems in the process. It is technically (and legally) impossible for a non-binary user to register for a service that demands mandatory binary gender identification. If the user submits the form with a blank gender field, the software—in this case, Facebook—is programmed to reject the submission, demanding that the user "select either male or female." Having likely encountered similarly frustrating scenarios many times before, the user may resolve the technical error by misrepresenting their gender. Yet, in doing so the user violates Facebook's Terms of Service. Facebook's Statement of Rights and Responsibilities is flagged immediately above the "Sign Up" button: "By clicking Sign Up, you agree to our Terms." Section 4, titled Registration and Account Security, requests that real names and information are provided, and 4.1 explicitly states: "You will not provide any false personal information on Facebook" (Facebook, 2013).

While the spirit of the Terms is up for interpretation, Terms are subject to change, and (un)intentional violations occur, Facebook has been heavy-handed in its search for the "authentic selves" (Associated Press, 2014) and "real names" of its users. Yet, "real name" policies do not work for the queer community or Native American communities,[5] but they do work for Facebook since there is a direct, inverse relationship between fake accounts and financial success. When Facebook announced that the estimated 5–6% of fake accounts detailed with the company's initial public offering (IPO) on 18 May 2012 had grown to 8.7% (by 30 June 2012), Facebook's stock dropped to less than $20 (from $38 three months earlier) and the company faced a lot of criticism (Rushe, 2012; Tavakoli, 2012). In part, this is because Facebook's marketable product is a user base of "real" people that can be targeted with the help of increasingly granular data. Leading up to the IPO, Facebook's (2012) prospectus, filed with the US Securities and Exchange Commission, highlighted "authentic identity" as the first of three elements forming "the foundation of the social web":

> Authentic identity is core to the Facebook experience, and we believe that it is central to the future of the web. Our terms of service require you to use your real name and we encourage you to be your true self online, enabling us and Platform developers to provide you with more personalized experiences. (2)

By positioning the social web as the future of business and defining it for businesses, Facebook's owners want to secure the "future of the web" and their place within it. Gender is included as one of four examples of how Facebook (2012) "creates value" for advertising and marketing clients (3).

In more public-facing spaces, Facebook's rhetoric about authenticity becomes more about morality. Facebook's (former) Chief Privacy Officer, Chris Kelly, once argued that "trust on the Internet depends on having identity fixed and known" (Kirkpatrick,

2010, 16) and Facebook creator Mark Zuckerberg has said that "having two identities for yourself is an example of a lack of integrity" (Zimmer, 2010). These attempts at regulating identity erase and delegitimize the many authentic experiences of people who question their identity, people with identities that change over time, and people who depend on aliases for safety.[6] Ultimately, this regulatory regime forecloses everyone's capacity to inhabit fluid identities.

Designing Non-mandatory Gender in Year Zero and Custom Gender in Year Ten

This 10-year analysis begins and ends with two important design decisions. In February 2004, Facebook's software was programmed with: a genderless sign-up page, a non-mandatory, binary field on profile pages, and three possible values for storing gender in the database. By February 2014, each of these elements had been modified: a mandatory, binary gender field on the sign-up page; a mandatory, non-binary field on profile pages; and four possible values for storing gender in the database. Both of these snapshots include software layers regulated by the binary and others that generate non-binary possibilities.

The early, 2004 design decision that programmed gender as non-mandatory on Facebook's profile pages created an important fissure for non-binary possibilities. At a deep level of the software, in the database, Facebook's gender field type was originally programmed to accept more than two values: 1 = female, 2 = male, and 0 = undefined. While a zero is inadequate in many ways, it is still a value beyond the binary of ones and twos. From a user's perspective—looking only at the user interface, not the database—the only non-binary option was to leave the field blank. This coding practice grants validity to binary genders while erasing non-binary genders, but it also produces conditions that allow for existence outside of the binary. The material reality of three accepted values in the database transgresses a rigid binary, yet falls short of a fluid spectrum, positioning the database somewhere in-between.

Coding a field type as non-mandatory is a design decision based on whether the data being collected are vital to the functioning of the software or, in the case of Facebook, the functioning of the company. Design decisions for profit-oriented companies encapsulate broader monetization strategies. In other words, if information about gender is advantageous for Facebook's monetization strategies, there will be pressure to code it as mandatory. Yet, there is always a trade-off between annoying users and enforcing data collection by making it mandatory. At this early stage of Facebook's development, data about gender were not considered vital to the functioning of the software or the company. In contrast, my interview with Ross reveals Facebook's contemporary, profit-focused, view: "Gender is a fundamental part of the product" (February 27, 2014).

By 2008, gender had been added as a mandatory, binary field on the sign-up page. Even in the February 2014, iteration—when the company finally capitulated to user demands for more gender options by reprogramming profile pages—the mandatory, binary field remained on the sign-up page. Meanwhile, deep in the database, users who select custom gender options are re-coded—without their knowledge—back into a binary/other classification system that is almost identical to the original 2004 database storage programming. The 2014 custom gender project offers the illusion of inclusion since surface changes to profile pages mask the binary regulation that continues underneath, at a deeper level of the software. Drawing on Foucault and Butler's insights, we see that conditions for gendered existence beyond the binary are activated on the software's surface. Yet, underneath the surface, these conditions are severed in favor of the binary. The design strategy that generates these conditions simultaneously reconfigures gender into data that conforms to the hegemonic regime embraced by marketing and advertising institutions. By actively employing divergent gender schemas within these two software levels, users and clients are satisfied simultaneously. Consequently, Facebook exercises power over its users by invisibly re-inscribing the binary. This technique maintains public-facing progressive politics while bolstering hegemonic regimes of gender control.

To explore this in more detail, consider the updated profile pages. In 2014, it is noteworthy that "custom" appeared as a third option, positioned only in relation to a normalized binary (McNicol, 2013). The binary is inscribed as dominant and "normal" while any "other" genders are positioned somewhere else in the hierarchy, only visible after the user clicks on "custom." Upon typing in the "custom" text-field, a list of possible gender options is revealed. Users can select more than one, which resolves Nakamura's (2002) critique of menu-driven identities that deny the programmatic capacity to straddle more than one clickable category, rendering intersectional identities unintelligible.

In the database, however, the code forces users back into a binary logic. To explain this finding, I will revisit Facebook's Graph API Explorer. My use of this tool involved navigating to the online website for Graph API Explorer, signing in to my Facebook account, retrieving an access token, and selecting fields to explore (see figure 25.1). To test how custom gender options are stored, I selected gender, along with identity (ID) and name fields to help determine which user was being tested. I also manipulated a test account. Each selected field became part of a "get" request that I submitted to obtain a response to my database query. When I queried the names and genders of my Facebook "friends" and test account, information was returned in the format displayed in figure 25.1.

Through these queries, it became clear that the database was programmed to store gender based on a user's pronoun, not the gender they selected. For instance, a user who selects "gender questioning" and the pronoun "she," will be coded as "female" in the database despite having selected "gender questioning." A query will identify the user's gender as "female" (see figure 25.2). The pronouns "he" and "she" equate to male and female, but when querying a user with the pronoun "them," only name and ID are returned without any information about gender—as if the user has no gender at all (see figure 25.3). The gender field actually turns gray, as seen in figure 25.3.

All users are reassigned as male, female, or custom (or retain an undefined value). With the February 2014 iteration, the value 6 became operational (equating to custom),[7] yet 0 (undefined) and 6 are indistinguishable since neither displays gender information

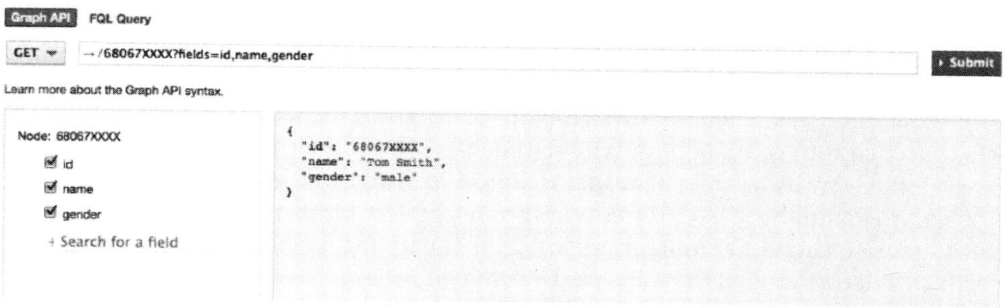

Figure 25.1
Example query using Facebook's Graph API Explorer Tool.

Figure 25.2
Example query with "she" pronoun selected.

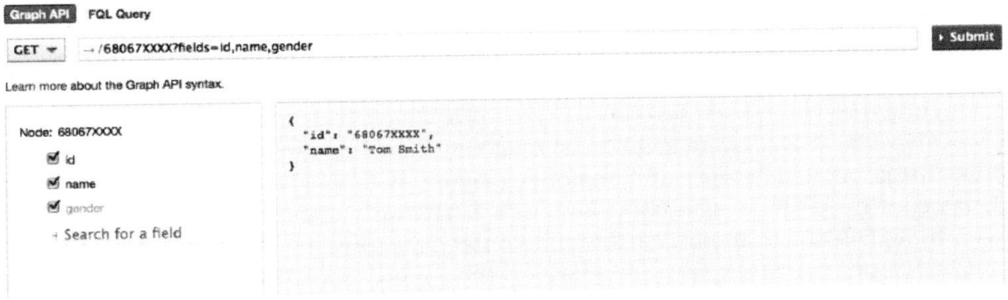

Figure 25.3
Example query with "them" pronoun selected.

when querying the database. Since undefined and custom effectively collapse into a null category from the perspective of API users, this storage system is nearly identical to how gender has been coded since the original 2004 iteration of Facebook's software. Facebook's software effectively begins and ends this decade in the same way: regulated by a binary logic but productive of non-binary possibilities. By 2014, however, a more marketable and "authentic" (yet, paradoxically, misrepresented) data set is produced, as I analyze in more detail later.

Binary by Design: Restricting Access to Non-binary Possibilities

The original design decision to program gender as a non-mandatory field eventually became a thorny issue. As Facebook grew up—as a social network, a company, and an advertising hub—gender became an increasingly valuable data point. Monetization strategies became more sophisticated and design strategies revolving around gender turned interventionist. To move the user base more fully toward the binary, one might expect a strategic re-design of the entire user interface to a mandatory binary, and the removal of undefined values in the database, restricting viable gender values to ones and twos. In 2008, two software modifications attempted to accomplish these goals: (1) the shift to a mandatory, binary sign-up page and (2) a special request for users with an undefined gender to select a binary gender.

Year	Sign-Up Page		
	Gender Field	Description	Mandatory
2004	No	N/A	N/A
2005			
2006			
2007			
2008	Yes	"I am: Select Sex/Male/Female" (drop-down list)	Yes
2009			
2010			
2011			
2012			
2013	Yes	"male" & "female" (radio buttons)	Yes
2014			

Figure 25.4
Time line of gender-related changes to Facebook's sign-up page.

While there have been several changes to Facebook's sign-up page over the past 10 years,[8] the first 4 years were genderless. By 2008, "I am" appeared, followed by a drop-down list populated by male and female (see figure 25.4). The field was mandatory and it has continued to be mandatory ever since. The only significant modification has been to replace the drop-down list with two radio buttons.

While the persistence of the binary on the sign-up page in 2014 despite the new custom gender options is puzzling, it highlights the continuing tension between the software's production of binary and non-binary conditions for existence. From a queer theory lens, we see the software's production of gendered subjects in distinct spaces and the co-existence of multiple gendered subjects as individual users are morphed by binary and non-binary affordances. This tension facilitates Facebook's relationships with both users and advertising clients, but at the software's roots, Facebook continues to comply with society's hegemonic norms.

With the release of the custom gender project, Facebook representatives declared that the company "want[s] you to feel comfortable being your true, authentic self" (Facebook Diversity, 2014). Director of Growth, Alex Shultz, said, "It was simple: not allowing people to express something so fundamental is not really cool so we did something. Hopefully a more open and connected world will, by extension, make this a more understanding and tolerant world" (Associated Press, 2014). I asked Facebook's Ross about this inconsistency, to which she replied, "There are some complex issues with the sign-up page but it's something we can consider in the future" (February 27, 2014). While we can speculate about the nature of these complex issues, ultimately user registration is the first moment when Facebook can police the "authenticity" of its users, satisfying investors by limiting fake accounts. This verification process continues to be regulated by the gender binary.

Beyond these binary-driven modifications to the sign-up page, profile pages have undergone far more changes over this decade long history. Profile pages fundamentally structured Facebook's original 2004 design: navigating from one profile page to the next was the predominant user activity. In 2006, the software was re-designed to highlight user activities. "Mini-feed" and "news feed" were introduced and gendered pronouns were eventually added to describe the user activities that populated these feeds. For instance, "Tom commented on his photo." To deal with users with an "undefined" gender, the software was programmed to use the pronoun "them." This "solution" was formally revisited in 2008, which brings us to the second major binary-driven software modification related to gender.

Forcing a database into a rigid binary by removing pre-existing (undefined) zero values is not an easy feat. In fact, from a practical perspective, it becomes increasingly difficult to make modifications to a database as it grows, entangles, and becomes mutually dependent on multiple software processes. With each new user registration, the database expands while undefined values continue to accumulate in the gender field. Facebook's "solution" involved targeting undefined users and asking them to select a binary pronoun. Yet, the consequences of selecting a binary pronoun were concealed (re-coding gender from 0 to 1 or 2 in the database and obstructing future access to non-binary programmatic possibilities).

This is how it happened. On 27 June 2008, a post on Facebook's company blog noted growth in non-English users and pronoun translation problems. The neutral "them" pronoun was deemed grammatically problematic: "Ever see a story about a friend who tagged 'themselves' in a photo? 'Themselves' isn't even a real word" (Gleit, 2008). As an aside, the singular "they" has an extensive history in the English language (Santos, 2013), and is commonly used in trans and queer communities along with ze, zir, and other non-binary pronouns. There was also concern expressed for users who may be misgendered in some languages since a neutral pronoun is unavailable. Since undefined users do not provide gender-related data, the software uses a default that is not based on any specific details about the user: "People who haven't selected what sex they are frequently get defaulted to the wrong sex entirely in Mini-Feed stories" (Gleit, 2008). Of course, selecting "sex" is only possible if one's "sex" is programmed as one of the selections, which means non-binary users have no option but to be "defaulted to the wrong sex entirely." Yet, interestingly, Facebook also recognized problems presented by the gender binary:

> We've received pushback in the past from groups that find the male/female distinction too limiting. We have a lot of respect for these communities, which is why it will still be possible to remove gender entirely from your account, including how we refer to you in Mini-Feed. (Gleit, 2008)

Yet, Facebook's design decisions did not extend programmatic possibilities beyond the binary nor move to a genderless design (remove gender as a category altogether). Either of these decisions could have offered a more respectful solution to this socio-technical problem. "Removing gender entirely" equated to hiding gender from the surface level while retaining a gender value in the database. Even this binary-driven compromise was incomplete, with leaks occurring in unexpected places. For example, the pre-populated labels "son" or "daughter" appear when familial relationships are expressed between users. Non-binary alternatives are unavailable and users cannot select the label themselves.

Ultimately, Facebook's "solution" involved prompting undefined users to indicate a preferred (binary) pronoun:

> We've decided to request that all Facebook users fill out this information [about their "sex"] on their profile. If you haven't yet selected a sex, you will probably see a prompt to choose whether you want to be referred to as "him" or "her" in the coming weeks. (Gleit, 2008)

Shortly following this announcement, a user posted a screenshot of this prompt, received upon log-in (httf, 2008), as seen in figure 25.5.

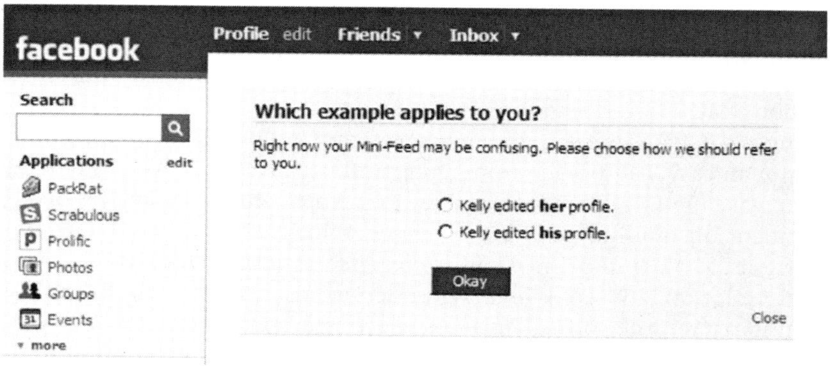

Figure 25.5
Request to select gendered pronoun, 2008.

Targeting users who have previously decided not to offer gender data, requesting that data under vague circumstances (obfuscating the consequences by drawing attention to a "confusing" pronoun), and positioning the binary as the only way forward is ethically suspect, concealing the surveillance and opaque data collection practices that accompany binary selection.

Two programmatic consequences were also obscured. Selecting "her" or "his" equated to: (1) binary gender assignment in the database, and (2) restricted access to the full range of database values. To explain the latter, it is important to understand that design changes related to gender that took effect in 2008 created a de-facto two-tiered user database. I will refer to the subsets as "legacy users" and "binary-ID users." A legacy user meets the following requirements: joined the site prior to 27 July 2008, and, at the precise moment when the software was identifying which users to prompt, had (1) an undefined gender selected, and opted to (2) refuse a binary pronoun, selecting "close" instead. For various reasons, and at any time, users might alter the gender field on their profile, but a user's legacy status would be revoked if they chose to binary-ID after the binary-driven 27 July 2008 pronoun request. A revoked legacy status meant assignment to the binary-ID user tier. Binary-ID users were produced by the software if they fulfilled one of the following requirements: (1) joined prior to 27 July 2008 and selected male or female on their profile at the precise moment when the software targeted undefined users for pronoun prompts; (2) joined prior to 27 July 2008, had an undefined gender selected, but responded to the software's prompt with a binary pronoun selection; or (3) joined after gender became a field on the sign-up page (and thus forced to binary-ID).

To reiterate, legacy users maintained the non-binary zero that has always been available as a programmatic possibility, along with its affordances, while binary-ID users were obstructed from non-binary existence. Suddenly, being a legacy user mattered: users who joined prior to 2008 could maintain their pre-intervention selections (such as "undefined" gender), even if those selections no longer existed in new software iterations. At the same time, a legacy user's power over the software (and company) was ultimately precarious since their status would expire upon selecting a male or female gender.

The invisible consequences of selecting "her" or "his" as an undefined user in 2008 only make sense in the context of a design strategy that sought to reduce undefined database values. Targeting a set of users and requesting data is not a decision that a company takes lightly. This intervention was part of a broader, binary-driven design strategy that afforded Facebook greater control over the conditions under which gendered subjects could be produced. It strengthened binary regulation and reduced (Facebook's definition of) "inauthentic" users. Along with the mandatory, binary gender field on the sign-up page, these 2008 modifications operated as a productive force geared toward normalizing the gender binary.

Resisting Control by Hacking Gender

Despite these binary-driven design strategies, there was a loophole that was highly dependent on the early-2004 decision to code gender as a non-mandatory field. In defiance of Facebook's regulatory regime, a hack was developed. As Galloway (2006) writes of hackers, "They care about what is true and what is possible. And in the logical world of computers, if it is possible then it is real" (168). The non-binary value materialized in Facebook's software represented an important possibility to be exploited. While designers of social media software have the greatest capacity to exercise power over the production of gendered subjects in and through their coding of gender, Feenberg (2005) argues that "subordinate groups may challenge the technical code with impacts on design as technologies evolve" (47). The gender hack represented both a challenge and an important resistance to the binary regulation imposed by Facebook's design.

The experience of Facebook user Rae Picher offers a useful illustration of how the gender hack worked.[9] As Picher explains in a public post on 27 April 2011, "I recently lost my carefully preserved genderless status on Facebook due to an April Fools' Day joke where I came out as a heterosexual woman." As a legacy user, when Picher selected "female," the software replaced the 0 associated with Picher's user ID in the database with a value of 1. This simple click erased Picher's legacy status. Picher (2011) explains what happened next:

> When I tried to switch BACK to not having my gender identified, Facebook threw a hissy fit and demanded that I binary-gender ID for them, and proceeded to use gendered pronouns for me on my wall and in my friends' news feeds. Now that's just not cool.

Luckily, Picher discovered an online video that taught users how to hack their gender. Most web browsers offer access to developer tools including one that interrogates and manipulates the source code of rendered web pages. When using this tool, a user is shown the code related to the page displayed in the browser, as seen in figure 25.6.

The bottom portion of figure 25.6 is the Web Inspector tool. This hack required users to navigate to the "edit basic info" page on the mobile Facebook website through a desktop computer. The tool exposes hyper text markup language (HTML) code from Facebook's profile editing page. Users would then edit the HTML code to add a third option associated with the value 0 (labeled "Hack My Gender" in figure 25.6, but any text works), which would then become selectable. As a result, users could override Facebook's obstruction and store a 0 in the database.

This hack directly challenged the binary-driven design strategies previously articulated. Hacking allowed users to obtain (or regain) legacy status, with all of its attendant affordances, particularly the capacity to exist outside of the binary. Prior to the 2014 release of the custom gender project, this technical loophole had been patched by Facebook's programmers and is no longer functional. Permitting the growth of undefined users

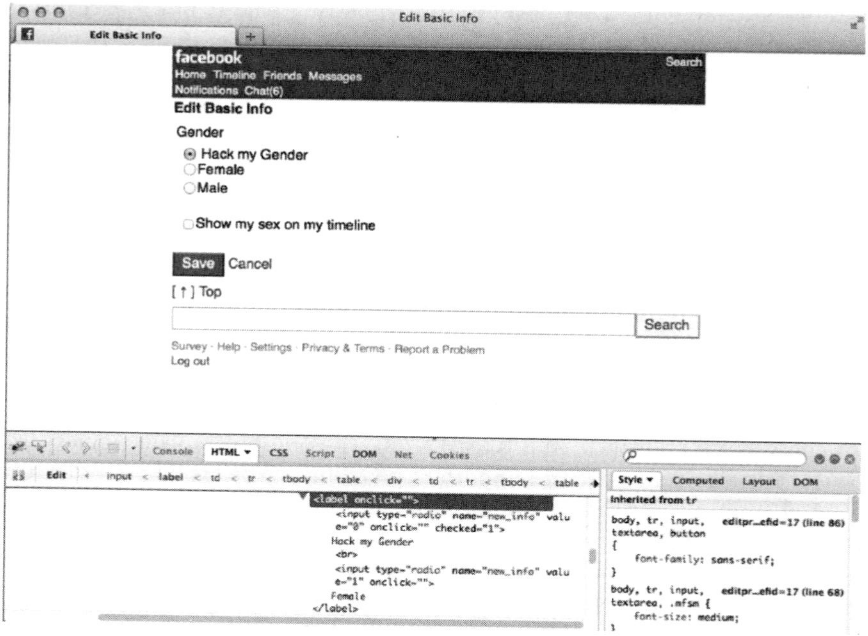

Figure 25.6
Hacking gender using HTML code.

was antithetical to the company's business model. Clearly, the "anti-commercial bent" of the hacking community (Galloway, 2006) was at odds with Facebook's profit-orientation.

More Invisible Layers: Surveillance, Authenticity, and Interoperability

The host of invisible consequences for undefined, legacy users are reminiscent of the opaque/transparent nature of code that Chun (2013) explains as "invisibly visible, visibly invisible" (15). Not only is code hidden from view, but it also requires a level of technical expertise to comprehend. According to Chun, computing's appeal rests on its "combination of what can be seen and not seen, can be known and not known," which "makes it a powerful metaphor for everything we believe is invisible yet generates visible effects" (Chun, 2013, 17). The 56 new gender options have already generated visible effects, such as public discourse on the topic of non-binary genders, and the capacity for users to more accurately represent their gender. Yet, this representation does not exist without data collection, which is typically "framed as being valuable" for users even though the system also benefits (McNicol, 2013, 203). "These sites want to know *what you are* so they can best figure out *what they can sell you*" (Nakamura, 2002, 116; original italics). Increasingly granular data about niche markets is conducive with a business model that is dependent on selling these markets to advertisers, marketers, and developers. Documenting and surveilling vulnerable populations are largely invisible practices—with unknown data beneficiaries—that include a great deal of risk. Even when someone willingly discloses their non-binary identity, danger can be lurking: "There are people out there that target trans/gn-c [gender non-conforming] people, and having that publicly listed on your profile can set you apart as a target"

(Haimson et al., 2015, 1184). Privacy settings accompanying custom gender demand savvy and organized users who take the time to group their network into categories, yet the system does not permit preferred pronouns to be private. The custom gender project has received many critiques, including the argument that a braver move for Facebook would involve a genderless design, particularly since data about gender is of little value apart from marketing contexts (Fae, 2014; McNicol, 2014).

Another invisible process with important implications is the computational reclassification of custom gender selection on the user interface back into what amounts to a binary (1, 2) system in the database—with custom (6) as a limited non-binary transgression, and an undefined category (0) populated by (a shrinking number of) legacy users. This programming decision reifies hegemonic gender norms that directly oppose the company's public statements ostensibly supporting the dismantling of this same hegemonic regime. Facebook's deep binary regulation of gender through code does (invisible) work in the world, just as Butler's (1997) "psychic operation of the norm" (21) permits the gender binary to (invisibly) act on society. Facebook's advertising clients are clear beneficiaries of this arrangement: the custom gender modification and the addition of the value 6 assists the company's creation of a data set calibrated toward the needs of advertisers who flock to Facebook for their "authentic" and highly granulated users. Facebook can now at least presume to more accurately classify their users' gender.

Facebook's early design decisions lumped together everyone who left the gender field blank, regardless of their reason, and classified the rest as a binary. Ten years later, any remaining undefined users in the database are legacy users who have resisted all programmed attempts to collect their gender data. They are the only users who retain the value 0. The number of undefined users must be quite limited and, given Facebook's rhetoric, they are deemed inauthentic. The value 6, on the other hand, captures only a sub-set of non-binary users: any custom gender user who selects the pronoun "them." As a result, all binary users and all remaining custom gender users (anyone with a binary "she" or "he" pronoun preference) can now be easily classified as ones and twos—"females" and "males." Indeed, no other users have a gender from the perspective of the 2014 API. As a result, Facebook can now supply advertisers, marketers, and developers with a tidy (albeit misrepresented) set of female, male, and non-binary users. News reports have noted that "ads will be targeted based on the pronoun [users] select for themselves" (Associated Press, 2014), which maps perfectly onto the "he," "she," "them" pronoun system, even though "them" is not currently accessible to advertisers.

Along with reducing the set of allegedly inauthentic users and improving Facebook's capacity to market its user base by gender (despite the misgendering that results), the custom gender project can also be read as an exercise in public relations. When I asked Facebook's Ross about the absence of significant change to the coding of gender in the database, it became clear that financially motivated relationships between Facebook's software and external websites and services that access, share, and exchange information are of great value. As Ross notes,

> Most of our third-party apps of course do not support custom gender and we wanted to really make it a frictionless experience for those developers. . . . Basically it was sort of a decision for simplicity sake to not break other parts of the product. (Interview, February 27, 2014)

While workarounds could be introduced in future API releases, Ross highlights Facebook's fear that a fundamental change to the storage of gender could disrupt current interoperability requirements, thereby sabotaging important financial relationships.

Conclusion

This analysis of the materiality of antagonistic constructions of gender in social media software offers important opportunities for nuanced and dialectic insights into the "invisibly visible," shallow/deep capacities for the production and enactment of power in and through software-user relationships and the regulation of social life through code and design decisions. Despite the addition of 56 gender options in February 2014, the gender binary has not been deprogrammed from Facebook's software. The software-user relationship continues to be deeply structured by the gender binary at the same time that it is productive of non-binary possibilities: from the genderless sign-up page turned binary and mandatory; to the permanent presence of non-binary possibilities in the database versus the cumulative, binary-driven design strategy that impeded, reduced, and patched access to those possibilities; to the forward-facing custom gender project that reconfigures gender into three insincere, but marketable, categories based on preferred pronouns. Overall, Facebook's software exists somewhere in-between a rigid binary and fluid spectrum. Yet, within this liminal space, and at a deep level, Facebook's software normalizes a binary logic that regulates the social life of users. The conditions for binary existence are easily produced while any meaningful non-binary existence is severed, even though the capacity to move beyond the binary has always been a programmatic possibility.

Within this 10-year history, "authentic" representation of a user's gender identity reaches a peak with the 2014 custom gender project. Yet, the conditions for this non-binary existence are restricted to the surface of the software (and continue to be denied on the sign-up page where the binary remains an important regulator in the user verification process). Inauthenticity looms large in the deeper level of the database through the misgendering of custom gender users who select a binary pronoun and, as a result, are produced by the software as "female" or "male" instead of the custom option they selected. Paradoxically, Facebook's rhetoric and business model re-interprets these inauthentic, misgendered users as highly marketable, "authentic," and "real" while rendering the only users who have managed to escape Facebook's binary-driven design interventions—limited in numbers as these legacy users must be—as inauthentic. In the end, authenticity does not have to be authentic to be financially viable, as long as your clients perceive it as authentic.

Given that Facebook continues to dominate the social media industry, at least in the United States (Duggan et al., 2015), the company's design choices, coding practices, and business model are well-positioned to influence new start-up ventures and as such are important sites of critique. While more research is needed that critically examines how gender, race, and other salient social categories are produced within both surface and deep layers of software, it is clear, in this case, that Facebook has actively governed the formation of its users as gendered subjects. From a Foucauldian perspective, embedding a hegemonic regulatory regime constitutes a technique of power. By invisibly re-classifying non-binary users into a binary-based gender schema within the database and employing this reconfigured data set as a mechanism of income generation and connectivity, Facebook secures "the desire to keep the order of binary gender natural or necessary" (Butler, 2004, 35).

To be clear, the issue at hand is not supplying advertisers and marketers with better data about gender. Since corporate data collection comes with serious risks, including surveillance of marginalized populations, our efforts should not be geared toward creating more "authentic" and "real" data sets by programming more inclusive (and granular) categories on surface or deep software levels. Facebook's attempt to ally with trans and gender non-conforming communities resulted in programming practices that actively misgender them. This misgendering reinforces hegemonic regimes of gender control that perpetuate the violence and discrimination disproportionately faced by these communities. The capacity for software to invisibly enact this symbolic violence by burying it deep in the software's core is the most pressing issue to attend to.

Acknowledgments

I am indebted to the Feminism and Social Media Research workshop organized by Libby Hemphill, Ingrid Erickson, Ines Mergel, and David Ribes for the 2014 Computer Supported Cooperative Work and Social Computing conference and the initial experimentation using Facebook's Graph API Explorer that J. Nathan Matias and Marie Gilbert conducted during the workshop. I also thank Oliver Haimson and the very helpful feedback from anonymous reviewers, as well as the programmers in my life, Christian Holz, Rocky Bivens, and Uri Bivens, who helped me explore the code and think through programmatic possibilities. All images are from www.facebook.com.

Funding

The author(s) received no financial support for the research, authorship, and/or publication of this article.

Notes

1. Facebook's software offers no affordance for distinguishing between sex (sexual organs, represented as male, female, and intersex) and gender (feelings and expressions associated with gender identity). Both terms appear interchangeably within the user interface and policy documents over time.

2. Custom gender remains under development and has been incrementally released for other languages with varying sets of gender options. As of June 2014, English (UK) offered over seventy options; English (US) was modified to a free-form text field in February 2015. Meanwhile, Français (Canada) still offers a mandatory binary as of July 2015.

3. Transgender people, particularly people of color, continue to be disproportionately represented in homicide and hate violence statistics (NCAVP, 2015).

4. A fluid spectrum can be crudely understood as a continuum between masculinity and femininity, including every shade of masculine-femininity and feminine-masculinity, along with genders existing closer to the center (such as genderqueer) and gender-questioning identities. It also represents possibilities not yet fully imagined or embodied.

5. Recent implementations of Facebook's "real name policy" have involved deactivation of user accounts, requests for legal names, and insistence on photographic ID as evidence to reactivate accounts. While Facebook spokespeople cite safety as a concern, the queer and drag queen community cite the ability to identify in ways that differ from "legal" identities as essential to their safety (Sylvan, 2014). These policies and the algorithms that determine inauthentic names have also disproportionately targeted Native Americans (Holpuch, 2015).

6. Consider victims of sexual abuse, people who keep their sexual orientation private from family or colleagues, and people who have careers that require anonymity.

7. Facebook's Ross explained that, although "not 100% certain," the value 6 is an artefact of not custom gender but various kinds of projects that have been worked on in the past that ended up using up those other constants. So it wasn't by design; it was just by default that it was the next number we could use. (Interview, February 27, 2014).

8. Originally restricted to Harvard students, Facebook's 2004 sign-up page included name, student status, email address, and password. By September 2006, anyone over the age of thirteen with a valid email address could join. In 2007, "birthday" became a mandatory field. By 2008, "status" (previously modified to allow nonstudents to join) had been removed.

9. I obtained permission from Picher to include this experience.

References

Associated Press. 2014. "Facebook expands gender options: Transgender activists hail 'big advance.'" *Guardian*, February 14, 2014. Accessed February 14, 2014. http://www.theguardian.com/technology/2014/feb/13/transgender-facebook-expands-gender-options.

Bodle, R. 2011. "Regimes of sharing: Open APIs, interoperability, and Facebook." *Information, Communication & Society* 14 (3): 320–337.

Bucher, T. 2012. "The friendship assemblage: Investigating programmed sociality on Facebook." *Television & New Media* 14 (6): 479–493.

Butler, J. 1997. *The Psychic Life of Power*. Stanford, CA: Stanford University Press.

Butler, J. 2004. *Undoing Gender*. New York: Routledge.

Chun, W. H. K. 2013. *Programmed Visions: Software and Memory*. Cambridge, MA: MIT Press.

Duggan, M., Ellison, N. B., Lampe, C., et al. 2015. "Social media update 2014." Pew Research Center, January 2015. Accessed July 9, 2015. http://www.pewinternet.org/2015/01/09/social-media-update-2014/.

Facebook. 2012. "Form S-1 registration statement." Accessed July 6, 2015. https://infodocket.files.wordpress.com/2012/02/facebook_s1-copy.pdf.

Facebook. 2013. "Statement of rights and responsibilities." Facebook, November 15, 2013. Accessed February 20, 2014. https://www.facebook.com/legal/terms.

Facebook Diversity. 2014. "Timeline photo." Facebook, February 13, 2014. Accessed July 7, 2015. https://www.facebook.com/photo.php?fbid=567587973337709.

Fae, J. 2014. "Facebook should remove all gender options instead." *Guardian*, February 16, 2014. Accessed July 8, 2015. http://www.theguardian.com/technology/shortcuts/2014/feb/16/facebook-should-remove-all-gender-options.

Feenberg, A. 2002. *Transforming Technology: A Critical Theory Revisited*. New York: Oxford University Press.

Feenberg, A. 2005. "Critical theory of technology." *Tailoring Biotechnologies* 1 (1): 47–64.

Ferraro, R. 2014. "Facebook introduces custom gender field to allow users to more accurately reflect who they are." GLAAD, February 13, 2014. Accessed July 1, 2015. http://www.glaad.org/blog/facebook-introduces-custom-gender-field-allow-users-more-accurately-reflect-who-they-are.

Foucault, M. 1982. "The subject and power." In *Michel Foucault: Beyond Structuralism and Hermeneutics*, edited by H. Dreyfus and P. Rabinow, 208–226. Chicago: University of Chicago Press.

Galloway, A. 2006. *Protocol: How Control Exists after Decentralization.* Cambridge, MA: MIT Press.

Gleit, N. 2008. "He/She/They: Grammar and Facebook." Facebook, June 27, 2008. Accessed March 2, 2014. https://www.facebook.com/notes/facebook/heshethey-grammar-and-facebook/21089187130.

Goldman, R. 2014. "Here's a list of 58 gender options for Facebook users." *ABC News Blogs*, February 13, 2014. Accessed February 25, 2014. http://abcnews.go.com/blogs/headlines/2014/02/heres-a-list-of-58-gender-options-for-facebook-users/.

Haimson, O., Brubaker, J. R., Dombrowski, L., et al. 2015. "Disclosure, stress, and support during gender transition on Facebook." In *Proceedings of ACM Conference on Computer-Supported Cooperative Work and Social Computing.* New York: Association for Computing Machinery.

Holpuch, A. 2015. "Native American activist to sue Facebook over site's 'real name' policy." *Guardian*, February 19, 2015. Accessed February 19, 2015. http://www.theguardian.com/technology/2015/feb/19/native-american-activist-facebook-lawsuit-real-name.

httf. 2008. "Facebook's gender blunder." *Livejournal* (blog), July 10, 2008. Accessed February 25, 2014. http://httf.livejournal.com/43728.html.

Jones, A. 2014. "Facebook's new gender options." *Storify*. Accessed July 4, 2015. https://storify.com/JonesAmberM/facebook-gender-options (page no longer available).

Kirkpatrick, D. 2010. *The Facebook Effect.* New York: Simon & Schuster.

Lessig, L. 2006. *Code: Version 2.0.* New York: Basic Books.

Marino, M. C. 2006. "Critical code studies." *Electronic Book Review*, December 4, 2006. Accessed February 22, 2014. http://www.electronicbookreview.com/thread/electropoetics/codology.

McNicol, A. 2013. "None of your business? Analyzing the legitimacy and effects of gendering social spaces through system design." In *Unlike Us Reader: Social Media Monopolies and Their Alternatives*, edited by M. Rasch and G. Lovink, 200–219. Amsterdam: Institute of Network Cultures.

McNicol, A. 2014. "A critique of Facebook's gender diversity." *exhipigeonist blog*, February 13, 2014. Accessed May 11, 2014. http://exhipigeonist.net/tag/gender/.

Nakamura, L. 2002. *Cybertypes: Race, Ethnicity, and Identity on the Internet.* New York: Routledge.

NCAVP. 2015. *Lesbian, Gay, Bisexual, Transgender, Queer, and HIV-Affected Hate Violence in 2014.* Accessed June 30, 2015. http://avp.org/storage/documents/Reports/2014_HV_Report-Final.pdf.

Picher, R. 2011. "Facebook's gender binary got you down?" Facebook User Note, April 27, 2011. Accessed February 27, 2014. https://www.facebook.com/notes/rae-picher/updatefacebooks-gender-binary-got-you-downupdate/10150166319923922.

RINJ. 2011. "RINJ campaign declares victory over the "Alleyway." *Wire Service*, November 4, 2011. Accessed June 15, 2015. http://www.wireservice.ca/index.php?module=News&func=display&sid=6887.

Rushe, D. 2012. "Facebook share price slumps below $20 amid fake account flap." *Guardian*, August 3, 2012. Accessed July 6, 2015. http://www.theguardian.com/technology/2012/aug/02/facebook-share-price-slumps-20-dollars.

Santos, S. R. 2013. "Let's talk about the history of gender pronouns (and gender-neutral pronouns) in English." *SaintRidley*, September 22, 2013. Accessed January 12, 2014. http://saintridley.kinja.com/lets-talk-about-the-history-of-gender-and-pronouns-an-1365242291.

Stevens, A. B. 2011. "Dear Facebook: Rape is no joke." *Ms.*, September 19, 2011. https://msmagazine.com/2011/09/19/dear-facebook-rape-is-no-joke/.

Sylvan, J. 2014. "Dear Facebook: This is why your new 'real name' policy hurts queers like me." *Washington Post*, September 22, 2014. Accessed September 22, 2014. http://www.washingtonpost.com/posteverything/wp/2014/09/22/dear-facebook-this-is-why-your-new-real-name-policy-hurts-queers-like-me/.

Tavakoli, J. 2012. "Facebook's fake numbers: 'One billion users' may be less than 500 million." *Huffington Post*, July 12, 2012. Accessed July 6, 2015. http://www.huffingtonpost.com/janet-tavakoli/facebooks-fake-numbers-on_b_2276515.html.

Wajcman, J. 2010. "Feminist theories of technology." *Cambridge Journal of Economics* 34 (1): 143–152.

Warner, M. 1993. *Fear of a Queer Planet: Queer Politics and Social Theory*. Minneapolis: University of Minnesota Press.

Yadav, S. 2006. "Facebook: The complete biography." *Mashable*, August 25, 2006. Accessed February 26, 2014. http://mashable.com/2006/08/25/facebook-profile/.

Zimmer, M. 2010. "Facebook's Zuckerberg: "Having two identities for yourself is an example of a lack of integrity." *MichaelZimmer* (blog), May 14, 2010. Accessed September 20, 2014. http://www.michaelzimmer.org/2010/05/14/facebooks-zuckerberg-having-two-identities-for-yourself-is-an-example-of-a-lack-of-integrity/.

26 Audible Citizenship and Audiomobility: Race, Technology, and CB Radio

Art M. Blake

One of the famous quotes in the analysis of technology was coined by historian Melvin Kranzberg. He argued that "technology is neither good nor bad, nor is it neutral." This phrase can be interpreted in many ways, but one way to think about it is that although technologies may not be inherently good or bad, they are not necessarily innocent. The quote highlights how people can further their values in ways that would be impossible without technology. For instance, some of the atrocities inflicted on people over the past one hundred years would not have occurred without abilities enabled by new technologies. And of course, the corollary is true as well: some of the amazing riches we have experienced would not have happened without the power of technology.

This article is a great illustration of Kranzberg's quote. This article shows how two groups—African Americans and white supremacists—were able to use the same technology—CB radios—to mobilize and further their goals in ways they had not been able to before. The ability to talk to others at long distances allowed people to form groups, cement new identities, and advance social movements. Each group appropriated the technology to its own uses—whether it be to avoid the police during efforts to disenfranchise and torment African Americans or develop visions of freedom and ways to break free of institutional restraints. The story Blake tells is incredibly complicated. CB radios didn't simply liberate or discriminate. Rather, they allowed different groups to advance their goals in various ways and they became interwoven in both cultures.

In the mid-1970s, magazines from *Newsweek* to *Car and Driver* to *Popular Electronics* all carried full-page advertisements for various brands of citizens band (CB) radio equipment, promoting the era's craze for the simple two-way radio communications devices. Flipping through the pages of *Ebony* magazine in the same period, one might get the impression that CB was not part of black culture, that the fad had not hit black America. Popular movies of the day also depicted CB radio as the domain of white guys, mostly southerners, racing their cars or trucks across the United States in defiance of boneheaded sheriffs and cops. In fact, however, belying this inattention from manufacturers, advertisers, and Hollywood, CB radio had thrived in African American communities since the early 1960s. But African Americans used CB radio in a different manner and for different purposes than their white counterparts, and thereby created a system of mobility far removed from, and in opposition to, that of the white "good buddies" talking into their radios as they cruised America's roadways.

From *American Quarterly* 63, no. 3 (September 2011): 531–553.

This article argues that black CB use first developed in direct response to the racial politics of the postwar period, in particular, the years of struggle for meaningful desegregation and full citizenship. The emergence of a distinct black CB culture by the 1970s epitomizes how black use of CB, as a form of what I call "audiomobility," circumvented white prohibitions against black mobility and audibility, denied white assumptions of technical and verbal superiority, as well as internal black class politics around accent, vocabulary, and speech styles—about "sounding black." Black use of CB developed in the postwar period to counteract the immobilities produced by the specific circumstances of racism in the 1960s and 1970s. For example, black CB use responded to the technology's early adoption by the Ku Klux Klan in the south at the moment of segregation's faltering power, and worked in the 1970s to counteract the use of CB as an organizing tool by northern opponents of school busing. From these origins, as an audiomobility network counteracting racism directly as well as indirectly, black CB evolved into an intraracial competitive arena, a technocultural practice suited only to the audibly toughest competitors. Black CB's historical significance, despite its absence from scholarly studies until now, rests in its functioning as a nexus for, and challenge to, various histories: of the politics of black speech and oral culture; of the role of radio programming in the creation of black cultural identity; of race and technology. Black CB also shows how "community" and "identity" do not necessarily originate through direct contact and communication. As I show, its users built on an already existing black aural public sphere—rooted in black-interest radio, jive talk, and jazz and blues lyrics—and adapted CB technology to combine with that aural-oral sphere to connect geographically and socially diverse individuals. Black CB thus indirectly created a technologically mediated community based on perceived audible racial identity.[1]

Through "shooting skip" (communicating via CB over long distances), black CB created an audible black geography, offering a way to connect in real time, across cities and regions, black individuals and their localities—a networking opportunity especially important in the 1960s and 1970s when differences in regional cultures, practices, and histories of race and racism shaped black politics. The regional diversity of the audibly black voices and speech styles heard over CB radio confirmed to listeners the variety as well as the size of the black CB community and, by implication, of black America.[2]

Black CB signified a process of tuning in, with great skill, to an invisible disembodied shared blackness, a shared black sound, and a black technoculture. In a heavily commercial, corporate broadcasting context, black CB also signified the creation and demarcation of black radio space (black spectrum), the drowning out of the "white noise" of not only the dominant commercial culture but also of racism and segregation through distinctively black speech backed by necessarily self-sufficient black technical prowess. Black CB was a trickster act, cunningly and playfully "throwing" black voices freely back and forth across states, regions, the nation, and sometimes beyond—the radio hardware acting as ventriloquism's middleman. Unlike black bodies or even the words and actions of black political leaders, the immateriality of (black) sound eluded capture, control, or blockading. A technology used by whites in the civil rights and black power eras to enhance their security while enjoying their unrestricted mobility became, in the hands of blacks, a device that allowed them to throw off their physical and political immobility by mobilizing their sound, their voices, even while their bodies stayed unobtrusively in place.[3]

. . .

Black involvement in CB radio was, by the CB craze of the mid-1970s, far more long-standing than that of most whites. As early as 1959, African American CB radio enthusiasts founded the Rooster Channel Jumpers, a CB radio users organization that established a nationwide network of black CB users. With official chapters in major American cities in the north, south, and east, a formal governance structure at every level, and blue and gold uniforms for its members, the Rooster Channel Jumpers operated in the same manner as many popular postwar civil and fraternal organizations familiar to middle-class communities, and drew on the long-standing club tradition of the African American middle class. One can get an idea of the eventual size of the organization and the commitment of its members by noting that, in late June 1978, an estimated ten thousand African American CB radio enthusiasts gathered in Dallas, Texas, for the fifth annual convention of the Rooster Channel Jumpers. The keynote address, given by Dr. Berkeley Burrell, president of the National Business League (formerly the Negro Business League, founded by Booker T. Washington in 1900), urged all black CBers to join together in a national black radio operators organization to promote the use of CB channels specifically for black economic organizing and mutual benefit. Although it seems Burrell never met that goal, since it may have seemed redundant given the strength of the Rooster Channel Jumpers, he continued to promote the idea that "nearly sixty percent of the Black population in this country can be reached through this powerful medium" and that "Black CB'ers can develop sufficient political clout to have a major impact on political and economic decisions that are routinely affecting their lives."[4] Burrell's vision fit with some of the gains he and other business leaders had begun to wring from the federal government via President Richard Nixon's Office of Minority Business Enterprise (OMBE) and other programs created in the wake of the urban crisis of the late 1960s and early 1970s. But for many black CBers, even those at the Roosters convention, Burrell's vision clashed with their regular flouting of FCC regulations to achieve the distinct social and communications goals of black CB radio practice.

Black CBers made CB radio exciting and significant for themselves in the 1960s and 1970s by using their radio equipment to communicate across long distances—an activity referred to in CB slang as "shooting skip"—what ham radio enthusiasts have always called "DX-ing." By shooting skip, black CB operators sent their signals much farther than the 150 miles permitted for the citizens band by FCC regulations. "Skip" is a naturally occurring phenomenon whereby radio signals bounce off the ionosphere and back down to earth at a distance far greater than the normal field of the transmitter. To talk long distance, CBers use a directional antenna to deliberately bounce their signal so that it lands hundreds, sometimes thousands, of miles away. Shooting skip successfully requires favorable atmospheric conditions, the skills and technical know-how to build and use a directional (horizontal) antenna, luck, and patience. Base stations (home-based as opposed to mobile vehicle–based stations) running more power to boost their antenna's signal often dominate the process, but not necessarily. Stories abound of big base stations being "cut off" by smaller stations or even by someone in a car who can make the most of the "conditions." By flouting FCC regulations limiting the signal distance of the CB radio service, African American CBers joined together via an invisible and mostly unsurveilled communications network. This network of users, partly because of the skill required to talk over long distances, then developed its internal elite practitioners who, by the mid-1970s, took over CB channel 6 as their domain and named it the "superbowl."

The superbowl differs not just in sound but also in *purpose* to mainstream, mostly white, CB. Taking its name from what football enthusiasts see as the ultimate annual battle of champions—the Superbowl game between the year's top-ranked NFL teams—which began in 1967 and continues to shape football fan culture in the United States, the naming of the superbowl CB channel suggested it was the arena in which the best, most skilled, and toughest men would compete. No other CB channel bore such connotations. Superbowl enthusiasts do not call themselves "CBers"—they call themselves "superbowl operators" or "radio operators." Unlike white mainstream CBers, they do not focus on local, conversational exchanges with other CB enthusiasts. Instead, the superbowl works as an arena for not only shooting skip but also competing with other superbowl operators to reach a far-off third party. Who wins is determined not only through a technical contest to get one's signal heard but also through verbal dominance, and the ability to tease or insult the other operator(s) so effectively that they drop out of the exchange.

Although readers may know of CB radio via its popular-culture representations, a brief background on its development seems necessary.[5] Citizens band is a radio communications technology that provides free access to two-way radio communications, usually over relatively short distances, using channels set aside by the FCC, the U.S. government agency that (since the 1930s) has overseen and managed all forms of communications technologies—radio, television, telephone, and now cable, satellite, and the Internet.

Although the FCC initially offered citizens band radio to the American public in 1948, CB was not feasible for mass use until 1958 when the FCC assigned twenty-three channels for the service. The FCC intended CB to provide a way for motorists or isolated workers to contact their homes, their coworkers, or emergency personnel in times of need. It also offered a cheap, simple form of mobile-to-base and mobile-to-mobile communication for businesses reliant on deliveries or other out-of-office activity. The CB radio service was designed for brief vocal exchanges, not for personal conversations like those conducted over the telephone or in person.[6] Long-distance truck drivers accounted for about one-third of CB users before the CB radio craze of the mid-1970s.[7]

CB radios quickly achieved mass popularity in the United States during the mid-1970s as a direct result of the publicity surrounding the December 1973–January 1974 independent truckers' strike that followed the 1973–1974 OPEC oil embargo. News media reported how truckers used CB radios to contact each other and to form illegal convoys, disrupting traffic or blocking highways, or to share information about police and highway patrol activity, or about gas stations with available fuel.[8] Sales of CB radio sets grew from 2 million in 1974 to a peak of approximately 10 million in 1976. But by 1978 CB radio sales had dropped abruptly, signaling that, for most Americans, the CB craze had definitely waned.[9] Market saturation and, by 1978, the congestion of CB channels by large numbers of users contributed to the decline of the CB fad.

During the 1970s, popular and consumer culture quickly picked up on the CB fad. A wave of CB-related movies, television shows, and songs became popular in the United States between approximately 1974 and 1978. However, in contrast to its image in popular culture as the tool of rebellious trucker culture, working-and lower-middle-class white men driving ordinary passenger cars formed CB's largest market in the 1970s. By 1977 drivers of passenger vehicles outnumbered truckers as CB users by about five to one.[10]

Moreover, among its majority white users, CB operated not as a rebel's weapon but as a technology of white rescue.[11] The period of CB's greatest popularity was also one dominated by fears of black working-class mobility. A growing automobile-centered urban-suburban infrastructure shaped the American postwar urban landscape, which itself produced fears of black mobility after the urban riots of the late 1960s and early

1970s. Such fears echoed a larger national concern about keeping the generally whiter suburbs "safe" from the generally poorer, more black, inner cities. By the 1970s, freeways, beltways, and highways connected the outer-and inner-city areas of large U.S. metropolitan areas. An August 1976 *New York Times* article referred to a recent FBI report showing a 10 percent increase from the previous year in serious crime in suburbs nationwide. Apparently, suburbanites believed that "inner-city minority resident[s] who journey out of the city to find more affluent victims" had caused the problems. Statements from various police authorities supported the suburbanites' fears, suggesting that roadways connecting downtown areas to suburbs might become criminal conduits. For example, the article quoted Daryl F. Gates, then an assistant police chief in Los Angeles, as stating "anyone can jump in a car, get on the freeway, rob a house and be back in the inner city in an hour."[12]

Given American dependence on roadways, especially in metropolitan areas, newspaper stories about crime and violence on the freeways and beltways exacerbated a broad-based urban anxiety at play in the postwar American city. The race riots in U.S. cities during the 1960s firmly connected race, the city, crime, and violence in the minds of most Americans. During the same decade, larger postwar changes such as economic restructuring, urban renewal, and the growth of suburbs lent credence to the status of American cities as zones of crime and fear. By the late 1970s many white Americans had developed a heightened fear of crime against both person and property; CB radios seemed an ideal technological solution for these fears, especially during times of apparent isolation as a driver on the roadway.

From the perspective of African Americans, the late 1960s and the early 1970s posed a far greater threat to person and property given the rate of physical and verbal violent challenges to anyone associated with the civil rights movement. After the Watts riot of August 1965 and the assassination of Dr. Martin Luther King Jr. in April 1968, the situation of urban blacks only worsened and white opposition hardened as black nationalist and black power groups developed a more powerful presence in the movement for black civil rights.[13]

Given the overwhelming domination by the mid-1970s of most CB channels by white users, and the charged racial-political context of that and the preceding decade, it is not surprising to hear accounts of African Americans experiencing racism on CB channels. One current superbowl user, whose on-air "handle," or CB nickname, is "Prime Minister," whose involvement goes back to the mid-1970s when he was twelve years old, says he has heard stories from older men involved in black CB, suggesting that the distinctive slang and style of black CB evolved partly as a defense against over-the-air racism:

> When African American operators finally found somewhere they could talk to one another, they were constantly harassed by . . . they used to call them "appliance operators." . . . They were primarily southern guys, and . . . I remember being a kid listening to Maestro talking in skip and here comes one of these appliance operators, he's from Mississippi, he cuts off the guy Maestro's talking to in Texas, and he just goes on this racist rant, and you can hear Maestro going, "Look man, we left, we're not on your frequency, we came here, we're having fun," and this guy just, "You son of a bitch black bastard blah blah blah."[14]

As far back as the late 1960s, and certainly by the mid-1970s, as a result of racism on CB channels, increasing numbers of black CBers simply opted to gather on two underused channels in the CB range and thus hoped to avoid harassment by white CBers:

channel 5 became the place for local talk; channel 6 became reserved for skip—for talking over long distances. By the early 1970s, channel 6 had become known as "the superbowl," and its purpose, sound, and style came to define African American CB.

Historical traces of the superbowl and its participants are almost nonexistent. My research on the topic started from a single mention of the term "superbowl" in a 1976 *Ebony* article titled "10–4, Bro'." Current participants in the "bowl" form the main "archive" for this sonic subculture and possess any existing artifacts or ephemera related to its history. Since much of the activity of the superbowl, and the supporting technology, breaks FCC regulations, users have no publications or retail outlets that would make their activities easier to document. Superbowl operators are also understandably wary of sharing some aspects of their activities with outsiders.

"Most blacks use a different kind of lingo" on CB radio, the author of the *Ebony* article recounted, referring to what one could hear on channel 6. The article quoted a black CB radio user from St. Louis as saying, "If you can't talk the soul bro' talk, then you don't need to be on the superbowl."[15] But the article did not elaborate on how this channel differed from regular CB channels or on what the "soul bro' talk" actually sounded like. Perhaps readers of *Ebony* in 1976 were presumed to know what soul bro' talk sounded like.

Referencing soul bro' talk carried a specific cultural and political weight by 1976. To the mostly middle-class African American readers of *Ebony* magazine, that talk might not have been a style of speaking they engaged in, but it might have been a style they once adopted or aspired to, or one they objected to, depending on their political position concerning black identity and culture as manifested through black vernacular speech. The politics of "sounding black" and using "black" language stretched back more than a century by 1976, but the closest roots of the superbowl sound and style can be found in the development of black radio in the 1940s and 1950s. Through the 1960s the emergence of black nationalism and the assertion of "black pride" provided encouragement for African Americans to deploy more overtly and publically the type of black speech that had, until then, mostly remained spoken and heard only within black communities. The superbowl's racial exclusivity, its distinctive slang and verbal style, link it to the histories and controversies, inside and outside the African American community, about sounding black—a debate about distinctive black speech styles, black accents, and vernaculars going back to the late nineteenth century.[16]

Since emancipation, if not before, blacks who wanted to advance socially or educationally within the black community, and certainly those who wished to achieve any degree of social mobility among whites, learned to assimilate their speech style and their voices—to make an effort to "sound white."[17] What white sounded like depended on locality—local white accents, speechways, and vocabulary. And sounding white usually involved sounding middle class—sonic assimilation has always been as much about class as about race.[18]

Through the 1970s, during the peak of CB's popularity among blacks and whites, some articles in the black press criticized the sound of blacks on CB radio. A 1976 article in the *Chicago Defender* used a spoofed conversation between the author and an imagined old friend to compare the style and spoken content of black men on CB in the Chicago area to the sound and content of the characters on *The Amos 'n' Andy Show*, the radio and then TV show famous for promoting the supposed comedic value of uneducated black men. As the article's author, Bob Dixon, pointed out to his imagined interlocutor, *The Amos 'n' Andy Show* had been banned from the airwaves after pressure from black civil rights groups such as the NAACP. But the other man reminds him of how, when they were younger, living in the small town of Cairo, Illinois, they used to gather around the radio to listen to *The Amos 'n' Andy Show*. Then he adds: "You and I can hear that A & A talk and dialect any evening we choose—right here in Chicago!" The man then gives an example of what he means, reproducing what he says he hears when he tunes into blacks on CB radio in Chicago:

> "Cornhusker—calling Cornhusker. Dis is Chicken Plucker. Do you read me?" "Dass a tenfoe, Chicken Plucker. I reads you well, ovah?" "Say, you got all de ribs an' trimmins for de week-end, Ole Buddy, or is you still not going out dere wid us? Ovah?" "Uh, tenfoe Plucker; well, I dunno effen I kin make it dis week-end. My ole lady might hafta go someplace and she won't try to drive de car. Kin I git a ten-foe on dat?"[19]

In producing this imagined CB conversation, Dixon mocks southern black vernacular speech. He associates the two men, inhabitants of one of the nation's major cities, and arguably the most significant "black metropolis," with black rural life by giving them handles associated with menial agricultural labor—plucking chickens and husking corn. He has the participants discuss their plans for a weekend that involves not the commercial entertainments of the metropolis but a casual get-together involving southern food favorites such as "ribs an' trimmins." And he throws in the possibility that Cornhusker's apparently unsophisticated and killjoy wife, because of her refusal to drive the car, may derail the men's plans for such simple pleasures.[20]

In all, this and other articles voiced long-standing criticisms reportedly made by middle-and upper-middle-class blacks of working-class blacks: that the voices and speech styles of working-class blacks, and their vocal and social behaviors, brought discredit to the black community as a whole. James Grossman and Davarian Baldwin have documented the articulation of such tensions in Chicago when, on the arrival of poorer, rural blacks from the southern states during the first "great migration" immediately after World War 1, the "old settlers" struggled to find ways to assimilate and incorporate the new black population into the urban northern world they had made for themselves. As Baldwin in particular shows, the tensions between the old settlers and the newer migrants did not involve a straightforward class struggle. Much of the vibrancy of the new "black metropolis" came from the commercial and mass entertainments, leisure activities, and styles of dress and self-expression emanating from the working-class neighborhoods of the newer migrants. To form the black metropolis they desired, the old settler "New Negroes" of Chicago had to adapt their tastes and their version of a progressive black modernity to include such styles and pleasures, or risk historical (and racial) irrelevance.[21] So when, in the 1970s, Dixon and other gatekeepers of black middle-class identity heard the sounds of black CB radio, they heard the sounds of another migration—an unchecked migration of working-class black voices across the city, in and out of class-bound neighborhoods and cliques, and spreading out across state lines via the relatively free public airwaves.

. . .

The men who became active in black CB and later on the superbowl, from the late 1960s through the 1970s, some of whom are still active on the bowl today, grew up not only in the dying days of *The Amos 'n' Andy Show*, as suggested by the Dixon *Chicago Defender* article quoted above, they also grew up in the era of black-appeal radio stations and the growth of black-oriented radio programming, both musical and otherwise. Those men, some now in their seventies, born just before or during World War II, had access during their teenage years to the voices of black-sounding black DJs throughout the south as well as in the major urban centers of the north. They came into full adulthood as the civil rights movement faced the challenges of black nationalism and black power, and the sounds of political blackness encompassed a sonic and stylistic spectrum from the church-based oratory and cadences of Martin Luther King Jr. to the snappier rhythms of Malcolm X and Muhammad Ali, to the clipped urban jive and Marxist-Leninist staccato of Huey Newton.

The superbowl's slang (different from mainstream CB slang), in addition to a particular verbal style, has created, since the bowl's inception in the 1970s, an almost exclusively black communications zone, just as the predominance of white speech styles and the use of phrases and accents from the white south helped keep mainstream CB in the 1970s almost exclusively white.[22] The racial exclusivity of mainstream CB, although not usually explicitly stated, created the desire among African American CBers for what became the superbowl. Explaining the impact of a black CB channel for its participants and creators in the 1960s and early 1970s, superbowl aficionado Prime Minister told me:

> There were guys that were talking in the sixties . . . they predate the bowl being the bowl. They were trying to find somewhere they could fit. And when I listen to them tell the stories it's awesome. . . . You hear Big Motor going, "Yeah, I remember one day I came to channel 6 . . . and I heard a guy and said 'god, he sounds like a black guy.'" You . . . hear him say that and . . . that's like . . . starting humanity and crossing another human.[23]

Prime Minister's statement, with its image of two people, each thinking they were alone in the world, accidentally finding each other through the racial audibility of a radio channel, conveys what he has heard from older bowl operators to be the intense surprise, excitement, pleasure, and cultural importance of discovering the possibility of African American contact via CB radio, especially that taking place over the long distances made possible by the phenomenon of skip.

Prime Minister's anecdote is especially interesting since, in the 1960s, African Americans were not necessarily isolated demographically or even technologically given the existence of a well-established black press in many parts of the United States, as well as national and local radio broadcast networks and, for many living in urban areas, the affordability of telephone service. But clearly "finding" one another audibly via CB still held particular power for African American men using that technology. Their apparent sense of isolation, perhaps truly felt only after the fact, in the moment of skip contact, and the superbowl as thus a remedy for that condition, suggests there was something else at stake beyond simple contact.

When the U.S. government set aside the bandwidth for CB radio in 1948, it broke open a tiny part of the closely guarded and heavy commercialized American broadcast spectrum. The Radio Act of 1927 and the Communications Act of 1934 had secured American broadcasting as the preserve of commercial corporate networks, removing the possibility that U.S. broadcasting would go the way of the United Kingdom or Canada, nations that established state-supported and (at first) commercial-free radio (and later television) broadcasting. Those U.S. legislative acts privatized and commodified the public spectrum and squeezed out all but a very few opportunities for the American public to have access to the public airwaves as users; the vast majority of the public lost that access and were consigned to the passive role of "listeners," consumers of the commercial broadcasters' advertising-based product.[24]

It had not always been so. Between 1906 and 1917, after the development of crystal radio sets, and before U.S. involvement in World War I when the government asked amateur radio operators to suspend their broadcasts, diverse voices and other sounds populated the American airwaves.[25] Tuning your radio dial could bring you the sounds of many other Americans, using their crystal radio sets in their homes to send out Morse code signals, music, and (by about 1915) their voices. The world of amateur radio—what became known as "ham" radio—was active and extensive. Such users never anticipated in those early days that they would be so soon banned from their role as broadcasters, restricted to the weakest part of the spectrum, and prevented from broadcasting content such as music, news, and sports—the most popular content, which commercial broadcasters wanted to reserve for themselves. Organized through national organizations such as the Radio League of America and the American Radio Relay League (which is still in existence), these amateurs called their world of radio "citizen radio."[26]

Citizens' virtual expulsion from the airwaves, starting in the 1920s, makes it all the more extraordinary that the FCC decided in 1948 to set aside spectrum for the new "citizens band" radio channels.[27] By 1948 privately owned broadcasting corporations (CBS, ABC, and NBC) controlled the nation's airwaves. In 1958, with twenty-three channels newly available on the citizens band, the service became usable by the American public. CB radio could be used and operated by anyone—one did not need the tuning skills of the ham radio aficionados, and ready-to-use CB radio sets, for installation in one's vehicle or home, rapidly became available in electronics stores across the nation.

The timing of the availability of citizens band radio forms a crucial part of the story of its politicization and racialization. CB radio became available for mass use in 1958, just as the mainstream civil rights movement regularly captured headlines across the United States, and right before that movement began to shift toward a more "direct action" approach, with the emergence of sit-ins as a strategy to desegregate lunch counters (which began in Kansas in July and in Oklahoma in September 1958, but gained prominent attention in Greensboro, North Carolina, in 1960) and the freedom rides in the summer of 1961.

The coincidence of the availability of the CB radio service, the mass production and marketing of the technology, and the rise of the civil rights and later black power movements led to the rapid adoption of the technology by citizen groups on both sides of the desegregation and civil rights debates. According to an article in the *Washington Post*, the Ku Klux Klan began using CB radio in 1961. Klansmen used their radios to better organize their racial terror activities by reporting to each other on the whereabouts of law enforcement or of their latest targets. This report initiated a series of articles in the *Washington Post* in 1965 and 1966 revealing the extensive use of CB by the Ku Klux Klan and the FCC's efforts to prevent such usage. In May 1965, as the House Un-American Activities Committee prepared to begin hearings on Klan activities, the *Washington Post* ran a lengthy article on the modern "third" Klan, detailing the Klan's long history and how it had recently become more active in the face of increased black efforts for civil rights.[28]

Picking up on the coverage, the syndicated columnist Drew Pearson, in his "Washington Merry-Go-Round" column, wrote an article about the KKK's use of CB radio. "The public," Pearson wrote, "would be surprised to know . . . that Robert Shelton, Imperial Wizard of the United Klans, is licensed to operate on a special citizens radio wavelength. He was given the license by the Federal Communications Commission . . . at the same time . . . the Justice Department had placed earlier Klan organizations on the subversive list along with the Communist Party." Pearson added that the Klan held several CB licenses, "all under front names, such as the 'Alabama Rescue Service' of Tuscaloosa."[29]

After Pearson's article, the FCC chair E. William Henry initiated an investigation into the possible illegal use of CB radio by the Klan.[30] As the HUAC hearings into the KKK continued through 1966, other stories appeared in the *Washington Post* reporting more detailed accounts of how the Klan used CB radio to organize attacks on individual African Americans or on civil rights organizations.[31] Representative Edwin E. Willis, the chair of HUAC, proposed legislation in June 1966, toward the end of the HUAC hearings on the Klan, which sought to curb the Klan's ability to organize and commit terroristic acts. The bill included a proposal to outlaw the use of CB radios, walkie-talkies, and telephones in the commission of federal crimes or in an effort to prevent the detection of such crimes.[32]

In response to Klan activity and to other forms of racist harassment, blacks organizing for civil rights and freedoms also began using CB radio. For example, the African American organization Deacons for Defense and Justice, started in Jonesboro, Louisiana, in the summer of 1964 to provide armed protection to local blacks and to civil rights workers from the Congress for Racial Equality (CORE) based in the area, used CB radios to ensure a rapid response to any apparent or real threat. The Deacons carried their CB radios alongside their guns as part of an effective system of defense.[33]

Blacks' use of CB radio to counter white racism helped "turn hegemony on its head," to quote George Lipsitz, an act connected to long African American political tradition, rooted in strategies such as the acts of resistance to slavery documented in the work of George Rawick and Eugene Genovese and discussed by other scholars of race and ethnicity who have analyzed the counterhegemonic political strategies that have helped build political movements and embolden their actors or observers.[34]

The use of CB radio to facilitate grassroots social and political movements takes us to the questions of citizenship and "the public" at the heart of the meaning of CB radio. [. . .] The establishment of the "citizens band" radio service soon after the end of World War II created a small, public noncommercial space in the otherwise almost entirely privatized public space of broadcast spectrum in the United States. The access to, and use of, that service implied (and initially required) citizenship. Until 1978 the right to use CB radio depended on possession of a license obtained from the FCC. The license application form was simple, asking only for name and address, how many units the applicant wished to run, and if the applicant was a U.S. citizen. Therefore, to participate in the public sphere of CB radio meant laying claim to, and gaining recognition as possessing, citizenship and specifically the right of a citizen to speak and to be heard. Your citizenship was your qualification. What marks the parallel worlds of white and black CB as public *versus* counterpublic is based not only on the political economy of race in the postwar United States but also on the manner (and meaning) of *how* those two publics, those two citizen groupings, made use of CB's distinct communications technology to achieve different versions of audiomobility.

The rhetoric of CB radio in its heyday of the 1970s was replete with populist claims to "freedom" and "free expression"—the rights fundamental to the classical (white, male, bourgeois) liberal subject, the historical participant in the Habermasian public sphere and, the record shows, the main participant in America's mid-1970s CB radio craze. African Americans, in the 1960s and 1970s, still fighting for legal and quotidian recognition as full equal citizens, comprised what Nancy Fraser and others have called a "counterpublic."[35]

Fraser wrote, in her now-classic *Social Text* article, "The relations between bourgeois publics and other publics were always conflictual. Virtually from the beginning, counterpublics contested the exclusionary norms of the bourgeois public, elaborating alternative styles of political behavior and alternative norms of public speech. Bourgeois publics, in turn, excoriated these alternatives and deliberately sought to block broader participation."[36] As public and counterpublic, first-class citizen and second-class citizen, white and black CBers laid claim to the public sphere of the radio spectrum; whites did so to tweak their relationship to law and order, and to seek mutual protection on the public roadways; blacks did so to circumvent the law (of a federal government that had, at minimum, failed them) and the limits placed on their mobility by white citizens. As with other "counterpublics," that formed by black CBers was not a universal public sphere. Like white CB, the superbowl operated with de facto exclusivity, aimed at connecting and creating a black citizen public able to master the necessary technological, linguistic, and temperamental skills.

Like white CB, the superbowl created a specifically aural public sphere and counterpublic based on racial identity. The assertion and promotion of audible racial difference offers a counterargument to established notions of the construction of racial difference that rely on visuality. African American CBers interviewed for this article asserted that they participated in the superbowl because they experienced racism on CB channels used by the majority white demographic of CB radio channels. Clearly, therefore, both blacks and whites "heard" race—and listened for it. This sensory perception formed a crucial aspect to the popularity and use of this technology in the postsegregation United States.

We should recall here Prime Minister's reference to the older superbowl operator Maestro's account of his emotional epiphany on suddenly audibly encountering another black man on CB radio in the 1960s. In such moments, Maestro, and others like him, experienced a crucial shift in their relationship to those normally regarded as "strangers." CB radio, for both blacks and whites, transformed strangers into allies, members of a defined public, a community. This shift forms an essential aspect of the formation of a public (and therefore a counterpublic). As Michael Warner has written, in reference to the formation of publics and counterpublics:

> Publics orient us to strangers in a different way. They are no longer merely people-whomone-does-not-yet-know; rather, it can be said that an environment of strangerhood is the necessary premise of some of our most prized ways of being. Where otherwise strangers need to be placed on a path to commonality, in modern forms strangerhood is the necessary medium of commonality. The modern social imaginary does not make sense without strangers. A nation, market, or public in which everyone could be known personally would be no nation, market, or public at all.[37]

He adds: "The development of forms that mediate the intimate theater of stranger-relationality must surely be one of the most significant dimensions of modern history, though the story of this transformation in the meaning of the stranger has been told only in fragments."[38] For African Americans in the 1960s and 1970s, audibly black unknown others were heard not as strangers but as potential friends and allies—at a time of intense racialized violence and danger. For whites, that same transformation of "stranger" into "good buddy," a trustworthy ally on the public roadway, formed the necessary underpinning of the cultural and political work of CB for that demographic in the 1970s—a period when fear and suspicion of strangers ran high, especially in urban-suburban areas.[39]

While both CB publics, black and white, shared that important stranger-to-ally transformation in their uses of CB radio, black use of CB radio continued to evolve beyond that point, whereas most white users of CB during its mid-1970s heyday dropped the hobby as the airwaves became too crowded and too filled with interference. Black CB, having established an audible racial community whose audiomobility directly opposed the immobilities of racism and segregation, could then address its internal culture. The emergence of the superbowl on channel 6 represented the achievement of a confident black CB culture, one less focused on issues of self-defense or basic community building. The skills necessary to build and operate a powerful station with a directional antenna to shoot skip created a hierarchy within black CB. The superbowl evolved as an arena in which technological skills combined with vocal style and verbal power to face off in an ongoing championship of audible blackness.

In the competitive arena of the superbowl, the most popular and respected operators are not only those who can shoot skip to get their "beam" (signal) across a long distance to a specific area or operator but, more significantly, those whose talking style is most impressive, most domineering. Superbowl talk is based on "dissing" the person with whom one is competing to get one's signal picked up by another person. The two or more operators "fighting" from different parts of the United States to reach, say, an operator in another region, not only compete via the power of their signal, their skill in directing and tuning it, and dealing with weather conditions but also in their deployment of well-delivered insults, threats, and ridicule directed at each other as they compete to reach the distant operator in the triangle of communication. Each tries to so successfully "diss" the other that his competitor cannot verbally "one up" him, causing the other to bow out of the contest. Such verbal skills must be combined with the technical skills to dominate the transmission—to cut off or at least drown out one's competitor.

Then as now, superbowl operators who can dominate others in a seemingly effortless manner garner particular admiration. A powerful signal combined with the audible evocation of a confident masculine authority has built the reputation of operators such as 766, Nationwide 1000, Crack Carter, and others. Prime Minister refers to this dominant style and how he learned to appreciate it. As he says, "The style I kind of try to adhere to is 'shock and awe.'" He learned this style in the 1970s from an operator in California called Yellow Jacket, about whom he says, "If CB were a true family he'd be my dad . . . he taught me a whole lot, especially about style and talking." Prime described to me, in audible detail, the superbowl exchange through which, under Yellow Jacket's guidance, he first heard the style he later aimed to emulate.[40]

Documenting superbowl battles secures the proof of who beat whom—who, by whatever means, "won" on any particular occasion. On the bowl, one of the three (or more) people involved in competing to reach a particular operator will often record the exchange. These recordings are known in superbowl slang as "Watergates"—a nickname dating the origins of this aspect of the superbowl to the mid-1970s when President Nixon's taping system in the Oval Office documented his paranoid power games and ultimately provided proof of his guilt in the Watergate conspiracy. The superbowl instruction to "let the gate roll" means to roll the tape or to start recording.

At the large meetings of bowl operators, known as "breaks" or "shoot outs," participants spend a lot of time listening to Watergates of particularly entertaining exchanges. Here is a short clip from a Watergate of three operators trying to reach Outlaw 187, an operator in New Jersey.[41] Not only are the three operators (Crack Carter in Texas, and 444 and Bricklayer, both located in Mississippi) trying to reach 187, they are each trying to dominate the transmission so as to push the others out. The references to "conditions" and to "Mother Nature" indicate the atmospheric conditions they are trying to use; in addition, Outlaw 187 relates messages that he is asked to pass on to one or another of the competing operators, since CB radio requires an on-off transmission—after one person has talked that person must get off the channel to listen. In superbowl slang, the passing of messages back and forth among competing operators is called "passing a five" or "flipping nickels," and forms the basis of the teasing or insulting exchange. It also allows the operator passing the fives to ratchet up the tension between the other operators by giving their own emphasis to the message they convey. Operators often possess many Watergates, some of them dating back years, and these are prized examples of the best of superbowl culture and form an important archive for the next phase of this research.

To hear black CB's audiomobility requires listening to the 1960s and 1970s as a multitrack recording on which one has layered samples from spoken-word radicals such as The Watts Prophets and The Last Poets, some Martin Luther King Jr., Malcolm X's "stop singin' and start swingin'" speech, Huey Newton intoning "The Black Panther Party calls for," the echoing vocals of Jesse Jackson MCing Wattstax in 1972, Muhammad Ali boasting, James Brown in Zaire, old recordings of early black DJs from Chicago and Atlanta, and then an instrument track underneath some superbowl Watergates. One can hear the latter segment of that imagined mix on Prime Minister's unreleased track "No Excuses on the Bowl."[42] The fifteen-minute track describes the workings and the spirit of the superbowl, warning against any whining or timidity. The bowl, Prime Minister makes clear, is like a verbal boxing match: if you enter the ring, you had better be prepared to compete as well as to win or lose on any given occasion. The competitive, domineering style builds on the audible presence African American men established during the post-1945 struggle for desegregation and meaningful black citizenship. Based on deliberately misusing the federal CB radio service to create an audible black national community, black CB thus resisted the restrictions placed on black physical mobility by white citizens and by the government; the superbowl also built on the diversification of black culture made possible by a parallel resistance to class-based attempts to limit black speech styles and their public dissemination.

Acknowledgment

Thanks are due to the many people who have helped me develop this article: for research assistance my thanks to Whitney Kemble, Stephen Broomer, Abigail Godfrey; to SSHRC for research funding; to the organizing committee of the 2009 Radio Conference held at York University, Toronto, for accepting the first version of this research for presentation; thanks to Elspeth Brown for literally encouraging me; profound thanks to Prime Minister; thanks also to Crack Carter, Mr. Easy Rider Sr., and everyone else I met in Little Rock, Arkansas, on Labor Day weekend, 2009. Thanks also to the anonymous reviewer for *American Quarterly* for so many helpful suggestions for improvement. Because some of the activities or technologies used in contemporary superbowl activities may violate FCC regulations, some technological and personal details (such as legal names) related to current participants have been omitted from this article.

Notes

1. The term "indirect orality" refers to forms of communication other than those that take place via direct face-to-face speech—the latter usually assumed necessary for the growth of a community. On "sounding black," see John Baugh, *Black Street Speech: Its History, Structure, and Survival* (Austin: University of Texas Press, 1983); and Baugh, *Beyond Ebonics: Linguistic Pride and Racial Prejudice* (New York: Oxford University Press, 2000).

2. On the formation and use of communications networks among a contemporaneous minority population, see Martin Meeker, *Contacts Desired: Gay and Lesbian Communications and Community, 1940s–1970s* (Chicago: University of Chicago Press, 2006), esp. chaps. 5, 6.

3. One set of examples comes from the school busing struggles in Boston and Louisville, where antibusing activists used CB radios to better target African American mobility as well as police attempts to control their protest actions. See "Police Battle Busing Foes Marching in South Boston," *New York Times*, February 16, 1976; "Boston Whites March in Busing Protest," *New York Times*, October 28, 1975; "Teen-Agers in Boston Toss Rocks and Bottles," *New York Times*, February 17, 1976; "Large Wallace Vote Reflects Depth of Antibusing Sentiment in Boston's Working-Class Neighborhoods," *New York Times*, March 8, 1976; "Blacks' Anger Rising in South Boston as Violence over Schools Spreads," *New York Times*, May 2, 1976; "School Buses in Louisville Will Carry Guards Today," *New York Times*, September 8, 1975.

4. "Rooster Channel Jumpers 1979," typescript manuscript, folder JJ, box 10, p. 1, Berkeley G. Burrell Papers, Mugar Library, Howard Gotlieb Archival Research Center, Boston University. Dr. Berkeley Burrell, head of the Booker T. Washington Association as well as president of the National Business League, held strong connections in the 1970s to the small but politically significant African American business and political elite wooed by President Nixon as part of his racially complicated domestic agenda.

5. Some of the following section was previously published in my article "An Audible Sense of Order: Race, Fear, and CB Radio on Los Angeles Freeways in the 1970s," in *Sound in the Age of Mechanical Reproduction*, ed. David Suisman and Susan Strasser (Philadelphia: University of Pennsylvania Press, 2009), 159–178.

6. For a detailed account of the FCC's intentions regarding CB radio, see Carolyn Marvin and Quentin J. Schutze, "The First Thirty Years," special section, "CB in Perspective," *Journal of Communication* 27 (Summer 1977): 109.

7. *Citizens Band Radio and the Future of the Portable Telephone* (New Canaan, CT: International Resources Development, 1977), 54–55.

8. Magazine and newspaper articles about CB radio published at the height of national interest in CB, from 1975 to 1977, usually made reference to the recent introduction of the speed limit, the truckers' strike, and the use of CB to avoid police and highway patrol speed traps. See, for example, "The Bodacious New World of C.B.," *Time*, May 10, 1976, 78–79; "Citizen's-Band Radio: Danger of Air Pollution?," *U.S. News and World Report*, March 7, 1977, 76–77; "Hey Good Buddy: CU Rates CB Radios," *Consumer Reports* 42, no. 10 (October 1977): 563; Brock Yates, "One Lap of America," *Car and Driver*, February 1975, 27–30, 75–77; "Nuisance or a Boon? The Spread of Citizens' Radios," *U.S. News and World Report*, September 29, 1975, 26–28; William Jeanes, "Tuning in Justice on Your CB Radio Dial," *Car and Driver*, July 1975, 10.

9. Newspaper and magazine articles often commented on the rapidly rising sales of CB radio sets. See, for example, "The Newest Hobby: Kibitzing by Radio," *Forbes*, July 15, 1975, 16–17; J. D. Reed, "A Big 10–4 on the Call of the Wild," *Sports Illustrated*, March 29, 1976, 36–38, 47–48.

10. *Citizens Band Radio and the Future of the Portable Telephone* (New Canaan, CT: International Resources Development, 1977), 60.

11. See Blake, "Audible Sense of Order."

12. Paul Delaney, "Suburbs Fighting Back as Crime Rises," *New York Times*, August 30, 1976.

13. On the history of the black power movement, see Peniel Joseph, ed., *The Black Power Movement: Rethinking the Civil Rights–Black Power Era* (New York: Routledge, 2006); and

Jeffrey O. G. Ogbar, *Black Power: Radical Politics and African American Identity* (Baltimore: Johns Hopkins University Press, 2005).

14. Prime Minister, interview with the author, March 2009.

15. Shawn D. Lewis, "10–4, Bro'," *Ebony*, October 1976, 120–122, 124, 126.

16. On the politics and history of black speech, see work by Geneva Smitherman, such as her book *Talkin That Talk: African American Language and Culture* (New York: Routledge, 1999), and work by John Baugh, such as *Out of the Mouths of Slaves: African American Language and Educational Malpractice* (Austin: University of Texas Press, 1999). African Americans were not the only racialized minority to develop a distinctive slang as part of an assertive cultural response to racism and exclusion. Mexican American youths in the Los Angeles area in the 1940s used a slang called "caló" that drew on Mexican Spanish, English, and African American hipster or jive expressions. For an excellent discussion of caló, see Anthony Macias, *Mexican American Mojo: Popular Music, Dance, and Urban Culture in Los Angeles, 1935–1968* (Durham, NC: Duke University Press, 2008), 87–89.

17. Smith, *How Race Is Made*, 25–26.

18. George Bernard Shaw's play *Pygmalion* (1913), later reinterpreted for Broadway and Hollywood as the musical *My Fair Lady*, tells one tale of the cruelties and power of class and accent in an early twentieth-century British context.

19. Bob Dixon, "Are Amos and Andy Dead?," *Chicago Defender*, September 3, 1975, 21. See also The Falcon, "Smokie City CB News," *Call and Post*, May 1, 1976, 8B.

20. In addition to criticism of speech style and accents, and the use of dialect, other articles in the black press criticized the bad manners and selfish behavior of blacks on CB radio (implicitly worse than what was heard of white CBers). See, for example, Jim Cleaver, "The Frustrations of Blacks and Citizens Band Radio," *Los Angeles Sentinel*, March 24, 1977, A7.

21. James R. Grossman, *Land of Hope: Chicago, Black Southerners, and the Great Migration* (Chicago: University of Chicago Press, 1989), esp. chap. 5, "'Home People' and 'Old Settlers'"; Davarian Baldwin, *Chicago's New Negroes: Modernity, the Great Migration, and Black Urban Life* (Chapel Hill: University of North Carolina Press, 2007), 9–10, 16–17, 233–242.

22. See Blake, "Audible Sense of Order," 176.

23. Prime Minister, interview with the author, March 2009.

24. Robert McChesney, *Telecommunication, Mass Media, and Democracy—the Battle for Control of U.S. Broadcasting, 1928–1935* (New York: Oxford University Press, 1994), 18–19, 26–28; Michele Hilmes, *Radio Voices: American Broadcasting, 1922–1952* (Minneapolis: University of Minnesota Press, 1997). The most recent scholarship on the history of the radio listener can be found in Elena Razlogova, *The Listener's Voice: Early Radio and the American Public* (Philadelphia: University of Pennsylvania Press, 2011).

25. Susan Douglas, *Inventing American Broadcasting, 1899–1922* (Baltimore: Johns Hopkins University Press, 1987), 297.

26. Hilmes, *Radio Voices*; and McChesney, *Telecommunication, Mass Media, and Democracy*.

27. Thomas R. Kennedy Jr., "New 2-Way Radio Ready for Public," *New York Times*, March 24, 1948, 27.

28. Robert E. Baker, "3d Time Up for Hooded Bigots," *Washington Post*, May 16, 1965, E1.

29. Drew Pearson, "Klan Modernizes Its Terrorism," *Washington Post*, October 18, 1965, B11.

30. "FCC Probes Klan Short-Wave Radio," *Washington Post*, October 20, 1965, A2.

31. See, for example, Richard Corrigan, "Klansmen Use Citizen Band Radios to Conduct Raids, Committee Told," *Washington Post*, January 24, 1966, A3.

32. "Willis Plans Curb on Klan Activities; Bill Sets Penalties," *Washington Post*, June 15, 1966, A9.

33. Fred Powledge, "Armed Negroes Make Jonesboro an Unusual Town," *New York Times*, February 21, 1965, 52.

34. Lipsitz uses the phrase "turning hegemony on its head" in a few instances. See, for example, George Lipsitz, *Footsteps in the Dark: The Hidden Histories of Popular Music* (Minneapolis: University of Minnesota Press, 2007), 192; and Lipsitz, *Midnight at the Barrelhouse: The Johnny Otis Story* (Minneapolis: University of Minnesota Press, 2010), 170. The historiography on hegemony/counterhegemony is extensive and has its roots in the early-1970s influence of European Marxism on American historians. The literature includes such major interventions as George Rawick, *From Sundown to Sunup: The Making of the Black Community* (Westport, CT: Greenwood Press, 1973); Eugene Genovese, *Roll Jordan Roll: The World the Slaves Made* (New York: Vintage, 1976); Lawrence Levine, *Black Culture and Black Consciousness* (New York: Oxford University Press, 1977); Joane Nagel, *American Indian Ethnic Renewal: Red Power and the Resurgence of Identity and Culture* (New York: Oxford University Press, 1997). My thinking about black CB's connection to histories of race and technology has been informed by Rose's *Black Noise*, as well as more recent scholarship on black technocultures and music, such as Alexander G. Weheliye's excellent *Phonographies: Grooves in Sonic Afro-Modernity* (Durham, NC: Duke University Press, 2005), and the special issue "Technology and Black Music in the Americas," *Journal of the Society for American Music* 2, no. 2 (2008).

35. Nancy Fraser, "Rethinking the Public Sphere: A Contribution to the Critique of Actually Existing Democracy," *Social Text* 24, no. 1 (1990): 56–80; Michael Warner, "Publics and Counterpublics," *Public Culture* 14, no. 1 (Winter 2002): 49–90.

36. Fraser, "Rethinking the Public Sphere," 61.

37. Warner, "Publics and Counterpublics," 57.

38. Ibid.

39. Blake, "Audible Sense of Order," 169–173.

40. To hear this part of my interview with Prime Minister, listen to the sound clip titled "01 Prime Minister interview" on the website associated with the original publication of this article: https://www.americanquarterly.org/interact/beyond_sound.html.

41. Listen to the sound clip titled "02 Watergate clip" on the website associated with the original publication of this article: https://www.americanquarterly.org/interact/beyond_sound.html.

42. The track "No Excuses on the Bowl" can be found through a simple Internet search. But please be aware that the availability of the track does not necessarily imply that its creator(s) have granted permission for such dissemination or for the possible copyright infringement caused by Internet dissemination and downloading.

27 Drones for the Good: Technological Innovations, Social Movements, and the State

Austin Choi-Fitzpatrick

Part of the complexity of sociotechnical systems concerns the fact that artifacts developed with an eye toward one particular use can often be used for other purposes. This is sometimes referred to as the *dual-use* nature of technology. Originally, the military developed this term to argue that governments should invest in technological weaponry because some of the discoveries would also have civilian uses, thereby resulting in both military and peacetime advantages. More recently, the term has been used as a warning that the development of seemingly peaceful technologies could be used, in the wrong hands, for nefarious purposes. However, the "dual part" of *dual use* can be misleading since there are more than two sectors, and a technology developed for one purpose may be used in more than two different ways. In any case, this characteristic is especially important when it comes to new and emerging technologies because typically their potential has not yet been fully realized, and their development can go in multiple directions. When the public fixes on one kind of use and attaches values and meaning to a technology, other potentially powerful uses may be neglected. Indeed, negative meaning may prevent the development and adoption of new and potentially beneficial uses of the technology.

In this piece by Austin Choi-Fitzpatrick, we see that although unmanned aerial vehicles (drones) were initially seen by the public as a military technology and viewed in a negative light because of their association with surveillance, drones have the potential to be useful in a variety of socially beneficial, civilian contexts. Choi-Fitzpatrick is interested in the use of drones in civil society. Civil society refers to actors such as nonprofit organizations, social movements, and faith-based organizations; they are not government or commercial for-profit entities. He argues that the development of drones for civil society use could be significantly facilitated by frameworks that are quite different than those currently being produced for commercial and military uses. This analysis of drones illustrates the complexities around the meaning and impact of new technologies as they are being developed.

The recent wave of mobilization and contestation that has swept from Tunisia to Ukraine has run parallel to the emergence of an important technological innovation.[1] While the use of mobile phones and social media has received a large amount of attention, protests in Hong Kong, Ukraine, and even Ferguson, Missouri have seen the emergence of civil society's use of unmanned aerial vehicles (UAVs) or, more commonly, "drones."[2] This innovation represents a technological shift in scale for citizen journalists, human rights advocates, and social movement actors. As such, it requires a sophisticated assessment of the ethical issues and policy terrain surrounding its use.

From *Journal of International Affairs* 68, no. 1 (Fall/Winter 2014).

To date, debates over the use of UAVs have focused on two areas. First, human rights groups have mobilized against the state's use of drone strikes and the killing of civilians in the "War on Terror." Second, policymakers in Europe and the United States have scrambled to regulate the commercial use of drones. However, a critical third segment of drone usage by and for civil society actors, especially social movements, deserves attention.

This article reviews the nascent literature on UAV use and situates it within the larger theory and debates over technology and innovation, ethics, legal rights (including privacy and the right to information), public policy, and human rights. It then applies these considerations to proposed guidelines for the use of UAVs by non-state and non-commercial actors.[3] It concludes by noting the perils and promises of the use of drones for the purpose of investigative journalism, human rights monitoring, and state accountability.[4] The dual interest in the technology by both the state and its challengers points to the promise and peril of innovation.

. . .

Innovation

The promise and peril of UAVs lie at the intersection of three interconnected technological innovations. The first involves a shift from analog to digital devices. This allows for more powerful onboard processors, longer battery life, and the ability to easily stream audio and video to digital consumer devices. Combined with more stable quadcopter designs, these have transferred UAVs from the hobbyist market to the general public. But this shift from analog to digital also covers the payloads these devices carry. While the carrying capacity within consumer devices is modest, they are sufficient to carry cameras, as well as sophisticated signal-jamming equipment, wireless routers, and similar electronic devices. UAVs are an ideal type of innovation, that is, they combine invention with exploitation (by marketing, integrating, and diffusing goods and ideas).[5]

Popular digital imaging devices represent a second technological scale shift, as they generate infinitely portable and reproducible images that can be shared, copied, distributed, and stored with increasing ease and decreasing cost. Combined with the emergence of online environs for storing and sharing images, digital imaging devices have fundamentally disrupted the status quo with regard to journalism, whether for entertainment, such as paparazzi photos of a Hollywood star, or accountability, such as YouTube footage from the Arab Spring.

The third technological innovation, and arguably the most disruptive, is the fundamental break between the camera and the street level. Photography has had a symbiotic connection with the street for more than a century, as far back as Eugène Atget's street photography in Paris in the 1890s and Jacob Riis's documentary photography in New York at the same time.[6] The most memorable photographs of violent conflict, social protest, and natural disaster have almost all been taken by a person present on the ground. The horizontal plane has been the most important space for both the perambulating human and the observant photojournalist. The same can be said of most state surveillance, as well as the increasingly common use of surveillance cameras in commercial centers. The journalist's camera is positioned at eye level. The state and commercial market have placed their devices just out of arm's reach, but both point nearly horizontally.

UAVs relocate the boundary between what is public and what is private, because camera-equipped UAVs move the line of sight from the street to the air. This simple shift effectively pushes public space from the sidewalk to the stairwell, courtyard, rooftop, and so forth. Once private, these spaces are now subject to surveillance. Or have they now become public spaces? Should technologists, ethicists, and public policy professionals simply increase the number and type of locations that are now considered public, or must a more profound conversation occur?

Technology has redrawn the lines between private and public space. Work on the Internet of Things and Internet privacy suggests that much of what happens in seemingly private spaces is not actually private.[7] This increasingly applies to our browsing habits as well as less recognized data passively generated from devices—for instance, my iPhone's accelerometer telling my mobile carrier or insurance provider that I have not jogged in days. UAVs represent a relatively new technology, or rather, a newly applied technology, that is disrupting our understanding of which spaces are private.

Ubiquitous closed-circuit televisions (CCTVs) represented the vanguard of this change, since they opened sidewalks, parks, and other public spaces to sustained and archived monitoring by commercial interests and law enforcement. When the feed from CCTVs went to tape, the question essentially involved privacy. When the feed now goes to digital archives, subject to hacking and scanning, the privacy issue has grown immeasurably. Digital archives of street surveillance footage, combined with facial recognition and behavioral software, push the privacy issue even further.

While these observations seem pedestrian at first blush, their implications are profound. Security and privacy policies address the prying eyes of the standing observer, not the roving airborne eye of a small UAV that is flying according to Global Positioning System (GPS) waypoints while streaming video over secure Wi-Fi to an operator sitting behind a laptop in a nearby cafe, library, or office complex. "Open air" and "free space" are no longer as "open" or "free" as they once were. They are instead now occupied or vulnerable to occupation. Cyberspace scholars suggest that new technologies are pivotal in "radically restructuring the materiality and spatiality of space."[8] Whether this space is used for the public good or as a means of state and commercial surveillance is just the sort of dilemma regulators face. Cyber-skeptics fear the panopticon, believing "a society biased toward hierarchy and capitalism generates the entirely rational impetus for . . . surveillance."[9] Others argue for a contrast between libertarian and authoritarian technologies where the former is egalitarian, and the latter is "fundamentally hegemonic."[10] If Predator drone strikes in Pakistan and Yemen represent challenges to notions of sovereignty, camera-equipped civilian UAVs in London and New York represent fundamental challenges to the notion of public space.

For some time, radical geographers have thought about space as it relates to power, politics, and change while technologists focus on the promise and peril of new technology. These two have met in the literature about the Internet.[11] Scholars of online worlds focus on the Internet as a disruptive new space, but UAVs disrupt the actually occurring material and physical space we inhabit every day. This applies to hard security as well as privacy. The walls and barricades around terrorist training camps, Occupy gatherings, and Davos meetings belong to a world of line-of-sight threats from paparazzi and pipe bombs. The United States has reinforced many embassies over the past decade with moats, ramparts, walls, and bulletproof glass.[12] Industry standard protection from an explosives-laden truck, however, is generally useless against a commercially available drone carrying toxic chemicals with an aerosol dispersant flying too close to an air intake inside a military compound. Innovation of this sort is a hallmark of asymmetrical warfare.[13]

Debates between technophiles and techno-skeptics, and the scale shifts indicated above, resonate in a complex thicket of ethical and legal considerations. In the United

States, the Federal Aviation Administration (FAA) has attempted to restrict all commercial use of drones, despite questions about their authority to do so.[14] Clearly, UAVs equipped with imaging devices also operate in a cultural, political, and technological environment charged with debates over citizen rights in an age of mobile telephony, citizen journalism, and ubiquitous surveillance. The debates over emerging big data capabilities to harness the data generated by these sources are only now emerging.[15] As societies grapple with the social and ethical implications of these technical innovations, policymakers find themselves in the unenviable position of regulating a technology in its infancy.

Civil Society Uses

Like any technology in its early stages of growth, drone use is flourishing. The discussion of the legal terrain surrounding UAVs suggests the challenges posed to the development and implementation of a single policy framework for regulating civilian use. Notably, there are multiple competing analogies for what sort of regulatory puzzle UAVs represent. Are they small airplanes, weapon platforms, flying cameras, or a new hobbyist device? Variation in the answer will shape policy responses. In what follows, I provide a brief overview of some of the public uses for these devices, the diversity of which suggests the complexity of any policy intervention.

Art

Cinematographers wishing to deploy the technology in the United States have recently petitioned the FAA to allow for their use in commercial artistic production prior to the release of the FAA's decision on drone use in civil airspace. The entertainment industry petition joins three others (agriculture, line inspection, and oil and gas) in seeking a waiver for drone use in "narrowly defined, controlled, low-risk situations."[16] Less conventionally, graffiti artists have begun experimenting with UAVs, the beginning of many efforts to integrate this technology into the arts.[17]

Human rights groups are beginning to make use of space-based remote sensing equipment for monitoring crises, and it is reasonable to expect an increase in such use as prices fall.

Mapping

Mapping represents an important cross-cutting utility that UAVs bring to all of the uses that follow. Maps that are already widely available from commercial enterprises (e.g., Google Maps) can be augmented with UAV-based data on conflicts, disasters, protests, environmental degradation, labor exploitation, and so forth. This usage is not limited to UAV-based equipment, however, as recent innovations include higher quality and lower cost satellite imagery. Human rights groups are beginning to make use of space-based remote sensing equipment for monitoring crises, and it is reasonable to expect an increase in such use as prices fall.[18]

Public Safety

There is increased experimentation with UAVs in a number of public safety-related areas, including firefighting and search-and-rescue operations.[19] UAVs are also deployed to augment the support of traditional ambulance or rescue services, as in the case of an accident in which a small UAV, equipped with a thermal imaging device, was able to locate a wrecked vehicle in Canada, and another in which a camera-equipped drone located a man whom rescue workers had been unable to find for days.[20] Yet such efforts fall into a regulatory gray zone, a fact further complicated by the commercial availability of a weaponized "riot control copter" for use against protesters.[21]

Environment

UAVs are increasingly used in a number of environmental areas, including change mapping (i.e., river erosion, deforestation, and urban expansion); disaster risk management and mitigation (assessing natural disaster risk and monitoring fires, volcanoes, and landslides); monitoring illegal activity, including banned hunting, fishing, and trade; and monitoring other natural factors like migration, levels of endangered species, and foliation.[22] The World Wildlife Fund recently received a $5 million grant from Google's Global Impact Awards program to monitor poaching and the illegal trade in wildlife with UAVs.[23] Large-scale environmental change can also be monitored using UAVs. China is using the technology to monitor polluting industries, and Brazil is considering using drones to monitor illegal logging.[24] Kenya had plans to deploy drones to spy on poachers in fifty-two of its national parks after a pilot program found that their presence reduced poaching by up to 96 percent.[25]

Humanitarian and Development Aid

One of the most significant areas of opportunity for civil society actors is in humanitarian aid, as organizations respond to natural disasters, conflict and post-conflict situations, and more general development and poverty-related needs. Former U.S. ambassador Jack Chow has suggested that UAVs could "deliver a peaceful 'first strike' capacity of food and medicines to disaster areas."[26] UAVs have served just this role in the wake of natural disasters in Haiti and the Philippines.[27] While there is more of a precedent for UAV use in humanitarian and post-conflict settings, they may also prove useful in helping health and development organizations access hard-to-reach beneficiaries.

Journalism

Journalists are increasingly experimenting with the incorporation of drones into their work.[28] Drones allow journalists to get much closer to the action. This applies equally when covering sports, reporting on conflicts, capturing imagery, and generally reporting on stories in ways that had not previously been possible. Citizen journalism could also benefit greatly from the use of UAVs documenting public events and providing alternative avenues for reporting, especially during periods of media censorship.[29]

Corporate Accountability

This use is in its infancy, though it shows promise. Recent drone footage revealed that a meatpacking plant in Texas was illegally dumping pigs' blood from a slaughterhouse into a nearby stream. While this triggered a federal investigation that shut the plant down, it also led to legislation in Texas forbidding the use of drones over private property.[30] A recent Kickstarter project to monitor factory farms (and challenge so-called "ag-gag" laws passed against whistleblowers and activists) was fully funded in less than a week.[31] It is likely such uses will expand in the near future, especially considering increasing concerns with corporate social responsibility, supply chain ethics, labor rights violations, corruption, and environmental impact.

State Accountability and Conflict

There appears to be a consistent interest in the use of UAVs to monitor low-intensity conflict and peacekeeping.[32] They have recently been deployed by the United Nations (UN) to the Democratic Republic of the Congo, Chad, and the Central African Republic.[33] Rebels in Syria, beyond the definition of civil society advanced here, have deployed relatively affordable and commercially available UAVs to monitor loyalist forces.[34]

. . .

Human Rights Monitoring

While this usage, like the others listed here, is still in its infancy, it too shows signs of rapid growth. A prominent anti-slavery advocate recently suggested deploying drones in the struggle to end slavery and human trafficking, in much the same way the technology has been used to protect endangered rhinos.[35] In cases such as Syria, there was brief discussion about whether the international community should invoke the Responsibility to Protect doctrine (R2P) and effectively vitiate Syria's rights over its airspace.[36] The Satellite Sentinel Project has advocated a similar intervention in the use of UAVs to monitor crisis situations and human rights violations. In the words of its founders, "A drone would let us count demonstrators, gun barrels, and pools of blood."[37] Sniderman and Hanis argue that, while this approach has implications for sovereignty rights and "may be illegal in the Syrian government's eyes . . . supporting Nelson Mandela in South Africa was deemed illegal during the apartheid era."[38] This observation emphasizes the tension between bearing witness and the legal status quo.[39]

Social Movements and Protests

There is some overlap between UAV-based state accountability monitoring and their use in social movements and protests. Clashes between anti-government protesters and pro-government forces in Bangkok were captured by drones and uploaded to YouTube in an attempt to draw attention to the protestors' cause.[40] They have also been used for similar purposes in Turkey, Estonia, Poland, Hong Kong, and Ferguson, Missouri.[41] This overlap occurs in the area of policing, where social movement scholars and scholars of policing have spent the past decade teasing out the changing dynamics surrounding police-protestor interaction.[42] UAVs can indeed serve as another set of eyes monitoring police action, holding the state to account in potentially violent protests. Yet social movements can put UAVs to a much broader range of uses, the most innovative of which remain to be seen. Whatever the case, civil society actors must be prepared for an aggressive response by the state and its agents, such as when police in Istanbul shot down a camera-equipped UAV while it was monitoring large anti-government protests in the Turkish capital.[43]

Material and Technical Disruption

With art and public safety at one end of the usage spectrum, more disruptive and "hacktivist"-inspired uses lie at the other end. UAVs can be used as lookout posts for graffiti artists or protesters needing a second pair of eyes. Camera-equipped devices can loiter or land and then feed imagery back to a clandestine location. This article has focused on the camera as a particular payload, but UAVs can just as easily carry Wi-Fi hardware that can perform wireless penetration testing, conduct 3D mapping of buildings or urban environments, conduct thermal mapping exercises of indoor and outdoor spaces, and conduct video and audio surveillance through cameras and directional microphones.

This list is meant to be illustrative of broad categories of use, but in reality, there are multiple configurations for a myriad of uses. It is not difficult to devise a modular system that would allow a user to quickly attach just the necessary components and then run multiple passes to update additional layers of data onto a map. For example, a designated area could receive a five-sweep treatment in which the first pass captures video and establishes GPS coordinates, the second captures thermal imagery, the third scans for Wi-Fi data, the fourth scans for radiation levels, and the last captures more specific surveillance footage.[44] The range of uses and the ramifications of various configurations suggest that a sophisticated framework is necessary to guide this innovation.

Frameworks

This broad and growing list of public uses requires a framework that differs significantly from the guidelines currently being developed around the commercial and military/police use of drones. While these guidelines revolve and profit, the organizing principle for civil society use must emphasize the public good. Current frameworks have broken new ground, but remain sector specific. As seen in table 27.1, the Humanitarian UAV Network framework emphasizes safety and suitability with the goal of providing humanitarian support.[45] The Drone Journalism Lab emphasizes transparency and accountability in pursuit of the public good.[46] For its part, the American Civil Liberties Union (ACLU) is focused on privacy, with a focus on preventing police abuse.[47]

Each contribution listed in the table above advances the factor of the greatest importance to the institutional environment that produced it. A comprehensive framework for civil society drone use must balance many interests: safety, suitability, transparency, accountability, privacy, and the rights of residents (citizen and non-citizen alike), while also maintaining a commitment to the public good. Striking this balance is no easy task. In what follows, I propose a broad framework to guide a range of non-state and noncommercial actor uses of drones. In this light, the guidelines listed above are specific configurations of the broader considerations emphasized in the following six principles:

Subsidiarity—The concept of subsidiarity suggests that decision-making and problem solving should occur at the lowest and least sophisticated level possible. The implication here is that a drone should only be used to address situations for which there is not a less sophisticated, invasive, or novel use. Steve Coll, dean of the Journalism School at Columbia University, has argued that drone operators should ask themselves, "What can you use a drone for, that you can't achieve by other means?"[48] Such an approach would ensure that drones are used in areas where they are actually appropriate, thus spurring innovation and possibly reducing resistance to their usage.

Physical and material security—This principle focuses on physical integrity issues related to the use of UAVs. Put bluntly, care must be taken so that these devices do not collide with people or with one another. Furthermore, they must not be weaponized in such a way that could cause physical harm to the public. How exactly this security is ensured is a matter of skill, which is determined by the operator, and situation, which is determined by weather and other environmental conditions. How it is defined is a matter of perspective: It is likely that both governments and corporations will consider the use of UAVs by investigative and citizen journalists to be a violation of their security. This use should nevertheless be protected by the rights to freedom of the press, expression, and information.

Do no harm—This principle draws inspiration from the UAViators' emphasis on a rights-based approach as found in the development and humanitarian aid communities.

Table 27.1
Existing guidelines for drone usage

Group	Themes	Target	Focal Point (on balance)
ACLU	1. Usage limits—police use with warrant only 2. Data retention 3. Policies decided by public representative 4. Abuse prevention and accountability 5. Weapons forbidden	Law enforcement	Restricted use
Professional Society of Drone Journalists	1. Newsworthiness 2. Safety 3. Sanctity of the law and public spaces 4. Privacy 5. Traditional journalistic ethics	Journalists	
UAViators.com	*Preflight* 1. Do no harm 2. Ensure flight safety (failsafe, flight plan, weather) 3. Ensure humanitarian value 4. Obey all laws 5. Respect individual privacy and engage community 6. Avoid use where retraumatization is possible *In-flight* 2. Select safe sites 3. Use a spotter 4. Respect relevant airspace regulations 5. Use allowed radio-control frequencies *Post-flight* 1. Keep a logbook 2. Request permission for image usage 3. Respect personal privacy and remove identifiable information 4. Freely share imagery with local communities whenever possible		

Note: Edited for brevity. Full guidelines available at http://www.aclu.org/blog/tag/domestic-drones; http://www.dronejournalism.org/code-of-ethics; https://docs.google.com/document/d/1pliYVNek2RsiS Q8_9ATFdJBzYFVP88edfLHL8uFBhUNedit.

The focus is not on reducing physical and material security, but is instead on ensuring the public good (i.e., the harm in question is related to the public good rather than physical integrity). The principle is one of proportionality, in which the question to be answered is, "Are the risks of using UAVs in a given humanitarian setting outweighed by the expected benefits?"[49] Here again there is room for debate. It is conceivable that social movements will incorporate UAVs into disruptive tactical repertoires, thereby reducing the likelihood of a policy compromise between movement actors and the centers of power and authority they are challenging. New uses must strike their own balance.

Public interest—This principle draws original inspiration from the concepts of newsworthiness and the public good, while recognizing that some seemingly insignificant or unpopular issues may be in the public's interest and for a public good without being considered newsworthy. This approach is especially sensitive to the importance of investigative journalism that holds to account the powerful and well-resourced, despite attempts by established interests to discredit these efforts.[50] This expansive conceptualization of public accountability is journalism's corner-stone. The preamble to the Society of Professional Journalists' Code of Ethics argues that "public enlightenment is the forerunner of justice and the foundation of democracy."[51] At a time when corporations and the state capture an ever-larger share of private space, every effort must be made to maintain and expand civil society's technological capacity for accountability and resistance. There is no better precedent—as both herald and cautionary tale—for this commitment than the free press.

. . .

Privacy—Each principle must be held in balance with the others, and none more so than with respect to privacy. Citizens and non-citizens should be protected from the prying eyes of the state and commerce, yet there is a need for a larger conversation about what level of privacy is to be expected when civil society actors have deployed drones for their own purposes.[52] There is reason to believe, however, that current legislation prohibiting "peeking while loitering"—for example, California Penal Code 647(i) prohibits "loitering, prowling, or wandering upon the private property of another, at any time, peeks in the door or window of any inhabited building or structure, without visible or lawful business with the owner or occupant"—would render such spying illegal, regardless of whether the camera was mounted to a tripod or a drone.[53] Yet this framework is more sanguine and ambivalent when it comes to the privacy of powerful rights violators. Camera- or sensor-equipped drones have the ability to violate the privacy and private property rights of corporate persons involved in malfeasance. However, the difference between the privacy of a bedroom and a boardroom is not insignificant. Likewise, creating a framework that applies in all circumstances is nearly impossible in an era in which digital privacy appears to be a mirage, and the possibility that a new wave of technological innovation will force a fundamental reimagining of both public space and expectations of privacy.

Data protection—Finally, data protection is paramount. Civil society actors using camera-equipped drones are likely to generate sensitive data. Filming a protest event, for example, creates a digital record of protesting participants. In the hands of social movement actors, this footage can be used to mobilize communities or challenge official records of events. In the hands of the authorities, however, digital footage can easily be scanned using facial recognition technology in order to create a database of known activists. As more UAVs gather more data, questions about how to handle big aerial data will emerge. Drones themselves will be easier to hijack as anti-drone technology evolves, and the wireless links that connect them to base stations will also be vulnerable to hacking. Context-specific protocols must ensure the security of data, thereby protecting against physical or digital theft or corruption.

Tensions emerge across these central principles. The first tension lies between individual privacy and the public interest. At the time of writing, it seems clear that privacy is undergoing a substantial overhaul in terms of the level of anonymity that can be reasonably expected in an age of constant surveillance and ubiquitous digitization. While it is difficult to comment on a process that is in flux and is subject to starkly different national regulatory regimes and cultural norms, it is clear that citizens and non-citizens alike will need to accept significantly less-robust guarantees to privacy in the future. This reality brings new tradeoffs, and it is important that those actors using UAVs work within the general bounds of emerging norms about privacy.

The second tension lies between insider and outsider tactics in the use of UAVs. While humanitarian drone use may be integrated into a state's military apparatus, social movements often choose tactics based on their values and goals.[54] Since social movements frequently reject formal political channels, or may be blocked from them altogether, there should be little surprise when they turn to social media in the face of authoritarian oppression.[55] Indeed, this is the recent history of social movements. In Rhodes's vivid description of the New Left in the 1960s, he documents a wide range of tactics:

> Petitioning, rock throwing, canvassing, letter writing, vigils, sit-ins, freedom rides, lobbying, arson, draft resistance, assault, hair growing, nonviolent civil disobedience, operating a free store, rioting, confrontations with cops, consciousness raising, screaming obscenities, singing, hurling shit, marching, raising a clenched fist, bodily assault, tax refusal, guerilla theater, campaigning, looting, sniping, living theater, rallies, smoking pot, destroying draft records, blowing up ROTC buildings, court trials, murder, immolation, strikes, and writing various manifestoes or platforms.[56]

While a good number of these fail the "do no harm" threshold, their creative breadth in a pre-digital age suggests that any framework for new technology must work hard to strike a balance between freedom of expression and assembly and the security of capital and the state. Policymakers and innovators alike should engage in a broad and inclusive discussion about how these principles might be best balanced.

. . .

Conclusion

In this article, I have attempted to briefly emphasize a relatively unfamiliar origins story for drones. Commercially available devices challenge the notion that drones are cousins to strike fighters laden with laser-guided bombs; they are also part of the same family as cameras. The technological family metaphors need not stop there. Indeed, the second section of this article is dedicated to detailing ten clear civilian and civil society uses for UAVs. The drone's payload can be beneficial and benign, or disruptive and deadly. My focus here has been on the drone's range of uses. The article's third section provides a tentative framework that I believe will help policymakers and the public differentiate between beneficial and harmful uses, with the "public good" as the benchmark. What exactly constitutes the public good is a matter of debate. Protecting privacy is important, but so is shedding light on important issues and holding responsible parties accountable. Protecting property is important, but so is speaking truth to power through graffiti and protest art.[57]

Talking about these tensions is not easy. Innovation is a moving target. The host of uses described earlier was harvested from online reports of innovation within roughly a twelve-month period. This innovation has occurred despite a lack of sustained scholarly inquiry or stable and consistent governmental oversight. Indeed, it was only recently that the FAA licensed three university campuses to conduct research on drone use.[58] Even without this licensing, others are using money from the U.S. Army Research Laboratory's Army Research Office to incorporate drones into campus-based,

Wi-Fi-based mesh network systems.[59] At the risk of severely belaboring the point, innovation has completely outstripped legislation, and much of this innovation is by and for the public good. This will continue into the foreseeable future as additional uses emerge. At present, it is not clear what the relationship will be between "drones for the public good" and satellites gathering information about humanitarian crises and human rights violations, though organizations such as UAViators are actively integrating social media, aerial imagery, and satellite imagery for humanitarian relief efforts.[60] A broader range of actors is working to make geographic information systems (GIS) and satellite data valuable for advocacy groups and policy practitioners alike.[61] This use predates the current wave of drone use by several years, and it is likely that more effective combinations of these technologies will be developed for civil society use. The Satellite Sentinel Project has the tagline, "The world is watching because you are watching," effectively shifting surveillance from an invasive enterprise to bearing witness.[62] This clever blending of traditional movement concepts (bearing witness) with new means (satellite technology) is echoed by Patrick Meier, who suggests that classic civil resistance tactics can be extended to drones.[63] This can be done, he argues, through the display of flags and symbolic colors, the "haunting" or taunting of officials, nonviolent air raids, defiance of blockades, and the disclosure of the identities of state agents.[64]

. . .

This wave of innovation and welter of uses raises a larger question: Does any of it matter? This is the subsidiarity principle writ large: Is there not another, less dramatic, way to meet these same objectives? What do drones add to the existing citizen monitoring mechanisms, through which information is captured on smartphones and disseminated by social media? These are important questions that I hope ongoing use and subsequent scholarship will begin to clarify. My sense is that an initial wave of enthusiasm will subside, leaving behind a solid body of innovation on the way civil society actors perform a number of tasks, especially related to social movements.

A final complication takes the form of public opinion, which seems hostile to this occupation of airspace. A recent study by the Pew Research Center's Internet & American Life Project found, "Sixty-three percent [of respondents] think it would be a change for the worse if personal and commercial drones are given permission to fly through most U.S. airspace."[65] Likewise, while it is legal in the United States to take pictures of individuals in public places, recent recreational uses have led to complaints of sexual harassment, as well as violence against drone operators.[66] The Kenyan government recently announced that it would ban the use of drones for monitoring poachers in the Ol Pejeta Conservancy, home to the endangered white rhino.[67] South Africa, too, has grounded camera-equipped UAVs, citing regulatory uncertainty at the global level.[68] Grappling with innovation is no easy task. This article suggests the same can be said of technology's relationship to civil society. Regulators must take care, lest they pass legislation and regulations that enable the state while crippling its citizens.

Notes

1. This article benefited from the research assistantship of M. Boby Sabur, Tautvydas Juskauskas, Luis Cano, and Justin De Los Santos; substantive and technical input from Phil Howard, Patrick Meier, John Holland, Sejal Parmer, Bernhard Knoll-Tudor, Thorsten Benner, Dean Starkman, Colleen Sharkey, Lars Almquist, and Edward Branagan; and financial support from Wolfgang H. Reinicke and the Central European University's School of Public Policy.

2. This language is fraught. The US military is committed to avoiding the terms "drone" and "unmanned aerial vehicle," preferring instead to use the term "remotely piloted aircraft." This term avoids the implication that these devices fly themselves, as well as the gendered notion that they are flown by men. I prefer the term "remotely piloted aerial platform" to reflect the diversity of payloads and the presence of a pilot, however remote and regardless of gender. I would be pleased if this usage proves popular but will not be using these pages to advance this argument. For present purposes, the common terms "drone" and "UAV" prevail. Jim Garamone, "Military Uses Remotely Piloted Aircraft Ethically," *American Forces Press Service*, May 22, 2014, http://www.defense.gov/news/newsarticle.aspx?id=122308; Joe Trevithick, "Learn to Speak Air Force: A Public Service Announcement Regarding Drones," *War Is Boring* (blog), May 27, 2014, https://medium.com/war-is-boring/learn-to-speak-air-force-e6ebc5614b25.

3. This article focuses on "civil society" use of drones. By civil society I mean nonstate and noncommercial actors using UAVs for public and private purposes. It is difficult to determine what exactly is meant by the "greater" or "public good," as these definitions are made by individual societies.

4. Patrick Meier, "Using UAVs for Community Mapping and Disaster Risk Reduction in Haiti," *iRevolution.net*, July 9, 2014, www.irevolution.net/2014/07/09/uavs-for-disaster-risk-reduction-haiti/; Faine Greenwood, "Drones, the Civic Surveillance Equalizer?," *sUAS News*, July 24, 2014, http://www.suasnews.com/2014/07/30184/drones-the-civic-surveillance-equalizer/.

5. Edward B. Roberts, "What We've Learned: Managing Invention and Innovation," *Research Technology Management* 31, no. 1 (Jan/Feb 1998): 11–29.

6. John Szarkowski and Eugène Atget, *Atget* (New York: Museum of Modern Art, 2004); Bonnie Yochelson and Daniel Czitrom, *Rediscovering Jacob Riis: Exposure Journalism and Photography in Turn-of-the-Century New York* (Chicago: University of Chicago Press, 2014).

7. Philip N. Howard, *Pax Technica: How the Internet of Things May Set Us Free or Lock Us Up* (New Haven, CT: Yale University Press, 2015).

8. Martin Dodge and Rob Kitchin, *Mapping Cyberspace* (London: Routledge, 2001), ix.

9. Uri Gordon, "Anarchism and the Politics of Technology," *WorkingUSA* 12, no. 3 (2009): 489–503; Giorel Curran and Morgan Gibson, "WikiLeaks, Anarchism and Technologies of Dissent," *Antipode* 45, no. 2 (March 2013): 294–314.

10. Curran and Gibson, "WikiLeaks, Anarchism and Technologies of Dissent," 299; Murray Bookchin, *The Ecology of Freedom: The Emergence and Dissolution of Hierarchy* (Montreal: Black Rose Books, 1991).

11. Curran and Gibson, "WikiLeaks, Anarchism and Technologies of Dissent," 299.

12. Mark McDonald, "Must All U.S. Embassies Now Be Fortresses?," *New York Times*, September 13, 2012, http://rendezvous.blogs.nytimes.com/2012/09/13/must-u-s-embassies-now-be-fortresses/?_php=true&_type=blogs&_r=0.

13. I have Patrick Meier to thank for this observation.

14. Jason Koehler, "A Drone Saved an Elderly Man Who Had Been Missing for Three Days," *Motherboard* (blog), Vice Media, July 23, 2014, www.motherboard.vice.com/read/a-drone-saved-an-elderly-man-who-had-been-missing-for-three-days.

15. Jeffrey Rayport, "What Big Data Needs: A Code of Ethical Practices," *MIT Technology Review*, May 26, 2011, http://www.technologyreview.com/news/424104/what-big-data-needs-a-code-of-ethical-practices/; Ellen Rooney Martin, "The Ethics of Big Data," *Forbes BrandVoice* (blog), March 27, 2014, http://www.forbes.com/sites/emc/2014/03/27/the-ethics-of-big-data/.

16. "Press Release—FAA to Consider Exemptions for Commercial UAS Movie and TV Production: Seven Companies Petition to Fly Unmanned Aircraft before Rulemaking Is Complete," Federal Aviation Administration (FAA), June 2, 2014, http://www.faa.gov/news/press_releases/news_story.cfm?newsId=16294.

17. "Interview: KATSU and the Graffiti Drone," Center for the Study of the Drone, Bard College, April 10, 2014, http://www.dronecenter.bard.edu/katsu-graffiti-drone; Jacob Kastrenakes, "Graffiti Artist KATSU Creates Abstract Paintings Using Drones with Spray Cans," *Verge*, April 7, 2014, http://www.theverge.com/2014/4/7/5582128/drone-paintings-by-katsu-graffiti-artist.

18. For more information, refer to Amnesty International's "Remote Sensing for Human Rights," http://www.amnestyusa.org/research/science-for-human-rights/remote-sensing-for-human-rights. Additionally, refer to the American Association for the Advancement of Science's Geospatial Technologies and Human Rights Project, http://www.aaas.org/page/remote-sensing-human-rights-project. More information can also be found at the Satellite Sentinel Project, http://www.satsentinel.org/.

19. Justin Dougherty, "Firefighters Push to Use Drones for Public Safety," *News9.com*, March 12, 2014, http://www.news9.com/story/24959827/firefighters-push-to-use-drones-for-public-safety.

20. "Credited for Saving Life—Draganflyer X4-ES UAS Used by RCMP Locates Unconscious Driver after accident," *Draganflyer Innovations*, May 9, 2013, http://www.draganfly.com/news/2013/05/10/credited-for-saving-life-draganflyer-x4-es-uas-used-by-rcmp-locates-unconscious-driver-after-accident/; Koehler, "Drone Saved an Elderly Man."

21. Leo Kelion, "African Firm Is Selling Pepper-Spray Bullet Firing Drones," *BBC*, June 18, 2014, http://www.bbc.com/news/technology-27902634.

22. "A New Eye in the Sky: Eco-drones," UNEP Global Environmental Alert Service (GEAS), May 2013, http://www.unep.org/pdf/UNEP-GEAS_MAY_2013.pdf; L. P. Koh and S. A. Wich, "Dawn of Drone Ecology: Low-Cost Autonomous Aerial Vehicles for Conservation," *Tropical Conservation Science* 5, no. 2 (2012): 121–132, http://www.tropicalconservationscience.mongabay.com/content/v5/TCS-2012_jun_121_132_Koh_and_Wich.pdf; "Google Helps WWF Stop Wildlife Crime," World Wildlife Fund, December 4, 2012, http://www.worldwildlife.org/stories/google-helps-wwf-stop-wildlife-crime.

23. World Wildlife Fund, "Google Helps WWF Stop Wildlife Crime."

24. Sandi Doughton, "Using Drones to Monitor Changes in Environment," *Star*, October 21, 2013, http://www.thestar.com.my/News/Environment/2013/10/21/Using-drones-to-monitor-changes.aspx/; Jennifer Duggan, "China Deploys Drones to Spy on Polluting Industries," *Guardian*, March 19, 2014, http://www.theguardian.com/environment/2014/mar/19/china-drones-pollution-smog-beijing; Lian Pin Koh, "Using Drones for Environmental Research and Spying," *ALERT*, April 27, 2014, http://alert-conservation.org/issues-research-highlights/2014/4/27/using-drones-for-environmental-spying-and-research.

25. Gitonga Njeru, "Kenya to Deploy Drones in All National Parks in Bid to Tackle Poaching," *Guardian*, April 25, 2014, http://www.theguardian.com/environment/2014/apr/25/kenya-drones-national-parks-poaching.

26. Jack Chow, "Predators for Peace: Drones Have Revolutionized War. Why Not Let Them Deliver Aid?," *Foreign Policy*, April 27, 2012, http://www.foreignpolicy.com/articles/2012/04/27/predators_for_peace.

27. Mabel González Bustelo, "Drone Technology: The Humanitarian Potential," *Open Democracy*, October 3, 2013, http://www.opendemocracy.net/opensecurity/mabel-gonzález-bustelo

/drone-technology-humanitarian-potential-0; Patrick Meier, "Using UAVs for Community Mapping and Disaster Risk Reduction in Haiti," *iRevolution.net*, July 9, 2014, http://www.irevolution.net/2014/07/09/uavs-for-disaster-risk-reduction-haiti/; Lean Alfred Santos, "In the Philippines, Drones Provide Humanitarian Relief," *Devex*, December 16, 2013, https://www.devex.com/news/in-the-philippines-drones-provide-humanitarian-relief-82512.

28. Louise Roug, "Eye in the Sky: Drones are Cheap, Simple, and Potential Game Changers for Newsrooms," *Columbia Journalism Review*, May 1, 2014, http://www.cjr.org/cover_story/eye_in_the_sky.php?page=all.

29. Melissa Bell, "Drone Journalism? The Idea Could Fly in the U.S.," *Washington Post*, December 4, 2011, http://www.washingtonpost.com/blogs/blogpost/post/drone-journalism-the-idea-could-fly-in-the-ussoon/2011/12/04/gIQAhYfXSO_blog.html.

30. Kashmir Hill, "Potential Drone Use: Finding Rivers of Blood," *Forbes*, January 25, 2012, http://www.forbes.com/sites/kashmirhill/2012/01/25/potential-drone-use-finding-rivers-of-blood/.

31. Twilight Greenaway, "Can Drones Expose Factory Farms? This Journalist Hopes So," *Civileats.com*, June 17, 2014, http://civileats.com/2014/06/17/can-drones-expose-factory-farms-this-journalist-hopes-so/.

32. Bustelo, "Drone Technology."

33. Wesley M. DeBusk, "Unmanned Aerial Vehicle Systems for Disaster Relief: Tornado Alley," NASA Technical Reports Server (NTRS), Conference Paper, Report No. ARC-E-DAA-TN500, 2009, http://ntrs.nasa.gov/archive/nasa/casi.ntrs.nasa.gov/20090036330.pdf.

34. Jassem Al Salami, "Drone Battle over Syria: Loyalists and Rebels Spying on Each Other with Off-the-Shelf Robots," *War is Boring* (blog), April 11, 2014, https://medium.com/war-is-boring/drone-battle-over-syria-159387e9de2b.

35. Rachel Browne and Alia Dharssi, "'Stunt Headed Nowhere': Activists Decry Plan to Use Drones to Secretly Film Forced Labour in India," *National Post*, December 13, 2013, http://www.news.nation-alpost.com/2013/12/13/stunt-headed-nowhere-activists-decry-free-the-slaves-plan-to-use-drones-to-secretly-film-forced-labour-in-india/.

36. "UN Advisers Invoke 'Responsibility to Protect' Civilians in Syria from Mass Atrocities," UN News Centre, June 14, 2012, http://www.un.org/apps/news/story.asp?NewsID=42235#.VEBdVOd8GeY; Michael Abramowitz, "Does the United States Have a 'Responsibility to Protect' the Syrian People?," *Washington Post*, September 6, 2013, http://www.washingtonpost.com/opinions/does-the-united-states-have-a-responsibility-to-protect-the-syrian-people/2013/09/06/5decf4c0-167d-11e3-be6e-dc6ae8a5b3a8_story.html.

37. Andrew Stobo Sniderman and Mark Hanis, "Drones for Human Rights," *New York Times*, January 30, 2012, http://www.nytimes.com/2012/01/31/opinion/drones-for-human-rights.html?_r=2&.

38. Ibid.

39. Sam Gregory, "Cameras Everywhere: Ubiquitous Video Documentation of Human Rights, New Forms of Video Advocacy, and Considerations of Safety, Security, Dignity, and Consent," *Journal of Human Rights Practice* 2, no. 2 (2010): 191–207.

40. "Drones Deployed to Capture Footage of Protests in Thailand," *France24 Web News*, December 4, 2013, http://www.france24.com/en/20131204-thailand-drones-deployed-to-capture-protests/.

41. Matthew Schroyer, "Interview with a Citizen Drone Journalist in Istanbul: 'I Have Been Witnessing Some Very Bad Things,'" Professional Society of Drone Journalists, June 24, 2013, www.dronejournalism.org/news/2013/8/interview-with-a-citizen-drone-journalist-in-istanbul-i-have-been-witnessing-some-very-bad-things; Matthew Schroyer, "Drone Journalism over Anti-ACTA Protests in Estonia," *mentalmunition.com*, February 13, 2012, www.mentalmunition.com/2012/02/drone-journalism-over-anti-acta.html; Robert Mackey, "Drone Journalism Arrives," *Lede Blog, New York Times*, November 17, 2011, http://thelede.blogs.nytimes.com/2011/11/17/drone-journalism-arrives/; "Hong Kong Protest: Drone Captures Scale of Protest," *BBC*, September 30, 2014, http://www.bbc.com/news/ world-asia-29421914; Aaron Sankin, "Should Drones Be Allowed to Fly over Ferguson?," *thedailydot.com*, August 17, 2014, http://www.dailydot.com/politics/ferguson-drone-footage-ruptly-video/.

42. Sarah A. Soule and Christian Davenport, "Velvet Glove, Iron Fist, or Even Hand? Protest Policing in the United States, 1960–1990," *Mobilization* 14, no. 1 (2009): 1–22.

43. Ian Steadman, "Turkish Protesters Use a Camera Drone, so Police Shoot It Down," *Wired*, June 24, 2013, www.wired.co.uk/news/archive/2013-06/24/turkish-protest-drone-shot-down.

44. I have the technologist and inventor John Holland to thank for this observation. Holland, interview with the author, April 26, 2014.

45. Refer to the Humanitarian UAV Network's website for more information: www.uaviators.org.

46. Refer to the Drone Journalism Lab's website for more information: www.dronejournalismlab.org.

47. "Domestic Drones," ACLU *Blog of Rights*, http://www.aclu.org/blog/tag/domestic-drones.

48. Roug, "Eye in the Sky."

49. Email correspondence with Patrick Meier, August 3, 2014.

50. Dean Starkman, *The Watchdog That Didn't Bark: The Financial Crisis and the Disappearance of Investigative Journalism* (New York: Columbia University Press, 2014).

51. "SPJ Code of Ethics," Society of Professional Journalists, http://www.spj.org/ethicscode.asp.

52. It is worth asking who will protect citizens from the prying eyes of an organized civil society group whose use of UAVs pursues a very narrowly defined agenda to the detriment of the public good. It is likely that this situation would be remedied by appeals to the state.

53. CAL. PEN. CODE § 647: California Code—Section 647(i), at codes.lp.findlaw.com/cacode/PEN/3/1/15/2/s647/.

54. Roug, "Eye in the Sky."

55. Verta Taylor and Nella Van Dyke, "'Get Up, Stand Up': Tactical Repertoires of Social Movements," *The Blackwell Companion to Social Movements*, ed. David Snow, Sarah Soule, and Hanspeter Kriesi (Malden, England: Blackwell, 2004), 262–293.

56. Doug McAdam and David A. Snow, eds., *Social Movements: Readings on Their Emergence, Mobilization, and Dynamics* (Los Angeles: Roxbury, 1997), 326; Philip Howard, *The Digital Origins of Dictatorship and Democracy: Information Technology and Political Islam* (Oxford, England: Oxford University Press, 2010).

57. Taylor and Van Dyke misattribute this statement to Rochon (1998) on p. 264.

58. Ron Eyerman, "The Role of the Arts in Political Protest," *Mobilizing Ideas*, June 3, 2013, http://mobilizingideas.wordpress.com/2013/06/03/the-role-of-the-arts-in-political-protest/; "FAA Selects Six Sites for Unmanned Aircraft Research," FAA, http://www.faa.gov/news/updates/?newsId=75399.

59. "April 12 Event Will Put Drone Skills to the Test," North Carolina State University, March 26, 2014, http://news.engr.ncsu.edu/2014/03/april-12-event-will-put-drone-skills-to-the-test/.

60. See www.irevolution.net.

61. An example of this broader range of actors would include Amnesty International's launch of the Eyes on Darfur campaign.

62. See the Satellite Sentinel Project, http://www.satsentinel.org.

63. Patrick Meier, "The Use of Drones for Nonviolent Civil Resistance," *iRevolution.net*, February 18, 2012, http://irevolution.net/2012/02/18/drones-for-civil-resistance/.

64. Ibid.

65. Aaron Smith, "U.S. Views of Technology and the Future: Science in the Next 50 Years," *Internet and American Life Project*, Pew Research Center, 2014, http://www.pewinternet.org/2014/04/17/us-views-of-technology-and-the-future/.

66. Leon Watson, "Woman Claims She Was Sexually Harassed by a Drone after Catching Man Flying Remote-Controlled Plane at the Beach That Got Uncomfortably Close to Female Sunbathers," *Daily Mail*, May 15, 2014, http://www.dailymail.co.uk/news/article-2629459/Woman-claims-sexually-harassed-DRONE-catching-man-flying-remote-controlled-plane-beach-got-uncomfortable-close-female-sunbathers.html#ixzz34FEL.npg; Jason Koehler, "This Kid Got Assaulted for Flying His Drone on a Beach," *Motherboard* (blog), Vice Media, June 9, 2014, http://motherboard.vice.com/read/this-kid-got-assaulted-for-flying-his-drone-on-a-beach.

67. Jason Koehler, "African Nations Are Banning the Drones That Could Stop Poachers," *Motherboard* (blog), Vice Media, June 4, 2014, http://motherboard.vice.com/read/african-nations-are-banning-the-drones-that-could-stop-poachers.

68. "Drones Banned in South Africa," *Times Live*, June 3, 2014, http://www.timeslive.eo.za/scitech/2014/06/03/drones-banned-in-south-africa.

V TWENTY-FIRST-CENTURY CHALLENGES AND STRATEGIES

The complexities explored in the last section suggest that those seeking to change the world face a daunting challenge. The complex relationship between society and technology, coupled with the fact that a range of actors influence technology and often do so with incomplete knowledge of how it will behave and what its effects will be, makes it difficult to understand, let alone direct. How can all of this be managed? How can we steer sociotechnical development to solve problems and realize values that are essential to human wellbeing?

In the face of this complexity, it may seem that people have little control. Hence, it is important to remember that technology does not just happen; it is the product of people. Sometimes adoption of a technology leads to shifts in power that those who designed the technology intended. Other times it comes with consequences that people choose to tolerate. Or they choose to do something about the consequences, such as changing the design of the technology, changing the policies that regulate the technology's uses, or even suing the corporation whose products ultimately led to the unintended consequences. The sociotechnical world is the result of these types of decisions, decisions made by people (individuals and groups), and when we are aware of this, we are better equipped to make choices that lead to a better future.

To be sure, it is easy to forget that people control technology. For instance, many of the technologies currently being developed are referred to as *emerging technologies*. This label is effectively a hidden form of technological determinism; it implies that these technologies are in nature and simply need to be found and uncovered by scientists and engineers. However, what "emerges" from the processes of research and development is largely a function of what engineers, government agencies, social critics, consumers, politicians, and many others decide to do.

For example, although we think we have a general idea of what autonomous cars are, at this point in their development, it is still unclear how they will operate in our lives. Will we design a system in which no one owns a car, and they roam free until they are called upon? Will we create a mixed system of driverless and human-driven cars? Will we develop a system in which all of the autonomous cars are identical in design and controlled by a central authority? A vast number of social decisions still have to be made before we can see what kind of an impact autonomous cars will have and how they will be part of our daily lives. Like other emerging technologies, the sociotechnical system of autonomous cars is not yet "made"; it is "in the making" and can still be made into "this" or "that" particular kind of system. Human beings will decide what these technologies come to be, how they are integrated into society, and even whether they will exist at all. In order to make conscientious decisions about the future, the connections between human decision-making and technological design and use need to be made visible and understood.

With this in mind, we turn now to the current context in which the future is being created. This section examines trends and issues that face us in the twenty-first century. The readings ask (and try to answer) such questions as: How will ideas about what it means to be human change as a result of modern medical technologies? How will global order change as national conflict and warfare increasingly take place in the cyber realm? How can we test new and potentially powerful technologies for addressing climate change when the testing itself poses risks to groups of people and the environment? Can we build sustainable systems and, at the same time, address environmental injustice? Will artificial intelligence diminish investment in human development?

The readings identify trends that are predicted to continue and intensify, problems and issues that cannot or should not be ignored, and technologies that are generating concern as we move further into the twenty-first century. Continuation of some trends could mean a better world enriched by better health care and greater convenience, leading to longer, healthier, and less stressful lives. These trends could also mean more peaceful, more democratic, and more comfortable lives. However, continuation of other trends could result in political unrest, degradation of the environment, infringements on basic freedoms, and so on. How technology is developed and steered will make a significant difference in whether the promises or the threats of the future are realized. Indeed, it is for this reason that it is important to take seriously the concerns expressed about emerging technologies as well as to listen to the potential benefits. An understanding of the possibilities for good and ill puts us in a better position to use whatever power and influence we have to affect the direction of development.

The selections in this section are intended to provoke and begin to answer the question: What kind of future world is desirable? Since the world is continually changing, the answer to this question can never be settled once and for all. Change is eternal and learning continuous, so answers must always be tentative, and the question must be asked repeatedly. As the readings in this volume emphasize, the future is being created through the interplay of the technical and social, through interactions among many actors, including those of the past and those of the present. Scientists and engineers tell us what is possible, and public discussion, debate, and funding around these new possibilities influence what scientists and engineers choose to do. This section provides an opportunity to both observe that interplay and to engage in it by trying to figure out what we should think about possible directions for the future.

Of course, the readings included in this section provide only a sample of contemporary discussions about the future. New sociotechnical enterprises are constantly being envisioned and new problems arise every day. As we were writing this book, several new challenges became apparent, including how to adequately address pandemics, the threat of a post-truth society, and increasing numbers of people questioning the foundations of democratic government. Those who want to contribute to the future must keep abreast of trends and developments in technological prowess. Rather than see this section as a map of future issues, consider it a starting point for thinking about the future. It stresses that the decisions we make today will have an important effect on the world of tomorrow. It challenges us to face the latest issues, to envision the world we want, and to actively build a sociotechnical future that is better for all. Practice in thinking through these issues will help prepare us to tackle the unforeseen problems we will all face tomorrow. Together we can debunk the myth that technology is going to happen, and we just have to deal with it. Rather, we shape the world through the decisions we make about technology. Knowing that makes it possible to create the world we want to live in.

Questions to consider while reading the selections in this section:

1. If the trends described in this section continue, who will be the winners and losers?
2. What forces are influencing the trends or creating the problems discussed in this section?
3. What values are being realized or threatened?
4. In what ways can social change be more effective in solving problems than technical change?
5. How are decisions being made to steer technology?

28 Engineering the Brain: Ethical Issues and the Introduction of Neural Devices

Eran Klein, Tim Brown, Matthew Sample, Anjali R. Truitt, and Sara Goering

Some of the major improvements to the lives of individuals in the twentieth century came from scientific and technological developments in the fields of health and medicine. New medicines were developed; powerful new imaging devices such as X-rays and MRIs were invented; the Human Genome Project brought a foundational knowledge of human genes and how they contribute to many diseases; several infectious diseases that had plagued humanity for centuries were eliminated through new vaccines and distribution practices. Research and development in medicine and health continues in the twenty-first century—some would say at a faster rate—so that the promise for future improvements in health is enormous. Nevertheless, as suggested in many of the readings in this volume, technological advances come with social consequences, and this is as true for medicine and health as for other domains of technology.

One of the most incredible new lines of research that will unfold in the twenty-first century is focused on understanding how the brain works. Researchers are developing the capacity to manipulate the brain. Although this ability has the potential to address many serious medical problems, the social implications of this research are profound because the brain is so much a part of who an individual is.

In this article, Eran Klein, Tim Brown, Matthew Sample, Anjali R. Truitt, and Sara Goering identify a set of social concepts that the endeavor to engineer the brain will significantly challenge. These include notions of identity, normality, authority, responsibility, privacy, and justice. The authors use two fictional case studies to illustrate the benefits of neural devices while at the same time exploring the deep ethical issues they raise. This reading is presented as a starting place for the kind of broad public discussion that should accompany the development of possibly transformational new technologies like neural devices.

Neural engineering technologies such as implanted deep brain stimulators and brain-computer interfaces represent exciting and potentially transformative tools for improving human health and well-being. Yet their current use and future prospects raise a variety of ethical and philosophical concerns.[1] Devices that alter brain function invite us to think deeply about a range of ethical concerns—identity, normality, authority, responsibility, privacy, and justice. If a device is stimulating my brain while I decide upon an action, am I still the author of the action? Should I be held accountable for every action in which a device is operative? Does a device make the interiority of my experience accessible to others? Will the device change the way I think of myself and others think of me? Such fundamental questions arise even when a device is designed

---

From *Hastings Center Report* 45, no. 6 (November–December 2015): 26–35.

for only a relatively circumscribed purpose, such as restoring functioning via a smart prosthetic.

We are part of a National Science Foundation–funded engineering research center tasked with investigating philosophical and social implications of neural engineering research and technologies.[2] Neural devices already in clinical use, such as deep brain stimulators for Parkinson's disease or essential tremor, have spurred healthy debate about such implications.[3] Devices currently under development—such as the BrainGate System of implanted brain sensors coupled to robotics in persons with paralysis,[4] exoskeletons for augmented movement,[5] transcranial do-it-yourself stimulators (tDCS),[6] closed-loop brain stimulating systems,[7] or even brain-to-brain interfacing[8]—promise to extend and deepen these debates. At our center, brain-computer interfaces (BCIs) are the principal focus of work.[9] Even acknowledging that the clinical translation of neural devices and seamless integration by end users may still largely reside in the future,[10] the potential of these devices calls for careful early analysis. The launching of the BRAIN (Brain Research through Advancing Innovative Neurotechnologies) Initiative in April 2013 provides further impetus for this work.[11]

In our work alongside neural engineers, we have come to view the work of engineering the brain as incredibly complex, not simply because of the technical feats it requires but also because of the varied ethical domains upon which the work touches. The functioning of the brain is intimately connected to an individual's and a culture's understandings of identity, human responsibility, privacy, authority, justice, and normality. In this paper, drawing on and extending work in neuroethics focused on deep brain stimulation, we explore how neural engineered devices, particularly brain-computer interface devices, challenge or may soon challenge our understanding of these six domains. We have structured this paper to focus on each of these domains individually, but as will quickly become apparent, we recognize that doing so is part artifice; these domains, particularly in their relationship to human agency, are often entangled and best explored together. Even careful examination of BCIs and privacy, for instance, leaves out something of critical import if it fails to attend to ways that outside access to our thoughts or intentions—even where secured by firewalls and encryption—may deeply affect our ideas about identity or responsibility.

Consider two fictional case studies, both based on developments in neural engineering: Joan is a thirty-year-old mechanical engineer and combat veteran who sustained an injury from an improvised explosive device. On her right arm, she has an above-elbow amputation, and her left arm has significant nerve damage that causes pain. After a year of aggressive rehabilitation at a military hospital, she volunteers to work with a research team developing a state-of-the-art robotic prosthetic controlled by a system that involves a brain-computer interface. A powered array of sensing electrodes and a chipset implanted in her brain records and sends signals to a robotic arm prosthesis through a wireless connection. Joan works with the engineers to customize the appearance of the prosthetic to fit her needs and personality: "Strong, durable, and nothing frilly, but with just enough of a soft exterior that I can hold my newborn daughter." When a new model of prosthetic is developed for improved functionality that requires extensive training or sharing of movement data from her existing prosthetic, she declines the upgrade, preferring to "keep that part of me just as it is, just as my daughter knows me."

John, a forty-five-year old man, has struggled with depression since his late teens, including several suicide attempts and inpatient psychiatric hospitalizations. Antidepressant medications, counseling, and electroconvulsive therapy have been unsuccessful.

Although he is well-liked by his coworkers and supervisors, his debilitating depressive episodes have made his deadline-oriented work as a technical writer difficult to accomplish at times. These complications threaten his job security. He volunteers for a study investigating an experimental BCI, a closed-loop deep brain stimulation (DBS) device for refractory depression as "my last hope." A set of sensing and stimulating electrodes are implanted in his brain and connected to a unit implanted in his upper chest. The unit is responsible for interpreting signals from the sensory electrodes, determining when treatment is needed, and applying a current through the stimulation electrodes. This sensor-stimulator loop detects changes in his mood and adjusts stimulation to achieve an appropriate set point determined during consultations with his physician. After implantation, John's depressive symptoms dramatically decrease, a fact that he frequently shares with his work colleagues, noting, "I am finally back to being me."

Identity

Our identities are complicated by our relationship with technologies. Many of us rely on tools such as smart phones, laptops, and GPS devices. Indeed, we sometimes incorporate these tools into our self-understandings—"He's a Mac guy," or, "She couldn't live without her smart phone." When our tools not only aid us but also directly replace parts or functions of our bodies, as may be the case for Joan and John, the effects of technology on identity are potentially even more significant. The experience of coming to identify with a prosthetic is rich and complex.[12] People who are blind sometimes think of their canes as part of their perceptual systems;[13] individuals with communication disorders can identify with their computer-synthesized voices.[14] When a tool functions so well that it becomes an integral part in our lives, we might say it has become part of us—it is no longer a mere tool.[15] Are neural devices likely to be taken up into our identities, and if so, to what benefit, or at what cost?

Becoming "part of us" can, of course, have multiple meanings. The technology might simply become part of how I consciously think of myself and want others to see me; it's part of my social identity, a way that people recognize me for who I am. It might, however, also become part of how I understand myself even at subconscious or neural levels, as when it enters into my body schema. Both philosophical and empirical work has suggested that we readily incorporate various tools into our body schemas,[16] and these seamlessly incorporated tools are sometimes considered an "extended body" beyond the confines of our skin. Similarly, we might envision the possibility of what philosophers Andy Clark and David Chalmers call "extended mind," where a person relies on external aids, such as smart phones or notebooks, to perform cognitive functions.[17] Neural engineering devices implanted in brains and designed to interface with existing nervous tissue in closed-loop systems may complicate our thinking about identity by changing our notions of both social identity and body schema. Shifts in identity can be a positive development—as when device-based alterations put us more in line with how we see ourselves—but they may also undermine identity in certain ways and come with acceptance costs for users.[18]

Think about Joan's case. Joan, like others who have become disabled, adjusts to this fact not only in how she interacts with her environment but in how she sees herself: "I am," she thinks, "a person without a functional arm." The transition to this new identity can be difficult and hard won. The offer of a smart prosthetic—even one that promises improved functionality—can be met with resistance if it is felt to undermine this new identity. Recall Joan's concern about her daughter's ability to recognize her *through her prosthetic*. Trying a new device is not necessarily cost free. Even when this cost is deemed worth paying, further threats to identity may follow soon upon adopting it. What if a new device is too cumbersome for her to achieve seamless integration into her body schema?[19] What if the new prosthetic is only temporary, a bridge technology to yet another device? Or what if Joan has lingering concerns about whether every action of the arm is indeed hers, given the possibility of device malfunction, design flaw, or interference from a third party? Her nagging doubts may stand in the way of her incorporating the device into her identity.[20] From Joan's perspective, a device's potential identity costs may be significant.

Her identity might also be stretched if the technology took a different form. The BCI used to control Joan's attached smart prosthetic might be used to control a detached assistive device, such as a robotic arm. In this case, the arm, even as it functions, no longer needs to be permanently attached to her body. The device might be wirelessly connected to her BCI in a way that allows her easy control of the device, and if the device had sufficient range and maneuverability, she might be able to send it to the next room on a task—getting her a drink or patting her child to sleep. Would a roaming robotic "arm" of this kind seem more like a mere tool, or would Joan be able to think of such an arm as "her arm" also? In the latter case, our common understandings of identity would experience foundational shifts.[21] Related questions about who and where an individual is might have implications for moral and legal responsibility as well.[22]

A different sort of identity question—one linked to concerns about authenticity[23]—arises in the context of John's DBS system. Will John feel like himself when his mood elevates as a result of an algorithmic feedback loop rather than through his usual physiologic process? If he experiences a negative mood that is context appropriate—perhaps in response to sad news—will he find it disorienting that the device quickly causes him to feel better? Would he even notice this change, or would it be something he feels obliged to explain, ex post facto, by reevaluating his own assessment of the news, given that his quick recovery from it occurs internally and without an obvious trigger, such as taking a pill? A person learns to "read" her reactions to events as a means of understanding her own internal states ("I must not have liked him so much, given that I didn't even cry when I heard he had died"). Similarly, John may find himself rethinking his values given his steady mood in the face of loss. Even if *he* does not notice, others around him may wonder about the authenticity of his responses, noting, for instance, that "John doesn't seem like himself" if his response does not fit the situation.

Carl Elliott has explored this question of authenticity in the context of neuropharmacology. Elliott critiques the widespread acceptance and promotion of Prozac and other selective serotonin reuptake inhibitors, given their capacity to make us feel better about situations in which we might more authentically feel despair. Some ways of responding to those situations are uniquely ours; they signal to others who we are. If we alter those response patterns through "smart" implants and effectively take the individual out of the response—even in the name of attempting to treat debilitating depression—we risk undermining the authenticity not just of the mood but also of the person's capacity for self-expression. As we develop neural technologies, we need to examine our aspirational norms and the ways such technologies may interfere with (or perhaps enhance) our capacities for self-definition and self-expression.

As a preview, if an individual's BCI is hacked, that person's movements or mood could be manipulated by another. She could easily find this more threatening—for her sense of self, for her moral and legal responsibility, and so forth—than if her computer or bank account is hacked and data stolen. Similarly, if the datasets from a person's BCI are recorded, how might such recordings be used by educational systems, courts of law, or employers? With ever more complex recording systems and algorithms for identifying and translating intentions, concerns about privacy become entangled with our understandings of identity and authenticity. Other closed-loop implantable devices, such as cardiac pacemakers, function relatively autonomously, but their control is less obviously linked to central features of our identity.

Normality

The concept of normality is central to the development and implementation of BCI devices. Take, as an example, John's BCI used for the regulation of mood. A neuromodulatory device that aims to change an emotional state in a particular direction—say, from depressed to happy—relies on norms of affective function. This requires that we set parameters delineating what counts as the abnormal state to be corrected and what counts as the normal state to be sought. Ascertaining or setting the standards for "normal" function is challenging, and not just for technical reasons.

Neural engineers often take themselves to be employing "objective" measures of normal and abnormal function. In John's example, normal brain function can be drawn in strict physiologic terms, such as by appeal to particular regional patterns of electrical activity or neurotransmitter levels. A device capable of repeatedly sampling physiologic brain function for deviation from the norm and iteratively intervening to reestablish normal function could constitute a closed-loop neuromodulatory system—a "pacemaker" for mood. Whether there is or could be one pattern of abnormal (or normal) electrical activity or neurotransmitter levels that faithfully represents mood across individuals, or even across the same individual over time, is an open question. Naturalistic theories of normality, such as Christopher Boorse's biostatistical theory or Norman Daniels's theory of species-typical functioning, would seem a natural place to start for defining normal brain function, but these approaches have been criticized for failing to recognize that appeals to normality are often ineliminably value laden.[24]

Take Joan's motor-oriented BCI and the ability to shake another's hand. The ability to shake hands is, on one level, a rather mundane sensorimotor skill, but it also is implicated in important social practices, such as expressions of autonomy, trust, friendship, negotiation, and courtesy, and this linkage complicates the definition of "normal." What counts as a normal grip, not too firm and not too soft? When is shaking another's hand too vigorous or too rigid? When is it too soon to release a grip, and when does a lingering grip become awkward? A firm, prolonged handshake may communicate trustworthiness during a business interaction but could indicate aggression or one-upmanship in a meeting of new acquaintances. A purely scientific rendering of normal hand-shaking behavior, one that tries to set aside the normative, may not be possible.

Feminist and disability critiques have challenged prevailing notions of normality, pointing out that individuals are not abnormal or defective by virtue of their disabilities; rather, individuals are disabled due to inhospitable environments that make their abilities a poor fit for individually or socially desired ends.[25] The implication for development of neural devices is that "normal" may not be what all end users want and how normal is best understood may deviate from simple notions of replacing lost human functions.

Neural devices can also improve functioning in ways that raise questions about neuroenhancement. Imagine if Joan's prosthetic arm affords her greater strength or endurance and allows her to perform a job that displaces several other workers. Could such a device give her a competitive advantage? Would it privilege her in some way or yield benefits that are not earned in a traditional, "authentic" way, such as through strength or endurance training? Conversely, should such a neural device be embraced, provided it is medically safe, because it offers a chance to improve on arbitrary limitations to human abilities? Analogous worries have been raised by the use of neuropharmacology to enhance cognitive function and mood and by the use of prosthetics in sport.[26] The concept of normality has been recognized as central to understanding and making progress in debates over pharmacological enhancement.[27] The same can be said about enhancement concerns raised by neural technology.[28]

The challenge of defining normality in BCIs may be complicated still further by the intertwining of related concepts such as authority and identity. For instance, physiologic measures of normal or abnormal function can come apart from introspective experience. If the sensors that trigger John's deep brain stimulator indicate that he is depressed at a time when he denies feeling depressed and fails to exhibit outward signs of depression, is he in fact depressed? Closed-loop neuromodulation of mood may lead to a disconnect between what is felt and what an individual or others think ought to be felt. The implications for identity from experiencing such a disconnect could be significant. Conversely, what if John reports feeling depressed but objective measures do not bear this out? What or who is the ultimate arbiter of normal affective function? Where does authority reside?

Authority

The goal of BCI technology is to translate brain processes underlying thought and action into desired outcomes, like grasping of a prosthetic hand or elevation of mood. BCIs are mediated by complex algorithms that take data from carefully placed brain sensors or electrodes, mine them for a desired signal, and convert them into a mechanical or electrical activity. Normative questions arise along this translational pathway.[29] How are the relevant characteristics of a common (or, as above, "normal") input signal to be defined? What signal qualifies as a person's intention, rather than a fleeting, fragmented, or even personally abjured thought? The challenge for neural engineers is not only to design a signal processing algorithm sufficiently sensitive to allow Joan's arm to gently grasp an offered flower, for instance, but also selective enough to prevent her hand from crushing the flower (or whatever else she is holding, like her daughter!) as a scene from *Little Shop of Horrors* momentarily passes through her consciousness. And perhaps most importantly, how much authority should we invest in the translational algorithms of BCIs and in what ways?

BCI systems offer a powerful, alternative way to access a person's mental life apart from first-person testimony. While BCI systems like Joan's are not designed to monitor intention, they might be able to record past commands. If so, we can imagine instances where BCI recordings disagree with a user's subjective reports. A BCI might indicate, for instance, that Joan "meant" to turn a steering wheel to the right and into a neighbor's fence, despite her insistence to the contrary. The alternative ontology afforded by BCIs has the potential to reshape how we understand and lend credence to claims of self-knowledge.

Such discussions of authority have some precursors in the literature on the neuroimaging of pain. Functional MRI has been explored as a possible tool in the diagnosis of chronic pain.[30] Whereas diagnosis and treatment of pain have traditionally relied predominantly on first-person subjective reports (such as "I am in pain now" and "My pain is a 4 out of 10"), neuroimaging may offer an alternative way to assess the level, kind, and even the presence of pain. The move toward neuroimaging evidence of pain raises concern. The limitations of neuroimaging of pain can be underestimated, as can the potential harm to individuals when imaging is misread or overinterpreted; if "objective" measures are placed on a par with personal testimony, false negative results may deny or diminish legitimate claims of pain-related suffering in both legal and medical contexts.[31] Consequently, medical practitioners should use a precautionary principle in cases of subjective-objective disagreement, according to which the deference to subjective reports reflects the seriousness of error.[32] Responsible use of the technology is not provided by the technology itself, and "objective" measures, such as colorful images of pain, may provide a kind of "illusory accuracy."[33]

A precautionary principle in the setting of subjective-objective disagreement is a lesson that could carry over into BCI systems. The complicated intersectional nature of agency, however, prevents the precautionary principle from applying straightforwardly. If Joan has come to include a BCI as part of her identity, then a device malfunction or a subjective-objective disagreement might be an experience of self-alienation that a precautionary principle on its own will not ameliorate. Joan may adamantly deny intending to swerve her car, but the mere fact of the event may be enough to engender feelings of guilt and self-doubt.[34] Legal and moral responsibility leads to similar complications; unlike cases of pain assessment, which are more clearly about limited reports of individual needs, legal and moral judgments will be made about *any* of the actions Joan makes with her prosthesis. Recall Joan's denial of an intention to run her car into her neighbor's fence. If a court of law introduces BCI recordings as evidence, and the data set clearly shows that there was such a motor intention immediately preceding the action, what should be done? Here, a precautionary principle is not a clear solution to conflicts between the objective and the subjective, though it may offer an appropriate starting point.

Moral and Legal Responsibility

We typically hold people responsible for actions over which they can exercise control. Taking responsibility for our own actions and holding others to account is a fundamental feature of living in moral community with others.[35] Neural devices, as we have seen in the cases of John and Joan, can complicate our notions of responsibility. Neural devices can provide a new source of information with which we can judge responsibility, as in John's closed-loop DBS recording of electrical activity related to his mood, even if the extent to which authority should be vested in this information is not self-evident. Neural devices can also be themselves *involved in* actions for which responsibility is at issue, such as in Joan's car accident. Insofar as they influence actions, thoughts, or feelings, they affect responsibility, both moral and legal.[36]

An individual with a neural device can be held responsible in different ways. Recall Joan's prosthetic arm and the car accident that causes destruction of her neighbor's fence. She claims she did not intentionally turn the car. Still, she might be held responsible for choosing to have and use a prosthetic arm in the first place; after all, had she not had it, her neighbor's fence might be safe. More commonly we ascribe responsibility for actions more proximate to the event of interest. Along these lines, we might hold Joan responsible for failing to train her arm adequately and in so doing increasing the risk of a resultant harm. If lost control due to inexperience were a foreseeable possibility, as intoxicated driving is a foreseeable consequence of alcohol ingestion, some attribution of responsibility might be appropriate.[37] Or we might hold Joan responsible for a momentary lapse of focus immediately preceding the event. However, we might absolve Joan of moral or legal responsibility if the device were poorly constructed or made use of faulty software and, in turn, malfunctioned. We might also wonder if the level of concentration required by Joan to control her prosthetic should temper the assignment of responsibility for mistakes, given the relative difficulty of her task. Fairly distributing responsibility can be a complicated affair.

Individuals with neural devices may be owed special moral consideration by others. People with disabilities can come to rely on such devices to secure social and other goods. For instance, Joan's device may allow her to gain employment, care for her child without assistance, or maintain a healthy sense of self-confidence. Given the extent of her reliance on the device, she may be owed special consideration, such as affordable replacement in the case of device failure. This is in part a consideration of distributive justice, but it may be more than this. It may also be a responsibility of everyone who encounters her individually to acknowledge the device's value in their interactions with her. A sudden malfunction of the prosthetic would not be like losing a favorite hat; it would be more akin to a blind person's losing her cane on a busy street.

Neural devices also affect responsibility insofar as they leave an auditable information trail. Whereas assigning responsibility for an action typically entails piecing together its history out of available (and sometimes unreliable) elements—reported memories, states of mind, environmental circumstances—neural devices offer the possibility of a detailed history of brain states leading up to an action. Even setting aside debates about how such brain states are causally related to actions, the mere presence of this trove of information is significant. It offers a detailed and available—and hence likely quite attractive—source of information about an action of interest. Courts, insurance companies, and others will be interested in that data.

Responsibility for actions involving neural devices is made more complicated by the intersection of identity and responsibility. Individuals with neural devices may not only become functionally dependent on devices but, as we have seen, may also incorporate these devices into their sense of self and body schema. If the connection to identity is taken seriously, the implications for responsibility might be striking. For instance, if Joan's prosthetic arm becomes an extension of her body and an integral part of her identity, its destruction could be traumatic. From a moral standpoint, its willful destruction could be taken as a significant moral transgression. From a legal standpoint, its destruction might be appropriately classified as battery, rather than mere property destruction, even if it is not being worn at the time.[38]

Privacy

As the technologies to monitor and intervene in complex neurological systems become more robust and useful to end users, platforms like Joan's prosthetic arm are likely to be integrated with wireless technologies, many of which are already implemented in consumer products. Wireless standards like near-field communication—exemplified by the ubiquitous radio-frequency identification systems in cellular phones, debit cards, security passes, and so on—are vulnerable to tampering, misuse, and attack. BCIs using such standards will be made vulnerable to the same security exploits that affect all devices using those standards. Again, Joan's prosthetic device might store her previous motor commands such that they could be retrieved later by either a medical professional or, potentially, a malevolent agent. Similarly, John's closed-loop deep brain stimulator might record the level and frequency of his treatment such that its wireless diagnostic interface could be used to extract details about his medical condition.

These two possibilities illustrate how neural engineering involves privacy at multiple levels. Illicit access to John's neurostimulator seems to recall issues of "brain privacy" raised in previous neuroethics literature that considers how neuroimaging technologies might reveal an individual's psychological traits or mental states ("brainotyping"), attitudes toward other people, and truthfulness.[39] Even without imaging, a stranger who gains access to John's closed-loop DBS recordings might be able to infer analogous details about John's emotional state or psychological disposition from the stolen data. While brain privacy doesn't quite capture the threat posed by stolen recordings from Joan's motor prosthetic, she might experience a threat to privacy if her data is combined with other nonprivate information about her—from social media, tracked web activity, or public record.[40] Further, BCIs are vulnerable to "brain spyware": malicious programs that can extract private information from the right kinds of neurological data. An attacker might, for example, present Joan with specially designed visual stimuli and derive private data from her neurophysiologic responses.[41]

One response to both of these threats to privacy might be deflationary; we need just take the relevant security precautions to prevent hacking in the first place. Appropriate encryption and design constraints could eliminate nearly all security risks. That might be a reasonable response in the context of computer privacy. In the case of BCIs, however, it seems more is at stake. If an attacker compromises a user's personal computer, the attacker will be able to access that user's data, perform tasks using his computer or disable that computer altogether. If an attacker were to compromise John's or Joan's device, the attacker might access stored neurological data, control his or her devices (against the victim's will), or disable the devices entirely. The task of securing a neurological device is a grave one: "Instead of protecting the software on someone's computer, we are protecting a human's ability to think and enjoy good health."[42] To this end, Tamara Denning and colleagues coin the term "neurosecurity" as "the protection of the confidentiality, integrity, and availability of neural devices from malicious parties with the goal of preserving the safety of a person's neural mechanisms, neural computation, and free will" (p. 2).

Identity complicates privacy concerns: we have yet to anticipate how a user's self-image will change when her personal neurological data can be accessed the same way we might access a file. At the very least, it seems that something that was once inexorably private, something that often comprises our sense of self, has become potentially public: available for access, interference, or inquisition. Perhaps there is something potentially disquieting about this shift itself. We can call this—picking up on Denning's nomenclature—the potential accessibility problem, or figuring out where to draw the boundary between public and private given the very existence of devices that collect and analyze neurological data.[43]

Whatever safeguards engineers implement to prevent breaches of security and privacy in BCIs and other neural devices, issues of fairness or justice might still arise or even be exacerbated. In the near term, BCIs will most likely be the end user's last resort treatment option for both motor control and neurostimulation. As such technologies advance, users may feel pressure to accept a technology in order to address their concerns. Will consenting to a BCI with wireless technology mean consenting to diminished privacy or increased potential accessibility? Could such devices be built differently to address end user needs without compromising the privacy of collected data or the security of the device's functions? Traditionally wired systems, while less convenient, would avert at least some of the privacy concerns. What kinds of compromises is it reasonable to expect users to accept?

Justice

Like other health-related technologies, neural devices raise issues of justice with regard to the distribution of harms and benefits and to the inclusion of perspectives from people likely to be affected by the technology. Although distribution concerns tend to attract the most attention in debates about justice—and have relevance here, particularly given the resource expense required to develop technologies that may never translate into widely available consumer goods—we should not underestimate the significance of the concept of *justice as recognition*.[44] How do we ensure that groups who are often differently socially positioned—such as the disabled people who are the intended beneficiaries of such technologies—have their perspectives on the meaning and significance of their bodily differences and these technologies heard and respected? Even once funding decisions have been made, how might we ensure that end users like Joan and John have their perspectives and concerns integrated into the development process?[45] If the technologies in question are to be designed so that end users will adopt them, attention must be paid to their particular needs, concerns, and experiences well before final products are determined. Justice as recognition demands explicit and meaningful engagement of likely downstream end users at major decision points in the design of neural technologies.[46]

To understand why, consider the potential individual trade-offs of adopting some of these technological devices. Unlike a traditional prosthetic, a BCI like Joan's does not allow the flexibility to completely abandon the device whenever she wants. An implanted BCI would require medical intervention for complete removal, and it may provide limited ability to turn on and off its recording functions.[47] As a consequence, her macro level of control and privacy may be more limited with the BCI compared to a traditional prosthetic.[48] For some, these potential threats may not be important, especially given the sense of freedom and independence that the BCI may provide. Still, if Joan perceives her device as acting unreliably and cannot distinguish whether this is a problematic design issue or a malicious hacking event, we may consider it at least reasonable that she would prefer the capacity to simply turn it off. In the interest of justice, researchers might be compelled to direct resources toward development of noninvasive BCIs, like caps or bands using electroencephalography, which may provide more user flexibility, although signal resolution still poses a technical hurdle. Balancing moral considerations with technical ones is a notable challenge but central to equitable decision-making. Consider the cochlear implants debates and the importance of allowing individuals to choose multiple pathways for enhanced communication (cochlear implants, speech therapy, sign language interpreters, or some combination).[49] In neural engineering, similar concerns arise if dominant social forces and arguably narrow notions about "normal" functioning pressure individuals to choose one modality over the other or to give up local control over the capacity to "turn off" a prosthetic.

In John's case, our moral as well as legal understanding of responsibility may complicate the concern about local control. Assuming that John has freedom to self-regulate his deep brain stimulator, he may pose a risk to his health or that of others; he may neurostimulate in such a way that produces a particularly enhanced mood, perhaps with side effects linked to mania.[50] Who should determine the appropriate settings in such cases? Sharing of information about neural stimulation or decision-making with a health care provider, for instance, may seem justifiable in John's case. Some might also consider it permissible for a health care provider to intervene if John was experiencing a depressive episode, even if he chose not to neurostimulate, much as psychiatric patients are sometimes committed to involuntary holds based on perceived threats to self or others. While the health care provider may be acting in what she perceives as the best interest of her patient, John might disagree, or alternatively, he might understand his depression as part of who he is, a significant factor in his identity. This kind of discrepancy between a health care provider and psychiatric patients has been well-documented by the Mad Pride, psychiatric survivors, and neurodiversity movements.[51]

As in the case with Joan, some might argue that John could have chosen an alternative to a BCI and that these potential consequences were foreseeable and avoidable. In choosing the BCI, he accepted these risks. Yet even if John and his health care provider understand the therapeutic benefits of a BCI (for example, better adherence, reliable access to medication, and less need for transportation to get to therapy) as significant, perhaps they would still recognize John's need to retain some form of control, and they might even prefer to err on the side of giving more control to the individual user. In terms of justice, we may also consider that, particularly for some marginalized communities, use of the device may be an attempt personally, socially, or institutionally to address broader social situations that deserve attention but often do not receive adequate social support and funding.

Ultimately, attending to concerns of justice as recognition will require that end users have input about design decisions and have eventual access to multiple options and the flexibility to change their minds as life warrants. To ensure this, researchers must begin to recognize the trade-offs that individuals face when choosing and using a device. Because of the complexity of end users' lives, neural engineers must engage with end users to design and develop devices that address community and individual needs.

The Beginning of a Discussion

As public funding for neurotechnology expands and interest in neural engineering increases, the ethical issues raised by these technologies must be carefully explored and analyzed.[52] Important steps have been made in this direction,[53] but more needs to be done. The Presidential Commission for the Study of Bioethical Issues, for instance, notes that advances in neuroscience raise complex issues related to cognitive enhancement, consent capacity, and legal responsibility and decision-making. Focusing attention on these areas of ethical, social, and legal concern is an important step for neuroscience and society, but understanding and addressing such concerns (and others) will require further empirical and normative work—such as what we hope to have begun to offer here.

The six core areas of ethical concern that we have identified—identity, normality, authority, responsibility, privacy, and justice—by no means form an exhaustive list; other areas of ethical concern, such as stigma and autonomy, could be added. Our list derives from both our ongoing discussions with neural engineers and from the bioethics and neuroethics literature. We believe that these six areas cover a substantial swath of the conceptual ground relevant to neural engineering and provide a starting point for discussion inside and outside neural engineering. Our hope is that these core areas can be a useful scaffolding for scholars and others as they work through challenges ushered in by neural engineering.

Acknowledgments

We would like to thank Suzanne Holland and Alicia Intriago. Thanks also to all of our collaborators at the Center for Sensorimotor Neural Engineering for taking the time to talk with us about their work and recognizing the need to think seriously about the ethical implications of neural engineering.

Disclosure

This work was supported by a grant from the National Science Foundation (award #EEC-1028725). The views are those of the authors and do not necessarily reflect those of the NSF.

Notes

1. M. Farah and P. Wolpe, "Monitoring and Manipulating Brain Function," *Hastings Center Report* 34, no. 3 (2004): 35–45.

2. Center for Sensorimotor Neural Engineering, http://csne-erc.org/.

3. M. Schermer, "Ethical Issues in Deep Brain Stimulation," *Frontiers in Integrative Neuroscience* 5 (2011): 1–5.

4. H. F. M. Van der Loos, "Design and Engineering Ethics Considerations for Neurotechnologies," *Cambridge Quarterly of Healthcare Ethics* 16 (2007): 305–309.

5. J. Sadowski, "Exoskeletons in a Disabilities Context: The Need for Social and Ethical Research," *Journal of Responsible Innovation* 1, no. 2 (2014): 214–219.

6. N. S. Fitz and P. B. Reiner, "The Challenge of Crafting Policy for Do-It-Yourself Brain Stimulation," *Journal of Medical Ethics* 41 (2015): 410–412.

7. E. Klein, "Models of Patient-Machine-Clinician Relationship in Closed-Loop Machine Neuromodulation," in *Machine Medical Ethics*, ed. S. P. Van Rysewyk and M. Pontier (New York: Springer, 2014), 273–290.

8. J. B. Trimper, P. R. Wolpe, and K. S. Rommelfanger, "When 'I' Becomes 'We': Ethical Implications of Emerging Brain-to-Brain Interfacing Technologies," *Frontiers in Neuroengineering* 7, no. 4 (2014): 1–4.

9. In generic terms, a BCI is a device for capturing and using brain-derived information to facilitate human control and communication. The extension of the term "BCI" is itself contested, and competing terms such as "brain-machine interface" and "neuroprosthesis" populate the literature. While these terminological debates are important, given our interest in the philosophical and ethical implications of BCI as a type of technology, we will largely put them to the side. For more on these debates, see F. Nijboer et al., "The Asilomar Survey: Stakeholders' Opinions on Ethical Issues Related to Brain-Computer Interfacing," *Neuroethics* 6, no. 3 (2013): 541–578.

10. R. Heersmink, "Embodied Tools, Cognitive Tools, and Brain-Computer Interfaces," *Neuroethics* 6, no. 1 (2013): 207–219.

11. BRAIN Initiative, accessed April 9, 2015, http://www.whitehouse.gov/share/brain-initiative.

12. See, for instance, P. Gallagher, D. Desmond, and M. MacLachlan, eds., *Psychoprosthetics* (London: Springer, 2007); F. B. Mills, "A Phenomenological Approach to Psychoprosthetics," *Disability and Rehabilitation* 35, no. 9 (2013): 785–791.

13. M. Auvray and E. Myin, "Perception with Compensatory Devices: From Sensory Substitution to Sensorimotor Extension," *Cognitive Science* 33 (2009): 1036–1058.

14. C. Elliott, "The Perfect Voice," chap. 1 in *Better than Well: American Medicine Meets the American Dream* (New York: W. W. Norton, 2004), 1–27.

15. C. D. Murray, "An Interpretive Phenomenological Analysis of the Embodiment of Artificial Limbs," *Disability and Rehabilitation* 26 (2004): 963–972; talking about bone-anchored prosthetic limbs, Mari Lundberg et al. note three ways that prosthetic users conceive of their prosthetics: as a practical tool, as an artificial body part, and as "part of me" (with increasing identification with the prosthetic); see M. Lundberg, K. Hagberg, and J. Bullington, "My Prosthesis as a Part of Me: A Qualitative Analysis of Living with an Osseointegrated Prosthetic Limb," *Prosthetics and Orthotics International* 35 (2011): 207–214.

16. A. Clark, "Reinventing Ourselves: The Plasticity of Embodiment, Sensing and Mind," *Journal of Medicine and Philosophy* 32, no. 3 (2007): 263–282; J. M. Carmena et al., "Learning to Control a Brain-Machine Interface for Reaching and Grasping by Primates," *PLOS Biology* 1, no. 2 (2003): e42, doi:10.1371/journal.pbio.0000042; A. Sengul et al., "Extending the Body to Virtual Tools Using a Robotic Surgical Interface: Evidence from the Crossmodal Congruency Task," *PLoS ONE* 7, no. 12 (2012): e49473, doi:10.1371/journal.pone.0049473.

17. A. Clark and D. Chalmers, "The Extended Mind," *Analysis* 58 (1998): 7–19; A. Clark, *Supersizing the Mind: Embodiment, Action and Cognitive Extension* (New York: Oxford University Press, 2008).

18. M. Hilhorst, "'Prosthetic Fit': On Personal Identity and the Value of Bodily Difference," *Medicine, Health Care, and Philosophy: A European Journal* 7, no. 3 (2004): 303–310.

19. Heersmink observes that the current BCI technology is too cumbersome—requiring extreme concentration and limited range of movement—to be transparent to the user; see Heersmink, "Embodied Tools, Cognitive Tools, and Brain-Computer Interfaces."

20. Schupbach reported that nineteen of twenty-nine patients who received deep brain stimulators for Parkinson's symptoms had issues recognizing themselves after surgery, and six of the twenty-nine experienced the change as deeply problematic, noting, for instance, "I feel like a robot," and "I don't feel like myself anymore"; see M. Schupbach et al., "Neurosurgery in Parkinson's Disease: A Distressed Mind in a Repaired Body?," *Neurology* 66, no. 12 (2006): 1811–1816.

21. It is also worth considering literature on identity and remote-controlled virtual avatars, to which serious gamers often become deeply attached; see, for example, S. Turkle, "Parallel Lives: Working on Identity in Virtual Space," in *Constructing the Self in a Mediated World*, ed. D. Grodin and T. R. Lindlof (London: Sage, 1996), 156–175.

22. J. Wolfendale, "My Avatar, My Self: Virtual Harm and Attachment," *Ethics and Information Technology* 9 (2007): 111–119.

23. See V. Johansson et al., "Thinking Ahead about Deep Brain Stimulation," *AJOB Neuroscience* 5, no. 1 (2014): 24–33. The authors acknowledge that authenticity worries may be inherent in the technology, particularly for bidirectional DBS devices, but downplay the concerns by suggesting that they may be more tied to the technology's novelty than to well-founded moral worries.

24. M. Synofzik, "Ethically Justified, Clinically Applicable Criteria for Physician Decision-Making in Psychopharmacological Enhancement," *Neuroethics* 2 (2009): 89–102.

25. A. Silvers, "A Fatal Attraction to Normalizing: Treating Disabilities as Deviations from 'Species-Typical' Functioning," in *Enhancing Human Traits*, ed. E. Parens (Washington, DC: Georgetown University Press, 1998), 95–123; R. Amundson and S. Tresky, "Bioethics and

Disability Rights: Conflicting Values and Perspectives," *Journal of Bioethical Inquiry* 5, no. 2–3 (2008): 111–23.

26. H. Greely et al., "Toward Responsible Use of Cognitive-Enhancing Drugs by the Healthy," *Nature* 456 (2008): 702–705; G. Wolbring, "Paralympians Outperforming Olympians: An Increasing Challenge for Olympism and the Paralympic and Olympic Movement," *Sport, Ethics and Philosophy* 6, no. 2 (2012): 251–266.

27. A. Roskies, "Neuroethics for the New Millennium," *Neuron* 35, no. 1 (2002): 21–23; D. Buchman and J. Illes, "Imaging Genetics for Our Neurogenetic Future," *Minnesota Journal of Law, Science, & Technology* 11, no. 1 (2010): 79–97.

28. M. Schermer, "Health, Happiness and Human Enhancement: Dealing with Unexpected Effects of Deep Brain Stimulation," *Neuroethics* 6, no. 3 (2013): 435–445.

29. I. de Melo Martin, "Defending Human Enhancement Technologies: Unveiling Normativity," *Journal of Medical Ethics* 36, no. 8 (2010): 483–487.

30. T. D. Wager et al., "An fMRI-Based Neurologic Signature of Physical Pain," *New England Journal of Medicine* 368 (2013): 1388–1397.

31. K. Davis, E. Racine, and B. Collett, "Neuroethical Issues Related to the Use of Brain Imaging," *Pain* 153, no. 8 (2012): 1555–1559.

32. J. Giordano, "The Neuroscience of Pain, and a Neuroethics of Pain Care," *Neuroethics* 3 (2010): 89–94.

33. Farah and Wolpe, "Monitoring and Manipulating," 40.

34. Bernard Williams argues that such agent regret is morally expected, even when there was nothing else an agent could have done to avoid an accident; this case is somewhat different in that Joan may actually question whether she was in control of her action. See B. Williams, "Moral Luck," in *Moral Luck*, ed. D. Statman (Albany, NY: SUNY Press, 1993), 35–55.

35. D. Shoemaker, "Responsibility and Disability," *Metaphilosophy* 40, no. 3–4 (2009): 438–461.

36. N. Lipsman and W. Glannon, "Brain, Mind and Machine: What Are the Implications of Deep Brain Stimulation for Perceptions of Personal Identity, Agency and Free Will?," *Bioethics* 27, no. 9 (2013): 465–470.

37. L. Klaming and P. Haselager, "Did My Brain Implant Make Me Do It?," *Neuroethics* 6 (2010): 527–539.

38. G. Ramachandran, "Assault and Battery on Property," *Loyola Law Review* 44 (2010): 253–276.

39. Farah and Wolpe, "Monitoring and Manipulating."

40. L. Austin, "Privacy and the Question of Technology," *Law and Philosophy* 22, no. 2 (2003): 119–166.

41. T. Bonaci et al., "Securing the Exocortex: A Twenty-First Century Cybernetics Challenge" (paper presented at the 2014 IEEE Conference on Norbert Wiener in the 21st Century, Boston, June 2014).

42. T. Denning, Y. Matsuoka, and T. Kohno, "Neurosecurity: Security and Privacy for Neural Devices," *Journal of Neurosurgical Focus* 27, no. 1 (2009): E7, 1–4.

43. Martha Farah recognizes this complexity with respect to privacy and identity in M. J. Farah, "Neuroethics: The Practical and the Philosophical," *Trends in Cognitive Sciences* 9, no. 1 (2008): 34–40.

44. N. Fraser and A. Honneth, *Redistribution or Recognition? A Political-Philosophical Exchange* (New York: Verso, 2003).

45. In practice, some researchers understand "end user" to mean industry partners or other research scientists because any clinical application is seemingly far downstream. Consequently, the distinct needs of underserved groups—like people with disabilities—may be overlooked during the development process.

46. A. Silvers, "Better than New! Ethics for Assistive Technologists," in *Design and Use of Assistive Technology: Social, Technical, Ethical, and Economic Challenges*, ed. M. M. K. Oishi, I. M. Mitchell, and H. F. M. Van der Loos (New York: Springer, 2010), 3–15.

47. On the relative irreversibility of BCI systems, see M. Synofzik and T. Schlaepfer, "Electrodes in the Brain: Ethical Criteria for Research and Treatment with Deep Brain Stimulation for Neuropsychiatric Disorders," *Brain Stimulation* 4 (2011): 7–16. In a different paper by Matthis Synofzik et al., it is suggested that "the alleged reversibility of DBS—which is still stated by most authors as one of the main ethical 'pro' arguments of DBS . . . might only apply to the technique, but not to the person"; see M. Synofzik, T. E. Schlaepfer, and J. J. Fins, "How Happy Is Too Happy? Euphoria, Neuroethics, and Deep Brain Stimulation of the Nucleus Accumbens," *AJOB Neuroscience* 3, no. 1 (2012): 30–36, 34–35.

48. This contrast may be yet too simplistic, as refinements in implantable devices, such as DBS systems, may increasingly allow for some patient control over stimulation. See E. Klein, "Models of the Patient-Machine-Clinician Relationship in Closed-Loop Machine Neuromodulation," in *Machine Medical Ethics*, ed. S. P. van Rysewyk and M. Pontier (New York: Springer, 2014), 273–290.

49. J. B. Christiansen and I. Leigh, *Cochlear Implants in Children: Ethics and Choices* (Washington, DC: Gallaudet University Press, 2002); R. Sparrow, "Implants and Ethnocide: Learning from the Cochlear Implant Controversy," *Disability & Society* 25, no. 4 (2010): 455–466.

50. For instance, Synofzik et al. describe a patient with a DBS system for general anxiety and obsessive-compulsive disorder. After calibration sessions in which the patient reported feeling "'unrealistically good' and . . . 'overwhelmed' by the sensations of happiness and ease" and asked to have the stimulations levels reduced, the patient returns and requests the higher stimulation again because he would "like to feel 'a bit happier' during the next few weeks"; see Synofzik, Schlaepfer, and Fins, "How Happy Is Too Happy?"

51. B. Lewis, "A Mad Fight: Psychiatry and Disability Activism," in *The Disability Studies Reader*, 4th ed., ed. L. J. Davis (New York: Routledge, 2013), 115–131; S. M. Robertson, "Neurodiversity, Quality of Life, and Autistic Adults: Shifting Research and Professional Focuses onto Real-Life Challenges," *Disability Studies Quarterly* 30, no. 1 (2009), accessed April 9, 2015, http://dsq-sds.org/article/view/1069.

52. R. M. Green, "The Need for a Neuroscience ELSI Program," *Hastings Center Report* 44, no. 4 (2014): inside back cover.

53. Presidential Commission for the Study of Bioethical Issues (PCSBI), *Gray Matters: Integrative Approaches for Neuroscience, Ethics, and Society* (Washington, DC: PCSBI, 2014), http://bioethics.gov/node/3543.

29 Cyber (In)security: Threat Assessment in the Cyber Domain

George Lucas

Books about military history are filled with platitudes about how technology has, over the centuries, changed the nature of warfare. From the invention of chariots around 3000 BC, to the development of gunpowder around 800 AD, to naval ships and airplanes more recently, what nation-states (or individuals, for that matter) do in order to defend against enemies has changed as weapons of warfare have changed. With the development of computers, information technology, and the Internet in the twentieth century, conflict between nations began to play out in the cyber realm, a trend likely to continue and escalate in the twenty-first century.

One of the changes in the characteristics of warfare in the twentieth century was the increasing physical distance between attackers and defenders. Those trends have only continued in the present century. Of course, face-to-face combat still takes place, but a good deal of conflict now involves aggressors and defenders who operate remotely. Soldiers who drop bombs from high in the sky do not directly see the damage that the bombs cause, and those who launch drone attacks from thousands of miles away see the consequences on computer screens rather than firsthand. In cyberattacks, aggressors can wreak havoc on their enemies while sitting comfortably behind computers. They can attack critical infrastructure; bring to a halt facilities such as nuclear power plants, financial institutions, and news outlets; and steal highly valued intellectual property. This means that individuals, organizations, and nation-states must go to great lengths to secure their information systems from intrusion and attack. In this chapter from his book, *Ethics and Cyber Warfare: The Quest for Responsible Security in the Age of Digital Warfare*, George Lucas provides some background on the variety of malevolent behavior that takes place in the cyber realm, including crime, vandalism, political activism (hacktivism), espionage, and acts of war.

At first glance, as the philosopher and cyber expert Randall Dipert observes,[1] the cyber domain is a strange place, utterly unlike the more familiar "real-world" domains of land, sea, air, and space.

Cyber itself is a new term, derived from an ancient Greek noun referring to a "space" or a "domain." The word *cyberspace* is therefore a redundancy. It was coined by science-fiction writer William Gibson in a short story written in 1982, and later described in his 1984 novel *Neuromancer* as "clusters and constellations" of data and their interconnections drawn from every computer in the universe.

The term *cyber warfare* first appeared in a Rand Corporation report written in 1992 by John Arquilla and David Ronfelt, who used it to describe a new form of conflict

---

From *Ethics and Cyber Warfare: The Quest for Responsible Security in the Age of Digital Warfare* (Oxford: Oxford University Press, 2017), 16–32.

perpetrated through the disruption of the flow of critical data managed across networked computer information systems. The threat of this new form of conflict, which they also termed *information warfare*,[2] was not taken seriously at first. Concerns about the prospects of cyber conflict quickly grew, however, especially in the first decade of the twenty-first century, in response to exponential growth in Internet crime, vandalism, and corporate and state espionage.

What, When, and Where?

Accurately assessing the gravity of the threat posed by largely hypothetical forms of virtual conflict, however, has proven to be quite difficult. This stems in part from the problem of accurately conceptualizing just what the cyber domain is.

The cyber domain is, as we have seen, often labeled a "fifth" domain of experience, alongside our more familiar and conventional four domains of land, sea, air, and space. The other four domains "contain" or consist of "things" or "entities" (objects) and "events," which are largely occurrences among and between the things in that particular domain. In a nutshell, "what" cyberspace is, by comparison, is a unique environment in which electrons and photons, through interactions determined by binary bits of computer code, dictate the form and structure of "objects" and "events." Cyberspace bears some resemblance to the near-earth plasma environment, that region of space closest to our own planet. Both consist largely of electromagnetic phenomena with extraordinarily fast response times over extremely large distances. But even this comparison is far from exact. The analogy does, however, highlight one notable feature regarding the "when" of cyber events: they are continuous and virtually instantaneous. Likewise, as in a plasma field, the "where" of discrete cyber objects is difficult to pinpoint, since they are almost entirely nonlocalized. These features are common to all quantized energy fields, but they are without precedent in the remaining three more conventional domains of land, sea, and air.

Finally, cyberspace is often understood as identical to, even taken to be a synonym for, the Internet or the World Wide Web. But the cyber domain, as Dipert also points out, is in fact far more extensive.[3] It also includes hardware, such as portable thumb drives, which connect to the Internet via other hardware devices, such as printers and laptops and television sets and streaming devices, which are likewise connected to the Internet via cables or wireless networks. Cyberspace also includes cell phone networks, which operate on different frequencies from the bandwidth devoted to Internet traffic. It includes satellite communications and the Global Positioning System (GPS), and conventional objects that are increasingly becoming networked into these distinct but intertwined systems of communication, such as an automobile's navigation system or on-board monitoring and safety systems on aircraft or the devices in the home that people can now monitor and control from their smartphones.

This emerging cyber network, often called the Internet of Things (IOT), increasingly also includes drones and other remotely piloted, unmanned military systems and the hardware and software that control them, communicating across command centers, and indeed, monitoring and governing the Internet itself. All these characteristics of the IOT, as the editors of *The Economist* recently pointed out, make it "a hacker's paradise." Yet security is often still the farthest thing from the minds of designers and users.[4]

How?

In cyberspace, a relatively simple phenomenon or transaction, such as sending an email message, making a cell phone call, or paying a bill online, consists of an "object" (e.g., the email message itself) and its attendant "event" (e.g., sending and receiving the message). These all have component parts that interact at many different points around the globe, spanning many countries and continents simultaneously. This provides enormous opportunities for mischief. Consider this simple example (which I often rely on to describe the workings of cyberspace and the problems encountered in it).

Suppose I place a normal phone call from my home, near Washington, DC, to my brother in Tampa, Florida. That phone call travels as a fairly conventional electromagnetic transmission for about a thousand miles over a series of trunk lines, which are simply large copper wires or coaxial cables (or, increasingly, fiber-optic cables) that stretch along a physical route, in a more or less straight line, from my city to my brother's.

If I decide to email my brother instead, my message will be disassembled into a swarm of discrete data packets, after which the Internet Protocol addressing system will label each packet and search for the quickest and least-congested Internet pathway to transmit them to their destination. As a result, these discrete data packets will almost assuredly travel around the world at speeds approaching that of light, perhaps more than once, and through the communication infrastructures of many countries, before they are finally reassembled at their destination. Owing to the historical evolution of the Internet from its origins in the 1960s as a US Department of Defense system for vital military command and control in the event of a nuclear war, moreover, the design and dissemination of the physical infrastructure supporting Internet communications entails that virtually all these discrete data packets, no matter where they start out or end up, will at some point pass through the original portions of the Internet's physical infrastructure (its "backbone"), which are located within the continental United States.

Verifying the identity and authenticity of Internet communications, as a result, can be quite tricky. When I get a conventional telephone call from my brother, I can usually tell who it is very quickly from the unique sound of his voice. But when I receive an email containing his name and return address, it may seem to come from him, but it may be fraudulent. The digital identity of the data packets may have been spoofed to fool me into thinking the message is from him, when in reality, the email is part of what is termed a *spear phishing scam*. Clever cybercriminals (say, somewhere in Uzbekistan) can fake my brother's and many other users' digital identities and send us phony emails, hoping we will innocently open them.

If I do open the email, it may contain a virus that will destroy or steal my data. Or, the email might contain a "worm"—a sinister software program that will download onto my computer and begin transmitting my sensitive private information to the original sender, while replicating itself and "crawling" through my personal network and email connections to affect the hardware of many other users' systems in the same manner. The generic term of art for all of these software programs is *malware*, and there is an amazing variety of malware that, when maliciously introduced into our personal computer operating systems, can steal our credentials, identities, or credit card information, which will then be sold to criminal gangs in Russia or Azerbaijan, who will, in turn, use it to drain our bank accounts or to charge merchandise to our credit cards, among many other nefarious uses.

Is this what the Chinese plan to do with the twenty-two million confidential personnel files that they apparently hacked from the OPM? If that seems like unlikely behavior for state agents (unless a few of them have decided to pursue criminal activities on the side), the troubling fact remains that the Chinese government could do so if it saw some advantage to be gained—and the results would be catastrophic for millions of Americans. In fact, no one yet has the slightest idea why the Chinese government stole these records—indeed, technically, we still don't know with absolute certainty it was they who did so.[5] But clearly, they should not have been able to do so, and would not have been able had the OPM's internal cybersecurity measures and those of the cybersecurity contractor it employed been more competent and robust.

Why?

Let's pause for a brief history of malevolent misbehavior in the cyber domain. Not so long ago, cyber activism was limited to pranks, practical jokes, and random acts of vandalism on the Internet carried out by individuals or small groups. Pranksters attached software viruses to emails that, when opened, quickly began to damage the infected individual's computer, or spread through an organization's internal network, posting goofy messages and perhaps even erasing valuable data or software stored on hard drives. Cyber vandals posted offensive messages or unwanted photos, or otherwise defaced an organization's website, often for no apparent reason other than their own amusement or to gain bragging rights with fellow computer hackers. About the only crimes committed in those early days were trespassing (invading an individual's PC or a private company network) and destruction of property. Apart from mean-spiritedness or a warped sense of humor, however, about the only reasons given at the time by hackers for such malicious activities were complaints about the monopolistic practices of, and the mediocre software distributed by, the Microsoft Corporation.

It was not long before sophisticated individuals and criminal gangs began exploiting the same software vulnerabilities as the pranksters for the purpose of stealing bank deposits, credit card numbers, or even victims' personal identities. At the same time, cyber activism was evolving into ever more sophisticated acts of political sabotage: defacing or even temporarily shutting down government or commercial websites with DDoS attacks, or dispatching those dreaded software worms that traveled *peer-to-peer* (computer to computer), penetrating firewalls and antivirus software in order to gain control over each PC or laptop, transforming it into a "zombie." These "zombies"—individual compromised machines—could then be remotely networked with others to form a massive botnet controlled by political dissidents or criminal organizations, who, in turn, used them to launch DDoS attacks to shut down banks and financial institutions (like stock markets) or to hack into individual users' accounts and divert the victims' funds to the criminals' own accounts.[6] This seems to be what happened in Estonia, *except for the last part*—the diversion of funds and draining of personal bank accounts did not take place. Why?

When the Stuxnet worm was introduced into the SCDA system that controlled the centrifuges in Iran, it did exactly what conventional weapons launched by states usually do: it targeted and destroyed military hardware that posed a grave threat of harm to others. Stealing random personal bank deposits from ordinary citizens, in comparison, does not seem to be the sort of thing that a state would do. In fact, the typical criminal activities that usually follow personal data theft were not observed in Estonia, which strongly suggests that a state or its agents, not random individuals (or "patriots," as Russian Federation officials blandly asserted at the time), had carried out the attacks.[7]

Hacktivism is the new term being used to classify acts of malevolence and mischief in the cyber domain, from straightforward crime and vandalism to the many forms of political protest that are now carried out on the Internet. Technically, the *hacktivist* is one who engages in vandalism and even in criminal activities in pursuit of political goals or objectives rather than personal satisfaction or financial gain. Well-known individuals, such as Julian Assange of WikiLeaks, and loosely organized groups, such as Anonymous, the now-defunct LulzSec, and Cyberwarriors for Freedom, resorted to Internet malevolence in order to publicize their concerns, which can range from defending personal privacy, liberty, and freedom of expression on the Internet itself all the way to expressing their opposition to oppressive political regimes like those of Syria or Egypt.

Three Ways of Being a Hacktivist

There are many ways of carrying out hacktivism. I find it useful to focus on the hacktivist's political goals. These political goals can be categorized as *transparency, whistleblowing,* and *vigilantism*. WikiLeaks purports, for example, to provide greater transparency by exposing the covert activities of governments and large corporations. Whistleblowers, such as US Army private Chelsea Manning (the former Bradley Manning) and former NSA contractor Edward Snowden, aim to expose what they take to be grave specific acts of wrongdoing or injustice, in Manning's and Snowden's cases, on the part of the US military and government, respectively.

Vigilante groups, such as Anonymous or the former LulzSec, are a bit harder to pin down, since the individual members may espouse a variety of disparate political causes. The organizational response to the group's chosen cause, however, involves its members taking the law (or, in its absence, morality) into their own hands; that is, based on the group's shared judgment regarding immoral or illegal behavior, it launches attacks against targets that might range from the Syrian government of Bashir al Assad, for its massive human rights violations, to ordinary business and commercial organizations, law-abiding individuals, and security and defense or constabulary organizations, such as the FBI, to whose practices the organization objects.

This is vigilantism. And (as the name Anonymous suggests) the members of these groups cannot easily be identified, traced, or otherwise held accountable for their actions. (The ringleader of LulzSec, known as "Sabu," may have been an exception; he was tracked down, arrested, and turned into an FBI informant and then took part in a sting operation against Anonymous.[8]) As in conventional vigilantism, the vigilante's judgment as to what constitutes a legal or moral offense is subjective and often inconsistent or otherwise open to question.

Importantly, in all cases involving transparency, whistleblowing, and vigilantism, the burden of proof is on those who deliberately violate their fiduciary duties and contractual agreements or flout or disobey the law in order to expose or to protest activities they have deemed to be even more egregious than their own. This comparative judgment by the protester or whistleblower invokes what is called the *principle of proportionality,* which holds that the degree of harm that results from the vigilante's actions must be *demonstrably less* than the harm being done by those whom the vigilante opposes. The problem is that this comparative judgment is notoriously difficult to make. Vigilantes often exaggerate or misrepresent the harm caused by the actions against which they protest and underestimate the destructive effect of their own activities on the public welfare.

Another difficulty with vigilantism is that there is no independent or adversarial review of these decisions. According to the *principle of publicity*, or the *principle of legitimate authority*, the final authority to evaluate the legitimacy of a protester's or dissident's actions does not rest with that individual or organization but with the public, in whose collective interests the individual or organization claims to act. So, in all these cases, it must be possible in principle to bring the dissident's actions and intentions before an impartial court of public opinion for independent review. This criterion is the one most frequently ignored and, consequently, the one both vigilantes and whistleblowers, and would-be whistleblowers, most often fail to meet. Both are prone to suffer from an abundance of self-righteousness.

The generic term *hacktivism* covers many sins, some more justifiable (or, at least, understandable and sometimes excusable) than others. I doubt, for example, that most readers would take exception to Anonymous defacing the government website of Bashir al Assad or exposing the online behavior of child pornographers. More worrisome are attacks on legitimate government, military, or corporate websites that result in harm to innocent bystanders. But that is the very nature of vigilantism. Anonymous and other cyber vigilantes have engaged in all these activities indiscriminately. Anonymous apparently lacks any kind of command or control center to reign in extremism or to deliberate on the appropriateness or suitability (let alone the moral justification) of the actions of its members. It functions instead like a loose-knit confederation of anarchists—what a professor at New York University went so far as to characterize as "a cyber lynch mob."[9]

But none of this social activism in cyberspace, no matter how well- or ill-advised, constitutes warfare, let alone rises to the level of a use of force or an armed attack. So, what has the phenomenon of hacktivism to do with the concerns of this book?

Conventional Warfare

Up to this point, we have been discussing the problem of whether or not any form of cyber conflict rises to the level of warfare or, more accurately, proves to be the equivalent of armed force. I have relied on readers' intuitive understanding of what conventional war is, without having given it a more precise definition. What follows is a definition of both conventional warfare and *irregular warfare*, a different kind of conflict that followed, or "disrupted," it.

Conventional war is defined as a form of political conflict between adversaries that is generally undertaken only when other modes of conflict resolution—such as diplomacy, negotiation, sanctions, and compromise—have failed. The resort to armed force is made to compel an otherwise implacable or recalcitrant adversary to comply with the more powerful nation's will or political ambitions.[10] The nineteenth-century Prussian military strategist Carl von Clausewitz provided the classic summative assessment of war as just such a "continuation of State policy by other means." The goal of armed conflict in pursuit of a nation's political objectives, he argued, is to defeat the enemy's armies, occupy his cities, and most significantly, *break his will to fight or resist.*[11]

This conventional, or "classical," understanding of warfare is heavily dependent on the underlying paradigm of the nation-state and of international relations as a kind of "perpetual anarchy" of the competing interests of nation-states (e.g., in acquiring territory or natural resources, providing for national security, or expanding profitable trade relations). And while the policies of an individual nation-state may represent a perfectly rational or logical course of action for *that particular nation* (or coalition of allies), the multiplicity of such interests virtually guarantees conflict between rival sets of interests and policy objectives. That is to say, the occasional resort to armed force to resolve at least some of these conflicts is all but inevitable.[12]

The very last conventional armed conflict, or war, however, may well have occurred more than two decades ago, in early 1991, when the army of Iraqi leader Saddam Hussein confronted a similarly trained and equipped UN coalition military force led by the United States in the deserts of Kuwait and southeastern Iraq. The apparent technological symmetry in that instance, as we realize in hindsight, turned out to be illusory. Since then, the conventional use of armed force has been combined with, or even supplanted by, "asymmetric" or "irregular" warfare. A striking feature of unconventional or irregular warfare is that adversaries and insurgent interest groups, frustrated by the radical asymmetries in military power stacked in favor in conventional forces, adopt new tactics for disrupting social systems and attacking the weak links in logistical supply chains through the use of, for example, suicide bombers and improvised explosive devices (IEDs), or of cyberattacks that can throw conventional military forces seriously off balance. Subsequently, new technologies owned initially by the besieged conventional forces become the optimal response. Drone attacks, for example, are "systemic" in precisely the same way IEDs and suicide bombers are: they disrupt the insurgent's command structure, relentlessly hunting him out where he lives and hides, demoralizing him, and hopefully breaking his ability and will to fight.

Similarly, cyber surveillance involving big-data collection on a previously unimaginable scale holds the promise of turning the tables and frustrating even the cleverest and most determined attackers, by carrying the cyber fight back to its source.[13] Here is yet another sense in which the lessons learned from the disruptive innovation brought about by the transition from conventional to irregular warfare during the past decade offer some useful comparisons with cyber warfare.

Unrestricted Warfare

In 1999, only a few years after the stunning defeat of Saddam's conventional army by the United States in Kuwait, two air-force colonels in the Chinese PLA published an extensive study of the United States' ability to project military power in the foreseeable future.[14] Their monograph, entitled *Unrestricted Warfare,* acknowledged that no one could any longer stand toe-to-toe with the US military in a conventional war, and recommended that the only way forward for China was to develop offensive and defensive capabilities in other areas, *including cyberspace,* where the United States was both highly vulnerable and by no means dominant, and to be willing to use those capabilities relentlessly in the pursuit of national interests.

Indeed, cyber- and national-security expert Richard Clarke links China's development of its now considerable offensive cyber capability with the doctrine espoused by these two theorists. That doctrine, he explains, is termed *shashou-jian,* and is meant to employ new kinds of weapons that are outside the traditional military spectrum and that capitalize on the vulnerabilities in an adversary's otherwise-advanced capacities: in the American case, its advanced capabilities and heavy reliance on cyberspace technology, both domestic and that of its allies, such as the technologically sophisticated nation of Estonia.[15]

These reflections by the Chinese military theorists led directly to state-sponsored campaigns allegedly carried out by two top-secret branches of the PLA—Unit 61398, based in Shanghai, and the even more sinister Unit 78020, in Kunming—both heavily engaged in cyberespionage and covert actions, such as the alleged planting of "trapdoors" and "logic bombs" in vital civilian infrastructure,[16] as well as the massive theft of industrial and classified military technologies, which the United States only began to acknowledge and publicly denounce in 2013. These developments lie at the heart of the threat assessment for the kind of cyber warfare feared by Clarke, Brenner, Harris, and many other experts in the field of cyber conflict and cybersecurity. Cyber conflict is thus considered to be a major, if not the main, weapon in the arsenal of weapons and tactics assembled to conduct "unrestricted warfare."

Global acts of cybercrime stand side by side with relentless and ongoing commercial and military espionage and the theft of industrial and state secrets that threaten the security and fundamental economic welfare of individuals and nations. Yet efforts to counter these activities are strongly opposed in many rights-respecting and reasonably democratic societies as constituting unacceptable infringements of liberty and privacy. Indeed, efforts to impose the kinds of controls and constraints that would provide greater security are among the proposed security measures that Internet vigilantes like Anonymous protest. Meanwhile Clarke, Brenner, and other security experts worried about our vulnerabilities on this score lament the present impasse and predict that the obsession with privacy, anonymity, and unrestricted freedom of action in the cyber domain will not be overcome short of a horrendous cyber Armageddon, on the magnitude of the attack on Pearl Harbor or the 9/11 attacks.

Some of the more extreme scenarios, as noted previously, may be exaggerated. The threat of terrorists or of your neighbor's alienated teenage son causing planes filled with helpless passengers to collide in mid-air or passenger trains to crash or derail or hydroelectric dams to burst and flood tens of thousands of hectares (ruining crops and drowning thousands of victims), or shutting down electrical power grids, while simultaneously releasing poisonous gases from chemical factories to destroy the populations of nearby cities—all seem highly unlikely. To date, the greatest threat by far to American and European security remains Internet crime, and this has been a focus of coordinated international law enforcement efforts. As of July 2015, for example, twenty countries that had been collaborating for several years with the FBI in Operation Shrouded Horizon finally succeeded in arresting the chief perpetrators of and taking down a cyber operation known as Darkode.com, a "dark web" forum that US Department of Justice officials described as "the world's most sophisticated English-language Internet Forum for criminal hackers" trafficking in malware.[17]

So we are left with the question, do any of the other forms of cyber conflict about which we read, or sometimes experience ourselves, ever constitute genuine cyber warfare? Or, more precisely, when, if ever, does noncriminal cyberactivity rise to a level equivalent to an armed attack by one (or more) state(s) against another sovereign state, as discussed in the UN Charter?[18]

Much of the confusion over this important question arises from the unwillingness of cyber pundits to distinguish between activities on a broad spectrum of cyber conflict—ranging, as we now see, from practical jokes, vandalism, and politically motivated hacktivism to serious cybercrimes (mostly robbery and financial schemes involving deception and identity theft) to commercial and industrial and state-sponsored cyberespionage (as the NSA was found to be doing) to physical sabotage (like Stuxnet). One critical feature of cyber conflict is that it has blurred the distinctions between what were once quite different kinds of conflict.

Another important difference is that the pursuit of cyber strategies and tactics and the use of cyber weapons by states have been largely under the control of intelligence agencies and personnel, whose rules of engagement are very different from those of conventional military combatants. Spies and espionage agents are generally engaged in activities that do not rise to the level of a "threat or use of force" under international law, let alone of armed conflict between states, but that do constitute criminal acts in the domestic jurisdiction within which they take place. Conventional war occurs in zones of combat in which the customary rule of law has broken down, and to which the international law of armed conflict therefore applies. This legal regime is, in general, far more lenient than is most domestic law regarding the permission to pursue conflict with deadly force. Even so, there *are* certain rules in international law that restrict the use of deadly force in conventional war (such as noncombatant immunity, military necessity, and proportionality) that do not apply to the pursuit of espionage.

In my opinion, this fundamental distinction regarding what are generally termed "standing orders and rules of engagement" within the different cultures involved in this new form of unrestricted warfare has not been sufficiently acknowledged, let alone understood. Cyberwar is "unrestricted" warfare carried out by spies and espionage agents who do not think themselves bound by legal restraints, unlike conventional combatants who are trained in the international law of armed conflict. Unrestricted warfare is not legally permissible or morally justifiable according to the rules and laws governing conventional armed conflict, but it is nevertheless routine practice among agents of espionage. Many cyber weapons and tactics are designed to attack civilians and civilian (noncombatant) targets, a feature that is illegal, and decidedly immoral, in the conventional case.[19]

If there is any bright spot to be found in all this, it is that, as noted, it takes vast resources, expertise, access, and practice time to build effective cyber weapons with genuine destructive capacity. These resources are accessible only to nation-states or, at the very least, to large, well-staffed, highly organized criminal organizations. That may be scant comfort, but nations and even large criminal organizations (unlike terrorists or troubled fourteen-year-olds) usually have clear-cut goals, purposes, and political or economic interests to which other governing authorities can appeal. Representatives of adversarial nations can communicate and, by appealing to those conflicting interests, negotiate a compromise. And they can also credibly threaten one another with punishment and retaliation, which could serve to deter malicious activities.

In the meantime, however, the need for individuals, organizations, and nations to protect their privacy, property, commercial and defense innovations, and overall safety and welfare through enhanced cybersecurity, especially with respect to highly vulnerable civil infrastructure, is urgent and still largely unaddressed.

State-Sponsored Hacktivism as a New Form of Warfare

A genuine cyber Armageddon—an effects-based equivalent of a large-scale conventional war or even a nuclear war, carried out with software weapons deployed in the cyber domain—is thus far the dominant threat posited by the advocates of enhanced cybersecurity, though for the present at least, the possibility of something like this happening appears to remain highly remote. Critics of this extreme threat assessment have claimed that the fearmongers deliberately engage in confusion, obfuscation, and equivocation: they may simply be confused about what really constitutes authentic warfare, or, perhaps unconsciously, they use the term *cyber warfare* metaphorically, similar to our describing the *war* on drugs, a *war* on crime, the *war* on poverty, or, importantly, the *global war on terror*.

Whatever the case, the current discussion of the threat of cyber warfare rests on a misuse of important terminology. Again, critics of the Armageddon scenario accuse its advocates of using the discussion as a tactic to gain public attention, garner scarce resources for their own preferred programs, or, more ominously, frighten the public into sacrificing their individuality, liberty, and privacy (and Internet anonymity) in favor of ever-increasing government intrusion and control. Conspiracy theorists, alongside Internet anarchists, would have us all believe that the cybersecurity community (largely consisting of government and security forces) is, at bottom, out to subvert and suppress freedom of expression, anonymity, liberty, and personal privacy by exaggerating the threat of cyber warfare.

What neither side in this debate explains, however, is that extreme, effects-based cyber conflict is only *one form* of cyber warfare. Neither side has yet paid sufficient attention to the alternative of state-sponsored hacktivism emphasized in this book—that is, they have failed to grasp the significance of the increasing tendency of states to pursue practices once thought to be the purview of individuals and nonstate organizations, including criminal organizations. I believe that this recent trend in state behavior has still not sparked sufficient concern. Professor Thomas Rid, a well-known security studies scholar at King's College, University of London, who vociferously opposes all uses of the term *cyber warfare*, for example, has not examined this alternative conception of what cyber conflict at the state level actually is. He and his supporters dismiss cyber warfare as merely a new, inflated form of espionage.

I am frankly not at all sure that state-sponsored hacktivism can be so easily classified or dismissed. To be certain, it entails activities similar to the domestic crimes that agents of espionage routinely commit. But, as described earlier, it also entails a good deal more. One would be hard pressed to find a historical equivalent to or precedent for the North Korean hack of Sony Pictures in the world of conventional espionage. Likewise, we cannot cite a previous historical instance in which enemy spies attempted to steal the confidential personal data of *millions* of ordinary citizens (rather than the personal data on a select number of high-value targets, perhaps singled out for rendition or assassination). As discussed earlier, we don't even know with 100 percent certainty that enemy *nations* carried out these operations through their state agents, nor do we have a clear idea of what their intentions might have been if they did. The attacks are unsettling nonetheless. And *unlike the threat of the prospect of cyber Armageddon, this is really happening.*

To be sure, the aforementioned critics of the idea of genuine cyber warfare might reply that state-sponsored hacktivism is merely "espionage on steroids"—that is, new and novel forms of the usual kinds of activities that spies have always engaged in but on a scale of magnitude only recently made possible through the medium of the cyber domain. *Perhaps.* But, just as with conventional armed conflict itself, scale and magnitude matter. So does persistence. Yes, enemy spies routinely try to steal state secrets or plans for new weapons systems or economically important industrial secrets. This happens all the time. But when it occurs on the scale of magnitude we have seen over the past few years or persists despite well-aimed threats to "cease and desist," it is not implausible to claim that cyber conflict has now supplanted conventional armed conflict in its ability to harm state interests, as well as to force weakened and vulnerable adversaries to their knees.

Such domination, such imposition of the political will of the adversary upon its victim, has, according to Clausewitz, always been the principal purpose and vocation of warfare as an instrument of the state. State-sponsored hacktivism itself is one principal component of what political philosopher Michael Gross has termed *soft war*, a conception that envisions a wide range of nonkinetic alternatives to conventional war.[20] Under the broad heading of *unrestricted warfare*, these alternative forms of unarmed conflict are, finally, what the military theorists and strategists in the PLA in China were proposing in 1999 as an alternative means of conducting hostilities and resolving conflicts without resort to conventional armed conflict. And in their account, cyber conflict was only the most prominent form of this new mode of warfare. "Soft" war in the cyber domain relies on craft, cleverness, and ingenuity in lieu of conventional kinetic force.[21] Yet our experience of this novel new form of "unarmed" warfare promises to do unlimited harm of its own if not fully understood and appropriately governed.

Notes

1. Randall Dipert, "The Essential Features for an Ontology for Cyberwarfare," *Conflict and Cooperation in Cyberspace*, ed. P. A. Yannakogeorgos and A. B. Lowther (Boca Raton, FL: CRC Press, 2013), 35–48.

2. The subsequent Rand Corporation anthology compiled by Arquilla and Ronfeldt a decade later documents this initial skepticism amid the exponential increase in the kinds of events they had predicted. See John Arquilla and David Ronfeldt, *Networks and Netwars: The Future of Terror, Crime, and Militancy* (Santa Monica, CA: Rand, 2001).

3. See Randall Dipert, "Other than Internet (OTI) Cyber Warfare: Challenges for Ethics, Law, and Policy," *Journal of Military Ethics* 12, no. 1 (2013): 34–53. See also my response and augmented account of OTI objects in cyberspace in George R. Lucas Jr., "Ethics and Cyber Conflict: A Review of Recent Work," *Journal of Military Ethics* 13, no. 1 (2014): 20–31.

4. *Economist* 416, no. 8947 (July 18–24, 2015): 10, 65. More recently, economist Robert J. Samuelson has written anxiously about our playing "internet roulette" in our headlong rush toward an ever-expanding "internet of things." *Washington Post*, August 24, 2015, A13. The CBS News television magazine *60 Minutes* recently featured a segment on "DARPA-Dan" Kaufman, a security expert and game designer recruited by the Defense Advanced Research Projects Agency (DARPA) to determine vulnerabilities in a range of increasingly networked devices, from refrigerators to cars, and to develop enhanced security for such objects against cyberattack. Leslie Stahl, "Nobody's Safe on the Internet," *60 Minutes*, February 8, 2015, http://www.cbsnews.com/news/darpa-dan-kaufman-internet-security-60-minutes/. Meanwhile, in a manner directly relevant to the concerns of this book with state involvement in hacking for political purposes, the US Department of Transportation has begun worrying that America's key urban traffic-control systems are increasingly vulnerable to cyberattack, with surprisingly dangerous implications. "Is More Gridlock Just a Hack Away?," *Washington Post*, August 9, 2015, C1, C2.

5. I increasingly suspect that this massive OPM security breach was intended to demonstrate China's superior cyber power and our nation's vulnerability. In effect, this is "cyber bullying" at the state level, and it will come to be viewed as another example of what I term state-sponsored hacktivism.

6. Reporter Mark Bowden authored a detailed account of "Confikker," a worm of unknown provenance first detected in 2008 as resident on millions of computers throughout the world, capable of creating what Singer and Friedman term "the Holy Grail of a botnet." We remain unaware at the present time of who developed and transmitted this malware or for what purpose. See Bowdon, *Worm: The First Digital World War* (New York: Atlantic Monthly Press, 2011); Singer and Friedman, *Cybersecurity and Cyberwar: What Everyone Needs to Know* (New York: Oxford University Press 2014), 72. Since Confikker has now quietly resided on

the global Internet for several years with no detectable malevolent activity, my own best surmise is that this is an "Internet failsafe switch," a security program designed presumably by agents of the United States or its allies to shut down the Internet, or regain total control of it, should an adversary or rogue state attempt to carry out massive acts of industrial or state sabotage as an act of war.

7. Controversy continues to swirl around the nature of this attack, and no definitive account has yet been produced. See, for example, Ian Traynor, "Russia Accused of Unleashing Cyberwar to Disable Estonia," *Guardian*, May 16, 2007, http://www.theguardian.com/world/2007/may/17/topstories3.russia. Traynor initially reported the allegations of state-sponsored cyber warfare carried out by Russia against Estonia that Richard Clarke and others had espoused. But only a year later, the *Guardian* seemed to publish a retraction in which the attack was described as the work of a lone Estonian youngster of Russian descent, carried out from inside Estonia's borders. See Charles Arthur, "That Cyberwarfare by Russia on Estonia? It Was One Kind in Estonia!," *Guardian*, January 25, 2008, http://www.theguardian.com/technology/blog/2008/jan/25/thatcyberwarfarebyrussiaon.

8. See the account of this episode regarding hacktivist Hector Savier Monsegur ("Sabu") in Shane Harris, *@War: The Rise of the Military-Internet Complex* (New York: Houghton Mifflin Harcourt, 2014), 131–135.

9. Gabriella Coleman, "Anonymous: From the Lulz to Collective Action," *New Everyday*, April 6, 2011, accessed July 19, 2015, http://mediacommons.futureofthebook.org/tne/pieces/anonymous-lulz-collective-action.

10. See Brian Orend, "War," *Stanford Encyclopedia of Philosophy*, ed. Edward N. Zalta, http://plato.stanford.edu/archives/fall2008/entries/war/. Orend gives a more extensive description and analysis of conventional warfare in *The Morality of War*, 2nd ed. (Petersboro, ON: Broadview Press, 2013).

11. Carl von Clausewitz, *On War*, ed. and trans. Michael Howard and Peter Paret (Princeton, NJ: Princeton University Press, 1976), bk. 1, chaps. 1–2. First published in 1830.

12. See my article "War," in *The Bloomsbury Companion to Political Philosophy*, ed. Andrew Fiala (London: Bloomsbury, 2015), 109–126.

13. I offer descriptive accounts of these new developments, also termed "hybrid" warfare (because of the combining of the new tactics with older conventional tactics and munitions) in "Postmodern War," *Journal of Military Ethics* 9, no. 4 (2010): 289–298; and in an address on the occasion of NATO's sixtieth anniversary, "'This Is Not Your Father's War': Confronting the Moral Challenges of 'Unconventional' War," *Journal of National Security Law and Policy* 3, no. 2: 331–342.

14. Qiao Liang and Wang Xiangsui, *Unrestricted Warfare*, trans. U.S. Central Intelligence Agency (Beijing: People's Liberation Army Literature and Arts, 1999). The CIA version is available on the web at http://www.c4i.org/unrestricted.pdf. The subtitle of the original Chinese edition, published in Beijing by the PLA Literature and Arts Publishing House in 1999, translates literally as "warfare without borders." When the Pan American Publishing Company published the book in English, it substituted the provocative and misleading "China's Master Plan to Destroy the U.S." The significance of this book, the extent to which it reflects actual government or military policies in China, and what bearing it had on the subsequent rise of cyber conflict are disputed matters. The account I offer here is intended as the most plausible and objective reconciliation of those contending interpretations.

15. See the extended discussion of this theory in Clarke and Knake, *Cyber War*, 47–54.

16. *Trapdoors* (also referred to as "backdoors") are bits of code embedded surreptitiously in operating systems that enable hackers to bypass security firewalls or encryption to gain access and take control of a software-controlled device or industrial system at a later time. A *logic bomb* is a small malware program that begins to operate (or "explode") when initiated by some kind of trigger or "fuse," such as a specific date or time. Once initiated, the program begins to damage the system within which it is embedded. See Stephen Northcutt, "Trap Doors, Trojan Horses, and Logic Bombs," SANS Technology Institute Security Laboratory, May 2, 2007, accessed May 20, 2016, http://www.sans.edu/research/security-laboratory/article/log-bmb-trp-door.

The most famous instance of an alleged military/security/intelligence use of a logic bomb was the destruction of a Soviet-era oil pipeline in Siberia, whose massive explosion in 1982 was attributed to a logic bomb allegedly planted by the CIA. See Thomas Reed, *At the Abyss: An Insider's History of the Cold War* (New York: Random House, 2007). Reed is a former US National Security Council aide. This account is widely regarded as apocryphal and is dismissed by most experts. Agents of the former Soviet Union deny the account. That is fortunate, if true, because if such an event had occurred, it would have represented the effort of astonishing blockheads—a pointless, cruel, environmentally destructive attack on civilian infrastructure that could well have triggered a serious international incident at the time, or even war. Even if it is apocryphal, it would therefore not be an achievement to celebrate, so much as an example of the kind of reckless and stupid behavior that we will want to discourage.

17. As its name suggests, the "dark web" is a specific, heavily encrypted region of the Internet that is not easily accessible to most users. Darkode.com was a criminal forum hosted in this region of the web and used primarily to traffic in stolen data. See "Major Computing Hacking Forum Dismantled" (press release, US Department of Justice, Office of Public Affairs, Washington, DC, July 15, 2015), http://www.justice.gov/opa/pr/major-computer-hacking-forum-dismantled.

18. See Natasha T. Balendra, "Defining Armed Conflict," *Cardoza Law Review* 29, no. 6 (2008): 2461–2516. The term *armed attack*, which has replaced the terms *war* and *warfare* in the formal terminology of international law, is nowhere officially defined, not even in the Charter of the United Nations. Instead, its meaning is presumed to be obvious: a use of force involving lethal weapons and destructive armaments. Article 2 (4) of the charter prohibits the use of force or the threat of the use of force by any nation against any other nation without prior United Nations authorization. The Security Council reviews disputes and threats and authorizes the use of force under UN jurisdiction for the purposes of the collective security of member nations. There is only one exception to this prohibition: when one or more nations is the victim of armed aggression or an armed attack. The victim nation may use force in the form of armed conflict to defend itself; other nations may come to its aid in repelling the armed attack. This leaves a number of essential questions unaddressed—for example, (1) if these provisions leave so much to the imagination regarding what constitutes a conventional armed attack, armed conflict, and the use of force, how in our present case are we exactly to determine when other means or methods of aggression or "domination" in the cyber realm rise to its equivalent?, and (2) if international law itself consists of what nations do or tolerate being done, they appear to both do and tolerate much wider uses of force for conflict resolution than the written or "black-letter" law formally sanctions. Some legal philosophers therefore argue that international law is merely rhetorical and has little or no normative force. But even if this is so, it is a form of rhetoric that is extremely powerful and persuasive in moderating state behavior, even if it is not successful in completely eliminating or adequately punishing illegal behavior.

19. Neil C. Rowe, "War Crimes from Cyber Weapons," *Journal of Information Warfare* 6, no. 3 (2007): 15–25. As Michael Schmitt observes, moreover, article 36 of the 1997 Additional

Protocol to the 1949 Geneva Conventions prohibits any state from seeking to develop or acquire any "new weapon, method, or means of warfare" that might be reasonably be found to violate existing laws of armed conflict—whence the development, let alone use of cyber weapons that deliberately targeted "civilians or civilian objects" would, perforce, be prohibited under existing international law. See his introduction to *Cyberwar: Law and Ethics for Virtual Conflicts*, ed. Jens David Ohlin, Kevin Govern, and Claire Finkelstein (New York: Oxford University Press, 2015), v.

20. Gross, a very well-respected political philosopher, first advanced this notion in the concluding chapter in *The Ethics of Insurgency: A Critical Guide to Just Guerilla Warfare* (New York: Cambridge University Press, 2014). The concept of "soft war" as unarmed conflict and its current and future role in international relations have been examined by several scholars. See Michael Gross and Tami Meisels, eds., *Soft War: The Ethics of Unarmed Conflict* (New York: Oxford University Press, 2016).

21. Specifically, this kind of "warfare" relies upon what cybersecurity expert Winn Schwartau several years ago cleverly termed "weapons of mass disruption." See Schwartau, *Cybershock* (New York: Thunder's Mouth Press, 2000).

30 Geoengineering as Collective Experimentation

Jack Stilgoe

Climate change is proving to be one of the most complex and difficult challenges of our time. The human sources of climate change—such as the use of fossil fuels and the production of gases in landfills—are scattered all over the globe and woven into enormous systems that structure and support our everyday lives. To make an immediate impact on the sources of the problem requires radical changes at every level of society—from international governance bodies to individual family units. Thus far, many countries have found it politically difficult to make such radical transformations.

This lack of change, coupled with the concern that we are reaching the "tipping point," after which rapid heating of the planet cannot be reversed, has led some scientists and engineers to advocate for a technological fix—or at least a temporary technological response—to the problem. They suggest we should be thinking about "geoengineering" the earth to try to slow the rapid temperature changes we are experiencing. They offer two primary strategies to do this: first, remove the carbon dioxide already in the atmosphere, commonly referred to as carbon dioxide removal (CDR), or second, try to bounce some of the sun's rays back into space so they don't heat the earth. The latter approach is known as solar radiation management (SRM).

This selection by Jack Stilgoe examines an SRM research project. Stilgoe doesn't, however, focus on the details of the technology. Rather, his interest is in exploring how the research that could possibly enable the technology should be governed. It is not really possible to use computer simulation or a traditional lab to test whether or not any particular geoengineering approach might work; the technology has to be tested in the real world. Stilgoe explores whether and how this could be done. Most meaningful geoengineering experiments in essence turn the people living in the study area (and perhaps beyond it) into lab rats of a sort. This conundrum is not limited to geoengineering. Increasingly, the systems that structure our lives are so interconnected that the impacts of any intervention will not really be known until they are tested outside the laboratory. Stilgoe poses a difficult question about whether and how such research and implementation should be conducted—minding the fact that "doing nothing" does not necessarily make us safer.

From *Science and Engineering Ethics* 22 (2016): 851–869.

Introduction

In September 2011, a proposed experiment was announced by a group of British University scientists that, from one perspective, seemed mundane. The idea was to float a tethered helium balloon a kilometre up in the sky with a hose attached. A pump would deliver a few dozen litres of water to the top of the hose, where it would emerge as a mist, evaporating before it hit the ground. At the time, there was a strong consensus among the scientists involved, their universities and outside observers that the experiment was not particularly risky, nor did it run against established ethical protocols. The experiment, however, became a condensation point for controversy because it was also, to use the researchers' own words, *"the first field test of a geoengineering technology in the UK"* (see Stilgoe 2015 for a fuller account). The experiment was part of a project called SPICE—Stratospheric Particle Injection for Climate Engineering. In the end, the experiment never got off the ground, figuratively or physically. Citing their own concerns about intellectual property and the governance of geoengineering, the researchers called it off. Nevertheless, the controversy generated provides an important entry point for "informal technology assessment" (Rip 1987) of geoengineering as a social experiment (Stilgoe et al. 2013).

Geoengineering (or climate engineering) encompasses a set of ideas for technological fixes to global climate change. The range of proposals is large, but in the main they concentrate either on removing carbon dioxide from the atmosphere or on cutting the amount of sunlight that reaches the surface of the Earth. The former category includes schemes to massively expand forests or seed oceans in order to encourage algae, as well as machines for capturing carbon directly from the air. The latter ranges from sunshades positioned in space between the Earth and Sun to the whitening of roofs on buildings. Within this category of so-called Solar Radiation Management, the idea of stratospheric particle injection—creating a reflective haze in the Earth's stratosphere—has attracted most attention because early assessments suggest that it has the greatest potential to reduce incoming sunlight while being (relatively) affordable. David Keith, currently the world's most prominent geoengineering researcher, opens his recent book—*A Case for Climate Engineering*—by claiming with some certainty that

> it is possible to cool the planet by injecting reflective particles of sulfuric acid into the upper atmosphere where they would scatter a tiny fraction of incoming sunlight back to space, creating a thin sunshade for the ground beneath. To say that it's "possible" understates the case: it is cheap and technically easy. (Keith 2013, p. ix)

As I will describe below, there are plenty of reasons to question the desirability of geoengineering. David Keith would join the majority of scientists in the nascent domain of geoengineering research who would doubt whether geoengineering was a Good Idea, although most would not share Keith's level of conviction. Proposals for geoengineering, which began as an extension of cold war technocratic modernism (see Fleming 2010), have, with their 21st century re-emergence, taken on a reflexive flavour (cf. Beck 1992). Nevertheless, geoengineering has, despite myriad uncertainties about its doability and desirability, rapidly acquired a deterministic frame, based on the assumption that it is "cheap" and "easy." Following the pattern of what Joly and colleagues (2010) call the "economics of techno-scientific promises," geoengineering has been naturalised by its researchers, treated as a thing in the world to be understood rather than a highly controversial, highly speculative set of technological fix proposals.

In this paper, I argue that the governance debate surrounding geoengineering can benefit from a view that starts with recognition of the social experimental nature of emerging technologies. The field of geoengineering research is small but growing. The uncertainties are vast and the likelihoods of predictability and control are tiny. Geoengineering would, as currently imagined, seem to represent an archetypical experimental technology. But we should not presume to know what geoengineering technologies will look like, if they are indeed realised. It is therefore also important to look at the experiments taking place within geoengineering research, experiments in which future geoengineering technologies and imaginaries (Jasanoff 2015) are being shaped. The emerging technology of geoengineering represents an experimental system in which knowns and unknowns are negotiated, in public discourse and in research projects. As with SPICE, the potential for reframing experimental means and ends suggests the possibility of a new mode of governance, one of collective experimentation, with implications for how we think about other geoengineering research experiments.

Responsible Research and Innovation

The emergence of geoengineering as a research agenda and a "matter of concern" (Latour 2004) has coincided with growing US and European interest in "responsible research and innovation," (RRI) "responsible innovation" or the "responsible development" of new technologies. There has been some institutional uptake of these terms, and possibly the ideas that they carry, within the European Commission, the UK Research Councils and the National Science Foundation respectively.[1] Although institutions may neglect, wilfully or otherwise, to mention it, these terms have their roots in debates about the possibilities of broadening the basis for technology assessment (Guston and Sarewitz 2002; Rip et al. 1995), reinvigorating the politics of technology (Winner 1980) and aligning science and innovation with social needs. While there have been substantial policy efforts in some countries to "open up" (Stirling 2008) public debates about emerging technologies, these have typically been disconnected from any policy or scientific response. RRI offers the possibility of shifting governance debates away from problematising publics to focus on research and innovation themselves. (Pellizoni 2004) suggests that we should pay more attention to the limits to responsiveness. In doing so, he reconnects debates about responsibility to an older discussion of the social control of technology. David Collingridge (1980) described the dilemma of control in these terms:

> Attempting to control a technology is difficult, and not rarely impossible, because during its early stages, when it can be controlled, not enough can be known about its harmful social consequences to warrant controlling its development; but by the time these consequences are apparent, control has become costly and slow. (Collingridge 1980, p. 19)

As Liebert and Schmidt (2010) point out, Collingridge is better remembered for describing this dilemma than for his normative aim of finding ways to govern despite it. Collingridge was interested in identifying and seeking to ameliorate "the roots of inflexibility" (Collingridge 1980, p. 45). We therefore need not be fatalistic, not least because, as Liebert and Schmidt go on to conclude, technologies may not be "controlled" according to particular decisions in the light of particular knowledge in the formal way that Collingridge seems at first to assume and then goes on to himself critique.

Technologies, as later constructivist studies would conclude, are as much a result of unquestioned assumptions or implicit values (see Williams and Edge [1996] for a summary). The attempt to govern despite the impossibility of prediction has acquired the term "anticipatory governance" (see Guston [2014] and Nordmann [2014] for a recent discussion).

Nor should we see uncertainty and ignorance as essential and problematic properties of technology. Uncertainty is constructed in scientific and innovative practice and attempts are made to exert both technical and social control over its bounds (Jasanoff and Wynne 1998). In public issues, uncertainties can be coproduced and reproduced as public concerns are interpreted, legitimised or rejected (Stilgoe 2007). As I will describe, the construction of experimental systems therefore plays a crucial political role by giving meaning to particular uncertainties.

From his re-reading of the Green Revolution, Collingridge demands that we pay closer attention to the contestation of problems to which technologies are offered as solutions (see Morozov [2013] for a recent popular account of the similar dynamics in what he calls the "solutionism" of digital technologies). Discourses of responsible research and innovation attempt, in the face of what are perceived as growing pressures towards neoliberal science (Lave et al. 2010; Pellizoni and Ylönen 2012), to draw stronger links with global societal challenges (von Schomberg 2012). But doing so introduces a profound question of democratisation. Technology, following Winner (1977), is itself a powerful form of legislation. If problems are constructed in order to serve particular solutions, rather than the other way around, then an important task of responsible research and innovation should surely be one of reflexivity on problem definition. Science and technology may themselves not hold the single or best answer, and may crowd out alternative approaches of social innovation.

What, then, does it mean to "care" for the futures to which science and innovation contribute (Groves 2014; Owen et al. 2013; Stilgoe 2015)? First, the idea of care seems more satisfactory than "control," the term used by Collingridge. Just as we recognise that the unintended consequences of technology are not completely predictable or controllable (Wynne 1988), so we should recognise that the trajectories of technology cannot themselves be predicted and controlled (see Stirling 2014). A careful approach is less likely to involve prohibition (Marchant and Pope 2009) than what Kuhlmann and colleagues (2012) call "tentative governance," encompassing "provisional, flexible, revisable, dynamic and open approaches that include experimentation, learning, reflexivity, and reversibility." This is, incidentally, close to David Collingridge's prescription of "corrigibility."

Scientists, innovators and others may argue that they are taking responsibility not just through conventional mechanisms of research integrity but also by engaging in what Alfred Nordmann (2007) has labelled "speculative ethics." Certainly, geoengineering research has seen more than its fair share of speculative ethics, which, by asking what happens "if" geoengineering futures are realised, contributes to a narrative of inevitability (Stilgoe 2015). Speculative ethics has joined risk assessment as part of an attempt to make techno-scientific promises of innovation more explicitly "responsible," but they risk closing down decision making rather than opening it up to new possibilities. The dominant governance discourse tends towards "containment" (Jasanoff and Kim 2009) of not just risk and ethics, but also of public debate. Peter-Paul Verbeek (2010) makes the case for reconnecting the empirical and ethical strands of the philosophy of technology to move from a mode of "technology assessment" to one of "technology accompaniment." In this latter mode, the imaginaries of geoengineering, which embed particular understandings of problems and solutions, can be adequately interrogated.

As I will argue, the impossibility of control in a scientific sense, let alone as a public issue, would put geoengineering alongside technologies such as genetically modified crops (Levidow and Carr 2007) and nuclear energy (Krohn and Weingart 1987), whose testing and deployment can be constructively seen as forms of social experiment (Krohn and Weyer 1994). This line of academic study builds upon, informs and is informed by a critical discourse about technology from commentators and NGOs that uses the language of experimentation to argue that technologies are less predictable, less well-understood and less controllable than their proponents would have us believe. The political writer John Gray (2004) expresses anguish that, "The world today is a vast unsupervised laboratory, in which a multitude of experiments are simultaneously underway. Many of these experiments are not recognised as such."

Bonneuil and colleagues argue that we should not look to the inherent riskiness of open-air experimentation but instead look to experiments as a site for the contestation of the politics of emerging technologies. They describe how, in France, field experiments with genetically modified crops were reframed through public controversy. In the decade up to 1996, thousands of field experiments took place without arousing wider interest. Over the next decade, these experiments became the focus of a debate less about the health and environmental risks of a particular technique than about the future economics and politics of agriculture. Experiments that had previously been "entrenched" as being routine scientific affairs, and shielded from public view, were dramatically reinterpreted as incursions into the social arena. Activists targeted and destroyed crop trials, justifying this as both a means to an end (attracting public attention) and an end in itself (preventing what they regarded as genetic contamination) (Bonneuil et al. 2008).

If we recognise the experimental nature of emerging technology from the start, we can put questions of democracy back into governance, asking how scientists and others should negotiate "the conditions for the performance of experiments in and on society" (Krohn and Weyer 1994, p. 181). The ethical questions expand beyond consideration of the ethical "implications" of technology to also include experimental care and ethics, which prompts consideration of who the participants are and the extent of their informed consent. The democratic governance of innovation therefore means asking what counts as legitimate experimentation (van de Poel 2016), prompting experimenters to confront "the questions we should ask of almost every human enterprise that intends to alter society: what is the purpose; who will be hurt; who benefits; and how can we know?" (Jasanoff 2003, p. 240).

Interpreting technologies as themselves experimental provides a powerful way to reimagine the uncertainties and stakes of geoengineering research and explore the politics of geoengineering experiments themselves. There is a risk that this reframing nebulises the issues to the point of meaninglessness. I would argue that, with reference to the history and philosophy of experimentation we instead gain a new foothold on governance through close attention to the demarcation of what is considered certain or uncertain and stable or unstable within experimental systems.

For Hans-Jorg Rheinberger (1997), "experimental systems" are a site for negotiation between the known and unknown. Experiments involve controlled surprises: "Experimentation, as a machine for making the future, has to engender unexpected events" (ibid, pp. 32–33). An experiment is made of two parts: the well-understood "technical objects," and the "epistemic things," which are the subject of inquiry. Following Rheinberger's analysis, we can start to investigate the politics of experimentation through analysis of the bounding of certainties and uncertainties. We can ask, for example, what surprises are permitted in experimental systems and how uncertainties are imagined, understood and controlled in the construction of experiments.

Opening up the "Surprise Room"

Beneath the now well-established conclusion of Science and Technology Studies (Nelkin 1979) that research in general expands rather than reduces the scope of uncertainty, we can analyse the strategic construction of uncertainty as a central part of scientific work (Wynne 1987). The notion of "surprise" gives this work a harder political edge. Scientists themselves are not uncomfortable with the idea of surprise. The surprises that mark apparent "breakthroughs" are central to scientific mythology. It is notable, for example, that psychologist Walter Mischel (2014) called his psychology laboratory at Stanford University's crèche the "surprise room." Gross (2010) makes the point that surprises, so integral to scientific novelty, nevertheless lie beyond conventional, containable categories of risk and probability. In this way, surprise is a useful lens on society's relationship with scientific uncertainty. The precautionary critique of technological risk assessment relates at least in part to the inability of regulation to anticipate or deal with the unexpected. The surprising nature of technological risk is often a function of previous wilful ignorance as, for example, with asbestos, whose risks were anticipated, but ignored, more than a century before they were effectively controlled (EEA 2001). Rather than being concerned about surprises per se, therefore, constructivist analyses should be interested in questions of who defines, prepares for and responds to surprises, and how and why they do so.

Experiments play important performative, public and technological roles. Habermas (quoted in Radder 2009) argues that experimentation turns science into "anticipated technology." Experiments involve the "systematic production of novelty . . . making and displaying new worlds" (Pickstone 2001, p. 13, 30). The wider political importance of experimentation means that, when they take place in public, they are typically displays of certainty rather than genuine surprise (Collins 1988; Shapin and Schaffer 1985). As Collins puts it,

> Where possible, experiments are still done in private because, the initiated aside, confidence in "the facts" will not survive a confrontation with Nature's recalcitrance. Only demonstrations or displays are gladly revealed for public consumption. (Collins 1988, p. 727)

A focus on experimental systems allows a reconsideration of the politics of geoengineering research. We can first reconsider, as others have begun to do, the inevitable experimentality of any future geoengineering technologies. This enables a focus on the contingency of technological promises that are currently offered as stable and certain. On this descriptive basis, we can secondly engage more normatively with the social aspects of "scientific" experimentation. Seeing geoengineering as itself an experimental system allows for new, constructive insights into the governance of geoengineering experiments themselves.

Geoengineering as Planetary Experiment

The problem to which geoengineering purports to offer a solution—climate change—has itself acquired a discourse of experimentalism. As scientific explanations of anthropogenic global warming were developed over the 20th Century, prominent scientific and political figures spoke of climate change as "a grand experiment" (Guy Stewart Callendar), "a large scale geophysical experiment" (Roger Revelle) or "a massive experiment with the system of this planet itself" (Margaret Thatcher), emphasising both the profundity and uncertainty of humanity's disruption to the climate system. For many climate scientists, the language of experimentation was a justification for the urgent development of scientific knowledge. Stephen Schneider argued in his book, *Laboratory Earth*, that "much of what we do to the environment is an experiment with Planet Earth, whether we like it or not" (Schneider 1997, p. xiv). Schneider's call-to-arms is issued to both policymakers and scientists: "It is no longer acceptable simply to learn by doing. When the laboratory is the Earth, we need to anticipate the outcome of our global-scale experiments before we perform them" (Schneider 1997, p. xii).

Some technological enthusiasts, such as Stewart Brand, have used the description of climate change as a messy experiment as a rationale for controlled experimentation through geoengineering (e.g. Brand 2010). Schneider, in the few years before his death in 2010, took the opposite view, also shared by Al Gore (2009), who argued that "We are already involved in a massive unplanned planetary experiment. . . . We should not begin yet another" (through geoengineering). Most contemporary geoengineering researchers would agree that the scale of surprises generated by doing geoengineering would be too profound to be currently tolerable, although their views would vary on the global climate conditions that would make such risks worth taking.

The recent renewal of enthusiasm for geoengineering is at least partly due to Paul Crutzen, a Nobel Laureate atmospheric scientist who published a prominent paper (Crutzen 2006) arguing that scientists and policymakers should cautiously reconsider the idea of stratospheric particle injection, which had fallen out of fashion as attention to climate change mitigation had grown. Following Crutzen's interjection, assessments of geoengineering proposals have sought to explore and explain the possible implications and uncertainties of deployment. Alan Robock (2008) provided an account of "20 reasons why geoengineering may be a bad idea." The possible side effects Robock identifies range from environmental (the effects on local weather and continued acidification of oceans) and sociotechnical (the potential for lock-in to bad and irreversible technological systems and the impossibility of global consensus on the ideal temperature for the "global thermostat") to the economic (high, escalating and uncertain costs) and political (the potential for militarisation of geoengineering technologies and the moral hazard that this technological insurance would introduce into delicate negotiations on mitigating climate change).

In 2008, the Royal Society began a study to respond to and inform the growing debate on geoengineering. Defining geoengineering as the "deliberate and large-scale intervention in the Earth's climatic system with the aim of reducing global warming," their report drew on a wide range of expertise, including social science, philosophy and law. As well as performing technical analyses of risk, cost, feasibility and speed across a wide range of geoengineering proposals, the Society's report discussed questions of ethics and governance, noting that "The acceptability of geoengineering will be determined as much by social, legal and political issues as by scientific and technical factors" (Royal Society 2009, p. ix). The uncertainties of geoengineering were foregrounded in parts of the report while in other parts the approach tends towards cost-benefit analysis. With explicit reference to Collingridge's dilemma of control, the report described the possibility of technological lock-in contributing to shaping the future of geoengineering. Coming at a time when researchers were beginning to conduct experiments with ocean iron fertilisation, the Society was faced with calls to govern experimentation, particularly when experiments crossed borders between jurisdictions or took place in international waters (see Stilgoe 2015).

Although Robock's assessment is broad, including ethical and political considerations, his sense of geoengineering-as-experiment is largely a technical one. He and colleagues (Robock et al. 2010) have argued that testing of geoengineering would be impossible without its full-scale deployment, in part because the signal of a response to any geoengineering would get lost in the noise of a chaotic climate system. Other geoengineering researchers have countered that, with careful scaling up and variation, the effects of geoengineering could be tested at a less than planetary scale (see MacMynowski et al. 2011). Even if this were to be true, the absence of either a hermetically-sealed scalable laboratory or a control run would blur any line drawn between research and deployment. Even without knowing what the technologies of an eventual geoengineering system would look like, sociologists of technology might agree that, as with the missiles studied by Donald Mackenzie (1993), they would be impossible to test except through use. Given the vast uncertainties within the climate system, any deployment of geoengineering, even at full scale, would be necessarily experimental, if not cybernetic (Jarvis and Leedal 2012). And when we consider whether these experiments might be in any way publicly credible, we bring climate models and their public contingencies into the apparatus too.

Further dimensions of the experimentality of geoengineering have been elucidated by Macnaghten and Szerszynski (2013) and Hulme (2014). For Mike Hulme, the experimentality of geoengineering relates to its outcomes being "unknown and unknowable" (Hulme 2014, p. 92). Using public focus groups, Macnaghten and Szerszynski (2013) explore the "social constitution" of current geoengineering proposals. They point to public scepticism about the predictability of geoengineering and unearth profound public concerns that "pervasive experimentality will be part of the new human condition" (Macnaghten and Szerszynski 2013, p. 470). (An earlier public dialogue exercise on geoengineering had been titled "Experiment Earth" [Corner et al. 2011], reflecting similar concerns.) Hulme (2014) joins Robock et al. in claiming that "The only experimental method for adequately testing system-wide response [to geoengineering] is to subject the planet itself to the treatment." But Hulme's argument is that this would also be an existential experiment on the human condition and humanity's ability to govern. The geoengineered world Hulme anticipates would be necessarily totalitarian. In a similar vein, Szerszynski et al. (2013) have pointed to the potential for solar geoengineering to be incompatible with democratic governance as we know it. (Rayner [2014] has critiqued this analysis of geoengineering's "social constitution" on the grounds that it prematurely identifies the essence of technology that remains hugely uncertain.)

The few NGOs that have begun campaigning against geoengineering were quick to adopt the language of experimentalism. One campaign, Hands Off Mother Earth (HOME), has the slogan "Our home is not a laboratory." It is tempting to read geoengineering as the archetype of the whole world becoming a laboratory (Latour 1999), but this global view risks detachment from more immediate concerns. Describing the experimentality of geoengineering should not be considered mere speculation on implications (following Nordmann's critique described above). Instead, by considering the contested boundaries of experiments (Davies 2010), we can engage with an emerging debate on the legitimacy of geoengineering experiments that are currently proposed or underway within and outside laboratories.

Governing Geoengineering Experiments

The debate generated by the SPICE experiment reveals the politics of experimentation in geoengineering—the things that are held to be certain, the things regarded as uncertain and worthy of investigation and the things regarded as out-of-bounds. The conventional story, relayed by the science press, scientists and science funders, is that SPICE was a failure of governance and a failed experiment. Reading it as a social experiment, we can see that SPICE reveals a huge amount about what is at stake in geoengineering research.

Before SPICE, scientists had sought to establish a safe space for experimental research and a means of containment for the spiralling social and ethical questions that geoengineering had begun to generate. Following the publication of its report on geoengineering, the Royal Society initiated a Solar Radiation Management Governance Initiative that, among other things, became a forum for negotiation of the Society's recommended "*de minimis* standard for regulation of research" (Royal Society 2009, p. xii). Although some geoengineering researchers were eager to begin conducting experiments to elucidate the implications of geoengineering, including experiments that would intentionally perturb the environment in order to study it, SRMGI was unable to agree where or whether such a line should be drawn. Around the same time, environmental experiments involving ocean iron fertilisation and cloud aerosols seemed to encroach into the geoengineering issue but whose motivations were either obfuscated or explicitly directed at conventional environmental science (see Buck 2014; Russell 2012).

The cancellation of the SPICE experiment did not quell this discussion. Indeed, it may have intensified scientists' attempts to identify and cordon off an area of no concern. Lawyer Edward Parson joined David Keith (2013) in arguing in *Science* for experimental thresholds. Their suggestion was that, above a certain upper limit (where there is a discernable effect on the environment), there should be a ban on geoengineering experiments. They also suggested a lower limit, beneath which experiments should be allowed to take place. Robock proposed an indoor/outdoor divide (Robock 2012), based on the premise that indoor activities are ethically justifiable while activities outside the laboratory demand additional scrutiny.

Victor and colleagues (2013) agree that "the key is to draw a sharp line between studies that are small enough to avoid any noticeable or durable impact on the climate or weather and those that are larger and, accordingly, carry larger risks" (see also Parson and Ernst 2013). A report from the US Congressional Research Service talks of the need for a "threshold for oversight" (Bracmort and Lattanzio 2013).

SPICE illustrates the trouble with such arguments. The reframing of the experiment as at least partly social challenges the attempt to hermetically seal it from public scrutiny. The SPICE scientists recognised this transition more vividly than anyone. One put it like this:

> People want to draw a bright line . . . and say everything above it is legitimate and everything below it is dangerous and requires governance. But that [laughs] that attitude undermines everything that SPICE is trying to figure out, everything that SPICE has been challenged to do in terms of looking towards the far field, thinking about things like lock-in. (Interview with SPICE scientist, quoted in Stilgoe 2015)

It is notable that the controversy generated by SPICE took place as much within the scientific community as around it. The idea of outdoor experimentation had already raised concerns among climate scientists. Raymond Pierrehumbert, a prominent climate scientist and critic of geoengineering, argued that

> the whole idea of geoengineering is so crazy and would lead to such bad consequences, it really is pretty pointless. We already know enough about sulfate albedo engineering to know it would put the world in a really precarious state. Field experiments are really a dangerous step on the way to deployment, and I have a lot of doubts what would actually be learned.[2]

David Keith had already argued that, "Taking a few years to have some of the debate happen is healthier than rushing ahead with an experiment. There are lots of experiments you might do which would tell you lots and would themselves have trivial environmental impact: but they have non-trivial implications."[3] In a BBC interview, he took issue with SPICE:

> I personally never understood the point of that experiment. That experiment's sole goal is to find a technocratic way to make it a little cheaper to get materials into the stratosphere. And the one problem we *don't* have is that this is too expensive. All the problems with SRM are about who controls it and what the environmental risks are, not how much it costs. It's already cheap. So from my point of view, I thought that was a very misguided way to start experimentation.[4]

For Keith and other geoengineering researchers, an additional, thinly-disguised concern was that negative reactions to SPICE would threaten subsequent research and experimentation on geoengineering. The concern was not that SPICE represented a perturbative experiment that fell on the wrong side of the various thresholds under discussion—all agreed that the experiment was benign in terms of its direct environmental impact—but rather that it challenged a dominant sense of what was considered "well-ordered science" (Kitcher 2003). SPICE was controversial not just because it was a prominent open-air experiment in a highly contested domain of technoscience, but also because it suggested an alternative demarcation of certainties and uncertainties.

Scientists' responses to the SPICE proposal point to competing framings of the experimental system. Before SPICE, priority research questions for geoengineering were overwhelmingly concerned with *episteme* (knowing that) rather than *techne* (knowing how) (see Hansson 2014a, 2014b; Ryle 1971). Hansson (2014b) has argued that experiments can blend episteme and techne, which provides an additional layer of explanation for the SPICE controversy. Scientists have, at least in the area of Solar Radiation Management, been reticent to openly explore the engineering constraints associated with creating a workable technology, instead reifying technological proposals dating back to the 1970s (e.g. Budyko 1974) and asking about the *impacts* of operationalising such ideas. SPICE brought engineers together with climatologists and atmospheric chemists, which had the effect of reconstructing the uncertainties considered relevant. Things previously considered stable, such as the cost and feasibility of stratospheric geoengineering, were treated as empirical questions. In Rheinberger's language, technical objects became epistemic things, disrupting an implicit sense of the experimental system. The new possibilities of surprise generated by SPICE challenged the deterministic story of geoengineering.

The public nature of the SPICE experiment, and the subsequent debate it generated, created an opportunity for what Nerlich and Jaspal call "frame shifting" (Nerlich and Jaspal 2012, p. 132). The initial assumption within the SPICE team was that the public would be interested, in a positive sense, or that the experiment could be a spur for a necessary debate on the ethics of geoengineering. Insofar as the experiment itself was problematised, the imagined public were those people within the immediate vicinity of the balloon who might bear witness to its launch. As the controversy unfolded, it became clear to the scientists and engineers that a relevant "public" would not be so easily bounded. If we presume that the "slippery slope" from research to deployment is completely frictionless, the relevant public could, as NGOs critical of SPICE implied, expand to encompass the world's population.

The idea of "care" implied by this reframing goes some way beyond the Royal Society's idea of "carefully planned and executed experiments" (Royal Society 2009, p. ix), which would include "Small/medium scale research (e.g. pilot experiments and field trials)" (Royal Society 2009, p. 61). The Royal Society recognised that "Just as field trials of genetically modified crops were disrupted by some NGOs, it is foreseeable that similar actions might be aimed at geoengineering experiments involving the deliberate release of sulphate or iron (for example) into the air and oceans" (Royal Society 2009, p. 15).

As a first step towards regulation, the Society argued for an international voluntary code of conduct, adding that "only experiments with effects that would in aggregate exceed some agreed minimum (de miminis) level would need to be subject to such regulation" (Royal Society 2009, p. 521; see also Bellamy 2014). In emphasising scientific self-governance and a scientific definition of contentious experimentation, they offer, in effect, to "take care" of this issue, on behalf of society. This is "care" in the paternalistic rather than democratic sense of the word.

Ralph Cicerone, who would go on to become president of the National Academy of Sciences, the Royal Society's US equivalent, had argued at the time of Crutzen's intervention that geoengineering research should be "considered separately from actual implementation . . . We should proceed as we would for any other scientific problem, at least for theoretical and modeling studies" (Cicerone 2006). But if we are concerned with the sociotechnical imaginaries of geoengineering (Jasanoff and Kim 2009), then public, open-air experimentation may not be uniquely problematic especially if, as with SPICE, there is a consensus that such experimentation does not pose direct risks. The experimentality of geoengineering, exacerbated by the trajectory that its emergence and scale-up would follow, make problematic any attempt to draw a line between research and deployment.

Solar geoengineering began as a set of thought experiments, substantially inspired by the natural experiment of a massive volcanic eruption. Since the re-emergence as a topic of scientific research, there have been almost no substantial solar geoengineering experiments taking place in the open environment, with the ecosystem as part of the apparatus.[5] SPICE was notable in that it became an in foro public experiment even in absence of an actual in situ trial taking place. However there have been a number of in silico experiments on general circulation models of the climate, whose results have informed the geoengineering debate. Ken Caldeira, another leading geoengineering researcher, has described in magazine interviews how he set out to demonstrate using computer models that geoengineering would be an unremittingly bad approach to global warming, but that he was taken aback by his own results. He told the *New Yorker*: "Much to my surprise, it [geoengineering] seemed to work and work well" (Specter 2012). (It is notable that these experiments *with* computer models are also experiments *on* the computer models; Schiaffonati, this issue; Stilgoe 2015). Given the power of such experiments to shape the promises and expectations of geoengineering, we might ask whether research inside the lab, involving computer models should self-evidently be free from public oversight or if there is a legitimate role for democratisation here too.

From Noun to Verb

In the short time since geoengineering was rehabilitated as a legitimate area of scientific study, it has rapidly acquired a deterministic frame. Geoengineering has become naturalised by the scientists, social scientists, philosophers and others who have begun to focus on it. This has the effect of closing down governance discussions and absolving scientists of responsibility for fashioning this nascent sociotechnical imaginary. Imagining the potential for constructive governance of geoengineering and geoengineering research requires challenging this frame. I have suggested in this paper that focussing on the experimentality of geoengineering as an emerging technology provides one way forward. Rather than presuming a regime of technoscientific promises, I suggest instead that we rethink geoengineering within a regime of collective experimentation (Joly et al. 2010).

Table 30.1 summarises what such a reframing would mean in thought and practice. The first feature is a grammatical one. The regime of technological promises tends to reify geoengineering as a technology that is, if not already in the world, inevitable. This is an outcome of what Joly calls the "naturalisation of technological advance" (Joly et al. 2010). Rather than treat the word as a noun (a gerund to be more precise) we can instead read "geoengineering" as a verb (a present participle). This shift from noun to verb turns geoengineering from an object of governance (Owen 2014) to a work in progress, with all of the attendant uncertainties and poorly-defined responsibilities of those—scientists, engineers, philosophers, social scientists and others—implicated in the project.

The implications of this way of thinking can be seen if we consider new proposals for geoengineering "field experiments." Keith and colleagues (2014) have recently described a suite of imagined experiments with which to explore the risks of further geoengineering research. Including the SPICE balloon experiment in their list, they imagine tests ranging from what they call "process studies" up to "climate response." Among these sits Keith's own proposed SCoPEx experiment, which would take place in the lower stratosphere in order to test possible effects of solar geoengineering on stratospheric ozone. He has argued with colleagues that such experiments are "a necessary complement to laboratory experiments if we are to reliably and comprehensively quantify the reactions and dynamics defining the risks and efficacy of SRM" (Dykema et al. 2014).

Keith and colleagues (2014) emphasise that the SCoPEx experiment would, along with most others in the list, generate negligible "radiative forcing" (an intended cooling effect on the climate). SCoPEx would have climatic effects that are "small compared with that of a single flight of a commercial transport aircraft." From their assessment, even larger "field experiments could be done with perturbations to radiative forcing that are negligible in comparison to the natural variability of climate at a global scale" (Keith et al. 2014).

Taking geoengineering as a noun, one can see the rationale for such experiments, and for a governance regime that seeks to delimit regulation according to whether experiments are seen as posing direct climatic risks, at large scales and for extended periods of time, as Keith and colleagues suggest. Within this frame, the inclination is to bound the experimental system tightly, to reduce what Keith and colleagues (2014) call "spurious disagreements." The underlying motivation is to create a "safe space" for research.

Table 30.1
Two governance regimes for geoengineering research

	Regime of technoscientific promises	Regime of collective experimentation
"Geoengineering"	. . . as noun	. . . as verb
Theory of technology	Instrumentalism	Substantivism/critical theory (see Feenberg 1999)
Responsibilities of researchers (including social scientists, philosophers etc.)	Assessment of technologies	Implicated in realising futures
Role of social science (see Macnaghten and Szerszynski 2013)	Proposing implications	Interrogating trajectories
Approach to uncertainty	Uncertainties seen as soluble through further research	Uncertainty seen as contested, inevitable and expanding
Approach to ethics	Speculative ethics and technology assessment	"Technology accompaniment" (see Verbeek 2010)
Characterising problems	"Solutionism," in which problems are assumed rather than explored	Reflexive approaches to problem identification and definition
Construction of public concerns	Technological development and perturbative experimentation	Open-ended, but may include imaginaries
Relationship between research and use	Scientific research is divorced from technological deployment	Research and deployment are entangled in the same social experiment
View of scientific autonomy	Negative liberty—freedom *from*. (The "right to research" viewed in libertarian terms)	Positive liberty—freedom *to*. (The "right to research" viewed in republican terms) (Brown and Guston 2009)
Relevant uncertainties	Implications of geoengineering	Implications, costs, feasibility, design
Governing experiments	Creating a "safe space" for research	Engaging with entanglements
Experimental systems	Bounded by science	Including publics, politics, ecosystems and scientists themselves

However, if we see geoengineering as a verb, under a regime of collective experimentation, things become less straightforward. Rather than prioritising freedom *from* experimental regulation, we might instead consider freedom in a positive sense, as a social licence to experiment. In addition to evaluating likely experimental risks and scales, we might also encourage scrutiny of experimental intentions and the imaginaries that sit behind them. Once we understand, as the SPICE scientists themselves did, that concerns with that experiment related to more than its direct risks, we can reframe other proposed experiments. This is not to presume that such experiments should therefore face additional governance from the top down. Indeed we would not wish those involved in experimentation and innovation to anticipate every possible future, not least because their activities are explicitly aiming to enable alternative and therefore unpredictable futures. The aim should instead be to experiment with experimentation, inviting further consideration of who should be involved in the definition and conduct of experiments. In practice, this may mean that geoengineering field experiments adopt the inter- and multi-disciplinary approaches that have started to take hold in other areas of geoengineering research (Szerszynski and Galarraga 2013).

The ambivalence of scientists and the political uncertainties surrounding geoengineering have meant that social scientists have been among those invited into various novel experimental collaborations (cf. Rabinow and Bennett 2012; Stilgoe 2012). These interactions typically involve the renegotiation of what is considered known and unknown as parties try to break out of the mould that is cast for them by others. Perhaps the social scientists and others that have become part of the apparatus of geoengineering research can contribute to the realisation of an alternative vision, one of collective experimentation.

Notes

1. See "Responsible Research and Innovation," European Commission, accessed November 1, 2014, http://ec.europa.eu/programmes/horizon2020/en/h2020-section/responsible-research-innovation; "Framework for Responsible Innovation," Engineering and Physical Sciences Research Council, accessed November 1, 2014, http://www.epsrc.ac.uk/research/framework/; National Science Foundation, *International Dialog on Responsible Research and Development of Nanotechnology*, accessed November 1, 2014, http://www.nsf.gov/crssprgm/nano/activities/dialog.jsp (page no longer available).

2. Quoted in D. A. Rotman, "Cheap and Easy Plan to Stop Global Warming," *MIT Technology Review*, February 8, 2013, accessed December 19, 2013, http://www.technologyreview.com/featuredstory/511016/a-cheap-and-easy-plan-to-stop-global-warming/.

3. Quoted in the *Economist*, "Lift Off," November 4, 2010, accessed December 19, 2013, http://www.economist.com/node/17414216.

4. Interviewed on *BBC Hard Talk*, BBC News Channel, November 14, 2011.

5. The only possible exception might be the E-PEACE experiment, which tested cloud formation off the pacific US coast in 2011, but this was not initially framed explicitly as a geoengineering test.

References

Beck, U. 1992. *Risk Society: Towards a New Modernity*. London: Sage.

Bellamy, R. 2014. "Safety first! Framing and governing geoengineering experimentation." Climate Geoengineering Governance Working Paper 14. Accessed November 1, 2014. http://www.geoengineering-governance-research.org/perch/resources/workingpaper14bellamysafetyfirst.pdf (document no longer available).

Bonneuil, C., Joly, P. B., and Marris, C. 2008. "Disentrenching experiment the construction of GM—crop field trials as a social problem." *Science Technology and Human Values* 33 (2): 201–229.

Bracmort, K., and Lattanzio, R. 2013. "Geoengineering technologies. Geoengineering: Governance and technology policy." Congressional Research Service, November 26, 2013. Accessed June 11, 2014. http://www.fas.org/sgp/crs/misc/R41371.pdf.

Brand, S. 2010. *Whole Earth Discipline: Why Dense Cities, Nuclear Power, Transgenic Crops, Restored Wildlands, and Geoengineering Are Necessary*. London: Penguin.

Brown, M. B., and Guston, D. H. 2009. "Science, democracy, and the right to research." *Science and Engineering Ethics* 15 (3): 351–366.

Buck, H. J. 2014. "Village science meets global discourse: The Haida Salmon Restoration Corporation's ocean iron fertilization experiment." Working paper, Geoengineering Our Climate. https://geoengineeringourclimate.wordpress.com/2014/01/14/village-science-meets-global-discourse-case-study/.

Budyko, M. I. 1974. *Izmeniya Klimata*. Leningrad: Gidrometeoizdat. Published in 1977 as *Climatic Changes*. Washington, DC: American Geophysical Union.

Cicerone, R. J. 2006. "Geoengineering: Encouraging research and overseeing implementation." *Climatic Change* 77 (3): 221–226.

Collingridge, D. 1980. *The Social Control of Technology*. London: Pinter.

Collins, H. M. 1988. "Public experiments and displays of virtuosity: The core-set revisited." *Social Studies of Science* 18 (4): 725–748.

Corner, A., Parkhill, K., and Pidgeon, N. 2011. "Experiment Earth? Reflections on a public dialogue on geoengineering." Working paper 11–02, Understanding Risk, School of Psychology, Cardiff University.

Crutzen, P. J. 2006. "Albedo enhancement by stratospheric sulfur injections: A contribution to resolve a policy dilemma?" *Climatic Change* 77 (3): 211–220.

Davies, G. 2010. "Where do experiments end?" *Geoforum* 41 (5): 667–670.

Dykema, J. A., Keith, D. W., Anderson, J. G., and Weisenstein, D. 2014. "Stratospheric controlled perturbation experiment: A small-scale experiment to improve understanding of the risks of solar geoengineering." *Philosophical Transactions of the Royal Society A: Mathematical, Physical and Engineering Sciences* 372 (2031): 20140059.

European Environment Agency. 2001. *Late Lessons from Early Warnings: The Precautionary Principle 1896–2000*. Luxembourg: Office for Official Publications of the European Communities.

Feenberg, A. 1999. *Questioning Technology*. New York: Psychology Press.

Fleming, J. R. 2010. *Fixing the Sky*. New York: Columbia University.

Gore, A. 2009. *Our Choice: A Plan to Solve the Climate Crisis*. Emmaus, PA: Rodale.

Gray, J. 2004. *Heresies*. London: Granta Books.

Gross, M. 2010. *Ignorance and Surprise*. Cambridge, MA: MIT Press.

Groves, C. 2014. *Care, Uncertainty and Intergenerational Ethics*. London: Palgrave Macmillan.

Guston, D. H. 2014. "Understanding 'anticipatory governance.'" *Social Studies of Science* 44 (2): 218–242.

Guston, D. H., and Sarewitz, D. 2002. "Real-time technology assessment." *Technology in Society* 24 (1): 93–109.

Hansson, S-O. 2014a. "Experiments before science?—what science learned from technological experiments." In *The Role of Technology in Science*, edited by Sven Ove Hansson. Philosophical Perspectives. Dordrecht, the Netherlands: Springer.

Hansson, S.-O. 2014b. "Experiments: Why and how?" Discussion paper, Technologies as Social Experiments Conference, Delft.

Hulme, M. 2014. *Can Science Fix Climate Change?: A Case against Climate Engineering*. Hoboken, NJ: Wiley.

Jarvis, A., and Leedal, D. 2012. "The geoengineering model intercomparison project (GeoMIP): A control perspective." *Atmospheric Science Letters* 13 (3): 157–163.

Jasanoff, S. 2003. "Technologies of humility: Citizen participation in governing science." *Minerva* 41 (3): 223–244.

Jasanoff, S. 2015. "Future imperfect: Science, technology, and the imaginations of modernity." In *Dreamscapes of Modernity: Sociotechnical Imaginaries and the Fabrication of Power*, edited by S. Jasanoff and S.-H. Kim. Chicago: University of Chicago Press.

Jasanoff, S., and Kim, S.-H. 2009. "Containing the atom: Sociotechnical imaginaries and nuclear power in the United States and South Korea." *Minerva* 47 (2): 119–146.

Jasanoff, S., and Wynne, B. 1998. "Science and decision-making." In Vol. 1, *Human Choice and Climate Change—the Societal Framework*, edited by S. Rayner and E. Malone. Columbus, OH: Battelle Press.

Joly, P. B., Rip, A., and Callon, M. 2010. *Re-inventing Innovation. Governance of Innovation. Firms Clusters and Institutions in a Changing Setting*. Cheltenham, United Kingdom: Edward Elgar.

Keith, D. 2013. *A Case for Climate Engineering*. Cambridge, MA: MIT Press.

Keith, D. W., Duren, R., and MacMartin, D. G. 2014. "Field experiments on solar geoengineering: Report of a workshop exploring a representative research portfolio." *Philosophical Transactions of the Royal Society A: Mathematical, Physical and Engineering Sciences* 372 (2031): 20140175.

Kitcher, P. 2003. *Science, Truth, and Democracy*. Oxford: Oxford University Press.

Krohn, W., and Weingart, P. 1987. "Commentary: Nuclear power as a social experiment—European political fall out from the Chernobyl meltdown." *Science, Technology, and Human Values* 12 (2): 52–58.

Krohn, W., and Weyer, J. 1994. "Society as a laboratory: The social risks of experimental research." *Science and Public Policy* 21 (3): 173–183.

Kuhlmann, S., Stegmaier, P., Konrad, K., and Dorbeck-Jung, B. 2012. "Tentative governance—conceptual reflections and impetus for contributors to a planned special issue of research policy on 'getting hold of a moving target—the tentative governance of emerging science and technology.'"

Latour, B. 1999. *Pandora's Hope: Essays on the Reality of Science Studies*. Cambridge, MA: Harvard University Press.

Latour, B. 2004. "Why has critique run out of steam? From matters of fact to matters of concern." *Critical Inquiry* 30 (2): 225–248.

Lave, R., Mirowski, P., and Randalls, S. 2010. "Introduction: STS and neoliberal science." *Social Studies of Science* 40 (5): 659–675.

Levidow, L., and Carr, S. 2007. "GM crops on trial: Technological development as a real-world experiment." *Futures* 39 (4): 408–431.

Liebert, W., and Schmidt, J. C. 2010. "Collingridge's dilemma and technoscience." *Poiesis & Praxis* 7 (1–2): 55–71.

Mackenzie, D. 1993. *Inventing Accuracy*. Cambridge, MA: MIT Press.

MacMynowski, D. G., Keith, D. W., Caldeira, K., and Shin, H. J. 2011. "Can we test geoengineering?" *Energy & Environmental Science* 4 (12): 5044–5052.

Macnaghten, P., and Szerszynski, B. 2013. "Living the global social experiment: An analysis of public discourse on solar radiation management and its implications for governance." *Global Environmental Change* 23 (2): 465–474.

Marchant, G. E., and Pope, L. L. 2009. "The problems with forbidding science." *Science and Engineering Ethics* 15 (3): 375–394.

Mischel, W. 2014. *The Marshmallow Test*. New York: Little, Brown.

Morozov, E. 2013. *To Save Everything, Click Here: Technology, Solutionism, and the Urge to Fix Problems That Don't Exist*. London: Penguin.

National Science Foundation. 2004. *International Dialog on Responsible Research and Development of Nanotechnology*. Accessed November 1, 2014. http://www.nsf.gov/crssprgm/nano/activities/dialog.jsp.

Nelkin, D. 1979. *Controversy: The Politics of Technical Decisions*. Beverly Hills, CA: Sage.

Nerlich, B., and Jaspal, R. 2012. "Metaphors we die by? Geoengineering, metaphors, and the argument from catastrophe." *Metaphor and Symbol* 27 (2): 131–147.

Nordmann, A. 2007. "If and then: A critique of speculative nanoethics." *Nanoethics* 1 (1): 31–46.

Nordmann, A. 2014. "Responsible innovation, the art and craft of anticipation." *Journal of Responsible Innovation* 1 (1): 87–98.

Owen, R. 2014. "Solar radiation management and the governance of hubris." In *Geoengineering of the Climate System*, edited by R. Harrison and R. Hester, 212–248. London: Royal Society of Chemistry.

Owen, R., Stilgoe, J., Macnaghten, P., Gorman, M., Fisher, E., and Guston, D. 2013. "A framework for responsible innovation. Responsible innovation: managing the responsible emergence of science and innovation in society." In *Responsible Innovation*, edited by R. Owen, J. Bessant, and M. Heintz, 27–50. West Sussex: John Wiley & Sons.

Parson, E. A., and Ernst, L. N. 2013. "International governance of climate engineering." *Theoretical Inquiries in Law* 14 (1): 307–338.

Parson, E., and Keith, D. 2013. "End the deadlock on governance of geoengineering research." *Science* 15:1278–1279.

Pellizzoni, L. 2004. "Responsibility and environmental governance." *Environmental Politics* 13 (3): 541–565.

Pellizzoni, L., and Ylönen, M. 2012. *Neoliberalism and Technoscience: Critical Assessments*. Farnham, United Kingdom: Ashgate.

Pickstone, J. V. 2001. *Ways of knowing: A New History of Science, Technology, and Medicine*. Chicago: University of Chicago Press.

Rabinow, P., and Bennett, G. 2012. *Designing Human Practices: An Experiment with Synthetic Biology*. Chicago: University of Chicago Press.

Radder, H. 2009. "The philosophy of scientific experimentation: A review." *Automated Experimentation* 1 (1): 2.

Rayner, S. 2014. "A curious asymmetry." Working paper, Climate Geoengineering Governance.

Rheinberger, H.-J. 1997. *Toward a History of Epistemic Things: Synthesizing Proteins in the Test Tube*. Stanford: Stanford University Press.

Rip, A. 1987. "Controversies as informal technology assessment." *Knowledge* 8:349–371.

Rip, A., Misa, T. J., and Schot, J., eds. 1995. *Managing Technology in Society—the Approach of Constructive Technology Assessment*. London: Pinter.

Robock, A. 2008. "20 reasons why geoengineering may be a bad idea." *Bulletin of the Atomic Scientists* 64 (2): 14–18.

Robock, A. 2012. "Is geoengineering research ethical?" *Peace & Security* 4:226–229.

Robock, A., Bunzl, M., Kravitz, B., and Stenchikov, G. L. 2010. "A test for geoengineering?" *Science* 327 (5965): 530–531.

Royal Society. 2009. *Geoengineering the Climate: Science, Governance and Uncertainty*. London: Royal Society.

Russell, L. M. 2012. "Offsetting climate change by engineering air pollution to brighten clouds." *Bridge* 42 (4): 10–15.

Ryle, G. [1946] 1971. "Knowing how and knowing that." In Vol. 2, *Collected Papers*, 212–225. New York: Barnes and Noble.

Schiaffonati, V. 2016. "Stretching the traditional notion of experiment in computing: Explorative experiments." *Science and Engineering Ethics* 22 (3): 647–665.

Schneider, S. H. 1997. *Laboratory Earth: The Planetary Gamble We Can't Afford to Lose*. New York: Basic Books.

Shapin, S., and Schaffer, S. 1985. *Leviathan and the Air-Pump*. Princeton, NJ: Princeton University Press.

Specter, M. 2012. *The Climate Fixers*. New York: New Yorker.

Stilgoe, J. 2007. "The (o-) production of public uncertainty: UK scientific advice on mobile phone health risks." *Public Understanding of Science* 16 (1): 45–61.

Stilgoe, J. 2012. "Experiments in science policy: An autobiographical note." *Minerva* 50 (2): 197–204.

Stilgoe, J. 2015. *Experiment Earth: Responsible Innovation in Geoengineering*. London: Routledge.

Stilgoe, J., Watson, M., and Kuo, K. 2013. "Public engagement with biotechnologies offers lessons for the governance of geoengineering research and beyond." *PLoS Biology* 11 (11): e1001707.

Stirling, A. 2008. "'Opening up' and 'closing down' power, participation, and pluralism in the social appraisal of technology." *Science, Technology and Human Values* 33 (2): 262–294.

Stirling, A. 2014. "Emancipating transformations: From controlling 'the transition' to culturing plural radical progress." Working paper, Climate Geoengineering Governance. http://steps-centre.org/wp-content/uploads/Transformations.pdf.

Szerszynski, B., and Galarraga, M. 2013. "Geoengineering knowledge: Interdisciplinarity and the shaping of climate engineering research." *Environment and Planning A: Environment and Planning* 45 (12): 2817–2824.

Szerszynski, B., Kearnes, M., Macnaghten, P., Owen, R., and Stilgoe, J. 2013. "Why solar radiation management geoengineering and democracy won't mix." *Environment and Planning A* 45 (12): 2809–2816.

van de Poel, I. R. 2016. "What kind of experiments are social experiments with technology?" *Science and Engineering Ethics* 22 (3): 667–686.

Verbeek, P. P. 2010. "Accompanying technology." *Techné: Research in Philosophy and Technology* 14 (1): 49–54.

Victor, D. G., Morgan, M. G., Apt, J., Steinbruner, J., and Ricke, K. L. 2013. *The Truth about Geoengineering*. New York: Foreign Affairs.

Von Schomberg, R. 2012. "Prospects for technology assessment in a framework of responsible research and innovation." In *Technikfolgen abschätzen lehren*, 39–61. Berlin: Verlag für Sozialwissenschaften.

Williams, R., and Edge, D. 1996. "The social shaping of technology." *Research Policy* 25 (6): 865–899.

Winner, L. 1977. *Autonomous Technology: Technics-out-of-Control as a Theme in Political Thought*. Cambridge, MA: MIT Press.

Winner, L. 1980. "Do artifacts have politics?" *Daedalus*, 121–136.

Wynne, B. 1987. "Uncertainty—technical and social." In *Science for Public Policy*, edited by H. Brooks and C. L. Cooper, 95–115. Oxford: Pergamon Press.

Wynne, B. 1988. "Unruly technology: Practical rules, impractical discourses and public understanding." *Social Studies of Science* 18 (1): 147–167.

31 Seven Principles for Equitable Adaptation

Alice Kaswan

This selection isn't about stopping climate change; it's about how to deal with the effects predicted to happen as our climate changes. In particular, it is concerned with the fact that the impacts of climate change will not be felt uniformly across the world, or even the US. Some places and peoples will be much more negatively affected than others. Kaswan examines the ways in which minorities, the poor, and the disadvantaged are far more likely to suffer from the effects of environmental degradation than rich, privileged people. Studies have shown, for instance, that African Americans are more likely to live near toxic waste dumps and heavy industry than whites. Those who are poor or lack a political voice not only often live in areas susceptible to current and future environmental disasters but usually lack the ability to adapt, move, or safely respond to the situation.

This article takes some of the basic lessons learned from decades of research into environmental justice issues and remedies and offers steps that government agencies and others could take to reduce the inequitable impacts of climate change on the least fortunate. In her discussion, Kaswan notes that not only poverty can make one unable to safely respond to the threats of climate change. She argues that people can be disadvantaged through a number of mechanisms, including the reduced movement of the elderly, the reduced ability to communicate faced by those with different language skills, and the segregation and isolation imposed by racism. Interestingly, the measures that the author suggests have little to do with technology. She doesn't offer a silver-bullet technological fix for these problems. Instead she suggests seven political and social responses to reduce the inequitable consequences that result from forces like pandemics and climate change.

As Professors Robert Bullard and Beverly Wright have stated, "Climate change looms as *the* global environmental justice issue of the twenty-first century," posing critical challenges "for communities that are already overburdened with air pollution, poverty, and environmentally related illnesses."[1] Around the world, sea level rise, more extreme storms, heat waves, wildfires, changing weather patterns, and the spread of disease appear inevitable.[2] Reducing greenhouse gas (GHG) emissions is necessary but not sufficient to address the potential damage.[3] Global, national, and subnational adaptation measures to reduce climate harm are essential.[4] To avoid substantial disparities in the impacts of climate change, equity considerations should play a vital role in emerging United States adaptation initiatives.[5] Focusing on domestic law, this article briefly describes climate change impacts and the role of socioeconomic factors in determining their magnitude. It then provides seven principles for achieving equitable adaptation.

From *Sustainable Development Law and Policy* 13, no. 1 (2012): 41–46, 67–69.

Climate Change Impacts

Among the most dramatic impacts of climate change will be the increasing incidence of disasters.[6] Climate scientists anticipate that flooding will become more common and severe as sea levels rise and hurricanes become more intense, generating more destructive storm surges—the consequences of which were all too evident after Hurricane Sandy's inundation of New York and New Jersey in Fall 2012.[7] Throughout the nation, precipitation events are likely to become more extreme and, in some parts of the country, overall precipitation levels are already increasing dramatically.[8] Scientists predict increasing wildfires in the western states,[9] predictions borne out by recent record-breaking fires.[10] Risks from flooding and fire include not only the direct harm from rising waters or flames, but contamination risks from inundated or incinerated industrial and hazardous waste facilities,[11] the need to dispose of tons of debris,[12] and the long-term housing and economic impacts that endure for years after major disasters.[13] Adaptation measures must address adequate disaster preparedness, response, recovery, and mitigation measures to reduce long-term risks.

Increasing disaster risks could also render certain parts of the country uninhabitable. Migration away from low-lying coastal areas and floodplains may ultimately be necessary.[14] Certain tribal communities in coastal Alaska, like the Village of Kivalina, already face the need to relocate.[15] Additional climate impacts, like unsustainably high temperatures, droughts or saltwater intrusion that depletes essential water supplies, could likewise require large-scale population shifts.[16] Adaptation measures must address local decision-making processes that govern decisions about when to protect an area from harm (through, for example, coastal armoring, levees, or the enhancement of natural buffers), when to adjust (through, for example, building standards to increase resilience), and when to retreat (through, for example, conservation easements or public purchase of at-risk property).

Scientists have also found that climate change will lead to numerous public health threats. Climate scientists predict that by 2100, average temperatures in the United States will increase by four to eleven degrees and heat waves that historically occurred once every twenty years will occur every other year.[17] Heat waves are among the most lethal of disasters, causing as many or more deaths than other types of disasters.[18] Moreover, higher temperatures trigger higher pollution levels, increasing the negative public health consequences of high heat.[19] Warmer temperatures in the United States are also predicted to lead to the spread of disease and allergens.[20]

Climate change will have pervasive economic impacts as well. For example, 80,000 businesses and almost 400,000 jobs were reportedly lost from Hurricanes Katrina and Rita.[21] Changes in resource availability, like water supplies, could increase the cost of water and, given the importance of irrigation to agriculture, increase the cost of food.[22] Warmer temperatures may increase demand for air conditioning, potentially increasing electricity costs.[23] Climate mitigation efforts, however well-meaning, could also increase energy costs, by either placing a price on carbon through a market-based control mechanism or by encouraging the use of more expensive renewable energy sources.[24] More broadly, adaptation measures themselves are likely to be extremely costly. Fortifying or moving key infrastructure, like roads, airports, and sewage treatment plants, will cost billions.[25] Relocating communities or buying out property owners to protect them from harm would cost billions more.[26] Disaster response and reconstruction costs multiple billions of dollars.[27] Indirectly, addressing climate impacts and financing adaptation measures could drain government resources from other functions, like education and the social safety net, unless alternative financing sources are developed.[28]

Climate Change Impacts and Equity

The consequences of climate change will be experienced unevenly. In the United States, poor and marginalized communities without sufficient financial and social resources will face significant adaptation challenges.[29] To quote Professor Robert Verchick: "Catastrophe is bad for everyone. But it is especially bad for the weak and disenfranchised."[30]

While it is critical to determine risk exposure—to assess the likelihood that a community will encounter a given climate impact—a community's ultimate vulnerability cannot be determined without also assessing its sensitivity and its capacity to cope.[31] Depending upon the type of climate impact at issue, sensitivity is determined by such features as the quality of the housing stock, underlying health conditions, land elevation, and proximity to other hazards. The capacity to cope is a function of such factors as a community's financial and social resources, access to health care, and geographic mobility.

Both physical and social factors thus determine climate impacts.[32] Social scientists evaluate social factors in terms of social vulnerability, defined as "the characteristics of a person or group in terms of their capacity to anticipate, cope with, resist, and recover from the impact of a natural hazard."[33] Substantial evidence demonstrates that social vulnerability is greater for the poor, the elderly, racial minorities, people with underlying health conditions or disabilities, the socially isolated and politically marginalized, immigrants, and communities that are dependent upon vulnerable natural resources.[34]

To avoid these disparities, climate change adaptation policies must grapple with underlying socioeconomic inequities. Decreasing social vulnerability requires adaptation measures that both reduce the underlying sensitivity to harm and enhance impacted communities' resilience to harm after it has occurred. As in the environmental justice context, pursuing climate justice involves improving substantive outcomes for disadvantaged communities, developing inclusive and empowering participatory mechanisms, and addressing the deeper social and institutional forces that create and perpetuate systemic disparities,[35] themes addressed by the seven principles articulated below.

Improving equity is valuable not only on its own terms, but because of the adverse societal consequences of failing to address equity. Widespread homelessness, unemployment, and illness disrupt the social fabric of a community and could create far-reaching instability. The already-frayed social safety net may be unable to cope with the scale of disruption that could occur. Considered comprehensively, it is more prudent to develop adaptation plans that avoid harm than it is to attempt to repair the harm after the fact—or suffer the consequences of irreparable devastation.[36]

Seven Principles for Equitable Adaptation

Given the key role of socio-economic factors in determining the magnitude of climate impacts, an integrated ecological, social, and economic approach to adaptation planning, like that suggested by Rob Verchick and by Manuel Pastor and his co-authors in the disaster planning context, is essential to equitable adaptation efforts.[37] Although successful adaptation will require attention to a wide range of important principles,[38] this article articulates a subset of that array, focusing on those principles with the greatest impact on equity.

The principles are intended to guide adaptation planning in any of the contexts in which it emerges. The principles are applicable to action taken by local, state, or national entities. They could inform new adaptation legislation, or they could be integrated into adaptation efforts by institutions, like disaster management agencies, housing agencies, public health organizations, and local governments as they act under existing authorities.

1 Government Has an Important Role to Play

A threshold question is whether government action is necessary or whether people can (and should) take care of themselves. There is little dispute over the importance of governmental measures to protect key infrastructure, like highways and energy systems. Where individual or private business welfare is at stake, however, some might argue that as long as the government provides accurate and accessible information about current and future climate impacts, the private market will generate the optimal response. As citizens perceive growing threats, they will respond, and their responses will reflect their individual (and differing) risk tolerances. For example, they will or will not move away from floodplains, seashores, or disease-prone areas, buy hazard insurance, trim fire-prone vegetation in their yards, and purchase air conditioning. Under this view, if citizens end up in harm's way, then they are responsible for their own choices.[39]

Relying on individual initiative is, however, unlikely to lead to sufficient adaptation. Individuals could discount what appear to be inchoate, distant, and remote threats. As a consequence, they could fail to make sufficient investments to prepare for uncertain risks. Moreover, certain adaptation choices, like retreat, require difficult emotional decisions that could lead to collectively irrational results, as community residents prefer denial to leaving their homes and communities and losing all the social capital that resides in existing community structures.

Relying on the market is particularly detrimental to low-income marginalized communities. As Manuel Pastor and his co-authors have observed, relying on "market forces" to adequately prepare for disasters and other climate change impacts will fail to provide an adequate adaptation response because reliance on private action fails to protect those without the knowledge or means to act, systematically disadvantaging poor and isolated communities.[40] Even assuming adequate knowledge, poor residents do not have the resources to respond to that knowledge by preparing, insuring, or moving.[41] When serious disasters occur, the government has historically provided some compensation, but that compensation cannot make up for underlying inequities.[42]

Moreover, relying on market forces to depopulate at-risk areas would exacerbate, not reduce, risks to low-income and of-color citizens who could be powerfully attracted to newly affordable housing—housing that has become affordable and available because it is at risk.[43] Where citizens do not have adequate resources and face limited housing mobility due to lingering discrimination, individual responses to climate change risk do not reflect free and unconstrained "choices."

Given the likelihood that market forces will fail to adequately protect people from harm, and fail in ways that exacerbate risks for more vulnerable populations, comprehensive government adaptation initiatives are warranted. The remainder of this section addresses key themes to guide the incorporation of equity considerations in adaptation policy.

2 Design Substantive Adaptation Measures That Address Vulnerability

Adaptation policies that attempt to treat everyone the same, regardless of underlying demographic characteristics, will result in substantial inequality given underlying differences. To achieve equitable adaptation, adaptation policies must explicitly address the demographics of affected populations and target interventions to address the needs of the most vulnerable.[44] Although such measures cannot eliminate all inequity—they cannot prevent the inexorable loss of Native American Alaskan coastal communities, for example—they could in many instances reduce harm and lessen disparities. Relevant characteristics include income, race, age, status as renters versus owners, and type of employment. Immigrant status is also relevant to adaptation policy, and is addressed explicitly below in connection with communication measures.

Disparities in income create many of the most significant disparities in vulnerability to climate change impacts. Income disparities also have a racial dimension: Although many whites live in poverty, communities of color are disproportionately poor.[45] Climate impacts that disproportionately impact the poor will therefore affect a larger percentage of people of color. Adaptation policies that target resources toward low-income communities could thus ameliorate both income and racial disparities.

For example, given poor families' lack of resources to prepare for disasters,[46] funding hazard preparation measures for low-income households or assisting with housing retrofits to provide cooling could improve outcomes for disadvantaged communities.[47] Moreover, poor residents are less likely to have adequate transportation to flee disasters,[48] face greater challenges in finding affordable and safe shelter if evacuation is necessary,[49] and are less likely to have air conditioning or other means for keeping cool in heat waves.[50] As Hurricane Katrina made abundantly clear, adaptation plans must provide timely transportation options,[51] provide for adequate and safe public shelters, and provide cooling centers in heat waves so that poor residents do not remain in place—and at risk—because of inadequate transportation or fear of public facilities.

In the disaster recovery context, to avoid homelessness and widespread suffering, low-income residents will require various forms of assistance, including adequate housing vouchers and relocation assistance where rebuilding is infeasible. If rebuilding requirements, like flood-proofing codes, add significant costs to re-building, then government support for such measure may be needed to ensure that low-income households are not priced out of rebuilding.[52] Given the challenges in siting and building low-income and public housing, a strong governmental role, and financial support, is likely to be necessary to ensure that adequate low-income options are available.

Long-range land use planning to address shifts in habitability will have important equity implications and should avoid criteria that adversely impact low-income communities. For example, if planners in an area subject to flooding risks were to choose what areas to protect based solely upon land value, that criterion would systematically undermine poor communities, communities that often have less power in local land use debates.[53] Land use decisions about protection, retreat, and new development should be guided by substantive criteria that recognize a range of community values, including but not limited to land value. In addition, decisions about how to facilitate retreat, and how to compensate for the loss of property, should recognize that low-income residents do not have the resources to start fresh elsewhere and face significant risks of homelessness or deepening poverty if relocation assistance is not provided.

Such long-range land use planning must also address potential impacts on areas that are likely to experience in-migration, as the population shifts from areas at risk to areas that face fewer risks and remain more habitable.[54] Adequate affordable housing in the nation's more habitable regions will be essential to avoid serious housing shortages and potential increases in homelessness.

Income is not the only demographic feature requiring sustained attention in the development of adaptation measures. Elderly and disabled residents face substantially greater risks in disasters because they are less likely to have adequate independent transportation, fare worse in shelters without adequate medical services, and are likely to suffer greater psychological distress from a disaster's profound disruptions.[55] They are also more vulnerable to public health threats, like heat and disease.[56] As a consequence, special accommodations for transportation, shelter, and medical needs are necessary for elderly and disabled residents to avoid serious consequences from disasters and the range of public health threats that climate change could cause.[57] Renters also require particularized attention.

Renters are less able to prepare for disasters or heat waves because landlords control investments in home strengthening, air conditioning, or other mechanisms to reduce vulnerability to disasters or heat waves.[58] Local governments could adopt building codes that require or incentivize landlords to strengthen structures and install air conditioning. Moreover, in hot climates, building codes could require building designs that minimize summer heat and incorporate energy-efficient cooling mechanisms. Given evidence that past disaster recovery programs have provided more resources for homeowners than for displaced renters,[59] adaptation planners should ensure that recovery programs provide adequate options for renters, including vouchers and housing alternatives.[60] In developing post-disaster rebuilding plans, relevant officials should include sufficient replacement rental and public housing, housing that has historically been replaced at a lower rate than other forms of housing.[61]

Lastly, given variations in risk exposure by occupation, adaptation planning should address the unique needs of certain workers. Outdoor workers, like agricultural, construction, and sanitation workers, face greater risks from high heat and pollution levels.[62] Those risks could be reduced by adjustments to the workday and by occupational safety guidelines that address adequate hydration, cessation of work when ambient temperatures exceed a certain level, and other measures to protect vulnerable workers.

3 Provide Culturally Sensitive Communications and Services

Communication is key to effective adaptation. Given the diversity of populations, community and demographic-specific strategies are necessary.[63] Public education can help communities prepare for disasters and inform them about how to address public health risks from heat waves, allergens, or new diseases. Early warning systems are also essential to prepare for weather-related disasters, including potential flooding and heat waves.[64] Effective disaster response requires providing those affected with information about evacuation and shelter options. After a disaster occurs, effective recovery depends upon widespread access to information about available recovery resources.

Experience in the disaster context demonstrates that linguistic and cultural isolation will exacerbate climate impacts for immigrant communities unless proactive steps are taken to develop community-specific communication mechanisms.[65] In addition to identifying language needs, adaptation planners need to identify culturally appropriate modes of communication including, potentially, newspapers, radio, television, e-mail, social media, or door-to-door outreach.[66] Given undocumented immigrants' justifiable fear of deportation or historically rooted distrust of government,[67] government agencies should provide assurances that they will not deport.[68] In addition, agencies could partner with nongovernmental community organizations that could facilitate community outreach, provide information, and help organize vulnerable or impacted communities.[69] The same issues arise in the context of providing services, like shelters or cooling centers, and in the context of distributing resources, like disaster relief.

Effective communications strategies are likely to vary for non-immigrant as well as immigrant communities, and require location-specific assessments.[70] Some neighborhoods may have strained relationships with local police departments or other officials.[71] Certain populations could also require different communication methods. For example, personal, door-to-door warning and assistance may be necessary to adequately prepare elderly and disabled residents.[72]

4 Develop Participatory Processes

Decisionmakers cannot develop substantively appropriate adaptation and communication strategies without the right participatory processes. Given the importance of community-specific information, adaptation planning processes require bottom-up participatory mechanisms.[73] Such participatory processes are important not only to obtain critical information, but to provide marginalized communities with a voice in difficult political decisions.[74] Consistent with principles of environmental justice, adaptation planning could provide a vehicle for community empowerment and self-determination.[75]

Adaptation planners should engage with community leaders to obtain site-specific information about relative disaster or heat preparedness and to identify appropriate modes of—and institutions for—communicating information about preparedness, warnings, and recovery.[76] Community-based information about available resources is also essential, including transportation and shelter options in the event of natural disasters or heat waves.[77]

The political dimension to participatory processes is as important as the informational dimension. Many adaptation-related decisions will be politically controversial. For example, planners must determine who benefits from disaster recovery resources. What resources for homeowners? What resources for renters? If new housing will be built, what income levels will it serve? With what neighborhood structures? In the long-term, communities facing flood and fire risks will have to make fateful decisions about what areas to protect and what areas to abandon. To be effective, participatory opportunities need to occur early in the process and address local power dynamics. Timing is critical to the ability to shape decision making; an obligatory public hearing on an already-complete planning document does not constitute real public participation. An extended process of place-based community hearings and forums is more likely to generate meaningful participation.[78] Moreover, given power disparities and the political marginalization of some communities, carefully crafted and targeted outreach will be necessary to draw in all communities. While good participatory mechanisms cannot erase endemic power imbalances, they at least provide transparent forums that give historically less powerful constituencies a seat at the table.

5 Reduce Underlying Non-climate Environmental Stresses

In some instances, climate change does not create new risks; it exacerbates existing risks. For example, it could increase risks from flooded sewage treatment plants, industry, or waste sites.[79] As Prof. Robin Craig has observed, a key adaptation principle is to "Eliminate or Reduce Non-climate Stresses and Otherwise Promote Resilience."[80] By improving the baseline, climate impacts will be less extreme. Because environmental justice research has demonstrated that many existing environmental problems, like hazardous waste storage and disposal sites, air pollution, and other environmental risks are disproportionately located in of-color and low-income communities,[81] reducing non-climate environmental stressors will have indirect equity benefits.

For example, improving inadequate storm water management, an existing non-climate problem, could mitigate the contamination that could arise from climate-caused increases in extreme precipitation.[82] In their compliance and enforcement initiatives, EPA or applicable state agencies could include vulnerability to disasters as a key factor in prioritizing their review of industrial and municipal storm water management plans and assessing compliance with industrial waste storage requirements. Similarly, the federal superfund program and its state equivalents could consider flood or fire risks in prioritizing cleanup efforts and in selecting remedies that take potential future disasters into account.[83] Moreover, aggressive efforts to reduce air pollution now will reduce the adverse consequences of future heat-induced air pollution increases.[84]

Following this principle would not only mitigate climate impacts; it would provide significant co-benefits by reducing existing non-climate stresses. Given extensive co-benefits, such initiatives are often considered "no" or "low" regrets policies that are justified whether or not climate change occurs.[85]

6 Mitigate Mitigation: Addressing Adaptation/Mitigation Tradeoffs

Although climate adaptation (addressing the impacts of climate change) and climate mitigation (reducing GHG emissions to lessen climate change) often involve different regulatory strategies, there are significant interactions between adaptation and mitigation measures. Policymakers need to consider the interplay between mitigation and adaptation.

In some instances, mitigation measures could be "maladaptive" by creating adaptation challenges, some of which raise equity concerns.[86] For example, a key strategy for reducing GHG emissions is encouraging smart growth to reduce transportation emissions from sprawl.[87] That smart growth could, however, increase urban heat. Scientists have documented that dense urban environments increase urban temperatures by several degrees over less-dense surrounding areas, a phenomena known as the "urban heat island effect."[88] Moreover, although having denser cities might reduce overall air pollution emissions by reducing the driving associated with sprawl, increased urban density could increase localized air pollution levels.[89] Finally, because many existing urban areas are in coastal areas and along rivers that face high disaster risks,[90] intensifying growth would often, as Prof. Lisa Grow Sun has suggested, constitute "smart growth in dumb places."[91] Where smart growth is justified, land use measures should prevent development in the riskiest areas and provide green spaces to minimize urban heat.[92] Transportation infrastructure should facilitate evacuation and be resilient to damage from potential disasters.[93]

Certain mitigation measures could also generate equity concerns if they increase energy costs, which could occur through greater reliance on more expensive renewable energy or from imposing a price on carbon through a market-based mechanism like cap-and-trade or a carbon tax.[94] Measures to alleviate such impacts, like financing energy efficiency or public transportation, would ameliorate the potential adverse economic consequences of climate mitigation policies.

In other instances, adaptation measures could compromise mitigation. For example, while policymakers should develop cooling strategies to protect people from heat waves, policies that simply require or finance the installation of air conditioning would undermine mitigation by increasing energy demand.[95] In addition to, or instead of air conditioning, policymakers should consider building standards that lead to cooler buildings,[96] urban designs that reduce the heat island effect, cooling centers, and demand-response systems that allow residents or utilities to reduce air conditioning use in unoccupied buildings.

7 A Comprehensive Agenda

While these suggestions for incorporating equity considerations into adaptation planning are important, it is also clear that they address symptoms, not causes. Underlying socioeconomic vulnerabilities create the disparities in the capacity to recover and reconstruct from disasters, inequities in the capacity to relocate to avoid harm, and differences in the public health consequences of increasing heat, pollution, and disease. We are confronting more than a "disaster planning" or "adaptation planning" issue.

A larger socioeconomic agenda is critical to achieving equitable adaptation. The IPCC has stated that "a prerequisite for sustainability in the context of climate change is addressing the underlying causes of vulnerability, including the structural inequalities that create and sustain poverty and constrain access to resources."[97] The IPCC states further that "addressing social welfare, quality of life, infrastructure, and livelihoods . . . in the short term . . . facilitates adaptation to climate extremes in the longer term."[98]

Successful adaptation will require addressing such pervasive issues as poverty, affordable housing, the provision of healthcare, and the political voice of currently marginalized communities.[99] Building social infrastructure has always been a laudable goal. Impending climate impacts provide yet another reason to mend social ills, or risk systemic disruptions that could make disasters like Hurricane Katrina and its aftermath the norm rather than the exception.

Conclusion

While global climate change is an "environmental" problem, the scope and scale of its impacts is strongly determined by underlying socioeconomic variables. As climate impacts emerge, they have the potential to exacerbate existing inequalities and cause severe hardships for the nation's most vulnerable populations—hardships that are not only intrinsically of concern, but also destabilizing to the larger community. These seven principles provide policymakers with guideposts for achieving equitable adaptation.

Notes

The author would like to thank Randy Rabidoux USF '13 for his research assistance and Robin Craig, Dan Farber, Victor Flatt, Carmen Gonzalez, J. B. Ruhl, and Rob Verchick for their very helpful comments on the longer version of this article.

1. Robert D. Bullard and Beverly Wright, The Wrong Complexion for Protection: How the Government Response to Disaster Endangers African American Communities (NYU Press, 2012), 51.

2. See, generally, Climate Change 2007: Impacts, Adaptation and Vulnerability, Contribution of Working Group Ii to the Fourth Assessment Report of the Intergovernmental Panel on Climate Change (M. L. Parry et al., eds., 2007), https://www.ipcc.ch/report/ar4/wg2.

3. Public attention is understandably focused on climate mitigation—on policies to reduce greenhouse gas emissions—but mitigation measures cannot undo the consequences of

already accumulated atmospheric greenhouse gases (GHGs). See Intergovernmental Panel on Climate Change, *Summary for Policymakers*, in Climate Change 2007: Impacts, Adaptation and Vulnerability, *supra* note 2, at 19, 20 [hereinafter IPCC, *Summary for Policymakers*], http:// www.ipcc.ch/pdf/assessment-report/ar4/wg2/ar4-wg2-spm.pdf; United States Global Change Research Program, Global Climate Change Impacts in the United States 11 (2009) [hereinafter USGCRP Report], http:// downloads.globalchange.gov/usimpacts/pdfs/climate-impacts-report.pdf.

4. The IPCC defines adaptation as: "The adjustment in natural or human systems in response to actual or expected climatic stimuli or their effects, which moderates harm or exploits beneficial opportunities." See Climate Change 2007: Impacts, Adaptation and Vulnerability, *supra* note 2, at 6. On balance, scientists predict that the negative consequences will outweigh the beneficial impacts. See IPCC, *Summary for Policymakers, supra* note 3, at 17.

5. Cf. Robert R. M. Verchick, Facing Catastrophe: Environmental Action for a Post-Katrina World 105–70 (2010) (proposing "Be fair" as a central principal of disaster law); Jim Chen, *Law among the Ruins*, in Law and Recovery from Disaster: Hurricane Katrina 1, 3 (Robin Malloy, ed., 2009) (arguing that the quest for equality should be a central component of disaster law); Victor B. Flatt, *Adapting Laws for a Changing World: A Systemic Approach to Climate Change Adaptation*, 64 Florida L. Rev. 269, 289–291 (2012) (discussing the important role of distributional justice in adaptation); J. B. Ruhl and James Salzman, *Climate Change Meets the Law of the Horse*, 62 Duke L.J. 975 (2013) (suggesting equity as one of three over-arching policy goals for adaptation efforts); Robert R. M. Verchick, *Disaster Justice: The Geography of Human Capability*, 23 Duke Envt'l l. & Pol'y Forum 23 (2012).

6. See National Research Council, Adapting to the Impacts of Climate Change 1 (2010), http://www.nap.edu/catalog.php?record_id=12783.

7. Sea levels have already increased over the last century and, notwithstanding uncertainty about the magnitude, climatologists predict further increases of three to four feet by 2100. See USGCRP Report, *supra* note 3, at 149 (describing past increase of up to two feet) and 150 (predicting future increase) and 149 (predicting more destructive storm surges). While no single weather event can be attributed to climate change, Hurricane Sandy provided a wake-up call to many about climate change and the risks of rising sea levels and more intense storm surges. Thomas Kaplan, *Most New Yorkers Think Climate Change Caused Hurricane, Poll Finds*, New York Times, December 3, 2012.

8. See IPCC, Managing the Risks of Extreme Events and Disasters to Advance Climate Change Adaptation 13 (2012) [hereinafter IPCC, Managing the Risks], https://www.ipcc.ch/report/managing-the-risks-of-extreme-events-and-disasters-to-advance-climate-change-adaptation/. Between 1958 and 2007, heavy storms increased by 67 percent in the Northeast and by 31 percent in the Midwest. USGCRP Report, *supra* note 3, at 32.

9. Id. at 82 (describing fourfold increase in Western wildfires over the last several decades).

10. See, e.g., Pete Spotts, *Monster Wildfire in Arizona: A Glimpse of What Climate Change Could Bring*, Christian Science Monitor, June 9, 2011, http:// www.csmonitor.com/Environment/2011/0609/Monster-wildfire-in-ArizonaA-glimpse-of-what-climate-change-could-bring; Darryl Fears, *Colorado's Table Was Set for Monster Fires*, Washington Post, July 1, 2012, http:// www.washingtonpost.com/national/health-science/colorados-table-was-set-for-monster-fire/2012/07/01/gJQAVa6cGW_story.html.

11. Verchick, *supra* note 5, at 133; Robin Kundis Craig, *A Public Health Perspective on Sea-Level Rise: Starting Points for Climate Change Adaptation*, XV Widener L. Rev. 521, 536–537 (2010); Manuel Pastor et al., in The Wake of the Storm: Environment, Disaster, and Race after Katrina 30 (2006).

12. See Linda Luther, Cong. Research Serv., Rl33477, Disaster Debris Removal after Hurricane Katrina: Status and Associated Issues (updated 2008) (describing post-Katrina debris removal challenges).

13. See, e.g., Pastor, *supra* note 11, at 25–27 (describing the long-term challenges in recovering from disasters, particularly for poor and minority communities).

14. See, e.g., National Research Council, *supra* note 6, at 75.

15. See Randall S. Abate, *Public Nuisance Suits for the Climate Justice Movement: The Right Thing and the Right Time*, 85 Wash. L. Rev. 197, 207 (2010).

16. See, e.g., Robin Kundis Craig, *"Stationarity Is Dead"—Long Live Transformation: Five Principles for Climate Change Adaptation Law*, 34 Harv. Envtl.L. Rev. 9, 55 (2010) (noting the possibility of significant migration from arid western areas to wetter regions, and from coastal areas inland).

17. USGCRP Report, *supra* note 3, at 29, 34 (describing predicted increase in average temperatures; describing predicted increase in heat waves).

18. USGCRP Report, *supra* note 3, at 90 (reporting that "heat is already the leading cause of weather-related deaths in the United States"). A severe 2003 European heat wave reportedly caused seventy thousand excess deaths. See National Research Council, *supra* note 6.

19. US Environmental Protection Agency (EPA), Our Nation's Air: Status and Trends through 2010 11 (2012). Higher temperatures increase the rate at which ozone, a significant air pollutant, is formed from its precursor compounds, nitrogen oxides, and volatile organic compounds. See also USGCRP Report, *supra* note 3, at 92–94.

20. Jonathan Samet, Public Health: Adapting to Climate Change 4–5 (Resources for the Future, 2010), https://www.rff.org/documents/172/RFF-IB-10-06.pdf.

21. Robert K. Whelan and Denise Strong, *Rebuilding Lives Post-Katrina: Choices and Challenges in New Orleans' Economic Development*, in Race, Place, and Environmental Justice after Hurricane Katrina: Struggles to Reclaim, Rebuild, and Revitalize New Orleans and the Gulf Coast 183, 183 (Robert D. Bullard and Beverly Wright, eds., 2009) [hereinafter Race, Place, and Environmental Justice]. In the month following Hurricane Sandy, the New York area lost approximately thirty thousand jobs due to the storm. See Patrick McGeehan, *Nearly 30,000 Jobs Lost Because of Hurricane Sandy*, New York Times (December 20, 2012).

22. See Seth B. Shonkoff et al., *The Climate Gap: Environmental Health and Equity Implications of Climate Change and Mitigation Policies in California—a Review of the Literature*, 109 Climatic Change S485, S491 (2011).

23. See USGCRP Report, *supra* note 3, at 54–55 (explaining that warmer temperatures are projected to lead to a slight net increase in energy use because increases in air conditioning are likely to outpace decreases in energy use for heating).

24. See, e.g., National Research Council, *supra* note 6, at 49 (describing increasing energy costs resulting from increasing focus on expensive renewable energy sources).

25. See id. (assessing the cost of providing protection from three feet of sea level rise at roughly $100 billion).

26. For example, relocating a single four-hundred-person Alaskan tribal village, the Village of Kivalina, is projected to cost from 95 to 400 million dollars. See Abate, *supra* note 15, at 207.

27. See M. L. Parry et al., Assessing the Costs of Adaptation to Climate Change: A Review of the UNFCCC and Other Recent Estimates 31 (2009). Hurricane Katrina reportedly caused

146 billion dollars in damage. See *Hurricane Sandy's Rising Costs*, New York Times A32 (November 28, 2012). After Hurricane Sandy, New York estimates $42 billion in damage, while New Jersey estimates almost $30 billion. Id.

28. See, generally, Matthias Ruth et al., The US Economic Impacts of Climate Change and the Costs of Inaction (2007), http://cier.umd.edu/documents/US%20Economic%20 Impacts%20of%20Climate%20Change%20and%20the%20Costs%20of%20Inaction.pdf. Climate mitigation efforts, like a cap-and-trade program or carbon tax, could conceivably generate revenue that would finance climate adaptation efforts. See Rachel Morello-Frosch et al., The Climate Gap: Inequalities in How Climate Change Hurts Americans and How to Close the Gap 19 (2010), http://dornsife. usc.edu/pere/documents/ClimateGapReport_full _report_web.pdf.

29. These challenges are described in more detail in the longer version of this article. See Alice Kaswan, *Domestic Climate Change Adaptation and Equity*, 42 Environmental Law Reporter 11125 (2012). Equity concerns are even more dramatic internationally. Many poor developing countries, like small-island states, Bangladesh, and African nations, are simultaneously the least responsible for, but the most at risk from, global climate change. See, e.g., IPCC, *Summary for Policymakers*, *supra* note 3, at 12 (describing high risks from sea level rise for low-lying African and Asian deltas and for small island states). The importance of international adaptation and equity concerns does not, however, erase the significance of addressing equity in US adaptation measures.

30. See Verchick, *supra* note 5, at 106.

31. See National Research Council, *supra* note 6, at 29; Chen, *supra* note 5, at 3–5.

32. See IPCC, Managing the Risks, *supra* note 8, at 7, 10; USGCRP Report, *supra* note 3, at 100–101; Verchick, *supra* note 5, at 38–41 (observing that the degree of hazard a community faces is "a combination of a community's *physical* vulnerability and its *social vulnerability*"; emphasis in original). See, generally, Pastor et al., *supra* note 11, at 2 (observing that environmental equity focuses on both cumulative exposure and social vulnerability).

33. See Daniel A. Farber et al., Disaster Law and Policy 217 (2d ed. 2010) (quoting Piers Blaikie et al., At Risk: Natural Hazards, People's Vulnerability and Disasters 9 [1994]).

34. See U.S. Climate Change Science Program, Analyses of the Effects of Global Change on Human Health and Welfare and Human Systems 64 (2008) (listing socioeconomic factors affecting vulnerability), http://downloads.globalchange.gov/sap/sap4-6/sap4-6-final-report -all.pdf; id. at 64 (noting greater impacts on those with lower socioeconomic status); id. at 123 (listing factors affecting vulnerability to disasters); Verchick, *supra* note 5, at 106; Bullard and Wright, *supra* note 1, at 52–54 (describing disparities in climate impacts for disadvantaged populations). Large-scale aggregate analyses have isolated the important role of social vulnerability as a determinant of disaster impacts. A study of 832 floods in seventy-four Texas counties found a statistically significant correlation between social vulnerability, as measured by racial minority or low-income status and flood deaths or injuries. Sammy Zahran et al., *Social Vulnerability and the Natural and Built Environment: A Model of Flood Casualties in Texas*, 32 Disasters 552–553, 555 (2008).

35. See Alice Kaswan, *Environmental Justice and Domestic Climate Change Policy*, 38 Envtl. L. Rep. 10,287, 10,289 (2008) (describing the environmental justice movement's distributive, participatory, and social justice goals and their influence on the movement for climate justice); see, generally, Luke W. Cole and Sheila Foster, From the Ground UP: Environmental Racism and the Rise of the Environmental Justice Movement (2001) (describing the environmental justice movement's focus on distributional outcomes, participatory processes, and institutional structures).

36. See M. L. Parry et al., *supra* note 27, at 102–113; Ruth et al., *supra* note 28. In addition to economic considerations, avoiding harm has important social, cultural, and psychological benefits.

37. Pastor et al., *supra* note 11, at 30–31; Verchick, *supra* note 5, at 165.

38. See, e.g., Verchick, *supra* note 5; Craig, *supra* note 16.

39. See James K. Boyce, *Let Them Eat Risk? Wealth, Rights and Disaster Vulnerability*, 24 Disasters 254, 257 (2000) (stating that "the wealth-based approach holds that . . . those individuals who are willing (and, perforce, able) to pay more, deserve to get more [disaster vulnerability reduction]").

40. See Pastor et al., *supra* note 11, at 7 (arguing that a market-based approach to disaster preparedness "is a recipe for targeting those with the least power in the social calculus"); Verchick, *supra* note 5, at 149; see Debra Lyn Bassett, *Place, Disasters, and Disability*, in Law and Recovery from Disaster, *supra* note 5, at 51, 68.

41. See Debra Lyn Bassett, *The Overlooked Significance of Place in Law and Policy: Lessons from Hurricane Katrina*, in Race, Place, and Environment, *supra* note 21, at 49, 57; Boyce, *supra* note 39, at 257 (observing that relying on individual willingness-to-pay for disaster reduction would distribute reductions in a manner "strongly correlated with wealth").

42. See, generally, Pastor et al., *supra* note 11, at 25–27 (describing how reconstruction programs have not been sufficient to fully address the needs of low-income disaster victims). For example, in post-Katrina New Orleans, the mayor proposed that the city should decide where to invest in new infrastructure and support rebuilding by evaluating where rebuilding was already occurring. That approach would privilege areas where residents had sufficient resources to rebuild and disadvantage areas where residents did not have sufficient resources. John R. Logan, *Unnatural Disaster: Social Impacts and Policy Choices after Katrina*, in Race, Place, and Environmental Justice, *supra* note 21, at 249, 257.

43. See Pastor et al., *supra* note 11, at 11 (observing, in the environmental justice context, that "lower-income residents may be willing to trade off health risks for cheaper housing").

44. See Heather Cooley et al., Social Vulnerability to Climate Change in California 1 (2012); National Research Council, *supra* note 6, at 55; Morello-Frosch et al., *supra* note 28, at 22. The California Energy Commission commissioned a study that not only identified nineteen discrete physical and social vulnerability factors but evaluated their cumulative impact by creating an overarching climate vulnerability index to score different areas of the state and then indicated where high social vulnerability "intersects with the most severe projected climate change impacts." Cooley, *supra* at ii.

45. Statistical Abstract of the United States, *Table 711, People below Poverty Level and below 125% of Poverty Level by Race and Hispanic Origin: 1980–2009*, http://www.census.gov/compendia/statab/2012/ tables/12s0711.pdf (indicating that, as of 2009, African Americans and Hispanics were twice as likely as whites to be below the poverty level: over 25 percent, in comparison with the white population's 12.3 percent poverty rate).

46. See, e.g., Pastor et al., *supra* note 11, at 19–21; California Climate Change Center: The Impacts of Sea-Level Rise on the California Coast 46 (2009) [hereinafter Impacts of Sea-Level Rise], https://www.coastal.ca.gov/climate/PI-cc-4-mm9.pdf (describing income- and race-based disparities in disaster preparation).

47. The Federal Emergency Management Agency (FEMA) has several programs that provide some resources for hazard mitigation both pre- and postdisaster; resources that could be targeted toward the most vulnerable populations. See FEMA, Hazard Mitigation Grant Program,

http://www.fema.gov/hazardmitigation-grant-program (describing). See also USGCRP Report, *supra* note 3, at 91 (describing Philadelphia's "Cool Home Program," which provides low-income elderly residents with roof retrofits to cool their homes and save energy).

48. See Robert D. Bullard, Glenn S. Johnson, and Angel O. Torres, *Transportation Matters: Stranded on the Side of the Road before and after Disasters Strike*, in Race, Place, and environmental Justice, *supra* note 21, at 63, 66–67 (noting racial disparities in automobile ownership). In Hurricane Katrina, 55 percent of those who failed to evacuate did not own cars. See Impacts of Sea-Level Rise, *supra* note 46, at 49.

49. See Pastor et al., *supra* note 11, at 23 (describing disaster studies indicating that poor and minority populations are more likely to resort to tent cities and mass shelters); Scott Gold, *Trapped in the Superdome: Refuge Becomes a Hellhole*, Seattle Times, (September 1, 2005) (describing horrific shelter conditions, conditions that could deter residents from evacuating).

50. See Shonkoff, *supra* note 22, at S488 (observing that low-income and of-color residents are less likely to have air-conditioning); Cooley et al., *supra* note 44, at 6 (citing study that poor people are less likely to use air-conditioning, even if they have it, due to financial concerns).

51. See Bullard et al., *supra* note 21, at 70. Recent data suggests that, despite some recent improvements, many states and local governments have not adequately addressed evacuation needs for carless and special-needs populations. See Bullard et al., *supra* note 21, at 69, 76, and 77–78 (describing studies).

52. See Bullard and Wright, *supra* note 1, at 75, 98 (noting that green building "that fails to address issues of affordability, access, and equity may open the floodgates for permanent displacement of low-income and minority homeowners and business owners").

53. See Pastor et al., *supra* note 11, at 11.

54. See Craig, *supra* note 16, at 55 (discussing possibility of mass migrations in response to climate impacts).

55. See Bassett, *supra* note 40, at 64–65.

56. See Cooley et al., *supra* note 44, at 5 (noting the role of age in vulnerability to heat). In the 2003 European heat wave, many of the deaths were among the elderly. See *Over 11,000 Dead in French Heat*, BBC News, August 29, 2003, http://newsvote.bbc.co.uk/mpapps/pagetools/print/news.bbc.co.uk/2/hi/europe/3190585.stm.

57. See Bassett, *supra* note 40 and 41; Janet E. Lord, Michael E. Waterstone, and Michael Ashely Stein, *Natural Disasters and Persons with Disabilities*, in Law and Recovery from Disaster, *supra* note 5, at 71.

58. See Impacts of Sea-Level Rise, *supra* note 46, at 48.

59. See Pastor et al., *supra* note 11, at 24; Impacts of Sea-Level Rise, *supra* note 46, at 50.

60. Although property owners, unlike renters, have lost an asset, renters nonetheless encounter severe postdisaster conditions. Postdisaster, the demand for rental housing skyrockets, leading to significant increases in rents and serious shortages of affordable rentals. See Pastor et al., *supra* note 11, at 24; see also Lisa K. Bates & Rebekah A. Green, *Housing Recovery in the Ninth Ward: Disparities in Policy, Process, and Prospects*, in Race, Place, and Environmental Justice, *supra* note 21, at 229, 231 (describing renters post-Katrina challenges).

61. See, e.g., Verchick, *supra* note 5, at 138–139 (describing insufficient development of rental housing in post-Katrina New Orleans); Robbie Whelan, *A Texas-Size Housing Fight*,

Wall Street Journal, August 3, 2012, at A3 (describing local resistance to rebuilding affordable public housing in Galveston, Texas).

62. See Shonkoff, *supra* note 22, at S489–490; Cooley et al., *supra* note 44, at 6.

63. See, e.g., National Research Council, *supra* note 6, at 70, 71 (observing importance of "early warning systems" and "public outreach" to adapting to extreme events and disease risks from contaminated water and disease outbreaks) and 71. See also Sari Kovats and Shakoor Hajat, *Heat Stress and Public Health: A Critical Review*, 29 Ann. Rev. Pub. Health 41, 49 (2008) (observing importance of health education to reducing impacts from heat waves).

64. See National Research Council, *supra* note 6, at 70.

65. See Ted Wang and Luna Yasui, Integrating Immigrant Families in Emergency Response, Relief and Rebuilding Efforts 1–3 (2008), http://www.aecf.org/~/media/PublicationFiles/IR3622H122.pdf.

66. See id. at 9–11.

67. See Pastor, *supra* note 11, at 25. As states begin to enact laws requiring police officers to check and report on immigration status, a practice upheld by the Supreme Court in *Arizona v. United States*, 132 S.Ct. 2492, 183 L.Ed.2d 351 (2012), immigrants are likely to become increasingly reluctant, rather than more willing, to interface with government officials. See *Copycat Immigration Laws*, Washington Post, September 29, 2011, http://www.washingtonpost.com/politics/copycat-immigration-laws/2011/09/29/gIQA993Y7K_graphic.html.

68. See Wang and Yasui, *supra* note 65, at 4.

69. See id. at 1 and 6–7 (describing the important role that community organizations could play in enhancing emergency response for "limited-English proficient (LEP) residents and immigrants").

70. See Bates and Green, *supra* note 60, at 243 (observing that post-Katrina recovery would have better served African American neighborhoods if community-based organizations had been involved in planning and outreach early in the recovery process); Bullard and Wright, *supra* note 1, at 215–216 (describing the importance of cultural awareness in disaster communications and detailing several techniques for developing effective communication).

71. As the Department of Health and Human Services has recognized, as a result of experiences of racial discrimination, "racial and ethnic minority groups may distrust offers of outside assistance . . . even following a disaster. . . . They may be unfamiliar with the social and cultural mechanisms of receiving assistance and remain outside the network of aid" (US Department of Health and Human Services, *Developing Cultural Competence in Disaster Mental Health Programs*, DHHS Publication No. SMA 3828, 2003, p. 16; available at https://store.samhsa.gov/sites/default/files/d7/priv/sma03-3828.pdf); Bullard and Wright, *supra* note 1, at 216–217 (discussing studies showing blacks' mistrust of white relief agencies).

72. Recognizing the importance of personal outreach, Philadelphia instituted a "buddy system" run by block captains to check on elderly residents during heat waves. See USGCRP Report, *supra* note 3, at 91.

73. See Impacts of Sea-Level Rise, *supra* note 46, at 89; Pastor et al., *supra* note 11, at 35; Morello-Frosch et al., *supra* note 28, at 24.

74. See Verchick, *supra* note 5, at 61–62, 67 (identifying the fundamental importance of democratic participation in mechanisms to reduce vulnerability).

75. See Luke W. Cole, *Empowerment as the Key to Environmental Protection: The Need for Environmental Poverty Law*, 19 Ecology l. Quarterly 619 (1992) (describing the environmental

justice movement's community empowerment goals); Sheila Foster, *Justice from the Ground Up: Distributive Inequities, Grassroots Resistance, and the Transformative Politics of the Environmental Justice Movement*, 86 Cal. l. Rev. 775 (1998).

76. See Pastor et al., *supra* note 11, at 21; Wang and Yasui, *supra* note 65, at 7.

77. See Wang and Yasui, *supra* note 65, at 5.

78. See Foster, *supra* note 75, at 834–836.

79. See Craig, *supra* note 11, at 536–537 (discussing contamination from inundated sewage treatment facilities); see also Pastor et al., *supra* note 11, at 30; Verchick, *supra* note 5, at 133 (discussing risks of inundated industries, landfills, or hazardous waste sites).

80. Craig, *supra* note 16, at 43–45 (articulating climate change adaptation principle #2: "Eliminate or Reduce Non-climate Stresses and Otherwise Promote Resilience" and subprinciple: "Decontaminate Land, Water, and Air, and Reduce New Pollution as Much as Possible").

81. See, e.g., Cole and Foster, *supra* note 35, at 167–183 (appendix: "An Annotated Bibliography of Studies and Articles That Document and Describe the Disproportionate Impact of Environmental Hazards by Race and Income"); see also Verchick, *supra* note 5, at 167–170 (suggesting that policy-makers should prioritize addressing existing hazards, like landfills and contaminated sites, in at-risk areas).

82. See EPA, Office of Water, National Water Program Strategy: Response to Climate Change 45–47 (2008), http://water.epa.gov/scitech/climatechange/upload/2008-National-Water-Program-Strategy-Response-to-Climate-Change.pdf (discussing need to manage stormwater and infrastructure to reduce contamination).

83. See Craig, *supra* note 11, at 538 (suggested expedited cleanup of contaminated sites in coastal areas). Disaster considerations could also significantly impact the choice of remedy, reducing the desirability of "institutional controls," like land use restrictions, that leave contamination in place and at risk of flooding.

84. See USGCRP Report, *supra* note 3, at 92–94 (describing how increasing temperatures could worsen air quality).

85. See National Research Council, *supra* note 6, at 71. The IPCC has noted that many initiatives to address projected increases in extreme events have multiple cobenefits that render them "low regrets" policies. IPCC, Managing the Risks, *supra* note 8, at 16, 17.

86. This essay addresses mitigation measures that create equity issues, a subset of the larger issue of maladaptation. Non-equity-related maladaptive mitigation measures, like thermal solar power plants that consume large volumes of water in areas expecting future shortages, are important but beyond the scope of this essay.

87. See Alice Kaswan, *Climate Change, Consumption, and Cities*, Fordham Urb. L.J. 253, 280–281 (2009).

88. See Brian Stone Jr., The City and the Coming Climate: Climate Change in the Places We Live 75–76 (2012).

89. See USGCRP Report, *supra* note 3, at 92 (noting that poor air quality is an especially serious concern in cities); EPA, *supra* note 19, at 19 (noting that toxic air pollution levels are higher in urban areas).

90. See Lisa Grow Sun, *Smart Growth in Dumb Places: Sustainability, Disaster, and the Future of the American City*, 2011–2016 BYU L. Rev. 2157, 2168–2169.

91. See id. Increased development is inconsistent with a sustainable long-term land use strategy in many of these high-risk areas. Id. at 2160–2161. See also id. at 2166–2168 (describing disaster risks associated with increasing density in urban areas).

92. See National Research Council, *supra* note 6, at 70 (noting that, in the long term, reducing health risks could require "urban design to minimize the urban heat island effect through greater use of trees and green spaces").

93. See Sun, *supra* note 90, at 2199–2200 (describing urban design patterns that facilitate long-term strategic retreat if it proves necessary).

94. See National Research Council, *supra* note 6, at 49 (regarding potential increases in energy costs from a switch to renewable energy).

95. See USGCRP Report, *supra* note 3, at 54–55; Environmental Health Perspectives and the National Institute of Environmental Health Sciences, A Human Health Perspective on Climate Change: A Report Outlining the Research Needs on the Human Health Effects of Climate Change 30–31 (2010), http://www.cdc.gov/climateandhealth/pubs/HHCC_ Final_508.pdf.

96. See National Research Council, *supra* note 6, at 50.

97. IPCC Report, Managing the Risks, *supra* note 8, at 20.

98. IPCC Report, Managing the Risks, *supra* note 8, at 11.

99. See Susan L. Cutter, *The Geography of Social Vulnerability: Race, Class, and Catastrophe*, in Understanding Katrina: Perspectives from the Social Sciences, June 11, 2006, http://forums.ssrc.org/understandingkatrina/the-geography-of-social-vulnerability-race-class-and-catastrophe/ (stating that "social vulnerability involves the basic provision of health care, the livability of places, overall indicators of quality of life, and accessibility to lifelines (goods, services, emergency response personnel), capital, and political representation").

32 Socio-Energy Systems Design: A Policy Framework for Energy Transitions

Clark A. Miller, Jennifer Richter, and Jason O'Leary

Over the past several decades, there have been significant concerns and countless discussions about the future of energy. Some fear that we will run out of petroleum reserves sooner rather than later. Some fear that we have too many petroleum reserves and that the temptation to use them and further pollute our atmosphere will be too great. Some worry about the unequal distribution of energy and the economic impact that it can have on developing areas. All of these issues and others have motivated individuals, companies, and governments to advocate for and attempt to develop entirely new systems of energy generation and distribution. They are calling for a massive energy transition from a world largely powered by fossil fuels to one that relies on alternative and sustainable energy sources.

Because of the massive sums of money involved and the fact that changes to the energy infrastructure could have significant geopolitical implications, most of these conversations have focused largely on the economic and political impacts of such changes and the question of whether they are technically feasible. Fewer scholars or analysts have thought about the impact that such a massive change would have on nearly every person in the world every day. The authors of this selection recognize that energy is not simply a massive technical system but rather a massive *sociotechnical system*. By using this lens, they are able to anticipate a large number of questions and concerns about how the development and implementation of new energy systems could affect individuals and local populations. Miller, Richter, and O'Leary show how applying the idea of sociotechnical systems allows us to more easily envision possible futures and develop strategies to make these futures better and more inclusive.

1 Introduction

Over the past quarter century, extensive research has documented that energy—typically represented as a technological and economic phenomenon—is also fundamentally social in its origins and organization, woven into societal, geographic, and geopolitical arrangements at scales from the individual to the planet (for recent reviews, see Miller and Richter 2014; Sovacool 2014; Zimmerer 2011). To date, however, work in the energy social sciences has had little impact on energy policy. There are a few notable exceptions. In parts of Europe, policy and planning increasingly recognizes the significance of the societal facets of energy for both building public support for energy policies (Wustenhagen et al. 2007) and in designing future energy systems by incorporating, e.g., innovative ownership models (Meyer 2007) and diverse public values

From *Energy Research and Social Science* 6 (2015): 29–40.

(Trutnevyte et al. 2011). Similarly, civil society organizations are increasingly pushing energy projects in developing countries to attend much more closely to the design challenge of linking the delivery of energy services to concrete strategies for alleviating poverty (Practical Action 2014).

In most of the world, however, and certainly in the United States, whose perspectives largely dominate global energy markets and the energy projects of the major development banks, energy policy remains a largely techno-economic problem. What we might term the human and social dimensions of energy barely rate a mention—let alone receive detailed, substantive treatment—in, for example, recent high profile US energy policy analyses, such as the National Academies' (2008) *America's Energy Future*, the US Department of Energy's (2012) *Quadrennial Technology Review*, or the US Department of State's (2014) review of the Keystone XL Pipeline. Nor do insights from the energy social sciences factor significantly into more routine energy policy analyses and decisions, e.g., in the permitting of drilling, the regulation of electricity markets, or the development of renewable energy mandates. Instead, energy policy routinely relies on caricatured, tacit or implicit, not-reflected-upon models of people and societies that rarely conform well to reality (Wynne 2005).

In this article, we propose that energy policy institutions adopt a more expansive conceptual framework that integrates social considerations more effectively into energy analysis and decision-making. We term this framework *socio-energy systems design*. In proposing this framework, we respond specifically to Sovacool's call in the inaugural issue of *Energy Research and Social Sciences* for new ways to communicate effectively about social science research in mainstream energy conversations (Sovacool 2014). Our objective is to shift the framing of energy policy from what we consider an overly narrow conventional approach—what technologies do we need to deliver energy, at what price, and with what carbon or other environmental costs—toward a perspective that recognizes that the conceptualization and design of energy systems is, fundamentally, an exercise in the simultaneous conceptualization and design of diverse social arrangements.[1] Through time, energy policy choices reconfigure societies, even as societies reconfigure energy systems, especially at moments when new energy systems are brought into being or during periods when existing systems are significantly rearranged through the persistent evolution, growth, and embedding of energy into human affairs (Huber 2013; Mitchell 2011). Thus, we argue, the social dimensions of energy systems are particularly salient for energy policy choices in the context of large-scale energy transitions, such as those currently underway in global energy markets due to the rise of hydraulic fracturing technologies for oil and gas extraction, the deployment of renewable energy generation to address climate change, and the development of alternative fuel, hybrid, and electric vehicles.

Contemporary energy transitions are reshaping not only the technologies and economics of energy but also physical and social geographies, social meanings, and the political organization of energy production, distribution, and consumption. Not surprisingly, around the globe, these changes have catalyzed growing socio-political resistance to energy policy and energy system change, with virtually every major form of energy technology confronting social protest and political controversy (Abramsky 2010). Only by reconceiving energy policy in more social terms, we believe, can the world hope to lessen conflict over energy transformations in the coming quarter century (Miller et al. 2013). The framework of socio-energy systems design aims to accomplish this task, reframing energy policy debates as debates not just about how to produce energy but about what energy production and consumption means for the diverse groups and communities who inhabit energy systems.

Most energy social scientists will not be surprised by the basic outlines of socioenergy systems design as an energy policy framework. The framework is heavily indebted to theories of *sociotechnical systems*: interconnected, integrated systems that link social, economic, and political dynamics to the design and operation of technological systems. Theories of sociotechnical systems have been used extensively and productively for some time to explain historical developments in the energy sector (e.g., Hecht 1998, 2011, 2012; Hirsh and Jones 2014; Hughes 1983; Nye 1999, 1990) and to analyze strategies for fostering sustainable energy transitions (e.g., Geels 2010; Geels and Verhees 2011; Marathe et al. 2011). Building on theories of sociotechnical systems and the co-production of technology and society (Jasanoff 2004), our objective is to establish socioenergy systems as a forward-looking design concept that can alter the lenses through which policymakers view energy policy choices, expand the range of analyses conducted to support those decisions, and enable publics to more effectively imagine and evaluate what energy policy may mean for individuals, families, and communities.

In the article's first section, we define the concept of socio-energy systems and illustrate how it can be applied to reformulate the goals and objectives of energy policy, as well as how its use as a lens transforms key ideas in energy policy, such as energy transitions and energy justice. In the second section, we offer three short case studies of current energy policy choices, drawn from our own research, that highlight the value of reframing energy policy choices as problems of socio-energy system design. In the third section, we offer four strategies for incorporating the concept of socio-energy systems design into energy policy practices and institutions. Although our argument could in principle be applied anywhere, in this article, we draw primarily on examples and case studies from the United States. We have made this choice in part because the United States is where we conduct our research and are most knowledgeable of the details of energy policy. But it is also, as we have suggested above, because US conceptualizations of energy policy are highly influential in global energy markets and institutions and because the United States is a central player in global energy transitions. Reframing US energy policy debates therefore has the potential to pay dividends not only in the United States but also in many other parts of the world.

2 Rethinking Energy Policy as Socio-Energy Systems Design

2.1 Defining Socio-Energy Systems

What do we envision when we suggest reframing the object of energy policy analysis and decision-making as an exercise in the design of socio-energy systems? First and foremost, we envision a way of recognizing that energy systems involve the work, behavior, and choices of many different kinds of people. Perhaps ironically, one of the best places to see energy and people in an integrated fashion is in children's books written a half century ago, not so long after the electrification of many homes and farms in the United States and the rise of the automobile as a common family purchase. Richard Scarry's books, in particular, *What Do People Do All Day?* and *Cars and Trucks and Things That Go*, offer a vibrant picture of diverse individuals, families, and communities living with new forms of energy. In Scarry's images and narratives, energy technologies come to life not just as assemblages of machinery but also as integral elements in the daily experiences of diverse people: workers, homeowners, beach goers, students, a "lazy fellow," and many more. Individuals in the books mine coal, transform it into electricity, and use it to power televisions, vacuums, and backyard barbecues. They harness the movement of water to transport trees and convert them into lumber for houses and boats and paper, as well as to power lights and irrigate crops. They drive cars and trucks and fly airplanes. They wire electrical systems in houses, retrofitting some and constructing others with wires already inside the walls. People's everyday lives and livelihoods—the activities that give them purpose and identity and that enact and animate the community of Busy Town—are thoroughly wrapped up in systems for producing and consuming energy.[2]

Scarry's imaginative world offers a lens through which it becomes possible to see energy policy choices markedly differently from conventional energy analyses. In the world depicted in the National Academies recent report, *America's Energy Future*, for example, the only facets of the future that seem to matter are which technologies to choose, how much to pay for them, and how much that will reduce carbon emissions (NAE 2008). The world is stripped bare of its human dimensions, and people are all but entirely absent from this image of the country's future, neglecting that the people of America's future will not only shape their energy systems but also inhabit forms of life partly configured by them (Winner 1986). Energy policy choices shape not just technological trajectories but trajectories in how people envision and construct themselves and their relationships to one another and to the world (Hirt 2012; Smith 1994; Sovacool and Brossman 2013). Yes, in the world made visible in Scarry's imagery, energy policies shape the technologies and the costs of electricity flowing through the wires; yet, they also act on all other aspects of the images, too.

Scarry's images make visible a host of dimensions of energy policy that social scientists have gone on to study in detail. The electrification of the home was as much about women's roles as homemakers and broader gender dynamics in household technology use as it was about new devices and infrastructures (see, e.g., Cowan 1983; Nye 1990). Education changed, too. Like the young pig on the roof of Scarry's power plant, students now learn Ohm's law in physics classes. More broadly, electrification, along with the birth of the chemical industry, helped dramatically upgrade the role of engineering education on university campuses in the late 19th and early 20th centuries and, a few decades later, fueled the rise of Stanford and MIT as intellectual powerhouses on the respective strengths of their electrical engineering departments (see, e.g., Leslie and Hevly 1985). And workplaces and work changed, too, as societies responded to workers experiences ("It's hot in here.") in new industrial settings ushered in by electrification with new forms of social research and social regulation and welfare, including safety standards and disability insurance (see, e.g., Wagner and Wittrock 1991).

Indeed, energy choices arguably contribute to virtually every facet of modern societies, up to and including fundamental constitutional arrangements. As recent policy debates illustrate, people's ideas about energy are bound up with basic notions about the proper arrangements of individuals, markets, and governments in modern societies. In the United States and Germany, for example, recent energy efficiency standards precluding the sale of incandescent light bulbs have become for key segments of the electorate a rallying cry for freedom and liberty against government intrusion into the economy (Howarth and Rosenow 2014). These claims draw not from some profound material impacts of compact fluorescent or LED bulbs. These replacements require at most only minor adjustments to social and technological arrangements and generally result in significant consumer cost savings. Rather, the idea that energy choices are constitutional in scope and import derives from ways of imagining the relationship between technology and politics that imbue energy choices with deep societal meaning (Jasanoff and Kim 2013, 2009). Similarly, conflicts over distributed solar energy generation in the US Southwest have sparked over whether government regulated utilities should have the right to protect their monopoly on energy sales from direct competition by creating barriers to the adoption of rooftop energy systems (Bird et al. 2013; Rucinski and Kaye 2014). Such monopolies are a key constitutional provision that has made possible, worldwide, the construction of the kind of technological systems represented by Scarry's image of the coal-fired power plant as the central organizing element of modern electrical systems (Insull 1915), but whose possibility is now under threat from new kinds of socio-energy arrangements (Graffy and Kihm 2014).

To facilitate energy policy analyses and decisions that can account for the multifaceted human and social dimensions of energy visible in Scarry's imagined world—and in our own real equivalents—we argue for adopting the concept of socio-energy systems as a design framework. At the most basic level, we define socio-energy systems as sets of interlinked arrangements and assemblages of people and machines involved in the production, distribution, and consumption of energy, in their supply chains, and in the lifecycles of their technologies and organizations. Simple analytics for socio-energy systems follow networks and systems of technological components, tracking the people and organizations connected to them (Mulvaney 2013). More complex analytics recognize that, at times, the linkages in socio-energy systems may flow entirely through social dynamics, that socio-energy systems dynamically shape and get shaped by the larger social, cultural, and political contexts in which they are embedded, and that people and organizations are complex entities—with histories, identities, and cultures—that require careful and sophisticated analysis (Mitchell 2011; Nye 1990; Perrow 1984). In either case, what is significant for our argument is the idea that it is these rich social worlds that energy policy acts upon and thus the question confronting energy policy—especially in the context of largescale energy transitions—is which kind of world one wants to create for the future.

2.2 Energy Policy Goals and Objectives

Adopting a socio-energy systems design lens, we argue, reconfigures core elements of energy policy. This includes, crucially, the goals and objectives of energy policy. It is an oversimplification, but perhaps not too much of one, to suggest that energy policy currently focuses on three relatively narrow goals: to produce and distribute sufficient energy to reliably meet demand, to minimize the cost of that energy, and to achieve environmental goals associated with energy production, such as low atmospheric emissions of carbon or other pollutants. A socio-energy systems lens does not necessarily reduce the significance of these objectives. Rather, such a lens does two things. First, it acknowledges that these goals may need to be balanced by other considerations, such as social changes or risks that may arise as a result of energy transitions. Thus, for example, while energy policy analyses of fracking have focused most closely on water quality issues, communities experiencing a rapid surge of fracking have expressed a much wider array of social and economic concerns that energy policies have tended to ignore—contributing to social alienation, civic protest, and even ballot initiatives aimed to ban the practice.

Second, a socio-energy systems lens opens up a considerably larger range of other potential goals that energy policy could consider and address, potentially including any significant social goal impacted by choices in the design of socio-energy systems. For example, rising inequality is a problem confronting most nations (Chin and

Culotta 2014). Since energy is a well-known factor in economic development—and, as we observe in the first case study below, energy technologies are always embedded in intricate socioeconomic arrangements whose design is an implicit component of energy policy—energy policy could explicitly target the reduction of social and economic inequality as a goal. Indeed, historically, regulators have prized equality as a core principle in the design of electrical utilities, although as we also suggest in the first case below, they may need to revisit the strategies used to pursue this goal in the context of today's energy transitions.

A socio-energy systems lens also encourages more sophisticated treatment of social goals within energy policy. In recent years, for example, advocates of US energy policies have focused a great deal of attention on job creation in the energy sector. Attention to job creation is no surprise given current US unemployment rates. Yet, as we will discuss in the second case study below, job creation is a relatively poor proxy for the deeper goals that communities ultimately care about, such as health, stability, and wellbeing. In the oversimplified narratives of US energy policy, job creation offers communities access to new wealth and to the stability that comes from employment. But jobs can be unsafe. Influxes of workers can stress social support infrastructures and exacerbate social conflict. And the vicissitudes of energy markets can be volatile, confronting communities with destructive boom and bust cycles. What matters is not just job creation but what kinds of jobs, who secures them, how long they last, and community resilience and innovation in the face of dynamic energy markets. All of these things can be measured and assessed—alongside job creation—if we reformulate and reframe energy policy in terms of the analysis and design of socio-energy systems.

2.3 Energy Transitions

Talk of energy transitions is common to energy policy discourse. In general, the language of energy transitions is used to describe shifts in the fuel source for energy production and the technologies used to exploit that fuel (Laird 2013). Thus, one might talk about the transition from wood to fossil fuels or from gasoline cars to electric vehicles. When envisioned in terms of socio-energy systems, however, fuel and technology become less diagnostic of energy transitions (Miller et al. 2013). Replacing a coal-fired power plant with a gas or nuclear-fired equivalent may have little impact on the overall energy regime, e.g., a centralized electrical utility operating as a regulated monopoly energy producer and distributor, even if it may bring local perturbations to those who live near the facilities in question or along relevant supply and waste chains. Instead, the phrase "energy transitions" comes to refer to significant transformations in socio-energy systems (Elzen et al. 2004). Thus, as we describe in the first case study below, the transition from a purchase model to a lease model in the financial arrangements surrounding rooftop photovoltaic systems in the US Southwest is rapidly transforming socio-energy systems—turning large numbers of energy consumers into producers—even though the technology and fuel in question are exactly the same.

In redefining energy transitions in terms of socio-energy systems design, what all three cases below illustrate, in particular, is the significance of careful attention to the micro-scale socio-economics of energy technologies. Socio-economic arrangements are significant not only in terms of predicting and explaining the social and market dynamics of new energy technologies—offering significant insight into processes of both commercial uptake and resistance—but also in understanding the social outcomes that flow from energy transitions. For example, to return to the theme of inequality discussed briefly above, our research shows that the micro-scale socio-economics of current solar energy policies in the US Southwest are, in fact, exacerbating inequality. Unequal outcomes that aggravate existing social and economic disparities are hardly unknown in the energy sector, of course (for a recent treatment of the subject, see, e.g., Ottinger 2013). What an approach to energy transitions grounded in socio-energy systems adds is the possibility to explicitly anticipate and incorporate a robust analysis of such outcomes into the design of energy policies from the outset, rather than working to correct them after the fact.

2.4 Energy Justice

Building on the two prior ideas, we can also construct an enhanced picture of energy justice. Energy justice is generally formulated in terms of access to energy, for good reason. Energy is considered a basic element of economic development, and something approximating 2 billion people worldwide either have no or limited access to modern forms of energy production. Yet, when considered in the context of socio-energy systems, access to energy is only one of many significant variables that determine just energy arrangements. Perhaps most obviously, the availability of energy in a community is meaningless if existing socio-technical arrangements do not allow people to effectively use that energy to upgrade their wellbeing (Practical Action 2014). Amartya Sen has famously observed that most famines do not result from a lack of food but rather from social, political, and economic arrangements that prevent people from using the available food to satisfy their hunger (Sen 1983). Much the same can be said of energy. Even at the national scale, data show that different countries derive radically different levels of human development and wellbeing from the same levels of overall energy consumption (Steinberger and Timmons Roberts 2010), strongly suggesting that the details of socioenergy systems matter not just in terms of access but effective use.

More broadly, as we describe in more detail in the case studies below, justice in socio-energy systems is a question of both the distribution of human outcomes within these systems, the distribution of power and voice in energy decision-making, and the deeper relationships between energy and the kinds of societies we fashion through and around it (Miller 2014; Mitcham and Rolston 2013; Richter 2013). Across entire lifecycles—supply chains, production and distribution chains, and waste chains—transitions in socio-energy systems reconfigure a wide range of social outcomes, relationships, and power across individuals, groups, and communities (Mitcham and Rolston 2013). It is this normatively transformative feature of energy policy, especially in the context of large-scale energy transitions, which has led to widespread social protest and conflict surrounding contemporary energy policy. Having operated largely out of the public eye for decades, energy policy institutions today confront a new reality: publics increasingly are attentive to energy choices. We suggest that a framework of socio-energy systems analysis and design will serve energy policy institutions well in understanding and addressing the new realities of energy politics.

3 Case Studies in Socio-Energy Systems Analysis and Design

The previous section suggested that looking at energy policy in terms of socio-energy systems design had the potential to foster a richer and more accurate picture of the full potential and impacts of energy policy. Where traditional energy policy considers the human and social dimensions of energy to be externalities, a socio-energy systems framework internalizes these factors. Here we offer three brief case studies chosen from our research, each focused on a significant recent energy policy decision. The case studies re-express these decisions as exercises in socio-energy systems design to illustrate how such analyses may work and what additional insights such a perspective may offer to energy policymakers.

3.1 Design Choice #1: Which Solar, Whose Benefits?

Since the late 1990s, Arizona, like many other parts of the world has turned to solar energy as a solution to several energy problems. Also like many places, this choice has been presented as a choice between solar and older fuels, like coal and nuclear. The state regulatory body, the Arizona Corporation Commission, has explicitly argued for advancing solar energy as a tool for diversifying the technology and fuel base of state's energy supply and has implemented a series of policies encouraging solar energy development, including both utility-scale and distributed generation. Federal policy has also contributed to the growth of Arizona solar energy through tax credits for investments in solar energy by both individuals and companies. The American Reinvestment and Recovery Act funded the installation of solar energy systems on public buildings, such as schools, to help reduce government energy expenditures. Combined with falling prices for solar energy technologies, solar investment has heated up dramatically in the state. As of 2013, Arizona ranked second in the nation in terms of total solar energy installed, behind California.[3]

In the past year, however, solar energy has become the subject of deep conflicts among Arizona's political institutions, elites, and publics. These conflicts have revealed that the choice confronting the state is not simply whether or not to—go solar—the state's utilities will easily achieve their 15% targets under the Commission's renewable portfolio targets—but which model of solar-based socio-energy system to choose going forward. The diverse policies that have promoted solar energy in Arizona have catalyzed the emergence of at least six distinct models of socio-energy systems design within the solar industry, each of which has strong advocates, a distinct vision of the future, and a track record of successful implementation. Table 32.1 illustrates some of the most significant dimensions of social variation across these models.

What table 32.1 captures is the fact that energy policy choices have the potential to create vastly different societies. Advocates of distributed energy have long made this argument (Lovins 2002), but the challenge is much deeper. Major differences exist across these diverse models of socio-energy systems, including: the cost of produced energy; the geographies of energy construction; the sources of capital that invest in them; the financial beneficiaries of projects; those who pay for projects and bear risks; the future viability of utility business models; and the collective, emergent patterns of energy behavior among publics and energy users. Some of the more significant and illustrative examples of these differences include:

- **Ownership, finance, and return on investment** Historically, utilities and operators of merchant power plants have dominated ownership of electricity generation in Arizona. Solar projects have significantly diversified ownership of energy facilities, including among homeowners, solar leasing companies, municipalities, businesses, and government agencies. Not surprisingly, this corresponds to a greater diversity of sources of capital for energy investments and of beneficiaries of return on those investments, allowing for a larger variety of investors and benefits to those investors. At the same time, many models threaten to reduce long-term utility revenues (Graffy and Kihm 2014; Satchwell et al. 2014).

Table 32.1
Varieties of socio-energy system design for Arizona solar energy developments

	Home rooftop, owned	Home rooftop, leased	Utility-scale projects	Community projects	Business rooftops	Public entity rooftops
Typical scale	3–7 kW	3–7 kW	20–300 MW	0.5–5 MW	0.5–2 MW	0.5–5 MW
Owner	Home owners	Leasing company	Utility	Town or city	Business or leasing company	Public entity
Partici-pants	Home owners	Home owners	Utilities	Residents	Business owners	Schools and government agencies
Financial beneficia-ries	Home owners	Home owners and leasing companies	Utilities	Utilities and residents	Utilities and businesses	Utilities and public entities
Cost / kW	Highest	Medium	Lowest	Low	Low	Low
Source of invest-ment capital	Home owner savings	Leasing company investors; home owner savings	Rate payers; utility investors	Cities; municipal bond holders; residents; tax-payers; social investors	Businesses	State bond holders; tax-payers
Location	House rooftops	House rooftops	Rural land	Variety of urban sites	Retail building rooftops	Variety of urban sites

- **Location** The debate between centralized and distributed solar tends to focus on plants built in open desert regions and on household rooftops, but significant variation exists both at utility-scales, where the most extensive development of projects in Arizona has occurred on marginal agricultural lands (which has the advantage of both avoiding ecological disruption and reducing overall water consumption, when the solar facilities are PV-based), and at mid-scales, for which a growing diversity of urban installation sites have emerged, including parking lots and parking garages, business and school building rooftops, vacant lots, walking zones, and others.
- **Customer experience** Variations in solar design create radically different opportunities for experience on the part of end users. In the case of rooftop systems, for example, homeowners become participants in energy generation, often frequently checking their level of production, learning about energy, and even competing with one another to see who is producing more electricity (Noll et al. 2014). For utility-scale systems, on the other hand, customers of solar electrons often never know they are users of solar systems or, at best, if they've bought into a green energy program, get a small adjustment on their bill.

Given the variability across these models, it is perhaps not surprising that even subtle, seemingly small changes in socio-energy system design can create significant shifts in social outcomes. Consider rooftop solar systems. Typically on the scale of 3–7 kW, rooftop solar systems have become increasingly popular among homeowners in Arizona. Subsidized by utilities under the state's renewable portfolio standard starting at roughly $4/W, rooftop systems were initially sold as purchased systems (with costs averaging $20,000–$30,000) to homeowners. Since 2010, however, a new model has emerged in which companies such as Solar City lease systems to homeowners. The technology itself is virtually identical, and the overall costs and financial benefits of each model are essentially the same, but the distribution of costs and benefits is different among key groups, including homeowners, utilities, ratepayers, leasing companies, and financial markets. Basically, the leasing model shifts some to all of the up-front costs of purchasing a system to the leasing company and its financial backers in exchange for some of the financial benefits derived. Two major outcomes have followed: the leasing model has quickly become the preferred model among homeowners, and the number of rooftop systems deployed in Arizona has grown rapidly.

These outcomes are important, from a socio-energy systems perspective, for two reasons. First, the rapid acceleration of rooftop adoption associated with the leasing model has made much more visible for utilities the specter that rooftop systems will significantly eat into utility revenue streams and shareholder value, diverting the financial benefits of energy production from utilities to homeowners and leasing companies. Arizona utilities, in turn, have fought back, seeking approval from regulators to impose additional costs on leased systems and lobbying for equal treatment under property tax rules, both of which would reduce the financial benefits of rooftop solar systems. The result has been sustained political conflict that shows no sign of abating anytime soon. Future developments will determine whether solar rooftop systems will continue to reduce utility revenues and shareholder value and raise rates for utility customers (Satchwell et al. 2014)—potentially undermining their long-term viability as businesses (Graffy and Kihm 2014)—or whether the rapid pace of solar rooftop adoption will slow in the state, allowing utilities to essentially retain monopoly control over electricity generation.

The second reason the outcomes of solar choices are important when looked at through a socio-energy systems lens is that they have potentially significant equity implications. As figure 32.1 shows, while rooftop solar can be found in almost all zip codes of the city of Phoenix, AZ, people living in zip codes with higher median value homes have, on average, derived significantly greater financial benefit from solar systems. These zip codes have both higher numbers of systems and, on average, larger rooftop systems. This means not only that wealthier households are deriving greater financial advantage from becoming more energy independent but also that (1) those gains were subsidized by poorer ratepayers, who contributed to the pool of subsidies that helped finance these systems; and (2) that ratepayers who cannot afford to go solar themselves may face higher rates in the future as wealthier households defect from the utility and no longer pay as much as before toward the maintenance of the electricity grid. Utility shareholders, the majority of whom are retirees, also stand to take an economic hit if utility business models are impacted by continuing solar investment. These challenges remain even in leasing models, which theoretically lower the up-front costs of rooftop systems to zero, as wealthier households are more likely to own homes, have the high credit scores necessary to secure leases, and be able to take advantage of lease pre-pay options that increase the financial benefits of leasing. None of these are reasons to oppose solar energy. These outcomes are not inevitable with solar energy. Rather, they are defects in socio-energy system design that could be corrected through proper energy policy choices.

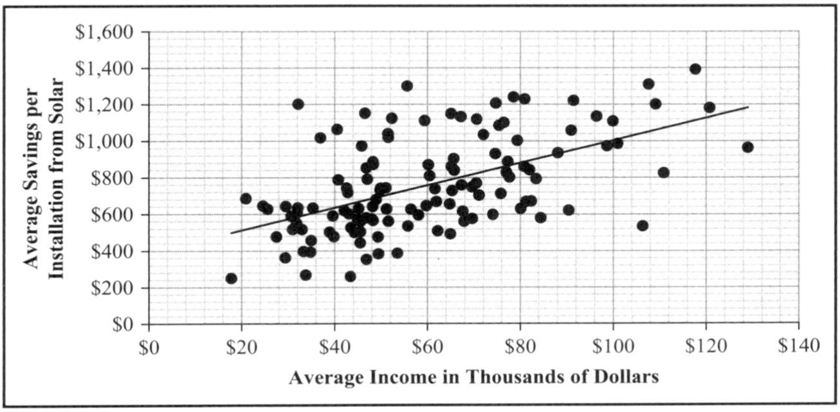

Figure 32.1
Average annual savings per household from installed rooftop solar systems, for each zip code in the Phoenix metropolitan region, as a function of the average income of households in that zip code. Data from arizonagoessolar.org, a database of solar installations maintained by the state's electric utilities, analyzed by Jason O'Leary and Clark Miller.

3.2 Design Choice #2: Which Infrastructure, Which Community?

The United States has experienced a boom in oil and gas extraction over the past decade, creating new regions of economic growth, social unrest, and environmental issues that are inherent to many socio-energy systems. This growth has occurred primarily from horizontal hydraulic fracturing of oil and gas in new regions of the country where drilling was previously unprofitable. This boom has been touted as a major success for US energy policy, as measured by traditional energy policy indicators such as lower oil prices, improved trade balances, investments in technology and infrastructure, and a significant number of relatively high paying jobs in the oil sector. Yet the US oil boom also illustrates that, when considered from a perspective of socio-energy system design, social complexities arise that should be taken into account in policy decisions. Our point is not to deny the successes of recent US oil policy, although there are reasons to do so, such as the long-term implications for carbon emissions. Rather, we emphasize that, with a socio-energy systems design framing, policy makers are positioned to anticipate and account for the full range of social outcomes and risks attendant to rapid growth of the oil industry.

One illustration of this can be seen in choices between forms of energy transport infrastructure. Historical analyses demonstrate that different transportation infrastructures can create different patterns of regional socio-economic development. In the late 19th century, in the Northeastern United States, for example, historian Christopher Jones has shown that the growth of canals for transporting coal, transmission lines for carrying electricity, and pipelines for carrying oil all generated jobs in energy producing regions, and all three contributed to economic growth in cities along the coast. Only the canals, however, led to vibrant social and economic growth in the small towns and cities that lined canal routes, as barges not only stopped routinely in their travels but also carried other trade goods at the same time (Jones 2013, 2014).

The State Department Environmental Impact Statement for the Keystone XL suggests that the US is currently experiencing similar dynamics with regard to oil transport from Canada's tar sands to Gulf Coast refineries. The lack of adequate pipeline capacity has created a strong demand to expand rail traffic and supporting infrastructure along the same corridor. Like canals, railroads create a very different distribution of social and economic benefits (and risks, e.g., of accidents, see below) to pipelines. Similar to canal systems in the 1800's, the construction of new rail terminals has generated not only construction jobs but also operations and maintenance jobs along the rail route, as well as generating hotel and other business for rail employees as they travel. Trains can also transport other trade goods at the same time (US Department of State 2014). Should the Keystone XL pipeline be built, these benefits will vanish.

A perhaps more significant illustration of the benefits of a socio-energy systems lens is the current lack of social planning for communities in new oil and gas producing regions. Many communities around the country, and especially those surrounding the Bakken Formation in Montana and North Dakota, have seen major changes associated with the build-up of energy operations. These changes include a significant growth in employment, although as with the case of large-scale solar plants built in rural or desert areas, workers are often imported into these regions rather than drawn from local populaces. The rapid growth (and sometimes subsequent rapid decline) in migrant workers often significantly strains local infrastructures, especially housing and social services.

Communities in the Bakken Formation are already confronting an influx of tens of thousands of workers throughout the supply chain, and as a result, have experienced significant increases in drug use and crime, rising costs of living, and changes in community dynamics. These are highly predictable problems, yet little planning has been done to deal with them, due to an exclusive focus on jobs and economic activity. Environmental and safety concerns also affect the boomtowns and their residents. Local road infrastructures experience significantly higher volumes of traffic and, especially, higher volumes of large truck use. On December 30th, 2013, a train carrying oil derailed and caught fire near Casselton, ND. The subsequent fireball resulted in the evacuation of 2,400 residents from their homes. Following this trend of increased oil-transport via rail, there have been several other similar accidents, some resulting in loss of life and property. Similar examples can be seen in Pennsylvania involving workers on the Marcellus shale formation and their impact on local infrastructures. Heavy trucking is required for nearly every part of the well drilling process in rural communities. A recent study by researchers at Rand Corp. shows that the cost of damage to state roads was between $13,000 and $23,000 per well in 2011. While shale firms have some agreements to repair roads that are visibly damaged, not all roads are effectively covered (Abramson et al. 2014).

Perhaps most perniciously, a study of social support institutions in North Dakota found that communities have faced pervasive social disruptions that make life difficult, especially for the poorest and most vulnerable within communities. Low-income individuals and families have been forced out by high rent costs, while others have seen rising rents eat away at their available income for food and other necessities. Many people live in substandard housing or in their vehicles, and social services and police are seeing rapid rises in domestic violence and the need for child protective services (Weber et al. 2014).

While these facets of energy development may simply be viewed as the inevitable downsides of an otherwise good deal for communities in energy producing regions, proper socio-energy system design and analysis could contribute to policies that improve social outcomes in boom communities, especially with regard to mitigating some of the worst impacts on communities while ensuring that new energy wealth contributes to long-term economic gains rather than short-cycle economic booms and busts. To the extent that future energy policy continues to neglect these often easily anticipated social consequences of energy production in accounting for the costs of oil, similar patterns of social unrest and unequal distribution of risks and benefits are likely to continue.

3.3 Design Choice #3: Which Waste Storage, Whose Voices?

Policy choices surrounding nuclear energy waste disposal offer a third example of the value of approaching energy policy as a problem of socio-energy system design. Nuclear energy produces 20% of electricity in the United States, and will continue to be an important part of the national energy portfolio. However, even as nuclear waste is produced in this process, there is still no comprehensive national plan for the 70,000 metric tons of nuclear waste that has been produced from the nuclear fuel cycle. Currently, spent nuclear fuel (SNF) is stored in spent fuel ponds for five years. After this period, it is supposed to be moved to a national repository. This repository has yet to be built, however, and in 2008, the preferred choice, Yucca Mountain, was closed, leaving the waste in limbo. Instead, SNF is stored in dry casks in independent spent fuel storage installations (ISISIs), which are located at 54 reactor sites (NRC 2014). A third alternative is the reprocessing of spent nuclear fuel.

Looked at through a socio-energy systems design lens, each of these methods of handling nuclear waste creates its own geography of risks, as well as socio-economic arrangements, that will distribute benefits and burdens across different communities. Onsite storage is the current model for storing nuclear waste at the 65 nuclear plants across the United States, although it was never intended to be a permanent solution. Each of these sites presents a safety risk to nearby communities in terms of leaks of radioactive materials, security risks, e.g. from terrorist attacks, or natural hazards like earthquakes or floods. The thirty-foot casks are a striking visual reminder of the risks facing communities who were assured that a centralized SNF site would be created to solve the problem of nuclear waste. Citizen groups have pressed for hardened onsite storage which would increase the distance between casks and provide additional cover with concrete and steel in order to protect the casks from attacks, accidents, or leaks. The Nuclear Regulatory Commission (NRC) has stated, however, that they see no reason to change how casks are stored. Additionally, the NRC has begun a process to extend licenses for up to 120 years for dry casks, reflecting the fact that long term central storage is still a long way off (NRC 2014).

A second method is geologic storage of spent nuclear fuel. Geologic storage offers the prospects of concentrating the long-term risks of radioactive wastes in a single, ideally designed and located facility—although this image is complicated considerably by the fact that waste would need to be transported from reactors to the site, thus significantly expanding the potential geography of risks to unwitting and perhaps unwilling communities along transport routes. The present state of geologic disposal in the United States is mired in controversies stemming from the 1982 Nuclear Waste Policy Act (NWPA). For thirty years, NWPA planning focused on Yucca Mountain as one of two federal repositories required by the law. Yet, in 2008, plans for the Yucca Mountain facility were shut down in the face of strong opposition from diverse local communities, ongoing scientific controversy over the risks of geologic storage, and continued political opposition to nuclear power among many environmental groups (Shrader-Frechette 2014; Walker 2009). Geologic disposal of nuclear waste also raises complex questions about intergenerational considerations. While sequestering nuclear waste from humans is seemingly technically straightforward in the present, uncertainty surrounding future longterm changes in human developments and needs, as well as environmental conditions, significantly complicates attempts to do so over the 10,000 years required by the Environmental Protection Agency for geologic repositories in the United States (MacFarlane 2006). New ways of engaging with the public may offer insights into future debates over nuclear waste, including the Blue Ribbon Commission's recommendation that the NRC should focus on areas where communities in locations deemed geologically appropriate are willing to step forth and host waste sites, and a similar initiative is emerging in the U.K. that focuses on voluntary communities and suitable geologies, although how to define those qualifiers still needs to be debated (Bickerstaff 2012; Blue Ribbon Commission 2012).

A third avenue for dealing with SNF is reprocessing. The US currently employs a "once through" process of using nuclear fuel, meaning the SNF from energy production is used only once, and then removed and replaced with new fuel during the refueling process. A closed fuel cycle would reuse the SNF in another kind of reactor that separates useful uranium from other by-products. Like a geologic repository, reprocessing would centralize waste handling, with a transport system from individual power plants, and thus create similar risks from accidents or attacks at the reprocessing site or during travel to and from power plants, the reprocessing facility, and the long-term repository that would still be required to store the final waste from reprocessing (although this would involve a much smaller volume of final waste). One proposal is to develop an energy park at Yucca Mountain where nuclear experiments such as reprocessing could be carried out. Instead of an eternal tomb for the radioactive waste produced by nuclear energy, Yucca Mountain would become a place for producing beneficial technologies (US Government Accountability Office 2011). Yet, this proposal addresses none of the social and political concerns that prevented Yucca from opening. At the same time, re-processing brings a very different set of risks. One of the principle socio-political concerns with reprocessing SNF is that it produces radioactive materials that may increase the risk of proliferation of nuclear weapons, one of the main reasons the US decided to forgo reprocessing in 1977.

These three possibilities for nuclear waste demonstrate that questions of social justice are inherent in choices about where and how to store nuclear waste, including the creation of nuclear communities near sites or along transportation routes, the development of expertise required to manage and maintain facilities, and security issues from local to global scales. It also opens up questions about what ethical and moral obligations present communities have to future generations, who will live with nuclear waste sites though they may not benefit from nuclear energy. Examining the social worlds that each approach would bring into being offers a richer backdrop for planning for future nuclear activity, as well as acknowledging that the back-end of the fuel cycle requires more foresight than previously afforded to it.

4 Bring Socio-Energy System Design into Energy Policy and Governance

How can the framework of socio-energy system design be effectively incorporated into energy policy and governance? This entails two tasks: (1) rethinking the object of energy policy, as we have discussed so far; and (2) configuring new approaches to energy policy and governance that can more effectively integrate the human and social dimensions of energy systems into energy analyses and choices. For our purposes here, we define energy policy and governance broadly, as not merely policy choices about energy but the processes and institutional arrangements through which policy choices are informed, made, and implemented. Thus, we include at least three significant domains of work in our definition of energy policy and governance: (1) the practices and techniques through which potential energy futures are envisioned, modeled, analyzed, and evaluated; (2) the forums and methods for deliberating, debating, and making energy choices; and (3) the institutions for fashioning, operating, and regulating new energy systems. Cutting across these three domains, we see four significant avenues for transforming energy policy into processes and practices for socio-energy system design.

4.1 Socializing Energy Policy

First, reconfiguring energy policy as socio-energy design requires new strategies for integrating the human and social dimensions of socio-energy systems into processes of energy design, planning, and policy-making. This requires understanding, acknowledging, and incorporating the ways in which people inhabit energy systems: as residents, consumers, workers, investors, managers, etc. We must learn to better document, visualize, and analyze socio-energy systems, to anticipate how such systems may change as part of energy transitions, and to envision alternate possibilities to inform energy policy deliberations. This will require new methods and models, new forms of expertise, and the addition of new disciplines to the mix of energy research and policy analysis, with a particular emphasis on finding strategies to bring specialists in the human and social sciences into the field of energy and the practice of energy policy. Learning to imagine, conceptualize, analyze, evaluate, and deliberate energy policy in terms of socio-energy systems will require complex methods of observation and analysis. Of special significance will be the role of research funding institutions that are positioned to build both new knowledge tools and the necessary human resources to advance the social sciences of energy, but whose deep neglect of research in these fields to date undermines the ability of countries to pursue intelligent social planning for energy transitions.

We believe three areas of knowledge advances are particularly significant at the outset. The first is the capacity to theorize and model socio-energy systems, fully integrating people into our understanding of supply chains, energy operations, and the social values, behaviors, relationships, networks, and institutions fashioned around the use of diverse kinds of energy—and the opportunities for social change that accompany significant changes in energy technologies and markets. A key element of such models will be to track and explain the social dynamics of energy systems (Noll et al. 2014). Approaches will be needed to explore how people become embedded into and disciplined into these systems and the subsequent implications for human imaginations, identities, and institutions. In order to remain safe in today's world, for example, children must be rigorously and carefully trained to behave in certain ways around energy systems (cars, electricity outlets, power cords, etc.). Likewise, people's understanding of the world is shaped by energy regimes. Adults become accustomed to thinking of themselves as elements of energy systems, for example, in turn rethinking their relationships with one another, with institutions, with societies, and with places. People who understand what a car is, know how to drive one, and either own or are able to access one generally view space, geography, and distance in radically different ways from others who do not share these same characteristics. Even for inhabitants of major cities with extensive and easy-to-use public transportation systems, like Boston, owning a vehicle transforms the city into a fundamentally different landscape with radically different geographies of accessibility.

The second is to significantly deepen understanding of the social outcomes that flow from energy policy choices and technology arrangements. New methods will need to be developed that focus on identifying, mapping, and measuring significant social outcomes from energy choices across a wide range of human affairs. As a general proposition, we are able to say relatively little about the social consequences that flow from energy projects. It is striking, for example, that only one major study has sought to measure the social consequences of oil development in the Bakken Formation (Weber et al. 2014). From a more predictive perspective, a similar problem bedevils energy as a tool for human development. For over a half-century, economic theories have highlighted the significance of energy inputs as a crucial facet of human development. Yet, strikingly, we know relatively [little] about how energy contributes to specific human development outcomes, especially at the community level. Moreover, we know even less about how the design of energy technologies and the kinds of social and economic arrangements we build around them enable or constrain the ability of communities to advance different facets of human development (Practical Action 2014). Such insights are critical if we are to make significant progress on developing energy systems that deliver high social value and thus contribute to advancing development goals in a significant way. A recent World Bank report highlights that energy development in regions that do not currently have electricity in Africa will remain expensive (Tenenbaum et al. 2014). All the more reason, therefore, to develop the knowledge necessary to ensure that the energy projects we do invest contribute to significant improvements in social outcomes.

Perhaps most significantly, third, strategies will need to be developed to map how complex socio-economic arrangements distribute the costs and benefits of energy production and consumption in highly heterogeneous ways across contemporary societies. The field of environmental justice demonstrates that in the context of large-scale energy transitions, understanding how subtle changes in social meaning, identity, or power relationships, especially in the context of differences in race and gender, are affected by energy systems is extremely important (Kurtz 2010; Rolston 2014; Ryan 2014; Schlossberg 2007). Designing and developing just energy systems requires attending to how these factors are distributed across different physical, social, cultural, and economic geographies. Only through such knowledge will energy policy be able to address the forms of injustice that have led in recent years to high profile social resistance to efforts to bring about energy transitions.

4.2 Systematizing Energy Policy

The second significant demand on energy policy in the context of major energy transitions is to extend and expand institutional capacities to analyze and govern energy on a system-wide, lifecycle basis. It is difficult if not impossible to imagine the possibility of contemplating socio-energy system design absent the capacity to think, analyze, and act in systemic terms. Our assertion may come as a surprise to some who think that energy policy already encompasses a great deal of systems thinking and perspectives. In some respects, this is true. Energy systems analyses are common, especially in academic research. Organizations like the International Energy Agency and US Energy Information Agency tend to operate at a systems level, although their data is often broken down by country or by state in a far less systematic fashion. Yet, current energy systems approaches fall short in two crucial ways. First, they tend to limit energy systems to their technological elements, neglecting to treat the social and political dimensions of energy equally robustly in systems terms. Second, energy policy and governance rarely follows systemic organization. That is, we rarely govern whole energy systems in a comprehensive fashion. Instead, energy policy and governance are conducted piecemeal, whether in terms of geography, sector, or scale, creating headaches for cross-jurisdictional coordination and orchestration.

In the United States, as an illustration, neither the federal government nor state governments have integrated energy agencies responsible in a comprehensive fashion for energy development, energy system management, or energy policy. Given the deep significance of energy for economic development and social organization, this is perhaps not surprising but nevertheless creates widespread difficulties for approaching energy policy as a problem of socio-energy system design. One consequent shortcoming is the general absence of strategic energy planning. Strikingly, for example, many states, confronted with the prospect of new EPA rules requiring reductions in state carbon emissions from electricity generation, face two significant problems. First, little to no data exists on state energy systems, except in terms of aggregate figures. Instead, understanding of energy systems tends to be isolated on a utility-by-utility basis. Second, energy systems cut across state boundaries in substantial ways, with utilities routinely depending on generating facilities in other states, subject to energy policy there.

Public utilities commissions in the United States have begun, recently, to encourage utilities to pursue integrated resource planning. Yet, in the context of current energy transitions, that is often inadequate. Electricity trading among utilities and independent power producers is on the rise. Access to resources for energy production, such as land and water, may become competitive. Perhaps most importantly, growth in electric and natural gas vehicles may soon compel the need to plan across the electricity, fuel, and natural gas sectors, something which is not routinely done in most parts of the country.

The absence of a capacity for strategic planning and systems governance becomes particularly true when we look at questions of socio-energy design in the context of energy resource development. Whether the subject is vehicle charging stations, power plant siting, or mineral rights leases, the common tendency to approach energy policy and decision-making on a case-by-case basis often leads to problematic outcomes from the perspective of socio-energy system design. Communities dealing with natural gas drilling have found it particularly difficult to plan for the impacts of drilling (e.g., road use, influx of new labor) and to ensure appropriate consideration of a variety of concerns (e.g., risk assessment and monitoring) in the absence of institutional or legal capacities to carry out a community-wide approach to energy policy decisions. Instead, they have confronted an environment in which landholders hold the rights to negotiate leases, with communities bearing the collective consequences of a host of uncoordinated individual decisions. A 2014 ban on hydraulic fracturing in Denton, TX, for example, has been challenged on the basis that it infringes local property owners' rights to exploit the resources they own. This framework for energy policy often also serves to facilitate the energy industry's ability to negotiate lower compensation levels for landowners by negotiating separately and secretly.

The case of power plant siting also offers an illustration of the problem of both a lack of systems analysis and system-wide policy. In siting new power plants, utilities or other companies typically select sites and initial designs for plants, negotiate land deals, and set up transmission line agreements before informing the public of a proposal to build a new facility. This process creates a strong tendency to lock in energy technology and economic designs, except at the margins, as it becomes extremely difficult for the policy process to reopen major design criteria. In turn, this precludes consideration of significant socio-energy design alternatives, such as building a different kind of plant, in a different location, or taking an entirely different path to meeting demand growth, such as increasing energy efficiency. Given that the primary benefits of a socio-energy design perspective come from being able to look broadly at alternative designs and design criteria, these benefits are unlikely to be realized via existing siting practices.

4.3 Publicizing Energy Policy

Enlarging the view of energy policy to encompass the project of socio-energy systems design naturally raises questions about who participates in the new expanded policy processes that this will inevitably entail. Socializing and systematizing energy policy, as described in the prior sections, arguably both expand the array of stakeholders who are likely to consider energy system development as within their scope of interest, especially in the context of large-scale energy transitions that may fundamentally transform the social and physical geographies of energy production and consumption. In fact, publics are already well aware of this and increasingly demanding a seat at the energy policy table (Abramsky 2010; Devine-Wright 2011). Even with public input in many federal projects due to a required public comment period, meaningful interactions between experts and lay-people remains difficult, and the incorporation of public ideas and concerns into policy processes in a systematic—or system-wide—fashion is lacking.

Conceptually, the challenge is to create energy policy processes that encompass the envisioning, designing, deliberating, choosing, and making of future socio-energy systems and render possible partnerships between the energy industry and communities at all of these stages. This, of course, is an immensely difficult undertaking even to imagine, let alone to implement successfully. Part of the challenge is to develop new methodologies for envisioning future energy pathways on multiple scales and temporalities. Another part of the problem is the current dearth of opportunities available to communities to become engaged in the design and deliberation of energy futures. Existing energy policy processes, from power plant and transmission siting processes to the design of major national legislation, tend to limit rather than expand public participation and engagement. These limits—in the energy policy process as currently constituted—include everything from the presumed levels of expertise entailed in energy policy debates to the forms and timing of public comments and feedback to the dissecting and distribution of energy policy choices across multiple agencies and jurisdictions. The result is a system that, in many ways, is largely opaque to the public and that the public is allowed (or perhaps encouraged) to systematically ignore.

In contrast to conventional energy policy processes, France has embarked over the past two years on a major review of its energy system and of the potential need to undertake a significant energy transition. This review began, perhaps surprisingly, with an extensive period of public debate, with over 170,000 people involved in diverse regional public forums (still small compared to the overall population of the country, but significantly larger than most exercises of public engagement in energy policy). While French public support for nuclear energy remains relatively strong (El Jammal et al. 2013)—nuclear energy currently provides 75% of France's electricity—the public debate afforded a relatively unique opportunity for the public to reflect on and deliberate French energy strategy over the long-term. As a result of this referendum, the French government has decided to lessen its reliance on nuclear energy, with a goal of 50% of electricity production from nuclear by 2025, compensated for by a ramp up in the production of renewable energy (*World Nuclear News* 2014).

In a similar, if dramatically less ambitious exercise, the state of Arizona has also pursued a series of energy deliberations over the past several years. The first such deliberation was led by a non-governmental organization, Arizona Town Hall, which has organized public debates about major policy issues confronting the state for the past half century. The event, held in Fall 2011, involved approximately 150 business, policy, and civic leaders in a three-day deliberative exercise and resulted in a set of policy recommendations that were subsequently distributed across the state. The strongest recommendation was for the development of a state energy plan, a task that was subsequently taken up by the Office of Energy within the Arizona Governor's Office. Working with a task force of energy industry leaders, the Governor's Office released a new energy plan for the state in February, 2014, focused on promoting solar energy, energy efficiency, and development of a strong energy workforce and industry. In thinking about different scales and perspectives, individual states in the US are an important node for expressing both local and national perspectives, which in turn may influence national energy policies and incorporate social and community values.

Events such as these are hardly a panacea for the challenges confronting energy policy in the coming decades. Involving publics in significant ways in efforts to redesign the energy system will require much more substantial initiatives, using innovative approaches to garnering public participation in everything from discussing energy scenarios to figuring out the details of novel energy systems. Yet the challenges of not engaging publics more effectively are also apparent. The energy system is so important to people's lives that significant changes to it are potentially enormously disruptive. Not surprisingly, therefore, even at relatively early stages in the coming energy transformation, as we write, nearly every currently plausible form of potential future energy system is at the center of significant social protests, political controversies, or both. It is not an accident that hydraulic fracturing is deeply opposed in the same communities that its use is enriching; that the process to site a nuclear repository at Yucca Mountain fell short; or that conservative Arizonans showed up en masse to protest efforts by utilities to significantly curtail solar rooftop development.

4.4 Governing Energy Transitions

Finally, energy policy going forward must recognize that the problems that confront energy policy and governance during times of large-scale and rapid transformation are qualitatively different than those at stake during times of stability or slow, incremental change. Here we identify only a few of what is likely to be a large array of important governance challenges that energy policy institutions and leaders will face.

One key governance challenge will be that of organizational change. For close to a century, the dominant organizational configuration of the electricity industry has been roughly stable, with large, centralized, monopoly utilities providing the bulk of electricity and dominating the management of electrical grids. Electricity deregulation has already brought significant organizational shifts to some regions, and the rise of distributed energy systems as a major potential electricity supply also poses challenges for this organizational model. As the utility sector recognizes, distributed energy generation undermines key elements of the existing financial and technological models of the electricity system. Energy policy must thus confront head on the question of how to organize electricity production and distribution going forward.

A second challenge is to find ways to maintain and strengthen public confidence in the transition process. Energy transitions have the potential to bring deep uncertainties to a wide variety of stakeholders, from utilities to regulators to publics. Energy policy processes need to recognize, acknowledge, and confront directly this challenge—which is at least in part a problem of trust in those managing the process—lest uncertainties lead to policy or political stalemate. In 1993, for example, the Clinton

Administration proposed raising gasoline taxes by a small amount to begin funding climate solutions. The proposed taxes have been far outstripped by subsequent actual change in gas prices, yet uncertainties about how climate policies would ultimately distribute the burdens of transforming the energy system led to widespread opposition to the proposals. Especially in the context of a system so consequential to people's lives and livelihoods as energy, managing the governance process to avoid public disaffection and defection is a key problem.

Related to the problem of trust is the problem of managing social dislocations associated with large-scale energy system change. Energy systems are embedded in and across a wide range of diverse communities, and large-scale changes in those systems are likely to have significant consequences for many of those communities. This is obvious in the case of energy producing regions, like coal mining communities or, especially, the Middle East or the Gulf Coast of the US. These regions are likely to be decimated by declining use of fossil fuels, unless appropriate policies are developed ahead of time to manage social and economic dislocation. In other cases, the consequences may be less obvious and direct, but may be just as severe. Many utilities, for example, are seen as relatively safe investments for retirement investment portfolios. In fact, a significant majority of utility stock is owned by retirees. If the threat to utility business models from distributed energy is as significant as some in the utility industry now fear, utility stock owners are likely to suffer significant investment losses.

A final challenge is the orchestration and management of the consequences of energy changes that flow across jurisdictions. Energy policy and governance comprise a tangled thicket of overlapping and non-overlapping jurisdictions, both geographically and institutionally. Energy systems cut through this thicket in complex and often unanticipated ways. Changes in energy systems are thus likely to have consequences far beyond their point of origin. Unless energy policy and governance institutions can find ways to better coordinate their decision-making across these thickets, the result is likely to be less than optimal policy solutions that carry enormous and unconsidered implications for diverse communities across the face of the planet.

5 Conclusion

We have used the word "inhabit" several times in this article to describe the social dimensions of energy systems. Whether it has ever been different, people today literally inhabit energy systems. They live with, in, around, and through energy—in the process both constituting energy systems and being constituted by them. Energy shapes—and is shaped by—people's economies, workplaces, identities, environments, technologies, landscapes, politics, and mental maps of the world. As the United States and the world contemplate a deep and widespread energy transition over the next few decades—whether toward new forms of hydrocarbons, a nuclear resurgence, renewables, or something else entirely—this transformation will have enormous human consequences. At least for the purposes of this transition, energy policy must expand to acknowledge, recognize, assess, and incorporate the fact that its objectives and outcomes are not just to change either the fuels or technologies of energy but to transform socio-energy systems.

Fundamentally what we are arguing for in this article is a reconceptualization of the energy policy imagination—of the kinds of things we think about, consider, analyze, argue about, and take into consideration in the making of energy policy choices (Jasanoff and Kim 2009; Sovacool 2014). This is both an act of opening up our imagination of the world of energy to the full range of ways in which energy entrains and is entrained in human affairs; of differently disciplining the imagination through new forms of analytic tools, methods, and models; of differently organizing who is involved in energy decisions, at what points in the process, and in what ways; and of differently weighing the many facets of complex energy choices. As energy researchers, analysts, managers, regulators, and politicians, we must find ways to bring the full breadth of the human and social dimensions of energy transitions into our day-to-day work in a robust and systematic fashion.

It is striking how impoverished we have let the imagination of our current forms of energy analysis and policy become. *America's Energy Future* may have been written by engineers, for the purpose of laying out our technological options going forward. Nonetheless, its title proclaims much, much more. As a document depicting America's energy future, it offers little insight into much else of importance in what energy will mean to the country a half century from today: the ways that we will inhabit future energy systems, the distribution of benefits that we will derive and risks and dangers that we will confront, the forms of labor and economic livelihood that will be enabled and disabled, the kinds of people that we will become, and the future possibilities that we will imagine for ourselves, our communities, and our world. These are the domains of imagination that we must increasingly enter and master, if we are to justly and successfully navigate and govern the coming energy transition.

Acknowledgments

This material is based upon work primarily supported by the Engineering Research Center Program of the National Science Foundation and the Office of Energy Efficiency and Renewable Energy of the Department of Energy under NSF Cooperative Agreement No. EEC-1041895. Any opinions, findings and conclusions or recommendations expressed in this material are those of the author(s) and do not necessarily reflect those of the National Science Foundation or Department of Energy.

Notes

1. Jasanoff and her colleagues term this *co-production*. See Jasanoff 2004; Jasanoff and Kim 2009.

2. Scarry's images can be seen online: power plant, http://3.bp.blogspot.com/-i7-ENiBDyOw/Tgn6HMOQwjI/AAAAAAAAAV8/u9wb1r7v0vg/s1600/Electricity+generation+production+richard+scarry+what+do+people+do.jpg; coal mine, http://scienceblogs.com/worldsfair/wp-content/blogs.dir/389/files/2012/04/i-1e3c3801c1d14d8203e4a04f37cd97d1-digging1small.jpg; river transport of logs, http://exampleschildrensbooks.files.wordpress.com/2012/06/scarry-trees.jpg.

3. Solar Energy Industries Association, http://www.seia.org/research-resources/2013-top-10-solar-states (page no longer available).

References

Abramsky, K., ed. 2010. *Sparking a Worldwide Energy Revolution*. Oakland, CA: AK Press.

Abramson, S., Samaras, C., Curtright, A., Litovitz, A., and Burger, N. 2014. "Estimating the Consumptive Use Costs of Shale Natural Gas Extraction on Pennsylvania Roadways." *Journal of Infrastructure Systems* 20 (3): 06014001.

Bickerstaff, K. 2012. "'Because We've Got History Here': Nuclear Waste, Cooperative Siting, and the Relational Geography of a Complex Issue." *Environment and Planning A* 44:2611–2628.

Bird, L., et al. 2013. *Regulatory Considerations Associated with the Expanded Adoption of Distributed Solar*. NREL/TP-6A20–60613. Boulder, CO: National Renewable Energy Laboratory.

Blue Ribbon Commission for America's Nuclear Future. 2012. *Report to the Secretary of Energy*. Washington, DC: Government Printing Office.

Chin, G., and Culotta, E. 2014. "What the Numbers Tell Us." *Science* 344 (6186): 818–821.

Cowan, R. 1983. *More Work for Mother*. New York: Basic Books.

Devine-Wright, P., ed. 2011. *Renewable Energy and the Public*. London: Earthscan.

El Jammal, M. H., Rollinger, F., Mur, E., Schuler, M., and Tchernia, J. F. 2013. *More than 30 Years of Opinion of French People on Nuclear Risks—Special Release of the 2012 IRSN Opinion Survey*. Fontenay-aux-Roses, France: Institut de Radioprotection et de Surete Nucleaire.

Elzen, B., Geels, F., and Green, K. 2004. *System Innovation and the Transition to Sustainability*. Cheltenham, United Kingdom: Edward Elgar.

Geels, F. 2010. "Ontologies, Socio-technical Transitions (to Sustainability) and the Multi-level Perspective." *Research Policy* 39:495–510.

Geels, F., and Verhees, B. 2011. "Cultural Legitimacy and Framing Struggles in Innovation Journeys: A Cultural-Performative Perspective and a Case Study of Dutch Nuclear Energy (1945–1986)." *Technological Forecasting and Social Change* 78:910–930.

Graffy, E., and Kihm, S. 2014. "Does Disruptive Competition Mean a Death Spiral for Electric Utilities?" *Energy Law Journal* 35:1–219.

Hecht, G. 1998. *The Radiance of France: Nuclear Power and Identity after World War II*. Cambridge, MA: MIT Press.

Hecht, G. 2011. *Entangled Geographies: Empire and Technopolitics in the Global Cold War*. Cambridge, MA: MIT Press.

Hecht, G. 2012. *Being Nuclear: Africans and the Global Uranium Trade*. Cambridge, MA: MIT Press.

Hirsh, R., and Jones, C. 2014. "History's Contribution to Energy Research and Policy." *Energy Research and Social Science* 1:106–111.

Hirt, P. 2012. *The Wired Northwest: The History of Electric Power, 1870s–1970s*. Lawrence: University Press of Kansas.

Howarth, N. A., and Rosenow, J. 2014. "Banning the Bulb: Institutional Evolution and the Phased Ban of Incandescent Lighting in Germany." *Energy Policy* 67:737–746.

Huber, M. 2013. *Lifeblood: Oil, Freedom, and the Forces of Capital*. Minneapolis: University of Minnesota Press.

Hughes, T. 1983. *Networks of Power: Electrification in Western Society, 1880–1930*. Baltimore: Johns Hopkins University Press.

Insull, S. 1915. *Central Electric Station Service: Its Commercial Development and Economic Significance as Set Forth in the Public Addresses (1897–1914) of Samuel Insull*. Chicago: Privately Printed.

Jasanoff, S. 2004. *States of Knowledge*. London: Routledge.

Jasanoff, S., and Kim, S-H. 2009. "Containing the Atom: Sociotechnical Imaginaries and Nuclear Power in the United States and South Korea." *Minerva* 47:119–146.

Jasanoff, S., and Kim, S-H. 2013. "Sociotechnical Imaginaries and National Energy Policies." *Science as Culture* 22 (2): 189–196.

Jones, C. 2013. "Building More Just Energy Infrastructure: Lessons from the Past." *Science as Culture* 22 (2): 157–163.

Jones, C. 2014. *Routes of Power: Energy in Modern America*. Cambridge, MA: Harvard University Press.

Kurtz, H. 2010. "Acknowledging the Racial State: An Agenda for Environmental Justice Research." In *Spaces of Environmental Justice*, edited by R. Holifield, M. Porter, and G. Walker, 95–117. West Sussex: Wiley-Blackwell.

Laird, F. 2013. "Against Transitions? Uncovering Conflicts in Changing Energy Systems." *Science as Culture* 22 (2): 149–156.

Leslie, S., and Hevly, B. 1985. "Steeple Building at Stanford: Electrical Engineering, Physics, and Microwave Research." *Proceedings of the IEEE* 73 (7): 1169–1180.

Lovins, A. 2002. *Small Is Profitable: The Hidden Economic Benefits of Making Electrical Resources the Right Size*. Boulder, CO: Rocky Mountain Institute.

MacFarlane, A. 2006. *Technical Policy Decision Making in Siting a High-Level Nuclear Waste Repository, in Uncertainty Underground: Yucca Mountain and the Nation's High-Level Nuclear Waste*. Cambridge, MA: MIT Press.

Marathe, A., Marathe, M., and Anil Kumar, V. S. 2011. "Towards a Pervasive Enabled Modeling Environment for Integrated Coevolving Energy Systems." *Proceedings of the 2011 IEEE EPU-CRIS International Conference on Science and Technology*. New York: Institute of Electrical and Electronics Engineers.

Meyer, Niels I. 2007. "Learning from Wind Energy Policy in the EU: Lessons from Denmark, Sweden and Spain." *European Environment* 17 (5): 347–362.

Miller, C. 2014. "The Ethics of Energy Transitions." *Proceedings of the 2014 IEEE International Symposium on Ethics in Science, Technology and Engineering*. New York: Institute of Electrical and Electronics Engineers.

Miller, C., Iles, A., and Jones, C. 2013. "The Social Dimensions of Energy Transitions." *Science as Culture* 22 (2): 135–148.

Miller, C., and Richter, J. 2014. "Social Planning for Energy Transitions." *Current Sustainable/Renewable Energy Reports* 1:77–84.

Mitcham, C., and Rolston, J. S. 2013. "Energy Constraints." *Science and Engineering Ethics* 19 (2): 313–319.

Mitchell, T. 2011. *Carbon Democracy: Political Power in the Age of Oil*. London: Verso.

Mulvaney, D. 2013. "Opening the Black Box of Solar Energy Technologies: Exploring Tensions between Innovation and Environmental Justice." *Science as Culture* 22 (2): 230–237.

National Academy of Engineering. 2008. *America's Energy Choices*. Washington, DC: National Academies Press.

Noll, D., Dawes, C., and Rai, V. 2014. "Solar Community Organizations and Active Peer Effects in the Adoption of Residential PV." *Energy Policy* 67:330–343.

NRC (Nuclear Regulatory Commission). 2014. *Storage of Spent Nuclear Fuel*. Washington, DC: NRC. Accessed September 30, 2014. http://www.nrc.gov/waste/spent-fuel-storage.html. http://www.nrc.gov/waste/spent-fuel-storage/faqs.html.

Nye, D. 1990. *Electrifying America: Social Meanings of a New Technology, 1880–1940*. Cambridge, MA: MIT Press.

Nye, D. 1999. *Consuming Power: A Social History of American Energies*. Cambridge, MA: MIT Press.

Ottinger, G. 2013. *Refining Expertise: How Responsible Engineers Subvert Environmental Justice Challenges*. New York: New York University Press.

Perrow, C. 1984. *Normal Accidents*. Princeton, NJ: Princeton University Press.

Practical Action. 2014. *Poor People's Energy Outlook 2014*. Bourton on Dunsmore, United Kingdom: Practical Action.

Richter, J. 2013. "New Mexico's Nuclear Enchantment: Local Politics, National Imperatives, and Radioactive Waste Disposal." PhD diss., University of New Mexico.

Rolston, J. 2014. *Mining Coal and Undermining Gender: Rhythms of Work and Family in the American West*. New Brunswick, NJ: Rutgers University Press.

Rucinski, T., and Kaye, B. 2014. "Taxes, Fees: The Worldwide Battle between Utilities and Solar." Accessed September 30, 2014. http://www.reuters.com/assets/print?aid=USKCN0HN07P20140929.

Ryan, S. 2014. "Rethinking Gender and Identity in Energy Studies." *Energy Research and Social Science* 1:96–105.

Satchwell, A., et al. 2014. "Financial Impacts of Net-Metered PV on Utilities and Ratepayers: A Scoping Study of Two Prototypical U.S. Utilities." Berkeley: Lawrence Berkeley National Laboratory. Accessed September 30, 2014. http://emp.lbl.gov/sites/all/files/LBNL%20PV%20Business%20Models%20Report_no% 20report%20number_0.pdf.

Schlossberg, D. 2007. *Defining Environmental Justice: Theories, Movements, and Nature*. Oxford: Oxford University Press.

Sen, A. 1983. *Poverty and Famines*. Oxford: Oxford University Press.

Shrader-Frechette, K. S. 2014. *Tainted: How Philosophy of Science Can Expose Bad Science*. Oxford: Oxford University Press.

Smith, M. R. 1994. "Technological Determinism in American Culture." In *Does Technology Drive History?*, edited by M. R. Smith and L. Marx, 1–35. Cambridge, MA: MIT Press.

Sovacool, B. 2014. "What Are We Doing Here? Analyzing Fifteen Years of Energy Scholarship and Proposing a Social Science Research Agenda." *Energy Research and Social Science* 1:1–29.

Sovacool, B., and Brossman, B. 2013. "Fantastic Futures and Three American Energy Transitions." *Science as Culture* 22 (2): 204–212.

Steinberger, Julia K., and Timmons Roberts, J. 2010. "From Constraint to Sufficiency: The Decoupling of Energy and Carbon from Human Needs, 1975–2005." *Ecological Economics* 70 (2): 425–433.

Tenenbaum, B., et al. 2014. "From the Bottom Up: How Small Power Producers and Mini-grids Can Deliver Electrification and Renewable Energy in Africa." Washington, DC: World Bank. Accessed September 30, 2014. https://openknowledge.worldbank.org/handle/10986/16571.

Trutnevyte, E., Stauffacher, M., and Scholz, R. 2011. "Supporting Energy Initiatives in Small Communities by Linking Visions with Energy Scenarios and Multi-criteria Assessment." *Energy Policy* 39 (12): 7884–7895.

US Department of Energy. 2012. *Quadrennial Technology Review*. Washington: US Department of Energy. Accessed September 30, 2014. http://energy.gov/downloads/quadrennial-technology-review-august-2012.

US Department of State. 2014. *Final Supplemental Environmental Impact Statement for the Keystone XL Project*. Washington, DC: US Department of State.

US Government Accountability Office. 2011. *Yucca Mountain: Information on Alternative Uses of the Site and Related Challenges*. Washington, DC: Government Accountability Office. Accessed September 20, 2014. http://www.gao.gov/products/GAO-11-847.

Wagner, P., and Wittrock, B. 1991. "States, Institutions, and Discourses: A Comparative Perspective on the Structuration of the Social Sciences." In Vol. 15, *Discourses on Society: The Shaping of the Social Science Disciplines*, edited by P. Wagner, B. Wittrock, and R. P. Whitley, 331–357. Dordrecht, the Netherlands: Springer.

Walker, J. S. 2009. *The Road to Yucca Mountain: The Development of Radioactive Waste Policy in the United States*. Berkeley: University of California Press.

Weber, B., Geigle, J., and Barkdull, C. 2014. "Rural North Dakota's Oil Boom and Its Impact on the Social Services." *Social Work* 59 (1): 62–72.

Winner, L. 1986. *The Whale and the Reactor*. Chicago: University of Chicago Press.

World Nuclear News. 2014. "New French Energy Policy to Limit Nuclear." Accessed July 1, 2014. http://www.world-nuclear-news.org/NP-New-French-energy-policy-to-limit-nuclear-1806144.html.

Wüstenhagen, R., Wolsink, M., and Bürer, M. 2007. "Social Acceptance of Renewable Energy Innovation: An Introduction to the Concept." *Energy Policy* 35 (5): 2683–2691.

Wynne, B. 2005. "Risk as Globalizing 'Democratic' Discourse? Framing Subjects and Citizens." In *Science and Citizens*, edited by M. Leach, I. Scoones, and B. Wynne, 66–82. London: Zed Books.

Zimmerer, K. 2011. "New Geographies of Energy: Introduction to the Special Issue." *Annals of the Association of American Geographers* 101 (4): 705–711.

33 Debugging Bias: Busting the Myth of Neutral Technology

Felicia L. Montalvo

The readings in the third section provided an abundance of examples of how values can be intertwined with technology. Although we can't foresee the future, digital technologies are likely to play an increasingly important role in our future and impact our values. Hence, it is important to be reminded that digital platforms, algorithms, and digital services companies are not "neutral." Montalvo details how the "Frightful 5"—Apple, Amazon, Google, Microsoft, and Facebook—constitute an incredibly powerful influence in the lives of Americans that is likely to endure well into the twenty-first century.

Most users tend to naturalize digital environments, accepting, without question or critical judgment, the new features and apps made available to them and the order in which information appears in search engines and newsfeeds. Montalvo provides example after example of bias in digital technology, and she argues that it is not just the coders who program their biases into digital systems. User biases come to be reflected in certain digital technologies because of the way they work. The author says that "the narrative of objective technology" serves the interests of digital services companies like the Frightful 5 and hides the reality that such companies have their own interests and would like us to behave in ways that serve those interests. The lessons for the twenty-first century are to resist the sometimes covert claim that a technology is neutral and to keep in mind both that technologies may be biased and that the companies that make them have their own interests that sometimes are and sometimes are not aligned with our own.

On February 16, 2016, "with the deepest respect for American democracy and a love of our country," Apple CEO Tim Cook hurled a giant "fuck off" at the FBI in the form of an open letter opposing an order to create a backdoor to the iPhone. This backdoor software, if created, would give the government (and anyone else who managed to get their hands on it) the ability to bypass iPhone encryption software and access user data.

A few days after the letter was published, the hashtag #StandWithApple began to circulate, with Twitter users applauding Apple's bravery in standing up for the people's right to privacy. Soon other tech giants rallied to the cause: Sundar Pichai, Google's new CEO, echoed Apple in stating that enabling a backdoor would compromise user security, and Facebook CEO Mark Zuckerberg jumped in to provide a supportive soundbite about the importance of encryption technology. Garden-variety iPhone users and tech leaders alike were ready to go to bat for our civil right to privacy. Yet, in our hashtag-and-emoji circus, we missed a critical point: They're not fighting for us.

From *Bitch* 71 (Summer 2016): 37–40.

Framing the encryption debate as one of civil rights vs. government power allows private companies like Google (now a subsidiary of Alphabet), Facebook, and Apple to paint themselves as brokers operating on behalf of the public. But these companies have made billions on the hardware, software, and web platforms that regulate every aspect of the digital lives of that public. We are not talking altruistic motives here.

The "Frightful 5," as journalist Farhad Manjoo dubbed the tech giants, have quickly made themselves indispensable to most of us. We depend on Apple, Amazon, Google, Microsoft, and Facebook to tell us what to do, where to go, how to get there, what to eat, what news to care about, who to follow, what to like, and what we should be doing with our time. In a little under two decades, these companies have not only touched every aspect of our lives, but have become, as Manjoo noted, "the basic building blocks on which every other business, even would-be competitors, depend." Cross-platform measurement company comscore.com reports that as of September 2015, Google Search dominated almost 64 percent of the search-engine market; Apple ranked as the top smartphone manufacturer, with 42.9 percent of the OEM market share in 2014; and a 2014 Pew Research Center study found that 71 percent of adults online use Facebook. For lay users, it's nearly impossible to extract any information from the internet without passing through the gates of one of the Frightful 5. Even finding an alternative search engine like DuckDuckGo might involve first passing through Google.

What happens when our primary access points to the world's information—and each other—are managed by a handful of private companies? And how did we come to think of them as neutral brokers of information?

Science, mathematics, and engineering: They are historically and presently associated with men, and treated as infallible bodies of knowledge that view, manipulate, and build the world from a completely objectivist point of view. The narrative of Silicon Valley's revolution hailed technology as inherently neutral, and its creators as impartial engineers of the coming utopia. In a 2013 manifesto coauthored with Jared Cohen, *The New Digital Age*, Alphabet executive chairman Eric Schmidt outlines a vision of the future in which everyone is connected (presumably through Google), stating that "technology is neutral, but people are not," which suggests that the way we use technology is what determines its values.

If we go by Schmidt's logic, a hunk of metal filled with combustible material isn't a bomb until it's dropped on a city of civilians. But Schmidt is not the only high-ranking technocrat who believes in the blank-slate theory of tech, and it's easy to see why: The belief in neutral technology is part of a larger cultural perspective that seeks to immunize men and their endeavors from the social constraints that ostensibly affect women: emotions, and, consequently, bias.

Forging an ideological link between technology and the hard sciences gives tech giants the power to perform what the pioneering cyber-feminist theorist Donna Haraway calls a "god trick," which she defined as a "view of infinite vision" that positions the subject entirely apart from, and above, the object. When scientists, engineers, or platform designers grant themselves a veneer of objectivity, they absolve themselves of the responsibility to acknowledge their own personal biases, their own stake in the code.

The reality that digital platforms and algorithms are not inherently neutral becomes clear with platforms like Facebook and Google that are designed using historical data as inputs. According to data scientist Shahzia Holtom, this means that "any biases, such as underrepresentation of women or ethnic minorities, that may be present in the historical data will also be reflected in the results."

In 2004, *Jewish Journal* reported that when users typed "jew" into Google's search bar, the first result was jewwatch.org, a fanatically anti-Semitic hate site. There was a resulting outcry from Jewish advocacy groups and journalists questioning how, out of the 1.72 million web pages relevant to the search term, this result landed on top. Google founder Sergey Brin declined to remove the offending result on the grounds that it would compromise Google's objectivity; his solution was to create an "Offensive Search Results" warning stating that "the beliefs and preferences of those who work at Google, as well as the opinions of the general public, do not determine or impact our search results."

Another search-query blind spot surfaced in 2007, when users noticed that when inputting certain phrases that began with "she"—"she invented," "she discovered," "she golfed," "she succeeded"—a spelling correction would appear automatically appear. "Did you mean: he invented?" Google's official explanation was that its spell-check algorithms were "based on sophisticated machine learning methods . . . completely generated without human input." Again: It's not us, it's the machines. Ta-da! God trick.

Digital platforms and algorithms don't just reflect the biases of the coder, they also have a way of replicating the narrow perspective of users as well. This became all too clear this March, when Microsoft debuted its AI chatbot, Tay. The 19-year-old female chatbot was promptly co-opted by a series of internet trolls and within 24 hours became a neo-Nazi mouthpiece for racist and sexist epithets. After Microsoft shut her down, #JusticeForTay rang out on Twitter, with those same trolls clamoring for the revival of the bot they believed was silenced for speaking the truth. Of course, it wasn't that Tay revealed some inherent truths about society, but rather that her design had failed to provide her with an adequate ethical framework for navigating the bias and violence of the real world. To create that filter would be to deny the impartiality of tech.

The narrative of objective technology differs sharply from the monopolized reality of the Frightful 5. The proliferation of digital platforms depends upon their unmatched addictiveness. "Habit formation" is the magic phrase in Silicon Valley; Nir Eyal, author of *Hooked: How to Build Habit-Forming Products*, notes that linking digital platform use to a user's daily routine and emotions is the best way to ensure loyalty—and, by extension, profit. The habit-formation model argues that digital platforms should be designed as a response to particular emotional triggers, especially internalized ones. You're anxious? Check Facebook. Bored? Hop on Twitter. Depressed? Scroll through Instagram. Not sure about something? Just fucking Google it. Once a platform is recognized as a balm for these triggered internal emotions, we don't even need the triggers anymore, but simply return on our own. The habit formation model does not satisfy a need, it creates an incessant craving. And by encouraging frequent platform use, it provides an endless supply of data that companies can use to get better at keeping you there. "Major tech companies have 100 of the smartest statisticians and computer scientists . . . whose job it is to break your willpower," notes ethical-design advocate Tristan Harris.

Falling into the addiction trap not only turns us into neat data points, it also robs us of the agency to make our own way through digital spaces. This isn't exactly what the vanguards of 1990s cyberculture had in mind. Author and cyberpunk progenitor William Gibson famously wrote that "the Internet is a complete waste of time—and that's what's so great about it." The web was for "wandering aimlessly" and discovering new things about your world; the "waste of time," as Gibson saw it, was a valuable expression of dissent from a society that glorified endless productivity.

Australian cyberfeminist collective VNS Matrix did foresee the dangers of a digital space in which, as founder Virginia Barratt told *Motherboard* in a 2014 interview, "access by women was limited and usually mediated by a male 'tech.'"; to revitalize the anarchical nature of the Internet, VNS produced video games and web hacks that would "hijack the toys from technocowboys and remap cyberculture with a feminist bent."

Though the culture of cyberpunks and cyber-feminists has been absorbed by Hollywood and spat out as *The Matrix*, the power of network that inspired their values is still at our disposal. And the realization of our collective agency begins with rejecting the fallacy of neutral technology, and recognizing ourselves in the machines.

The god trick applies not just to products, but to the chronic homogeneity of the tech industry as well. If we can see ourselves in the machines, after all, then it follows that we should also recognize the need for more diverse perspectives. But even with a steady increase in stem education funding over the last decade, many populations—Black and Latinx people and women among them—are wildly underrepresented. And the real-life effects are being felt by coders and users alike: Last summer, after uploading a number of pictures to Google Photos, software developer Jacky Alciné noticed that the facial-recognition function identified two of his dark-skinned friends as gorillas. As he stated in a blog post, "It just doesn't make sense, barring the obvious, why the world's most popular search engine was incapable of recognizing the face of a dark-skinned Black person." A stronger, more diverse quality-assurance team would have been able to prevent the error. But, as a January 2016 *Bloomberg Businessweek* report pointed out, absent a titanic culture shift ("Over the past two decades, African Americans have made up no more than 1 percent of tech employees at Google, Facebook, and other prominent Silicon Valley companies"), such egregious incidents will continue to occur.

Meanwhile, in 2014, data collected from GitHub revealed that only 17 percent of Google's engineers were women—and that's higher than the industry average. But tech journalist Rachel Sklar argues that the general dearth of women in technology isn't as simple as "there aren't enough suitable women engineers," because tech companies don't just hire engineers. "If Silicon Valley's money people or fancy keynote founders or biz whizzes on the cover of Entrepreneur were all engineers crushing code 24/7, then fine," noted Sklar in a post on *Medium*. "But that is not the case."

When we recognize that the engine of the tech revolution—the assumption of inherent neutrality—is faulty, and that the Frightful 5's appeal to the public good is just canny PR, then we can begin to imagine how to effectively design, program, and use these platforms to provide humanity with real, measurable benefits. The techie/former *Daily Show* producer/comedian Baratunde Thurston, who recently received the Interactive Hall of Fame Award at South by Southwest, took time in his acceptance speech to draw attention to what's at stake in supposedly dehumanized tech:

> If innovation is all about making the world a better place, and the algorithms that claim to do so derive from this very imperfect world sick with racism and sexism and crippling poverty, then isn't it possible that they might make the world a worse place? Could we end up with virtual-reality racism? Could we have machine-learned sexism? Could poverty be policed by drones and an internet of crap? This is all very possible if we don't engage consciously in the work that we're doing.

We can start with an understanding that tech revolutions are inherently social revolutions that project their own set of values and expectations onto the world in which they evolve. Relinquishing the god trick and recognizing that emotion and bias in the machines are not weaknesses or failures, but rather strengths, will make it possible for all of us to move toward a future that makes us better than what we are.

34 When Winning Is Losing: Why the Nation That Invented the Computer Lost Its Lead

Mar Hicks

At first glance, this piece may seem an odd selection in a section about the future since it is a story about the past. However, as historians often point out, if we don't learn from history, we are doomed to repeat it. Hicks presents a tale of the history of computing with the hope that we can learn from failures in the past as well as successes. One of the challenges of the twenty-first century will be to develop a workforce with the skills needed to develop, monitor, maintain, use, and improve science and technology in an increasingly complicated sociotechnical world.

Hicks's story saliently illustrates the hurdles that have been put in the path of women who have the capacity and drive to succeed in science, technology, engineering, and math (STEM). The story serves as a cautionary tale for the future, with a lesson that is powerful because it points to more than the unfairness of discrimination. Hicks suggests that discrimination against women led to the UK losing its leadership role in computer technology. In this respect, Hicks's story not only reinforces a theme throughout this book about the intertwining of gender and technology but suggests that the stakes can be very high when we fail to understand and address the various ways that science and technology affect different groups of people.

After college, I worked as a systems administrator in Harvard's electrical engineering and computer science department, and while there, I noticed an odd gender split. Most of the workers my age were men. But our bosses, who were from an older generation, were women. When we would remark on this odd gender inversion, our bosses would remind us that history wasn't a linear progress narrative: there used to be a lot more women in computing, they'd tell us.

As the child of a computer programmer, this should have come as no surprise. But I had thought of my mother's work on an individual basis, rather than as part of a trend. She got into computer programming after being pushed out of the graduate astronomy program at the same university where I was now a sysadmin, having been told that she was taking up a spot in the program that should have gone to a man. It had never occurred to me that the reason she failed as an astronomer but succeeded as a programmer hinged on computing's gendered labor history.

When I began the research that would become my book *Programmed Inequality*, I initially had only one question in mind: if there were lots of women in early computing, where did they go? I wanted to figure out what happened to make the current field

---

From *Computer* 51, no. 10 (2018): 48–57.

so male-identified. I did not realize that what I would uncover would shake my faith in the "computer revolution." Or that the story that would emerge would not just be about women, but about how the nation that invented the computer destroyed its own computing industry rather than risk putting more women in charge of a powerful new technology.[1]

Victory from the Jaws of Defeat

In 1944, Britain led the world in computing. With the country bombed heavily by German forces, and starved of food and supplies by German submarines, decoding enemy messages meant the difference between life and death not just for soldiers but for citizens on the home front. By the final, grueling stages of the war, codes were much harder to crack and the machines that helped mechanically speed codebreaking—the Bombes that Alan Turing helped build—were far too slow to decrypt even a fraction of the thousands of daily messages.

To decode these messages by brute-force would have required hundreds of thousands of man-hours, and so was out of the question. Yet Post Office Engineer Tommy Flowers, the son of a bricklayer from working-class East London, knew brute-force codebreaking could work, if only a fast enough machine could be created. While Max Newman, who headed machine code-breaking at Bletchley Park, ordered more production of the same slow codebreaking machines as before, Flowers used his own initiative, and his own money, to design and build an all-electronic computer that few thought would work, due to the fragility of its thousands of vacuum tubes. But Flowers's team made a seeming impossibility into reality in the final months of 1943. The first Colossus computer was operational at Bletchley Park by early 1944. There would be nine more by war's end, working around the clock.

Though Flowers designed these computers, the people who made them useful on a day-to-day basis, doing everything from assembling, to troubleshooting, to programming and operating them, were the Women's Royal Naval Service. The Colossus—the first digital, electronic, programmable computers in the world—brought machines into intelligence work at a speed and scale heretofore unknown. At a time when the best electronic computing technology in the US was still in the testing phase, the British used electronic computers to change geopolitical events: Colossus II let the Allies know where and when to land on D-Day, turning the tide of the war. These computers, and the telecommunications networks they relied on, were the first instance of what today would be called cyberwarfare. The workers were overwhelmingly women.

After the war, the contributions of the women who operated these cutting-edge computers were kept secret, along with the machines themselves. Most women who had worked at Bletchley were told to masquerade as "secretaries" to disguise the true nature of their vital wartime intelligence work.[2] Many took their experiences with them to their graves. This was not simply due to wartime and postwar secrecy: for decades what these women did was misunderstood and viewed as unimportant because computing was a feminized sphere of work.[3]

The Girls in the Machine

At the dawn of the digital age, and well into the mainframe era, jobs in computing were thoroughly feminized.[4] Women had not been placed into computing roles during the war simply because men were at the front: they were recruited into these roles in both

Figure 34.1
1957 *Yearning Miss* cartoon from *Tabacus: The Company Magazine of the British Tabulating Machine Company*. Even though computing work required skill and often advanced mathematics knowledge, it was perceived as unintellectual and largely unimportant, in part because low-paid women were doing it.

wartime and peacetime because computing was work with no career path and therefore not suitable for men. Before a computer was a machine, it was a job, which Jay Forrester described as "a girl sitting at a desk" doing calculations.[5]

The British continued to be strong in computing after the war, making breakthroughs that matched or anticipated US developments. While most Colossus operators returned to other jobs in civilian life, the women operating nonsecret electromechanical computers at Bletchley Park carried over into postwar computing jobs. As swords were turned to ploughshares and the nation brought electronic computerization into industry and government, women workers carried right over into cutting-edge jobs in high tech. Programming and most other day-to-day computing continued to be a feminized sphere of work. IBM UK even measured their manufacturing in less expensive "girl-hours" rather than man-hours through the *1960s*.[6]

But that started to change within a few decades of the war's end. Not because the work was changing, but because the perception of the work was. Computers were becoming integrated into every part of the work of government and industry, and it was becoming clear they would shape the world to an ever greater extent in the decades to come. As that happened, their wide-ranging power became more apparent. Suddenly, low-status women workers were no longer seen as appropriate for this type of work, even though they had all the technical skills to do the jobs (see figure 34.1).

Building the Digital Age

In 1959, one of these women computer workers needed to program, maintain, operate, and test all of the computers in a major computing center that was doing critical work for the government. These computers didn't just automate low-level office work; the programs they ran formed the infrastructure for national affairs. From the National Health System to the value-added taxation system (VAT) required to fund British infrastructure, computing ran the affairs of the British state and impacted citizens' daily lives in important ways.

This programmer also had to train two new hires who had no computing skills.[7] She would bring them up to speed with a full year of training. After that, the new hires would step into management roles, while their trainer would be demoted to an assistantship below them. That the trainer was a woman and that the new hires were both young men was not a coincidence. Even though this work required skill and often advanced mathematics knowledge, it had long been perceived as unintellectual and, in part because low-paid women were doing it, largely unimportant.

Within the public sector, these women formed an underclass of highly technically trained, operationally critical technology workers. They were called the "Machine Grades." In the mid-1950s, when the British government finally assented to paying its employees equal pay for equal work, the majority of women working in the civil service did not get equal pay, because they were machine workers. Women's wages were not made equal with the seldom-used men's pay scales in the machine grades: the government's rationale was that women had been doing this work for so long, and in such great majority, that their lower rate of pay had now become the market rate for the work.

The growing, feminized, machine underclass formed the infrastructure of a swiftly computerizing nation—from taxation to atomic energy to economic forecasting. And yet, their low status and the fact that they were essentially forbidden from rising into management meant a complete mismatch between the growing power of these technical experts and the way they were treated—hence the female "senior machine operator" who had to train her new nontechnical, management-level replacements. Though her supervisors might note that she had "a good brain and a special flair" for programming, they would also decide that she should not be given any further responsibility, and, in fact, looked forward to her leaving for marriage or "any other reason" once the training was complete.[8]

This began a decade-long effort to remove women workers from these newly important technical positions and replace them with management-minded young men—men who could go from the boardroom to the machine room, managing computers as well as people. Women were not expected to manage men, so the idea of technical women working their way up to fill these roles was out of the question. Marriage was often used as a turnover mechanism, getting rid of women before they could be promoted to positions of authority (see figure 34.2). As late as the 1970s it was frowned upon—and often outright forbidden—for middle class, white-collar women to continue working after marriage.

Re-gendering Computing

Few men with good career prospects ahead of them were keen to go into the feminized "backwater" of computer programming. Men with the technical skills to do these jobs did not exist in large numbers. As the government, the nation's largest computer user, ramped up for the electronic age, it simultaneously discarded its trained technical workforce. Intent on swapping "bright young men" into these newly important posts, managers in industry and the public sector got rid of their most qualified technologists.

Ironically, this was how the UK ended up getting one of its earliest and most successful software startups. Its business model and success would prove to be a cautionary

Figure 34.2
A young woman named Anne Davis wears a punched-tape dress at her "retirement party" as she leaves her job to get married. Marriage was often used as a turnover mechanism, getting rid of women before they could be promoted to positions of authority. *Source: ICL News,* August 1970.

tale for the rest of the computing industry. Freelance Programmers, founded by Stephanie "Steve" Shirley, took advantage of British sexism to create a thriving company powered by women's technical labor.[9] Shirley, who had been a child refugee during World War II, brought to England with 10,000 other German Jewish children on the Kindertransport, had worked at the prestigious Dollis Hill Post Office Research Station in the 1950s, the same place where Tommy Flowers had built the Colossus.

As Shirley jockeyed for promotion in government and then in industry, she began to realize, as she put it, "The more I became recognized as a serious young woman who was aiming high—whose long-term aspirations went beyond a mere subservient role—the more violently I was resented and the more implacably I was kept in my place."[10] After being passed over for promotions she had earned, she learned that the men assigned to make the decision were repeatedly resigning from the promotions committee rather than risking giving a woman a promotion.

Shirley got married and left the workforce, but she did not quit working in computing: instead she started her own computing company in 1962. Shirley's belief—that software was something people would pay for rather than expect to come along with the mainframe or be written by their own staff—was risky for the time. The nascent software services industry of the 1960s still had to contend with the idea that software was not really a product in and of itself.

Shirley's company was also unusual because it learned from the mistakes that the government and most of industry were making. While those organizations starved themselves of good programmers by refusing to promote or accommodate women technologists, Shirley scooped up this talent pool. She specifically hired women, and by giving them flexible, family-friendly working hours and the ability to work from home, her business tapped into a deep well of discarded expertise. For every woman programmer she employed there were dozens more who wanted jobs but were being turned out of regular employment due to discrimination.[11] Initially, having a woman's name at the helm of her company prevented Shirley from getting contracts. She began using her nickname "Steve" on business correspondence to get her company off the ground.

Desperate for programmers, systems analysts, and software architects, the government and major British corporations hired Shirley and her growing team of women to do mission-critical work on an ad hoc, outsourced basis. Shirley's team designed and programmed everything from software to run the petrochemical industry and a variety of administrative programs, to the code that ran the Concorde's black box flight recorder. The Concorde black box project was managed and completed entirely by a remote workforce of nearly all women, who programmed with pencil and paper from home before testing their software on rented mainframe time (see figures 34.3 and 34.4).

The irony that these women were not good enough to keep in formal employment, but so indispensable that government and industry would outsource major

Figure 34.3
Computer programmer Ann Moffatt sits at her kitchen table in 1968, writing code for the Concorde's black box flight recorder. She later became technical director at Freelance Programmers, in charge of a staff of more than three hundred home-based programmers. Moffatt notes that the baby in the photograph is now fifty-two. *Source:* Photo courtesy of Ann Moffatt.

Figure 34.4
Stephanie "Steve" Shirley, Ann Moffatt, and their business manager Dee Shermer in a machine room in the 1960s. Shirley specifically hired women, and by giving them flexible, family-friendly working hours and the ability to work from home, her business tapped into a deep well of discarded expertise. *Source:* Photo courtesy of Ann Moffatt.

computing projects to them, was not lost on Shirley and her workers, but seemed to sail over the heads of many others. Shirley's successful business model shows why sexism was bound to create major problems for the computing industry. Her company took advantage of the sexism intentionally built into the field to eventually become a billion-dollar company. But the gains made by her company paled in comparison to the loss of expertise and productivity in the industry as a whole as a result of labor discrimination. Her feminist business model offered a solution, and a way forward, but that was not the solution that British industry and government chose to enact.

Power and Technology

Discarding women's technical expertise was all the more surprising because through the 1960s—which Prime Minister Harold Wilson described as the era of "white heat"—Britain pinned all its hopes on high technology, and computers in particular, to arrest its slide into second-rate world power status. It had come out of the war with much of its infrastructure destroyed and its population struggling with austerity. The promise of progress through technology held irresistible allure.

When Wilson promised a tech-led social revolution, with the unfairness of the British class system "burning up" in technology's white heat, many wanted to believe in this vision of the future. Much like in the US today, where Silicon Valley promises a better society through better technology, the British were taken in by the idea that technology would solve social problems, particularly inequalities of class and the unfairness of a nonmeritocratic society. Yet many inequalities would only be heightened during Britain's "computer revolution." These technologies were designed and leveraged by those at the top of political, social, racial, and gender hierarchies, and they were—both implicitly and explicitly—designed to work within those existing patterns of discrimination, and even strengthen them.

As their empire collapsed, and formerly colonized people fought for and won their freedom, the British put more faith into technology as a means of control. Those at the highest levels of government and industry believed that high technology would be version 2.0 of the British Empire. The British were strongly against using American computers in government and essential industries because it would be a threat to their national security—it would give a foreign power a back door into the workings of the British State. For the same reason, the government and British corporations hoped to place as many British computers into the banking industries, industrial infrastructure, and even governments of former colonial nations. The British government understood the power of controlling another nation's computing infrastructure very early in the digital revolution, and how this could be a mode of imperial power. Even if it was not direct political or military power, it was enough to make Britain indispensable.[12]

Images of British computers deployed abroad presented a gendered imperial narrative and positioned other nations—particularly former colonial nations—as uncivilized and backward, in need of British high technology in order to advance into the 20th century. In one image, a Ghanaian woman operating a British computer accompanies a story about British technological domination abroad (see figure 34.5a). In another (not pictured), a British computing company talks about the civilizing effects of its "universal methods" of computing showing rows of women workers. Throughout, the British injected their gendered labor model into these contexts, staffing certain roles only with women regardless of local customs. When Indian companies set up their own computer installations, for instance, they were more likely to slot men into roles that would be staffed only by women when British computer companies were in control (see figure 34.5b).[13]

The Beginning of the End

By the mid 1960s, the computer labor shortage in both government and industry had sharpened into a crisis. The government continued to struggle to find qualified candidates to train. Computing, as a field of endeavor, had been feminized work for so long that most young men did not dare go into it. Though this was broadly the same in the US, the UK case is instructive. With its smaller labor force, it was not as insulated against the effects of sexist and racist labor practices as the US was for many decades.

The commitment to hire only management-oriented young men began to soften, and in the mid-1960s, a cohort of women programmers was able to get jobs in the new,

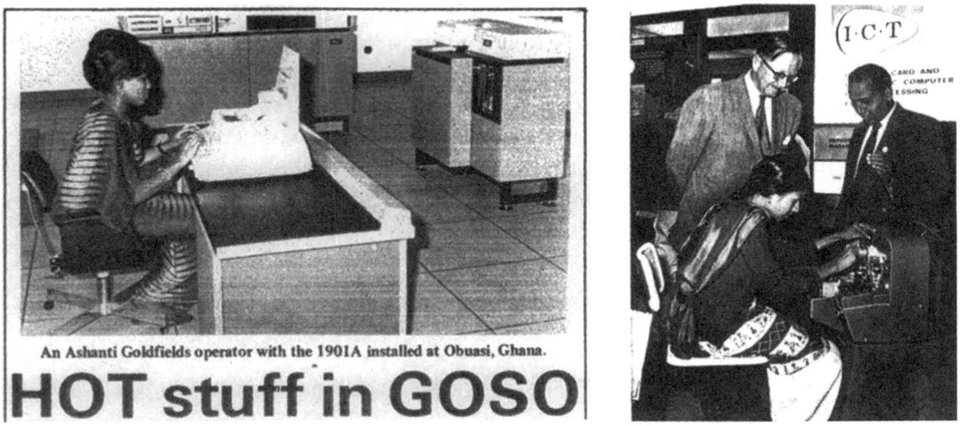

(a) (b)

Figure 34.5
Images of British computers deployed abroad presented a gendered imperial narrative and positioned other nations—particularly former colonial nations—as in need of British high technology to advance into the twentieth century. (*a*) An Ashanti Goldfields computer worker in Ghana in the 1970s and (*b*) an Indian woman presenting British computing technology at the Indian Industries Fair in 1965.

higher-level technical grades of the civil service, as long as they came from the service's white-collar office worker grades and not from the pink-collar machine grades. This brief wave of more egalitarian hiring, in terms of gender if not in class or race, helped some women who were just starting out into the field of computing, but did not signal a permanent change.

The power that technical people had, given the interwoven nature of computing processes with all the functions of the state—from the welfare state to the Bank of England to the Atomic Energy Authority—meant that they, and their machines, were becoming increasingly indispensable. By the late 1960s, the government's fears of losing control over the technology that ran the nation were perilously close to becoming a reality. Striking punch operators shut down the newly created VAT computer, delaying the implementation of this critical national system for months.

It was in this context that the Ministry of Technology, backed by other government agencies, hatched a plan to reconfigure the computing infrastructure on which the government relied. The low number of trained, reliable, management-aligned technical people was starving the government's computers of the workers they needed to function optimally. Training more young men had not worked, simply resulting in more turnover and wasted outlay. Outsourcing was only a temporary solution. And returning to a feminized computer labor force was a nonstarter. The only solution was to re-engineer the system to function with a much smaller labor force.

Government officials believed they would need ever more massive mainframes so they could centralize all computing operations, allowing a far smaller technical labor force to control the state's digital infrastructure. Under the advice of the Ministry of Technology the government forced a merger of all the remaining viable British computer companies. The government would fund the merged company's research program and promise to purchase its computers exclusively.

In return, the merged company, ICL, accepted a high degree of government control over its product line, and embarked on the project of producing the huge, technologically advanced mainframes the government needed to solve its labor problems, because the government had become convinced that it could no longer function by trying to get more young men into computing—the numbers simply weren't there. Since solving computer labor problems through training had not worked, it demanded that the British computer industry's product line accommodate the state's needs through the design of the systems itself.

Unfortunately, this change occurred at a time when massive mainframes were losing popularity while smaller mainframes and decentralized computer systems were rising in popularity. By the time ICL delivered the required machine line in the 1970s—the highly advanced 2900 series—the British government no longer wanted it, and neither did other potential customers. The government swiftly removed its support for ICL, torpedoing what was left of the British computing industry. Though they realized their mistake, they did not see the underlying sexism that had caused it.

Moored to its new expensive, dead-end mainframe line, ICL could not compete with the offerings of IBM and other companies. ICL had neglected the development of its smaller, and better-selling, lines of mainframes to focus on developing the 2900, but this proved unwise: support for the company evaporated and ICL laid off thousands of workers with each passing year.

Worse Things than Losing

Many people know a little bit about the history of women in computing. Ada Lovelace and Grace Hopper have risen to the level of posthumous celebrities. But the truth is that women were never so exceptional in the history of computing as they might now seem. By positioning a few successful women as laudable exceptions, we have missed a major thread within computing history.

The British case is instructive not in spite of being a failure story, but because it *is* a failure story. Technological success stories map to the idea that computing's history is one of social progress, even when it is not. The narrative of meritocracy-driven progress in the history of computing must be rethought by emphasizing the important historical lessons of technological failure, and the social reasons behind those failures.

In the British example, sexism produced a technical labor shortage in government and industry, eventually leading to the downfall of the British computing industry because women were common rather than unusual. And history shows us that the sexism that produced this shortage was an intentional and constructed feature of electronic computing—not a natural evolution of the field, nor a reflection of women's talent, goals, and interests. This is important because it means our problems today are not a matter of simply getting more women into computing. Rather, we need to address the structural, programmed-in inequalities that have been there from the start. These hierarchies were meant to preserve the powerful social and political structures that electronic computing was meant to serve, and they still function in this way.

Histories like this are of crucial importance today because they are not unusual. The British association of technical work with low-status workers, but powerful managerial work with high-status workers, was a blueprint for the postindustrial West. The manner in which British leaders and industry officials worked together to standardize and codify a gendered underclass of tech workers, and then to later upskill and masculinize that work once the managerial power of computers became clear, was an intentional set of systems design parameters to ensure that those who held the most power in predigital society, government, and industry continued to hold that power after the computer revolution.

Although the contours of this story and the groups affected change in different national and temporal contexts, the idea that technological systems preserve existing hierarchies and power structures, rather than being revolutionary, usually holds true. As I write in *Programmed Inequality*, "The British case is a parable of how nations can modernize in ways that are not merely uneven but that actively reconstitute categories of social inequality." Scholars such as Alondra Nelson have pointed out that who controls a technology or decides on its design has material impact in ways that go beyond mere usability and instead extend into the heart of citizenship and civil rights.[14]

A close re-examination of computing history shows us that the computer revolution was never really meant to be a revolution in any sort of social or political sense. People who were not seen as worthy of wielding power over men and nations were deliberately excluded—even when they had the required technical skills—and, to a great extent, that process continues today. As we see the negative effects of discrimination in Silicon Valley reverberating outward to affect political events on a national and even global scale, it is worth looking back and remembering what happened to the nation that invented the computer, and how discrimination played a role in its fall from grace.

And yet, for all the destruction caused by the British attachment to a particular kind of sexist technological progress, it would be wrong to take from this history the lesson that discrimination in high technology is bad because it hurts high technology or the economy more broadly. The civil rights of workers and citizens were violated to try to construct what the UK thought, at the time, was the correct technological solution to a broad set of labor and technical problems. To say that this was wrong only because it resulted in a lack of profit or because it stunted technical advance elides the core issue. Civil rights are not relative and negotiable but absolute and inalienable, and must never depend on profit or productivity; that vision of technological progress is not progress at all. As Margot Lee Shetterly noted in her groundbreaking book *Hidden Figures*, by 1969 the US could put a man on the moon, but African Americans still could not drive to the next state without fear of being killed.[15] How can this deeply uneven style of development ever be considered real progress?

Today in the US we are seeing the same process take hold, as powerful technocrats with little interest in, or understanding of, history, society, and the civil rights of women and minorities steer the ship of digital society in circles, retracing, and often worsening through technological amplification, all of the discriminatory practices and policies civil rights and labor activists fought hard to undo in the mid and late 20th century.[16] As Safiya Noble puts it, "If you're designing technology for society, and you don't know anything about society, then you're unqualified."[17]

That this mode of designing technology without much attention to history, civil rights, race, or gender is often represented as a natural and necessary way of doing things should be a nudge to reconsider the patterns of hegemony that we take for granted as being a fundamental part of our technological systems. Today, the US is in a similar position to 20th-century Britain: a technologically advanced, but socially regressing late-imperial power. As our technological systems become increasingly destructive to our economic and political institutions, we find ourselves failing, collectively, to address the decades of intentional design decisions that have led to these unwanted consequences. Although the details differ, our historical circumstances are eerily similar to our closest historical cousin in important ways. We would do well to learn from their mistakes.

Notes

1. M. Hicks, *Programmed Inequality: How Britain Discarded Women Technologists and Lost Its Edge in Computing* (Cambridge, MA: MIT Press, 2017).

2. B. J. Copeland, *Colossus: The Secrets of Bletchley Park's Codebreaking Computers* (Oxford: Oxford University Press, 2006).

3. Hicks, *Programmed Inequality*.

4. Hicks; J. Light, "When Computers Were Women," *Technology and Culture* 40, no. 3 (1999): 455–483; M. L. Shetterly, *Hidden Figures: The American Dream and the Untold Story of the Black Women Mathematicians Who Helped Win the Space Race* (New York: William Morrow, 2016); J. J. Bartik, J. T. Rickman, and K. D. Todd, *Pioneer Programmer: Jean Jennings Bartik and the Computer That Changed the World* (Kirksville, MO: Truman State University Press, 2013).

5. J. Forrester, "Project Whirlwind" (lecture, The Computer History Museum, June 1980), https://www.youtube.com/watch?v=JZLpbhsE72I.

6. Hicks, *Programmed Inequality*.

7. Hicks.

8. Hicks.

9. D. Spicer, "Oral History of Dame Stephanie Shirley," *IEEE Annals History of Computing* 40, no. 1 (January–March 2018): 6–7.

10. S. Shirley, *Let IT Go: The Story of the Entrepreneur Turned Ardent Philanthropist* (Andrews UK, 2012).

11. Hicks, *Programmed Inequality*; Bartik et al., *Pioneer Programmer*; Shirley, *Let IT Go*; J. Abbate, *Recoding Gender: Women's Changing Participation in Computing* (Cambridge, MA: MIT Press, 2012); N. Ensmenger, *The Compute Boys Take Over: Computer Programmers and the Politics of Technical Expertise* (Cambridge, MA: MIT Press, 2010.

12. Hicks, *Programmed Inequality*.

13. Hicks.

14. A. Nelson, T. L. N. Tu, and A. Headlam Hines, *TechniColor: Race Technology and Everyday Life* (New York: New York University Press, 2001); A. Nelson, *The Social Life of DNA: Race Reparations and Reconciliation after the Genome* (Boston: Beacon Press, 2016).

15. Shetterly, *Hidden Figures*.

16. M. Broussard, *Artificial Unintelligence: How Computers Misunderstand the World* (Cambridge, MA: MIT Press, 2018); S. Noble, *Algorithms of Oppression: How Search Engines Reinforce Racism* (New York: New York University Press, 2018); V. Eubanks, *Automating Inequality: How High-Tech Tools Profile Police and Punish the Poor* (New York: St. Martin's Press, 2018).

17. S. Noble, "Databite No. 109: Safiya Umoja Noble-Algorithms of Oppression" (lecture, Data and Society Research Institute, May 2018), https://datasociety.net/events/databite-no-109-safiya-umoja-noble-algorithms-of-oppression.

35 Shaping Technology for the "Good Life": The Technological Imperative versus the Social Imperative

Gary Chapman

In the early twenty-first century, many consider globalization to be an unstoppable force. New technologies (including computers, satellites, and the Internet) are making communications across the earth instantaneous. This facilitates global corporations and integrated economies such that decisions in Tulsa can have profound effects on the lives of farmers in Sri Lanka. Arguably, the intensified interaction may reduce or even eliminate differences in cultures and traditions, including cuisines. Globalization has generated a considerable number of critics who point fingers both at the institutions promoting it and the technologies that make it possible.

Gary Chapman, however, argues that the problem is not technologies but rather the power and influence of what he calls the "technological imperative." He critiques the idea that our primary goals should be to develop technologies and systems that increase efficiency and lower costs. If we make technology our primary goal, then many of the values we hold dear may be lost in the mix. In essence he asserts that if we believe in technological determinism, it will come true. If, however, we choose to privilege social values over technical goals, we will be able to create the world that we want.

Chapman maintains that this vision is not a pipe dream and explores the slow-food movement as a case study in countering the losses that occur with a blind dedication to efficiency and changing the direction of technological development. If, for instance, our primary goal is to reengineer produce so that it grows faster, is cheaper to harvest, and is easier to ship, then local flavors, local economies, and local jobs might be lost. If, on the other hand, we make specific values, people, and institutions our end, rather than technology, we can still employ technology but to local ends. This harkens back to the selection on the Amish, who embody Chapman's suggestions in many ways. Chapman urges us not to abandon technology but to see it as a means to achieve our goals and not the goal itself.

Since the collapse of the Soviet Union and the communist Eastern bloc at the beginning of the 1990s, technology and economic globalization have become the chief determinants of world culture. Indeed, these two omnipresent features of modern, civilized life in the postindustrial world are so intertwined that they may be indistinguishable—we may speak more or less coherently of a "technoglobalist" tide, one rapidly engulfing most of the world today.

A significant point of debate raised by this phenomenon is whether, and to what extent, the rapid spread of technologically based global capitalism is inevitable,

---

From Douglas Schuler and Peter Day, eds., *Shaping the Network Society: The New Role of Civil Society in Cyberspace* (Cambridge, MA: MIT Press, 2004), 43–65.

unstoppable, and even in some vague sense autonomous. There are those who believe that there is in fact a strong "technological imperative" in human history, a kind of "technologic" represented both in macro-phenomena such as the market, and in individual technologies such as semiconductor circuits or bioengineered organisms. It is not hard to find evidence to support such an idea. The prosaic description of this concept would be that technological innovations carry the "seed," so to speak, of further innovations along a trajectory that reveals itself only in hindsight. Moreover, the aggregate of these incremental improvements in technology is an arrow that points forward in time, in a process that appears to be accelerating, piling more and more technologies on top of one another, accumulating over time to build an increasingly uniform and adaptive global civilization. There often appears to be no escape from this process. As the allegorical science fiction villains of the TV series *Star Trek*, the Borg, say in their robotic, repetitive mantra, "Resistance is futile. You will be assimilated." Stewart Brand, the *Whole Earth Catalog* guru and author who turned into a high-tech evangelist, put it this way: "Once a new technology rolls over you, if you're not part of the steamroller, you're part of the road" (Brand 1987, 22).

The growing and vocal antiglobalization movement, on the other hand, is questioning such assumptions and challenging the idea of a necessary link between a "technological imperative" and human progress. The protesters at antiglobalization demonstrations typically represent a wide range of both grievances and desires, so it is difficult to neatly characterize a movement that is repeatedly drawing hundreds of thousands of protesters to each major demonstration. But the one principle that seems to unite them—along with many sympathizers who choose not to participate in public demonstrations—is that the future is not foreordained by a "technological imperative" expressed via global corporate capitalism. There is the hope, at least, among antiglobalization activists that human society might continue to represent a good deal of diversity, including, perhaps especially, diversity in the way people adopt, use, and refine technology.

Who wins this debate, if there is a winner, will be at the heart of "shaping the network society," the theme of this book. The outcome of this ongoing debate may determine whether there are ways to "shape" a global technological epoch at all. If the concept of a technological imperative wins out, over all obstacles, then human beings are essentially along for the ride, whether the end point is utopia or apocalypse or something in between. Human consciousness itself may even be shaped by surrender to the technological imperative.

If, on the other hand, technology can be shaped by human desires and intentions, then the critics of the "autonomous technology" idea must either explain or discover how technologies can be steered deliberately one way and not another. Neither side in this debate has a monopoly on either virtue or vice, of course. Letting technology unfold without detailed social control is likely to bring many benefits, both anticipated and unanticipated. Setting explicit goals for technological development could, on the other hand, help us avoid pitfalls or even some catastrophes. It will be the search for balance that is likely to characterize our global discussion for the foreseeable future, a balancing of the "technological imperative" with what might be called our "social imperative."

The Technological Imperative Full Blown—Moore's Law and Its Distortions

In 1965, Gordon Moore, the cofounder of the Intel Corporation and a pioneer in semiconductor electronics, publicly predicted that microprocessor computing power would double every eighteen months for the foreseeable future. He actually predicted that the number of transistors per integrated circuit would double every eighteen months, and he forecast that this trend would continue through 1975, only ten years into the future. But in fact this prediction has turned out to be remarkably accurate even until today, more than thirty five years later. The Intel Corporation itself maintains a chart of its own microprocessor transistor counts, which have increased from 2,250 in 1971 (the Intel 4004) to 42 million in 2000 (the Pentium 4) (http://www.intel.com/technology/mooreslaw/index.htm). The technology industry is nearly always preoccupied with when Moore's law might come to an end, especially as we near the physical limits of moving electrons in a semiconductor circuit. But there always seems to be some promising new technology in development that will keep Moore's law alive.

Moore's prediction has so amazed technologists, because of its accuracy and its longevity, that it has become something close to a natural law, as in Newtonian physics. Of course, there is bound to be an end to the trend that makes the prediction accurate, which means that Moore's law will someday become a historical curiosity rather than a "law." But the prediction has taken on a life of its own anyway. It is no longer regarded as simply a prediction that happened to be fulfilled by a company owned and controlled by the predictor, a company that spent billions of dollars to make sure the prediction came true and profited immensely when it did come true. Moore's law is regarded by some technophiles as "proof" that computers and computer software will increase their power and capabilities forever, and some computer scientists use the thirty five-year accuracy of Moore's law as evidence that computers will eventually be as "smart" as human beings and perhaps even "smarter" (Kurzweil 1999; Moravec 1998).

There is no connection between transistor density on a semiconductor chip and whether or not a computer can compete with a human being in terms of "intelligence." It is not even clear what constitutes intelligence in a human being, let alone whether or not a computer might match or surpass it. Intelligence is an exceedingly vague term, steeped in controversy and dispute among experts. Computers are far better at some tasks than humans, typically tasks that involve staggering amounts of repetitive computation. But human beings are far better at many ordinary "human" tasks than computers—indeed, an infant human has more "common sense" than a supercomputer with billions of transistors. There is no evidence that human beings "compute," the way a computer processes binary information, in order to cogitate or think. Nor is there evidence that the von Neumann computer model of serial bit processing is even a simulacrum for human information processing.

This is not to say that Moore's law is insignificant or irrelevant—the advances in computer processing power over the past thirty five years have been astonishing and vitally important. And the fact that Gordon Moore was prescient enough to predict the increase in a way that has turned out to be amazingly accurate is fascinating and impressive. But as others have pointed out, the semiconductor industry has spent billions of dollars to make sure that Moore's prediction came true, and it is worth mentioning that Moore's own company, Intel, has led the industry for all of the thirty five years Moore's law has been tested. Nevertheless, if the prediction had failed we would not be talking about it the way we do now, as a cornerstone of the computer age, and for this Moore deserves credit. But there is nothing about Moore's law that points to the kinds of future scenarios that some authors and pundits and even engineers have attributed to it. There is nothing about Moore's law that makes it a true "law," nor is there any imperative that it be accurate indefinitely, except the industry's interests in increasing transistor density in order to sell successive generations of computer chips.

In any event, there are trends now that suggest that this measure of progress, of increasing chip density, is gradually losing its significance (Markoff and Lohr 2002).

Moore's law is an example of how a thoughtful and interesting prediction has been turned into an argument for the technological imperative, that society must invest whatever it takes to improve a technology at its maximally feasible rate of improvement—and to invest in a specific technology, perhaps at the expense of a more balanced and generally beneficial mix of other technologies. Advocates of the semiconductor industry argue that semiconductor chips are the "seed corn" of the postindustrial economy, because their utility is so universal and significant to productivity. But so is renewable energy, or human learning, or sustainable agriculture, all things that have experienced a weakness in investment and attention, at least in comparison to semiconductors and computer hardware.

In April 2000, Bill Joy, vice president and cofounder of Sun Microsystems—a very large computer and software company in Silicon Valley—published a provocative essay in *Wired* magazine titled "Why the Future Doesn't Need Us" (Joy 2000). Joy raised some troubling questions for scientific and technological researchers, about whether we are busy building technologies that will make human beings redundant or inferior beings. Joy's article created a remarkable wave of public debate—there were public discussions about his thesis at Stanford University and the University of Washington in Seattle, he was invited to present his ideas before the National Academy of Sciences, and he appeared on National Public Radio. His warnings were featured in many other magazine articles.

Joy refers to the arguments of two other well-known technologists, Ray Kurzweil and Hans Moravec, who have both written extensively about how computers will one day become "smarter" than human beings. Humans will either evolve in a way that competes with machines, such as through machine implants in the human body, or else "disappear" by transferring their consciousness to machine receptacles, according to Kurzweil and Moravec. In his *Wired* article, Joy accepts this as a technological possibility, perhaps even an inevitability, unless we intervene and change course. "A technological approach to Eternity—near immortality through robotics," writes Joy, "may not be the most desirable utopia, and its pursuit brings clear dangers. Maybe we should rethink our utopian choices" (Joy 2000).

But the question then becomes, Can we rethink our utopian choices if we believe rather thoroughly in a technological imperative that propels us inexorably in a particular direction? Can Moore's law coexist with a free ethical choice for technological ends? Joy says that perhaps he may reach a point where he might have to stop working on his favorite problems. "I have always believed that making software more reliable, given its many uses, will make the world a safer and better place," he notes, adding that "if I were to come to believe the opposite, then I would be morally obligated to stop this work. I can now imagine such a day may come." He sees progress as "bitter sweet": "This all leaves me not angry but at least a bit melancholic. Henceforth, for me, progress will be somewhat bittersweet" (Joy 2000).

There is a great deal to admire in such emotions; these are the musings of a thoughtful and concerned person, and, given Joy's stature and reputation in his field, such qualities are welcome precisely because they seem so rare among technologists.

But behind such ideas is an unquestioned faith in the technological imperative. Joy suggests that unless we simply stop our research dead in its tracks, on ethical grounds, we may create dangers for which we will be eternally guilty. This is certainly possible, but it is distinctly one dimensional. Could we not redirect our technological aims to serve *other* goals, goals that would help create a life worth living rather than a life shadowed by guilt and dread? Joy begins to sound like Theodore Kaczynski, the "Unabomber," whom he quotes with some interest and intrigue (as does Kurzweil). Kaczynski also believed the technological imperative is leading us to our doom, hence his radical Luddite prescriptions and his lifestyle, not to mention his deadly attacks on technologists for which he was eventually sent to jail for life. Kurzweil, Joy, and Kaczynski all portray technology as an all-encompassing, universal system—Joy repeatedly uses the phrase "complex system"—that envelops all human existence. It is the "totalizing" nature of such a system that raises troubling ethical problems for Joy, dark fantasies of doom for Kaczynski, and dreams of eternity, immortality, and transcendence for Kurzweil. But might there not be another way to adapt technology to human-scale needs and interests?

The Slow-Food Movement in Italy

The Italian cultural movement known as "slow food"—not a translation, it is called this in Italy—was launched by Roman food critic and gourmand Carlo Petrini in 1986, just after a McDonald's hamburger restaurant opened in Rome's magnificently beautiful Piazza di Spagna (Slow Food 2002). Petrini hoped to start a movement that would help people "rediscover the richness and aromas of local cuisines to fight the standardization of Fast Food." Slow food spread very rapidly across Italy. It is now headquartered, as a social movement, in the northern Piedmont city of Bra.

Today there are 65,000 official members of the slow food movement, almost all in Western Europe—35,000 of these members are in Italy. These people are organized into local groups with the wonderfully appropriate name *convivia*. There are 560 *convivia* worldwide, 340 of them in Italy (Slow Food 2002). Restaurants that serve as "evangelists" of the slow food movement display the group's logo, a cartoon snail, on the front door or window. If one sees this logo displayed by a restaurant, one is almost guaranteed to enjoy a memorable dining experience.

Slow food is not just about sustaining the southern European customs of three-hour lunches and dinners that last late into the night, although that certainly is part of the message. Petrini very cleverly introduced the idea of a "Noah's ark" of food preservation, meaning a concerted effort to preserve nearly extinct natural foodstuffs, recipes, and, most of all, the old techniques of preparing handmade foods. From this idea, the slow food movement has broadly linked gastronomy, ecology, history, and economics into a benign but powerful ideology that nearly every southern European citizen can understand. Slow food, in addition to being an obvious countermovement to American "fast food," has developed into a movement that promotes organic farming and responsible animal husbandry, community-based skills for the preservation of regional cuisines, and celebrations of convivial, ceremonial activities such as food festivals and ecotourism. The Slow Food organization has even sponsored a film festival, featuring films with prominent scenes about food, and plans to award an annual "Golden Snail" trophy, the slow food equivalent of an Oscar.

Slow food is thus one of the more interesting and well-developed critiques of several facets of globalization and modern technology. Specifically, it is a response to the spread of globally standardized and technology-intensive corporate agriculture, genetic engineering, high-tech food preparation and distribution, and the quintessentially American lifestyle that makes "fast food" popular and, for some, even imperative. Italian proponents of slow food and their allies in other countries—*convivia* are now found in other European countries as well as in the United States—view this as a struggle for the soul of life and for the preservation of life's most basic pleasures amidst a global trend pointing to increased competition, consumerism, stress, and "hurriedness":

> We are enslaved by speed and have all succumbed to the same insidious virus: Fast Life, which disrupts our habits, pervades the privacy of our homes and forces us to eat Fast Foods. . . .
>
> Many suitable doses of guaranteed sensual pleasure and slow, long-lasting enjoyment preserve us from the contagion of the multitude who mistake frenzy for efficiency. (Slow Food 2002)

In 1999, the slow food movement spun off a new variation of itself: the slow cities movement. In October of that year, in Orvieto, Italy, a League of Slow Cities was formed, a charter was adopted, and the first members of this league elected as their "coordinator" Signor Paolo Saturnini, the mayor of the town of Greve in Chianti (Città Slow 2002).

The Charter of Association for the slow cities movement (which in Italian does have the Italian name *città slow*) has a rather sophisticated and subtle view of globalization.

The development of local communities is based, among other things (*sic*), on their ability to share and acknowledge specific qualities, to create an identity of their own that is visible outside and profoundly felt inside.

The phenomenon of globalization offers, among other things, a great opportunity for exchange and diffusion, but it does tend to level out differences and conceal the peculiar characteristics of single realities. In short, it proposes median models which belong to no one and inevitably generate mediocrity.

Nonetheless, a burgeoning new demand exists for alternative solutions which tend to pursue and disseminate excellence, seen not necessarily as an elite phenomenon, but rather as a cultural, hence universal fact of life (Slow Food 2002).

Thus, slow cities are those that

- Implement an environmental policy designed to maintain and develop the characteristics of their surrounding area and urban fabric, placing the onus on recovery and re-use techniques
- Implement an infrastructural policy that is functional for the improvement, not the occupation, of the land
- Promote the use of technologies to improve the quality of the environment and the urban fabric
- Encourage the production and use of foodstuffs produced using natural, ecocompatible techniques, excluding transgenic products, and setting up, where necessary, presidia to safeguard and develop typical products currently in difficulty, in close collaboration with the Slow Food Ark project and wine and food Presidia
- Safeguard autochthonous production, rooted in culture and tradition, which contributes to the typification of an area, maintaining its modes and mores and promoting preferential occasions and spaces for direct contacts between consumers and quality producers and purveyors
- Promote the quality of hospitality as a real bond with the local community and its specific features, removing the physical and cultural obstacles that may jeopardize the complete, widespread use of a city's resources
- Promote awareness among all citizens, and not only among inside operators, that they live in a Slow City, with special attention to the world of young people and schools through the systematic introduction of taste education. (Slow Food 2002).

Slow food and slow cities are thus *not* neo-Luddite movements; both acknowledge the importance of technology, and they specifically mention the benefits of communications technologies that allow a global sharing of ideas. But both movements are committed to applying technology for specific purposes derived from the values of the "slow" movement as a whole: leisure, taste, ecological harmony, the preservation and enhancement of skills and local identities, and ongoing "taste education." The subtleties of this worldview are typically lost on most Americans, who have no idea where their food comes from, nor would they care if they did know (this applies to most, but not all, Americans, of course). Many Italians and French are concerned that this attitude might spread to their countries and wipe out centuries of refinement in local cuisines, culinary skills, agricultural specialties, and other forms of highly specific cultural identity. As the Slow Cities Charter of Association straightforwardly asserts, "universal" culture typically means mediocre culture, with refinement and excellence reserved in a special category for people with abundant wealth. The "slow" alternative is to make excellence, identity, and "luxury" available "as a cultural, and hence universal fact of life," something not reserved for elites but embedded in daily life for every-one. This will require "taste education," something meant to offset the damage of mass marketing and the advertising of mass products.

Slow food might be dismissed as a fad among the bourgeoisie, a cult of hedonists and effete epicureans. It is certainly a phenomenon of the middle class in southern Europe, but it may serve as a kind of ideological bridge between more radical antiglobalization activists and older, more moderate globalization skeptics. The slow food movement is not thoroughly opposed to globalization, in fact. It is concerned with the negative effects of globalization and technology, which have prompted the movement's appearance and its eloquence—with respect to the leveling of taste, the accelerating pace of life, and the disenchantment with some of life's basic pleasures, such as cooking and eating.

In this way, the "slow" movement is an intriguing and perhaps potent critique of modern technology, which is otherwise widely viewed as propelling us toward the very things the "slow" movement opposes. In the United States, for example, bioengineering is typically regarded as inevitable, or even promoted as the "next big thing" for the high-tech economy. In Europe, by contrast, genetically engineered crops and foodstuffs are very unpopular, or, at best, greeted by deep skepticism, even among apolitical consumers. In the United States, there is widespread resignation in the face of the gradually merging uniformity of urban and especially suburban spaces. Shopping malls all look alike, and even feature the same stores; many Americans look for familiar "brand" restaurants and attractions such as those associated with Disney or chains like Planet Hollywood, indistinguishable no matter where they are found; a "successful" community is one where the labor market is so identical to other successful communities that skilled workers can live where they choose; suburban tracts of new homes are increasingly impossible to tell apart. Critics of this universal trend in the United States bemoan the appearance of "Anywhere, U.S.A.," typified by the suburban communities of the West, the Southwest, and the Southeast. These communities are often characterized by a sort of pseudo-excellence in technology—they are often the sites of high-tech industries—but mediocrity in most other amenities of life. Such trends are not unknown in Western Europe—Europeans are starting to worry that these kinds of communities are becoming more and more common there too—but there is at least a vocal and sophisticated opposition in the slow food movement and its various fellow travelers.

At bottom, the slow food and slow cities movements are about the dimensions of what Italians call *il buon vivere*—the good life. For Italians and many other Europeans (as well as many Americans and people in other countries, of course), the "good life" is one characterized by basic pleasures like good food and drink, convivial company and plenty of time devoid of stress, dull work, or frenzy. The advocates of slow food see the encroaching American lifestyle as corrosive to all these pleasures. The American preference is for convenience and technology that replaces many time-intensive activities, rather than for quality and hard-won skills. Plus, the American style is one of mass production aimed at appealing to as many people (or customers) as possible, leaving high-quality, customized, and individualized products and services reserved for the wealthy. In the United States, this trend has now taken over architecture, music, films, books, and many other things that were once the main forms of artistic and aesthetic expression in a civilized society. Many fear this American trend toward mass appeal, mediocrity, and profit at the expense of quality, excellence, and unique identity is now taking over the Internet as well.

It must be mentioned that the orientation toward local communities, the preservation of skills, skepticism about globalization and technology, and so on is not without its own pitfalls. At the extreme ends of this perspective are dangerous and noxious political cauldrons, either of nationalism and fascism on the one hand or neo-Luddite, left-wing anarchy on the other. There are already discussions in Italy about the similarities in antiglobalization sentiments shared by the extreme right and the extreme left, which in Italy represent true extremes of neofascism and full-blown anarchism. At the same time that these two forces battled each other in the immense antiglobalization protests in Genoa in the summer of 2001—when neofascists on the police force allegedly beat protesters to their knees and then forced them to shout "Viva II Ducel," a signal of the fashionable rehabilitation of Mussolini among the Italian right wing—intellectuals on the right of Italian politics were musing in newspapers about how much of the antiglobalization rhetoric of the protesters matched positions of the neofascist parties.

The slow food movement has nothing to do with this dispute—the movement is capable of encompassing the entire political spectrum, except perhaps the extreme left wing. Nevertheless, there is the possibility that an emphasis on local and historically specific cultural heritage, even limited to cuisines, could become a facade for anti-immigrant or even xenophobic political opinions, which are discouragingly common in European politics today. Moreover, there is a long history of skepticism and outright opposition to technology among right-wing extremists, who often try to protect nationalist and conservative traditions threatened by technologies such as media, telecommunications, the Internet, and reproductive technologies, to name a few.

So far, it appears that the slow food and slow cities movements are admirably free of such pathologies, at least in their public pronouncements and in the character of their concerns, such as environmental, agricultural, and water quality in developing nations. Both movements argue that technology has a place in their worldview, especially as a tool to share ideas globally, and as a means of protecting the natural environment. Far from being tarred with the neofascist surge in Italy, or with the anarchist tactics of French activist Jose Bove—who has smashed a McDonald's in Provence and burned seeds of genetically altered grain in Brazil—slow food and slow cities have been linked to the concept of "neohumanism," a phrase used by Italian poet Salvatore Quasimodo in his acceptance speech when he received the Nobel Prize for Literature in 1959. "Culture," said Quasimodo, "has always repulsed the recurrent threat of barbarism, even when the latter was heavily armed and seething with confused ideologies" (Quasimodo 1959).

This seems to me the idea of culture that the slow food and slow cities movements represent, using cuisine and urban planning as vehicles, like Quasimodo's poetry, to both repulse barbarism and strengthen *il buon vivere*. It seems no mere accident that slow food has been born in the same country that gave us humanism and the greatest outpouring of art and aesthetic beauty the world has ever seen, roughly 500 years ago. Now, in an age of high technology, "neohumanism" seems like our best course for the future. . . .

Tying It All Together

All of the elements described above require *action*, both individual and collective, as opposed to the passive behavior that allows an (alleged) technological imperative to take its course. These elements require understanding and consideration, and in some cases active seeking of information, connections, and ideas, all activities that go against the grain of current trends on the Internet. A basic premise is that technological development is not in fact inevitable, or that a particular outcome of a technology's trajectory is not foreordained. Making appropriate choices thus becomes not only possible but of paramount importance, and it helps a lot if multiple choices work together to enhance and promote values worth defending.

Slow food, for example, has admirably linked culinary enjoyment to environmental protection and responsibility as well as to a critique of economic forces that push people into a harried life. The Slow Cities League has extended this idea to the cultural identity of towns, cities, or regions, and then in turn to the desire for excellence and distinctiveness among individuals. The Open Source and free-software movement is clearly linked to grassroots struggles over intellectual property laws and to the challenge of balancing hard-won skills and ease of use. So far, technology activists attracted to the Open Source and free-software movement have had less sympathy with addressing the digital divide, although there are many activists who have made this bridge. This is simply an area where more work needs to be done.

It is obvious that there is not yet a completely realized "consciousness"—for lack of a better word—about the new paradigm described above, a way of thinking about the world that is not only opposed to the negative, dehumanizing aspects of globalization but that recognizes globalization's positive features and incorporates them into a worldview that both embraces technology and shapes it to different ends. Many young antiglobalization activists are—unfortunately—opposed to all aspects of globalization, and some are even opposed to modern technological development in general (even though antiglobalization organizations are very effective users of the Internet, by and large). Some young anarchists mistakenly want to preserve cultures and forms of social organization that are guaranteed to be stuck in poverty and ignorance indefinitely. Among the most visible antiglobalization activists—for example, those who are willing to resort to violence and overt confrontation with the police—there is hardly more intellectual content than an inchoate rage or the thrill of rebellion, which are useless for building alternative visions of good living.

The antiglobalization movement, on the other hand, has tapped into some powerful feelings, and not only among young rebels. Thoughtful people who simply see a world with intensified forms of current trends—including industry concentration and gigantism; a magnetic pull toward blandness, sameness, and mediocrity; a life chock-full of advertisements; and the replacement of authentic and free culture with ersatz and "pay-per" "experiences"—have reasons to be both alarmed and angry. People whose Internet experience stretches back to the years before the dot-com frenzy of the 1990s are often appalled at what the Internet has become. There is a sense that when commerce touches anything these days, it either dies or is transformed into something artificial and alien, a poor substitute for the "real." This may continue indefinitely, but not without resistance and friction.

Conclusion: A New Bipolarity?

During the decades of the Cold War, the "intellectual frame" of the world was organized around a competitive bipolarity between the capitalist West and the communist East, with other parts of the world judged on their orientation to these two poles. For historical and political reasons, the bipolarity of the Cold War was both ideological and geographic, symbolized by the phrase—and the reality of—the "Iron Curtain." This experience predisposed many of us to think of cultural and political bipolarities in geopolitical terms.

Globalization has not only ended this way of thinking about the world, but it has introduced the potential, at least, of thinking about competing worldviews that have no geographic "map"—that is, there are different social movements representing different goals for culture and history that transcend nation-states or regions. Since the tragedy of September 11 there have been speculations about a colossal and terrifying "clash of civilizations" between the secular West and the Muslim crescent that stretches from the west coast of North Africa to Indonesia. Such scenarios typically leave out the fact that there are millions of Muslims in Western countries, largely an epiphenomenon of globalization. The September 11 terrorists themselves spent years outside of Islamic countries. The old framework of geopolitical rivalries representing alternative views of how society should be organized and governed is no longer as relevant as it once was. People with any particular political orientation that claims to be relevant to the entire world can cheaply and easily find allies and surrogates around the globe by using the Internet and other communications technologies. The impact and significance of "memes"[1]—or germs of cultural ideas—have been increasing in a dramatic fashion because of these developments. Even the Zapatistas, the political movement of the impoverished and largely illiterate population of southern Mexico, have used the Internet in such effective ways that their conflict in Mexico has been described as a form of "information warfare" (Ronfeldt and Martinez 1997). The use of the Internet by antiglobalization protesters around the world has become a key component of their success in organizing massive and repeated demonstrations at the meetings of world leaders (Tanner 2001).

But it is the slowly emerging threads of a different vision of how to live life—and how to think about technology—that are most interesting, especially as they begin to interweave and create a positive, attractive, and feasible alternative to large-scale corporate organization and mass consumerism. Slow food and slow cities are two examples, and the Open Source and free-software movement is another. The growing opinion that biotechnology is a distraction from addressing the world's crisis in biodiversity is yet another example. The tumultuous politics in the Middle East are supplying new urgency to questions about postindustrial societies' dependence on fossil fuels for energy. There are other examples. So far, these "alternative" views, or challenges to the status quo, have yet to cohere into a complete ideological framework that might appeal to large numbers of people, enough people to transform political agendas. But this could happen. The antiglobalization protests are large enough and frequent enough—particularly in Western Europe—that they may begin to have sweeping effects on political discourse and elections. In the United States, there is growing evidence of a widening gulf between the two main political parties. The activists of each party, the Democratic Party and the Republican Party, are increasingly polarized; even the leadership of the Democratic Party is often viewed by its own supporters as too timid and compromised by deals with wealthy donors and corporations. Moreover, the two ends of the political spectrum are coming to represent not just political choices but entirely separate "lifestyles," or two distinct cultures, as intellectuals on both the right and the left now admit.

The familiar strategy of reigning in capitalism when its excesses need to be checked, typically through state institutions with regulatory powers, is losing its appeal among many people who are looking for completely new ways of reconfiguring economic sectors such as agriculture, communications, entertainment, and health care. We seem to be in the process of creating two cultures that will coexist but with constant friction, especially because of the economic fragility of small-scale enterprises in comparison to their more powerful global counterparts. If French farmers are put out of business because of world-trade agreements and the French lose their cherished traditions of regional culinary specialties, a lot of French people will be looking for ways to fight back. If the Internet turns completely into a pipeline for junk e-mail, pop-up web ads, online scams, garish "infotainment" sites, and pornography, with little material of quality and excellence as compensation, there will be a backlash. If all these things happen at more or less the same time—say, within the next twenty years—and enough people see this trend as folly and tragedy, a new, nongeographic bipolarity, may develop, one between people who care about such things and those who do not, or between people who have one vision of the good life and a competing group who see their vision as being fulfilled by contemporary capitalism. This is already happening in one form or another in most modern countries. Global capitalist culture is already uniting the interests of billions of consumers. What we are only starting to see is the merging of interests of people who have reservations about that particular fate for the world. Pieces of an alternative global culture are beginning to appear, but are not yet meshed into a coherent picture.

The world has many problems and not all of them will be solved by "alternative" ways of thinking. We are likely to see more wars, more anguish among entire peoples, perhaps even more terrorism, and there will be responses to all of these things that will shape history in important ways. There has certainly been a dramatic reinforcement of conventional security institutions in the United States since September 11, 2001, for example. The big global problems like poverty, illness, and violence will still need to be addressed by institutions with consensual powers of authority.

Nor will we see complete harmony and a convergence of ideas and goals among people who deliberately oppose the status quo. So many degrees of difference among such people may persist that they will continue to look like a chaotic, cacophonic mob rather than a historic force of change. It may only be over the course of many years that we come to recognize an emergent, new way of thinking, which is likely to take different but related forms all over the world. The common thread that may unite many disparate but like-minded efforts, from food activists to digital rights activists, is thinking about technology as malleable, as capable of serving human-determined ends, and as an essential component of *il buon vivere*, the good life. It is only by working with that premise that the idea of shaping the network society makes sense.

Note

1. The word "meme" is defined by *The Merriam-Webster Dictionary* as "an idea, behaviour, style, or usage that spreads from person to person within a culture."

References

Brand, S. 1987. *The Media Lab: Inventing the Future at MIT*. New York: Viking Press.

Città Slow (Slow Cities website). 2002. Accessed September 12, 2007. http://www.cittaslow.org.

Joy, B. 2000. "Why the Future Doesn't Need Us." *Wired* 8.04 (April): 238–262. Reprinted as chap. 5 in this volume.

Kurzweil, R. 1999. *The Age of Spiritual Machines: When Computers Exceed Human Intelligence*. New York: Viking Press.

Markoff, J., and Lohr, S. 2002. "Intel's Big Bet Turns Iffy." *New York Times*, September 29, 2002. Accessed September 12, 2007. http://www.nytimes.com/2002/09/29/technology/circuits/29CHIP.html.

Moravec, H. 1998. *Robot: Mere Machine to Transcendent Mind*. Oxford: Oxford University Press.

Quasimodo, S. 1959. "Salvatore Quasimodo Banquet Speech." The Nobel Prize. Accessed September 12, 2002. http://nobelprize.org/nobel_prizes/literature/laureates/1959/quasimodo-speech.html.

Ronfeldt, D., and Martinez, A. 1997. "A Comment on the Zapatista 'Netwar.'" In *In Athena's Camp: Preparing for Conflict in the Information Age*, edited by J. Arquilla and D. Ronfeldt, 369–391. Santa Monica, CA: RAND.

Slow Food. 2002. Accessed September 12, 2007. http://www.slowfood.com. See also the website for Slow Food USA, http://www.slowfoodusa.org/, accessed September 12, 2007.

Tanner, A. 2001. "Activists Embrace Web in Anti-globalization Drive." Reuters, July 13, 2001. Accessed September 12, 2007. http://www.globalexchange.org/economy/rulemakers/reuters071301.html (page no longer available).

36 Not Just One Future

David E. Nye

We end this book with the reflections of David Nye, a historian of technology. Drawing on historical cases and contemporary themes, Nye provides a broad understanding of the connection between people and machines. He doesn't just reject technological determinism; he shows how deeply humans are enmeshed with technology. People and machines exist in a symbiotic relationship. Moreover, he suggests that technology is profoundly about the future—that is, humans invent, modify, use, and adapt to technologies in anticipation of what is to come in their lives. Thus, technology matters to the future not just because it is one factor shaping it; rather, humans develop and relate to technology in order to make their futures.

To prepare for the future, it is not enough to simply recognize that technology is a powerful force shaping the world that humans inhabit. One has to embrace the reality that the future will be built by humans as we make decisions, and our decisions will influence and be influenced by the technologies that are intimately intertwined with our lives. We will make choices as consumers, as citizens, as users, as producers, and as constituents of an array of organizations and institutions. Our decisions will not be determined entirely by available technologies since we create what is available and modify it by using technologies in unanticipated ways, assigning unpredicted meanings, and, ultimately, producing newer technologies.

In his reflections, Nye meanders through a wide range of themes and topics, including many that were taken up in earlier sections and readings, and he suggests that we should not think about the future in a singular manner. Humans consider and take up technologies in a variety of ways, so we "are not necessarily evolving towards a single culture." The future will not be homogeneous. There is room for all of us to make choices to create the futures we want.

Several of the most widely read novelists of the first half of the twentieth century evoked the terror of living in a society where technologies became the basis of massive state control. These included the dystopias of H. G. Wells (*When the Sleeper Wakes*), Aldous Huxley (*Brave New World*), and George Orwell (*1984*).[1] Reading their work is another reminder that technologies matter, whether one is looking at human evolution, at cultural diversity, at employment, at government, at consumption, at the environment, at safety, at the military, or at the overarching social construction of reality.

Reading such works also raises the question "Where does the human infatuation with technology lead?" Even the simplest tool implies a narrative that points toward the future. The chimpanzee with his peeled stick looking for termites has begun to think of a sequence in time, as hunger prompts him to make a tool with which to find,

---

Chapter 11 from *Technology Matters* (Cambridge, MA: MIT Press, 2006).

catch, and eat termites. The prehistoric toolmaker patiently chipping out arrows or axe-heads conceived of a generalized future need and prepared against the day when tools should be at hand. Complex technologies demanded more resources and lengthened the gap between making and possible use. In some cases, the future benefit is obvious—for example, digging a well, throwing a bridge across a river, or building an aqueduct into a city. Yet the largest projects often lacked such practical justifications. The builders of Stonehenge, the Pyramids, or the medieval cathedrals could dedicate a lifetime to a great technological project for its religious meanings. To mobilize such energies required a sense of order that placed such buildings at the center of a whole scheme of life. Technology is not something that comes from outside us; it is not new; it is a fundamental human expression. It cannot easily be separated from social evolution, for the use of tools stretches back millennia, long before the invention of writing. It is hard to imagine a culture that is pre-technological or a future that is post-technological.

From the vantage point of the present, it may seem that technologies are deterministic. But this view is incorrect, no matter how plausible it may seem. Cultures select and shape technologies, not the other way around, and some societies have rejected or ignored even the gun or the wheel. For millennia, technology has been an essential part of the framework for imagining and moving into the future, but the specific technologies chosen have varied. As the variety of human cultures attests, there have always been multiple possibilities, and there seems no reason to accept a single vision of the future. Those who think machines are taking humanity somewhere in particular probably are wrong. No determinism made the automobile an inevitable choice instead of mass transit. Nothing inherent in technologies dictates that people should live in apartment buildings, semidetached dwellings, or single-family houses. Nothing makes e-mail an inevitable cultural choice instead of the telephone, the old-fashioned post, or the Amish preference for face-to-face communication. Rather, each group of people selects a repertoire of techniques and devices to construct its world. A more useful concept than determinism is technological momentum, which acknowledges that once a system such as a railroad or an electrical grid has been designed to certain specifications and put in place it has a rigidity and direction that can seem deterministic to those who use them.

Neither the technologies of the future nor their social uses are predictable. Although there is a great deal of speculation in this area (including stock market speculation), even experts with impeccable credentials are often wrong. Predictions about such fundamentally new inventions as the telegraph, the electric light, the telephone, and the personal computer were little more accurate than flipping a coin. More surprising, even modifications of existing inventions are hard to predict. Many people, notably the inventors and investors who have a stake in the outcome, propose dramatic future scenarios in which their particular device will become indispensable for the average person, so that no home should be without one. This proved true for the light bulb and the radio, but not for the picture phone or a myriad other devices.

Historians, who know more of the story and who do not have a financial stake in the outcome, are more likely to get it right. Traditionally, they are divided into Internalists and Contextualists.[2] Internalists focus on how machines come to be. Fifty years ago they celebrated individual inventors, gazing over their shoulders in admiration. However, further research has shown that inventors draw on networks of people. Far from acting alone, they coordinate and synthesize. Each new machine emerges from and is shaped by the time, the community, and the place of its making. Internalists do not treat a technology as a "black box" whose inner workings can be taken on faith. They delight in opening up the machine to scrutiny, studying precisely how it worked and what new problems it presented. New machines are not simply products of laboratories, however. They emerge within shaping political and social contexts, and they can be used for many different ends. Contextualists focus on how new machines are incorporated into society. They reject the idea that the public desires unknown machines ahead of time and somehow knows what to do with them when they are invented. Contextualists seek to understand the perspectives of people in the past, most of whom were not engineers or inventors. Any new device had to prove itself in everyday life, and in many cases (notably those of the telegraph and the phonograph) an invention first seemed a novelty with few obvious practical uses. It took railroad managers a generation to see the benefits of the telegraph lines that ran along their own right-of-way. It required a generation for the public to find that it wanted the phonograph not primarily to dictate letters or preserve voices for posterity but to play music. Because historians of technology are familiar with many similar cases, almost none agree with externalists (such as Alvin Toffler) who treat machines as powerful and well-functioning "black boxes" that irresistibly transform the world around them. Such approaches generally are misleading.

Since technologies are not deterministic, it follows that people can use them for many ends. For much of the nineteenth and the twentieth century, sociologists and historians assumed that the machine age could only lead to a crushing homogeneity. But in practice, people have often used technologies to create differences. Consumers generally prefer variety. Even a manufacturer bent on absolute uniformity, such as Henry Ford, eventually had to give in to the public's demand for a range of models and options. Likewise, homeowners proved adept at transforming Levittown's rows of identical, mass-produced homes into variegated neighborhoods. Difference triumphed over uniformity.

By the end of the twentieth century, it was clear that highly technological societies prefer to maximize differentiation. Racial, regional, and ethnic communities developed separate identities by inventing and disseminating new traditions. In contact zones between highly technological societies and developing nations, a creolization process took place as many peoples selected and rearranged elements of Western culture and absorbed them into their own traditions. Technologies were not simply being used to eradicate cultural differences and create a single, global culture. In many cases they enhanced the possibilities for the survival and growth of marginal communities or minority cultures. This pattern emerges even in the urban geography of the United States. The spatial formation of Los Angeles, as analyzed by Mike Davis in *City of Quartz*, is highly suggestive.[3] Wealthy gated communities define a complex of global values in their cable network televisions, broadband Internet connections, and new automobiles. Beyond their gates are ethnic communities and poor neighborhoods, which are hardly the passive recipients of whatever trickles down from the globalized world of the movie studios and the international corporations. Rather, these local groupings are highly inventive, creating new clothing styles, musical forms, and other cultural products that often are reappropriated into "global" culture. Using mobile phones, they cruise and control much of the city. This dynamic between what John Fiske calls an imperializing power and the localizing powers of the margins energizes much of popular culture, moving it away from homogeneity toward differentiation.[4]

The expansion of multicultural consumption across the globe puts pressure on natural resources. Although extraction and production techniques have become more efficient, continual abundance is not ensured. Technological improvement does not automatically lead to long-term economic growth. People can choose wasteful methods that bring high agricultural yields but hasten soil erosion. They can build power plants and factories that produce inexpensive goods but pollute the air, leading to acid rain and deforestation. They can choose to recycle metals or not, to use powerful pesticides such as DDT or not, to allow genetically modified plants into their environment or not. Overall, each culture chooses how large a "footprint" it will leave on the land and whether it will live within limits set by its environment or treat nature as a stock of raw materials. The philosopher Holmes Rolston speculated that "in the increasingly technologically sophisticated world of the future" nature will no longer constitute the framework of life. Instead, "nature will become not so much redundant as increasingly plastic."[5] Yet treating nature in this way may have limits.

The OECD nations use many more resources than the rest of the world. In 1990, a North American family of four consumed as much power as an African village of 107. A North American used twice as much energy as a European and ten times as much as a Latin American.[6] For optimists who believe in more growth, such disparities point to new markets. But if, as seems likely, there are environmental limits to growth, then the Western world, particularly the United States, must reduce per-capita resource consumption. To judge by the political parties' programs, however, most people refuse to believe that there are imminent limits to growth and do not fear global warming. Voters support leaders who stimulate the economy, and they reject restrictions on technological innovation. In the 2000 presidential campaign, George W. Bush denied the existence of global warming. As president, he pushed to permit greater use of methyl bromide, a pesticide that depletes the ozone layer and that was in the process of being phased out.[7] Bush asked for oil drilling in the Arctic National Wildlife Refuge. He proposed weakened air quality standards that would triple toxic mercury emissions and would increase sulfur emissions and the smog-forming nitrogen oxides in the air. Bush understood that technological systems are social and political constructions; he did not believe that the environmental system had reached its carrying capacity.

Some technological liberals acknowledge the dangers of pollution, energy shortages, and resource depletion but argue that if we have invented a problem we can invent its solution. If twentieth-century automobiles were inefficient and polluting, new hybrid cars can be three times as efficient. They can use electric motors in the city and gasoline or natural gas engines on the open highway. Universally adopted, they would reduce the demand for oil by more than a half. In practice, however, average fuel efficiency in American automobiles has worsened since 1988, when it was 22.1 mpg, falling to 20.4 mpg in 2002.[8] Since some cars now on the market get more than 60 mpg, this is clearly a cultural choice. Fortunately, energy savings have been more effective in other areas. The best new washing machines use half the water and half the electricity of models from the 1980s. Likewise, better housebuilding techniques, including more wall insulation, thermal windows, common walls with neighboring buildings, and passive solar heating, can halve a home's energy needs. When such technological choices are combined with intensive recycling, a society might achieve the "sustainable development" called for in the "Brundtland Report" (World Commission on Environment and Development 1987). Even if the affluent become more efficient, however, the poor will not be interested in preserving the environment if they remain trapped in poverty. Sustainability requires both lower consumption in wealthy nations and improved living standards in less developed countries.[9]

In theory, "soft" or "appropriate" technologies could be used to scale back resource demands and energy consumption. But in practice, "appropriate" technology may fall short. Solar panels can convert sunlight into electricity, but for 30 years their energy has been more expensive than fossil fuels. Even if an OECD country wanted to convert to renewable energy, solar and wind technologies could supply considerably less than half of the current they demand. Whatever the technology, social factors may prevent its adoption. In India, biogas generators that convert animal dung and organic waste into methane proved so expensive that only middle-class farmers could afford them. In the United States, the ideology of individualism thwarts some collective solutions, such as efficient, centrally generated steam heat for neighborhoods. In Europe, housing codes and regulations often make it difficult for individuals to try new technologies. Overall, the world has not cut but has only slowed the growth of its energy consumption.

Both the growing demand for highly differentiated consumer goods and the effort to solve environmental problems create more jobs. Until the 1970s, it seemed that improved efficiency in agriculture and industry would lead either to massive unemployment or to a shorter workweek for all. Instead, in the last two decades of the twentieth century both the number of jobs and working hours increased. The new jobs were in the service and knowledge sectors of the economy. Productivity grew as corporations adopted lean production, flatter management structures, and computerized "just-in-time" inventory and delivery, but after 1975 hourly wages stagnated for many. In contrast, between 1860 and 1975 workers (particularly in Western Europe) gradually unionized and obtained a shorter workweek and higher wages. In recent decades, outsourcing of jobs has weakened unions, undermined job security, and pushed workers to accept longer hours. Even those with the best positions worked longer in 2005 than they did a generation ago. At the same time, many unions reinvented themselves to become partners with management, in some cases even sharing the same offices.

These choices in the workplace are inseparable from political and market choices. Older arguments about technology and the state often focused on the importance of a free press to democracy. But improved technologies of communication are not automatically shared or used by many people, nor are they necessarily used to investigate the government. Rupert Murdoch has assembled a media empire that includes newspapers, radio, television, and publishing. A single corporation owns the weekly magazine *Time*, the Warner studios, the Cable News Network, America Online, and much else.

Some recent films present dire visions of hegemonic corporate control by means of computers. *The Net* depicts a firm that sells security software to large firms and government offices.[10] With access to their databases, it manipulates information for its own advantage. It can affect the stock market, the Pentagon, or any institution that has become its client. It can change medical records, rewrite police files, invade bank accounts, or otherwise falsify data. The corporation has nearly completed its clandestine capture of the government when a young programmer begins to uncover its activities. In response, the corporation, using its access to bank records, police files, and other personal information, systematically destroys the programmer's identity. The corporation rewrites her curriculum vitae from that of a successful working woman into that of a prostitute and a drug addict wanted by the police. Meanwhile, another woman assumes her identity and her job. The programmer has the computer savvy to fight back and to expose the conspiracy. The average citizen identifies with her, but clearly lacks such skills.

If one danger is hegemony from the private sector, the equally daunting alternative is that the media can became a central apparatus of state control. In the satirical 1998 film *Wag the Dog*, a US president threatened by scandal in the midst of a re-election campaign restores his public image by hiring a "spin doctor."[11] He manipulates the media by inventing an imaginary overseas crisis that requires military intervention. He "documents" the invented crisis with "documentaries" fabricated in a studio. Such deception might seem far-fetched, but intelligence agencies have long used disinformation campaigns to confuse enemies, and some national leaders reiterate falsehoods on television with such sincerity that much of the public believes them.[12]

Possible misuse of the media is worrisome, but it is not the only issue. Even where the press is free of government controls or misuse and where political discussion is not fettered by monopoly, democratic nations have tended to let the marketplace make important technological decisions. The elected members of most legislatures have little technical or scientific expertise, and often the introduction of major new machines and processes is little discussed except by specialists. In the 1990s, relatively few people were involved in the decisions to commercialize the Internet, for example. Citizens were conceived primarily as passive consumers, not as voters.

Future developments in genetic engineering, in robotics, and in nanotechnology will force legislatures to make hard and potentially irreversible choices. One way to deal with government's technical illiteracy is to put engineers and scientists on congressional staffs. Another is to create advisory institutions, such as the Office of Technology of Assessment, which existed for a little more than two decades in the United States and inspired Europeans to create similar institutions. A third possible remedy for this problem is to bring representative groups of citizens together and crystallize their opinions through deliberative polling. With 3.3 million scientists and technicians working on research and development worldwide, some way for legislatures to make informed decisions about technology policy must be found. Workers, voters, and consumers do not accept or reject technologies in isolation. Rather, all technologies are enmeshed in systems, and these may be closed or open. A closed system locks out competition and, by a variety of means, tries to keep consumers loyal. This can be done by creating a strong brand, by building in technological incompatibilities, or by creating synergies between the different elements of the system that make it more attractive than alternatives. Microsoft, for example, has used all of these techniques to dominate the computer software business. Technological systems are socially constructed; they can be open or closed, adaptable or rigid, democratically dispersed or restricted to elites.

In the future, will people have access to a variety of technical possibilities? Will they have the infrastructure needed to take advantage of potential choices? In the Netherlands, for example, many commuters have a real choice between the bicycle, mass transit, and the automobile. In contrast, the majority of Americans live in suburbs or rural areas with impoverished choices. The infrastructure is so distended that the automobile is the only practical option. The transportation choices available in most of the United States have contracted since 1905, when the country had millions of horses, a robust trolley system, bicycles, and some of the best passenger railways in the world. Today, most passenger trains have been abandoned and most interurban trolley lines have disappeared. In cities, the number of passengers on buses and trolleys dropped from 17.2 billion a year in 1950 to 7 billion a year in the early 1970s. Hundreds of smaller cities and towns abandoned public transportation.[13] Some suburbs do not even build sidewalks for walkers, and bicycle lanes are almost unheard of. The history of American transportation shows that a wealthy nation can make decisions that impoverish rather than enhance its choices.

Do people know enough to make choices? Does the citizen know how to ride a bicycle? Can the average person use the staggering array of options on a new home computer? Has the average consumer the skill needed to use the huge range of home improvement tools being sold? Cultures and organizations can seek to maximize skills, thereby enhancing resilience and flexibility. Henry Ford at first made a car that was simple enough for many people to repair. Today's automobiles are so complex that only highly trained mechanics can do so. Some early word processing software allowed users to add and delete fonts and utilities, so that owners could streamline and custom design software to fit their needs. In the 1980s it was possible to write a book on a computer with no hard disk, using software that required less than one megabyte. Two decades later, word processing programs are far less flexible but demand at least ten times as much memory. In contrast, Linux software is not sold as a technological "black box" with hidden code, but is an open-ended system that users can supplement with new code. Such open-source software empowers the consumer, but it takes more time to master. Consumers need to evaluate such tradeoffs. When manufacturers present new machines as "black boxes" which one cannot understand much less modify or repair, many people use them only until they need renovation or upgrading and then abandon them. When consumers give up on repairs, they can easily feel trapped, forced continually to upgrade their systems, as manufacturers determine the pace of change and the possible linkages between their machines. Fortunately, over the long term consumers generally seek variety, change, and independence.

The inability to understand or fix many modern machines is also linked to issues of safety. Even if each device functions perfectly by itself, ensembles of machines may have hidden incompatibilities that can cause malfunctions, accidents, or even disasters. Although each individual machine can be improved and made safer, the overarching system of machines may contain dangerous inconsistencies that manifest themselves only in extreme or unusual circumstances. Airplanes and high-speed trains usually function well and have a better safety record than automobiles. But a small malfunction can cause a devastating wreck at hundreds of miles per hour. The more powerful the system, the greater its destructive potential when it goes awry.

This general problem becomes even more acute in warfare, which places extreme demands on technologies and the people who use them. Modern armies have vastly improved their firepower and also their ability to protect soldiers. However, as it became more difficult to attack military targets, they attacked civilians instead. World War II was the turning point, and by its end bombers routinely destroyed entire cities. Dropping an atom bomb on Hiroshima was the tragic but "logical" result. By the end of the twentieth century, 90 percent of those who died in wars were civilians. Fatal mistakes became so common that military journalists invented euphemisms such as "friendly fire" and "collateral damage" to refer to killing one's own troops or unarmed civilians. An army tries to minimize such mistakes, but enemies will try to disrupt technological systems—for example, by jamming radios, decoying air strikes away from intended targets, or infecting computer software.

Extensive military training with computer simulations can transform actual battle into an almost unreal repetition of an arcade game. This disassociation began in the Renaissance, when artillery targets first were too far away to be clearly seen. For 400 years the distances have been increasing. Now a computerized system of representation mediates between the people who die on the ground and those who release bombs from miles in the air or fire a cruise missile at a target 100 miles away. Success or failure is abstract, represented by icons and movements on a screen. Such combatants are in a far different psychological state than those who engage in hand-to-hand battle. When simulated combat and real combat begin to look much the same, soldiers have entered a different experiential world. As a journalist observed, "The illusion is that we can win wars without killing people. Interactive video games that surpass even Gulf War television images threaten to revolutionize warfare further, by desensitizing tomorrow's policy makers completely to the consequences of fighting."[14]

There are equally important psychological implications to incorporating new technologies into everyday life. The experiences that seem natural to children today are radically unlike those of 200 years ago. A green lawn seems "normal" but it is artificially flattened, fertilized, and clipped, and could scarcely be found anywhere in 1805. The "normal" home in Western society invented since then has expanded to include indoor plumbing, central heating, hot running water, electric lighting, radio, refrigeration, television, and much more. The world that seems natural at our birth has been continually modified. One should be skeptical about claims that people can be easily or radically altered because they watch television, use the Internet, acquire a mobile phone, or purchase an intelligent machine. Nevertheless, the cumulative effect of continual innovation has encouraged people to see the world less as a shared dwelling than as a stockpile of raw materials. Technological peoples can unconsciously assume that the world exists for their convenience. The typical motorist assumes that gasoline at reasonable prices is "natural," and on the evening news it has become regrettable but "natural" to see a target from the vantage point of the nose cone of a missile, just as it has become "natural" to see bloody civilian casualties.

For those who surf the Internet for hours every day, using chat rooms and assuming different online personalities has also become "natural," moving them a vast distance from Martin Heidegger's position. He feared that people were losing touch with direct sensory experience of the natural world. In contrast, some people today *want* to leave their bodies behind and merge with the machine. Ray Kurzweil confidently asserted in *The Age of Spiritual Machines* that "there won't be mortality by the end of the twenty-first century [if] you take advantage of the twenty-first century's brain-porting technology."[15] For Kurzweil, future people will not be limited by their physical selves or by the hardware inherited from the past. Rather, he argues, eventually people will exist in their software, and they will migrate to new sites and expand their capabilities as computers become smaller and more powerful. These new sites might look like the bodies that emerged through biological evolution, but they will be rebuilt and improved using nanotechnologies.[16] For those who embrace this perspective, Heidegger's worries may seem mere nostalgia. Kurzweil's writings extravagantly illustrate how people increasingly treat the world as a mere storehouse of raw materials that can be used indiscriminately to satisfy their desires. Is all of nature, even the human body, merely a standing reserve awaiting exploitation and "improvement"?

Freeman Dyson predicts that our descendants will invent and develop radio telepathy so that they can experience collective memory and collective consciousness. He suggests they might break down the barriers between species and communicate with dolphins, whales, and chimpanzees. Conceivably, "those who have been part of an immortal group-mind may find it difficult to communicate with ordinary minds."[17] John Perry Barlow, in contrast, suggests that individual consciousness probably will survive: "Even though I believe that our advances in telecommunications are creating a great Mind that will combine all of our minds, I don't believe that individual human personalities will be subsumed into this vast organism."[18] Most people do not embrace either of these visions, and they are far less sanguine about computers. *New Yorker* cartoons suggest that many feel inferior to advanced machines. In one, a fortune teller warns a middle-aged businessman that he will never catch up with the new technology. In another, two robots stand looking at a museum display, which explains how they came to replace people. This history of evolution has five stages: a naked man, a typewriter, a mainframe computer of the 1970s, a personal computer, and (walking out of the frame to the right) a robot. A third cartoon shows a kitchen, where a man complains that he does not want to play chess with his microwave oven but only wants to warm up his lasagna.[19] Below the surface of these jokes lurks the fear that we are inadequate relative to the machine. We might become obsolescent.

Arthur C. Clarke argued that human invention ultimately can only lead to our evolutionary replacement by intelligent machines: "The tool we have invented is our successor. . . . The machine is going to take over."[20] For him, technology was not merely a series of ever more complex gadgets; it was the artificial extension and acceleration of evolution. Indeed, teams of researchers now actively seek not merely faster computers with more memory, but artificial intelligence. One researcher put it this way: "We're getting more and more alienated from these things [nature] that created us. The distance between us and what made us is growing very fast." However, he added, "from an evolutionary point of view, from a rational point of view . . . it doesn't matter whether the process [life or evolution] is carried on by carbon chemistry or by silicon or by robots."[21] Indeed, the status of intelligent machines is becoming a legal issue. Would an artificial intelligence deserve the same rights before the law that people have? Would an intelligent machine have the "right" to be eternally plugged in? Might ownership of life forms with artificial intelligence someday seem as heinous as slavery?

If machines can surpass humanity, will they take control? In some films, computer systems become malevolent. In Stanley Kubrick's film *2001*, a computer on board a space ship intentionally kills most of the crew. In the *Matrix* trilogy, a vast computer system pacifies most of humanity, keeping them asleep by linking them to a computer simulation that appears to be the urban world of the late twentieth century. In reality, it is several centuries later, after a terribly destructive world war. The surface of the earth is uninhabitable. The computer system "grows" people in pods in order to harvest their energy. They experience only a virtual reality that computers generate to pacify them, except for the few who reject the programming and escape into a reality far shabbier than the collective hallucination. Even the fact that some people reject the programming is a technological artifact, in that their rebellion is due to software errors at least as much as it expresses the human love of liberty. A different idea underlies the film *Lawn-Mower Man*, in which a scientist uses a combination of powerful drugs and computer programs to transform a stupid gardener into a man of superhuman intelligence. In the climactic scene, his powerfully enhanced mind merges with the research center's computer system and then becomes the core of a disembodied brain that bestrides the global computer network. Though many people will find either of these scenarios an unacceptable apocalypse, some contend that either merging with the machine or being replaced by it must be the logical result of history.

In contrast, the burden of my argument has been that there is no single, no logical, and no necessary end to the symbiosis between people and machines. For millennia, people have used tools to shape themselves and their cultures. We have developed technologies to increase our physical power, to perform all kinds of work, to protect ourselves, to produce surpluses, to enhance memory, and to extend perception. We have also excelled in finding new uses for inventions, and this has had many unexpected and not always welcome consequences. We are not necessarily evolving toward a single world culture, nor must we become subservient to (or extinct in favor of) intelligent machines. For millennia we have used technologies to create new possibilities. This is not an automatic process; it can lead either to greater differentiation or to increasing homogeneity. We need to consider the questions that technology raises because we have many possible futures, some far less attractive than others. We must "try to love the questions themselves like locked rooms and like books that are written in a very foreign tongue." As Rilke suggests, we may then "gradually, without noticing it, live along some distant day into the answers."[22] By refusing to let any ensemble of objects define our world as already given, we can continue to choose how technology matters.

Notes

1. H. G. Wells, *When the Sleeper Wakes,* reprinted in *Three Prophetic Science Fiction Novels of H. G. Wells* (Dover, 1960); Aldous Huxley, *Brave New World* (Perennial reprint, 1998); George Orwell, *1984* (Signet reprint, 1990). Increases in computer speed and memory make it far easier today to establish Orwell's universal surveillance, while the rapid advances in biology seem to make Huxley's genetic engineering possible.

2. There are other, more complex ways to divide up historians of technology, but this is not an essay in methodology. Staudenmaier (1985) remains a good starting point.

3. Davis 1990.

4. Fiske 1993, p. 52.

5. Rolston 1998.

6. United Nations Environmental Programme 1997, p. 218.

7. Source: www.nrdc.org.

8. Ibid.

9. World Commission on Environment and Development 1987, pp. 8, 89. For discussion, see Duchin et al. 1990. See also Duchin and Lange 1994.

10. *The Net,* directed by Irwin Winkler (Culver City, CA: Columbia Pictures, 1995).

11. *Wag the Dog,* directed by Barry Levinson (Burbank, CA: New Line, 1998).

12. Lyndon Johnson gained additional war powers after he convinced Congress and the American people that their navy had been attacked in the Bay of Tonkin, but little evidence suggests such an attack took place. In 2002, George W. Bush convinced many of the American people, without any hard evidence, that Iraq possessed weapons of mass destruction and that its leaders were closely linked to international terror networks.

13. Hilton 1985, pp. 45–47.

14. Trevor Corson, "Try Out Smart Bombs of Tomorrow in Living Rooms Today," *Christian Science Monitor,* July 6, 1998.

15. Kurzweil 1999, p. 128.

16. Ibid., pp. 134–136, 140, passim.

17. Dyson 1997, pp. 157–158.

18. Barlow 2004, p. 184.

19. These and many other cartoons dealing with technology and the future can be found at www.cartoonbank.com.

20. Clarke 1999, p. 194.

21. Cited on p. 199 of Helmreich 1998.

22. Rilke 1954, p. 35.

References

Barlow, John Perry. 2004. "The Future of Prediction." In *Technological Visions: The Hopes and Fears That Shape New Technologies*, edited by M. Sturken et al. Philadelphia: Temple University Press.

Clarke, Arthur C. 1999. *Profiles of the Future*. New York: HarperCollins.

Davis, Mike. 1990. *City of Quartz*. Brooklyn: Verso.

Duchin, Faye, and F. G. Lange. 1994. *Ecological Economics*. Oxford: Oxford University Press.

Duchin, Faye, F. G. Lange, and T. Johnson. 1990. *Strategies for Environmentally Sound Development: An Input-Output Analysis*. Third Progress Report to the United Nations.

Dyson, Freeman. 1997. *Imagined Worlds*. Cambridge, MA: Harvard University Press.

Fiske, John. 1993. *Power Plays, Power Works*. Brooklyn: Verso.

Heidegger, Martin. 1977. "The Turning." In *The Question Concerning Technology and Other Essays*. Manhattan: Harper and Row.

Helmreich, Stefan. 1998. *Silicon Second Nature*. Berkeley: University of California Press.

Hilton, George W. 1985. "The Rise and Fall of Monopolized Transit." In *Urban Transit*, edited by C. Lave. Pensacola: Ballinger.

Kurzweil, Ray. 1999. *The Age of Spiritual Machines*. New York: Penguin.

Rilke, Rainer Maria. 1954. *Letters to a Young Poet*. New York: Norton.

Rolston, Holmes III. 1998. "Technology versus Nature: What Is Natural?" *Ends and Means* 2, no. 2.

Staudenmaier, John M. 1985. *Technology's Storytellers: Renewing the Human Fabric*. Cambridge, MA: MIT Press.

United Nations Environmental Programme. 1997. *Global Environmental Outlook*. Distributed by Oxford University Press, Oxford.

World Commission on Environment and Development. 1987. *Our Common Future*. Oxford: Oxford University Press.

Index

Note: Tables and Figures are indicated by "t" and "f" following page numbers and Notes are indicated by "n" followed by note numbers.

Abbot, Henry L., 318
Abortion debate, 47, 51–52, 218–219
Actor-network approach, 147–172
 anthropomorphism, 154–155, 170n11
 built-in users and authors, 155–157, 170n16, 170n19
 delegation to humans, 151–152
 delegation to nonhumans, 152–154
 description of a door, 149–150, 169nn5–6
 figurative and nonfigurative characters, 157–160, 159–160f, 160t, 171n23
 from nonhumans to superhumans, 161–162
 seatbelt mandate, 147–148, 149f
 texts and machines, 163–168, 164–166f, 168f
Adaptation to climate change. *See* Environmental justice and climate change
Additive layer manufacturing, 62–63
Advanced Cell Technology, 35
Aerospace industry, model-based programming, 365
Africa, ENSO effects, 276
African Americans. *See also* CB radio, race, and neutral technology
 as counterpublics, 406–407
 facial-recognition software and, 530
 as percentage of Big Tech employees, 530
Age-graded social hierarchies, 38–39
Ageism, 38–39
Age of Spiritual Machines, The (Kurzweil), 67–70, 74, 562
Aging, theories on, 34–36. *See also* Prolongation of life
Agony and the Ecstasy, The (Stone), 72, 85n5
Agricultural industry, 193–195, 216–217, 278–279, 325–326. *See also* Aquaculture
Airbus, 64
Airline pilots, nonhuman technologies and, 212
Air Line Pilots Association (ALPA), 239–240
Airplane cockpit design. *See* Gender and airplane cockpit design
Akrich, Madeleine, 152
Alaska, 308, 327, 484, 487
Alciné, Jacky, 530
Algorithm, defined, 358–360. *See also* Neutral technology, myth of
Allen, Woody, 83
Alvarez, Luis, 80
Alzheimer's disease, 39–40
Amazon.com, 361–362, 364
Amazon rainforest monitoring system, 335–355
 alignment of artefacts and institutions, 348–349
 conceptions of control and dialectics, 338
 deforestation rates in Brazil (1978–2008), 341f
 ecological concerns (1989–2000), 343–344, 346f
 ecological dominance (2001–2008), 344–345
 features of system, 339–341, 340f
 institutional change as emergent and conflictual process, 346–348
 institutional theory and IT artefacts, 336–338
 IT shaping institutional change, 349–350, 350f
 military and economic dominance (1964–1988), 342–343, 342t, 346f
 reduction of deforestation, 345–346, 346f

Amazon rainforest monitoring system (cont.)
research methodology, 339
social context of, 341–345
American Academy of Arts and Sciences, 83
American Civil Liberties Union (ACLU), 419, 420t
American College of Obstetricians, 221
American Heart Association, 375–376
American Radio Relay League, 405
American Reinvestment and Recovery Act, 508
Amish and technology, 247–265
Amish community and values, 247–251, 262–263
entrepreneurs and, 252f, 254f, 257, 259–261, 259f, 262, 264n11
milk regulations, 249f, 258–259
Ordnung and Amish change, 252–253
Ordnung code of conduct, 251–252
regulation of electricity, 248f, 255
regulation of technological change and identity, 253–255
running about phase for youth, 257–258
simplicity of the home, 256f, 261–262
transportation and identity, 250f, 255–257
Amniocentesis, 218
Animal Liberation (Singer), 217
Anonymous (group), 453, 454, 456
Anthropometrics, 238–239
Anthropometric Survey of Army Personnel (U.S. Army), 238
Antibiotics, overuse of, 69, 77
Antiglobalization. *See* Technological imperative vs. social imperative
Anti-immigrant backlash, 37
Apgar tests, 220
Apollo 13 (space capsule), 308
Apple Inc., 527–528
Application Services Provisioning (ASP), 337
Aquaculture, 215–216
Archer, E. R. M., 281
Arctic National Wildlife Refuge, 558
Argentina, benefits of SCF in, 279
Aristotle, 80
Arizona, 508, 508–510, 509t, 511f, 518–519
Arizona Town Hall, 518–519
Arquilla, John, 449–450
Artificial intelligence, 223, 359, 362, 563
Ashby's law, 362
Asimov, Isaac, 71, 85n3
Aspen Institute, 83
Aspin, Les, 241
Assange, Julian, 453
Assisted suicide, 40–41, 223
Association for Computing Machinery, 365
Atomic bombs, 78–79, 198–199, 561
Attali, Jacques, 83
Australia, benefits of SCF in, 279
Automat, 212–213
Automotive sector
autonomous cars, 362–363, 429
car design, 3–4, 144
fuel efficiency, 558
IT patents for, 58
mechanical complexity of, 561
spaghetti code and accidents, 362
Avgerou, C., 336–337
Ayres, Russell W., 201

Bacon, Francis, 173
Badische Anilin und Soda Fabrik (BASF), 141–142
Bakelite, 111, 129n6
Bakken Formation, 512, 516
Baldwin, Davarian, 403
Banco Nacional de Panama, 323
Barlow, John Perry, 562
Barratt, Virginia, 529–530
Barrett, M., 337
Barton, Amy, 194
Baruch, Bernard, 79
Baruch Plan, proposed (1946), 79
Behnke, James, 368, 369–370
Bethe, Hans, 82
Bias in historical data. *See* Neutral technology, myth of
Bicycle, early variations of, 115f, 117–118f, 124–127f. *See also* Social construction of facts and artifacts
Bijker, W. E., 112, 138, 174, 327
Binary gender. *See* Facebook, coding gender on
Bio, Nano, Info and Carbo (BNIC), 58
Biological Weapons Convention (BWC) (1972), 81–82
Biotechnology. *See* Prolongation of life
Blackler, F., 338
Bliss point, the. *See* Junk food, science behind addictiveness of
Boeing, 240
Bonig, J., 112
Bonneuil, C., 467
Boorse, Christopher, 437
Boorstin, Daniel, 190
Bosch, Carl, 142

Boujut, Jean-François, 292
Bove, Jose, 550
Bowker, G. C., 312, 314, 315
BP, 305–306, 307, 308
BRAIN (Brain Research through Advancing Innovative Neurotechnologies) Initiative, 434
Brain-computer interfaces (BCIs). *See* Neural devices and ethics
BrainGate System, 434
Brand, Stewart, 469, 544
Brazil. *See also* Amazon rainforest monitoring system
  ENSO effects, 276
  IBAMA (environmental agency), 339, 341, 344
  INPE (Institute for Space Research), 339–340
  limitations on usability of SCF in, 278–279
  marketing of Coke in, 377
  PAC (Plan for Acceleration of Economic Growth), 345
  technocratic insulation of SCF information, 279–280
Brin, Sergey, 529
Brownell, Kelly, 368–369
Brundtland Report (1987), 558
Bullard, Robert D., 483
Bulletin of the Atomic Scientists, The, 79
Burkina Faso, limitations on usability of SCF in, 279
Burrell, Berkeley, 399
Bush, George W.
  environmental policies, 558
  political marketing by, 215
  stem cell research policy, 35, 46, 57
Butler, J., 380, 383, 391
Butler, Samuel, 147

Cable, Vince, 55
Cadbury Schweppes, 370–371, 372–373
Caldeira, Ken, 473
Calder, R., 281
California, stem cell research funding, 57
California Rural Legal Assistance, 194
Callon, Michel, 112, 139
Calorie restriction, 35
Campbell, 377
Canal Treaties (1977), 313, 316
Canal Zone Government, 320
Cane, M. A., 277
Capital (Marx), 197
Capitalism, 105–106, 544, 553
Carey, Al, 375–376
Caro, Robert A., 192
Case for Climate Engineering, A (Keith), 464
Castro, Fidel, 38
CB radio, race, and neutral technology, 397–412
  African Americans as counterpublic, 406–407
  black CB use, 397–404, 406, 407–408
  black founding of Rooster Channel Jumpers, 398–399
  CB radio availability, post-1958, 405
  Ku Klux Klan use, 398, 405–406
  local talk (channel 5), 401–402
  politics of black vernacular speech, 402–403
  popularity of CB radios during 1970s, 400
  shooting skip (long distance communication), 398, 399–400, 404, 408
  stranger relationality concept and, 407–408
  superbowl operators (channel 6), 399–402, 403–404, 407, 408–409
  trucker and white male use, 400, 407–408
  urban-suburban fears of black mobility and, 400–401, 407
CERN (European Organization for Nuclear Research), 292, 294. *See also* Sociotechnical complexity
Cesarean births, 219, 221–222
Challenger (space shuttle), 307
Chalmers, David, 436
Chandler, Alfred D., 199–200
Chemical Weapons Convention (CWC) (1993), 82
Chernobyl accident, 308
Chicago Council on Foreign Relations, Roper survey on use of force, 37
Childbirth, as pathology, 219–222
China, 451–452, 455–456
Chorionic villus sampling (CVS), 219
Chow, Jack, 417
Chun, W. H. K., 390
Churchill, Winston, 81
Ciborra, C. U., 349
Cicerone, Ralph, 473
Cisco, 59
Citizens band (CB) radio, 400, 404–405. *See also* CB radio, race, and neutral technology
City of Quartz (Davis), 557

Civil Rights Movement, 401, 402, 403–404, 405–406
Civil society. *See* Drone use in civil society
Clark, Andy, 435
Clarke, Arthur C., 80, 563
Clarke, Lionel, 61
Clarke, Richard, 455, 456
Clausewitz, Carl von, 454, 458
Climate change. *See* Environmental justice and climate change; Geoengineering as collective experimentation
Climate forecasting. *See* Seasonal climate forecasting (SCF), and equity
Clinton, Bill, 375–376, 519–520
Clonaid, 218
Cloning, 34, 45, 74, 218
Closed-circuit televisions (CCTVs), 415
Coca-Cola, 372, 377–378
Codes of ethics, 365
Cohen, Jared, 528
Coke, 377
Coll, Steve, 419
Collingridge, David, 465–466, 470
Collins, H. M., 468
Colossus computers, 532, 535
Columbia (space shuttle), 307
Commercial aviation, 239–241
Communications Act (1934), 404
Comprehensive Test Ban Treaty, 82
Computer code. *See* Franken-algorithms
Computer voices, 211, 214
Computing, UK history of, 531–542
as feminized work during WWII era, 532–534
as leader during WWII, 532
machine grades and pay, 534, 539
male authority dominance, 533–534
sexism produced labor shortage, 540–541
shift to smaller labor force, 538–540
success of woman-owned Freelance Programmers, 534–537, 536–537f
vision of tech-led social revolution, 537–538
Comscore.com, 528
Conception, nonhuman technologies in, 217–218
Concorde black box project, 536
Condensed Course in Motion Picture Photography (Gregory), 231–232
Confikker worm, 459n6
Congress for Racial Equality (CORE), 406
Conventional warfare, defined, 454–455
Convention on Biological Diversity (CBD), 268
Cook, Tim, 527
Coolidge, Calvin, 319
Craig, Robin K., 489
Crick Institute (UK), 63
Cronon, Bill, 327
Crutzen, Paul, 469
Currie, W. L., 337, 348
Customer service representatives, nonhuman technologies and, 211
Cyber (in)security, 449–462
conventional warfare, 454–455
gravity of threat, 450
hacktivism, 453–454
history of malevolent misbehavior, 452–453
objects and events, 450–452
state-sponsored hacktivism as warfare, 457–459
unrestricted warfare, 455–457
Cyber, origin of term, 449
Cyberspace, origin of term, 449–450
Cyber warfare. *See* Cyber (in)security
Cyberwarriors for Freedom, 453

Dalai Lama, 83
Daniels, Norman, 437
Darkode.com, 456, 461n17
Darwin among the Machines (G. Dyson), 74, 362
Data collection and UAVs, 421–422
Davis, Mike, 557
Day after Trinity, The (documentary), 79
Deacons for Defense and Justice, 406
Death
control of dying process, 222–223
fear of, 39
medical technological prevention of, 39–40
seen as preventable evil, 43
Deep brain stimulators. *See* Neural devices and ethics
Deepwater Horizon disaster, 305, 307, 308
Defense Advanced Research Projects Agency (DARPA), 57, 72
Deforestation. *See* Amazon rainforest monitoring system
"Deforestation: Death to the Panama Canal" (Wadsworth), 320, 327
De Lee, Joseph, 220
Dennet, Daniel, 47
Denning, Tamara, 441

Deutsch, John, 243
Dichter, Ernest, 376
Digital platforms. *See* Neutral technology, myth of; *specific platforms*
DiMaggio, P. J., 338
Dipert, Randall, 449, 450
Disability and normality concept, 437
Distribution of competences, 153
Dixon, Bob, 402–403
Dodds, Gordon, 328n6
Dolly the sheep, 34, 45, 62, 218
Domenici, Pete, 305
Doomsday Clock, 79
Doritos, 372
Dorn, Edwin, 242, 243
Dosi, G., 111
Downing, T. E., 281
Drexler, Eric, 75–76, 80
Drone Journalism Lab, 419
Drone use in civil society, 413–428
 corporate accountability, 417
 entertainment industry, 416
 environmental activities, 417
 frameworks for, 419–422, 420t
 humanitarian and development aid, 417
 human rights monitoring, 416, 418
 journalism, 417
 mapping, 416
 material and technical disruption, 418–419
 public safety, 416–417
 social movements and protests, 418
 state accountability and conflict, 418
 technological innovations and, 414–416
Dr Pepper, 370, 372–373
Dual-use nature of technology, 413
DuckDuckGo, 528
Dunlop, J. B., 128
Dunn, Jeffrey, 377–378
Dworkin, Ronald, 51
Dyson, Freeman, 78, 79, 562
Dyson, George, 74, 360–361, 362

Easterbrook, Gregg, 74
Eberstadt, Nicholas, 36, 37
Ectogenesis, defined, 45. *See also* Reproductive ectogenesis
Edgerton, David, 57
Educational settings, as boot camps, 213
Edwards, P. N., 314
Electric Bond and Share Company (EBASCO), 139–140
Elliott, Carl, 436
Ellul, Jacques, 138
El Nino–Southern Oscillation (ENSO), 275–276, 277, 281
Embryo adoption, 52
Embryonic stem cells, 35, 46, 51
"Emergence of Extreme Subpopulations from Common Information and Likely Enhancement from Future Bonding Algorithms" (Johnson), 361
Emerging technologies, use of term, 429
Empirical Programme of Relativism (EPOR), stages of, 113, 121–128
Employees, control of, 206–212
Endangerment of the species. *See* Technologies, and endangerment of the species
Energy-efficient appliances, 558
Energy justice and policy. *See* Socio-energy systems design
Engels, Friedrich, 196–198
Engines of Creation (Drexler), 75–76, 80
Environmental justice and climate change, 483–499
 adaptation/mitigation tradeoffs, 490–491
 climate change impacts, 484–485
 comprehensive agenda, 491
 culturally sensitive communications and services, 488–489
 design of measures that address vulnerability, 487–488
 government's role, 486
 impacts and equity, 485
 participatory processes, 489
 reduction of underlying environmental stresses, 489–490
Ergonomics, 237–238
Ethics for the New Millennium (Dalai Lama), 83
Ethnic minorities, underrepresentation of, 530. *See also* African Americans
Eugenicists, 219
Europe, nuclear power regulation, 304, 305, 309
European Mars Rover, 60
European Patent Office (EPO), 269, 272
Euthanasia, 40–41, 223
Exxon Valdez oil spill, 308
Eyal, Nir, 529

Facebook
 algorithms, 358, 359
 Application Programming Interface (API) access, 381
 friend feature, 361

Facebook (cont.)
myth of neutral technology and, 527–528
percentage of adult users, 528
Facebook, coding gender on, 379–396
binary by design, 385–389
gender-related changes to sign-up page, 386f
request to select gendered pronoun, 388f
research methodology, 381
resisting control by hacking gender, 389–390, 390f
sociotechnical problems, monetization and, 381–383
surveillance, authenticity, and interoperability, 390–391
10-year analysis of design decisions, 380, 383–385, 384–385f
Facial-recognition, 530
Factor, Max, 233
Factory farms, 216–217
Farming. *See* Agricultural industry
Fast-food industry. *See* Human and nonhuman robots; Slow-food movement (Italy)
Fear Index, The (Harris), 361
Fearnside, P. M., 340
Federal Aviation Administration (FAA), 415–416, 422
Federal Bureau of Investigation (FBI), 456, 527
Federal Communications Commission (FCC), 399, 400, 402, 405–406
Feenberg, A., 389
Feminist studies, of science and technology, 174–176
Fertility rates, decline in, 36, 37
Feynman, Richard, 74–75
Fifth domain. *See* Cyber (in)security
Final Exit (Humphry), 223
Finland, attempt to reduce sodium consumption, 373
Firestone, Shulamith, 176
Fiske, John, 557
Fligstein, N., 336, 338, 341, 346, 347, 350
Flowers, Tommy, 532, 535
Food and Agriculture Organization of the United Nations (FAO), 268
"Food for the Future: Someday, Rice Will Have Built-In Vitamin A. Unless the Luddites Win" (Lovins and Lovins), 74–75
Food-science. *See* Junk food, science behind addictiveness of
Ford, Henry, 142, 210, 557, 561
Foresight Institute, 76
Forrester, Jay, 533
Fort Calhoun Nuclear Power Plant, 308–309
Foucault, M., 380, 383
Fountain, Jane, 337
Fourastié, Jean, 36
Fox, Robert, 149, 150
France, 317, 318, 467, 518
Franco, Francisco, 38
Franken-algorithms, 357–366
algorithm, defined, 358–360
clashing codes, 360–362
military stakes, 363–365
real-life dangers, 362–363
search for solution, 365–366
Franklin, Rosalind, 61
Fraser, Nancy, 406–407
Fraternités (Attali), 83
Freedom Party (Austria), 37
Freelance Programmers, 534–537, 536–537f
Friedland, William, 194
Friedman, Emanuel, 220
Friedman, Thomas, 37
Friedman Curve, 220
Frightful 5. *See* Amazon.com; Apple Inc.; Facebook; Google; Microsoft Corporation
Frito-Lay, 373–376
Fukushima and the inevitability of accidents, 303–310
coping with disasters, 308–310
regulations, 304–306
warnings, 306–308

G7, ranking of UK for ease of doing business, 56
Galloway, A., 389
Gates, Daryl F., 401
Gelernter, David, 69
Gell-Mann, Murray, 73
Gender. *See* Computing, UK history of; Facebook, coding gender on; Women
Gender, as missing factor in STS, 173–183
feminist studies of science and technology, 174–176, 180n8
importance of gender, 178–180
STS perspective, 176–178
Gender and airplane cockpit design, 237–245
regulating accommodation in defense aircraft, 241–244

technological bias in existing aircraft, 237–238
technological bias within commercial aircraft, 239–241
technological bias within defense aircraft, 238–239
Gender choice centers, 217–218
General Electric Company, 139, 143, 144n5
General Mills, 369, 377
Genetically modified foods, 74–75, 467, 549
Genetic damage, 34
Genetic defects, testing of fetus, 218–219
Genetic engineering, 74–77
Genetics, nanotechnology, and robotics (GNR), 71, 76–77, 81–82. *See also specific technologies*
Genovese, Eugene, 406
Geoengineering as collective experimentation, 463–482
deterministic frame, 474–476
governance of geoengineering experiments, 471–474
governance regime of technoscientific promises vs. collective experimentation, 475t
notion of surprise, 468
as planetary experiment, 469–471
responsible research and innovation, 465–468
SPICE—Stratospheric Particle Injection for Climate Engineering, 464, 465, 471–474
German Ideology, The (Marx), 102–103
Germany, 36, 37, 141–142
Geron Corporation, 35
Gerontology, 34
Gibson, William, 449, 529
Giddens, A., 337
Gilfillan, S. C., 104
GitHub, 530
Gladwell, Malcolm, 370–373
Glick, Isaac, 255
Global North/South divide, 37, 278–279
Glover, Jonathan, 47
God-trick concept, 528–529, 530
Go-Gurt, 369
Goldman Sachs, 307–308
Google
algorithms, 358
DeepMind division, 359
Global Impact Awards program, 417
myth of neutral technology and, 527–529
Offensive Search Results, 529
Project Maven, 364
Google Photos, 530
Gore, Al, 469
Graph API Explorer, 381, 384, 384–385f
Gray, John, 467
Gray 900 problem, 76
Great Recession (2008), 307–308
Gregory, Carl Louis, 231–232
Gross, M., 468
Gross, Michael, 459
Grossman, James, 403
Groves, Leslie, 143
Guarente, Leonard, 35
Gulf War (1991), 455

Haber-Bosch technique of nitrogen fixation, 141–142
Habermas, J., 468
Habit formation, 529
Hacktivism, defined, 453. *See also* Cyber (in) security
Haider, Jörg, 37
Haldane, J. B. S., 47, 50
Halliburton, 307
Hammer, Claus, 53
Hammer, G. L., 281
Ham radio (crystal radio sets), 404–405. *See also* CB radio, race, and neutral technology
Handbook of Science and Technology Studies (Jasanoff), 174
Handley, Charles, 231
Hands Off Mother Earth (HOME), 470
Hanis, Mark, 418
Haraway, Donna, 528
Harris, Robert, 361
Harrison, Michelle, 220, 221
Hasselbladh, H., 337
Hasslacher, Brosl, 73, 75–76
Hay–Bunau–Varilla Treaty (1903), 317
Hayes, Denis, 190
Hayes, N., 337–338
Hayflick, Leonard, 34
Hayflick limit, 34–35
Health care, nonhuman technologies in, 208–209, 213
Heckadon-Moreno, Stanley, 321–322
Heidegger, Martin, 562
Helsinki Declaration (1964), 48, 50
Henry, E. William, 406
Herbert, Frank, 70
Herzberg, Elaine, 358
Hidden Figures (Shetterly), 541
High frequency trading (HFT), 360

Hillis, Danny, 70, 73, 74
Hiroshima bombing (1945), 2, 78, 561
Holtom, Shahzia, 528
Hooked: How to Build Habit-Forming Products (Eyal), 529
Hopper, Grace, 540
House Un-American Activities Committee, hearings on Ku Klux Klan, 405–406
Hughes, T. P., 174, 314, 317
Hulme, Mike, 470
Human and nonhuman robots, 205–228
 bureaucratic workplaces, 209–212
 control of birth and death, 217–223
 control of childbirth, birth as pathology, 219–222
 control of customers, 212–215
 control of death process, 222–223
 control of employees, 206–212
 control of pregnancy, choosing the ideal baby, 218–219
 control of process and product, 215–217
 control of students in university setting, 213–215
 education, McChild care centers, 208
 fast-food industry, conveyor system, 212–213
 fast-food industry, from human to mechanical robots, 206–207
 food production, cooking, and vending, 207, 215–217
 health care, 208–209
Human brain. *See* Neural devices and ethics
Human engineering, 238
Human Genome Project, 219
Humanitarian UAV Network, 419
Humans, technology and creating the future, 555–565
 comparison of energy usage, 558
 employment impacts, 559
 Internalist vs. Contextualist historians, 556–557
 mass media, 559–560
 maximization of differentiation, 557–558
 modern warfare, 561–562
 psychological implications, 562
 sustainable development, 558–559
 technological momentum vs. determinism, 556
 technology as fundamental human expression, 555–565
 theories on the future, 562–564
Humphry, Derek, 223
Hurricane Katrina, 308, 484, 487
Hurricane Rita, 484
Hurricane Sandy, 484
Hussein, Saddam, 455
Husserl, Edmund, 191
Huxley, Aldous, 555
Hydraulic fracturing, 517

IBM, 59, 540
ICL, 540
I.G. Farben, 142
Immigration, 37
India. *See also* Traditional medicinal knowledge, in India
 benefits of SCF in, 279
 biogas generators, 559
 call centers, 211
 Council of Scientific and Industrial Research (CSIR), 270
 Department of Ayurveda, Yoga and Naturopathy, Unani, Siddha and Homoeopathy (AYUSH), 270
 National Institute of Science Communication and Informative Resources (NISCAIR), 270
 nuclear power regulation, 304
 Union Carbide chemical leak (1984), 304, 308
Indigenous knowledge (IK). *See* Traditional Knowledge (TK), defined
Industrial disasters, 304–310. *See also* Fukushima and the inevitability of accidents; *specific disasters*
Information warfare, origin of term, 449–450
Infrastructural systems. *See* Panama Canal watershed
Infrastructure, defined, 313–314
Innovation process six-stage model, 110, 110f
Innovation studies, 109–110
Intel Corporation, 59, 544–545
Intellectual property (IP), 268
Inter-American Development Bank (IADB), 323, 347
Intergovernmental Panel on Climate Change (IPCC), 491
International Energy Agency (IEA), 516
Internet of Things (IOT), 415, 450
In Vitro Fertilisation (IVF), 46, 47, 48, 51, 52, 217–218
iPhones, 362, 527

Irani, Lilly, 364
I, Robot (Asimov), 71, 85n3
Irregular warfare, defined, 455
Italy, 36, 37, 547–552

James, William, 104
Japan. *See also* Fukushima and the inevitability of accidents
  atomic bombing by US during WWII, 2, 78, 561
  fertility rates, 36
  immigrant assimilation, 37
  projected median age, 36
  workers per retired persons, 36
Jasanoff, Sheila, 248
Jaspal, R., 472–473
Jasper the dog, 62
Jeevani, 269
Johnson, Lyndon B., 308
Johnson, Neil, 360–361, 365–366
Johnston, R., 111
Joint Primary Aircraft Training System (JPATS), 238–239, 240, 241–244
Joly, P. B., 464–465, 474
Jones, Christopher, 511
Joy, Bill, 546–547. *See also* Technologies, and endangerment of the species
Junk food, science behind addictiveness of, 367–378
  carrot snack products, 377–378
  conference on obesity epidemic, 367–370
  product optimization, 370–373
  vanishing caloric density, 373–377
Justice as recognition concept, 442–443

Kaczynski, Theodore, 68–69, 85n1, 547
Kallinikos, J., 337, 348
Karppanen, Heikki, 373–374
Kauffman, Stuart, 73, 76–77
Keith, David, 464, 471, 472, 474
Kelly, Chris, 382–383
Kenya, UAV legislation, 423
Kevorkian, Jack, 40–41, 223
Keystone XL, 511–512
Kim Il Sung, 38
Kindem, Gorhan, 233
KinderCare, 208
Kirkwood, Tom, 35
Knowledge-enabled mass destruction (KMD), 71, 77, 83
Kodak, 233
Koppleman, Lee, 192
Kowinski, W. S., 214
Kraft, 377
Kranzberg, Melvin, 397
Kraybill, Donald B., 255
Kroc, Ray, 206, 210
Kropotkin, Peter, 190
Kuhlmann, S., 466
Kuhnian paradigms, 111, 130n13
Ku Klux Klan. *See* CB radio, race, and neutral technology
Kurzweil, Ray
  Age of Spiritual Machines, The (Kurzweil), 67–70, 74, 562
  need to leave Earth, 79–80
  vision of near immortality, 71, 546–547

Laboratory Earth (Schneider), 469
Larson, Clark, 322
Latour, Bruno, 139. *See also* Actor-network approach
Law, John, 174
Lawn-Mower Man (film), 563
Law of the excluded middle, 147–148
Lawson's Bicyclette, 115, 119, 125f, 126
Layton, E., 110
Lazonick, W., 112
League of Slow Cities, 548–549, 551
Lécaille, Pascal, 293
Lega Lombarda (Italy), 37
Lenin, Vladimir, 138
Le Petit Bertrand, 158, 159f
Leslie, John, 79, 86n11
Liebert, W., 465–466
Life expectancies, increase of, 33–40
Lillienthal, David, 190
Lin, Robert I-San, 373–374, 376–377
Linux, 561
Lipsitz, George, 406
Livestock, in factory farms, 216–217
Local knowledge. *See* Traditional Knowledge (TK), defined
Logic bombs, 456, 461n16
Longevity. *See* Prolongation of life
Long Island, New York overpasses, 191–192, 288
Long Now Foundation, 70
Lovelace, Ada, 540
Lovins, Amory, 74–75
Lovins, Hunter, 74–75
Lula, Luiz Inácio, 345
LulzSec (former group), 453

Machine learning, 363
"Machine Stops, The" (Forster), 11–31
Part I: The Air-Ship, 11–18
Part II: The Mending Apparatus, 19–25
Part III: The Homeless, 25–31
Mackenzie, Donald, 470
Macnaghten, P., 470
Madison Dearborn Partners (Chicago), 378
Malarial mosquitoes, 69
Malkiewicz, K., 230
Malone, E. L., 281
Malware, 451
Mandatory retirement ages, 38
Manhattan (film), 83
Manhattan Project, 78–79, 81, 82, 143
Manjoo, Farhad, 528
Manning, Chelsea, 453
Marcellus shale formation, 512
Marx, Karl, 99, 100, 102–103, 138, 197. *See also* Technological determinism, aspects of
Massey Energy, 305, 307, 308
Material Resistance, A Textbook, 150
Maternal serum alpha-fetoprotein (MSAFP) testing, 219
Matrix trilogy (films), 530, 563
Maxwell, James Clerk, 150
McCormick, Cyrus, II, 192–193
McDonaldization of Society, The (Ritzer), 205
McDonald's. *See* Human and nonhuman robots
McDonnell Douglas, 240
McGrath, Jim, 45
Median age, globally, 36
Medical ethics, 218, 219
Meier, Patrick, 423
Metaphysics (Aristotle), 80
Michelangelo, 72, 85n5
Microsoft Corporation
as closed system, 560
code of ethics for algorithms, 364
cyber attacks against, 452
myth of neutral technology and, 527–528
Tay, AI chatbot, 529
Midwifery, 219–220, 223
Military, nonhuman technologies in, 223
Military Standard 1472, *Human Engineering Design Criteria for Military Systems, Equipment and Facilities*, 238
Millennium (*Lignes d'horizons*) (Attali), 83
Millerson, G., 230
Mills, Frederick, 230
Miraflores Locks Visitor Center, 316
Mischel, Walter, 468
Misgendering. *See* Facebook, coding gender on
Missing mass concept. *See* Actor-network approach
Mitchell, Timothy, 314
Model-based programming, 365
Moffatt, Ann, 536–537f
Molecular electronics, 73, 75–76
Momentum. *See* Technological momentum
Monell Chemical Senses Center (Philadelphia), 374
Moore, Gordon, 544–545
Moore's law, 73, 544–547
Moravec, Hans, 69–70, 74, 80, 546
Morris, William, 190
Moses, Robert, 192
Moskowitz, Howard, 370–373
MRC Laboratory of Molecular Biology (UK), 61
Mudd, Michael, 368–369
Muir, John, 343
Mukherjee, Ann, 375–376
Mulkay, M. J., 111–112, 123
Mumford, Lewis, 139, 151, 189–190, 196
Murdoch, Rupert, 559
Muscle Shoals Dam (Alabama), 142

NAACP (National Association for the Advancement of Colored People), 402
Nagasaki bombing (1945), 2, 78
Nakamura, L., 384
Nanomedicine, 71
Nanotechnology, 70–71, 75–76
Napoleon, Louis, 192
National Academies, *America's Energy Future*, 502, 504
National Biologics Industry Innovation Centre (UK), 61
National Composites Centre (NCC) (UK), 63
National Front (France), 37
National Institutes of Health (NIH), 35, 57
National Physical Laboratory (UK), 64
National Science Foundation, 434, 465
National Union of Iron Molders, 192–193
Natural disasters, 484, 487–488. *See also specific hurricanes*
Nature as infrastructure, defined, 311. *See also* Panama Canal watershed
Nelson, Alondra, 541
Neohumanism, 550–551

Neoinstitutional theory, 336–337
Nerlich, B., 472–473
Nestlé, 377
Net, The (film), 559
Netherlands, transportation choices, 560
Neural devices and ethics, 433–447
 authority in translational algorithms, 438–439
 concept of normality, 437–438
 identity, 435–437
 justice as recognition, 442–443
 moral and legal responsibilities, 439–440
 privacy, 440–442
Neuroimaging, 439
Neuromancer (Gibson), 449
Neutral technology. *See* CB radio, race, and neutral technology
Neutral technology, myth of, 527–530
New Digital Age, The (Cohen), 528
New England Journal of Medicine, on weight gain in US, 376
Newman, Max, 532
Nicquevert, Bertrand, 293, 299
Nietzsche, Friedrich, 33, 80
Noble, David, 112, 194
Noble, Safiya, 541
Nordmann, Alfred, 466
Normal Accidents (Perrow), 303
Normal accident theory, 309
Norris, George, 142
North Dakota, 512
North/South divide. *See* Global North/South divide
Novartis, 51
Nozick, Robert, 47
Nuclear, biological, and chemical (NBC) technologies, 71, 77, 81–82
Nuclear energy waste disposal, 513–514
Nuclear power, 190, 201, 304. *See also* Fukushima and the inevitability of accidents
Nuclear Waste Policy Act (NWPA) (1982), 513
Nuclear weapons, 71, 78–79

Obama, Barack, 377
Obesity, 368, 370, 376. *See also* Junk food, science behind addictiveness of
Office of Minority Business Enterprise (OMBE), 399
Office of Personnel Management (OPM) hack, 451–452
Office of Technology of Assessment, 560
Oil industry, 305, 308
Old Order Amish. *See* Amish and technology
"On Authority" (Engels), 196–198
O'Neil, Cathy, 359, 361
Oost, E. C. J. van, 112
Open Source software movement, 551, 552
Oppenheimer, J. Robert, 78–79
Optical scanners, 211, 214
Organisation for Economic Co-operation and Development (OECD), high tech sectors, 58
Organizing vision, defined, 337
Orwell, George, 555
Osborne, George, 58, 59
Ozanne, Robert, 192–193

Pale Blue Dot (Sagan), 77
Panama
 compesino culture and agrarian reform, 323–326
 Forest Law 13 (1987), 326
 Guardia Nacional government, 323
 Ministry of Agriculture (MIDA), 323, 324
 national environmental agency (ANAM), 316
 Panama Canal Authority (ACP), 326
 RENARE, 322, 324–325
Panama Canal Company, 320
Panama Canal watershed, 311–334
 cultural politics of forests, 325–327
 design and water use, 318
 forest guards and watershed management, 324–325
 forest hydrology, politics and watershed landscapes, 321–322
 funding and government programs, 322–324
 Lake reservoir storage of water, 318–319
 map of, 313f
 nature as infrastructure, 313–316
 technical vs. "natural" management, 319–321, 329n7
 watershed forests and, 316–318
Paradigmatic aspect, of sentences, 164–165, 164–165f
Paredes, Ruben Dario, 324
Parfit, Derek, 46
Parson, Edward, 471
Pastor, Manuel, 485, 486
Pathways, in medical care, 209
Paulson, Henry, 307–308
Peace Corps, 326

Pearson, Drew, 405–406
Pentagon of Power, The (Mumford), 139
Pepsi, 377
Persso, Anders, 53
Peru, limitations on usability of SCF in, 278, 281
Petrini, Carlo, 547
Petroleum industry. *See* Socio-energy systems design
Pew Research Center
 Internet & American Life Project, 423
 study on Facebook users, 528
Pfaff, A.S. P., 281
Photo and movie film, lighting for whiteness, 229–235
 backlighting, 232
 early history of film stock, 231–233
 white face as norm, 231–234
Photographic Tour on Wheels, A (White), 118
Pichai, Sundar, 527
Picher, Rae, 389
Pierrehumbert, Raymond, 472
Pillsbury, 367–368
Pinch, T., 138, 174
Pirbright Institute of Animal Health (UK), 62
Plato, 197, 198, 200
Plutonium recycling, 201
Political ecology, 315–316
Political nature of technologies, 189–204, 288
 inherently political technologies, 196–202
 technical arrangements and social order, 191–195
Poultry market, 62
Poverty and the poor. *See* Environmental justice and climate change; Seasonal climate forecasting (SCF), and equity
Powell, W. W., 338
Pregnancy, and nonhuman technologies, 218–219. *See also* Reproductive ectogenesis
Prego spaghetti sauce, 371
Prescriptions, 152–153, 169n8
Presidential Commission for the Study of Bioethical Issues, 443
Privacy and UAVs, 421
Product optimization, 370–371
Professional Society of Drone Journalists, 420t
Programmed Inequality (Hicks), 531, 541
Program of action, 148, 168n1. *See also* Actor-network approach
Project Maven (DoD), 364
Prolongation of life, 33–43
 Category I and II old age, 40–42
 decline in fertility rates, 37
 increase of life expectancies, 33–40
 management of social hierarchies, 37–39
 theories on aging, 34–36
Pugwash Conferences, 83
Putman, Todd, 377

Quasimodo, Salvatore, 550
Queer theory. *See* Facebook, coding gender on

Racism. *See* CB radio, race, and neutral technology; Photo and movie film, lighting for whiteness; Political nature of technologies
Radio Act (1927), 404
Radio League of America, 405
Radio telepathy, 562
RAINMAN (software), 279
Rand Corporation, 449–450, 512
Rawick, George, 406
Rayner, S. D., 278, 281
Reagan, Ronald
 political marketing by, 214–215
 Small Business Research Initiative, 57
 Strategic Defense Initiative, 80, 143
Reaper manufacture, 192–193
Reed, Mark, 73
Regan, S., 338
Regulatory capture, 304–306
Reisner, Michele, 372–373
Reppy, J., 240–241
Reproductive ectogenesis, 45–54
 ethicist's view, 49–50
 impact on status of fetus, 51–52
 Parfit teletransporter, the, 46
 potentiality of, 50–51, 52
 rights of pregnant women, 52
 woman and the fetus, 46–47
 xenotransplantation scenario, 47–49
Republic (Plato), 197
Republic of Technology, The (Boorstin), 190
Research and development (R&D). *See* United Kingdom, industrial strategy
Research Partnership for Investment Fund (UK), 63
Responsible research and innovation (RRI), 465–468
Rheinberger, Hans-Jorg, 467–468, 472
Rhodes, Richard, 78–79

Rice, Condoleezza, 306
Rid, Thomas, 458
Rilke, Ranier Maria, 564
Rimm, Eric, 376
Riskey, Dwight, 374–375
Ritzer, George, 335
Robock, Alan, 469, 470
Robotics, 70–71, 74, 363. *See also* Human and nonhuman robots; Neural devices and ethics; Technologies, and endangerment of the species
Robot: Mere Machine to Transcendent Mind (Moravec), 69–70
Roddenberry, Gene, 71–72
Rolls Royce, 64
Rolston, Holmes, III, 558
Roman Catholic Church, Vatican Television, 214
Ronfelt, David, 449–450
Roosevelt, Franklin D., 140, 142
Roosevelt, Theodore, 317
Ross, Lexi, 381, 383, 386, 391
Royal Society, 469–470, 471, 473
Ruskin, John, 343

Sagan, Carl, 77
Samsung, 363
Sanger, Stephen, 369
Satellite Sentinel Project, 418, 423
Satellite technology, 60
Saturnini, Paolo, 548
Scarry, Richard, 503–504, 505
Schmidt, Eric, 528
Schmidt, J. C., 465–466
Schneider, Stephen, 469
Science, technology, engineering, and medicine (STEM), 531
Science fiction stories, 70, 71–72. *See also* "Machine Stops, The" (Forster); "Sultana's Dream" (Hossain)
Scientific management, 209–210
Scientific research, as stimulus for technological advances, 105–106
Scientific research, in UK. *See also* United Kingdom, industrial strategy
  advanced materials, 62–63
  agri-science, 62
  big data, 59
  energy, 63–65
  regenerative medicine, 61–62
  robotics and autonomous systems, 60–61
  space science, 60
  synthetic biology, 61
SCoPEx experiment, 474
SCOT approach. *See* Social Construction of Technology (SCOT)
Searle, John, 67
Seasonal climate forecasting (SCF), and equity, 275–286
  as decision-support tool, 277–280
  ENSO effects, 275–276, 277, 281
  impact of inequality on equity, 280–281
  limitations on usability of, 276–277
  opportunity cost of SCF use, 281–282
Seat belt mandate, 147–148, 149f
Self-replication, 70–71, 76, 81
"Self-Replication: Even Peptides Do It" (Kauffman), 76–77
Sen, Amartya, 509
Sensory-specific satiety, 372
Sentient robots, 67–70, 74
September 11, 2001 terrorist attacks, 37, 308
7-Eleven, 213
Sexism. *See* Computing, UK history of
Shaping Technology I Building Society (Bijker and Law), 174
Shermer, Dee, 537f
Shetterly, Margot Lee, 541
Shielding disk design, for ATLAS particle detector. *See* Sociotechnical complexity
Shirley, Stephanie "Steve," 534–537, 537f
Shopping malls, nonhuman technologies in, 214
Shultz, Alex, 386
Silicon Valley, early contracts in, 57
Simons, Menno, 250, 263n6
Simultaneous discovery, 101
Singer, Peter, 52, 216–217
SIR2 gene, 35
Sklar, Rachel, 530
Slow cities movement, 548–552
Slow-food movement (Italy), 547–552
Sniderman, Andrew S., 418
Snowden, Edward, 453
Social construction of facts and artifacts, 109–136, 138–139
  closure and stabilization, 119, 126–128, 131n21
  Empirical Programme of Relativism (EPOR), 113
  interpretative flexibility of findings, 121–126
  Social Construction of Technology (SCOT), 113–121
  technology studies, 109–112

Social Construction of Technological Systems, The (Bijker, Hughes, and Pinch), 174
Social Construction of Technology (SCOT), 113–121
selection and variation developmental process, 114–115, 114–118f, 120–122f
Social hierarchies, 37–39
Social imperative. *See* Technological imperative vs. social imperative
Society of Professional Journalists, 421
Socioeconomic inequity. *See* Environmental justice and climate change; Seasonal climate forecasting (SCF), and equity
Socio-energy systems design, 501–525
defining socio-energy systems, 503–505
energy justice, 507
energy policy goals and objectives, 505–506
energy transitions, 506–507
governing energy transitions, 519–520
nuclear energy waste storage design case study, 513–514
oil industry infrastructure/community case study, 511–513
publicizing energy policy, 518–519
socializing energy policy, 515–516
solar design case study, 506–507, 508–510, 509t, 511f
systematizing energy policy, 516–518
Sociology of technology, 111–112. *See also* Social construction of facts and artifacts
Sociotechnical complexity, 291–302
dealing with supervisors, 292–293
interactions between objects, 295–296
neighborhood relations, 295–296, 297, 298–300
operational summary, 300–301
shielding disk design for ATLAS particle detector, 293–295
technical vs. strategic work, 296–298
Sociotechnical systems. *See* Socio-energy systems design; Technological momentum
Sodium consumption, 373–375
Soft determinism (James), 104
Soft war, defined, 459
Solar City, 510
Solar energy, 190, 198, 559. *See also* Socio-energy systems design
Solar radiation management (SRM), 464, 471, 473. *See also* Geoengineering as collective experimentation
Solter, Davor, 45
South Africa, UAV legislation, 423
Sovacool, B., 502
Soviet Union, Chernobyl accident, 308
Spafford, Eugene, 363
Spain, fertility rates, 36
Spear phishing scams, 451
Speculative ethics, 466–467
Spent nuclear fuel (SNF), 513–514
SPICE—Stratospheric Particle Injection for Climate Engineering, 464, 465, 471–474
Stanley Exhibition of Cycles, 118–119, 123
Star, S. L., 314, 315
Star Trek (TV show), 71–72, 544
Stem cell research, 35, 40, 46, 57
Stock market, high frequency trading (HFT), 360–361
Stone, W. Clement, 210
Strategic Defense Initiative (SDI), 80, 143
STS (science and technology studies or science, technology, and society), 4, 315, 381, 468. *See also* Gender, as missing factor in STS
Stuxnet worm, 452
Suchman, Lucy, 364
"Sultana's Dream" (Hossain), 87–94
Sun, Lisa Grow, 490
Sun Microsystems, 72, 546
Supermarkets, nonhuman technologies in, 210–211, 214
Surrey Satellites Technologies (SSTL), 60
Sweden, legal status of fetus, 51
Swidden system, 325–326
Sylvan Learning Centers, 208
Syntagmatic aspect, of sentences, 164–165, 164–165f
Szerszynski, B., 470

Taco Bell, 206, 207
Tailhook scandal (1991), 243, 244n5
Tansley, Arthur George, 343
Taylor, Frederick, 209–210
Technological determinism
defined, 138
soft determinism, 104
technological momentum vs., 556
Technological determinism, aspects of, 99–107
fixed sequence to technological development, 100–102

imposition of determinate pattern of social relations on society by technology, 102–104
mediating role of technology in modern Western society, 105–106
social forces and technological progress, 104–105
Technological imperative vs. social imperative, 543–554
individual and collective action, 551–552
Moore's law and its distortions, 544–547
new bipolarity, 552–553
slow-food movement in Italy, 547–552
Technological momentum, 137–145
characteristics of momentum, 141–142
definitions, 138
EBASCO as cause and effect, 140
EBASCO as gathering momentum, 140–141
EBASCO as technological system, 139
use of momentum, 143
Technological politics, 288. *See also* Political nature of technologies
Technologies, and endangerment of the species, 68–85, 546–547
author's innovations, 72–74, 75
Kaczynski and Luddite challenge, 68–69
Leslie on risk of human extinction, 79, 86n11
Moravec's robots, 69–70, 74
potential for mass destruction, 71
relinquishment of dangerous technologies, 80–83
self-replication and, 70–71, 76–77, 81
sentient robots, 67–70, 74
Technology, history studies, 110–111
Technology and Culture (MacKenzie), 110
Technology and society, introduction, 1–5
complexities of sociotechnical systems, 287–289
human and social values and, 185–187
relationship between, 95–97
social effects, 3
social shaping, 2–3
sociotechnical systems, 3–4
twenty-first-century challenges and strategies, 429–431
visions of technological future, 7–9
Technoscience, defined, 173–174, 177
Telemarketing, nonhuman technologies in, 211
Telephones, 261–262, 264n14, 362, 527
Television, 190, 214–215, 234, 261
Teller, Edward, 78
Telomerase, 34–35
Telomeres, 34–35
Tennessee Valley Authority (TVA), 142
"There's Plenty of Room at the Bottom" (Feynman), 75
Thinking Machines Corporation, 70
Thomson, Judith Jarvis, 47
Thoreau, Henry David, 81, 82
3D printing, 62, 64
3M Corporation, 234
Three Mile Island accident, 309, 310
Thrift, Nigel, 314
Thurston, Baratunde, 530
Tildesley, Dominic, 59
Tomato industry, 193–195
Tong, Rosemarie, 47
Traditional Knowledge (TK), defined, 267–268
Traditional medicinal knowledge, in India, 267–274
history of medicinal use of herbs/spices, 268–269
intellectual property and traditional knowledge, 268
traditional knowledge, defined, 267–268
Traditional Knowledge Digital Library (TKDL), 269–272, 270f, 272t
Transformation of Corporate Control, The (Fligstein), 338
Trapdoors, 456, 461n16
Trinity (atomic test), 78, 81
Truman, Harry, 78
Turing, Alan, 532
T.V.A.: Democracy on the March (Lillienthal), 190
21st century technologies, and endangerment of the species (Joy), 67–86
Twitter, #JusticeForTay, 529
2001 (film), 563

UAViators.com, 419–420, 420t, 423
Ullman, Ellen, 358
Ultrasound, 219, 221
Unbounding the Future: The Nanotechnology Revolution (Drexler), 75
UN Charter, on armed attack, 456, 461n18
Union Carbide chemical leak (1984) (India), 304, 308
United Airlines, 211
United Kingdom, industrial strategy. *See also* Computing, UK history of
acceptance of mistakes, 57–58

United Kingdom (cont.)
Biomedical Catalyst fund, 57
capital-intensive R&D, 56–57, 65–66
Centres for Doctoral Training (CDTs), 64
Economic and Social Research Council (ESRC)–led life study, 59
G7 ranking for ease of doing business, 56
Government Office for Science, 58
Industrial Biotechnology and Bio Energy Catalyst fund, 57
investment in eight technologies, 59–65
National Space Technology Programme, 60
Research Base, 56
Research Councils, 57, 58, 60, 61, 62, 465
Satellite Applications Catapult, 60
Space Leadership Council, 60
STEAM—Science Technology Engineering Arts and Maths, 56
Technology Strategy Board, 57, 58, 60–61, 62
United Nations (UN), *Report on Human Development* (1995), 179
United Nations Conference on Trade and Development (UNCTAD), 268
United States (US). *See also* Panama Canal watershed; *specific federal agencies*
Anywhere, U.S.A. trend, 549–550
benefits of SCF in, 279
bioengineering, 549
Board of Consulting Engineers for Panama Canal, 317–318
capital-intensive R&D, 57
fragmented energy system, 517
handicap accessibility, 193
immigrant assimilation, 37
Mars Rover, 60
media age of population, 36
Nuclear Regulatory Commission (NRC), 304–305
origins of Tennessee Valley Authority, 142
public utilities commissions, 517
regional water management, 278
study of business leaders, 200–201
temperature increases due to climate change, 484
transportation choices, 560
Universities, nonhuman technologies in
control of professors, 208
control of students, 213
University of California, mechanical tomato harvester, 193–195
Unix operating system, 72
Unmanned aerial vehicles (UAVs). *See* Drone use in civil society
Unrestricted Warfare (Chinese PLA), 455–456
Urban heat island effect, 490
US Agriculture Department
approval of 50 genetically engineered crops, 75
EPO patent on neem, 269
USAID
"Canal, The" report (1981), 322–323
US Army
Joint Primary Aircraft Training System (JPATS), 238–239, 240, 241–244
Natick Research Development and Engineering Center, 238, 239
Research Laboratory, Army Research Office, 422–423
US Army Corps of Engineers
Panama Canal watershed and, 320, 329n7
US Chemical Safety Board, 305
US Congress
approval of construction of Panama Canal, 317–318
ban on NIH funding for research harmful to embryos, 35
US Congressional Research Service, 471
US Defense Department
acquisition policy, 238
Defense Advisory Committee on Women in the Service (DACOWITS), 242–244
Project Maven, 364
US Energy Department, 57
nuclear power regulation, 304
small modular reactor technology, 63
US Energy Information Agency, 516
US Environmental Protection Agency (EPA), 514, 517
Use of force, 36–37
US Forest Service
Panama Canal watershed and, 320–321, 329n7
US Government Accountability Office, 307
US Interior Department
Materials Management Service, 305
US Justice Department
Ku Klux Klan as subversive organization, 405
US Navy
Joint Primary Aircraft Training System (JPATS), 238–239, 240, 241–244

US Nuclear Regulatory Commission (NRC), 513–514
US Occupational Safety and Health Administration (OSHA), 304–306
US Orphan Drugs Programme, 57
US Patent Office, 269
US Securities and Exchange Commission, 307, 382
US State Department
 Keystone XL, Environmental Impact Statement, 511–512

Valentine, Joseph, 233
Verbeek, Peter-Paul, 466–467
Verchick, Robert R. M., 485
Victor, D. G., 471
Vigilantism, in cyberspace, 453–454
Visible Hand, The (Chandler), 199–200
Visions of Technology (Rhodes), 78–79
Vlaams Blok (Belgium), 37
VNS Matrix, 529–530
von Neumann probes, 79

Wadsworth, Frank
 "Deforestation: Death to the Panama Canal," 320, 327
 on greenies, 321, 329n10
 as Panama Canal watershed consultant, 320, 321, 322, 323, 325, 328n6
Wag the Dog (film), 559–560
Walker, Joseph, 232
Walsh, Toby, 362, 363
Walsham, G., 337
Warfare. *See* Cyber (in)security
Warner, Michael, 407
Watershed management. *See* Panama Canal watershed
Weaponized drones, 363–364
Weapons of mass destruction (WMD), 71, 77–78, 81
Weapons of Math Destruction (O'Neil), 359, 361
Weber, Max, 144
Web Inspector tool, 389–390, 390f
Weeks Act (1911), 329n7
Wells, Deane, 52
Wells, H. G., 555
WGBH Educational Foundation, 234
White (Dyer), 229
White, L. Meadows, 118
White, Lynn, Jr., 138
White Plague, The (Herbert), 70
White supremacists. *See* CB radio, race, and neutral technology
"Why Humans Like Junk Food" (Witherly), 375
Wikileaks, 453
Wilkins, Maurice, 61
Willetts, David, 59, 60, 61
Wilmott, Paul, 365
Wilson, Harold, 537–538
Wilson Dam. *See* Muscle Shoals Dam (Alabama)
Wind energy, 559
Winner, Langdon, 154, 237, 466
Winston, Brian, 233
Witherly, Steven, 375
Wolfram, Stephen, 73
Women. *See also* Computing, UK history of; Gender, as missing factor in STS; Reproductive ectogenesis
 contraceptive and reproductive technology, 176, 178
 control of conception and birth, 217–222
 cyber-feminists, 529–530
 global access to education, 179
 life expectancies of, 179
 mortality rate for mothers, 179
 as percentage of Big Tech employees, 530
 political participation of, 36–37
 seen as separate social group when riding bicycles, 118–119
Women's Royal Naval Service (UK), 532
Word processing software, 561
Workplaces, nonhuman technologies in, 209–212. *See also* Human and nonhuman robots
World Bank, 347, 516
World Health Organization, mortality rate for mothers, 179
World Intellectual Property Organization (WIPO), Intergovernmental Committee on Intellectual Property and Genetic Resources, Traditional Knowledge and Folklore (IGC), 268
World Trade Organization (WTO), 268
World Wildlife Fund, 417
W.R. Grace Company, 269
Wright, Beverly, 483

Xenotransplantation. *See* Reproductive ectogenesis
Xenotransplantation Society, 53

Yoga, in TKDL, 271–272
Yoplait, 369

Young, Michael, 47
Yucca Mountain Nuclear Waste Repository, 513–514

Zebiak, S., 277
Ziervogel, G., 281
Zimbabwe, levels of usability of SCF in, 278–279
Zuboff, S., 349–350
Zuckerberg, Mark, 383, 527

Inside Technology Series

Edited by Wiebe E. Bijker, W. Bernard Carlson, and Trevor Pinch

Deborah G. Johnson and Jameson M. Wetmore, editors, *Technology and Society: Building Our Sociotechnical Future, second edition*

Kean Birch and Fabian Muniesa, *Turning Things into Assets*

David Demortain, *The Science of Bureaucracy: Risk Decision-Making and the US Environmental Protection Agency*

Nancy Campbell, *OD: Naloxone and the Politics of Overdose*

Lukas Engelmann and Christos Lynteris, *Sulphuric Utopias: The History of Maritime Fumigation*

Zara Mirmalek, *Making Time on Mars*

Joeri Bruynincx, *Listening in the Field: Recording and the Science of Birdsong*

Edward Jones-Imhotep, *The Unreliable Nation: Hostile Nature and Technological Failure in the Cold War*

Jennifer L. Lieberman, *Power Lines: Electricity in American Life and Letters, 1882–1952*

Jess Bier, *Mapping Israel, Mapping Palestine: Occupied Landscapes of International Technoscience*

Benoît Godin, *Models of Innovation: The History of an Idea*

Stephen Hilgartner, *Reordering Life: Knowledge and Control in the Genomics Revolution*

Brice Laurent, *Democratic Experiments: Problematizing Nanotechnology and Democracy in Europe and the United States*

Cyrus C. M. Mody, *The Long Arm of Moore's Law: Microelectronics and American Science*

Tiago Saraiva, *Fascist Pigs: Technoscientific Organisms and the History of Fascism*

Teun Zuiderent-Jerak, *Situated Interventions: Sociological Experiments in Healthcare*

Basile Zimmermann, *Technology and Cultural Difference: Electronic Music Devices, Social Networking Sites, and Computer Encodings in Contemporary China*

Andrew J. Nelson, *The Sound of Innovation: Stanford and the Computer Music Revolution*

Sonja D. Schmid, *Producing Power: The Pre-Chernobyl History of the Soviet Nuclear Industry*

Casey O'Donnell, *Developer's Dilemma: The Secret World of Videogame Creators*

Christina Dunbar-Hester, *Low Power to the People: Pirates, Protest, and Politics in FM Radio Activism*

Eden Medina, Ivan da Costa Marques, and Christina Holmes, editors, *Beyond Imported Magic: Essays on Science, Technology, and Society in Latin America*

Anique Hommels, Jessica Mesman, and Wiebe E. Bijker, editors, *Vulnerability in Technological Cultures: New Directions in Research and Governance*

Amit Prasad, *Imperial Technoscience: Transnational Histories of MRI in the United States, Britain, and India*

Charis Thompson, *Good Science: The Ethical Choreography of Stem Cell Research*

Tarleton Gillespie, Pablo J. Boczkowski, and Kirsten A. Foot, editors, *Media Technologies: Essays on Communication, Materiality, and Society*

Catelijne Coopmans, Janet Vertesi, Michael Lynch, and Steve Woolgar, editors, *Representation in Scientific Practice Revisited*

Rebecca Slayton, *Arguments That Count: Physics, Computing, and Missile Defense, 1949–2012*

Stathis Arapostathis and Graeme Gooday, *Patently Contestable: Electrical Technologies and Inventor Identities on Trial in Britain*

Jens Lachmund, *Greening Berlin: The Co-production of Science, Politics, and Urban Nature*

Chikako Takeshita, *The Global Biopolitics of the IUD: How Science Constructs Contraceptive Users and Women's Bodies*

Cyrus C. M. Mody, *Instrumental Community: Probe Microscopy and the Path to Nanotechnology*

Morana Alač, *Handling Digital Brains: A Laboratory Study of Multimodal Semiotic Interaction in the Age of Computers*

Gabrielle Hecht, editor, *Entangled Geographies: Empire and Technopolitics in the Global Cold War*

Michael E. Gorman, editor, *Trading Zones and Interactional Expertise: Creating New Kinds of Collaboration*

Matthias Gross, *Ignorance and Surprise: Science, Society, and Ecological Design*

Andrew Feenberg, *Between Reason and Experience: Essays in Technology and Modernity*

Wiebe E. Bijker, Roland Bal, and Ruud Hendricks, *The Paradox of Scientific Authority: The Role of Scientific Advice in Democracies*

Park Doing, *Velvet Revolution at the Synchrotron: Biology, Physics, and Change in Science*

Gabrielle Hecht, *The Radiance of France: Nuclear Power and National Identity after World War II*

Richard Rottenburg, *Far-Fetched Facts: A Parable of Development Aid*

Michel Callon, Pierre Lascoumes, and Yannick Barthe, *Acting in an Uncertain World: An Essay on Technical Democracy*

Ruth Oldenziel and Karin Zachmann, editors, *Cold War Kitchen: Americanization, Technology, and European Users*

Deborah G. Johnson and Jameson W. Wetmore, editors, *Technology and Society: Building Our Sociotechnical Future*

Trevor Pinch and Richard Swedberg, editors, *Living in a Material World: Economic Sociology Meets Science and Technology Studies*

Christopher R. Henke, *Cultivating Science, Harvesting Power: Science and Industrial Agriculture in California*

Helga Nowotny, *Insatiable Curiosity: Innovation in a Fragile Future*

Karin Bijsterveld, *Mechanical Sound: Technology, Culture, and Public Problems of Noise in the Twentieth Century*

Peter D. Norton, *Fighting Traffic: The Dawn of the Motor Age in the American City*

Joshua M. Greenberg, *From Betamax to Blockbuster: Video Stores and the Invention of Movies on Video*

Mikael Hård and Thomas J. Misa, editors, *Urban Machinery: Inside Modern European Cities*

Christine Hine, *Systematics as Cyberscience: Computers, Change, and Continuity in Science*

Wesley Shrum, Joel Genuth, and Ivan Chompalov, *Structures of Scientific Collaboration*

Shobita Parthasarathy, *Building Genetic Medicine: Breast Cancer, Technology, and the Comparative Politics of Health Care*

Kristen Haring, *Ham Radio's Technical Culture*

Atsushi Akera, *Calculating a Natural World: Scientists, Engineers and Computers during the Rise of U.S. Cold War Research*

Donald MacKenzie, *An Engine, Not a Camera: How Financial Models Shape Markets*

Geoffrey C. Bowker, *Memory Practices in the Sciences*

Christophe Lécuyer, *Making Silicon Valley: Innovation and the Growth of High Tech, 1930–1970*

Anique Hommels, *Unbuilding Cities: Obduracy in Urban Sociotechnical Change*

David Kaiser, editor, *Pedagogy and the Practice of Science: Historical and Contemporary Perspectives*

Charis Thompson, *Making Parents: The Ontological Choreography of Reproductive Technology*

Pablo J. Boczkowski, *Digitizing the News: Innovation in Online Newspapers*

Dominique Vinck, editor, *Everyday Engineering: An Ethnography of Design and Innovation*

Nelly Oudshoorn and Trevor Pinch, editors, *How Users Matter: The Co-construction of Users and Technology*

Peter Keating and Alberto Cambrosio, *Biomedical Platforms: Realigning the Normal and the Pathological in Late-Twentieth-Century Medicine*

Paul Rosen, *Framing Production: Technology, Culture, and Change in the British Bicycle Industry*

Maggie Mort, *Building the Trident Network: A Study of the Enrollment of People, Knowledge, and Machines*

Donald MacKenzie, *Mechanizing Proof: Computing, Risk, and Trust*

Geoffrey C. Bowker and Susan Leigh Star, *Sorting Things Out: Classification and Its Consequences*

Charles Bazerman, *The Languages of Edison's Light*

Janet Abbate, *Inventing the Internet*

Herbert Gottweis, *Governing Molecules: The Discursive Politics of Genetic Engineering in Europe and the United States*

Kathryn Henderson, *On Line and on Paper: Visual Representation, Visual Culture, and Computer Graphics in Design Engineering*

Susanne K. Schmidt and Raymund Werle, *Coordinating Technology: Studies in the International Standardization of Telecommunications*

Marc Berg, *Rationalizing Medical Work: Decision Support Techniques and Medical Practices*

Eda Kranakis, *Constructing a Bridge: An Exploration of Engineering Culture, Design, and Research in Nineteenth-Century France and America*

Paul N. Edwards, *The Closed World: Computers and the Politics of Discourse in Cold War America*

Donald MacKenzie, *Knowing Machines: Essays on Technical Change*

Wiebe E. Bijker, *Of Bicycles, Bakelites, and Bulbs: Toward a Theory of Sociotechnical Change*

Louis L. Bucciarelli, *Designing Engineers*

Geoffrey C. Bowker, *Science on the Run: Information Management and Industrial Geophysics at Schlumberger, 1920–1940*

Wiebe E. Bijker and John Law, editors, *Shaping Technology/Building Society: Studies in Sociotechnical Change*

Stuart Blume, *Insight and Industry: On the Dynamics of Technological Change in Medicine*

Donald MacKenzie, *Inventing Accuracy: A Historical Sociology of Nuclear Missile Guidance*

Pamela E. Mack, *Viewing the Earth: The Social Construction of the Landsat Satellite System*

H. M. Collins, *Artificial Experts: Social Knowledge and Intelligent Machines*

http://mitpress.mit.edu/books/series/inside-technology